UNDERSTANDING SPACE WEATHER AND THE PHYSICS BEHIND IT

Delores J. Knipp
USAF Academy, Emeritus

Editors:
Marilyn McQuade
USAF Academy

Doug Kirkpatrick
USAF Academy

Boston   Burr Ridge, IL   Dubuque, IA   New York   San Francisco   St. Louis
Bangkok   Bogotá   Caracas   Lisbon   London   Madrid
Mexico City   Milan   New Delhi   Seoul   Singapore   Sydney   Taipei   Toronto

The McGraw·Hill Companies

Copyright © 2011 by The McGraw-Hill Companies, Inc. All rights reserved. Printed in the United States of America. Except as permitted under the United States Copyright Act of 1976, no part of this publication may be reproduced or distributed in any form or by any means, or stored in a data base retrieval system, without prior written permission of the publisher.

Understanding Space Weather and the Physics Behind It

1 2 3 4 5 6 7 8 9 0   QVR QVR   13 12 11

ISBN-13: 978-0-07-340890-3
ISBN-10: 0-07-340890-5

Learning Solutions Manager: Darlene Bahr
Production Editor: Lynn Nagel
Cover Design: Mary Tostanoski
Front Cover Photo Credits:
  • *Polar Orbiting Environmental Satellite (POES) courtesy of National Oceanic and Atmospheric Administration (NOAA)*
  • *Radiation Belt (Purple graphic to right of title) courtesy of Center for Integrated Space Weather Modeling (CISM)*
  • *Magnetosphere Visualization (Overlying Coronal mass ejection and above satellite) courtesy of Center for Integrated Space Weather Modeling (CISM), Sponsored by the National Science Foundation*
  • *Coronal Mass Ejection courtesy of LASCO C2, Solar and Heliospheric Observatory (SOHO) (operated by ESA and NASA)*
  • *Alaska Aurora courtesy of Joshua Strang, USAF*
  • *Artist's rendition of the Heliospheric Current Sheet courtesy of NASA artist Werner Heil*
Back Cover Photo Credits:
  • *Communication/Navigation Outage Forecasting System (C/NOFS) satellite courtesy of USAF*
  • *Visible Sun with Sunspots courtesy of MDI consortium (co-sponsored by the European Space Agency [ESA] and NASA)*
Printer/Binder: Quad/Graphics Versailles

Understanding Space Weather and the Physics Behind It

Preface

UNIT 1. SPACE WEATHER AND ITS PHYSICS

Chapter 1 Space Is a Place…with Weather 1

    1.1 Introduction 2
        1.1.1 Space Is Not Empty! 2
        1.1.2 The Realms of Terrestrial and Space Weather 6

    1.2 Where and Why Does Space Weather Occur? 9
        1.2.1 Other Stars Affect the Near-space Environment 9
        1.2.2 The Sun Affects Space Weather 10
        1.2.3 Earth's Magnetosphere Results from and Contributes to Space Weather 16
        1.2.4 Space Weather Dominates Earth's Upper Atmosphere 19
        1.2.5 Meteors and Space Dust Affect the Space Environment 22
        1.2.6 Space Weather Effects in Earth's Lower Atmosphere and at Earth's Surface 23

    1.3 Tales of Two Storms 25
        1.3.1 A Brief History of Space Weather Observations 25
        1.3.2 1859: First Observation of a Solar Flare and Associated Earth Disturbance 26
        1.3.3 2003: Extreme Space Weather in the 21st Century 29
        1.3.4 Space Weather Storm Scales 32

    1.4 The Lorentz Force 36

Chapter 2 Space Is a Place…with Energy 43

    2.1 Introduction to Energy in Space 44
        2.1.1 The Physical Concept of Energy in a System 44
        2.1.2 Forms and Transformations of Energy 50
        2.1.3 Energy Conservation and Entropy 61
        2.1.4 Energy at High Speeds 62

	2.2	Energy Interaction within the Space Environment	65
		2.2.1 Energy Transfer via Particles and Fields	65
		2.3.1 Electromagnetic Energy Transitions: Electromagnetic Waves and Photons	69
	2.3	Characterizing Electromagnetic (EM) Radiation	75
		2.3.1 Blackbody Radiation	75
		2.3.2 Discrete Line Radiation and Absorption	81
		2.3.3 Radiation from Other Sources	86
Chapter 3	**The Quiescent Sun and Its Interaction with Earth's Atmosphere**		**93**
	3.1	Introduction to the Quiescent Sun and its Interior	94
		3.1.1 Our Local Magnetic Star	94
		3.1.2 The Stratified Sun	98
		3.1.3 The Convective Sun	104
	3.2	Regions of the Solar Atmosphere	109
		3.2.1 The Sun's Lower Atmosphere	109
		3.2.2 The Sun's Upper Atmosphere	114
	3.3	Characteristics of the Quiescent Solar Magnetic Field	121
		3.3.1 Multi-scale Solar Magnetic Fields	121
		3.3.2 Typical Solar Cycle Field Variations	123
	3.4	Basic Physics of Atmospheres	128
		3.4.1 Law of Atmospheres	128
		3.4.2 Absorption Interactions in an Atmosphere	131
Chapter 4	**Space Is a Place...with Fields and Currents**		**149**
	4.1	Introduction to Fields	150
		4.1.1 The Field Concept	150
		4.1.2 Static Electric and Magnetic Fields	156
		4.1.3 Maxwell's Equations	159
		4.1.4 Electric, Magnetic, and Electromagnetic Wave Energy Densities (Pressure)	169
	4.2	Electric Currents and Conductivity	173
		4.2.1 Charge Conservation, Current Continuity, and Current Density	173
		4.2.2 Ohm's Law and Electrical Conductivity	177
	4.3	Magnetic Behavior in the Space Environment	180
		4.3.1 Frozen-in Magnetic Flux	180
		4.3.2 Electromagnetic Characteristics of Dynamos	182

	4.4	Currents and Electric Fields in Near-space	185
		4.4.1 Electric Currents in the Near-Earth Environment	185
		4.4.2 Electric Fields in Space	189

Chapter 5 — The Quiescent Solar Wind: Conduit for Space Weather — 199

	5.1	The Quiescent Solar Wind	200
		5.1.1 Solar Wind Plasma Characteristics	200
		5.1.2 The Supersonic and Super-Alfvénic Solar Wind	208
		5.1.3 Interplanetary Magnetic Field Characteristics	216
		5.1.4 The Current Sheet and Magnetic Sectors	220
	5.2	Solar Wind Interactions in the Heliosphere	227
		5.2.1 Solar Wind Deceleration at a Magnetized Barrier	227
		5.2.2 Solar Wind Beyond the Inner Planets	234

Chapter 6 — Space is a Place...with Plasma — 243

	6.1	Plasma Characteristics and Behaviors	244
		6.1.1 Qualitative Plasma Behavior	244
		6.1.2 Defining Characteristics of a Plasma	248
	6.2	Single-particle Dynamics—I	255
		6.2.1 Single-particle Accelerated Motion	255
		6.2.2 Single-particle Unaccelerated Motion	259
	6.3	Thermal Plasma Fluid Dynamics	268
		6.3.1 Maxwell-Boltzmann Distribution and the Saha Equation	268
		6.3.2 Energy Density in Plasma	271
	6.4	Magnetized Plasma Fluid Dynamics	274
		6.4.1 The Basics of Magnetohydrodynamics (MHD)	274
		6.4.2 Applying Magnetohydrodynamics (MHD)	281
	6.5	Elementary Plasma Waves	284
		6.5.1 Wave Propagation	284
		6.5.2 Plasma Oscillations and Waves	285

Chapter 7 — Earth's Quiescent Magnetosphere: Its Role in the Space Environment and Weather — 293

	7.1	Earth's Geomagnetic Field	294
		7.1.1 Geomagnetic Field Basics	294
		7.1.2 Coordinate Systems for Earth's Magnetic Domain	299
	7.2	Geomagnetic Structures and Charged Particle Populations	305
		7.2.1 Earth's Magnetized Geospace	305
		7.2.2 Inner Magnetosphere	308
		7.2.3 Connections to Distant Regions	311

	7.3	Single Particle Dynamics—II	316
		7.3.1 Gyration and Bounce	316
		7.3.2 Longitudinal Drift	321
	7.4	Quiescent Magnetospheric Processes	324
		7.4.1 Forming the Dayside Magnetopause	324
		7.4.2 Mass, Energy, and Momentum Transfer to the Magnetosphere	327
		7.4.3 Trapping Processes in the Plasmasphere	336

Chapter 8 Earth's Quiescent Atmosphere and Its Role in the Space Environment and Weather — 347

	8.1	Earth's Neutral Atmosphere	348
	8.2	Earth's Ionized Atmosphere	358
		8.2.1 Transient Luminous Events	358
		8.2.2 Ionization as a Function of Height	360
		8.2.3 The Ionosphere Characterized by Latitude and Layers	370
	8.3	Radio Wave Propagation in the Ionosphere	375
		8.3.1 Operational Aspects of High-frequency (HF) Radio Propagation	375
		8.3.2 The Physics of Ionospheric Radio Wave Propagation	378
	8.4	Ionospheric Interactions with Other Regimes	386
		8.4.1 Quiescent High-latitude Ionospheric Connections to the Magnetosphere	386
		8.4.2 Ionospheric Mass Outflows	392

Chapter 9 The Active Sun and Other Stars: Sources of Space Weather — 399

	9.1	The Solar Cycle and Its Roots	400
		9.1.1 The Solar Dynamo and Its Supporting Motions	400
		9.1.2 Active Regions and Their Components	405
	9.2	Active Sun Emissions	414
		9.2.1 Energy Paths of Magnetic Reconfiguration	414
		9.2.2 Flares and Radio Bursts	415
		9.2.3 Coronal Mass Ejections (CMEs)	425
		9.2.4 Solar Energetic Particles	433
	9.3	Space Weather Effects from Other Stars	436
		9.3.1 Low- and High-energy Particles from the Cosmos	436
		9.3.2 High-energy Photons from the Cosmos	441

UNIT 2. ACTIVE SPACE WEATHER AND ITS PHYSICS

Chapter 10 The Active Interplanetary Medium: Conduit for Space Weather — 445

- 10.1 Disturbances in the Quasi-stationary Solar Wind — 446
  - 10.1.1 Quasi-stationary Structures in the Slow Solar Wind — 446
  - 10.1.2 Co-rotating Structures in the Solar Wind — 450
- 10.2 Transient Disturbances in the Solar Wind — 457
  - 10.2.1 Characteristics of Transients in the Interplanetary Medium — 457
  - 10.2.2 Geoeffectiveness of Transients in the Interplanetary Medium — 462
- 10.3 Solar Energetic Particles from Interplanetary Shocks — 473
  - 10.3.1 Traveling Shocks and Solar Energetic Particles — 473
  - 10.3.2 Shock Acceleration Mechanisms — 480

Chapter 11 The Disturbed Magnetosphere and Linkages Above and Below — 491

- 11.1 Describing Geomagnetic Storms — 492
  - 11.1.1 The Interplanetary Medium and Magnetic Storms — 492
  - 11.1.2 The Dst Index and Its Relation to the Ring Current and Storm Phases — 493
- 11.2 Large-scale Magnetic Disturbances — 499
  - 11.2.1 Power from the Solar Wind — 499
  - 11.2.2 Energy Dissipation Processes with Links to the Ionosphere — 502
  - 11.2.3 Unloading Energy to the Trapping Regions — 508
- 11.3 Field and Current Coupling during Magnetic Storms — 522
  - 11.3.1 High-latitude Magnetosphere-Ionosphere Coupling — 522
  - 11.3.2 Mid- and Low-latitude Magnetosphere Shielding — 529

Chapter 12 Space Weather Disturbances in Earth's Atmosphere — 535

- 12.1 Upper-atmosphere Disturbance Drivers — 536
  - 12.1.1 Characteristic Solar Cycle Behavior — 536
  - 12.1.2 Flare and Fast CME Energy Response — 538
  - 12.1.3 Magnetospheric Energy Sources — 542
  - 12.1.4 Poynting's Theorem, Poynting Flux, and Joule Heat — 551
- 12.2 Thermospheric Disturbance Effects — 558
  - 12.2.1 Internal Disturbances: Thermospheric Tides and Winds — 558
  - 12.2.2 External Disturbances From High Latitudes — 559
  - 12.2.3 Outflowing, Upwelling, and Composition Changes in the Thermosphere — 562

	12.3	Ionospheric Storm Effects	569
		12.3.1 Ionospheric Polar Cap Absorption Events	569
		12.3.2 Ionospheric Storms and Disturbance Features	572

UNIT 3. IMPACTS AND EFFECTS OF SPACE WEATHER AND SPACE ENVIRONMENT

Chapter 13 Near-Earth Is a Place...with Susceptible Hardware and Humans — 589

	13.1	Damage and Impacts from Particles and Photons	590
		13.1.1 Particle Radiation Environment	591
		13.1.2 Energetic Particle Radiation Environment for Humans and Hardware	600
		13.1.3 Energetic Plasma, Photon, and Neutral Atmosphere Effects on Hardware	614
		13.1.4 Satellite Drag	620
	13.2	Damage and Impacts Associated with the Meteor and Artificial Debris Environment	625
		13.2.1 The Natural Meteor Environment	625
		13.2.2 Artificial Space Debris Environment	629
	13.3	Hardware Damage and Impacts Associated with Field Variations	633
		13.3.1 High-energy Electrons	633
		13.3.2 Geomagnetic Field Interactions with Satellites	636
		13.3.3 Geomagnetic Field Interactions at the Ground	638
	13.4	Surveying the Impact of Space Weather on Systems—I	644

Chapter 14 Effects of Space Weather and Space Environment on Signals and Systems — 653

	14.1	Background and Solar-driven Effects	654
		14.1.1 Background Ionosphere Effects	654
		14.1.2 Prompt Signal Effects	656
		14.1.3 Impulsive Solar Events Generating Polar Cap Absorption (PCA)	663
		14.1.4 Geomagnetic Storms and Seasonal Effects on Signals	665
	14.2	Non-solar Effects on Signal Propagation	677
		14.2.1 Meteor Effects	677
		14.2.2 Magnetar Effects	678
		14.2.3 High-altitude Nuclear Signal Effects	681
	14.3	Surveying the Impact of Space Weather on Systems—II	684

Appendix A	NOAA Space Weather Scales	689
Appendix B	Electromagnetic Waves and the Speed of Light	693
Appendix C	Vector Identities	695
Appendix D		697
Index		699

Preface

Space weather affects so much of our modern society that we need a good explanation of where it comes from, what it is, and how it affects what we do every day. High-altitude airliners, national power grids, and radio frequency communications are a few of the advanced-technology systems affected by solar flares, coronal mass ejections, radio bursts, and other space weather phenomena. Local and national planners, as well as system designers and builders, must account for the possible disruptions and interference caused by electromagnetic waves and charged particles that daily spew from our Sun and arrive from deep space. Even personal devices, such as Global Positioning System signals, react to solar emissions, which sometimes degrade service for a time. Space weather is everywhere and we need to know about it, so Dr. Delores Knipp wrote this book.

In it, she describes what we know about the Sun and its processes, the heliosphere, where the planets intercept solar emissions, and our local tear-drop shaped magnetosphere that expands and contracts during the solar cycle. She explains the concepts simply yet thoroughly, using math as needed, so that students may gain a deeper understanding of the causes of and reactions to solar ejections. Hundreds of diagrams depict the processes deep inside our Sun and around and within our planet's affected volume of space. Plots of sensor measurements show trends and cycles that help scientists describe and predict the solar phases. This fascinating and imaginative work gives life to a relatively little-known yet ever-present field that is vital to our safe existence.

We acknowledge the funding support for this lengthy effort from many organizations, including the US Air Force Academy, the Air Force Space and Missile Center, Air Force Space Command, NASA Headquarters, Goddard Space Flight Center, the Office of the Undersecretary of Defense for Acquisition, the National Reconnaissance Office, the National Security Space Office, the Air Force Rocket Laboratory, and the National Polar Orbiting Environmental Satellite Office. We appreciate their recognition of how valuable a space weather fundamentals textbook can be.

We recognize the following space weather organizations that supported this writing with substantial technical material: the US Air Force Academy Department of Physics, the Air Force Research Laboratory, the Air Force Weather Agency, the Naval Post Graduate School, the Air Force Institute of Technology, the National Science Foundation's Center for Integrated Space Weather Modeling, the NOAA Space Weather Prediction Center, Air Force Space Command, US Army Space Command, the National Aeronautics and Space Administration, the European Space Agency, the University Corporation for Atmospheric Research, the NASA Community Coordinated Modeling Center, the Aerospace Engineering Science Department at the University of Colorado, the Atmospheric Science Department at the University of Missouri, and Space Environment Technologies. Support from staff members of the High Altitude Observatory (HAO) at the National Center for Atmospheric Research and use of the HAO library were crucial in finishing the manuscript. Through many contacts, personal and electronic, we gathered portions of their expertise that helped describe the complexities of space weather. Their valuable contributions made the technical content relevant and state-of-the-art.

We received leadership and moral support for this unique textbook from Col Rex Kiziah, Physics Department Head, and Col Marty France, Astronautics Department Head, both at the US Air Force Academy. And without the encouragement and backing of Dr. Wiley Larson, we'd never have started writing. His unwavering support and timely nudges kept us going from start to finish. From the Physics Department, writing and reviewing assistance came from numerous faculty members. Students at many levels reviewed elements of the material. Others who reviewed and added valuable technical insight include Dr. James Head (Brig Gen., retired), Dr. Geoff McHarg, Dr. Heidi Fearn, Dr. Ryan Haaland, Dr. Heidi Mauk, Dr. Gabriel Font-Rodriguez, Ms. Jeanie Ferguson, and the many chapter contributors. Keeping the project funded took the combined efforts of Dr. Larson and Col Timothy Lawrence from the Air Force Academy. We appreciate the diligent service provided by all of these parties.

The author thanks her very patient husband, David Berens, and family for all of their support through many iterations of the text.

The book layout and format (and some graphics) are the detailed work of Anita Shute, who reworked chapters tirelessly at the request of the author and editors. All that's right with the book layout belongs to her. Our graphics artist is Mary Tostanoski, who created dozens of stunning original figures and updated figures on a moment's notice throughout our writing journey. Her colorful artwork brings concepts to life in every chapter. The involvement of the space physics community is reflected in the hundreds of graphics provided by individuals who love this discipline.

We made every effort to eliminate mathematical and factual errors, but we may have missed a few. We invite (and encourage) readers to send us comments and constructive criticism, so we can collect improvements for the next printing. We'll update it someday and include the corrections we have on hand.

We also recognize that space weather analysis and reporting are ever-changing fields and some of the material in this book will eventually be overcome by events and new knowledge. Science and technology never stand still, and neither will we. We look forward to the challenge of keeping up with the moving targets and making this book relevant for years to come.

All the figures are used with permission from the authors or owners.

With this volume on space weather, we hope we've captured and presented a large portion of what affects our world and the worlds around us. We hope our readers will find it well-written and easy to understand and retain. Our sincere desire is for many readers to get excited about this topic, study it more, and become space weather enthusiasts who make the world better and more livable. We wish all of our readers the greatest success in their space weather learning and doing.

September 2010

Dr. Delores Knipp	Dr. Doug Kirkpatrick	Marilyn McQuade
Author	Technical Editor	Technical Editor

Department of Astronautics
United States Air Force Academy
USAF Academy, Colorado 80840
Voice: 719-333-4110   FAX: 719-333-3723

Department of Physics
United States Air Force Academy
USAF Academy, Colorado 80840
Voice: 719-333-3510   FAX: 719-333-3182

Space Is a Place…with Weather

1

UNIT 1. SPACE WEATHER AND ITS PHYSICS

Contributions by Edward Cliver and William Murtagh

You should already know about…

- The metric system of units (SI) and scientific notation
- Vector notation and cross products
- The basics of kinematic and circular motion
- Basic concepts and units of force, energy, and momentum
- Newton's Laws of Motion
- Newton's Law of Gravitation
- Coulomb's Law for charged particles
- Three states of matter—solid, liquid, gas
- The Sun's and most planets' magnetic fields

In this chapter you will learn about…

- The constituents of the space environment
- Relative distance scales of the solar radius, Earth radius, and astronomical units
- A fourth state of matter—plasma
- The nature of space weather and where it originates and occurs
- Important effects of space weather storms on the near-Earth environment
- Periods of solar storminess and quiet, known as the solar cycle
- The Sun's extended, magnetized atmosphere
- Great space weather storms of the last three centuries
- Space weather storm scales
- The Lorentz force

Outline

1.1 Introduction
    1.1.1 Space Is Not Empty!
    1.1.2 The Realms of Terrestrial and Space Weather

1.2 Where and Why Does Space Weather Occur?
    1.2.1 Other Stars Affect the Near-space Environment
    1.2.2 The Sun Affects Space Weather
    1.2.3 Earth's Magnetosphere Results from and Contributes to Space Weather
    1.2.4 Space Weather Dominates Earth's Upper Atmosphere
    1.2.5 Meteors and Space Dust Affect the Space Environment
    1.2.6 Space Weather Effects in Earth's Lower Atmosphere and at Earth's Surface

1.3 Tales of Two Storms
    1.3.1 A Brief History of Space Weather Observations
    1.3.2 1859: First Observation of a Solar Flare and Associated Earth Disturbance
    1.3.3 2003: Extreme Space Weather in the 21st Century
    1.3.4 Space Weather Storm Scales

1.4 The Lorentz Force

1.1 Introduction

Modern aviators operate in a terrestrial atmosphere, about which much is known. Despite this knowledge, some flights divert because of bad weather. On a seasonal basis, airport ramps full of aircraft are evacuated from a hurricane's path. Terrestrial weather is monitored twenty-four hours a day, seven days a week. But what about the space environment, home to critical technology systems and billions of dollars in space assets? Who monitors it? What should space professionals know about it? What actions mitigate its effects on technology systems?

1.1.1 Space Is Not Empty!

Objectives: After reading this section, you should be able to...

♦ Define the terms space weather, space climate, and space environment
♦ Distinguish phases of matter: solid, liquid, gas, and plasma

Space is Not Benign

Effects of the space environment are generally beyond sight and touch. These effects occur in the realms of radio communication, spacecraft operations, high-altitude flight, extended power grids, and even computer microchip manufacturing. A significant part of civil and military operations now relies on the "good" behavior of the natural space environment. When the environment turns hostile, space operators need to know. Present and future aerospace leaders need space-weather awareness and knowledge of potential impacts. The following vignette from a space weather storm in late 2003 illustrates its effects on today's civil and military operations.

> *Immediate Effects. In mid-October 2003 solar cycle 23 was declining and space weather forecasters had a respite from the solar activity of the previous three years. But the calm was not to last. Remote observations indicated a brewing storm on the back side of the Sun. On October 22 an ugly, tangled web of magnetic field rotated into view (see images in Sec. 1.3.3). Spasms in the field hurled cannon balls of solar atmosphere into the interplanetary medium. (Figure 1-1 shows such a blast on October 26, 2003.) Within minutes of the explosion, energetic-particle levels at Earth were so high that NASA officials directed International Space Station (ISS) astronauts to take precautionary shelter. Airlines rerouted polar flights to avoid the high radiation levels and communication blackout areas—at a cost of $10,000 to $100,000 per flight.*

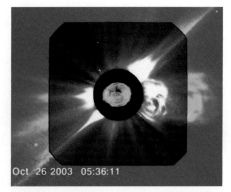

Fig. 1-1. Mass and Energy Flowing from the Sun on October 26, 2003, as Viewed by Three Instruments on the Solar and Heliospheric Observatory (SOHO) Spacecraft. Here we show a composite image with evidence of a mass ejection. The ejected mass associated with this event was roughly equal to the mass of a small mountain on Earth. The composite image is an overlay of several SOHO observations. At the center, in green, is an extreme ultraviolet image of the Sun that replaces the very bright solar disk and atmosphere, which have been masked to a distance of two solar radii (dark circle). Bright loops in the red rectangle beyond the mask show the mass ejection as it first departs the Sun. A subsequent image shows the ejection expanding into interplanetary space (to the right in the blue rectangle). The scope of the image is ~16 solar radii in the horizontal. The entire mass ejection structure is three-dimensional and extends into the foreground of the images. *(Courtesy of ESA/NASA – the SOHO Program)*

Short-term. Less than 24 hours after the fury on the Sun, utility companies reported storm-induced currents over northern Europe, resulting in transformer problems and a subsequent regional blackout. On October 28, the density of free electrons in the ionosphere on Earth's sunlit side jumped by 25%. High-frequency radio communication was disrupted. Climbers on Mount Everest could not signal base operations. Satellite navigation was compromised in some areas. Auroras were visible in Houston, Texas, one of the more southern cities in the United States.

Long-term and Ongoing. One month later the same solar-active region sent another blast toward Earth. Students at a western US college, who were finishing a precision-mapping project in a geography course, found that satellite-derived position errors went from 5 meters to 75 meters in just a few minutes. Over subsequent months, electric utility operators discovered serious damage deep inside power transformers. The cost to repair: tens of millions of dollars. Space was not benign!

Defining Space Weather, Space Climate, and Space Environment

Once generally viewed as a subset of aeronomy, astronomy, or perhaps astronautics, day-to-day interactions between Earth and the constituents of interplanetary space are now commonly referred to as *space weather* or *space environment interactions*. The science that attempts to explain variations in these interactions is *space physics*. Space weather describes the conditions in space that affect space-based and ground-based technology systems as well as Earth and its inhabitants. It is a consequence of the behavior of the Sun and other stars, as well as the nature of Earth's magnetic field and atmosphere, and our location in the solar system. *Space climate* depends on long-term solar emissions and the their effect on the Sun's extended atmosphere and bodies in its atmosphere, such as Earth. As a practical matter, space weather often means a disturbed situation, whereas space environment often refers to typical or background conditions and sometimes to the rise and decline of solar activity. Space weather is created by electromagnetic energy from the Sun and by the Sun's out-flowing atmosphere that streams by Earth (and all planets) at tremendous speeds. Earth's magnetic field and upper atmosphere store, dissipate, and redirect some of this energy. The vast majority of energy and mass involved in space weather is from the Sun, but distant stars and closer objects, such as meteoroids and electrified layers of Earth's atmosphere, play a role too.

Space Environment Volume and Matter

Table 1-1 lists some measures of distance we use often in this text. For practical purposes we consider the region of space within 200 Earth radii (200 R_E = 1.28 × 10^9 m) as "near" Earth. This is huge and could hold over 8 million Earths, but compared to the solar system, this volume is comparatively inconsequential. We often call this region simply the space environment. Does this mean Earth's atmosphere is part of the space environment? Most definitely it is, but for the most part, we confine our discussions to atmospheric interactions occurring above 50 km. We leave the atmosphere below that level (99.9% of atmospheric mass) in the capable hands of meteorologists. However, circumstances exist in which space weather processes spawn activity in the lower atmosphere and, occasionally, all the way to the surface. So when necessary, we claim even the lower atmosphere as space weather territory.

The Sun is a huge, dense swirl of plasma, held together by its own gravity. It radiates away its excess thermonuclear energy as rivers of photons and charged particles. The charged particles then form a tenuous plasma that sweeps through interplanetary space. Thus, an understanding of the space environment hinges on understanding plasmas. The material of space comes in several states: solid, liquid, gas, and plasma (Fig. 1-2). The latter state, one in which a gas is so energized that its electrons and ions have separated, dominates space interactions. A very thin soup of plasma and energy constrains and interacts with near-Earth space creating space climate, space weather, and space storms. Modeling this plasma environment remains a challenge.

Although the particles in plasma have electric charges, the plasma at the macroscopic level remains electrically neutral. The ions and free electrons, though not formally bound to each other, exist in a common hot "soup." The electrons are attracted to the ions, but they are usually traveling so fast they don't recombine with ions. Electrons normally travel faster than the ions because the heat energy moves their smaller masses more readily. Such a complex set of interactions makes plasma very dynamic. If plasma cools, the electrons slow and may recombine with positive ions to form a gas. On a macroscopic scale, plasma is very fluid, similar to a gas. However, unlike a gas, it exhibits one property other forms of matter rarely exhibit—it strongly interacts with electromagnetic fields. This property sets plasma apart from other matter. Additionally, the collective motions of charged particles generate magnetic fields that, in turn, govern the charged particle motions. (Sec. 1.4 and Chapter 4.)

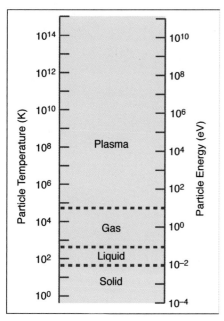

Fig. 1-2. Phases of Matter. Here we show phases of matter as a function of average particle energy in units of Kelvin (K) and electron volts (eV). The boundaries between the phases are approximate.

Table 1-1. Some Distances in Space and the Space Environment. This table lists approximate distances from the Sun and from Earth.

Sun-centered	Earth-centered
Distance to galactic center 2.47×10^{20} m = 1.65×10^9 AU	Distance to solar wind monitor* 1.50×10^9 m = 2.35×10^2 R_E
Distance to nearest star 4.02×10^{16} m = 2.68×10^5 AU	Distance to the Moon 3.84×10^8 m = 60.3 R_E
Distance to heliopause* 2.3×10^{13} m = 1.53×10^2 AU	Distance to bow shock* 9.60×10^7 m = 15 R_E
Sun-Pluto distance 5.91×10^{12} m = 39.4 AU	Distance to dayside magnetopause* 6.38×10^7 m = 10 R_E
Sun-Earth distance 1.50×10^{11} m = 1 AU	Distance to geosynchronous orbit 4.20×10^7 m = 6.6 R_E
Solar radius 6.96×10^8 m = 4.64×10^{-3} AU	Earth radius 6.38×10^6 m = 1 R_E

* Terms to be defined later in the chapter.

Pause for Inquiry 1-1

Consider the energy scales in Fig. 1-2. For subatomic particles an alternative to the Kelvin temperature scale is an energy scale in electron volts (eV). Approximately how many Kelvins of temperature are associated with each electron volt (eV)? *(Answer at the chapter's end)*

Focus Box 1.1: The States of Matter: Solid, Liquid, Gas, and Plasma

A large number of particles in a group exhibit macroscopic (large-scale) physical properties that depend on temperature and pressure. We model matter in three standard states—solid, liquid, and gas. More complete models with additional states are needed for space physics applications. In this section we compare the states familiar to us with the states that describe most of the universe beyond our planet.

Charged particles of gas, when behaving collectively, are *plasma*. Plasmas exhibit rather complex behavior compared to non-ionized gases. This behavior complicates the understanding and forecasting of space weather and space environmental effects. Figure 1-3 illustrates how the states of matter change as a function of temperature. The discussion below highlights similarities and differences between these states.

Solids. Low-energy molecules or atoms are bound to each other in such a way that they resist compression and shear. Solids may have organized crystalline structures or they may have more random arrangements of molecules. Only an external force can change the shape of the solid. On a macroscopic scale, a large enough solid possesses a well-defined surface. Ice, cold steel, and rocks are examples of solids we encounter daily.

Liquids. Energy added to some solids breaks material bonds, allowing them to become liquid. Liquid molecules or atoms are able to shear or slide around each other, but they are still bound to the general vicinity by inter-molecular forces. On a macroscopic scale, a liquid possesses a well-defined, but deformable surface. Everyday examples include water, alcohol, oil, and mercury.

Gases. With the addition of enough energy, a liquid evaporates into a gas. The added energy allows some, if not all, of the molecules to overcome intermolecular forces and escape each other. The weak forces between particles permit gases to have indefinite shapes and volumes. A gas under pressure compresses more readily than a liquid and expands to fill its container, whereas a liquid occupies a minimum portion of the container. On a macroscopic scale a gas possesses no surface. Air, water vapor, carbon monoxide, and gasoline fumes are gas examples.

Plasma. When matter, particularly gases, gains significant energy, outer electrons are stripped easily from parent atoms. An atom with one or more missing electrons is an ion, which is now positively charged. Thus, a heated gas becomes ionized to form plasma. Plasma consists of clouds of positively charged ions together with their negatively charged (free) electrons. For reasons we discuss later, magnetic fields are often used to confine high-energy plasmas. Examples of plasma include electrical sparks, lightning, the Sun, and stars. In fact, almost all of the material in the visible universe is in the form of plasma.

We may classify the states discussed above by temperature as shown in Fig. 1-2. We also classify the states with weak inter-molecular forces (fluids) by flow behavior. Fluids are matter that flow: liquids, gases, and plasmas.

Later in this text, we describe and quantify the behavior of plasmas, especially as that behavior relates to varying electric and magnetic fields.

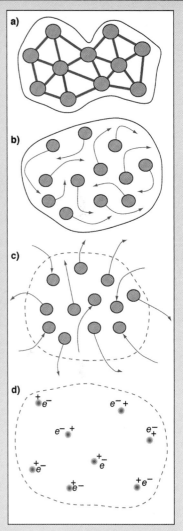

Fig. 1-3. a) **Solid Matter.** Solids usually have structures that deform only by external forces. b) **Liquid Matter.** Liquid molecules slide around each other, yet are still bound by inter-molecular forces. c) **Gaseous Matter.** Gas molecules are not bound by intermolecular forces, so they expand to fill their container. d) **Plasma Matter.** Plasma consists of clouds of ions mixed with but independent from their electrons. *(Adapted from the AF Weather Agency)*

1.1.2 The Realms of Terrestrial and Space Weather

Objectives: After reading this section, you should be able to...

♦ Distinguish regions of Earth's atmosphere and magnetosphere
♦ Associate temperature, altitude, and dominant characteristics with each region

Earth and its environs exist in a volume carved from the heliosphere—the Sun's extended atmosphere. When we survey the bodies of the solar system, we find that those with atmospheres share many common traits. The governing physics in these atmospheres produces several different regimes. Earth's atmosphere contains all of them. Other planets have a subset of these. Three of these regimes are primarily associated with terrestrial weather, but may experience solar cycle influences.

Below 100 km, gravity, pressure gradients, and Earth's rotation dominate the dynamics of Earth's atmosphere and drive terrestrial weather. Particle collisions in the relatively dense neutral gas prevent or disrupt strong interactions with Earth's magnetic field. Infrared (IR), visible, and ultraviolet (UV) solar radiation heat these regions.

- The troposphere, between 0 and 15 km, is a turbulent weather region marked by decreasing temperature with altitude. The Sun's radiation warms Earth's surface, allowing it to emit long-wave radiation that warms the base of the layer. Water vapor transports energy across latitude and longitude. Outgoing longwave IR radiation cools the regions above the surface, setting up an atmosphere prone to extreme mixing and overturning.

- The stratosphere, between 15 km–50 km, is a region where solar radiation creates ozone, which in turn shares its energy by heating the local atmosphere. This warm layer resting on top of the cool upper troposphere creates a stable atmosphere marked by increasing temperatures.

- In the mesosphere, between 50 km–80 km, the constituents of the atmosphere are relatively poor absorbers of solar radiation. This region experiences rapid cooling with altitude and is dynamically unstable. In the 1990s, scientists learned that forms of electricity related to lightning were jetting through the mesosphere to the edge of space. The stratosphere and mesosphere are now being linked to solar cycle variations in UV radiation. Stratospheric circulation changes may influence long-term tropospheric weather. Scientists do not yet know the strength of these links to climate variation.

Above 80 km, electric and magnetic forces become more influential as a small fraction of the gas becomes ionized. Beyond 1000 km, magnetic forces, rotation, and pressure gradients drive the dynamics of space weather in the highly ionized regions of near-space. Extreme ultraviolet (EUV) and X-ray radiation heat these regions.

- In the thermosphere, between 80 km–1000 km, short-wave radiation from the Sun interacts with the tenuous constituents and heats individual atoms and molecules, resulting in a hot atmosphere with increasing-to-steady temperatures that may exceed 1000 K. The lower thermosphere is home to

thin layers of ablated meteoric materials. Neutral grains are part of the thermosphere. When ionized by sunlight or particle impact, the grains form a portion of the ionosphere.
- The ionosphere is imbedded within the thermosphere generally between 80 km–1000 km. The short-wave radiation from the Sun removes electrons from parent atoms, creating distinct layers of ionization. In these layers electric and magnetic fields control ion motion.
- The exosphere is the extension of the atmosphere above 300 km in which individual particles have so much energy and engage in so few collisions that they stand a good chance of escaping Earth's gravitational hold. This region contributes particles to the plasma-dominated regions of near-space.

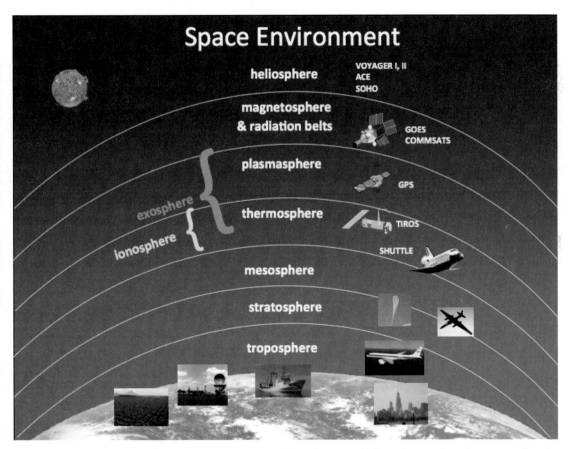

Fig. 1-4. Atmospheric Layers. The relative locations of Earth's atmospheric and magnetic regimes, as well as the heliosphere. Image is not to scale. (ACE is Advanced Composition Explorer; SOHO is Solar and Heliospheric Observatory; GOES is Geostationary Operational Environmental Satellite; COMMSATS is communication satellites; GPS is Global Positioning System; TIROS is Television Infrared Observation Satellite.) *(Modified from a graphic provided by the AF Weather Agency)*

The following "spheres" are absent from planets lacking a magnetic field. When present, they tend to be distorted in shape. These regions are the most influenced by space weather events.

- The plasmasphere is that part of the exosphere from 1 R_E– ~5 R_E dominated by Earth's magnetic field and Earth's rotational dynamics
- The magnetosphere, including the inner radiation belts, is the near-space region dominated by Earth's magnetic field. Electric currents allow distant regions of the magnetosphere to communicate with each other. The magnetosphere is more influenced by the heliosphere than by Earth's rotation. Its shape is very asymmetric, extending from roughly 10 R_E on the Sunward side to 100s of R_E on the nightside (Sec. 1.2.3). Beyond the magnetosphere the Sun's magnetic field and plasma dominate.

In summary, the lower atmosphere (0 km–80 km) is dominated by collisions and the gas dynamics of neutral constituents. In the upper atmosphere plasma dynamics becomes more important. Beyond the exosphere, magnetic fields, electric fields, and plasma dynamics are in control.

EXAMPLE 1.1

Gravitational Example:

- *Problem Statement:* Determine the gravitational acceleration at one solar radius (i.e., at the solar surface). Compare this value to gravitational acceleration at Earth's surface.
- *Relevant Equation or Concept:* Newton's Law of Universal Gravitation.
- *Given:* The Sun's mass is 2×10^{30} kg. The solar radius is 7×10^8 m.
- *Solution:* The gravitational force is

$$F_g = ma = -mg_S = -m\frac{GM_S}{R_S^2}$$

The minus sign tells us the acceleration is directed inward (opposite the direction of $+r$).

- Solving for the magnitude of g_s yields:

$$g_s = \frac{(6.67 \times 10^{-11} \text{N} \cdot \text{m}^2/\text{kg}^2)(2 \times 10^{30} \text{ kg})}{(7 \times 10^8 \text{m})^2} = 272 \text{ m/s}^2$$

- *Interpretation:* g_s is approximately 27 times that of Earth's gravitational acceleration (9.8 m/s^2).

Follow-on Exercise: Determine the acceleration from gravity 1000 km above Earth's surface. Provide a fractional comparison to Earth's surface value.

At what distance from the Sun's center is the Sun's gravitational acceleration equal to the gravitational acceleration at Earth's surface?

1.2 Where and Why Does Space Weather Occur?

This section briefly describes the primary agents and phenomena in space weather and the space environment system. We start at the galactic level and work toward Earth's surface. We explain each of the space weather elements more thoroughly later in the text. Here we simply develop a vocabulary that enables us to describe space weather effects in the intervening chapters.

1.2.1 Other Stars Affect the Near-space Environment

Objectives: After reading this section, you should be able to...

- Recognize that stars other than the Sun create space environmental disturbances
- Contrast the solar cycle intervals in which galactic cosmic rays and solar energetic particles are most likely to influence the space environment

Stars other than our Sun are so distant that they seem unlikely to influence Earth's near-space environment. However, violent supernova explosions (stellar death) in our own galaxy and in distant galaxies accelerate material to near light speed. During such events, vast amounts of electrical energy are suddenly freed. In the tumult, electrons are stripped from their parent atoms. The remaining heavy nuclei travel for years through interstellar space. These nucleonic remnants, known as *galactic cosmic rays (GCRs)*, arrive at Earth from all directions with extreme energies.

Cosmic ray deposition at Earth is strongest during solar minimum when the Sun's extended magnetic field provides the least shielding. Because of their high energy, they penetrate Earth's magnetic field, large amounts of spacecraft shielding, and in some cases Earth's atmosphere. On their way they affect satellite solar panels, clouds, aircraft avionics, and even manufacturing processes on the ground. They are particularly dangerous to humans and hardware in space. Numerous spacecraft malfunctions have been associated with these ever-present particles.

On rare occasions, upheavals on other stars, resulting in violent gamma and X-ray bursts, have disturbed Earth's upper atmosphere to the point that global radio-wave propagation was impacted. Scientists named the sources of these impulsive stellar bursts, *magnetars*.

In March 1979, widely separated Russian and US spacecraft detectors were disrupted by an energy front moving through the solar system. Radiation counts rose from 100 per second to 200,000 per second. Nearly 20 years later, a satisfactory explanation was finally formulated: the spacecraft had been hit by bursts of energy from a neutron star from the Large Magellanic Cloud, a sister galaxy to our own Milky Way [Kouveliotou et al., 2003]. At Earth's surface, humans remained blissfully unaware. Our atmosphere had protected us. Similarly, our atmosphere also protects us from particle outbursts from our own star, the Sun, that tend to occur during the active phase of the solar magnetic cycle.

1.2.2 The Sun Affects Space Weather

> **Objectives: After reading this section, you should be able to...**
>
> ♦ Describe energy releases from the Sun that create space weather
> ♦ Appreciate the distance and time scales involved in space weather
> ♦ Explain the terms: heliosphere, solar wind, and bowshock

The Nature of the Sun and Solar Dynamics

The Sun is a nuclear furnace that, each second, converts millions of tons of hydrogen to helium in the solar core. In the process, a tiny fraction of the mass belonging to the hydrogen atoms transforms into pure energy. The reservoir of energy creates temperature gradients that force an outward-directed flow of energy. If the energy flowed out of the Sun without further interaction with matter, we could end this book here. However, the high-energy photons created with each nuclear reaction are captured by matter, emitted, and recaptured and reemitted billions of times on their path out of the Sun. With each capture, they give a bit of energy to their surroundings. By the time the photons reach the solar surface, most of them have degraded to the lower-energy photons that we sense as yellow visible light. The visible light is only a small part of the total energy radiated from the Sun.

Unlike Earth, the Sun does not rotate as a solid body. Rather, it rotates fastest at the equator (~25 Earth-days per rotation) and slowest toward the poles (over 32 Earth-days per rotation). This process is *differential rotation* and is one of the fundamental drivers of space weather. Further, the outer layers of the Sun's fluid interior convect (boil) in a way similar to hot water in a pot. The solar plasma motions twist and wind the Sun's magnetic field into "islands" known as sunspots, whose field strength is hundreds or even thousands of times stronger than Earth's magnetic field. Thus, sunspots are compact storage zones for the Sun's magnetic energy. They usually appear in groups and are part of a large magnetic structure known as an *active region* that often extends through the entire solar atmosphere. The differential twisting of the solar plasma causes relative motion of one sunspot with respect to a neighboring sunspot and sets the stage for explosive releases of energy.

Observations of sunspots over many years revealed that the number of sunspots changes periodically. In some years, as many as 250 sunspots are observed per day, while in other years, such as 2009, long stretches of days go by with zero sunspots. This phenomenon is the *sunspot cycle* and we illustrate it in Fig. 1-5. The plot shows that the number of sunspots varies with an average period of about 11.4 years. Some cycles last only eight years, while others are as long as 15 years. A period of no or few observed sunspots is a *sunspot minimum*. Conversely, a period with the most sunspots is a *sunspot maximum*. The minima and maxima are separated by rising phases and declining phases. Beginning with the minimum that occurred around 1755, sunspot cycles have been numbered. The cycle with its maximum in the year 2000 was cycle number 23.

The Sun releases energy in a variety of ways. The most obvious way is its visible radiant energy. Because of the huge number of interactions between energy and solar matter, we see a relatively constant background of electromagnetic radiation released across a broad range of wavelengths called the *solar spectrum*. At 1 AU the Sun provides 1366 watts of power per square meter. This power ultimately energizes our atmosphere.

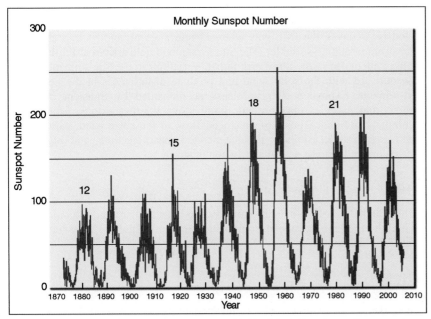

Fig. 1-5. Sunspot Values for Solar Cycles 12–23. Selected cycle numbers are shown above several spikes. *(Data provided by the National Geophysical Data Center.)*

Pause for Inquiry 1-2

Consider Fig. 1-5. Spacecraft have been in and above the atmosphere for about how many solar cycles? *Hint: You may need to look up the launch date for the Sputnik satellite.*

Pause for Inquiry 1-3

In which phase of the solar cycle were you born? Rise, Maximum, Decline, or Minimum?

Space-based observations also reveal a quiet background outflow of tenuous and highly ionized plasma (100,000 K) called the *solar wind*. On average, the solar wind blows past Earth at a tremendous speed of ~400 km/s (about 1,000,000 mph in Earth's vicinity), which is about ten times faster than sound speed in the rarified plasma. This directed (ram) flow of the solar wind is a form of kinetic energy. The random, omni-directional motion of the 100,000 K plasma particles is thermal-kinetic energy. Physicists associate this phenomenon with temperature, which we describe in the next chapter. The solar wind consists of particles, mostly ionized atoms of hydrogen and helium and traces of oxygen, carbon, iron, and other elements. In Earth's vicinity (1 *astronomical unit (AU)* from the Sun), each cubic meter of space typically contains 5×10^6 positive ions and a roughly equal number of free electrons. This particle content is very low compared to the 10^{25} molecules in each cubic meter of air at sea level. The radially out-flowing solar wind carries and stretches the Sun's magnetic field into space. In its stretched form, the solar field is the *interplanetary magnetic field*

(IMF). The stretching may be so severe that it is impossible to trace both ends of the field lines to their roots in the Sun. As the Sun rotates (the mean rotation period is 27 days), the IMF gets wrapped into a spiral, similar to a stream of water from a rotating sprinkler. The Sun's magnetic field arrives at Earth with an angle of about 45° to the Sun-Earth radial. Variations in the Sun's magnetic field flow outward with the solar wind and produce disturbances in the near-Earth environment. Explosive energy release accompanied by transient bursts of photons and particles may produce large space weather storms. Storms also accompany prolonged gusts of enhanced flow in the solar wind, called *high-speed streams*. Occasionally the magnetized mass ejections and high-speed streams merge to create a particularly potent storm.

> **Pause for Inquiry 1-4**
>
> The 100,000 K solar wind constantly sweeps past a solar wind monitoring satellite called the Advanced Composition Explorer Satellite. Why doesn't this satellite melt?

> **Pause for Inquiry 1-5**
>
> Would an astronaut doing extra vehicular activity get blown away by the 400 km/s solar wind?

Most of the impulsive disturbances occur when energy stored in solar magnetic fields converts to other forms. Three basic types of impulsive emissions exist:

- *Solar flares*—intense bursts of radiative energy across the entire electromagnetic spectrum, with the largest burst enhancements in the X-ray, extreme ultraviolet (EUV), and radio portions of the spectrum.
- *Solar energetic particles (SEPs)*—protons ejected with relativistic speeds near a flare site, or particles accelerated by a shock from the explosion site pushing into the solar wind. Some energetic particles reach Earth in as little as 20 minutes after a solar eruption.
- *Coronal mass ejections (CMEs)*—huge parcels of the Sun's atmosphere accelerated (ejected) into interplanetary space (Figs. 1-1, 1-6, 1-18b, and 1-19b). The parcels carry threads of the Sun's magnetic field into space as well. Much of the ejected plasma is from the Sun's upper atmosphere (corona).

We describe these emissions in future chapters and provide short case histories of two particularly notable events later in this chapter. Table 1-2 shows an impact grid for the important forms of the Sun's emissions. Even "quiet Sun" emissions produce geomagnetic storms.

The Solar Environment

Because the Sun is so close, we are able to study its behavior in detail. We are also benefactors (or victims) of its output. In our galaxy, the Sun is a middle-aged, yellow star, located around 150 million kilometers [1 AU] away from Earth—about 8.3 light-minutes away. We rarely think of the Sun as an entity with an extended atmosphere, but it has one, and it is large indeed, extending well beyond the planetary orbits to a distance of 120 AU–160 AU (Table 1-1).

1.2 Where and Why Does Space Weather Occur?

Table 1-2. Simplified Classifications of Space Weather Storm Effects. Here we list various forms of the Sun's emissions and their characteristics.

Quiet Sun Emissions	Time to Arrive at Earth	Storm Impact
Photons from ~5800 K surface	8 min	Normal conditions
Minimal solar energetic particles	several hours	Normal conditions
Solar wind plasma	100 hr	Normal conditions
• With strong magnetic field	60 hr–100 hr	Geomagnetic storm
• With high speed	30 hr	Geomagnetic storm
Disturbed Solar Emissions	**Time to Arrive at Earth**	**Storm Impact**
Solar flare photons (X-ray-radio)	8 min	Radio blackout
Burst of solar energetic particles	15 min–several hr	Radiation storm
Coronal mass ejection	20 hr–120 hr	Geomagnetic storm

EXAMPLE 1.2

Mass Example

- *Problem Statement:* How massive is a CME? Observations suggest the amount of material blown out of the Sun's atmosphere during a CME ranges from 10^{12}–10^{14} kg. Let's compare this to the mass of an ancient monolithic outcropping of rock in the Australian Outback—Mt. Uluru, formerly known as Ayer's Rock.

- *Relevant Equations or Concepts:* Definition of density and volume of a cylinder

- *Given:* The monolith is made of akrose sandstone, whose density (ρ) is ~2700 kg/m³. Assume the total depth of the structure is 1 km and the radius is ~1275 m.

- *Assume:* Uniform density.

  Mass = ρV

- *Solution:* We model the structure as a cylinder of sandstone. The volume of a cylinder is

$$V = \pi r^2 h$$

where
r = radius of the base of the cylinder [m]
h = height [m]

The mass contained within Mt. Uluru is the density of the rock multiplied by the volume of the mountain.

$$M = (2700 \text{ kg/m}^3) \pi (1275 \text{ m})^2 \, 1000 \text{ m} = 1.38 \times 10^{13} \text{ kg}$$

- *Interpretation and Implications:* This value is in the range of mass estimates for a CME. During solar maximum, the Sun sheds about three CMEs a day, in addition to its normal mass loss to the solar wind of about 8.6×10^{13} kg/day and the mass lost through nuclear fusion.

Fig. 1-6. **Mt. Uluru in the Australian Outback.** The mass in the rock monolith is comparable to that of a coronal mass ejection. *(Courtesy of the Commonwealth of Australia, Geoscience Australia website, Significant Rock Features, 2009)*

Follow-on Exercise 1: The largest modern aircraft carriers have a mass of about 8.5×10^7 kg. How many carrier mass-equivalents are carried into the interplanetary medium by a typical CME?

Follow-on Exercise 2: If the Sun ejects one CME a day, how long does it take to eject the mass equivalent of Earth's atmosphere?

Chapter 1 Space Is a Place...with Weather

Focus Box 1.2: The Size of Things in the Solar System

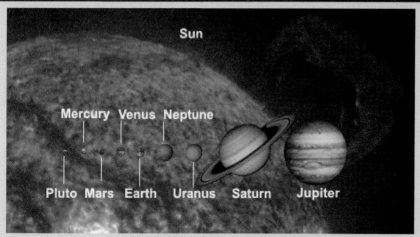

Fig. 1-7. **Solar System.** Here we show the relative sizes of the planets in the solar system. *(Courtesy of NASA)*

Most space textbooks provide a drawing of the Sun and its planetary system that is not to scale. To give a sense of the scaling of our solar system, we imagine the Sun to be the size (diameter) of a 0.30 m (12 in) pie plate, which is also the approximate size of an adult male foot. If we were to put the plate down and march about 110 feet beyond it we would reach Earth's orbit at one AU. (We recommend trying this in a long building with 12-in floor tiles.) In diameter, Earth is less than 1/100th the size of the Sun, about the size of a small pencil eraser in this scheme. To give more perspective on the relative sizes of these bodies, we note that the entire Earth-Moon system would fit inside the radius of the Sun (Fig. 1-7). We know the Sun is large and distant, but why does it matter? Because of the tremendous scale involved, forecasting which solar events will be geo-effective (Earth-disturbing) is very challenging.

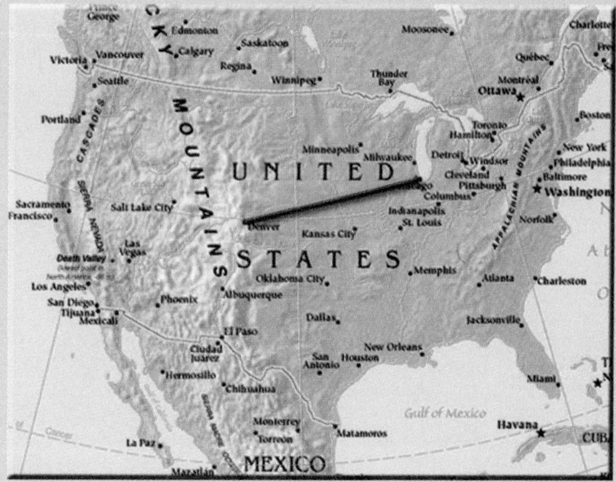

To put the entire heliosphere on a reasonable scale we imagine the Sun to be the size of a sports stadium. The distance separating the Sun and the near edge of the Sun's influence (the heliopause) would be approximately the distance between Denver and Chicago (Fig. 1-8).

For a perspective on our entire solar system, we consider the artist's rendering of our heliosphere in Fig. 1-9. In that image the small yellow dot encloses the entire planetary system of the Sun.

Fig. 1-8. **Distance between the Sun and the Near Heliopause If the Sun Is the Size of a Sports Stadium.** The red line represents the distance from the Sun to the near edge of the heliopause (153 AU).

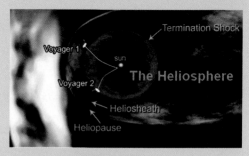

Fig. 1-9. **The Heliosphere beyond the Near-space Environment.** Here we show the contents of the heliosphere, including the solar system (yellow dot), the termination shock (~100 AU), the heliosheath (defined below), the heliopause, and the bow shock that hits the heliosphere from interstellar space. *(Courtesy of NASA)*

Heliosphere: The Extended Solar Atmosphere

The *heliosphere* is that region of space dominated by the Sun's extended atmosphere. The heliosphere is essentially a bubble or cavity in interstellar space produced by the solar wind. Figure 1-10 shows such a bubble forming around a distant star. The bubble forms because the Sun's extended atmosphere is permeated by a magnetic field that is roughly in the form of a dipole. When a magnetic dipole is immersed in flowing plasma from an outside source, in this case the interstellar medium, then it deforms into a tear-drop shape as in Figs. 1-10 and 1-11. The outer surface where the heliosphere meets the interstellar medium is the *heliopause*. Here the term "pause" indicates a break or discontinuity. The location of the discontinuity in the solar magnetic field depends on the phase of the solar cycle. When the Sun is more active, the boundary is further from the Sun because of the higher pressure generated by the solar magnetic field. Although a few cosmic rays and electrically neutral atoms from interstellar space penetrate this region, virtually all of the material in the heliosphere emanates from the Sun. The solar wind plasma creates structure within the heliosphere based on relative speed. At some distance from the Sun (well beyond the orbits of the major and minor planets), the solar wind slows to meet the gases in the interstellar medium. At such a location, the solar wind speed becomes subsonic as it goes through a *termination shock*. Beyond the termination shock, the solar field and plasma form the exterior heliosphere (*heliosheath*). The heliosheath and termination shock are turbulent regions that scatter some charged particles and energize others. Thus, these regions are of interest for systems that are vulnerable to cosmic rays. The Voyager 1 spacecraft was in the vicinity of the termination shock in late 2004 and by the end of 2009 was traveling through the heliosheath.

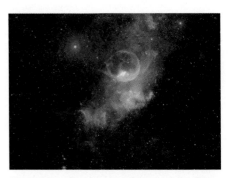

Fig. 1-10. Our Sun Is a Star; and the Stars Are Suns. Here we show a Hubble Telescope view of the Bubble Nebula, a distant stellar system with a stellar-sphere and an associated stellar wind in excess of 2000 km/s. *(Courtesy of the National Optical Astronomy Observatory)*

Pause for Inquiry 1-6

Using the 12-inch scale described in Focus Box 1.2, how many feet separate the Sun and Uranus? How many feet separate the Sun and the near heliopause?

Pause for Inquiry 1-7

How long does it take a typical solar wind parcel to get to Neptune?

Pause for Inquiry 1-8

What is the speed of a high-speed stream that reaches Earth in 30 hours?

Pause for Inquiry 1-9

At appropriate increments on the scale below, place the locations of Saturn, Uranus, the heliopause and Alpha Centauri. The letter S indicates the Sun.

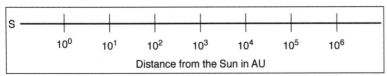

Inner Heliosphere. In the inner heliosphere the solar wind is supersonic. All planets behave as obstacles to the supersonic solar wind. A perturbation wave (*bow shock*), forms upstream of each planet as a means of slowing and diverting the solar wind around the planet. Smaller solar system bodies such as moons and asteroids generally don't have bowshocks. For a bowshock to form, the body must have a magnetosphere, an ionosphere, an atmosphere, or some combination thereof. Most moons and asteroids don't qualify because they are too small to create or retain such regions.

Earth's bowshock forms ~15 R_E upstream from Earth. The solar wind rapidly decelerates and turns turbulent as it passes through the bowshock to come in contact with Earth's magnetic domain. For many years, scientists had a goal of positioning a spacecraft upwind of the bowshock to monitor the incoming solar wind for analysis and forecasting purposes. In fulfillment of that goal, the Advanced Composition Explorer (ACE) spacecraft has been Sunward of the shock, about 230 R_E upstream from Earth, since 1997.

> **Pause for Inquiry 1-10**
>
> About how long does a photon from the Sun take to get to the edge of the heliosphere nearest Earth?

1.2.3 Earth's Magnetosphere Results from and Contributes to Space Weather

> **Objectives: After reading this section, you should be able to...**
> - Explain the term "magnetosphere"
> - Describe the shape of the magnetosphere
> - Distinguish between inner and outer magnetosphere regions
> - Appreciate the similarities between the heliosphere and the magnetosphere

Regions of the Magnetosphere

In this section we describe the similarity between Earth's magnetosphere and the heliosphere. This similarity is not an accident. Magnetic fields immersed in moving, conducting fluids exhibit self-similar behavior over many scales. Earth's magnetic field would resemble a magnetic dipole at large distances from Earth, if the solar wind didn't distort it into a bullet-shaped magnetosphere (Fig. 1-11). Field lines distant from Earth become increasingly stretched into a *magnetotail* that extends well beyond the lunar orbit (60.3 R_E). The distortion occurs because of field-line, current, and plasma interactions. As we discuss in Chaps. 7 and 11 an adequate description of the magnetosphere must include these interactions.

Outer Magnetosphere. Now we bring our tour closer to home. Just Earthward of the bowshock is a zone of disordered solar wind created by the turbulent braking of solar wind flow (Fig. 1-12). In this region, called the *magnetosheath*, flow energy converts into turbulent eddies and ultimately into random thermal motion of the solar wind constituents. Closer to Earth, the boundary separating the frothy and weakly magnetized solar wind from Earth's magnetic domain is the

1.2 Where and Why Does Space Weather Occur?

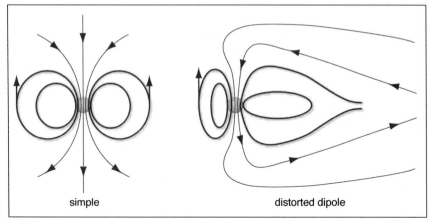

Fig. 1-11. The Distortion of Earth's Magnetic Field by the Solar Wind. On the left is a simple dipole magnetic field configuration. On the right is Earth's magnetic field as distorted by an external flow. Earth's north magnetic pole is currently near the south geographic pole and vice versa.

magnetopause, which is where the magnetosphere meets the interplanetary medium. The boundary is a fluid and magnetic membrane that flexes in the dynamic solar wind. Generally the nose of the magnetopause is located at ~10 R_E. Inside the magnetopause is that region of the space environment dominated by Earth's magnetic field; the *magnetosphere*. The local magnetospheric structure mirrors the structure of the heliosphere-heliopause system. This similarity is also no accident. Systems with differing plasma and magnetic characteristics attempt to maintain their own identity. They do so by forming boundaries made of flowing charged particles (currents). If one system of magnetized plasma flows against another, the boundary often forms a bullet or tear-drop shape.

Magnetospheric plasma originates in Earth's upper atmosphere and in the solar wind. Although much of the magnetosphere is a near-vacuum by Earth standards, its high plasma density, compared to interplanetary space, allows energy in the solar wind to drive electric currents and set plasma in motion within the magnetosphere and Earth's upper atmosphere. We know two important effects of this behavior: 1) solar wind disturbances energize plasma even within Earth's protective magnetic shield, and 2) the solar wind induces magnetospheric flows that are disconnected from Earth's rotation.

Magnetotail. Contained within the volume of the magnetotail are regions of relatively low and high plasma densities. The *magnetotail lobes*, one in each hemisphere, have relatively low-density, cool plasma (Fig. 1-12). The *plasmasheet* within the central magnetosphere has a denser and more energetic plasma population. During periods of geomagnetic storms, this plasma invades the orbit of geosynchronous satellites (6.6 R_E) and bathes the satellites in a soup of relatively energetic ions and electrons. This plasma bath leads to potentially lethal space weather impacts for instruments on these satellites.

Inner Magnetosphere. Close to Earth (within 5 R_E), Earth's magnetic field resembles a dipole associated with a giant bar magnet. The relatively strong dipole field lines near Earth act as a trap for energetic plasma. In 1958 an instrument onboard *Explorer 1*, built by James Van Allen, measured unexpectedly high values of ionizing radiation coming from a region within the dipolar magnetic field lines. Today we call this region of near-space the *radiation belts* or sometimes the *Van Allen belts*. The belts are doughnut-shaped and centered on

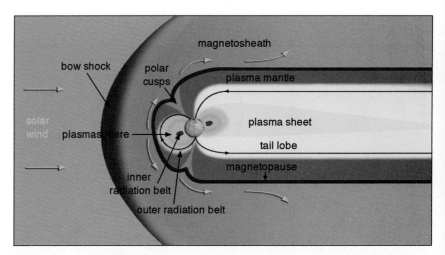

Fig. 1-12. Earth's Magnetosphere. Here we show a cross section of the magnetosphere viewed from a dusk location (above the terminator [day-night line] on Earth's surface, moving toward darkness). The bowshock typically forms about 15 R_E upstream of Earth and has a curved shape. Shocked solar wind plasma is shown in brown. The magnetopause is shown in rust. Just inside the magnetopause is the plasma mantle (lavender). This region is a mix of solar wind and Earth plasma. Plasmas that have access to the most distant regions of the magnetosphere are shown in light blue. The plasmasheet bridges the outer and inner magnetosphere. It is an inner magnetosphere feature at high latitudes. The radiation belts (red and aqua blue) overlap the plasmasphere. The lower-latitude and lower-energy region that bridges the inner magnetosphere and upper atmosphere is the plasmasphere (green). *(Adapted from Patricia Reiff at Rice University)*

Earth's magnetic equator. Spacecraft that operate in or near the radiation belts must carry extra shielding to prevent damage to instruments and components from energetic particles trapped in the belts.

Occupying much of the same region of space as the radiation belts is a region of higher density but significantly lower temperature plasma, called the *plasmasphere*. This plasma consists of hydrogen ions (protons) and electrons from Earth's upper atmosphere. The *plasmapause* (sharp outer edge of the plasmasphere) is usually 4 R_E–6 R_E from Earth's center (19,000 km–32,000 km above the surface). Inside the plasmapause, the plasma rotates with Earth. The inner edge of the plasmasphere, which intermixes with Earth's upper atmosphere, is the altitude at which protons replace oxygen as the dominant species in the atmospheric plasma. This edge usually occurs at about 1000 km altitude. The plasmasphere serves as a vital link between Earth's magnetic domain and Earth's atmosphere.

Pause for Inquiry 1-11

Could we use the Moon as a base for full-time magnetospheric monitoring? To answer the question indicate the locations of the Moon in the matrix below. (See Table 1-1 for distances.)

Hint: Consider the location of the Moon with respect to the Sun during the various lunar phases.

	Inside or Outside of the Magnetosphere?	Sunward, Anti-Sunward, or at Earth's Dawn-Dusk Line?
Full Moon		
New Moon		
Half Moon		

Pause for Inquiry 1-12

Return to Fig. 1-4 and put the distances to the regime boundaries in Earth radii on the left-hand side of the diagram. For the heliosphere use a distance scale in AU.

1.2.4 Space Weather Dominates Earth's Upper Atmosphere

Objectives: After reading this section, you should be able to...

- Explain classification schemes for layers of the atmosphere
- Understand the relationship between the thermosphere, ionosphere, and plasmasphere
- List the mechanisms for creating the ionosphere

The Edge of Space

Earth's upper atmosphere occupies a strategic position—the edge of space. For our purposes, it is a portion of the satellite (and rocket) operational arena, a source of matter for the regions of space outside the sensible atmosphere, and a protective blanket shielding the regions below from many energetic particles and photons. Many astronautical engineers designate the edge of space at about 130 km, because that is the lowest altitude at which a spacecraft can successfully make one full orbit. For our purposes, we designate the region at or above 50 km as near-space, because interesting space-physics-related effects occur at this level and above. This shell of molecules, atoms, and ions, extending down to 50 km, interacts with (1) the Sun by photons, (2) the magnetosphere through plasma and electromagnetic interactions, and (3) the atmosphere below via waves and gravity.

Driven by energy and momentum from above and below, the upper atmosphere's behavior is rather complex. It is best organized in four categories: mixing, temperature, retention, and degree of ionization. Each category in Fig. 1-13 contains the word "sphere." In this instance, the term is at least geometrically appropriate. Close to Earth, gravity becomes a dominating force, and much of the material in the upper atmosphere is stratified and constrained to a spherical volume about Earth.

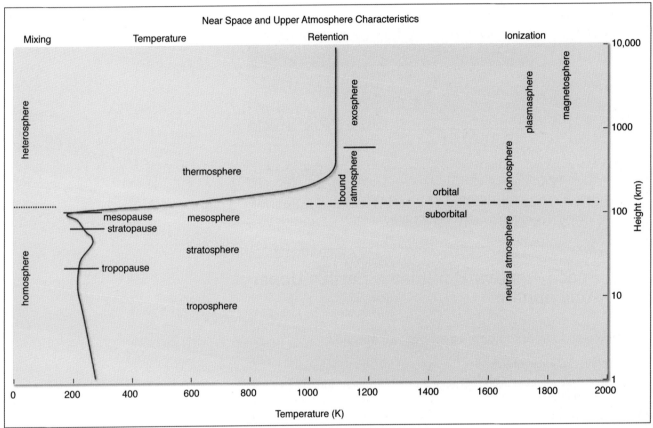

Fig. 1-13.   **Classifications for Earth's Atmosphere.** Here we show the layers of Earth's atmosphere and their characteristics. The central curve shows temperature variation with height. We note that the vertical scale is logarithmic. We define "homosphere" and "heterosphere" below on this page. [After Hargreaves, 1992]

Atmospheric Classification

Mixing. When collisions are numerous, the atmosphere is well mixed. Strong mixing leads to a homogenous ratio of atmospheric constituents. Between Earth's surface and 100 km, the chemical mix of atmospheric constituents is roughly 78% N_2, 21% O_2, and 1% other. The region below 100 km is the *homosphere*. Above 100 km however, the lower density of particles leads to fewer collisions. Therefore, atoms and molecules tend to stratify, with more massive particles remaining at relatively low altitudes, and the less massive ones diffusing to higher levels. The stratified region is the *heterosphere* (as in heterogeneous). A significant consequence of this stratification is the high relative abundance of atomic oxygen above 300 km where many low-Earth orbiting satellites orbit. Atomic oxygen is exceedingly reactive. It attacks spacecraft components and reduces mission lifetime (Chap. 3).

1.2 Where and Why Does Space Weather Occur?

Temperature. Figure 1-13 shows the full temperature range of Earth's thermal envelope. We provide a general discussion of the atmospheric temperature structure in Sec. 1.1.2. Between 100 and 400 km chemical species with an affinity for absorbing solar radiation are present. Though the density of these species is low, solar photons are abundant, meaning individual particles in this region, called the *thermosphere*, share in a wealth of solar energy. With high energy per particle available, temperatures rise rapidly. Exospheric temperatures exceed 1000 K, on average, over the solar cycle. As solar emissions vary with the solar cycle, so do thermospheric temperatures and densities. When thermospheric densities vary, low-Earth orbits are perturbed by variations in atmospheric drag. Sudden temperature changes created by geomagnetic storms deposit energy in the polar regions. The energy is partially redistributed in plumes of outflowing plasma and traveling density waves. These waves propagate to the equator and beyond, producing ripples in the plasma and irregularities in radio wave propagation.

Retention. Earth's gravitational force is reduced (proportional to $1/r^2$) in the most-distant atmospheric region, the *exosphere*. Plasma particles heated by various solar and magnetospheric processes fight against gravity. Typically only the lighter atmospheric elements occupy this upper domain. Some of these particles may gain sufficient energy to achieve escape velocity from the atmosphere. Their ballistic trajectories allow them to exit to the plasmasphere and magnetosphere. Below about 500 km, frequent collisions and gravity thwart attempts at escape, and particles are retained unless energized by extreme space weather storm effects. The lower boundary of this region, known as the *exobase*, is usually near 500 km but may be lower during higher solar activity. The upper reaches of the exosphere blend into the plasmasphere at 10,000 km.

Ionization. Within the thermosphere is perhaps the most dynamic of the upper atmosphere regions: the ionosphere. Shortwave solar electromagnetic radiation heats and excites atoms and molecules in Earth's upper atmosphere. It also rips molecules apart and tears electrons from some fraction of their parent particles. The free electrons and positive ions then form several weakly ionized layers of plasma. These ionized layers constitute the *ionosphere*. Because of the *diurnal* (day-night) nature of incident solar radiation, the ionization is more extensive during the day than during the night.

Solar photons are not the only culprit in ionization. Energetic particles arriving from outer space as cosmic rays and from the nightside magnetosphere also have a role in creating the high-latitude ionosphere. Ionization and its distribution largely control the propagation of radio waves. Communication and navigation signals may be severely degraded during ionospheric storms.

Pause for Inquiry 1-13

The mesosphere is shown to be in the suborbital regime, so satellites cannot stay in orbit there. It does not contain sufficient mass to support aircraft flight. List some other ways we might investigate the mesosphere.

Focus Box 1.3: The Ionosphere

The ionosphere serves as a high-altitude reflector for short-wave broadcasting and long-range communication. During a solar flare, the enhanced X-ray radiation from the Sun causes the electron density in the lowest layer of the ionosphere to increase considerably. This increase results in a *radio blackout*—the immediate and large-scale absorption (rather than reflection) of high-frequency radio waves and subsequent disruption of short-wave communication over Earth's sunlit hemisphere. In addition, beginning tens of hours after a solar flare, the global electron density in the upper levels of the ionosphere may undergo substantial variations because of the interaction of the disturbed solar wind with the near-space environment. These variations last for several days after disturbance onset. At high latitudes, a significant fraction of the ionization is produced by charged particles that have been dumped from the magnetosphere during times of magnetic storms and also by charged particles from the Sun that have been diverted to these latitudes by the geomagnetic field.

Even during quiescent times, instabilites arise in the ionospheric plasma. These instabilities cause the initially (relatively) homogeneous plasma to develop density turbulence, typically magnetic field-aligned, with scale sizes on the order of centimeters to hundreds of kilometers. Electromagnetic waves propagating through this turbulence are scattered, resulting in *scintillation* (twinkling) of signal sources, analogous to the twinkling of starlight by density turbulences in the atmosphere. These scintillations lead to strong disturbances in the radio band up to GHz frequencies, thereby disrupting communications and some navigation systems, such as the Global Positioning System signals. In addition, the scattering often blinds radar tracking (e.g., over-the-horizon radars) and disrupts—or sometimes improves—communications. Currently scientists are trying to find ways to use these scintillation and scattering effects as a diagnostic tool for space weather.

1.2.5 Meteors and Space Dust Affect the Space Environment

Objective: After reading this section, you should be able to...

♦ Describe the impacts of meteors on the space environment

As Earth orbits the Sun, it constantly intercepts the remains of old cometary tails and space dust. Earth intercepts about 10^7 to 10^9 kg of meteoric material each year. This material ablates as it falls into the atmosphere and creates small ionization trails in its wake. The elongated paraboloid of ionized air is many kilometers long. We call it a *meteor trail* (meteor echo). Occurring at an atmospheric height of about 85 km–105 km, this ionized trail of meteor debris is capable of reflecting radio waves from transmitters on Earth. Meteor trail reflections are brief, however. As the trail rapidly diffuses into the surrounding air, it quickly loses its ability to reflect radio waves, causing most reflections to last less than one second. Occasionally, a large meteor may create a trail capable of reflecting radio waves for several minutes. During times of enhanced meteor activity, some forms of communication may improve. On the other hand, radar beams may experience anomalous reflection known as radar clutter. In the worst cases, some radar systems are rendered temporarily ineffective. Meteoroids and meteors also represent a collisional hazard to spacecraft. Even tiny fragments endanger spacecraft components because they strike at such high speeds.

1.2.6 Space Weather Effects in Earth's Lower Atmosphere and at Earth's Surface

Objectives: After reading this section, you should be able to...

♦ Explain the space weather significance of solar energetic particles
♦ Know that solar energetic particles penetrate through several layers of atmosphere
♦ Recognize that some space weather effects may be observed at the ground

Galactic cosmic rays and their Sun-generated cousins, *solar energetic particles (SEPs)*, penetrate Earth's atmosphere. In the stratosphere and upper troposphere, these particles create radiation exposure to avionics, flight crews, and passengers on high-latitude flights. The European Council has recommended that aircrews be treated as personnel receiving occupational radiation exposure. As such, European crewmembers' annual radiation doses must be monitored [Jansen et al., 2000]. Energetic particles (cosmic rays) and their by-products may influence cloud formation by changing the ionization state of cloud condensation nuclei. Solar cycle variations of cloud cover are an active area of research [Tinsley, 2000]. Energetic particles have been identified as a source of soft errors in stored data on computer systems. Computer systems that operate in high, mountainous regions or where the magnetic field is somewhat weak are slightly more susceptible to energetic particle damage.

Extreme SEP events, associated with very energetic solar flares, are sensed in the troposphere and at the ground. The most energetic of these particles create chemical compounds in the atmosphere that subsequently drift to the ground. Snow in the polar cap ice sheets collects the particles and becomes a recording device for the great solar energetic particle events associated with the largest solar flares. Approximately 125 ground level SEP events have been identified from core samples in the polar ice for the interval 1561–1950 [McCracken et al., 2001].

Electrons energized in Earth's magnetosphere also penetrate into the stratosphere. In the upper regions of the stratosphere, energetic electrons alter ozone behavior and create short-term reductions in ozone density. Scientists are attempting to determine if a solar cycle variation of upper-level ozone exists. Researchers are certain a link exists between stratospheric winds and the solar cycle [Labitzke and van Loon, 1987]. Variations in solar ultraviolet radiation on solar cycle time scales appear to alter the amount of ozone and the distribution of ozone heating between 30 and 50 km in Earth's atmosphere. In turn, this heating affects the stratospheric circulation. Scientists are actively investigating the link between altered stratospheric circulation and the location of the Northern Hemisphere winter storm tracks.

Earth's inhabitants are rather comfortably shielded from almost all direct shortwave solar radiation and solar wind effects. However, space weather disturbances of the magnetic field and plasma in the magnetosphere and upper atmosphere propagate to the surface and even to the ocean floor. As we discuss shortly, surface magnetic field disturbances provided one of the first historical clues about what we now call space weather. Space weather effects at the ground include magnetically induced electrical surges in power lines that disrupt power grids and similar effects in unprotected long-distance communication lines on land or under the sea. An extreme example of this resulted in the 1989

Hydro Québec power outage. Even fiber optic systems can be compromised if their signal-amplifying equipment is powered by long metallic wires.

Our broad description of the space environment in this chapter emphasizes that particles, fields, and photons figure predominantly in space weather. Next we describe two of the largest space weather events of recent history. Many of the phenomena we mention above reappear in these short case studies.

EXAMPLE 1.3

Kinetic Energy

- *Problem Statement:* Determine the kinetic energy of a 1 gram meteoroid traveling at a speed of 30 km/s.
- *Relevant Equation or Concept:* Definition of kinetic energy
- *Given:* Mass of 1 gram (= 0.001 kg) and speed of 30 km/s.
- *Assume:* Forces are balanced, resulting in no energy addition to the system.

- *Solution:* $KE = \frac{1}{2}mv^2$

  $KE = 0.5 \times (0.001 \text{ kg}) (30 \times 10^3 \text{ m/s})^2 = 4.5 \times 10^5 \text{ J}$

- *Interpretation:* This energy is huge for such a small object. Most non-armored systems (including humans) will suffer catastrophic damage from the rapid delivery of only 10^4 J. Thus, we see why collisions with small pieces of meteoroids or space debris are of concern to space operators.

Follow-on Exercise 1: Determine the kinetic energy of a small car traveling at 60 mph and compare your results to the meteoroid above.

Follow-on Exercise 2: The Sun emits approximately 10^{31} particles, most of which are protons, into the heliosphere each second. If the average speed of these particles is 450 km/s, what is the solar kinetic energy loss each second? Is this loss more or less than the emission of solar radiative energy each second?

1.3 Tales of Two Storms

1.3.1 A Brief History of Space Weather Observations

Objectives: After reading this section, you should be able to...

♦ Understand that solar disturbances have a long observational history
♦ Understand that space weather is an emerging science because of society's reliance on technology

To the casual observer, the Sun's appearance changes only with weather conditions and seasons. However, for centuries more discerning Sun watchers have reported hints of a different kind of variability—dark regions on the Sun called sunspots. Bone inscriptions from China and other records from Asia suggest the ancients knew about sunspots 3000 years ago [Zhentao, 1990]. Records from Greece indicate that irrigation workers used sunspots to determine agricultural irrigation levels as early as 300 BCE. Sunspots the apparent size of hen's eggs are tallied in Korean literature. Reviews of European history suggest that in 1607, Johannes Kepler may have observed a sunspot but mistook it for the planet Mercury. Johannes Fabricius, a German student, wrote a short manuscript about sunspots in 1611, but society took little note. Other observers reported spots as well but assumed they were solar satellites.

Galileo Galilei was the first western scientist to enter into full public discourse on the matter. Based on observations made as early as 1611 with a newly invented telescope, Galileo began to summarize his observations of the less-than-perfect Sun. He made numerous illustrations of the sunspots, such as shown in Fig. 1-14. In 1616 he circulated a controversial manuscript entitled, "On Sunspots." His assertion of a spotted Sun as the central body of our local planetary system was contrary to the accepted political and religious views of his time. State and religious censors toiled to prevent his ideas from reaching the masses. For a time they succeeded. Fortunately, what Galileo could not publish was slowly revealed to his contemporaries and successors via the telescope. We now know that *sunspots* are relatively dark regions near the Sun's surface harboring intense magnetic fields. They are visible indicators that the Sun is anything but constant.

Because the Sun was (and is) dangerously bright to view, sunspot tracking was fraught with difficulties. Telescopic image projection, which Galileo pioneered, was the only safe mode for sunspot observation, but this method did not get consistent results. Not until a German scientist named S. Heinrich Schwabe (1844) published accounts of a roughly 11-year periodicity in sunspot numbers did these dark regions become astronomically significant. Schwabe's work was subsequently publicized by a more prominent scientist, Baron Alexander von Humboldt, in 1851. The next year Edward Sabine and others linked magnetic perturbations on the ground with the sunspot cycle. Shortly thereafter Richard Carrington (1859) noted a rhythmic 11-year development and relocation of sunspots from mid- to low-solar latitudes. Edward Maunder formalized the spatial description of sunspot development in the early 1900s with his famous butterfly diagram, the modern version of which is shown in Fig. 1-15. The diagram shows that sunspots develop at relatively high heliographic latitudes early in the sunspot cycle. As time progresses the spots appear at progressively lower latitudes.

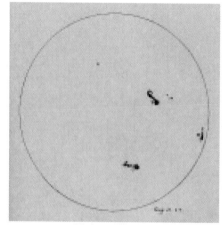

Fig. 1-14. Galileo Sunspot. Here we show Galileo's sunspot drawing from 17 June, 1613 *(Courtesy of The Galileo Project, Rice University)*

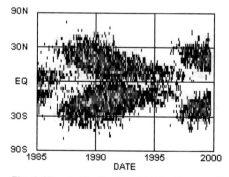

Fig. 1-15. Latitudinal Distribution of Sunspots for Solar Cycle 22. At the beginning of a new sunspot cycle the first sunspots form at mid-latitudes. With each subsequent solar rotation new sunspots form at slightly lower latitudes. Near the end of the cycle, sunspots form near, but not on, the solar equator. *(Courtesy of David Hathaway at NASA)*

Solar activity effects on technology were reported within eight years of installing the first telegraph lines in England. During a spring 1847 magnetic storm, a telegraph operator noted that strong alternating current deflections occurred on the telegraph instrument as a brilliant auroral light appeared in the sky [Barlow, 1849]. On September 1, 1859, the first observation of an explosive release of solar energy known as a solar flare was verified by independent observers, Richard Carrington and Richard Hodgson. Within a day, telegraph operators in Europe and America were astounded to find their devices operating on energy from the aurora rather than batteries. What we now know as space weather effects were reported in scientific journals within a few months.

Five solar cycles later, in May of 1921, telegraphic equipment was again so energized that it set fire to a New York relay station. The trail of disruption continued with new technologies. Radars being developed in the 1940s were jammed by solar bursts, radio transmissions on submarine cables were disrupted in the late 1950s, and long land-lines experienced induced current effects in the 1970s and 1980s. Numerous satellite glitches and some failures have been associated with space weather activity since the advent of the space age.

1.3.2 1859: First Observation of a Solar Flare and Associated Earth Disturbance

Objective: After reading this section, you should be able to...

♦ Describe events associated with the first reported technology impacts of space weather

How did humans become aware of the link between solar activity and Earth's magnetism? A tantalizing clue occurred on September 1, 1859 with the first observation of a solar flare. On that late summer day, two British amateur astronomers were independently monitoring sunspots from separate locations on the northern and southern outskirts of London. At 11:18 AM Greenwich Mean Time (now referred to as *Coordinated Universal Time [UTC]*), Richard Carrington noticed an extraordinary sparkle in two rapidly brightening patches of light near the middle of a sunspot group he was studying (indicated by A and B in Fig. 1-16). The area remained illuminated for about five minutes, with the intense light moving to positions C and D by 11:23 AM. This unusual event was also observed by Richard Hodgson. Both men presented their observations at the November 1859 meeting of the Royal Astronomical Society.

> *Hodgson reported: "While observing a group of sunspots on the 1st of September, I was suddenly surprised at the appearance of a very bright star of light, much brighter than the Sun's surface, most dazzling to the protected eye, illuminating the upper edges of the adjacent spots and streaks, not unlike in effect the edging of the clouds at sunset; the rays extended in all directions; and the centre might be compared to the dazzling brilliancy of the bright star Lyrae when seen in a large telescope with low power. It lasted for some five minutes and disappeared instantaneously about 11:25 AM...The magnetic instruments at Kew (magnetic observatory) were simultaneously disturbed to a great extent."*

This amazing set of observations represents the first clear description of a solar flare, corresponding to a sudden and intense heating of solar plasma. Only the largest flares are bright enough to be seen in visible light. Thus, we know that this was an extraordinary event.

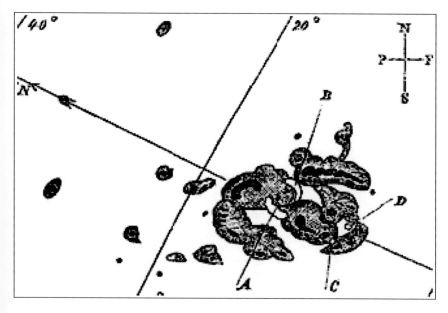

Fig. 1-16. Reproduction of a Drawing by Richard Carrington. Here we show the location of the flare he observed while making a drawing of an active region. Points A, B, C, and D were locations of extremely bright light. [Carrington, 1860]

Figure 1-17 shows the magnetic deflection observed coincident with the white-light flare. This magnetic signature, called a *geomagnetic crochet* (because of its hook-like appearance), is associated with a type of severe sudden ionospheric disturbance. The crochet indicates an electric current disturbance associated with the rapid, but short-lived, ionization event in Earth's upper atmosphere. Eight minutes after extreme flares on the Sun, Earth's upper atmosphere is subject to overwhelming changes in ionization and rapid increases of currents on Earth's dayside. The rapidly changing current systems create magnetic deflections at the ground.

Carrington and Hodgson learned that nearby magnetic monitoring instruments registered strong disturbances at about the same time, but neither pressed to associate the solar emission with the magnetic disturbance. Balfour Stewart, Director of the Kew Observatory where the magnetic observations were made, strongly suspected the association [1861] but had no theory to explain it. Only after E. Maunder [1905] presented convincing evidence of the Sun's 27-day rotation period in the geomagnetic record was a solar-geomagnetic link accepted. In fact, the full explanation of what is now known as the "Carrington Storm" was not published until scientists were able to link solar emissions, geomagnetism, and radio wave disturbances in the ionosphere [Bartels, 1937].

No doubt one of the reasons more than 75 years passed before scientists explained this event was the appearance of another and yet larger geomagnetic disturbance only 18 hours after the first crochet. As seen in the right side of Fig. 1-17, the great geomagnetic storm beginning early on 2 September drove the Kew magnetometer trace off-scale. Geomagnetic storms usually commence 60 to 140 hours after a well-positioned solar flare. This delay is typical for associated mass ejections to travel from the Sun to Earth at speeds of 300 km/s–700 km/s.

The 18-hour transit time for the Carrington storm indicates the mass ejection was hurtling through the interplanetary medium at well over 2000 km/s, the fastest in modern times.

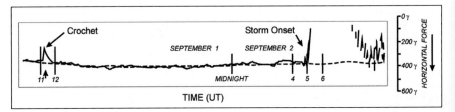

Fig. 1-17. **The Kew Observatory Magnetometer Trace from September 1–2, 1859.** The crochet was recorded at observatories in Earth's sunlit sector. The subsequent disturbance on 2 September was the recording of what was, arguably, one of the largest geomagnetic storms of modern times. We note the recording goes off scale just after 0500 UTC. [Cliver, 2006]

Did anyone else notice this disturbance? Emphatically, yes. Carlowicz and Lopez [2002] and Odenwald and Green [2008] provide an excellent accounting of the 1859 space weather events. The storm was accompanied by a great low-latitude aurora visible from Santiago (Chile), Honolulu (Hawaii), and Wakayama (Japan). Astonished readers in the southern US could read newspapers by auroral light alone. In France, telegraphic connections were disrupted as sparks flew from long transmission lines. In the US, telegraphs worked without batteries. Some telegraph lines ran exclusively on currents associated with the aurora. One hapless telegrapher was shocked by his own equipment. In Washington D.C., telegraph operator Frederick Royce reported:

> "During the auroral display, I was calling Richmond and had one hand on the iron plate…Happening to lean towards the sounder, my forehead grazed the ground wire. Immediately, I received a very severe electric shock…An old man who was sitting facing me said that he saw a spark of fire jump from my forehead to the sounder." [Loomis, 1859]

Very recently this storm has regained notoriety as the most intense solar energetic particle (SEP) event of the last four centuries. Over the interval 1561 to 1950, the largest single deposition of chemicals created by solar protons has been convincingly linked to the Carrington event [McCracken et al., 2001]. Obviously aircraft and spacecraft were not available to receive the radiation dose likely created by this solar blast. However, scientists are keenly interested in trying to recreate this event in models and simulations so that we may understand worst-case scenarios for current technology.

Next we describe a more recent storm, one less powerful than the Carrington Storm, but one in which our technology systems were exposed and vulnerable.

1.3.3 2003: Extreme Space Weather in the 21st Century

Objectives: After reading this section, you should be able to...

- Describe events associated with a modern space weather storm
- Understand remotely sensed images of the space environment

We fast forward to the 21st century. During mid-October 2003, the Sun was remarkably spotless. Solar particles and photons were in a quiescent state. But lurking on the Sun's farside was a sunspot larger than Jupiter. By October 22, the colossal region was rotating into view and spewing bursts of radiation and high-energy particles. Additional spots appeared, clustering in three active regions numbered 10484, 10486, and 10488. Approximately 3% of the Sun's visible surface harbored ominous sunspots with contorted and intertwined magnetic fields (Fig. 1-18).

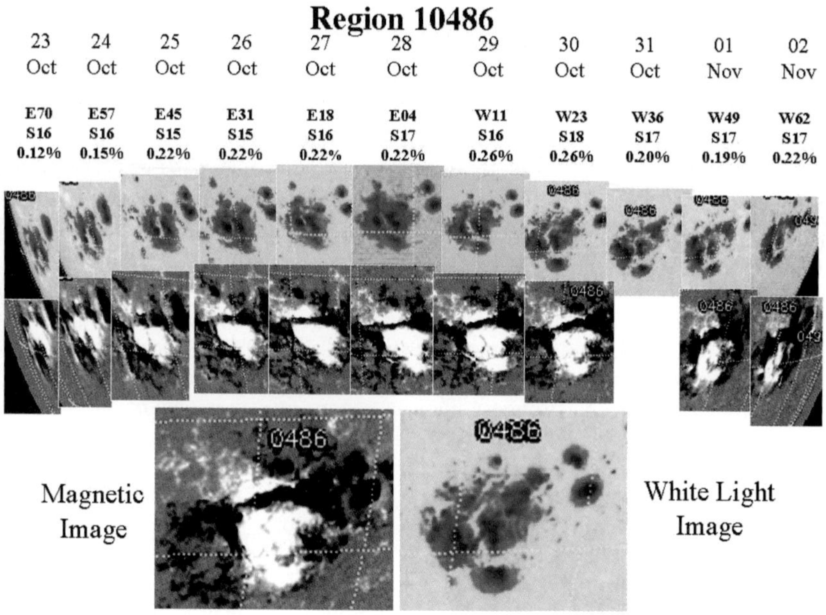

Fig. 1-18. White Light and Magnetic Imagery. Here we show the solar disk transit of super Region 10486. The dates of the individual images are shown at the top. Below the dates are the longitudes of the image and the percentage of how much of the solar disk that was covered by the sunspot. The white light images are false-colored to blue to improve contrast. In the magnetic images, black is inward-directed magnetic field lines and white is outward-directed magnetic field lines. *(Courtesy of Big Bear Solar Observatory)*

Space weather forecasters worldwide engaged in the most active and demanding solar-activity forecasting epoch in years. During the three-week period in late October to early November, the National Oceanic and Atmospheric Administration (NOAA) Space Weather Prediction Center (SWPC) staff issued over 250 solar energetic event watches, warnings, and alerts—in the previous two months they had issued only two warnings. This outbreak occurred well past the solar cycle 23 peak in April 2000. While late-cycle active periods have occurred in the past, the extreme level of activity during this stage of the solar cycle was unusual. Seventeen major flares erupted from these spots between October 19 and November 5, 2003. Many of these flares had associated radiation storms, including a severe storm that started on October 29 (the Halloween Storm). Geomagnetic storm periods were observed on 12 of the 20 days, with two storms reaching the extreme level on October 29 and 30.

Numerous anomalies were reported by deep space missions and by satellites at all altitudes. The Goddard Space Flight Center, Space Science Mission Operations Team indicated that approximately 60% of Earth- and space-science missions were impacted—25% of the missions shut down instruments or took other protective actions. The orbital level of the International Space Station (ISS) dropped at twice its normal rate because of enhanced atmospheric drag. Most industries vulnerable to space weather experienced some degree of impact to their operations. A magnetic crochet, the hallmark magnetic signature from the 1859 Carrington storm, was reported by many ground magnetic observatories. Power transformers in South Africa's ESKOM power grid experienced rapid and extreme heating of their internal cooling oil, ultimately forcing the replacement of millions of dollars worth of equipment. Solar images and solar activity stories were reported in newspapers around the world, making solar flares a household term. These chains of events are a subset of the space weather impacts originating in these solar regions.

Although solar flares are still categorized by their white light output, a more sensitive measure is now acquired by spacecraft observing X-ray emissions above Earth's atmosphere. The output is categorized in 5 levels, with the top category indicating extreme (X) levels of X-ray emissions. On October 28 at 1101 UTC, a powerful X-level flare occurred near the central meridian (Fig. 1-19a). This flare produced emissions across the electromagnetic spectrum and one of the largest radio bursts ever recorded. The flare region was the origin for a very fast (2125 km/s [4.75 million mph]) coronal mass ejection (Fig 1-19b) and a severe radiation storm. The ejected mass arrived at Earth in just 19 hours (one of the fastest arrivals on record). The CME impacted Earth's magnetic field and produced an extreme geomagnetic storm, lasting 27 hours.

Subsequently, on November 4, another extreme (X-level) flare (Fig. 1-20a) from this same region so disturbed Earth's dayside ionosphere that normal high-frequency radio communication was not possible. The space-borne X-ray instrument observing the event became saturated for 12 minutes during this flare. The flare was one of the strongest since spacecraft measurements began in 1976. Despite the record size of the disturbance, its location near the west limb of the Sun limited its impact at Earth but enhanced its interaction in the direction of Saturn. The associated CME traveled away from Earth (Fig. 1-20b). As a result, the subsequent radiation storm and magnetic storm at Earth only reached moderate and minor levels, respectively. However, the Cassini spacecraft, near Saturn, recorded alarming levels of high-energy particles.

Based on the information above, we ask if space weather has gotten worse in the new millennium. We doubt that disturbances have worsened. Rather we depend more on technologies that are vulnerable to space weather. In the chapters that follow, we describe the how's and why's of these vulnerabilities.

However, before we move to these topics we describe more about the impacts of the space weather and space environment. Next, we consider the space weather storm scales developed by the US National Oceanic and Atmospheric Administration (NOAA).

1.3 Tales of Two Storms

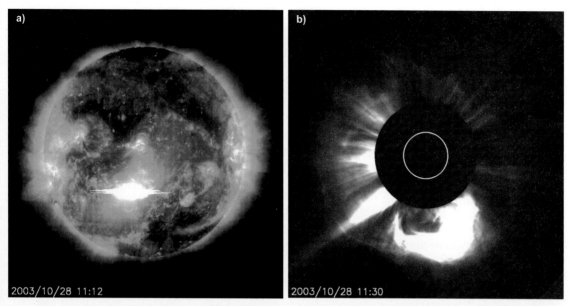

Fig. 1-19. The X17 Flare of October 28, 2003. These images come from two instruments on the Solar and Heliospheric Observatory (SOHO) Spacecraft. **a)** This extreme ultraviolet image shows a flare located below and just left of center. **b)** The bright emission that surrounds the Sun (the halo), indicates that the ejected mass is aligned with Earth. Less than 24 hours after the X17 flare, the same active region produced another major flare. *(Courtesy of ESA/NASA – the SOHO Program)*

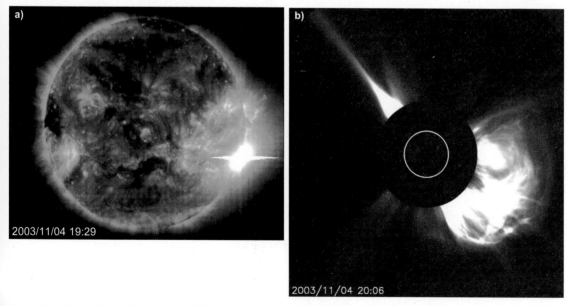

Fig. 1-20. Solar Flare of November 2003. a) On the left, we show the Geostationary Operational Environmental Satellite (GOES) Solar X-ray image of the November 2003, X28 (estimated) flare. This flare, which developed on the west limb of the Sun, was one of the largest X-ray flares ever recorded. **b)** On the right is the Solar and Heliospheric Observatory (SOHO) image of the mass ejection. We note that the bright emission is confined to the right side of the image, indicating that the bulk of the material is not Earth-directed. *(Courtesy of ESA/NASA – the SOHO Program)*

1.3.4 Space Weather Storm Scales

Objective: After reading this section, you should be able to...

♦ Distinguish between storm scales for radio blackouts, solar radiation storms, and geomagnetic storms

In Table 1-2 we list disturbed solar emissions that affect human and technology systems. These emissions—photons, energetic particles, and fields—operate on three different time scales and create disturbances in three areas of the near-Earth space environment. Table 1-3 shows an abridged version of the NOAA Space Weather Storm Scales, listing only the effects for Strong Storm level (Level 3) in each category. A full listing of effects is in Appendix A.

Table 1-3. Abridged NOAA Space Weather Storm Scales. Here we list only the effects, physical measures, and average frequency for storm level 3. Appendix A has descriptions of other Radio (R), Solar Radiation (S), and Geomagnetic (G) storm levels.

Category		Effect	Physical Measure	Average Frequency (1 cycle = 11 yrs)
Scale 1-5	Descriptor and (Location)	Duration of event influences severity of effects Some events last more than one day	Type of observations	Number of events when flux level was met (number of storm days)
R 3	Strong Radio Storm (R) (Dayside)	*High Frequency (HF) Radio:* Wide area blackout of HF radio communication, loss of radio contact for ~1 hour on Earth's sunlit side *Navigation:* Low-frequency navigation signals degraded for about an hour	GOES X-ray peak brightness by class and by flux* X1 and (10^{-4})	175 per cycle (140 days per cycle)
S 3	Strong Solar Radiation Storm (S) (Polar)	*Biological:* Radiation hazard avoidance recommended for astronauts on extra vehicular activity (EVA). Passengers and crew in high-flying aircraft at high latitudes may be exposed to radiation risk. *Satellite operations:* Single-event upsets, noise in imaging systems, and slight reduction of efficiency in solar panels are likely *Other systems:* Degraded HF radio propagation through the polar regions and navigation position errors likely†	Flux level of $\geq$10 MeV particles (ions)‡ 10^3	10 per cycle
G 3	Strong Geomagnetic Storm (G) (Auroral, Predominantly Nightside)	*Power systems:* Voltage corrections may be required, and false alarms triggered on some protection devices *Spacecraft operations:* Surface charging may occur on satellite components, drag may increase on low-Earth-orbit satellites, and corrections may be needed for orientation problems *Other systems:* Intermittent satellite navigation and low-frequency radio navigation problems may occur, HF radio may be intermittent, and aurora has been seen at geomagnetic latitudes of 50° or less	Planetary Index Kp = 7 (Derived from ground magnetometers)**	200 per cycle (130 days per cycle)

* Flux, measured in the 0.1–0.8 nm range, in W/m². The scale level is based on this measure, but other physical measures are also considered

† Other frequencies may be affected.

‡ Flux levels are 5 min averages. Flux in particles per unit time (second), per unit solid angle (steradians), per unit area (square centimeter) [#/(s · sr · cm²)]. The scale level is based on this measure, but other physical measures are also considered. High-energy particle measurements (>100 MeV) are a better indicator of radiation risk to passengers and crews. Pregnant women are particularly susceptible.

** We describe the Kp Index in Focus Box 1.5.

Associated with solar flares are X-ray photons that interact with Earth's ionosphere and change the radio propagation characteristics on Earth's dayside for minutes to hours after the beginning of the flare event. Under the worst conditions high-frequency radio communications on the dayside are blocked for a few hours. The physical measure for such effects currently comes from NOAA's Geosynchronous Operational Environmental Satellites (GOES). Solar flaring also disturbs plasma in the Sun's outer atmosphere, generating disturbances at radio frequencies. Some but not all X-ray events are accompanied by solar radio bursts. We list X-ray and radio effects in Table 1-3, Radio (R) Scale.

Some solar flares and many CME shock fronts energize particles to relativistic speeds and thus have the ability to produce ionizing radiation. The Solar Radiation (S) scale indicates that these particles have biological, satellite operations, and communication effects. They quickly gain access to geomagnetic field lines that link to Earth's polar regions. As these particles decelerate in collisions with Earth's upper atmosphere they ionize atoms and molecules, creating regions of atmosphere that absorb radio waves—*polar cap absorption (PCA)* events. They also deliver dangerous amounts of energy to humans and satellite components in the near-Earth environment. Arriving sometimes as quickly as 20 minutes after a solar disturbance, these particles sometimes create polar radio blackouts that last for days and cause hardware and human effects that compromise system integrity. Radiation storms occur less frequently than other types of storms, but they are more difficult to forecast.

Coronal mass ejections are a facet of the magnetic processes that create solar flares. These bubbles of magnetism and mass sometimes intercept Earth. If they do, then their speed and orientation determine how effective they are in creating aurora, energized particles, and current systems that affect technology systems and hardware such as satellites and power grids. The Geomagnetic (G) Storm scale deals with these effects. Geomagnetic storm effects generally begin more than a day after the causal event on the Sun.

In Fig. 1-21 we summarize the timing of space environmental disturbances and the actions that forecasting agencies take when events are likely or detected. Radio waves used by communication and navigation systems are subject to effects from multiple sources of solar disturbances.

Pause for Inquiry 1-14

Use the knowledge you've gained thus far to put the following in order of occurrence from first to last.

Auroral display
Coronal mass ejection
Flare
Power grid disruption
Radio blackout
High-energy particles at Earth

Pause for Inquiry 1-15

In addition to Fig. 1-22, consult Table 1-3 and Appendix A. What level of geomagnetic storm is indicated in Fig. 1-22? Justify your answer.

Fig. 1-21. The Time Scales of Solar Effects. Eight minutes after a flare erupts from the Sun, the first blast of extreme ultraviolet (EUV) and X-ray light increases the ionospheric density and causes HF communication loss. Thirty to 1000 minutes later, energetic particles may arrive. One- to-four days later, the CME passes and energizes the magnetosphere and ionosphere, affecting navigation systems and radio communications. The geometric signs indicate actions that forecasting agencies take when an event is anticipated or observed. Similar to the realm of terrestrial weather, watches are issued when conditions support the development of potentially hazardous situations. Alerts are issued when hazardous situations are foreseen with enough time for operators to take preventative action. Warnings are issued when an event is in progress. *(Courtesy of NOAA's Space Weather Prediction Center)*

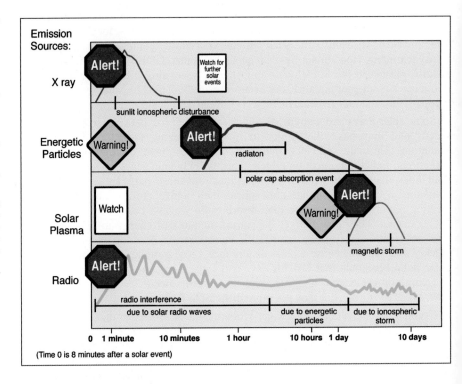

Focus Box 1.4: Terrestrial and Space Weather Comparisons

At this point we describe how terrestrial weather and space weather compare. We do this throughout the text, but here is a synopsis.

Terrestrial climate, in the grand scheme, has its ultimate source in the Sun. Variations and perturbations on this background (which we call terrestrial weather) are really driven by internal instabilities within Earth's ocean-atmosphere system. Terrestrial weather is not, in general, externally forced. Further, the time scale for large (*synoptic* or continental-scale) weather systems is on the order of days. In terms of observations, the terrestrial weather system is quite well measured. World-wide weather stations and satellites efficiently feed data into terrestrial forecasting models. The models that deal primarily with the neutral atmosphere have been in development for several decades.

Space climate (environment) also has its ultimate source primarily in the Sun. Space weather, on the other hand, is mostly driven by external forces. The Sun, the interplanetary medium, and the interstellar medium all enter directly into the forcing of space weather. The time scales for space weather are often very short, allowing the entire global space weather system to be rearranged in a matter of minutes. Unfortunately, observations in the space environment are quite sparse, so inputs for space weather models are relatively few. Exacerbating this problem is the relative youth of most space weather models that must account for ionized and neutral particles.

Governing equations for terrestrial weather are generally the fluid form of the conservation laws, plus appropriate constitutive equations. For space weather we use these equations plus Maxwell's equations, as they apply to plasmas.

As a discipline, space weather is in its infancy. We see in the following chapters that space weather needs to mature rapidly to keep pace with risks posed to space assets and technologies. This text is an important step in jumpstarting technical understanding of space weather.

Focus Box 1.5: Describing Space Weather Disturbances at Earth

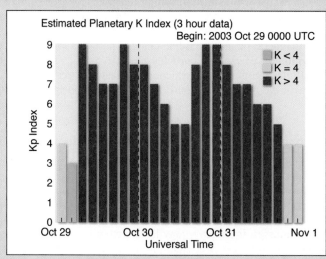

Fig. 1-22. Kp Values for 29-31 October, 2003. A value is reported for each three-hour interval of a UTC day. *(Courtesy of NOAA's Space Weather Prediction Center)*

Here we highlight the Kp index, an indicator of electric currents flowing in the ionosphere. Introduced by Julius Bartels in 1949, the index is determined by measuring abnormal deflections in Earth's magnetic field at the surface caused by these currents. Measurements for computing the index are made at a number of worldwide magnetic observatories (ground magnetometers) at high- and mid-latitudes. The Kp-index scale ranges from 0 to 9 and is directly related to the maximum amount of fluctuation (relative to a quiet day) in the geomagnetic field over a three-hour interval.

Similar to other physical indices that must cover vast energy ranges, the Kp index is logarithmic. Values exceeding 7 are unusual. The events of late October, 2003 (Fig. 1-22) show Kp levels of 9. NOAA categorizes any Kp value greater than 4 as a storm event. The Kp index is a general geomagnetic disturbance index—other scales address different forms of space weather disturbances.

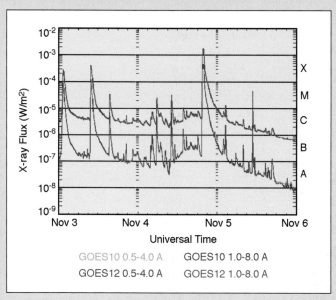

Fig. 1-23. GOES X-ray Flux for November 3–5, 2003. An X-ray flux value is shown for each five-minute interval. *(Courtesy of NOAA's Space Weather Prediction Center)*

Operational information about current solar activity comes from the X-ray sensors on the GOES satellites. A GOES X-ray flux plot (Fig. 1-23) contains five-minute averages of solar X-ray emissions integrated over the visible solar disk, in the 0.1 nm–0.8 nm (1 angstrom–8 angstrom) and 0.4 nm–0.05 nm (0.5 angstrom–4.0 angstrom) passbands. Normally, the plots include data from primary and secondary GOES satellites.

The scale is logarithmic on the left side of Fig. 1-23. On the right side is the corresponding alphabetic rating, which is converted to the R-storm scale we show in Table 1-3 and Appendix A. Levels A and B are considered quiet levels. The minor storm level starts at M1 (1×10^{-5} W/m^2). The NOAA Space Weather Prediction Center alerts are issued at the M5 (5×10^{-5} W/m^2) and X1 (1×10^{-4} W/m^2) levels, based upon one-minute data. Large X-ray bursts cause short wave fades for HF propagation paths through the sunlit hemisphere. Some large flares are accompanied by strong solar radio bursts that may interfere with satellite downlinks. This flare, which developed on the west limb of the Sun, was one of the largest X-ray flares ever recorded.

1.4 The Lorentz Force

On a macroscopic scale, plasma is a fluid, similar to a gas. However, unlike a non-ionized gas, plasma exhibits a strong interaction with electromagnetic fields. This property sets plasma apart from other types of matter.

In the presence of magnetic fields, individual charged particles follow curved paths. The collective motions of charged particles generate magnetic fields that, in turn, govern the particles' motions. The tendency for plasma to generate forces that feed back to control the plasma is of particular interest to space physicists and space operators.

> **Objective: After reading this section, you should be able to...**
>
> ♦ Describe the motion of a charged particle in an electric or a magnetic field

The simplified version of Newton's Second Law ($\Sigma \boldsymbol{F} = m\boldsymbol{a}$) applies in non-relativistic situations to individual particles and to collections of particles and to charged and uncharged masses. In our everyday lives we tend to be more aware of forces acting on uncharged bodies. For example we know the sum of gravity and lift ($\Sigma \boldsymbol{F} = -m\boldsymbol{g} + \boldsymbol{L}$) allows aircraft to stay aloft. In space weather applications, we often ignore gravity because its influence is small. However we consider other forces acting on charged particles, individually or collectively. Here we provide applications of the Second Law for charged-particle and plasma behaviors. The *electric (Coulomb) force* for charged particles is

$$\boldsymbol{F}_E = q\boldsymbol{E} \qquad (1\text{-}1)$$

where

$\boldsymbol{F}_E$ = electric (Coulomb) force vector [N]
q = charge on a particle [C]
$\boldsymbol{E}$ = electric field vector [V/m]

This force accelerates charged particles along electric field lines. That is, $\Sigma \boldsymbol{F} = m\boldsymbol{a} = q\boldsymbol{E}$.

If a magnetic field (*B*) rather than an electric field is present and the charged particle is moving, we describe the force on the particle as the *magnetic Lorentz force*:

$$\boldsymbol{F}_B = q(\boldsymbol{v} \times \boldsymbol{B}) \qquad (1\text{-}2)$$

where

$\boldsymbol{F}_B$ = magnetic Lorentz force vector [N]
$\boldsymbol{v}$ = particle velocity vector [m/s]
$\boldsymbol{B}$ = magnetic field vector [T]

Of particular interest to space physicists is the Magnetic Lorentz force's ability to change the direction of motion of charged particles without performing any work on them. By virtue of the cross product, this force acts perpendicular to $\boldsymbol{v}$ and $\boldsymbol{B}$ and creates a centripetal (toward the center) acceleration, but adds no motion along the direction of the force.

In the space environment we often find both fields acting on charged particles, creating a combined force called the generalized Lorentz force

$$\boldsymbol{F} = q\boldsymbol{E} + q\boldsymbol{v} \times \boldsymbol{B} = q(\boldsymbol{E} + \boldsymbol{v} \times \boldsymbol{B}) \qquad (1\text{-}3)$$

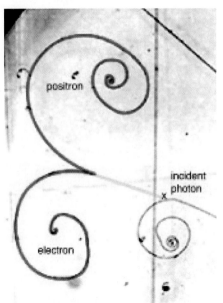

Fig. 1-24. Helical Tracks of Two Charged Particles. Here we show an electron and positron, created from the decay of an energetic photon. Their motions are guided by the presence of a magnetic field. *(Courtesy of the European Organization for Nuclear Research [CERN])*

Pause for Inquiry 1-16

What must be the direction of the magnetic field for the particle trajectories shown in Fig. 1-24?

The resultant motion for the charged particles is a circular or spiral path, as shown in Fig. 1-24, depending on the orientation of B and E.

We develop some initial ideas about solar wind interaction with the magnetopause using the Lorentz force concept. We assume that solar particles flow past the magnetosphere as suggested by the flow lines in the cross section of the magnetosphere viewed from above the north geographic pole in Fig. 1-25. Most of the solar wind particles skim past the magnetosphere's flanks. At the magnetopause a few wander in and feel the influence of Earth's magnetic field. Protons on the dawn side accelerate inward for part of their gyro orbit, while electrons accelerate outward. On the dusk side a slight tendency exists for electrons to accelerate into Earth's magnetic field while the protons accelerate outward. The net result is a charge separation with protons accumulating inside the magnetosphere on the dawn side and electrons accumulating on the dusk side. Within the magnetosphere the charge build-up creates a dawn-to-dusk electric field (E). As we learn in subsequent chapters, this electric field and enhancements in it, caused by a process called magnetic merging, lay the foundation for energy exchange between the solar wind and the near-Earth system.

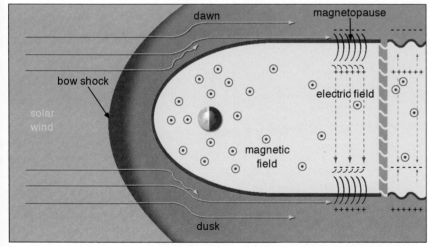

Fig. 1-25. Dynamo Action Viewed in the Equatorial Cross Section of the Magnetosphere. The Sun is to the left. Earth's magnetic field (dark gray circles with dots) points outward from the page. At the magnetotail flanks, charged particles from the solar wind are captured and separated by Earth's magnetic field. Charge separation is caused by the trapping of charged solar wind particles that complete only partial gyrations at the magnetopause. Thus, passage of the solar wind plasma generates a dawn-to-dusk electric field in the interior magnetotail and a dusk-to-dawn electric field at the magnetotail flanks. Only a few trapped particles are needed to set up a measurable electric field. This charge separation process acts as a dynamo (generator). In subsequent chapters we learn about a more efficient dynamo process that appears during solar wind disturbances. The diagram is not to scale.

> **Pause for Inquiry 1-17**
>
> Verify that the generalized Lorentz force has units of newtons [N].

> **Pause for Inquiry 1-18**
>
> When the only force acting on a charged particle is the $q(\mathbf{v} \times \mathbf{B})$ force, which of the following changes?
>
> a) The particle's mass
> b) The particle's momentum
> c) The particle's energy
> d) None of these change

EXAMPLE 1.4

Charged Particle Motion Caused by an Electric Field—Part 1

- *Problem Statement:* During space weather storms, oxygen ions and electrons in Earth's high-latitude, upper atmosphere encounter strong electric fields. Describe the force from an outward-directed electric field (anti-parallel to the magnetic field in the northern hemisphere) on a low-energy oxygen ion in Earth's distant upper atmosphere. Assume the ion is located in Earth's north polar region (Fig. 1-26).
- *Given:* The charge on the ion is 1.60×10^{-19} C.

  The magnetic field in the north polar region is directed downward, toward Earth's surface.
- *Assume:* An electric field strength of 1 mV/m directed away from Earth and that the ion is located in Earth's north polar region.
- *Solution:* The Electric Coulomb force causing the ion to accelerate along the electric field line is $\mathbf{F}_E = q\mathbf{E}$.

  The magnitude of the force on the ion is $\mathbf{F}_E = qE = 1.60 \times 10^{-19}$ C $\times$ 0.001 V/m $= 1.60 \times 10^{-22}$ N

  The direction of the force on the ion is away from Earth.
- *Interpretation and Implications:* A net Electric Coulomb force accelerates charged particles in the near-space environment. During some space weather events, large quantities of ionized oxygen atoms escape from Earth's atmosphere and enter regions of near-space, subsequently modifying the electrical environment of satellites.

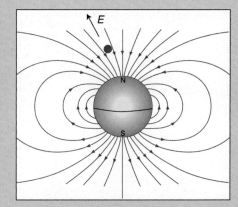

Fig. 1-26. Earth's Magnetic Dipole. Here we show the magnetic field lines leaving Earth's north polar region and entering the south polar region. The red dot is the approximate location for the example problem.

Follow-on Exercise 1: Will the ion's gyration about the field line be clockwise or counterclockwise?

Follow-on Exercise 2: Assume the gyration speed of the oxygen ion about the magnetic field is 1200 m/s. What is the Lorentz force on the ion if the local magnetic field strength is 7500 nT?

EXAMPLE 1.5

Charged Particle Motion in a Magnetic Field

- *Problem Statement:* Determine the radius of gyration of a proton at the magnetosphere flank.
- *Concept:* Radius of gyration.
- *Given:* Magnetic field strength at the magnetosphere flank = 20 nT, proton perpendicular speed = 300 km/s, proton mass = 1.67×10^{-27} kg, and proton charge = 1.6×10^{-19} C.
- *Assume:* A uniform magnetic field strength and the charged particle motion is perpendicular to the magnetic field.

- *Solution:* The force experienced by the proton is centripetal in nature. Set the centripetal force equal to the magnitude of the magnetic Lorentz force and solve for the radius of the circular motion.

$$|F_B| = \left|\left(-\frac{mv^2}{r}\hat{\mathbf{r}}\right)\right| = |q\mathbf{v} \times \mathbf{B}|$$

$$|r| = \frac{mv}{qB}$$

$$r = \frac{(1.67 \times 10^{-27} \text{kg})(3 \times 10^5 \text{m/s})}{(1.6 \times 10^{-19} \text{C})(20 \times 10^{-9} \text{T})} \approx 1.56 \times 10^5 \text{ m} = 156 \text{ km}$$

- *Interpretation and Implications:* Solar wind protons skimming past Earth's magnetic flanks may be trapped in gyro-motion about the magnetopause field lines with a motion scale size of a few hundred kilometers. The flank magnetopause layer is several gyro radii thick, on the order of 500 km–1000 km.

Follow-on Exercise 1: Determine the gyroradius of a solar wind electron with characteristics similar to that of the proton.
Follow-on Exercise 2: Determine the gyrofrequency and period of the proton and electron.

Summary

Humans probably have observed elements of space weather for thousands of years. Over many decades, observations revealed that sunspots are associated with magnetic fields on the Sun and with aurora at Earth.

In this chapter we explain the space environment. The Sun's influence extends to nearly 100 times the Sun-Earth distance. Within this volume of space the Sun's radiation and outer atmosphere influence all planets, forming magnetospheres around many and ionospheres around all. Our local ionosphere and magnetosphere are the sites of considerable energy exchange with the solar emissions. To date most space weather has been identified with these spheres of influence. We are currently learning that the energy does not stop at the ionosphere, but rather extends into the atmosphere and all the way to the ground.

In 1859, a particularly bright and long-lasting solar flare was followed shortly by a great magnetic storm at Earth. Investigations into the associations between sunspots, solar flares, and magnetic storms began in earnest with that event and have continued ever since. Much of what space physicists have learned was employed to forecast the timing and likely effects of the large space weather storms of solar cycle 23. Effects from these storms inhibited communication, caused some satellites to go into protective safe mode, and permanently damaged others.

Electric and magnetic forces play a large role in energizing space weather events. These forces rapidly put charged matter into motion. Unlike most terrestrial weather storms that develop over many hours to days, space storms develop over many minutes to hours.

Chapter 1 Space Is a Place…with Weather

Key Words

active region
astronomical unit (AU)
bow shock
Coordinated Universal Time (UTC)
coronal mass ejection (CME)
differential rotation
diurnal
electric (Coulomb) force
exobase
galactic cosmic rays (GCRs)
gas
generalized Lorentz force
geomagnetic crochet
heliopause
heliosheath
heliosphere
heterosphere
high-speed streams
homosphere
interplanetary magnetic field (IMF)
ionosphere
liquids
magnetars
magnetic Lorentz force
magnetopause
magnetosheath
magnetosphere
magnetotail
magnetotail lobes
meteor trail
plasma
plasmapause
plasmasheet
plasmasphere
polar cap absorption (PCA)
radiation belts
radio blackout
scintillation
solar energetic particles (SEPs)
solar flares
solar spectrum
solar wind
solid
space climate
space environment interactions
space physics
space weather
sunspot cycle
sunspot maximum
sunspot minimum
sunspots
synoptic
termination shock
thermosphere
Van Allen belts

Notation

Electric Coulomb force: $F_E = qE$
Magnetic Lorentz force: $F_B = q(v \times B)$
Generalized Lorentz force: $F_L = qE + qv \times B = q(E + v \times B)$

Answers to Pause for Inquiry Questions

PFI 1-1: An electron-volt is associated with $\sim 10^4$ K

PFI 1-2: ~5 solar cycles

PFI 1-3: Various individual answers

PFI 1-4: The solar wind particles have high speeds, but the density is very low. Therefore little heat energy is available for exchange with satellites in the solar wind.

PFI 1-5: Similarly, little momentum is exchanged with an astronaut, so the force that could accelerate an astronaut is miniscule.

PFI 1-6: Uranus is 19.2 AU from the Sun × 110 ft/AU = 2112 ft. Voyager 1 is 100 AU from the Sun × 110 ft/AU = 11,000 ft = 2.08 mi.

PFI 1-7: $(4.5 \times 10^{12}$ m$)/(4 \times 10^5$ m/s$) = 1.125 \times 10^7$ s ≈ 4.33 mo

PFI 1-8: $(1.5 \times 10^{11}$ m$) / (30$ hr $\times 3600$ s/hr$) = 1.4 \times 10^6$ m/s ≈ 1400 km/s

PFI 1-9: Saturn: ~10 AU; Uranus: ~20 AU; Heliosphere boundary: ~150 AU; Alpha Centauri: 2.7×10^5 AU

PFI 1-10: $(2.3 \times 10^{13}$ m$) / (2.998 \times 10^8$ m/s$) \approx 0.89$ d

PFI 1-11: No. The moon moves in and out of the solar wind. When it is not in the solar wind it is likely to be in the magnetosphere. It's inside the magnetosphere when it's anti-Sunward. It's outside when it's Sunward. It's outside when it's at the dawn and dusk lines.

PFI 1-12: Troposphere: $0\ R_E - 0.002\ R_E$
Stratosphere: $0.002\ R_E - 0.008\ R_E$
Mesosphere: $0.008\ R_E - 0.013\ R_E$
Thermosphere: $0.013\ R_E - 0.16\ R_E$
Ionosphere: $0.016\ R_E - 0.16\ R_E$
Exosphere: $> 0.05\ R_E$
Plasmasphere: $1\ R_E - 6\ R_E$
Magnetosphere: $10\ R_E - 100s\ R_E$
Heliosphere: 0 AU – 150 AU

PFI 1-13: Balloon or sounding rocket.

PFI 1-14: *Flares and coronal mass ejections* (may be simultaneous) but the flare is usually observed first.

Radio blackout—results from flare and develops on the dayside eight minutes after flare and lasts for minutes to hours.

High-energy particles at Earth—results from flare and/or CME, first particles can arrive as early as 20–25 minutes after flare, if they are traveling at ~1/3 light speed. Polar radio blackouts may develop shortly after arrival of high-energy particles.

Auroral displays—develops with arrival of CME at Earth, usually 2–3 days after flare.

Power grid disruption—develops a few minutes to hours after arrival of CME.

PFI 1-15: Storm level is G-5

PFI 1-16: *B* is out of the page

PFI 1-17: $[(C \times N/C) + (C \times m/s \times N/(C\ m/s))] = N$

PFI 1-18: b) Magnetic Lorentz force changes the momentum by changing the particle's direction.

References

Bartels, Julius. 1937. Solar eruptions and their ionospheric effects—A classical observation and its new interpretation. *Terrestrial Magnetism and Atmospheric Electricity*. Vol. 42, No. 3. American Geophysical Union, Washington, DC.

Carlowicz, Michael and Ramon Lopez. 2002. *Storms From the Sun: The Emerging Science of Space Weather*. John Henry Press, Washington DC.

Jansen, Frank, Risto Pirjola, and Rene Fevre. 2000. *Space Weather: Hazard to Earth?* Swiss Re Publishing, Zurich, Switzerland.

Kouveliotou, Chryssa, Robert Duncan, and Christopher Thompson. 2003. Magnetars. *Scientific American*. February. Nature America. New York, NY.

Labitzke, Karin and Harry van Loon. 1988. Associations between the 11-year solar cycle, the quasi-biennial oscillation, and the atmosphere: I. the troposphere and stratosphere in the northern winter. *Journal of Atmospheric and Terrestrial Physics*. Vol. 50.

McCracken, Ken G., Gisela A. M. Dreschhoff, Edward J. Zeller, Don F. Smart, and Margaret A. Shea. 2001. Solar cosmic ray events for the period 1561–1994, 1, Identification in polar ice, 1561–1950. *Journal Geophysical Research*, 106. American Geophysical Union. Washington, DC.

Odenwald, Sten and James Green. 2008. Bracing the Satellite Infrastructure for a Solar Superstorm. *Scientific American*. Vol. 299. Nature America. New York, NY.

Tinsley, Brian. 2000. Influence of Solar Wind on the Global Electric Circuit and Inferred Effects on Cloud Microphysics, Temperature, and Dynamics in the Troposphere. *Space Science Reviews*. Vol. 94. Springer. Dordrecht, Netherlands.

Zhentao, Xu. 1990. Solar Observations in Ancient China and Solar Variability. *Philosophical Transactions of the Royal Society of London*. Vol. 330, Issue 1615. Royal Society Publishing. London, England.

Figure References

Carrington, Richard C. 1860. Description of a singular appearance seen in the Sun on September 1, 1859. *Monthly Notices of the Royal Astronomical Society*. Vol. 20. Royal Astronomical Society. London, England.

Cliver, Edward W. 2006. The 1859 Space Weather Event: Then and Now. *Advances in Space Research*. Vol. 38, Issue 2. Elsevier. Amsterdam, Netherlands.

Hargreaves, John K. 1992. *The Solar Terrestrial Environment.* Cambridge, UK: Cambridge Press, Cambridge, UK.

Further Reading

Freeman, John W. 2001. *Storms in Space.* Cambridge Press.

Siscoe, George. 2007. Space Weather Forecasting Historically Viewed through the Lens of Meteorology. *Space Weather – Physics and Effects.* Volker Bothmer and Ionnis Daglis, editors. London, England: Praxis Publishers.

Tribble, Alan C. 2003. *The Space Environment Implications for Spacecraft Design.* Princeton University Press.

Tsurutani, Bruce T., Walter D. Gonzalez, Gurbax S. Lakhina, and Sobhana Alex. 2003. The extreme magnetic storm of September 1–2, 1859. *Journal of Geophysical Research,* 108, No. A7. American Geophysical Union. Washington, DC.

US Department of Commerce, National Oceanic and Atmospheric Administration, Service Assessment: Intense Space Weather Storms, October 19–November 07, 2003, National Weather Service, Silver Spring, Maryland, 2004.

Historical References

Barlow, William H. 1849. On the Spontaneous Electrical Currents Observed in the Wires of the Electrical Telegraph. *Philosophical Transactions of the Royal Society of London.* Vol. 139. Royal Society Publishing. London, England.

Carrington, Richard C. 1859. On the distribution of the solar spots in latitude since the beginning of the year 1854; with a map. *Monthly Notices of the Royal Astronomical Society.* Vol. 19. Royal Astronomical Society. London, England.

Hodgson, Richard. 1860. On a curious appearance seen in the Sun. *Monthly Notices of the Royal Astronomical Society.* Vol. 20. Royal Astronomical Society. London, England.

Loomis, Elias. 1859. The Great Auroral Exhibition of August 28th to September 4th, 1859. *American Journal of Science and Arts.* Vol. 78. Published by Benjamin Silliman.

Maunder, Edward W. 1905. Magnetic disturbances, as recorded at the Royal Observatory, Greenwich, and their association with sunspots. *Monthly Notices of the Royal Astronomical Society,* Vol. 65. Royal Astronomical Society. London, England.

Sabine, Edward. 1852. On periodical laws discoverable in the mean effects of the larger magnetic disturbances. *Philosophical Transactions of the Royal Society of London.* Vol. 142. Royal Astronomical Society. London, England.

Schwabe, S. Heinrich. 1844. Solar observations during 1843. *Astronomische Nachrichten.* Vol. 21, No. 495.

Stewart, Balfour. 1861. On the great magnetic disturbance which extended from August 28 to September 7, 1859, as recorded by photography at Kew Observatory. *Philosophical Transactions of the Royal Society of London.* Vol. 151. Royal Society of London. London, England.

Von Humboldt, Alexander. 1851. *Kosmos: Entwurf einer physichen Weltbeschreibung.* Vol. 3, No. 2. Cotta, Stuttgart.

Space Is a Place...with Energy

2

UNIT 1. SPACE WEATHER AND ITS PHYSICS

Contributions by C. Lon Enloe, M. Geoff McHarg, and Brian Patterson

You should already know about...

- Dot products, cross products, and integration
- Newton's Laws of Motion
- How force relates to momentum, pressure, and impulse
- The Ideal Gas Law
- Conservation of energy
- Atomic and molecular masses described in universal atomic mass units
- The relations between voltage differences and electric fields
- The origin and general properties of mechanical waves and the Doppler effect
- The radiative model of the hydrogen atom

In this chapter you will learn about...

- Transformation, storage, and distribution of energy in the space environment
- Particle, photon, and field interactions in the space environment
- Various ways to categorize energy content
- Sources and properties of electromagnetic energy
- Pressure and pressure changes
- The Planck radiation function and its relation to blackbody radiation
- The Stefan-Boltzmann Law and Wien's Law
- Discrete line spectra
- Spectroscopy and spectroscopic notation
- Bremsstrahlung, synchrotron, and plasma oscillation radiation

Outline

2.1 **Introduction to Energy in Space**
    2.1.1 The Physical Concept of Energy in a System
    2.1.2 Forms and Transformations of Energy
    2.1.3 Energy Conservation and Entropy
    2.1.4 Energy at High Speeds

2.2 **Energy Interaction within the Space Environment**
    2.2.1 Energy Transfer via Particles and Fields
    2.2.2 Electromagnetic Energy Transitions: Electromagnetic Waves and Photons

2.3 **Characterizing Electromagnetic (EM) Radiation**
    2.3.1 Blackbody Radiation
    2.3.2 Discrete Line Radiation and Absorption
    2.3.3 Radiation from Other Sources

Chapter 2 Space Is a Place...with Energy

2.1 Introduction to Energy in Space

Here we define energy, describe the forms in which it appears, and provide insight into how nature transforms, stores, and distributes it in the space environment. Advanced readers should review the example problems in Secs. 2.1 and 2.2 and then proceed to Sec. 2.3

2.1.1 The Physical Concept of Energy in a System

Objectives: After reading this section, you should be able to...

- Define work, heat, and energy and explain their relationships
- Distinguish between open, closed, and isolated systems
- Distinguish between energy and power
- Calculate values of energy and power in the space environment

Energy is the currency that nature uses to transact its business. We care about energy because every aspect of space weather is constrained by energy and its transformation or transfer within a system.

Energy in Systems

Energy is a complex concept, so we begin with a definition. Energy is the capacity of a system to perform work or transfer heat. The basic international system (SI) unit of energy is the joule [J] or $kg \cdot m^2/s^2$. Figure 2-1 illustrates energy relationships between a system and its environment.

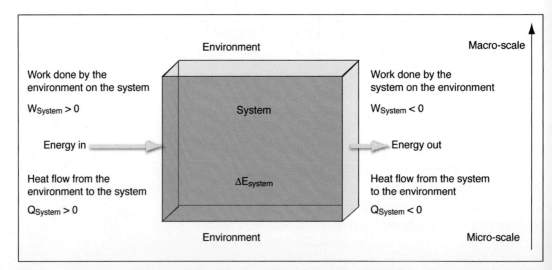

Fig. 2-1. Schematic of Energy Relationships in a Closed System. Energy is a scalar. The arrows represent the magnitude of the scalar. When work (W) is done on the system or heat (Q) flows to the system, energy enters the system. Energy exits the system via heat flow to the environment or work done on the environment.

2.1 Introduction to Energy in Space

All matter has energy, but not all energy is associated with matter. Energy exists in the forms of photons, fields, and the rest energy of particles. Forces act via particles, photons, and fields to transfer energy to other objects or fields. The basic transfer processes are work and heat.

Energy is conserved. Objects pass their energy to other objects, photons, or fields. Energy changes form within a system, but it is not created or destroyed. Only forces outside (external) to the system change its energy. When internal forces act, they change the form of energy within the system.

The energy of a whole system is the sum of the various forms of energy in each of its parts. We usually define a system as all matter and energy within a volume of space. An *open system* allows mass and energy exchange between the system and the environment. In contrast, a *closed system* does not allow mass transfer. An *isolated system* is even more restrictive, allowing neither mass nor energy exchange.

Pause for Inquiry 2-1

Enter one of the following terms in the blanks below: open, closed, or isolated.

Our discussions in Chap. 1 reveal that energy does in fact enter our geosphere. Our geosphere is clearly not a(n) _____ system. The magnetic nature of Earth's domain effectively deflects all but about 1%–2% of the mass delivered by the solar wind. Knowing this we describe the undisturbed magnetosphere as approximately a(n) _____ system. When the solar wind and interplanetary magnetic field merge with Earth's magnetosphere during storm time, our geospace becomes a(n) _____ system.

Work–Mechanical, Macroscale Interactions

Along with energy, we need to understand the concepts of *work* and *heat*. As defined in the physical sciences, work (W) describes the transfer of energy to or from a system by application of an outside force. We think of this force as an identifiable push or pull that creates a macro-scale, mechanical interaction. As an example, the exhaust gas from a rocket engine pushes a spacecraft forward.

A component of the external force must be parallel to the displacement for positive work to occur (Fig. 2-2). If the force is perpendicular to the displacement or if no displacement occurs, then no work is done.

$$\Delta E = W = \int F_{external} \cdot d\mathbf{r} = \int F_{external} dr \cos\theta \quad (2\text{-}1)$$

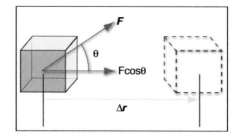

Fig. 2-2. **Force on an Object.** The force (**F**) at angle (θ) does work on the object as it moves through a distance (Δr).

where
- ΔE = energy change [J]
- W = work done [J]
- $F_{external}$ = applied external force vector [N]
- $d\mathbf{r}$ = infinitesimal displacement vector [m]
- θ = angle between the force and displacement vectors [rad or °]

The dot between the two vectors indicates a vector product involving only those components of force parallel to the system displacement. It produces a scalar quantity called work that quantifies the energy change. The integral sign

45

means to sum small segments of the force-times-distance product contributing to the total work. In applying Eq. (2-1) we assume the net force is constant over each small distance segment (d**r**) but may be different from one segment to the next.

Pause for Inquiry 2-2

In the three illustrations below where d**r** is the distance vector and **F** is the force vector, determine the sign of the work done on a system: Work > 0, = 0, or < 0.

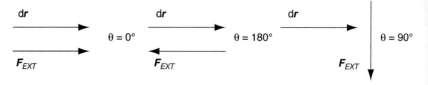

Pause for Inquiry 2-3

Suppose a 1000-kg satellite, located one R_E above Earth, completes a circular orbit. How much work does gravity do over the course of the orbit? Which figure in Pause for Inquiry 2-2 best describes this situation?

Heat—Thermal, Microscale Interactions

Energy transfer to or from a system on a microscopic level as a result of temperature difference is *heat* and is represented by Q. Heating occurs via countless atomic or molecular collisions. We don't try to track the individual pushes and pulls; rather we measure the energy change inside the system by macroscopic means such as temperature or phase change. As an example, radiative energy (photons) from the Sun energizes matter in Earth's upper atmosphere. Matter thus exposed to the Sun's radiative emissions gains energy that results in a temperature increase and often increased ionization. At Earth the temperature rise in the upper atmosphere causes the atmosphere to expand and increases drag forces on low-Earth orbiting satellites. Additionally, the ionization changes the chemistry of the upper atmosphere.

If we quantify heat incrementally as dQ, then we can investigate the heat changes in terms of associated temperature changes or the ability of a heated fluid system to perform work by changing its volume. Heat change in a mass of fluid is given by

$$dQ = mc_v dT + p dV \qquad (2\text{-}2)$$

where

m = mass of the material [kg]

c_v = specific heat at a constant volume [J/K kg]

dT = temperature change [K]

p = pressure [J/m³]

dV = volume change [m³]

Equation (2-2) is one form of the *First Law of Thermodynamics*, which is really a statement of energy conservation. Heat (energy on the microscale) never disappears. It transforms as temperature change, volume change, or both.

Pause for Inquiry 2-4

Suppose we isolate a certain amount of gas with a real or imaginary boundary so that it does not exchange heat with its surroundings. If the volume does work on its surroundings by expanding, where does the energy come from for the expansion?

Power—The Rate of Energy Exchange

Often we are as interested in the rate of energy change as we are in the amount of energy. *Power* is the rate at which work (heat) is produced or dissipated. Its units are J/s or watts [W].

$$P = \frac{dE}{dt} = \frac{d(W+Q)}{dt} \tag{2-3}$$

where
- P = system power [W]
- E = system energy [J]
- t = time [s]
- W = work on or from the system [J]
- Q = heat transfer to or from the system [J]

Power values tell us how rapidly energy is exchanged between a system and its environment. In our Earth-bound interactions, energy values range from a few joules to several million joules. Similarly, power values range from a few watts to several million watts.

Tables 2-1 and 2-2 compare energy and power values. Focus Box 2.1 plots the energy versus power of common and headline-making events. A large solar eruption is one of the most powerful and energetic events in our solar system.

Table 2-1. **System Energy Values.** Here we show typical values for ordinary and notable events.

System	Energy
Raising this textbook from the floor to a desk	~10 J
The daily nutritional energy requirement for a healthy adult human	~8×10^6 J
The daily consumption of a large commercial satellite	~3×10^8 J
Energy of a one megaton hydrogen bomb	~6×10^{15} J
Annual human global energy production	~3×10^{20} J
Large solar flare release	~10^{25} J

Table 2-2. **System Power Values.** Here we list power levels for various items or events.

System	Power
Hair dryer	~10^3 W
Power consumption of a large commercial satellite	~4×10^3 W
Catastrophic power to humans and most hardware	~10^4 W
Power from a large coal or nuclear power plant	~10^9 W
Power of a large solar flare	~10^{22} W
Power of a 1 megaton hydrogen bomb	~6×10^{24} W
Total solar power output (luminosity)	~4×10^{26} W

Pause for Inquiry 2-5

Where would a moderate hurricane lasting for five days fit on Fig. 2-3? Where would a 1500 kg car traveling at highway speeds that stops by crashing into a concrete barrier over a 3-second interval fit on the chart?

EXAMPLE 2.1

Energy in a Flare

- *Problem Statement:* How energetic is a solar flare? Observations indicate the October 28, 2003 solar disturbance released 10^{25} J of radiant energy during about two hours. What fraction of the solar luminosity (3.81×10^{26} J/s) is this?
- *Relevant Equations or Concepts:* Definition of Energy and Power
- *Given:* 10^{25} J of radiant energy released during two hours

- *Solution:* Power is energy change per unit time: Power = dE/dt
  The average radiant power from the flare was 10^{25} J /7200 s = 1.39×10^{21} J/s. Thus the fractional luminosity is

  $(1.39 \times 10^{21}$ J/s$)/(3.81 \times 10^{26}$ W$) = 3.6 \times 10^{-6}$

- *Interpretation:* The flare output was about 3 millionths (3×10^{-6}) of the total solar luminosity. However, in the X-ray portion of the Sun's spectrum, the flare was several thousand times above the background level (Fig. 1-23).

Follow-on Exercise: Compare this result to the average rate of work done to accelerate the associated mass ejection to a speed of 2000 km/s. Use the time interval of two hours. (Recall that you calculated the mass of this ejection at the end of Chap. 1.)

Focus Box 2.1: Energy Versus Power

How do space weather storms compare to events that occur in our daily lives or events that grab headlines? Further, should we compare energy or power? Figure 2-3 addresses these questions by plotting energy versus power for events such as marathons, thunderstorms, earthquakes, etc. Some of them release lots of energy for a short time and have high power output, while others release lower energy for a longer time and have lower power output.

In general, more powerful natural events make headlines, but they cannot occur without a large reservoir of energy. Solar eruptions are the most energetic events in our solar system. They release some of the Sun's store of magnetic energy. Fortunately Earth receives only a tiny fraction of this energy and power. Global magnetic storms, resulting from episodic solar emissions, are at the high end of powerful terrestrial-based events (pipeline electric currents, enhanced auroral displays, transformer meltdowns, etc.). The important thing to recognize about global magnetic storms is that they occur several times a year and influence the entire space environment.

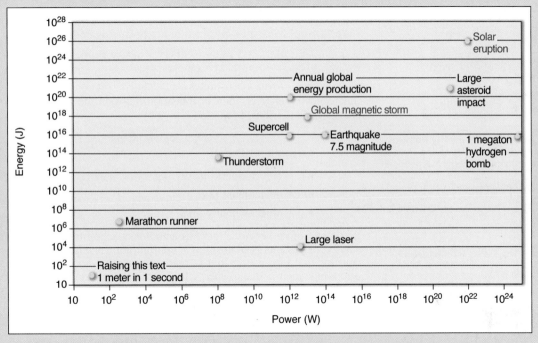

Fig. 2-3. Energy versus Power for Various Events. Here we plot some events in our lives and processes in the solar system. The plot is logarithmic on both axes, allowing us to plot data that span many orders of magnitude. Some events release huge amounts of energy but for a short time so their power output is higher than other events that last longer and have a lower energy output. Space weather events are shown in red. *(Data Courtesy of Steve Hill at NOAA, Space Weather Prediction Center)*

2.1.2 Forms and Transformations of Energy

Objectives: After reading this section, you should be able to...

♦ Compare and contrast kinetic and potential energy
♦ Explain the concept of internal energy
♦ Relate temperature to the speed of individual molecules and atoms
♦ Explain why units of electron volts (eV) and joules (J) are useful in space physics

We pursue a two-fold interest in energy: 1) How do external forces transfer energy to or from a system? That is, how does a system do work? and 2) How do forces internal to the system transform energy?

Mechanical Energy—System Energy on the Macro-scale

Figure 2-4 illustrates energy pathways inside a system. The most fundamental forms of energy are kinetic and potential energy. The sum of these, on the macroscopic level, constitutes *mechanical energy*. In physical systems mechanical energy is often in transition between the two basic forms. *Kinetic energy (KE)* is energy of motion

$$KE = (1/2)mv^2 \quad (2\text{-}4)$$

where

KE = object kinetic energy [kg·m^2/s^2 = J]
m = object mass [kg]
v = object velocity [m/s]

It comes from a net force that has acted or is acting to create motion. *Potential energy (PE)* is energy of position or configuration and is created and unleashed by conservative forces as matter is rearranged within a system. Energetic systems often trade their potential energy for kinetic energy and vice versa.

Particles accelerate as forces act and kinetic energy changes. The *work-kinetic energy theorem* says that the change in kinetic energy equals the net work done on the system by external forces. We write this as

$$\Delta KE = KE_{final} - KE_{initial} = \int \mathbf{F} \cdot d\mathbf{r} = W_{net} \quad (2\text{-}5)$$

where

ΔKE = kinetic energy change [J]
$\mathbf{F}$ = force vector applied to the system [N]
$d\mathbf{r}$ = system displacement vector [m]
W_{net} = net work done on the system [J]

Figure 2-5 shows the track of a pendulum as it moves under the influence of gravity. As the pendulum accelerates vertically downward from rest, gravity does positive work on it. The pendulum's initial potential energy transforms into kinetic energy in Earth's gravitational field. When the pendulum returns to it original height, the released kinetic energy converts back into potential energy. This cyclical process is associated with a conservative force—one that gives back the energy to its original form.

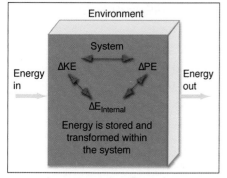

Fig. 2-4. Schematic of Energy Pathways Inside a System. According to the Second Law of Thermodynamics, energy that goes into heat and internal forms cannot completely transform into macroscopic potential and kinetic energy.

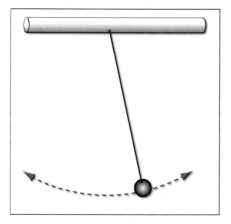

Fig. 2-5. Mechanical Energy Is Conserved. The total mechanical energy, the sum of kinetic (*KE*) and potential (*PE*) energy, is constant in a conservative field. We show this with a simple swinging ball on a string. At the bottom of the arc, the ball's speed is greatest and height is lowest, hence *KE* is at a maximum and *PE* is at a minimum. As the ball rises to the top of the arc, *KE* trades for *PE*, until the ball stops momentarily at the top, where *PE* is at a maximum and *KE* is zero.

If we ignore non-conservative forces such as air drag and string friction at the attachment point, then the object arrives at the base of the arc with an amount of kinetic energy equal to that which it lost from the gravitational potential form. Subsequently the pendulum returns to its original height with all of the potential energy restored, but kinetic energy is zero (it stops, so $v = 0$). In this example the gravitational force "conserves" the energy, releasing it first as kinetic energy and then restoring it to potential energy. The change in potential energy is exactly equal to the negative work done by gravity in raising the object.

$$\Delta PE = -\int \mathbf{F} \bullet \mathrm{d}\mathbf{r} = -\int F \mathrm{d}r \cos\theta \qquad (2\text{-}6a)$$

$$\Delta PE = -W \qquad (2\text{-}6b)$$

where
ΔPE = potential energy change [J]

Often $\mathbf{F}$ is the gravity force vector, but more generally it's any conservative force vector. The minus sign is important. Equations (2-6a) and (2-6b) say that to increase potential energy, energy in the form of work must come from elsewhere in the system. This transfer process is the essence of energy conservation. Stored energy has the potential to transfer back to kinetic energy (Examples 2.2 and 2.3). Many forms of potential energy exist, because nature has many types of conservative forces. We list several forms of potential energy in Table 2-3. The give and take among the two fundamental forms of energy allows us to make a more general statement about mechanical energy conservation

$$-\Delta KE = \Delta PE \qquad (2\text{-}7a)$$

Restating this in terms of the change in mechanical energy (ΔE) gives us

$$\Delta E = 0 = \Delta KE + \Delta PE \qquad (2\text{-}7b)$$

Table 2-3. Categories and Examples of Energy. Here we list several forms of kinetic and potential energy.

Kinetic Energy of Motion	Potential Energy Stored in the Configuration of Mass or Fields
Motion Movement of mass from one place to another or about an axis of rotation	*Gravitational* Energy of mass in a gravitational field
Electricity Organized motion of charged particles	*Electrical* Energy of charged mass held in or bound by an electric field
Thermal Macroscopic motions of particles internal to a substance: Translational (heat), rotational, vibrational; and torsional	*Chemical and Latent* Energy stored in the electronic bonds of atoms or molecules. Also referred to as internal energy
Mechanical Waves Oscillatory motion of mass	*Stored Mechanical Energy* Created by an application of a force
	Nuclear Energy Energy released by mass conversion, stored in the nuclear bonds of nucleons
Combined Kinetic-Potential Forms of Energy	
Radiant Energy: Electromagnetic energy propagating via field oscillations	
Magnetic Energy: Energy in the magnetic field created by moving charged particles	

EXAMPLE 2.2

Mechanical Energy Conversion

- **Background:** Cool dense tubes of mass, called filaments, may be supported in the solar atmosphere by magnetic tension force for days to weeks (Fig. 2-6). The filaments may span distances of up to a quarter of a solar radius. Although magnetic field reconfiguration is generally thought to be the driver behind most solar flares, a small number of flares may be created when mass suspended above the Sun suddenly drains to the surface. These are Hyder flares, after Charles Hyder, who suggested a mechanism for their creation.

- **Problem Statement:** Show that the gravitational potential energy released by the sudden downfall of about half of the material in a solar filament could power a major solar flare. For estimation purposes use a filament radius of 500 km, centered at a height of 2000 km. Also assume that the density of filament material is 10^{-3} kg/m^3 (Figs. 2-6 and 2-7).

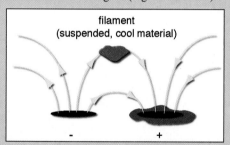

Fig. 2-6. Material Suspended in Filaments Formed by Magnetic Arches. The diagram shows a cross section of structures in the Sun's atmosphere that can support masses of cool plasma against the force of gravity in the Sun's lower atmosphere. The fields, shown in yellow, may be anchored in sunspots.

- **Relevant Concepts:** Work and potential energy, volume of a tube (cylinder)

- **Given:** Mass density, $\rho = 10^{-3}$ kg/m^3, tube length = 1.75×10^8 m, solar mass is 2×10^{30} kg, tube radius = 500 km.

- **Assume:** The magnetic tension force and gravity acting on the tube create a static equilibrium that can be disrupted by shock waves and other forms of disturbances. The gravitational acceleration is constant at the altitudes in question and equals the value of 272 m/s^2, as determined in Ex. 1.1.

- **Solution:** Determine the amount of work done by the magnetic tension force to lift the filament material shown in Fig. 2-6 and 2-7. The magnetic tension force does work on the tube that is equal and opposite to the work done by the gravitational force. The shock wave allows the stored energy to be converted to large-scale kinetic energy and then to heat and radiant energy.

One half of the mass in the tube is

(0.5)(tube volume)(density of tube material)

$$(0.5)(\pi r_{tube}^2 l_{tube})\rho =$$

$$(0.5)(3.14)(5 \times 10^5 \text{m})^2 (1.75 \times 10^8 \text{ m})$$

$$(10^{-3} \text{kg/m}^3) = 6.87 \times 10^{16} \text{kg}$$

The work done by the magnetic tension force is

$$\int F \cdot dr = \int F_{g\,Sun} \cdot dr = -m_{tube} g_{Sun} h$$

$$= (6.87 \times 10^{16} \text{kg})(272 \cdot \text{m/s}^2)(2 \times 10^6 \text{m}) \approx 3.74 \times 10^{25} \text{ J}$$

This energy could be transformed to kinetic energy if the filament were to fall.

- **Interpretation:** Table 2-1 indicates that the energy release in a large solar flare is $\sim 10^{25}$ J, so the downfall of a fraction of the filament material could provide a significant part of the energy radiated in a solar flare. Some of the remaining energy could propel the other half of the mass into space. Scientists believe that most flare energy comes from a process called magnetic merging that converts magnetic potential energy (rather than gravitational potential energy) to radiation. In some cases both energy sources may be tapped.

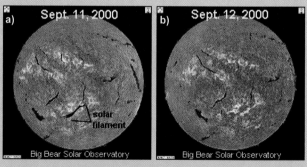

Fig. 2-7. Solar Images Viewed Through a Hydrogen-alpha Filter at Big Bear Solar Observatory. a) On Sept 11, 2000 many dark linear filaments are present. The cooler material in these filaments appear dark against the hot background. b) One of the filaments near the central meridian is absent in the Sept 12th image. During the interval between the two exposures, the filament collapsed, spawning a powerful X-ray solar flare and a powerful coronal mass ejection. *(Courtesy of the Big Bear Solar Observatory/New Jersey Institute of Technology)*

Follow-on Exercise: Estimate the filament material's speed as it nears the solar surface.

EXAMPLE 2.3

Escape Speed

- *Problem Statement:* Determine the escape speed of an object that is attracted to a central gravitational body.
- *Relevant Concept:* Conservation of mechanical energy
- *Given:* Universal gravitational constant (G = 6.67×10^{-11} N·m²/kg²)
- *Assume:* The universe is our system. The soon-to-escape object (m) is at the surface of the other object ($R_{surface}$).
- *Solution:* The gravitational potential energy associated with any pair of masses (M and m) separated by a distance (r) is given by

$$PE = -\frac{GMm}{r}$$

where

PE = gravitational potential energy [J]
G = universal gravitational constant (6.67×10^{-11} N·m²/kg²)
M = mass of the central body [kg]
m = escaping object's mass [kg]
r = distance between the centers of mass [m]

The negative sign indicates the masses are gravitationally attracted to each other.

The total mechanical energy is

$$E_{Total} = KE + PE = \frac{1}{2}mv^2 - \frac{GMm}{r} = \text{constant}$$

To break the gravitational binding, the object needs to separate to a distance, $r = \infty$. Arriving at that distance with minimum energy implies that the final speed (v_f) will be zero. We write the energy conservation equation as

$$\frac{1}{2}mv_i^2 - \frac{GMm}{R_{surface}} = 0 - \frac{GMm}{\infty}$$

where the right side equals zero.

$$\frac{1}{2}mv_i^2 = +\frac{GMm}{R_{surface}}$$

The initial speed (v_i) a body must have to escape is then

$$v_i = v_{escape} = \sqrt{\frac{2GM}{R_{surface}}}$$

- *Interpretation:* A very useful application of energy conservation is evident in determining: 1) the speed an object must have to escape the gravitational attraction of a central body, and 2) the distance the object must travel to convert the initial kinetic energy completely into potential energy ($v_f = 0$ m/s). This case is purely hypothetical, but it serves as a useful limiting case for spaceflight.

Follow-on Exercise 1: Determine the value of escape speed from the Sun. Use the data provided in Chap. 3, Table 3.1.
Follow-on Exercise 2: What initial speed did the Voyager 1 spacecraft need at 1 AU to escape the solar system?

Internal Energy—System Energy on the Microscale

Microscale Kinetic Energy. Interaction between point masses in a system produces potential and kinetic energy on the microscale

$$E_{internal} = KE_{internal} + PE_{internal}$$

Tracking the energy for each atom in a gas or plasma would be overwhelming, so for the kinetic portion of the energy we use statistics. Figure 2-8 shows the statistical distribution of speeds of individual gas particles for selected temperatures. The distribution is a mathematical function, called a *Maxwell-Boltzmann function* (Sec. 6.3). The distribution represents nature's tendency to spread energy among a population of particles. Most particles have speeds near the center (maximum) of the distribution. Small but finite probabilities exist for individual atoms having very small or very large speeds. Cooler (slower) particle populations have less energy to distribute and fractionally less energy to distribute at the high end of the distribution. The speed distribution is asymmetric because some particles gain high speeds after collisions with other particles, thus populating the right side of the distribution. This tendency is more pronounced for higher temperatures.

Fig. 2-8. Maxwell-Boltzmann Distribution of Particle Speeds for Gases of Different Temperatures. Each distribution curve has the same number of particles. Lower temperature particles have a narrower range of speeds. As the particles' temperatures increase, their speeds spread, and a few particles have very large speeds. We describe the function for creating these curves in Sec. 6.3.

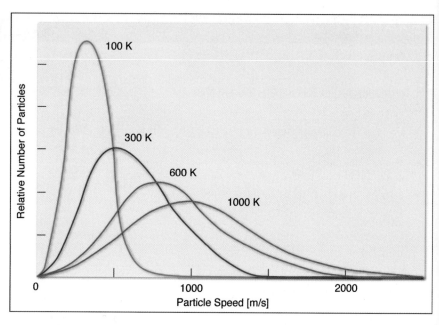

Figure 2-8 implies a relationship between temperature and the particle speed distribution in a gas. In fact, the *root-mean-square speed* (v_{rms}) is related to temperature by

$$\frac{1}{2}mv_{rms}^2 = \frac{3}{2}k_B T \quad (2\text{-}8a)$$

where

m = mass of an atom [kg]

$v_{rms} = \sqrt{v_x^2 + v_y^2 + v_z^2} = \sqrt{v^2}$ = root-mean-square speed [m/s]

k_B = Boltzmann's constant (1.38×10^{-23} J/K)

T = temperature [K]

Equation (2-8a) says that the macroscale variable we call temperature is a statistical measure of a gas's internal kinetic energy. Translational microscale energy is always present in a gas with a temperature above absolute zero. The translational energy increases with rising temperature. Atomic gas particles have motion in three dimensions. Each dimension contributes $(1/2)k_B T$ of energy for a total of $(3/2)k_B T$. For estimation purposes we commonly refer to the energy of a gas or plasma as simply $k_B T$, and ignore the constant in front of the equation.

For multi-particle gases additional forms of thermal motion are often excited. We extend Eq. (2-8a) to account for these additional motions by adding appropriate terms and units of $k_B T$.

$$KE_{internal} = KE_{translation} + KE_{rotation} + KE_{vibration} + KE_{bending}$$

$$= \frac{3}{2}k_B T + \frac{2}{2}k_B T + \frac{1}{2}k_B T + \frac{1}{2}k_B T = \frac{7k_B T}{2} \quad (2\text{-}8b)$$

For a polyatomic material, rotation about the common center of mass adds $(1/2)k_B T$ of energy apiece for the degrees of freedom associated with the latitudinal (θ) and longitudinal (φ) motions. The total contribution is $(2/2)k_B T$ to rotational energy. Vibration and bending each add $(1/2)k_B T$ to the energy budget. Equation (2-8b) indicates that polyatomic gases have multiple ways of distributing heat; that is, they have high heat capacity by virtue of the different types of small-scale motions they exhibit.

Usually we are interested in the energy of an ensemble of particles. To determine the ensemble energy we multiply the individual particle energy by the number of particles (N). To determine the energy density we multiple it by n, the number of particles per unit volume.

Pause for Inquiry 2-6

Convert translational speed to rotational speed to show that rotational kinetic energy is written as $KE_{rotation} = (1/2) mr^2 \omega^2$, where r is the distance of a mass from the axis of rotation and ω is the angular rotation rate.

EXAMPLE 2.4

Thermal Energy

- **Problem Statement:** Determine the root-mean-square speed of an atomic oxygen ion in the thermosphere.
- **Relevant Concept and Equations:** Temperature-energy relation, $\frac{1}{2}mv_{rms}^2 = \frac{3}{2}k_B T$
- **Given:** Mass of the oxygen atom in universal atomic mass units is 16 amu.
- **Assume:** Temperature of the thermospheric gas is 1000 K.
- **Solution:** The average random motion of the ion is related to its temperature by

$$KE = 1/2\ mv^2 = 3/2\ k_B T.$$ Where k_B is the Boltzmann constant, 1.38×10^{-23} J/K

The speed of the oxygen ion is

$$v = \sqrt{\frac{3k_B T}{m}} = \sqrt{\frac{3(1.38 \times 10^{-23} \text{J/K})(1000\text{K})}{(16\text{amu})(1.67 \times 10^{-27}\text{kg/amu})}} = 1245 \text{ m/s}$$

- **Interpretation:** Under rather typical conditions, oxygen atoms in Earth's upper atmosphere move with speeds in excess of 1 km/s. Furthermore, because this value is only an average and some of the atoms move much faster or much slower, speeds of 200 m/s to 2000 m/s are possible. As a point of reference, a good marathon runner moves at ~4 m/s.

Follow-on Exercise 1: Determine the root-mean-square speed for a proton in the corona. See Table 3-1 for the coronal temperature ranges.

Follow-on Exercise 2: Determine the thermal energy in a cubic meter of solar wind where the average particle temperature is 10^5 K and the particle density is 5 particles/cm^3.

Pause for Inquiry 2-7

Compare the root-mean-square speed for oxygen in the thermosphere with Earth's escape speed (Example 2.4) and explain why Earth retains an atmosphere.

We express energy, work, and heat in units of joules (J). However, because particle densities in space are very low—many orders of magnitude less than the air we breathe—space scientists often use a smaller energy measure, called the electron volt (eV), that is appropriate for the tiny particle masses involved. This measure equals the loss of electrical potential energy or the gain of kinetic energy for an electron moving through a potential difference of one volt. An eV relates to the joule by

$$1 \text{ eV} = 1.602 \times 10^{-19} \text{ J}$$

An electric field acting over a distance creates an electric potential difference. Symbolically this is

$$\Delta V = -\int \boldsymbol{E} \bullet d\boldsymbol{r} \qquad (2\text{-}9)$$

where

ΔV = voltage difference [V]

$\boldsymbol{E}$ = electric field intensity vector [V/m]

$d\boldsymbol{r}$ = distance vector over which the electric field acts [m]

If a charged particle interacts with, or experiences, a voltage (potential) difference, the particle will accelerate. Mathematically, we write this relationship as the product of the charge (q) and the voltage difference (ΔV)

$$q\Delta V = -q\int \boldsymbol{E} \bullet d\boldsymbol{r} = -\int q\boldsymbol{E} \bullet d\boldsymbol{r} = -\int \boldsymbol{F}_E \bullet d\boldsymbol{r} \qquad (2\text{-}10)$$

where

$q\Delta V$ = potential energy acquired by the charged particle [J]

$\boldsymbol{F}_E$ = electric force vector [N]

By Eq. (2-6), we can substitute ΔPE for the last term in the above equation. Therefore we write

$$\Delta PE = -W = q\Delta V - \Delta KE \qquad (2\text{-}11)$$

If the charge is that of an electron (–e), and the voltage difference is an increase of one volt, then the energy change is one electron-volt (eV).

$$\Delta KE = -\Delta PE = -q(\Delta V) = -(-e)(\Delta V) = +1 \text{ eV}$$

In all energy equations we may legitimately substitute [eV] for [J]. Table 2-4 provides other useful conversions for the joule.

Microscale Potential Energy. Microscopic potential energy also comes in many forms, such as latent heat energy, electronic energy, electrical energy, nuclear energy, etc. Here we describe two forms of microscale energy: binding electronic energy and nuclear energy.

Binding Electronic Energy and Ionization Energy. When we describe neutral matter, we often ignore the energy that binds electrons to the nucleus. On the microscale this energy is important, especially if we want to know how much energy is required for ionization. The energy that must be supplied to separate the ion-electron pair could come from thermally driven or high-energy impact collisions, from short-wavelength photons, or even from radio frequency oscillations. Typically the energy added to the system is not exactly equal to that

2.1 Introduction to Energy in Space

Pause for Inquiry 2-8

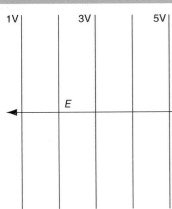

The diagram shows lines of equal potential. An electric field (E), shown by the arrow, is associated with the voltage difference. The voltage values are shown at the top.

Determine in which direction a proton would move to have its kinetic energy increased.

a) up   b) down   c) right   d) left

In which direction would an electron move to increase its kinetic energy?

a) up   b) down   c) right   d) left

Table 2-4. Energy Conversions. Here we list a few energy values with their joule equivalents.

1 erg	$= 10^{-7}$ J	1 British Thermal Unit	$= 1.054 \times 10^3$ J
1 thermo-chemical calorie	$= 4.184$ J	1 kilowatt-hour	$= 3.6 \times 10^6$ J
1 food calorie (1 kilocalorie)	$= 4184$ J	1 megaton	$= 4.18 \times 10^{15}$ J

required for ionization. In that case, the particles involved in the reaction carry away the excess energy as kinetic energy or store it, if they reach an excited state. Example 2.5 explores the energy needed to ionize hydrogen.

Nuclear Energy. The solar core and Earth-bound nuclear reactors draw energy from changes in the structure of the atomic nuclei that release potential energy. By fusion, stellar hydrogen nuclei combine to form helium nuclei. In a nuclear reactor, uranium or plutonium nuclei split apart in a fission process. The energy released in these reactions comes from rearranging the nucleons (protons and neutrons) into a more stable state. To achieve this lower-energy state, a small amount of the matter in the reaction is converted to energy. How does this conversion happen? The answer is that matter is a form of energy. This concept is quantified as Einstein's famous physics formula

$$\Delta E = \Delta m c^2 \qquad (2\text{-}12)$$

where

ΔE = energy change [J]

Δm = mass change [kg]

c = light speed (2.998×10^8 m/s) in a vacuum

The energy stored intrinsically in a piece of matter at rest equals its mass times light speed squared. The quantity, mc^2 is the *rest mass energy*, and the Δ indicates a change in this energy. The entire equation is a statement of mass-energy conservation.

Chapter 2 Space Is a Place...with Energy

EXAMPLE 2.5

Thermal and Ionization Energy

- *Problem Statement:* Determine the temperature that a collection of hydrogen atoms must have for most of the atoms to become ionized by thermal collisions.
- *Relevant Concepts and Equation:* Thermal energy and energy conversions.
- *Given:* For a hydrogen atom, the ionization energy is $E \approx 13.6$ eV.
- *Assume:* Ionization occurs because of thermally driven collisions.
- *Solution:* The temperature associated with this (average kinetic) energy is therefore

$$(3/2)k_B T = (13.6 \text{ eV})(1.6 \times 10^{-19} \text{ J/eV})$$

Solving for temperature, T, we get

$$T = \frac{2(13.6)(1.6 \times 10^{-19} \text{ J/eV})}{3(1.38 \times 10^{-23} \text{ J/K})} = 1.05 \times 10^5 \text{ K}$$

- *Interpretation:* The temperature required for thermal ionization of hydrogen atoms is very high. Temperatures in the Sun's interior and far outer atmosphere are an order of magnitude hotter than this. Thus, we know most hydrogen atoms in these regions are ionized. We recall from our previous discussion of speed distributions that even at lower temperatures some of the atoms have high speeds and hence, sufficient energy to become ionized. However because of collisions they won't maintain an ionized state for very long.

Follow-on Exercise: Based on the ionization temperature for hydrogen, would you expect much of the hydrogen at the Sun's surface to ionize?

EXAMPLE 2.6

Fusion

- *Problem Statement:* During proton-proton fusion events in the solar core, the net energy gained per conversion cycle is 26.7 million electron volts (MeV). What is the mass lost during this reaction? What percentage of the mass of the four reacting protons is lost?
- *Relevant Concept:* Rest mass and energy are equivalent.
- *Given:* $\Delta E = 26.7$ MeV
- *Assume:* Only nuclear forces act. The mass of one proton is 1.007276 amu.
- *Solution:* Mass-energy is conserved.

$$\Delta E = \Delta mc^2 = 26.7 \text{ MeV}$$
$$\Delta mc^2 = 26.7 \text{ MeV} \times 1.6 \times 10^{-19} \text{ J/eV} = 4.272 \times 10^{-12} \text{ J}$$
$$\Delta m = 4.272 \times 10^{-12} \text{ J}/(2.998 \times 10^8 \text{ m/s})^2 = 4.75 \times 10^{-29} \text{ kg}$$

Each hydrogen nucleus has the mass of one proton, and four protons interact to form the helium nucleus:

$$\frac{\text{mass loss}}{\text{original mass}} = \frac{4.75 \times 10^{-29} \text{ kg}}{4(1.007276 \text{ amu}) \times 1.66 \times 10^{-27} \text{ kg/amu}} = 0.007 = 0.7\%$$

- *Interpretation:* A tiny fraction of the mass involved in a fusion reaction is missing at the end. It has been converted to energy.

Follow-on Exercise 1: Assume you could extract all of the energy stored in the matter of a 0.025 kg coin. How much energy would you have? Compare this to the electrical energy needed to run an average city for 1 week.

Follow-on Exercise 2: How many fusion reactions must occur each second to keep the Sun shining with its current luminosity?

2.1 Introduction to Energy in Space

Focus Box 2.2: Solar Nuclear Reactions

Solar energy is generated primarily by reactions involving the strong nuclear force and the electromagnetic force. The most important reaction in the solar core is the *proton-proton (PP) chain reaction*, which has three branches. Here we explore the dominant branch, PP I. Depending on model assumptions, this branch is responsible for 70–90% of the solar nuclear reactions. It consists of multiple steps shown schematically in Fig. 2-9. The average time for each major step is quite different. At core temperatures only one proton in 100 million has enough energy to fuse during a collision. The reaction rate is so low that an average proton requires more than a billion years to find a suitable "hot" partner with which to collide in a successful fusion event. The Sun is only about 4.5 billion years old, so most of its protons have not yet found fusion partners. The intermediate steps take less than ten seconds, and the last step takes one million years. If it takes this long, how is it possible for the Sun to shine? We remember that the number of protons in the Sun is exceptionally large, so that the overall number of reactions is large, although the individual probability of reaction is small.

Fusion reactions in the Sun began only when gravitational compression caused the solar core temperature to exceed 10^6 K. High-speed hydrogen nuclei collide with each other, ultimately fusing into helium. The resulting helium atom is not as massive (by a tiny amount) as the four protons (plus two ambient electrons) that combined to create it. The difference in mass (mass defect) is released as energy, according to $\Delta E = \Delta mc^2$. Altogether 0.7% of the mass of four protons appears as energy.

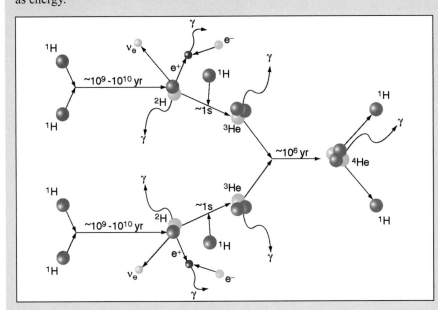

Fig. 2-9. The Proton-proton (PP I) Chain. The proton-proton chain reaction starts with four protons interacting, as shown. The process moves from left to right, releasing positrons, neutrinos, and gamma rays, before ending in a helium nucleus. Each reaction produces a total of 26.7 MeV of energy. Although the individual probability of a given particle finding a partner and interacting with it is small, the entire reaction chain is sustained by the immense numbers of particles available for interaction.

Here are the details of the process. Two protons (^{1}H) interact to create a deuterium nucleus (^{2}H), a positron (e^+), an electron neutrino (ν_e), and a gamma-ray photon (γ) with an energy of 0.42 million electron volts (MeV). The neutrino flies through the Sun and out into the universe at nearly light speed, producing very little interaction with surrounding matter.

$$2[^1H + {}^1H] \rightarrow 2[^2H + e^+ + \nu_e + \gamma] \quad \text{where energy out} = 2[\gamma = 0.42 \text{ MeV}] \quad (1)$$

The positron takes only a second to interact with an ambient electron (e^-), emitting two gamma rays (2γ).

$$2[e^+ + e^-] \rightarrow 2[2\gamma] \quad \text{where energy out} = 2[2\gamma = 1.022 \text{ MeV}] \quad (2)$$

The chain continues with the next reaction. A hydrogen atom (^{1}H) immediately combines with the deuterium nucleus, creating a helium-3 nucleus (^{3}He) and a gamma ray.

$$2[^2H + {}^1H \rightarrow {}^3He + \gamma] \quad \text{where energy out} = 2[\gamma = 5.49 \text{ MeV}] \quad (3)$$

Focus Box 2.2: Solar Nuclear Reactions *(Continued)*

In a subsequent interaction, two helium-3 nuclei eventually combine, creating a helium-4 nucleus (alpha particle, two protons and a photon).

$$[^3\text{He} + {}^3\text{He} \rightarrow {}^4\text{He} + 2{}^1\text{H} + \gamma] \qquad \text{where energy out} = \gamma = 12.86 \text{ MeV} \qquad (4)$$

Ultimately six protons and two electrons interact, producing one helium nucleus, two electron-neutrinos, two protons, and 26.7 MeV of energy. Line (5) summarizes the reaction.

$$[6{}^1\text{H} + 2e^- \rightarrow {}^4\text{He} + 2\nu_e + 2{}^1\text{H} + 26.7 \text{ MeV}] \qquad (5)$$

Theories developed in the 1930s by Hans Bethe and others proposed that every fusion reaction should produce a chargeless reaction by-product called a *neutrino*. Neutrinos and photons exit the Sun, but on vastly different time scales. Unlike photons that require millions of years to escape the solar interior, neutrinos leave the Sun within seconds of their creation. If we could sense every solar neutrino, the sheer number of them would blind us. Billions of neutrinos pass through an area the size of a fingernail every second. However, these ghostly particles ignore matter and most pass through the entire Earth without an interaction. Only the highest energy neutrinos are observable with current technology. Scientists expect to observe hundreds per day, but the best capture facilities observe only about 10 neutrinos per day. Investigations continue. For now, we have the results shown in Fig. 2-10: a fuzzy view of the tiniest particles associated with solar nuclear reactions.

To gain information about the deep solar interior, scientists study neutrinos, the extremely low-mass particles generated by fusion reactions. These particles are key figures in verifying our understanding of nuclear physics and are our primary means of peering into the Sun's heart. However, even demonstrating that these stealthy particles have mass was a major undertaking that yielded a definitive answer only in 1998 [Fukuda et al., 1998]. After the Super-Kamiokande experiment in Japan provided data showing that some types of neutrinos do have mass, further experiments conducted at the Sudbury Neutrino Observatory in Ontario, Canada, showed that neutrinos change appearance as they travel through space [Ahmad et al., 2002]. (They have a natural "cloaking" capability). These discoveries help explain the dearth of observations, but even these experiments capture fewer neutrinos than scientists think they should. Besides having no charge and only the tiniest mass, which makes them notoriously difficult to detect, these particles seem to disappear occasionally. The best hope of capturing any of them comes in focusing on a rare side branch of the proton-proton chain (PP III has <1% occurrence frequency) that should produce a particularly energetic neutrino. If we measure the neutrino flux from this reaction, then we can make progress in verifying the reaction rates of all of the chains.

Fig. 2-10. A Neutrino Image of the Sun. This view of the Sun is from the Super-Kamiokande (Super-K) experiment in Japan. Brighter colors represent a larger flux of neutrinos. Three hundred days of data were collected to produce this neutrino image of the Sun. The picture covers a significant fraction of the sky; however most of the neutrinos are coming from the vicinity of the solar core. *(Courtesy of NASA and Robert Svoboda at Louisiana State University)*

2.1.3 Energy Conservation and Entropy

> **Objectives:** After reading this section, you should be able to...
>
> - Define entropy
> - Relate the Second Law of Thermodynamics to the ability of a system to do mechanical work

The First Law of Thermodynamics states that energy is neither created nor destroyed. Energy that enters a system must be balanced by energy stored in the system and energy exiting the system as shown in Fig. 2-4. The key to solving many energy conservation problems is defining the system. We use a few system examples and describe conservation of mechanical energy for conservative forces earlier in the chapter. Next, we give a thought example that is a little broader, including non-conservative forces resulting in dissipative work ($W_{dissipative}$).

$$E_{total\ initial} = E_{total\ final} = \text{constant}$$

Yet, often we hear of "energy loss." Can energy loss happen? Energy loss usually refers to energy lost from the system to the environment. Accounting for the energy transferring to the environment solves the apparent loss problem.

Suppose a meteor collides with a satellite. The meteor, for a short time, exerts an external force on the satellite, doing work on the satellite and causing it to move faster (if hit from behind) and into a higher orbit, or causing some part of the satellite to "give" (deform) (Fig. 2-11). The satellite probably also gains internal energy—for example, its temperature could rise and it could generate sound waves. The sum of the newly acquired motional energy of the satellite on the large- and micro-scale, and the energy stored in deformation of its components resulting from the collision, equals the amount of energy transferred to the satellite by the meteor.

The energy loss for the meteor is energy gain for the satellite, and we account for all of it by defining the system to consist of the meteor and the satellite. Yet we suspect that something has been lost in this process. We have a net loss of high-quality energy that efficiently does work, and the quality of the energy will never return to the meteor or the meteor-plus-satellite system. In some engineering applications the fact that the energy is no longer available to do work, but instead is available as heat is called energy loss or energy dissipation. The forces involved in such degradations of energy are called *non-conservative forces*.

In most human endeavors (water heaters, car engines, refrigerators, etc.), heat energy or energy stored in random forms is less valuable energy because we don't have easy ways to extract work from it. Nevertheless, these energy forms remain a part of the overall energy budget. This inability to extract useful work, even as energy is conserved, is the basis for entropy.

Entropy (S) is a measure of disorder in a system. The *Second Law of Thermodynamics* says that nature tends to create a disordered state. Energy available for useful work decreases as entropy increases. This rule is part of the Second Law, as illustrated in Fig. 2-12. A corollary to this law is that low-temperature (low-quality energy) systems do not transfer net energy to higher-temperature systems. When the Sun unleashes some of its magnetic energy stored in a complex active region, part of the energy goes into flaring processes that ultimately heat the solar atmosphere. Some of the energy is retained in its

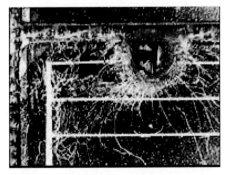

Fig. 2-11. Damage in a Hubble Telescope Solar Array. This damage came from a space impact. The picture was taken by Shuttle astronauts who rendezvoused with the orbiting telescope to make repairs. The hole size is 2.5 mm. *(Courtesy of NASA)*

high-quality magnetic form as an ejected bubble of coronal mass and magnetic field. But much of that energy ultimately dissipates as heat. Coronal mass ejections (CMEs) heat the solar wind and the magnetospheres and atmospheres of planets with which they collide. Focus Box 2.3 describes how CMEs increase entropy.

Pause for Inquiry 2-9

A satellite in low-Earth orbit is acted on by a drag force. In the process the satellite system's

a) Entropy increases

b) Mechanical energy increases

c) Potential energy increases

d) All of the above

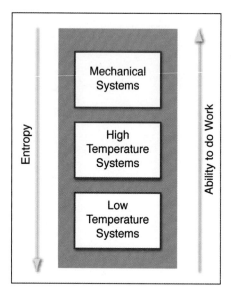

Fig. 2-12. Work in a Mechanical System. When driven by conservative forces, mechanical systems have their greatest ability to do work. But their operations have to be maintained against the tendency toward higher entropy. They are, therefore, the most costly to operate because they need continuous high-quality energy input.

2.1.4 Energy at High Speeds

Objectives: After reading this section, you should be able to...

♦ Calculate the kinetic energy of mass moving at high speeds

♦ Define rest energy

Thus far, our formulation for kinetic energy has been rather simple and quite adequate for mass moving at low speed: $KE = 1/2 mv^2$. In the early 20th century, Albert Einstein realized this approximation was valid for low-speed masses, but we needed a much more general formulation for high speeds. Further, he recognized that the formula failed to account for the energy that is inherent in every mass. The latter situation is rectified by accounting for rest mass energy as

$$E_0 = mc^2$$

This variant of Eq. (2-12) indicates that mass can be converted to energy. The rest energy may become available to do work. (Reference our discussion of nuclear energy in Sec. 2.1.2 and Focus Box 2.2.)

To properly account for the energy in systems whose speed is a significant fraction of light speed, we introduce Einstein's modification of the kinetic energy formula:

$$KE = \left(\frac{1}{\sqrt{1 - \frac{v^2}{c^2}}} - 1 \right) mc^2 \quad (2\text{-}13)$$

When the speed is much less than light speed ($v \ll c$), we approximate the first term in the brackets as the first two terms of a binomial expansion: $(1 + 1/2(v^2/c^2) + ...)$, and so recover our usual equation for kinetic energy:

2.1 Introduction to Energy in Space

$$KE \approx mc^2 + \frac{1}{2}\frac{v^2}{c^2}mc^2 - mc^2 = \frac{1}{2}mv^2$$

Assuming no change in potential energy occurs, the total energy of a moving particle consists of its rest mass and kinetic energy.

$$E_{Total} = mc^2 + KE = \frac{mc^2}{\sqrt{1-\frac{v^2}{c^2}}} = \gamma mc^2 \quad (2\text{-}14)$$

Where the relativistic factor $\gamma = \dfrac{1}{\sqrt{1-\dfrac{v^2}{c^2}}}$

The γ used in this equation should not be confused with the same Greek symbol used for gamma-ray photons.

By extension, the total energy of a system containing potential energy is the rest mass energy plus the potential energy. High-energy systems have more inertia and require a proportionally larger amount of energy to accelerate. They also deliver proportionally more energy to a system with which they interact (Example 2.7 and Focus Box 2.3).

EXAMPLE 2.7

Relativistic Kinetic Energy of Solar Particles

- *Problem Statement:* Determine the change in kinetic energy of a relativistic solar proton that stops in a satellite's solar panel during a solar proton event. Assume it approaches the panel with a speed of 0.33c. Compare the relativistic result with the non-relativistic formula result.
- *Key Concept:* Relativistic kinetic energy
- *Given:* Mass of a proton = 1.67×10^{-27} kg
- *Assume:* Final kinetic energy equals zero.
- *Solution:*

$$KE_i = \left(\frac{1}{\sqrt{1-\frac{v^2}{c^2}}} - 1\right)mc^2 = \left(\frac{1}{\sqrt{1-\frac{(0.33c)^2}{c^2}}} - 1\right)mc^2 = 0.059mc^2$$

$$W = \Delta KE = 0 - (0.059)(1.67 \times 10^{-27}\text{kg})(9 \times 10^{16}\text{m}^2/\text{s}^2)$$
$$= -8.87 \times 10^{-12}\text{ J} = -5.54 \times 10^7 \text{eV} \approx -55.4 \text{ MeV}$$

Inappropriately using the non-relativistic formula results in
$KE_f = 0$
$KE_i = 1/2\ mv^2$

$$W = \Delta KE = 0 - (0.5)(1.67 \times 10^{-27}\text{kg})(9.80 \times 10^{15}\text{m}^2/\text{s}^2)$$
$$= -8.18 \times 10^{-12}\text{ J} = -5.11 \times 10^7 \text{eV} \sim -50 \text{ MeV}$$

- *Interpretation:* The relativistic formulation calculates an impact with ten percent more energy change (work done). Spacecraft instrument designers must account for this result when developing materials for operating in the space environment.

Follow-on Exercise: Relativistic electrons are rather common near geosynchronous orbit. What is the speed of a 1 MeV electron? How do the rest energy and kinetic energy of such a particle compare?

Chapter 2 Space Is a Place...with Energy

Focus Box 2.3: Solar Energetic Particles

The growing number of speckles in Fig. 2-13 resulted from relativistic particles decelerating in the telescope tube of the POLAR spacecraft. The particles were accelerated near the Sun by the leading edge of an Earth-directed CME on July 14, 2000. They experienced multiple accelerations depending on their charge and the orientation of the electric and magnetic fields. After repeated interactions some particles flew toward Earth at approximately one-third light speed. If these particles encounter materials inside of a sensor, they decelerate. During the deceleration, the particles' energy is converted to high-energy photons that subsequently blind the sensor. Sudden increases in the flux of energetic particles are major space weather events that are currently not well forecast.

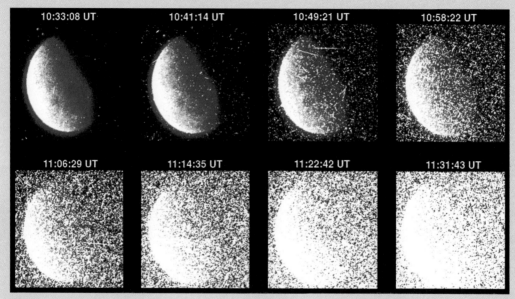

Fig. 2-13. The POLAR Spacecraft View of Energetic Particles Arriving at Earth on July 14, 2000. Image progression is to the right. In the first image, Earth's dayside is visible as a reflected glow of ultraviolet energy at 130.4 nm. The amount of reflected energy decreases at high latitudes, eventually giving way to darkness. Six minutes after the initial image, a "fuzz" appears in the images. The fuzz results from energized solar particles striking the POLAR camera and creating photons that interfere with viewing. The POLAR extreme ultraviolet sensor was overwhelmed by stray photons within a half hour of the event development. It remained out of use for several hours. *(Courtesy of NASA)*

As we discuss in this and subsequent chapters, energetic systems tend to disorder. However, the disorder may be most efficiently achieved by creating structures such as magnetically active regions on the Sun that produce fast CMEs that in turn energize individual particles that move quickly into space to disperse energy.

2.2 Energy Interaction within the Space Environment

2.2.1 Energy Transfer via Particles and Fields

Objectives: After reading this section, you should be able to...

- Compare and contrast energy transfer processes via particles and fields
- Relate the concept of energy density to pressure

Energy Transfer and Transport via Particles and Fields

On the macro scale, charged and neutral masses in bulk, collisional, and oscillatory motions have the ability to do work, transport energy, and exert pressure. Static gravitational, electric, and magnetic fields exert pressure and store energy in potential form. If these fields become dynamic (change with time), they generate waves that transport energy. The entries in Table 2-3 suggest several ways to store and transport energy.

Micro-scale energy transport processes are usually invisible, and the pathways are many. Scientists simplify these processes into three forms: conduction, convection, and radiation. *Conduction* and *convection* are temperature difference processes mediated by particle interactions. *Radiative processes* account for the transformation of thermal energy into photons and vice versa, thus enabling subsequent interactions at light speed.

Photons exert pressure. When a photon is absorbed, electromagnetic forces do work on or heat the absorbing material. We describe these processes in more depth in Sec. 2.2.2. In the context of heat energy, radiation is really a conversion of heat energy to photons of light, and the subsequent transport of those photons. Radiative cooling is important at the Sun's surface, where thermal energy escapes from convection cells, thus allowing the material at the Sun's surface to cool and sink to lower levels.

Particle Interactions

Conduction is the diffusion of thermal energy from hotter matter to cooler matter. Metals are excellent conductors of thermal energy, whereas wood and plastic are not; they are insulators. A small fraction of the energy transfer process in the Sun's upper atmosphere results from electron heat conduction.

Convection is heat transfer by the large-scale movement of hot or cold portions of a fluid. Convection is usually the dominant form of heat transfer in liquids and gases and occurs only in fluids. The rising bubbles in a pot of boiling water are examples of convective motion. We commonly distinguish two types of convection: *free convection*, in which natural forces (gravity and buoyancy forces) drive the fluid movement and *forced convection*, where a fan or other stirrer drives the fluid. Free convection is the dominant energy transfer process in the outer third of the Sun's interior. Forced convection occurs in the magnetosphere when the solar wind forces a stirring of magnetic fields and plasma.

Radiation refers to energy traveling as photons or electromagnetic (EM) waves. When dealing with radiant energy we either use the particle (photon) model of light or the wave (EM) model. Certain measurements and applications are described more easily by one model or the other. When EM waves encounter matter, they transform energy via radiation. The energy absorbed by matter usually produces *photothermal energy* (heat) in the object, whereby the photon's energy converts to vibrations of the molecules. These vibrations are called *phonons* (heat energy). As an example, the Sun-facing sides of satellites and planets are heated by photon interactions that subsequently transfer energy to non-sunlit regions of the satellite. Another possibility is a *photoelectric transfer* that converts photons to kinetic energy of conduction electrons (electrical energy). Solar energy, converted to electric potential energy, is stored in batteries. Solar panels produce electrical energy to power satellites using photoelectric transfer. *Photochemical processes* cause chemical changes that effectively store the energy. For example, in the upper atmosphere, compounds containing nitrogen and oxygen change their properties during solar storms and auroral disturbances. Photons with sufficient energy cause *photoionization* that subsequently contributes to any of the other photo-processes.

Field Interactions

Fields are one of nature's ways of storing and exchanging energy without physical contact. They exert pressure. There are several types:

Gravitational Field. When a falling object accelerates in a gravitational field, its potential energy converts to kinetic energy. Similarly, when an object is lifted, the gravitational field stores the lifter's energy as potential energy in the Earth-object system. *Example:* The descent of the Space Shuttle in Earth's gravitational field converts gravitational potential energy into large-scale kinetic energy, heat, and sound waves.

Electric Field. An electric field causes an electrical force that accelerates charged particles and converts potential energy into kinetic energy or vice versa. Charged particles possess potential energy in the presence of an electric field similar to the potential energy of a mass in a gravitational field. Further, charged particles interact and transfer energy via the electric and magnetic fields they create. *Example:* An electric current in a conductor moves charges and causes molecules to vibrate, converting electrical potential energy into kinetic and thermal energy. Microscale electric fields create friction in solids and viscosity in fluids.

Magnetic Fields. Static magnetic fields do no work on charged particles, because the magnetic field always acts perpendicular to the direction of motion of charged particles (without a parallel component between force and distance, work is zero). Magnetic fields do, however, exert pressure and store energy in their configuration. For example, as the solar wind flows past Earth's magnetosphere Earth's field lines become stretched. A stretched dipole field stores more energy than an unstretched field. Changing magnetic fields create electric fields that change charged particles' kinetic energy, and thus do work on them. *Example:* This concept gives us electromagnetic dynamos in a variety of space environment processes and Earth-based power generation systems.

Strong Nuclear Fields. When the fields of the strong and weak nuclear forces do work, some nuclei reach an excited state and some subatomic particles convert a portion of their mass to energy. There are two basic types of conversion: fusion (when nuclei combine) and fission (when nuclei split).

We delve further into field-energy relations in Chap. 4.

2.2 Energy Interaction within the Space Environment

Energy Density and Pressure

A unifying concept for energy storage and transport via particles, photons, and fields is energy density, or energy in a volume available to do work. The higher the energy density the greater the pressure (force per unit area) on the surroundings. In Table 2-5 we provide examples of several different forms of energy density associated with particles, fields, and waves.

Thermal pressure gives us a sense of warmth by transferring heat from air to our skin. Sometimes the transfer goes the other way and we feel cold. We know that the fast flow of water from a fire hose has energy and transfers it to other objects. Similarly, the fast flow of the solar wind transfers energy to objects in its path, in particular Earth's magnetosphere. Both of these flows have energy density that we call ram pressure. Pressure created by fields and waves relates to the squared magnitude of the field inside the volume. When magnetic fields are enhanced within the solar atmosphere, the energy density may exceed a threshold for stability. The result is a coronal mass (and magnetic) ejection. Wave intensity (power flux (I)), in the form of waves traveling at speed v, delivers energy to a volume to be stored as energy density (I/v), or in the case of EM energy (I/c). We say more about energy flux in the next section.

> **Pause for Inquiry 2-10**
>
> Verify that each formula in Table 2-5 has units of J/m^3 and N/m^2 (pressure).

Table 2-5. Forms of Energy Density. Here we list forms of energy density important to the space environment. (Units for **B** are T or $N/(A \cdot m)$).

Pressure (J/m^3)	Formula	Comments
Thermal	nk_BT	n = number of particles per unit volume k_B = Boltzmann constant (1.38×10^{-23} J/K) T = temperature
Ram or Dynamic	$1/2 \rho v^2$	ρ = mass density (kg/m^3) v = flow speed (m/s)
Magnetic Field	$B^2/2\mu_0$	B = magnetic field strength (T) μ_0 = permeability constant (1.26×10^{-6} N/A^2)
Electric Field	$\varepsilon_0 E^2/2$	E = electric field strength (V/m) ε_0 = permittivity constant (8.85×10^{-12} $C^2/(N \cdot m^2)$)
Electromagnetic Wave	$(1/2)(\varepsilon_0 E^2 + B^2/\mu_0)$	Average amplitude of electric and magnetic field
Radiation	I/c	I = electromagnetic wave intensity (W/m^2) c = light speed (2.998×10^8 m/s)

Focus Box 2.4: Energy Density in Thermal Fluids

Energy density in a static fluid is related to the pressure (force per unit area) created by particle collisions. To develop this idea we use a fluid-in-a-box approach; however the concept is applicable to other geometries and other media. We consider the one-dimensional behavior of a relatively low-density gas inside a cubic container as in Fig. 2-14. Along the y-axis, a particle (such as an electron, atom, or molecule) with velocity v_1 strikes the perfectly reflecting right face of the cube and rebounds with velocity v_2. If the particle has mass (m), then the change in the particle's momentum (mv) is

$$\Delta(m\mathbf{v}) = -2mv_y\hat{\mathbf{y}}$$

According to Newton's Second and Third Laws, this momentum change corresponds to an impulse ($F\Delta t\hat{\mathbf{y}}$) exerted by the particle on the face of the cube.

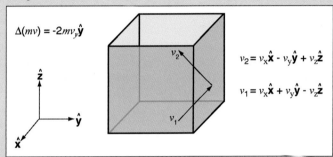

Fig. 2-14. A Particle Exchanging Momentum with the Wall of a Container. The particle with velocity v_1, impacts the box wall and rebounds with velocity v_2. Its momentum change occurs because the particle changes direction as it bounces off the wall.

If, on average, half of the particles in the container are moving to the right at speed v_y, and the particle density is n, then the number of particles per unit time striking the right cube wall with area A is $1/2 n v_y A$. The resulting *pressure* (P), on that face is

$$P = \frac{F}{A} = \frac{(\#particles\ striking\ area/unit\ time)(impulse/particle)}{area} = nmv_y^2$$

Not all particles in the container have speed v_y, so we replace v_y^2 with its average value $(v_y^2)_{av}$ and express the pressure on the wall in terms of the average kinetic energy $(1/2 mv_y^2)_{av}$ associated with motion along the y axis

$$P = \frac{1}{2}(2nmv_y^2)_{av}$$

Dividing and multiplying by a factor of 2 allows us to relate pressure, random kinetic energy, and temperature using the one-dimensional version of Eq. (2-8a):

$$\frac{1}{2}k_B T = \left(\frac{1}{2}mv_y^2\right)_{av}$$

Substitution shows that pressure is a function of temperature and particle density. Pressure, as described in the *Ideal Gas Law*, is

$$P = nk_B T$$

In an ideal gas, particles are well mixed. The speeds of the particles along the various directions are the same, so the pressure is independent of direction.

We extend this idea of pressure to a fully ionized thermal plasma. Pressure in such a fluid is given by $P = 2nk_B T$. The factor of 2 comes from the fact that electrons, free of their parent atoms, also exert particle pressure. With equal numbers of electrons and positive ions, the thermal energy density in a fully ionized plasma is twice that of a non-ionized gas. Intuitively this makes sense. We know that extra energy is required for the ionization, so it must be present in the system.

EXAMPLE 2.8

Thermal and Dynamic Energy Density in the Solar Wind

- *Problem Statement:* The solar wind is a hot, tenuous, magnetized flow of plasma. Compare the relative contributions of thermal and dynamic pressure in the average solar wind.

- *Assume:* The solar wind consists of equal numbers of hydrogen ions and electrons with similar temperatures.

- *Given:* Average solar wind density = 5 ions per cubic centimeter. Average solar wind speed = 400 km/s. Average solar wind ion temperature = 10^5 K.

- *Solution:* Thermal pressure = nk_BT; Dynamic pressure = $1/2 \rho v^2$

Thermal energy density (pressure) = nk_BT for both species: electrons and ions

$2nk_BT = 2 \times (5 \text{ particles/cm}^3) \times (100 \text{ cm})^3/\text{m}^3 \times (1.38 \times 10^{-23} \text{ J/K}) \times 10^5 \text{ K}$
$= 1.4 \times 10^{-11} \text{ J/m}^3 = 1.4 \times 10^{-11} \text{ Pa} = \approx 0.01 \text{ nPa}$

Dynamic energy density (pressure) = $1/2 \rho v^2 = 1/2 n_i m_i v^2 + 1/2 n_e m_e v^2$
$= (1/2)(2n_i)(m_i + m_e) v^2 \approx (n_i)(m_i) v^2$
$= (5 \times 10^6 \text{ ions/m}^3)(1.67 \times 10^{-27} \text{ kg/m}^3 \text{ ion})(4 \times 10^5 \text{ m/s})^2$
$= 1.37 \times 10^{-9} \text{ Pa} \approx 1 \text{ nPa}$

- *Interpretation:* The dynamic (flow) pressure in the solar wind is approximately 100 times the thermal pressure. This thermal pressure relates directly to the supersonic nature of the solar wind.

Follow on Exercise: Compute the magnetic pressure in the solar wind and compare it to the results above. The average magnetic field in the solar wind is 5 nT.

2.2.2 Electromagnetic Energy Transitions: Electromagnetic Waves and Photons

Objectives: After reading this section, you should be able to...

- Use basic wave terminology to describe electromagnetic waves
- Calculate the energy of a photon associated with an electromagnetic wave
- Characterize the electromagnetic spectrum in terms of wavelength and frequency
- Calculate the Doppler shift of an electromagnetic wave
- Calculate the power flux of electromagnetic radiation

Energy transfer and transformation via electromagnetic waves and photons is a dominant element of space weather and the space environment.

Electromagnetic Radiation

Electromagnetic (EM) radiation is energy in transit that has no material aspects. Radiant energy moves through space at light speed without the aid of matter. Electromagnetic radiation also travels through most media at lower speed. Indeed light travels through different materials at speeds that indicate the degree of interaction between the electromagnetic wave and the material. This speed is always less than light speed in a vacuum. Examples include EM waves from accelerating conduction electrons in a radio transmitter antenna and energy level transitions of electrons in an atom.

We know that electric and magnetic fields store energy. When charged particles accelerate they create an EM wave with electric and magnetic components that transport energy from one point to another. This wave transports energy via oscillating electric and magnetic fields. We sometimes describe radiant energy as a stream of particle-like objects called *photons*. The connection between wave-like and particle-like descriptions of EM radiation generally merges in the discipline of quantum mechanics (an accepted theory of atomic structure). In either case, the radiation originates from EM oscillators.

In the photon model we quantify energy as

$$E = hf \quad (2\text{-}15)$$

where

E = photon energy [J or eV]
h = Planck's constant (6.626×10^{-34} J·s = 4.136×10^{-15} eV·s)
f = wave frequency [hertz (cycle/s)]

Photons move at light speed (in a vacuum), regardless of their energy or origin.

In the wave model, higher frequencies are associated with higher energy. The EM wave *amplitude (A)* describes the greatest height or depth of excursion from the equilibrium value. For EM waves the electric field [V/m] and magnetic field [T] oscillate, similar to mechanical waves on the surface of a pond that move outward from their source in all directions. The crests of water waves form circles, and circular troughs follow them. Thus we see alternating concentric crests and troughs. Our eyes tend to follow the moving crests, which we call *wave fronts*. Perpendicular to the wave fronts defined by the crests, that is, in the radial direction of the circular crests, are *rays* that show the direction that a portion of the wave is moving. Electromagnetic waves exhibit similar wave-like characteristics but travel in three dimensions. We illustrate various wave characteristics in Fig. 2-15.

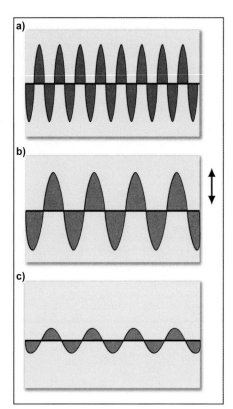

Fig. 2-15. Wave Propagation. Here we compare characteristics of different waves. **a)** a wave propagating to the right as indicated by the arrow. **b)** a wave with a wavelength twice the value of the wave in a). We measure the wavelength from crest to crest (arrow above the waves). The wave amplitude is indicated by the arrow along the right side of the waves. **c)** a wave with an amplitude one third of the value in b).

For water and EM waves, we call the displacement of the crest positive and the displacement of the trough negative, so that the undisturbed position is zero. The distance between successive crests or troughs is the *wavelength (λ)*, while the number of complete wave cycles passing a fixed point per second is the wave *frequency (f)*. The *phase velocity* of the waves is the distance they travel per unit time, which is just the length of each wave cycle multiplied by the number of waves passing a fixed point per unit time. Symbolically, we write phase speed as

$$v = \lambda f \quad (2\text{-}16a)$$

for mechanical (water or sound) waves, and

$$c = \lambda f \quad (2\text{-}16b)$$

for electromagnetic waves

where

v = phase velocity of a mechanical wave [m/s]
λ = wavelength [m]
c = phase velocity (light speed) of an electromagnetic wave (2.998×10^8 m/s in a vacuum)

(Some texts use ν (the Greek letter nu) instead of *f* for frequency.)

Although all EM waves propagate in the same way and with the same speed (light speed) in empty space, for convenience we divide the EM wave spectrum into regions according to wavelength. If we arrange the entire sequence such that the shortest wavelengths are at one end and the longest are at the other end, then we produce the *electromagnetic spectrum*, as shown in Fig. 2-16. The spectrum covers just fewer than 20 orders of magnitude in frequency and wavelength. Toward the short-wavelength end is the very limited portion to which our eyes are sensitive, called the *visible spectrum* (visible light or just light). The physiological response of the eye to the various frequencies of the visible spectrum results in what we perceive as color.

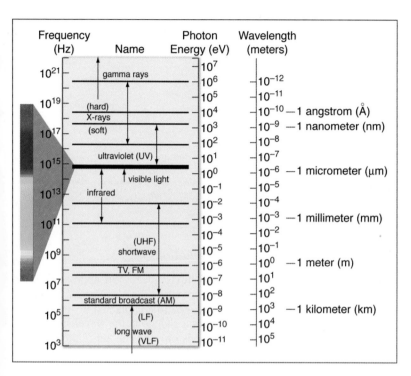

Fig. 2-16. The Spectrum of Electromagnetic Radiation. Here we show the range of frequencies and wavelengths for electromagnetic waves. The color bar on the left is an expansion of the visible light band. We label different regions in the full spectrum not because of any intrinsic difference in the radiation, but because nature has different ways of generating radiation depending on its wavelength. Further, detection and biological systems interact differently with EM radiation from different regions of the spectrum. (UHF is ultra-high frequency; TV is television; FM is frequency modulation; AM is amplitude modulation; LF is low frequency; VLF is very-low frequency.)

Short wavelengths in the visible spectrum are violet, with progressively longer wavelengths producing the response we identify as indigo, blue, green, yellow, orange, and red. Visible light has wavelengths between approximately 350×10^{-9} m and 700×10^{-9} m, corresponding to frequencies between 8.5×10^{14} Hz and 4.3×10^{14} Hz.

Gamma rays and X rays have sufficient energy to penetrate human tissue and the shields of sensitive equipment on spacecraft. Extreme ultraviolet (EUV) light and ultraviolet (UV) light pack enough energy to ionize matter in the upper atmosphere and cause physical damage to human skin and eyes. Radio waves and the radiant heat we feel at a distance from a campfire are also forms of electromagnetic radiation (long wavelength in the infrared band and lower) that human eyes can't detect.

For historical reasons, some units of wavelength measurements are more convenient than others for describing different regions of the EM spectrum. An *angstrom (Å)* is 10^{-10} m. Angstroms are often used to quantify visible and shorter wavelengths, while infrared wavelengths are generally expressed in micrometers (1 micrometer [µm] = 10^4 Å = 10^{-6} m). In this text, we describe wavelength in meters with the appropriate prefixes (centi-, micro-, nano-, etc.), but some diagrams and images display wavelengths in other units. The hertz (Hz or cycle per second) is the unit for measuring frequency for all EM radiation.

EXAMPLE 2.9

Photon Energy

- *Problem Statement:* Compare the energy of a solar X-ray photon at 10 nm with that of a solar radio photon at 10.7 cm.
- *Relevant Concepts:* Energy and electromagnetic spectrum
- *Given:* $\lambda_{x\text{-}ray} = 10$ nm, $\lambda_{radio} = 10.7$ cm
- *Assume:* Nothing

- *Solution:*

$$E = hf = h\frac{c}{\lambda}$$

$$\frac{E_{10nm}}{E_{10.7cm}} = \frac{\left(\frac{1}{10 \times 10^{-9}\,m}\right)}{\left(\frac{1}{1.07 \times 10^{-1}\,m}\right)} = 1.07 \times 10^7$$

- *Interpretation:* An X-ray photon has more than 10 million times the energy of a radio photon.

Follow-on Exercise: Compare the energy of a 1-angstrom photon with that of a 1-micrometer photon. To what spectral regions do these photons correspond?

Doppler Effect for Electromagnetic Radiation

The sonic effect heard as the varying pitch (frequency) of approaching and retreating train whistles is the phenomenon known as the *Doppler effect*, named for Christian Doppler, the Austrian physicist who first explained it. An observer moving relative to a wave source experiences a change in the wavelength of the wave. The size of the observed effect depends only on the net relative motion along the line of sight between the source and the observer. The wave source may be moving, or the observer may be moving, or both. For EM waves the Doppler effect is evident when detected radiation has a shorter wavelength (blue-shift) as the source and observer approach each other or a longer wavelength (red-shift) as they recede from each other. The amount of wavelength change is proportional to the relative speed along the line between the source and observer.

We let $\Delta\lambda$ be the shift in wavelength equal to the difference between the measured wavelength (λ_m) and the wavelength (λ) in the absence of relative motion ($\Delta\lambda = \lambda_m - \lambda$). Then the formula relating the Doppler shift for light to the relative speed is

$$\Delta\lambda/\lambda = dv_{rel}/c \qquad (2\text{-}17)$$

where

$\Delta\lambda = \lambda_m - \lambda$ Doppler wavelength shift [m]

v_{rel} = relative line-of-sight velocity between the observer and the light source (astronomers call it the radial velocity [m/s])

The wavelength shift is greater for longer-wavelength radiation. We measure the Doppler displacement if we know the wavelength when relative motion is absent. Fortunately, this unperturbed wavelength is relatively easy to determine if we look for line emissions from common elements such as hydrogen, as we describe in the next section.

We consider the wavelength shift positive if the wavelength shifts to the red (distance increasing) and negative if it shifts to the blue (distance decreasing). Measuring the Doppler shift of specific wavelengths emitted by stars (emission lines) helps us determine the velocity of stellar recession or approach. In the case of nearby or large objects, we measure rotational rates because the edges of an object, which move toward and away from the viewer most rapidly, experience opposite Doppler shifts. Figure 2-17 shows this type of observation for the Sun.

Intensity

Our ability to quantify the energetic process associated with electromagnetic energy relies on the properties and mathematical description of waves. The power transported across a unit area by a wave is its *intensity (energy flux), I*. In general, intensity is proportional to the square of the wave amplitude.

$$I = kA^2 \quad (2\text{-}18)$$

where

- I = wave intensity (also called radiance for EM waves) [W/m^2]
- k = proportionality constant that depends on the wave type [W/m^4]
- A = wave amplitude whose units depend on the wave type [m]

Fig. 2-17. A Full-disk Dopplergram. Because the Sun rotates, one side (left side) comes toward our Doppler instrument and the other goes away. The figure is a map of the speed of the Sun's surface as measured by the SOHO/MDI instrument. The colors indicate motion toward the instrument (dark) or away (lighter). Superimposed on the dominant left-right rotational gradient are smaller-scale features. These features are caused by solar phenomena including granulation and supergranulation, which we describe in Chap. 3. *(Courtesy of Edward J. Rhodes at NASA)*

Figures 2-15 b and c show waves of amplitude A and $1/3A$. The intensity of the lower-amplitude wave is one ninth the intensity of the higher-amplitude wave. Table 2-6 gives several examples of wave intensity values.

Table 2-6. Intensity of Wave Energy. Here we list some wave sources and their energies.

Source	Intensity or Radiance (W/m^2)
Bright aurora	1.5×10^{-7}
Solar X-ray flare (0.05 – 0.8 nm) at 1 AU	$10^{-4} - 10^{-7}$
Sound—normal conversation at 1 meter	10^{-4}
Sound—threshold of pain	1
100 Watt light bulb	~8
Sunlight at 1 AU summed over all wavelengths	1366

The surface area illuminated by an expanding sphere of light (or a portion of it) increases as the square of the distance from the light source. However, in the absence of processes that transform EM energy to other forms, the total amount of EM energy is conserved as the light sphere expands outward from the source. Therefore, the light intensity (power per area) must decrease with the square of the distance from the source. We show this relationship between observed intensity and distance in Fig. 2-18 where I varies inversely as the square of the distance (r) from the light source.

$$I_{final} \times Area_{final} = I_{initial} \times Area_{initial}$$

$$I_{final}(4\pi r_{final}^2) = I_{initial}(4\pi r_{initial}^2) \quad (2\text{-}19)$$

$$I_{initial}(r_{initial}^2)/(r_{final}^2) = I_{final}$$

Fig. 2-18. Intensity Decreasing with Distance from a Constant Power Source. If we were to surround the diagram with a sphere of surface area $4\pi r^2$, we would capture all of the radiated energy and find it to be the same value as that going through a sphere surrounding only the light bulb. Thus, we know that energy is conserved.

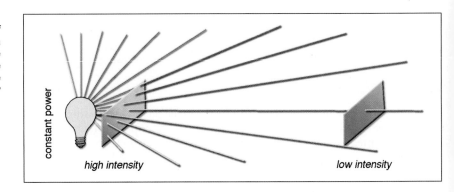

EXAMPLE 2.10

Doppler Shift

- *Problem Statement:* The measured shift of an absorption line at 500 nm in the spectrum of a star is –0.05 nm. Determine the star's radial velocity relative to Earth.
- *Relevant Concept:* Doppler Shift
- *Given:* $\Delta\lambda = -0.05$ nm
- *Assume:* Line of sight motion

- *Solution:* From the Doppler formula we find the radial velocity of the star relative to Earth using
$$v = (\Delta\lambda/\lambda)c = (-0.05 \text{ nm}/5 \times 10^2 \text{ nm})(2.998 \times 10^8 \text{ m/s}) = -30 \text{ km/s}$$
where the negative sign denotes a velocity of approach. If the algebraic sign of the velocity had been positive (that is, the wavelength shift had been positive), the star would be receding from Earth.

Follow-on Exercise: Assume a CME traveling towards Earth generates radio frequency emissions at 75 MHz. If the CME is Earth-directed with a speed of 2000 km/s what is the Doppler shift of the signal?

EXAMPLE 2.11

Inverse Square Reduction of Energy

- *Problem Statement:* Suppose that at 1 AU from the Sun, the apparent brightness of the radiation over 1 m² of surface area is 1 unit. What will the brightness be at 2 and 5 AU?
- *Relevant Concept:* Intensity
- *Given:* The distance from the source is 2 and 5 AU.
- *Assume:* No absorption of EM energy

- *Solution:* Use the inverse square law ($1/r^2$) to find the ratio of the original intensity at 1 AU distance to the intensities at the requested distances. At 2 AU, each square kilometer receives $(1/2)^2$ or 1/4 of a unit of illumination. At 5 AU, the brightness will be $(1/5)^2$ or 1/25 of a unit of illumination.
- *Interpretation:* The Sun's intensity decreases significantly with distance. At 1 AU, Earth receives an adequate intensity to sustain life, but at Jupiter (5 AU), the intensity is much too low for life as we know it.

Follow-on Exercise 1: Determine the intensity of solar radiation at the heliopause.

Follow-on Exercise 2: Determine the intensity of a solar flare that develops near the Sun's surface. (Use data from Table 2-6.)

2.3 Characterizing Electromagnetic (EM) Radiation

Electromagnetic energy plays a fundamental role in the space environment and space weather. We investigate the origins of this energy in more detail here.

2.3.1 Blackbody Radiation

Objectives: After reading this section, you should be able to...

- Compare and contrast blackbody radiation with other forms of radiation
- Recognize the representation of a blackbody curve on linear and logarithmic scales
- Use Wien's Law to determine the wavelength peak radiative output from a blackbody
- Use the Stefan-Boltzmann Law to calculate power flux from a blackbody
- Relate the Planck radiation function to Wien's Law and the Stefan-Boltzmann Law
- Relate Planck's Radiation Law to blackbodies

Blackbody Radiation

Blackbody continuum radiation is created by the constant random motion of electromagnetic oscillators in relatively dense and opaque matter. An ensemble of accelerating particles in suitably dense solids, liquids, and gases supports oscillations at countless quantized frequencies. All objects with temperatures above 0 K continuously emit EM waves over a continuum of wavelengths. The source of this radiation is the thermal energy of the matter, that is, the microscopic movement of the object's EM oscillators: electrons, ions, atoms, and molecules. *Blackbody radiators* are equally good absorbers and emitters. Continuum blackbody radiation from the Sun is predominantly in the ultraviolet-to-infrared (UV-IR) wavelengths.

A blackbody's temperature determines the amount of radiant power (# photons/s) emitted at any given wavelength. In Figs. 2-19 and 2-20 we plot the radiated energy flux versus wavelength from blackbodies at several temperatures. The curves follow *Planck's Radiation Law* that relates radiance of the body in a given wavelength interval to its temperature (Focus Box 2.5). From the peak of each curve the radiance decreases steeply, but not symmetrically. Emissions at either extreme become less and less likely. The shape of the curves is similar to those in Fig. 2-8. This similarity is not an accident. One interpretation is that the Planck curve represents the distribution of all possible photon energies in a sample of matter. The energy for thermalized photons is distributed in the most probable manner, consistent with their specified energy. That is, for a given temperature a most likely energy value exists for the photons—the peak of the curve.

The Planck function (Eq. (2-23) in Focus Box 2.5) prescribes two distinguishing aspects of blackbodies: the wavelength at which the curve peaks (λ_{max}) and the total power radiated (area under the curve derived from integrating across all wavelengths). Each curve in Figs. 2-19 and 2-20 has a distinct peak whose location tends to shorter wavelengths for higher temperatures. Differentiating the Planck function produces *Wien's Displacement Law*, which specifies the wavelength of maximum radiation.

$$\lambda_{max} T = \alpha \qquad (2\text{-}20)$$

where

λ_{max} = wavelength for maximum amount of radiation [m]
T = object temperature [K]
α = constant (2.898×10^{-3} m·K)

For relatively low temperatures, such as room temperature, most of the blackbody spectrum consists of infrared (IR) wavelengths that are invisible to the human eye. For example, an ordinary object sitting on a desk appears not to radiate (although it does), because most of its radiation is at IR wavelengths. At higher temperatures, the spectral peak is at shorter wavelengths. When an electric stove burner heats, it eventually reaches a temperature at which its blackbody emissions overlap the region of visible light and produce a red glow.

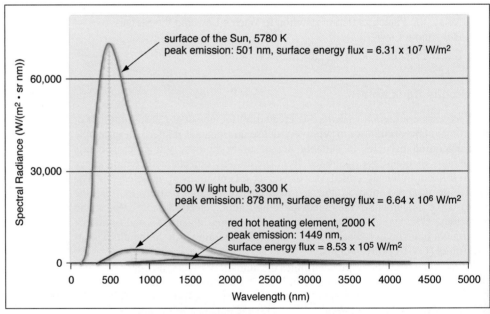

Fig. 2-19. Spectral Power Distribution (Planck Curves) from the Surfaces of Blackbodies at Three Temperatures. Peak radiation for hotter bodies is at shorter wavelengths. Hotter objects radiate more energy at all wavelengths. The power flux for each temperature is the total area under the corresponding curve. Black-body radiance covers several orders of magnitude for objects that astronomers and space physicists study. Therefore we often plot Planck radiation curves in log-log form as shown in Fig. 2-20. Note that on the linear scale in this figure the values for the red hot heating element are not even visible.

Integrating the Planck equation (Eq. (2-23)) in Focus Box 2.5 over all wavelengths produces the total radiance $R(T)$ emitted to space from a unit area of the body. This value is given by the *Stefan-Boltzmann Law*.

$$R(T) = \int_0^\infty R(\lambda, T) d\lambda = \varepsilon_\lambda \sigma T^4 \qquad (2\text{-}21)$$

where

$R(T)$ = energy flux (radiance) over all wavelengths [W/m²]
ε_λ = emissivity of radiant efficiency [unitless]
σ = Stefan-Boltzmann constant (5.67×10^{-8} W/(m²·K⁴))

For all blackbodies ε equals one. Bodies with ε less than one are graybody radiators. When a body radiates uniformly from its entire surface, the total energy flux is luminosity, given by

$$L = \varepsilon \sigma A T^4 \qquad (2\text{-}22)$$

where

L = luminosity [W]
A = radiating area [m²]

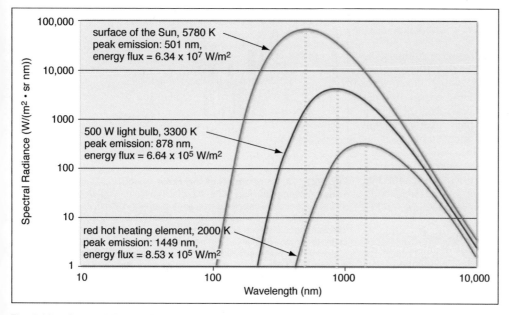

Fig. 2-20. Spectral Energy Flux Versus Wavelength for Three Temperatures. The dotted lines show the peak of the individual curves with the emission maximums for each temperature. This data is the same as shown in Fig. 2-19, but we created the plot in log-log format. This format allows us to view energy levels that are much lower on the vertical scale than in Fig. 2-19. These levels don't make large contributions to the total energy—they are tens to thousands of times smaller than energies in the upper tiers—but their variations often have important effects.

Focus Box 2.5: Planck Radiation Law

The formula that quantifies the radiated energy flux at any particular wavelength for the curves in Figs. 2-19 and 2-20 is the Planck Radiation Law

$$R(\lambda, T)d\lambda = \frac{2\pi hc^2}{\lambda^5}\left(\frac{1}{e^{(hc/\lambda k_B T)} - 1}\right)d\lambda = \frac{c_1}{\lambda^5}\left(\frac{1}{e^{(c_2/\lambda T)} - 1}\right)d\lambda \tag{2-23}$$

where
- R = radiated energy flux per wavelength [W/(m²·nm)]
- λ = wavelength [m]
- T = temperature [K]
- h = Planck's constant (6.626×10^{-34} J·s = 4.136×10^{-15} eV·s)
- c = light speed (2.998×10^8 m/s)
- c_1 = 3.74×10^{-16} W·m²
- c_2 = 1.44×10^{-2} m·K

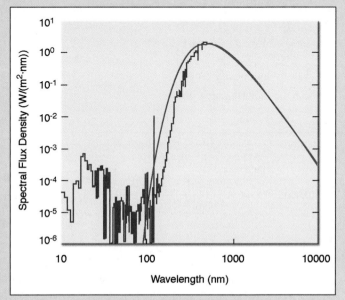

Fig. 2-21. A Comparison of the Solar Spectrum. This plot shows solar radiation (blue line) at the top of Earth's atmosphere, along with a plot of a blackbody spectrum (red line) in log-log format. The Sun emits more radiation in the waveband short of 100 nm than does a blackbody. Between about 100 nm and 140 nm, the Sun is a graybody radiator; and in these regions ε_λ is less than one. The vertical axis values are scaled to account for the energy intensity reduction in its transit from the Sun's surface to Earth, i.e., $(R_s/AU)^2$. *(Courtesy of Kent Tobiska at Space Environment Technologies)*

The radiance function is a product of two terms. The $1/\lambda^5$ term represents the reduced energy flux per unit wavelength at long wavelengths, while the inverse exponential term represents the reduced energy flux per unit wavelength at shorter wavelengths. Thus, at a given temperature the probability of a high level of photon energy approaches zero for very long and very short wavelengths. Differentiation of the function for this curve produces Eq. (2-20) and reveals that the peak height of the curve is at a specific wavelength that depends only on the temperature of the blackbody. Integration of the radiance function produces Eq. (2-21), that too is a function of temperature only.

Often a physical body is not a perfect blackbody radiator, so Eq. (2-21) includes *spectral emissivity (ε_λ)* that specifies the degree to which an object behaves as a blackbody at a given wavelength. Where $\varepsilon_\lambda < 1$, the Planck curve (now a graybody curve) achieves only a fraction of the height of the blackbody curve. Some materials behave as blackbody radiators at certain wavelengths but have quite different spectral characteristics at other wavelengths. They could have jagged curves similar to the blue curve of Fig. 2-21. Most of the radiators we deal with in space have emissivities close to one. Technically, they are graybodies, but we often treat them as blackbodies.

If we compare the spectrum of solar emissions at the top of Earth's atmosphere to a theoretical blackbody curve, we find the solar spectrum best fits to a 5770 K blackbody with its maximum radiation at 503 nm, which is in the yellow portion of the visible spectrum we associate with sunlight. However, a 5770 K curve is a poor fit at short and very-long wavelengths. This poor fit is caused by absorption of radiation in the Sun's overlying layers and by additional, highly variable emission components in the Sun's upper atmosphere.

For the Sun, the value of ε_λ is nearly unity at wavelengths between 300 nm and 10^6 nm. In the visible and infrared wavelengths, where the Sun emits most of its energy, it behaves as a blackbody. However, in the shortest wavelength regime (X ray and EUV) and long wavelength regime (microwave and radio), the Sun is not a blackbody emitter. Instead it emits much more radiation than a blackbody. The excess long wavelength emissions come from plasma oscillations that we describe in subsequent sections. Most of the shortwave energy from the quiet Sun comes from specific emission lines in the Sun's upper atmosphere. At a few thousand kilometers above the Sun's surface, the million-K corona sustains a shroud of highly ionized gases whose emissions create the EUV and X-ray region of the solar spectrum. We discuss this topic in the next section.

2.3 Characterizing Electromagnetic (EM) Radiation

EXAMPLE 2.12

Luminosity

- *Problem Statement:* At 1 AU the solar irradiance is 1366 W/m^2. Use this value to determine the solar radiant output (luminosity) at the Sun's surface.
- *Relevant Concepts:* Solar luminosity and radiant intensity
- *Given:* The solar radius is 7×10^8 m. One AU is 1.49×10^{11} m

- *Assume:* No absorption of energy between the Sun and Earth
- *Solution:* The total power flowing through a sphere at 1 AU is
$$E = 4\pi r^2 I = 4\pi (1.49 \times 10^{11} \text{ m})^2 (1366 \text{ W/m}^2) = 3.81 \times 10^{26} \text{ W}$$
- *Interpretation:* Our Sun's radiative output is 3.81×10^{26} W, corresponding to 1366 W/m^2 at Earth's distance from the Sun.

Follow-on Exercise: Show that a room temperature object has a radiance of more than 400 W/m^2.

Pause for Inquiry 2-11

Explain why an activated heating element on an electric stove sometimes glows and sometimes doesn't.

Pause for Inquiry 2-12

Why do stars cooler than the Sun appear red?

Pause for Inquiry 2-13

Refer to Fig. 2-21.
In what wavelength band does the Sun best approximate a blackbody?
In what wavelength band does the Sun best approximate a graybody?
In what wavelength bands is the Sun a non-thermal radiator?

Pause for Inquiry 2-14

Show that the maximum energy flux in Fig. 2-21 is of the correct order of magnitude to be consistent with a $1/r^2$ decrease as energy transits from the Sun to 1 AU.

The amount of energy radiated by ordinary objects is significant, even though we usually don't notice it. For example, an object at room temperature, 24° C (297 K), radiates about 450 W/m^2 from its surface (Follow-on Exercise 2.12). If this radiance level is maintained, why doesn't everything grow cold quickly and why don't we feel this radiation? In fact, if an object is suddenly placed in outer space, far away from any strong energy input, then the object will grow cold quite rapidly by radiating its heat away. Normally, however, an object is completely surrounded by other objects of the same temperature (such as air), and these objects also radiate energy at the same rate. Most objects reach thermal equilibrium with their surroundings. Thus, absorbed radiation from surrounding objects balances energy lost by radiation. We don't feel the effects of the radiation because of this balance, unless we happen to stand between objects that have different temperatures. For example, if we stand next to a wall that was just warmed by the Sun and the Sun sets, we will feel the warmth from the wall as EM energy in transit.

Figures 2-19 and 2-20 reveal that thermal bodies radiate across a continuum of wavelengths. When viewing a continuum of radiation, a phenomenon called dispersion makes it possible to separate and identify the different wavelengths present in the spectrum. For example, when light in the visible portion of the spectrum passes through a glass prism, it bends through a range of angles as shown in Fig. 2-22. The difference in each wavelength's bending is *dispersion*. In practice, we often use a diffraction grating in place of the glass prism to create the separation because gratings have higher resolving power and other useful characteristics, but the results are the same.

The dispersed spectrum may be photographed to produce a *spectrograph*. Also the spectrum may be manipulated digitally with a sensitive semiconductor called a charge-coupled device (CCD). A CCD converts light into electrical current, where the current becomes a digital camera's equivalent of film. If the photographic plate or CCD is replaced by an eyepiece, the instrument is called a *spectroscope*.

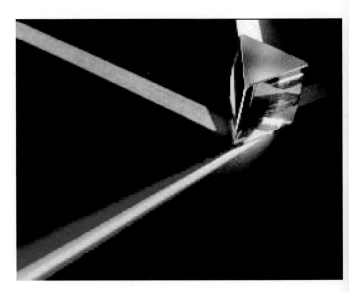

Fig. 2-22. Light Dispersion through a Prism. White light passing through a glass prism disperses into its constituent colors. *(Courtesy of Adam Hart-Davis)*

2.3.2 Discrete Line Radiation and Absorption

> **Objectives:** After reading this section, you should be able to...
> - Explain the origin of discrete-line radiation
> - Distinguish between emission and absorption lines and continuum emissions
> - Calculate energy associated with emitted or absorbed light wavelengths

Early in the twentieth century, scientists learned that electrons in any atom possess only certain, discrete amounts of energy. Each of these allowed amounts of energy is an electron energy level (*energy state*). When an electron moves between two allowed states, the atom emits or absorbs radiation at specific frequencies.

To illustrate transitions from excited states, we use an energy level diagram for hydrogen, as in Fig. 2-23. These energy states are similar for all like atoms (one valence electron), but are different for other kinds of atoms. The lowest level is the *ground state* of the atom and is the most stable state. The higher levels represent the atom's excited states. The length of time that an atom remains in the excited state is the *lifetime* of the state. This time is as short as a nanosecond for some states and as long as years for others.

Line spectra are created in low-density gas when electrons in individual atoms radiate at specific wavelengths during transition between bound energy levels, or during transition from free (ionized) to bound states. Absorption lines arise from the inverse process. The molecular analogy to line spectra is *band spectra*. Many lines and bands are observed in the X-ray through IR portions of the solar spectrum.

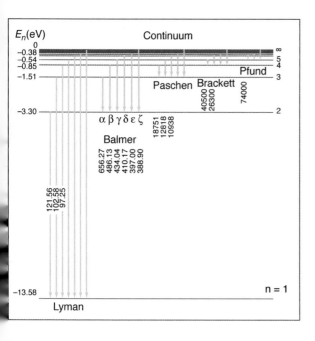

Fig. 2-23. Line Spectra and Energy Level Diagram of the Hydrogen Atom. Here we show many of the possible energy state transitions. The numbers on the left side are energy differences in electron volts (the ground state has a value of −13.58 eV). The numbers down the right side of the diagram are primary quantum numbers (n) designating the electron levels. The values on the vertical lines correspond to the wavelengths in nanometers of the emitted radiation for several transitions. When 13.58 eV of energy is supplied to a neutral hydrogen atom, an electron at the $n = 1$ level transitions to the $n = \infty$ level (leaves the atom) and we say the atom is ionized. The Greek letters associated with the Balmer Series are an historical artifact.

When an electron in an excited state returns to a lower state, it emits a photon. The equation relating the change in energy between the two levels and the frequency of the emitted photon is

$$\Delta E = E_f - E_i = hf = h(c/\lambda) \qquad (2\text{-}24)$$

where E_f and E_i are the energies of the final and initial electron states.

The lowest orbital level (n) associated with a transition defines the series floor. For example, the Lyman series has its floor at $n=1$. Lyman-α radiation (interesting to those who study EUV radiation behavior in the solar and Earth atmospheres) is generated by the $n=2 \rightarrow n=1$ transition and is associated with 121.6 nm radiation. The primary hydrogen Balmer (H-α) radiation is generated by the $n=3 \rightarrow n=2$ transition in the visible wavelength regime at 656.3 nm (visible red light). This H-α radiation is used for monitoring solar activity.

The various parts of Fig. 2-24 help distinguish between blackbody continuum radiation and the line spectra associated with electronic transitions. As we discuss in Sec. 2.3.1, blackbody continuum radiation (Fig. 2-24a) arises from countless random oscillations of the matter in the body. If the body also has a hotter tenuous gas component, such as the Sun's atmosphere, the atoms in the non-collisional regime may radiate a discrete set of wavelengths as in Fig. 2-24b. The wavelengths are characteristic of the particular constituent atoms or molecules in the gas and are thus a powerful tool for determining the composition of the radiating material. Emission lines are most easily viewed from an optically-thin gas without intervening matter that could absorb the radiation. In contrast, if continuum radiation passes through an intervening layer of cool gas, the atoms or molecules of the cool gas absorb a discrete set of wavelengths, leading to a series of absorption lines (Fig. 2-24c). Figure 2-24d depicts the possible emission situation for a blackbody emitting radiation through an intervening cool tenuous gas that depletes the radiative flux. Added to the continuum emission is radiation emitted by electronic transition from individual excited atoms or molecules that are not in thermodynamic equilibrium with their surroundings.

The Sun's blackbody emissions pass through layers of solar atmosphere that absorb part of the outbound radiation. The top row of Fig. 2-25a shows the solar visible blackbody emission spectra plus the dark absorption *Fraunhofer lines*. These lines are named after German physicist Joseph Von Fraunhofer, who investigated the Sun's spectrum and discovered the dark lines in 1814. The photon deficit develops when intervening gas in the solar atmosphere absorbs some of the outbound blackbody radiation and re-emits a portion of it in directions that do not coincide with the line of sight.

Discrete emissions are not limited to those created by electronic transitions. When atoms join to create diatomic or polyatomic molecules, the atoms may rotate about a common center of mass, oscillate to and fro as tiny spring-mass systems, or bend in response to thermal jostling. The associated motions create emissions at discrete (quantized) wavelengths. These motions require relatively little energy to excite them and are usually at much longer wavelengths than those of electronic transitions. Airglow from Earth's upper atmosphere is in the IR portion of the spectrum as are some emissions from the solar chromosphere.

Focus Box 2.6: Spectral Notation

Scientists use several designations for spectral lines and transitions on energy level diagrams. Some of the designations are rooted in the energy transitions allowed under the rules of quantum mechanics. Atomic systems can be quantified with four numbers called the quantum numbers. The letters n and l represent two of them. The major energy level is n, while l designates the small splitting of the major levels associated with electron angular momentum effects. The remaining quantum states are represented by the magnetic quantum number m; and for poly-electronic atoms, the spin quantum number m_s.

For hydrogen, the energy associated with the principal quantum number (n) is given by a formula made famous by Niels Bohr, a Danish scientist who was a leader in quantum physics:

$$E_n = 13.6 \text{ eV}/n^2$$

Spectral lines (emissions) are associated with transition or change in energy between levels for hydrogen (Z = 1) and other simple atoms. We calculate the energy change as

$$\Delta E = 13.6 \text{eV}(Z^2)\left(\frac{1}{n_f^2} - \frac{1}{n_i^2}\right), \text{ where } Z = \text{\# protons} \qquad (2\text{-}25)$$

We obtain the emission wavelength by applying $\lambda = hc/\Delta E$, where $h = 4.136 \times 10^{-15}$ eV·s.

Each of the series of lines in Fig. 2-23 represents electron transitions to the same final energy level (n) for that series. Figure 2-23 shows the Lyman, Balmer, Paschen, Brackett, and Pfund series, with notations of H-α, H-β, H-γ, etc. for the lines of the Balmer Series. Similarly, L-α, L-β, L-γ, etc., represent the three lowest energy transitions for the Lyman series. Emissions at H-α, H-β, and H-γ wavelengths are present in the spectrum labeled H in Fig. 2-25. More massive elements, such as iron (Fe), undergo many transitions yielding many lines, each influenced by the valence state, location in the crystal structure, and other characteristics. Several visible and ultraviolet iron line emissions are present in the spectrum labeled Fe in Fig 2-25.

For multi-electron systems, the electron orbits are often shown as shells. The lowest shell (closest to the nucleus) is the K shell. Further out are the L, M, N, O, etc. shells. A transition from the L shell to the K shell produces a spectral line that is called the K-line of the substance.

Solar spectroscopists use another notation for neutral and ionized atoms that focuses on the electron most available for excitation. In a neutral atom it's the outermost electron. For example, neutral calcium is Ca I. For a singly ionized calcium atom (Ca^+) the outermost electron is absent, but its second electron is available for excitation, so we refer to it as Ca II, and its ion as Ca^{++}.

Fig. 2-24. Radiation in Continuum, Emission, and Absorption Processes. a) Here we show a radiance-versus-wavelength plot for a blackbody radiator. **b)** This plot is a radiance versus wavelength diagram for line emissions, where the line is broadened by Doppler shifts of the moving emitters. **c)** This radiance versus wavelength plot for line absorption shows where the line is broadened by Doppler shifts of the moving emitters. **d)** This plot shows a combination of continuum, line emission, and line absorption.

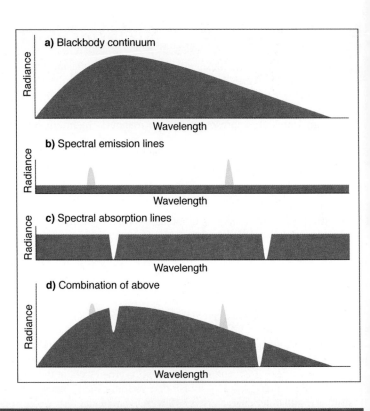

EXAMPLE 2.13

Balmer H-α Emission

- *Problem Statement:* Verify that a Balmer H-α emission corresponds to ~656 nm wavelength light
- *Relevant Concept:* Quantized emissions from electronic transitions in hydrogen atoms and energy levels. Energy-wavelength relationship for light.
- *Given:* Emission is from the H-α line
- *Assume:* An isolated atom
- *Solution:*

$$\Delta E = 13.6 \text{ eV}(Z^2)\left(\frac{1}{n_f^2} - \frac{1}{n_i^2}\right)$$

$$\lambda = hc/\Delta E$$

$\Delta E = 13.6 \text{ eV}(1)^2[(1/(2)^2 - 1/(3)^2] = (13.6 \text{ eV})(0.139) = 1.889 \text{ eV}$

$\lambda = hc/\Delta E = (4.136 \times 10^{-15} \text{ eV} \cdot \text{s})(2.998 \times 10^8 \text{ m/s})/(1.889 \text{ eV})$
$= 656.4 \text{ nm}$

- *Interpretation and Implication:* The $n=3$ to $n=2$ transition in an isolated hydrogen atom produces visible light in the red portion of the spectrum. The Sun is predominantly hydrogen. In regions of the Sun where hydrogen is excited thermally to the $n=3$ level, a significant emission should exist in this line. The appropriate temperatures for this thermal excitation are typically found in the solar atmospheric layers and above the surface—the photosphere and chromosphere.

Pause for Inquiry 2-15

Two bright emission lines, one blue and one red are present in the hydrogen spectrum in the middle of Fig. 2-25. To which lines in the Balmer series of Fig. 2-23 do these correspond?

2.3 Characterizing Electromagnetic (EM) Radiation

Pause for Inquiry 2-16

Carbon monoxide (CO) molecules are present in tiny amounts in the solar chromosphere. One of the allowed rotational transitions decreases the rotational energy by 4.75×10^{-4} eV. What is the associated wavelength of this emission?

Dissociation of this molecule requires 11.1 eV. What temperature of gas would provide enough thermal energy to cause dissociation by collision?

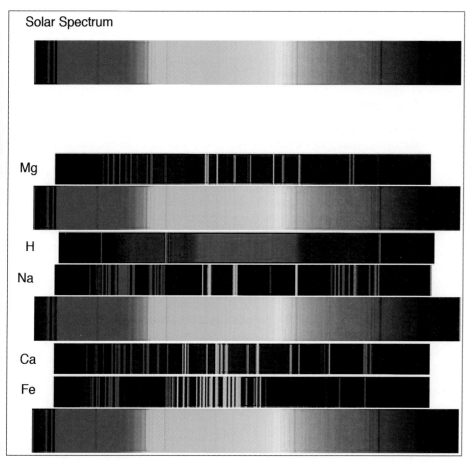

Fig. 2-25. Spectrograms. The top spectrum is the visible blackbody solar radiation. The dropouts in intensity (vertical dark lines) represent wavelengths of energy that have been absorbed as the blackbody radiation passes through the Sun's atmosphere. Individual emission lines from various elements (magnesium-Mg, hydrogen-H, sodium-Na, calcium-Ca, and iron-Fe) are shown in the lower part of the image. The solar spectrum is repeated for reference below the images of the individual elements, Mg, H, Na, Ca, and Fe. The emission lines correspond to the dropouts in the solar spectrum, meaning that these elements absorb certain wavelengths in the Sun's atmosphere. *(Courtesy of NASA)*

2.3.3 Radiation from Other Sources

Objective: After reading this section, you should be able to...

- Describe the physical origins of plasma waves, bremsstrahlung, and synchrotron radiation

The Sun and other stars emit electromagnetic (EM) radiation ranging from very energetic gamma rays to low-energy radio waves. We observe distinct forms of radiation with different physical origins. In this section we describe the origins of non-blackbody EM radiation that is present at a background level but may be enhanced during times of solar disturbance: synchrotron radiation, plasma wave radiation, and bremsstrahlung radiation. Each of these radiations comes from accelerating charged masses.

Plasma Wave Radiation

When disturbances in the form of EM waves or fast-moving matter travel through plasma, they excite the relatively mobile electrons, causing them to oscillate about the more massive ions (Fig. 2-26). As the electrons attempt to return to their equilibrium positions, they behave as tiny spring-mass systems, with the accelerating electrons constantly overshooting their undisturbed position. The motion generates horizontally propagating EM waves with a frequency determined by the local electron density. When these propagating waves encounter electron density irregularities in the background plasma, they are scattered in all directions and so are detectable outside of the plasma. In effect the electrons extract energy from the disturbances and reradiate the energy at a frequency determined by the plasma density. Because of the varying electron density in the solar atmosphere, the radiation forms a continuum of radiation in the UV through radio portions of the spectrum. Earth-based radars search for these emissions, especially those arising from fast-moving disturbances such as coronal mass ejections.

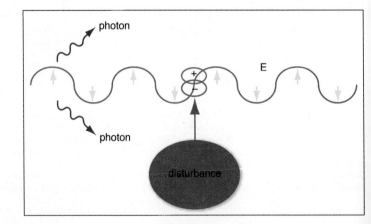

Fig. 2-26. Radiation from Plasma Oscillations. Electrons oscillate about their equilibrium positions and emit photons after being disturbed by a passing electromagnetic wave.

Bremsstrahlung

An important source of X-ray radiation in space comes from the sudden deceleration of high-speed electrons. When an electron slows as it approaches another charged particle or deflects in an electric or magnetic field, it radiates EM energy (Fig. 2-27). This energy is *bremsstrahlung* (German for braking radiation). Bremsstrahlung is the common process by which laboratories (or doctors and dentists) generate X rays. The electron beam produces radiation with a continuous spectrum of wavelengths down to a minimum wavelength given by $\lambda_{min} = hc/(E_{max})$. Here, E_{max} is the original electron energy. Bremsstrahlung is possible over the entire EM spectrum, but X-ray bremsstrahlung dominates the space-environment disturbances. During solar flares, bremsstrahlung at X-ray and even γ-ray wavelengths becomes very intense. During unusually quiet solar wind conditions, solar wind electrons have direct access to Earth's polar regions. Charged particles with sufficient energy may slam into Earth's upper atmosphere and ionize particles, creating a bremsstrahlung glow of X rays. Satellite X-ray sensors record backscattered X-ray events when this happens.

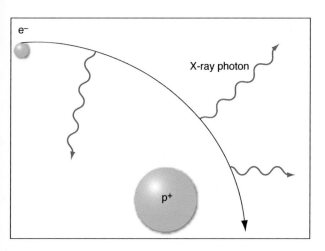

Fig. 2-27. An Electron Accelerating in the Presence of an Ion. As the electron's path curves it emits a continuum of photons.

EXAMPLE 2.14

Bremsstrahlung Emission

- *Problem Statement:* Determine the minimum wavelength of bremsstrahlung emission from a beam of 30 keV electrons.
- *Relevant Concepts:* Minimum wavelength of emission comes from the electron decelerating in a single energetic interaction
- *Given:* 30 keV electron beam
- *Assume:* Only bremsstrahlung interactions

- *Solution:* $\lambda_{min} = hc/(E_{max})$
  $\lambda = (4.136 \times 10^{-15} \text{ eV s})(2.998 \times 10^8 \text{ m/s})/(30{,}000 \text{ eV})$
  $= 0.04 \text{ nm}$
- *Interpretation and Implication:* This emission is in the X-ray portion of the spectrum. This wavelength of light is observed in the solar corona. Longer wavelengths are possible if the deceleration occurs in smaller increments, which is often the case, so bremsstrahlung is frequently observed as a continuum from the shortest wavelengths to much longer wavelengths.

Follow-on Exercise: Beams of 10 eV–1000 eV solar wind electrons sometimes travel with high-speed solar wind streams. What is the minimum bremsstrahlung wavelength that could be emitted if the electrons are stopped by a detector?

Synchrotron and Cyclotron Radiation

Charged particles spiral around magnetic fields. Because the particles continuously change direction, they are accelerating and emitting radiation (Fig. 2-28). Synchrotron emissions come from electrons traveling at nearly light speed in the presence of a magnetic field. Cyclotron radiation is the same process but from less energetic motions. The longer an electron is in the magnetic field, the more energy it loses. As a result, the electron makes a wider spiral around the magnetic field, and emits electromagnetic radiation at a longer wavelength. To maintain synchrotron radiation, a continual supply of relativistic electrons is necessary. These electrons are usually supplied by very powerful energy sources such as supernova remnants or quasars. Synchrotron radiation with sufficient energy to be sensed at Earth appears only sporadically as radio emissions from flare activity on the Sun.

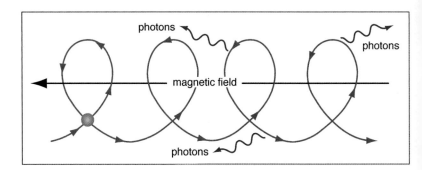

Fig. 2-28. An Electron Accelerating as It Gyrates around a Magnetic Field Line. These photons most often have a frequency in the radio portion of the electromagnetic spectrum.

At this point we have enough information about energy to proceed with studying the particles and photon emissions from the quiescent Sun in Chap. 3. We continue our investigation of energy stored and transported by fields in Chap. 4.

Summary

Energy is nature's currency for transacting business. It is at the heart of space weather as a conserved quantity that we trace from nuclear reactions in the Sun's core to Earth's magnetosphere and atmosphere. These manifestations of kinetic and potential energy act on macro- and micro-scale levels. Heat is microscale kinetic energy. Fields store energy on both scales. Radiation is energy in motion and we also quantify it on both scales.

Energy conservation is a key concept that helps us track energy in the space environment. Over-all energy is conserved, but it may enter and exit various subsystems of the space environment. Rapid entry or exit leads to powerful space weather events. Nature tends toward a state of lower energy organization (higher entropy), but the process of getting to the lower state often involves creating highly organized energy structures that store and then rapidly transform energy. Sunspots and active regions that we describe in the next chapter qualify as such entities.

Further sub-categories of energy are useful for describing the way energy interacts with matter or propagates as waves. The agents of energy transfer are particles, fields, and photons. The processes of energy transfer are bulk motion, cyclic motion, conduction, convection, and radiation. Depending on the nature of the interaction, we may describe energy in a per-time form (power), a power-per-area form (energy flux), or a per-volume form (energy density). Electromagnetic energy does not require matter to propagate.

Planck's Law, Wien's Displacement Law, and the Stefan-Boltzmann Law describe the EM energy from perfectly radiating bodies. We use spectroscopy to study radiative emissions and to learn about the composition, temperature, and behavior of radiating bodies.

Energy leaves the Sun in the form of photons, energized particles, and stretched magnetic field configurations. In the Sun's atmosphere, hydrogen, helium, and a host of minor heavier elements in various states of ionization imprint their signatures upon the solar spectrum. Absorption lines deplete the Sun's surface radiation, whereas hot ionized gases create thousands of emission lines and a few weak continua. The lines and continua raise the EUV flux of photons by many orders of magnitude above that of a ~6000 K blackbody. Electromagnetic oscillations from sources other than blackbodies are important to space weather. Bremsstrahlung and plasma oscillations contribute to non-blackbody radiation from the quiescent Sun.

Chapter 2 Space Is a Place...with Energy

Key Words

amplitude (A)
angstrom (Å)
band spectra
blackbody continuum radiation
blackbody radiators
bremsstrahlung
closed system
conduction
convection
dispersion
Doppler effect
electromagnetic (EM) radiation
electromagnetic spectrum
energy
energy state
entropy (S)
First Law of Thermodynamics
forced convection
Fraunhofer lines
free convection
frequency (f)
ground state
heat
Ideal Gas Law
intensity (energy flux), I
isolated system
kinetic energy (KE)
lifetime
line spectra
Maxwell-Boltzmann function
mechanical energy
neutrino
non-conservative forces
open system
phase velocity
phonons
photochemical processes
photoelectric transfer
photoionization
photons
photothermal energy
Planck's Radiation Law
potential energy (PE)
power
proton-proton (PP) chain
radiation
radiative processes
rays
rest mass energy
root-mean-square speed (v_{rms})
Second Law of Thermodynamics
spectral emissivity (ε_λ)
spectrograph
spectroscope
Stefan-Boltzmann Law
visible spectrum
wave fronts
wavelength (λ)
Wien's Displacement Law
work
work-kinetic energy theorem

Notation

Relation between work, energy, and force

$$\Delta E = W = \int F_{external} \cdot d\mathbf{r} = \int F_{external} dr \cos\theta$$

Energy change in a mass of fluid $dQ = mc_v dT + p\, dV$

Definition of power $P = \dfrac{dE}{dt} = \dfrac{d(W+Q)}{dt}$

Definition of non-relativistic kinetic energy $KE = (1/2)mv^2$

Relation between work and kinetic energy

$$\Delta KE = KE_{final} - KE_{initial} = \int \mathbf{F} \cdot d\mathbf{r} = W_{net}$$

Change in potential energy

$$\Delta PE = -\int \mathbf{F} \cdot d\mathbf{r} = -\int F\, dr \cos\theta$$

$$\Delta PE = -W$$

Conservation of mechanical energy in the absence of dissipative forces $-\Delta KE = \Delta PE$

$\Delta E = 0 = \Delta KE + \Delta PE$

Relation between kinetic energy and temperature

$$\tfrac{1}{2}mv_{rms}^2 = \tfrac{3}{2}k_B T$$

Components of internal kinetic energy $KE_{internal} = KE_{translation} + KE_{rotation} + KE_{vibration} + KE_{bending}$

$$= \tfrac{3}{2}k_B T + \tfrac{2}{2}k_B T + \tfrac{1}{2}k_B T + \tfrac{1}{2}k_B T = \dfrac{7k_B T}{2}$$

Voltage difference $\Delta V = -\int E \cdot dr$

Relation between electric fields and potential energy

$$q\Delta V = -q\int E \cdot dr = -\int qE \cdot dr = -\int F_E \cdot dr$$

$\Delta PE = -W = q\Delta V - \Delta KE$

Rest mass energy $\Delta E = \Delta mc^2$

Relativistic kinetic energy $KE = \left(\dfrac{1}{\sqrt{1-\dfrac{v^2}{c^2}}} - 1\right)mc^2$

Total energy of a moving particle in the absence of a conservative force field

$$E_{Total} = mc^2 + KE = \dfrac{mc^2}{\sqrt{1-\dfrac{v^2}{c^2}}} = \gamma mc^2$$

Electromagnetic wave energy $E = hf$
General wave speed relation $v = \lambda f$
Electromagnetic wave speed relation $c = \lambda f$
Doppler shift $\Delta\lambda/\lambda = dv_{rel}/c$

Intensity (energy flux) related to wave amplitude $I = kA^2$
Conservation of energy (Intensity formulation)

$$I_{final} \times Area_{final} = I_{initial} \times Area_{initial}$$

Wavelength of maximum blackbody emission $\lambda_{max} T = \alpha$

Stefan-Boltzmann Law $R(T) = \int_0^\infty R(\lambda, T)d\lambda = \varepsilon_\lambda \sigma T^4$

Luminosity $L = \varepsilon \sigma A T^4$

Planck blackbody radiation equation

$$R(\lambda, T)d\lambda = \dfrac{2\pi hc^2}{\lambda^5}\left(\dfrac{1}{e^{(hc/\lambda k_B T)} - 1}\right)d\lambda = \dfrac{c_1}{\lambda^5}\left(\dfrac{1}{e^{(c_2/\lambda T)} - 1}\right)d\lambda$$

Photon emission energy (from one of a particle's excited states) $\Delta E = E_f - E_i = hf = h(c/\lambda)$

Energy change between levels for simple atoms

$$\Delta E = 13.6\,eV(Z^2)\left(\dfrac{1}{n_f^2} - \dfrac{1}{n_i^2}\right), \text{ where } Z = \# \text{ protons}$$

Answers to Pause for Inquiry Questions

PFI 2-1: Isolated, closed, open

PFI 2-2: a) $W > 0$, b) $W < 0$, c) $W = 0$

PFI 2-3: $W = 0$, the gravitational force is perpendicular to the displacement vector. The right-most illustration best represents the situation.

PFI 2-4: The energy for expansion comes from the internal energy of the gas. This expansion means the gas will cool as it expands.

PFI 2-5: $\sim 2 \times 10^{18}$ J energy dissipated assuming a five-day hurricane with power $\sim 5 \times 10^{12}$ W. Car: ~830 kJ, 280 kW

PFI 2-6: Incremental arc length $dl = rd\theta$.

Angular speed $= \omega \equiv d\theta/dt$.

Translational speed $v = dl/dt = rd\theta/dt = r\omega$.

Therefore, $KE = 1/2\, mv^2 = 1/2\, mr^2\omega^2$.

PFI 2-7: From Ex. 2.4 the root mean square speed for oxygen in the thermosphere is ~1.2 km/s. Escape speed from Earth is ~11.2 km/s. Very few oxygen particles, even in the high-speed tail of the Boltzmann distribution, have a speed 10 times greater than average. The oxygen atoms do not have enough kinetic energy to overcome gravitational potential energy. Therefore Earth retains most of its oxygen.

PFI 2-8: Ion-left; electron-right

PFI 2-9: c. Entropy increases.

PFI 2-10: The units for $nk_B T$ are $(\#/m^3)(J/K)(K) = (J/m^3)$. The units for $(1/2)\rho v^2$ are $(kg/m^3)(m^2/s^2) = (J/m^3)$. The units for $(B^2/2\mu_0$ are $(N/A \cdot m)^2/(N/A^2) = (N/m^2) = (J/m^3)$. Others are left to the reader.

PFI 2-11: When a heating element is really hot, a small portion of its emitted spectrum overlaps with the visible wavelengths. If the element is merely warm, then the energy is almost exclusively in the infrared portion of the spectrum.

PFI 2-12: Cool stars emit more radiation at longer wavelengths that coincide with the red portion of the visible spectrum.

PFI 2-13: Use Fig. 2-21. The best blackbody approximation is at 700 nm to > 10,000 nm. Gray body approximation is at 200 nm to 700 nm. Non-thermal emitter < 200 nm.

PFI 2-14: The peak radiance (power spectral flux density) from the Sun shown in Fig. 2-21 is ~7 $\times 10^4$ W/(m$^2 \cdot$ nm). Multiply this value by the ratio of the squared surface areas of the solar surface and 1 AU, $4\pi(7 \times 10^8$ m$^2)^2 / 4\pi(1.5 \times 10^{11}$ m$^2)^2$ to arrive at ~1.5 W/(m$^2 \cdot$ nm).

PFI 2-15: 656.3 nm is Balmer H-α. 486.1 nm is Balmer H-β.

PFI 2-16: $\lambda = hc/\Delta E = (4.136 \times 10^{-15}$ eV s)$(2.998 \times 10^8$ m/s)/(4.75 eV) = 261.2 nm

$hf = \Delta E$; $f = \Delta E/h = (4.24 \times 10^{-20}$ J)/$(6.63 \times 10^{-34}$ J/s) = 6.39×10^{13} Hz

$\Delta E \sim k_B T$; (11.1 eV) $(1.6 \times 10^{-19}$ J) = 1.78×10^{-18} J; $T = (1.78 \times 10^{-18}$ J)/$(1.38 \times 10^{-23}$ J/K) $\approx 1.3 \times 10^5$ K

References

Ahmad, Qazi R., et al. 2002. *Direct Evidence for Neutrino Flavor Transformation from Neutral-Current Interactions in the Sudbury Neutrino Observatory.* Physical Review Letters. Vol. 89. American Physical Society. College Park, MD.

Fukuda, Y., et al. 1998. *Measurements of the Solar Neutrino Flux from Super-Kamiokande's First 300 Days: The Super-Kamiokande Collaboration.* Physical Review Letters. Vol. 81. American Physical Society. College Park, MD.

Further Reading

Burnell, S. Jocelyn Bell, Simon Green, Barrie Jones, Mark Jones, Robert Lambourne, and John Zarnecki. 2004. *An Introduction to the Sun and Stars.* Edited by Simon Green and Mike Jones. Cambridge University Press.

Enloe, C. Lon, Elizabeth Garnett, Jonathon Miles, and Stephen Swanson. 2001. *Physical Science: What the Technology Professional Needs to Know.* John Wiley and Sons.

Fleagle, Robert G. and Joost A. Bussinger. 1980. *An Introduction to Atmospheric Physics.* 2nd Edition. Academic Press.

Serway, Raymond. 1997. *Principles of Physics.* 2nd Edition. Saunders College Publishing.

Schroeder, Daniel V. 1999. *An Introduction to Thermal Physics.* Addison Wesley Longman.

Wallace, John M. and Peter V. Hobbs. 2006. *Atmospheric Science, An Introductory Survey.* 2nd Edition.

Wolfson, Richard and Jay M. Pasachoff. 1999. *Physics with Modern Physics.* 3rd Edition. Addison Wesley.

The Quiescent Sun and Its Interaction with Earth's Atmosphere

3

UNIT 1. SPACE WEATHER AND ITS PHYSICS

Contributions by C. Lon Enloe and Charles Lindsey

You should already know about...

- Using exponential and logarithmic functions
- How to describe angular motion
- The Ideal Gas Law
- Conservation of momentum
- Earth's atmosphere and its layers (Chap. 1)
- Energy (Chap. 2)
- Spectral emissions (Chap. 2)

In this chapter you will learn about...

- The composition and structure of the quiet Sun and its atmosphere
- The transport of energy from the solar core to the solar atmosphere
- Solar differential rotation and circulation
- Thermal and magnetic buoyancy
- The solar dynamo as the origin of the Sun's magnetic field
- The global and synoptic scale features of the Sun's magnetic field
- The 11-year variation in solar features
- Remote observations of the Sun
- Optical depth in exponential atmospheres
- The structure of Earth's quiescent, ionized atmosphere
- Impacts of the quiescent Sun on Earth's atmosphere

Outline

3.1 **Introduction to the Quiescent Sun and its Interior**
    3.1.1 Our Local Magnetic Star
    3.1.2 The Stratified Sun
    3.1.3 The Convective Sun

3.2 **Regions of the Solar Atmosphere**
    3.2.1 The Sun's Lower Atmosphere
    3.2.2 The Sun's Upper Atmosphere

3.3 **Characteristics of the Quiescent Solar Magnetic Field**
    3.3.1 Multi-scale Solar Magnetic Fields
    3.3.2 Typical Solar Cycle Field Variations

3.4 **Basic Physics of Atmospheres**
    3.4.1 Law of Atmospheres
    3.4.2 Absorption Interactions in an Atmosphere

Chapter 3 The Quiescent Sun and Its Interaction with Earth's Atmosphere

3.1 Introduction to the Quiescent Sun and its Interior

3.1.1 Our Local Magnetic Star

Objectives: After reading this section, you should be able to...
- State the primary constituents of the Sun and their relative abundances
- Identify the regions of the Sun's atmosphere that exhibit the most variability
- Describe the rotation rate at various locations in and on the Sun
- State the rotation rate in terms of period and frequency
- Explain what insights helioseismology provides

Energy and Composition. Each second, millions of tons of hydrogen nuclei fuse to form heavier nuclei in the solar core. As we discuss in Chap. 2, with each reaction a small amount of mass converts to energy that ultimately arrives at the solar surface and atmosphere. From there, much of it is emitted across a broad range of wavelengths as the solar luminosity. In the solar atmosphere a small fraction of the energy goes to accelerate particles that ultimately become part of the solar wind.

To understand the impacts of the quiescent Sun on the space environment, we study the solar background emissions. The Sun is a modest-sized, relatively stable star that continuously emits electromagnetic radiation and magnetized plasma. The radiative flux from the Sun's lower, comparatively cool atmosphere is steady over time to within approximately 0.1%. A more variable flux (at times greater than 10,000% variation) of X-ray, extreme ultraviolet (EUV), and radio emissions comes from the Sun's multi-million-degree outer atmosphere. Figure 1-23 shows the X-ray variability. The total radiative output from this region of the solar spectrum is less than 2% of the total emission, but the impact at Earth and other planets is profound. Table 3-1 summarizes some of the solar characteristics.

Table 3-1. **The Sun.** Here we list some observational facts and computer model results.

Solar Properties	
Mass = 1.99×10^{30} kg	Age = 4.6×10^9 yr
Luminosity = 3.86×10^{26} W	Core pressure = 2×10^{16} Pa
Surface gravity = 272 m/s^2	Core temperature = 1.5×10^7 K
Escape velocity at surface = 618 km/s	Core density = 1.6×10^5 kg/m^3
Effective blackbody temperature = 5780 K	Photon escape time = $10^5 - 10^6$ yr
Surface magnetic field strength = 10^{-4} T	Radius, R_S = 6.96×10^8 m or ~111 R_E
Mean Sun-Earth distance = 1.50×10^{11} m = 215 solar radii = 1 AU	Coronal temperature $1-2 \times 10^6$ K
Atomic composition: ~ 92% H, ~7.8% He, 0.2% other elements (C, N, O....)	Mean density = 1.4×10^3 kg/m^3
Average rotation period = 27 days, $\omega = 2.7 \times 10^{-6}$ rad/s	Mass loss rate = 10^9 kg/s
Equatorial inclination to Earth's orbital plane = 7°	

As the Sun developed, it collected cosmic remnants of supernova explosions (the remains of extremely massive dying stars) and now has a composition similar to the background universe. On average, about 92% of the atoms in the

3.1 Introduction to the Quiescent Sun and its Interior

Sun are hydrogen and 7.8% are helium. The rest are heavier elements, such as lithium, carbon, oxygen, and iron. In the solar center nuclear reactions convert hydrogen to helium. Hence the relative abundance of helium is higher than the other heavier elements. About 60 elements are present in the Sun. Although the minor elements constitute only a tiny fraction of the solar mass, they contribute substantially to the line spectra of the solar atmosphere. The spectra yield important information about temperature, density, solar structure, and activity.

Rotation. Galileo's early observations and drawings revealed that the Sun rotates in a counterclockwise direction (left to right as we view it from Earth). The rotation developed because of the conservation of angular momentum during the stellar formation. The Sun continues to rotate today, but its rotation rate has slowed, by perhaps a factor of ten, during its lifetime. In the early stages of solar formation, the Sun may have transferred angular momentum to gas jetting from the polar regions. Currently, the Sun's magnetic field is the primary culprit in the deceleration. We recall that by extending an arm or leg, a spinning skater rotates more slowly. As the solar wind draws out the Sun's magnetic field, the field exerts a torque that slows the solar rotation. Younger stars with spectral characteristics similar to our Sun (G-class to astronomers) rotate every three to ten days, compared to the 27-day rotation period for our Sun.

The Sun is a gaseous body. Its rotation rate differs with latitude and depth. Shearing and stretching in the outermost portion of the Sun concentrate energy in magnetic patches that generate space weather storms (Secs. 3.1.3 and 3.3 and Chap. 8). Figure 3-1a illustrates some of the characteristics of rotational motion inside the Sun, including the marked tendency for the outer equatorial portions of the Sun to rotate faster than the poles (Example 3.1). This *differential rotation* produces a large-scale shearing motion that redistributes angular momentum by turbulent eddies. As we see in Example 3.1, surface rotational period is a function of solar latitude, varying from about 25 days at the equator to 33 days at the poles.

Figure 3-1b shows that the Sun's interior has about the same rotational frequency as the mid-latitude surface. Below about 0.7 solar radii (R_S) a change occurs from differential rotation to a regime where all latitudes of the deep Sun's layers rotate as a solid body. Where differential rotation meets solid-body rotation, we find a narrow layer of rotational shear, called the tachocline, thought to be the factory for magnetic field variations that produce the solar cycle. Near the equator a strong forward shear develops between the fast moving exterior and the more slowly spinning interior. Conversely, a backward shear develops in the polar regions. A combination of heating and cooling, shearing, and circulating produces the Sun's magnetic cycles.

Pause for Inquiry 3-1

Use Fig. 3-1b to determine the surface rotation period (in days) at 45° latitude.

Scientists determine the Sun's interior rotation rate by analyzing patterns of surface oscillations caused by internal waves (Focus Box 3.1). The analysis produces a helioseismic record of motion that helps us understand the physical conditions below the visible solar surface (In Greek "helio" means Sun.) The rotation influences the effective speed of the sound waves within the Sun, so scientists use that knowledge to determine the rotation rate at different depths within the Sun.

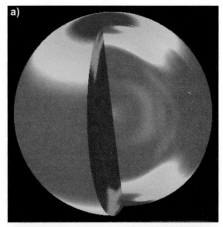

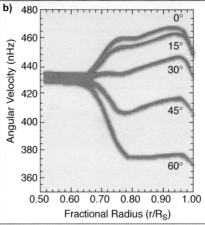

Fig. 3-1. a) **Global Visualization of Internal Differential Rotation of the Sun.** The visualization is based on data from the Global Oscillation Network Group (GONG) observatories and Solar and Heliospheric Observatory (SOHO) spacecraft. Red indicates faster rotation, blue is slower. The inner Sun rotates as a nearly solid body. *(Courtesy of ESA/ NASA – the SOHO Program)* b) **The Angular Velocity Plotted as a Function of Radius for Several Latitudes.** All results are based on data from the Michelson Doppler Imager (MDI) instrument aboard the SOHO spacecraft, averaged over 144 days. [Adapted from Thompson et al., 2003]

EXAMPLE 3.1

(Based on rotation model from Snodgrass and Ulrich [1990])

Differential Rotation

- *Problem Statement:* Determine the angular advance of a parcel at the solar equator with respect to a parcel at ±40°. Use this information to determine the advance of an equatorial parcel during one half of a solar cycle.

- *Relevant Concept:* Differential rotation.

- *Given:* We model the differential rotation of the solar surface with a series of sine functions of the form $\omega = a_0 + a_1 \sin^2\theta + a_2 \sin^4\theta + \ldots$ where ω is the rotation rate in °/day and θ is the solar latitude.

- *Assume:* The rotation model is $\omega = 14.3 + (-2.4)\sin^2\theta + (-1.8)\sin^4\theta$ [°/day].

- *Solution:* For 40° north or south of the equator, the equation above gives a rotation rate, ω of

  $\omega = 14.3 + (-2.4)\sin^2(40°) + (-1.8)\sin^4(40°) = 13.0$ [°/day]

  resulting in a rotational period of 360°/(13.0°/day) = 27.6 days.

  Using the same equation, we find for the equator

  $\omega = 14.3 + (-2.4)\sin^2(0°) + (-1.8)\sin^4(0°) = 14.3$ °/day

  resulting in a rotation period of 360°/(14.3°/day) = 25.1 days.

  At 40° the photosphere rotates 365 day/(27.6 day/rot) = 13.2 rotations in a year.

  At the equator the photosphere rotates 365 day/(25.1 day/rot) = 14.5 rotations in a year.

  The total advance of an equatorial parcel during one-half of a solar cycle is 1.3 rotations/yr × 5.5 yr = 7.15 rotations.

- *Interpretation and Implications:* Thus, in one year, the equator moves 14.5 − 13.2 = 1.3 times around, relative to the region at 40°—enough to cause quite a tangle in the 5.5 years from a sunspot minimum to a sunspot maximum. Figure 3-9 illustrates this situation.

Follow-on Exercise: If the interior Sun rotates at the rate of 13.2°/day, how far does a parcel at the equatorial surface advance with respect to an interior parcel each 27 days? How far does a parcel at the polar surface fall behind with respect to interior parcel each 27 days? Give your answer in degrees and radians.

Focus Box 3.1: Sensing the Solar Interior

Helioseismology is the study of the Sun's interior from observations of the vibrations of its surface. In a manner similar to the way physicians use sonograms to check on the health of soft tissue inside the human body, scientists use sound waves reverberating through the Sun to investigate the solar interior and even the far side of the Sun. The Sun acts as a resonant cavity for acoustic waves generated mainly by the Sun's convective motions (Fig. 3-2).

Acoustic waves involve compression and rarefaction of the plasma. By subtracting two images of the Sun's surface taken minutes apart, scientists observe alternating patches in brightness that result from heating and cooling in response to acoustic vibrations of the interior, called solar oscillations. Most of the detectable oscillatory power is in the 2 mHz–5 mHz range associated with horizontal wavelengths of greater than 8×10^3 km. These standing waves (a whole number of waves must "fit" into the solar circumference) cause the Sun to resonate in a manner similar to a bell.

In the Sun some sound wave fronts propagate at an angle with respect to the radial direction. The wave front speed is proportional to the square root of the temperature ($T^{1/2}$). Because the solar temperature increases dramatically toward the core, the lower part of the wave front moves faster. Therefore, the acoustic wave refracts and bends back toward the surface (Fig. 3-3).

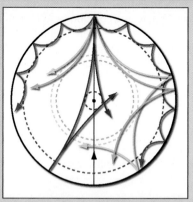

Fig. 3-2. Solar Cross Section Showing Sound Waves Refracting Inside. Various waves reflect and refract off the Sun's upper layers. In the process, they cause solar oscillations and affect the solar seismology. [Adapted from Christensen-Dalsgaard, 2002]

Focus Box 3.1: Sensing the Solar Interior (Continued)

Near the surface, the temperature decreases rapidly. A given temperature is associated with a cutoff frequency below which acoustic waves can't propagate. If waves have a frequency higher than the cutoff, they penetrate upward into the solar atmosphere and dissipate, heating the atmospheric gas and plasma. Waves below the cutoff frequency reflect back into the solar interior and become trapped inside the Sun, causing entire sections of the Sun to repeatedly oscillate (resonate). The waves that penetrate deep into the solar interior last long enough to propagate around the entire solar circumference. This circumnavigation takes several hours. Such low-frequency waves create a resonance that lasts for days or weeks.

Helioseismology has many applications. Scientists estimate the solar rotation rate as a function of its radius (Fig. 3-1b), as well as composition, temperature, and motions within the Sun. They also determine the radial location of forces that drive large-scale plasma motions. They have even devised a method to locate sunspot groups on the far side of the Sun. This development has tremendous potential as a space weather forecasting tool. Figure 3-4 is a whole-Sun map of magnetic regions obtained from helioseismic sounding. Dark regions indicate where a magnetic field has accumulated. Data from the far side tend to be noisier and less certain than data on the Earthward side of the Sun. Nonetheless, when magnetic concentrations are intense on the Sun's far side, forecasters get an early warning of potential activity.

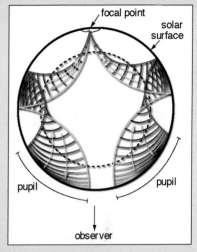

Fig. 3-3. A Cross Section of Waves Bouncing in the Solar Interior. The dotted circle indicates the base of the solar convective zone. Wavefronts from a far-side point source (focal point) with intervals of 286 s (~3.5 mHz) reflect once from the solar surface (right and left sides) and arrive in a pupil on the Sun's near surface (toward observer). Waves from the focal point (green arrows) are reconstructed in the image from the surface disturbance they create on the near surface. The computer simulation demonstrates how reconstructed waves travel a reverse route (yellow arrows) to create nearly identical waves converging into the focal point, contributing to and pin-pointing the local disturbance. Because of symmetry constraints, the type of holography illustrated here is only practical for regions within approximately 50° of the far-side central meridian (antipode). A variation of this wave configuration detects active regions near the solar limb. *(Courtesy of Doug Braun and Charlie Lindsey at North West Research Associates)*

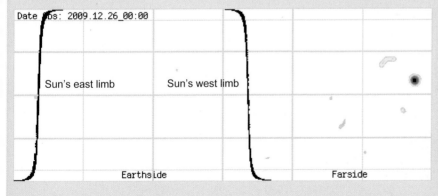

Fig. 3-4. Whole-Sun Magnetic Map. The map extends from the solar south pole to the north pole in an equal-area projection. The solar equator and each 30° of latitude and each 60° of longitude are marked. Here the distortion in creating a flat map from a spherical surface is minimized at about 27° latitude. This location is near the latitude of active regions in the first half of the cycle, so active regions are least distorted during that time. The dark curves separate the visible (Earthward) side of the Sun from the hidden (far side) of the Sun. See Focus Box 3.2 for a discussion of solar coordinates. The data in the map were taken at the same time) *(Courtesy of the Solar Oscillations Investigation (SOI) at Stanford University)*

Chapter 3 The Quiescent Sun and Its Interaction with Earth's Atmosphere

3.1.2 The Stratified Sun

Objectives: After reading this section, you should be able to...

- Apply momentum balance to the motion of parcels in the solar interior
- Describe the energy transport mechanisms in the core and radiative zone
- Describe the link between the tachocline and regions above and below it
- Describe basic solar coordinates

Astronomers believe most quasi-steady-state stars are radially stratified. The stratification produces concentric regions dominated by different physical processes. In this section, we briefly describe the Sun from the inside out.

Core

Figure 3-5 shows regions of the solar interior in terms of their contribution to energy transport. In the solar *core* ($r < 0.25\ R_S$), the weight of the overlying mass produces enormous pressure and high temperatures. Theory indicates that core temperatures exceed 1.5×10^7 K, and the core pressure is $\sim 2 \times 10^{16}$ Pa (Fig. 3-6). The temperature is so high that nuclear reactions fuse hydrogen into helium, as we describe in Chap. 2, and occasionally fuse helium into carbon. The energy created by nuclear reactions begins largely as gamma (γ) rays (~MeV photons). These photons immediately downgrade to hard X rays (> 10 keV) by interactions with electrons and ions in the core, but remain trapped inside the Sun for millions of years, scattering off core material and slowly diffusing outward (Fig. 3-7).

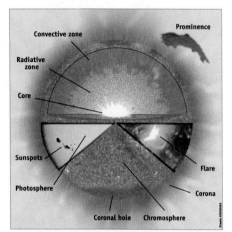

Fig. 3-5. Basic Processes and Conditions in the Sun. Here we show the Sun's interior, highlighting its core, convective zone, and temperature. *(Courtesy of ESA/NASA – the Solar and Heliospheric Observatory (SOHO) Program)*

Fig. 3-6. Temperature and Pressure Model of the Solar Interior. Deep in the interior the temperature and pressure decrease slowly with height. In the outer, convective zone the temperature decreases rapidly with height, leading to convective overturning. *(Data courtesy of Jørgen Christensen-Dalsgaard)*

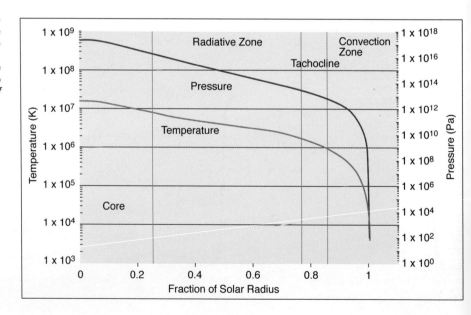

Radiative Zone

Radiative Diffusion. Surrounding the core is a huge region extending to approximately 0.7 R_S, in which photon-energy transport dominates (Fig. 3-7). In this *radiative zone* the temperature decreases with increasing radial distance as heat flows outward through successively cooler overlying layers. As the hard X-ray photons meander outward they interact with ions and free electrons and are absorbed and re-emitted many times, each time at a somewhat lower energy and in a different direction—even back toward the core. This process of *radiative diffusion* consists of a large number of discrete steps in a random walk pattern. As a result of these multiple interactions, the average energy of the photons gradually decreases to soft X rays as they diffuse outward. By the time the energy reaches the upper portions of the radiative zone, it is in the form of extreme ultraviolet (EUV) radiation. Because of the torturous photon path, their trip from the core to the solar surface (which could be done in a vacuum in two seconds) takes most photons hundreds of thousands of years to complete. The energy we see in the sky today was created by a nuclear reaction that occurred a very long time ago. We won't be alive to observe the photons that were created in recent years.

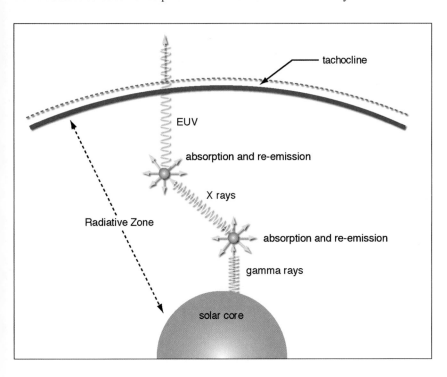

Fig. 3-7. The Radiative Zone. The absorption and re-emission process occurs countless times as radiant energy slowly diffuses to the top of the radiative zone. *(Adapted from the AF Weather Agency)*

Hydrostatic Balance in the Solar Interior. The core and radiative zone exhibit *hydrostatic equilibrium*. Gravitational attraction, acting inward, balances an outward-directed change in the pressure force (pressure gradient force). We understand this balance in terms of a conservation principle for a variable x that is written as

$$\frac{dx}{dt} = Sources - Sinks$$

Chapter 3 The Quiescent Sun and Its Interaction with Earth's Atmosphere

If the quantity x is momentum ($m\mathbf{v}$), then external forces are the sources and sinks of the rate of momentum change. The conservation equation becomes

$$\frac{d(m\mathbf{v})}{dt} = \text{Sum of external forces} \tag{3-1a}$$

where

m = particle mass [kg]
$\mathbf{v}$ = particle velocity vector [m/s]
t = time [s]

For fluids we write Eq. (3-1a) in terms of mass density and force density:

$$\rho\frac{d\mathbf{v}}{dt} = \rho\mathbf{a} = \Sigma\mathbf{F}_{ext}/volume \tag{3-1b}$$

where

ρ = mass density [kg/m^3]
$\mathbf{a}$ = particles' collective acceleration vector [m/s^2]

In the vertical (radial) direction the force densities are associated with gravity and the pressure gradient force

$$\rho\frac{d\mathbf{v}}{dt} = -\frac{GM\rho}{r^2}\hat{\mathbf{r}} - \frac{dP}{dr}\hat{\mathbf{r}} = \Sigma\mathbf{F}_{ext}/volume \tag{3-2}$$

where

G = universal gravitational constant (6.67×10^{-11} [N·m^2/kg^2])
M = mass within a sphere of radius, r [kg]
r = radial distance, positive outward [m]
P = pressure [Pa or N/m^2]

For *hydrostatic equilibrium* the net force density equals zero, so Eq. (3-2) simplifies to

$$\frac{dP}{dr}\hat{\mathbf{r}} = -\frac{GM\rho}{r^2}\hat{\mathbf{r}} \tag{3-3}$$

$$= -\rho g\,\hat{\mathbf{r}}$$

where

$$g = \frac{GM}{r^2} = \text{local gravitational acceleration [m/s}^2\text{]}$$

Equation (3-3) is the statement for hydrostatic equilibrium where the pressure gradient force per unit mass equals the gravity force per unit mass in magnitude but is in the opposite direction.

Pause for Inquiry 3-2

Verify that each element of Eq. (3-3) has units of force/volume [N/m^3] and pressure/distance [Pa/m].

Tachocline

Hot, highly massive, short-lived stars have very dominant radiative zones that move their core-generated energy to their surfaces almost exclusively by radiative processes. In contrast, stars less massive than 0.3 solar masses are fully convective, transporting their energy outward by bulk plasma motion.

Long-lived, cool stars such as the Sun transport energy using both processes. Scientists believe these stars need a thin layer at the top of the radiative zone that acts as an interface region for the two energy transport processes. For differentially rotating stars this interface zone is the *tachocline*. This thin region, shown as the bright red shell in Fig. 3-8, allows the relatively calm and stratified conditions of the radiative zone to give way to the turbulent fluid motions overhead.

Changes in flow velocities across the tachocline stretch and enhance magnetic field lines generated by the moving plasma. Thus the tachocline, with a width less than 0.1 R_S, is likely to be the region in which the Sun generates, stores, and amplifies magnetic fields. Because the region sits atop the comparatively stable radiative zone, the magnetic storage may be long-term, perhaps years or decades. Turbulence from the overlying convective zone accentuates the magnetic field in localized regions. Some researchers believe the localized concentrations of magnetic field in the tachocline are the root fields for the intense magnetic activity during the maximum phase of the solar cycle.

Sheared flow at the tachocline deforms pole-to-pole (*poloidal*) *magnetic fields* into concentrated, azimuthally stretched fields called *toroidal magnetic fields* (Figs. 3-9 a and b). The concentration of the magnetic field generates high pressure that in turn pushes plasma out of the pressurized zone, creating low mass density bubbles. Parcels with low mass density are buoyant. Bubbling motions that we describe in the next section help to buoy some of the ropes of concentrated field to the surface where they may form sunspots (Figs. 3-9 c and d). The generating and cycling of magnetic flux is the *solar dynamo process*. This process has multiple scales, the largest of which links the tachocline, through the unstable convection region that overlies it, to the solar atmosphere.

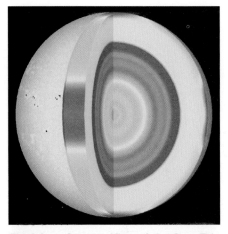

Fig. 3-8. Cutaway View of the Sun. This depiction, based on Solar and Heliospheric Observatory (SOHO) data, shows the tachocline in bright red. SOHO instruments measured higher than expected sound speed, and a rapid change in the Sun's rotation speed, between the faster-turning outer region and the slower interior. This thin shear layer may generate intense magnetism in a layer about 0.1 R_S thick. For simplicity, smaller scale velocity variations are not shown. *(Courtesy of ESA/NASA – the SOHO Program. Data imaged by Alexander Kosovichev at Stanford University)*

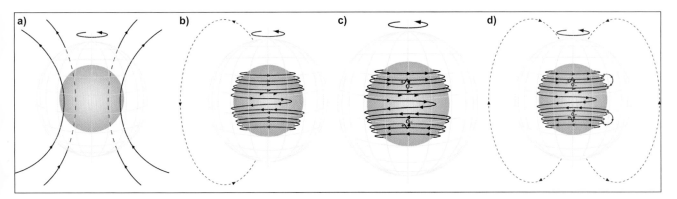

Fig. 3-9. Solar Dynamo Process. This sequence of diagrams depicts qualitatively the succession of processes that lead to cyclic magnetic field changes about the equator. The inner semi-opaque sphere represents the Sun's radiative core/tachocline and the blue mesh is the solar surface. In between is the solar convective zone where a dynamo resides. **a) Shearing of the Poloidal Field by the Sun's Differential Rotation near the Convective Zone Base.** The Sun rotates faster at the equator than at the poles. **b) Toroidal Field Produced by This Shearing.** Over the course of many solar rotations the field wraps around the Sun as we discuss in Example 3.1. The result is a sheared toroidal field. **c) Strong Toroidal Field.** When the toroidal field is strong enough, buoyant loops rise to the surface, twisting as they rise because of the rotational influence. Sunspots form bipolar regions (+ and − field orientation) at the base of the magnetic loops. **d) Flux Emergence.** Additional flux emerges with an orientation opposite that of the previously existing poloidal fields and spreads in latitude and longitude from decaying spots. [Adapted from Babcock, 1961 and Dikpati and Gilman, 2006]

Chapter 3 The Quiescent Sun and Its Interaction with Earth's Atmosphere

Near the tachocline the solar plasma cools to ~5 × 10⁶ K. At larger radial distances the temperature drops further, allowing trace amounts of carbon, oxygen, calcium, and iron ions to combine with free electrons. At about 0.86 R_S the temperature becomes sufficiently low for a significant fraction of the free electrons to slow enough for hydrogen or helium nuclei to recapture them. Photons are more easily absorbed by the resultant neutrals. Photon absorption impedes radiative transfer, making the plasma more opaque. Increased opacity causes the strong temperature changes with height (Fig. 3-6), helping to create thermal instability in the Sun's outer envelope.

> **Pause for Inquiry 3-3**
>
> Solar scientists often use millions of meters (Mm or megameters) as a unit of measure. Determine the thickness of the tachocline and of the convective zone in Mm.

EXAMPLE 3.2

Pressure Gradients

- *Problem Statement:* Determine the value of the pressure gradient at the tachocline and compare it to the vertical pressure gradient in Earth's atmosphere near sea level. Consider that ~95% of the solar mass is interior to the tachocline.
- *Relevant Concept:* Hydrostatic equilibrium forces must balance to create hydrostatic equilibrium.
- *Given:* Sun's mass = 2 × 10³⁰ kg, the tachocline is at 0.7 R_s, and the density at the tachocline = 250 kg/m³.
- *Assume:* The tachocline region is in hydrostatic equilibrium.
- *Solution:* For the tachocline

$$0 = -\frac{GM\rho}{r^2}\hat{r} - \frac{dP}{dr}\hat{r}$$

$$\frac{dP}{dr}\hat{r} = -\frac{GM\rho}{r^2}\hat{r}$$

$$= -\frac{(6.67 \times 10^{-11}\,\mathrm{Nm^2/kg^2})(0.95)(2 \times 10^{30}\,\mathrm{kg})(250\,\mathrm{kg/m^3})}{(0.7 \times 7 \times 10^8\,\mathrm{m})^2}\hat{r}$$

$$= -132{,}000\,\mathrm{Pa/m}\,\hat{r}$$

For terrestrial sea level

$$0 = -\frac{GM\rho}{r^2}\hat{r} - \frac{dP}{dr}\hat{r}$$

$$\frac{dP}{dr}\hat{r} = -\frac{GM\rho}{r^2}\hat{r}$$

$$= -\frac{(6.67 \times 10^{-11}\,\mathrm{Nm^2/kg^2})(5.97 \times 10^{24}\,\mathrm{kg})(1\,\mathrm{kg/m^3})}{(6378 \times 10^3\,\mathrm{m})^2}\hat{r}$$

$$= -9.79\,\mathrm{Pa/m}\,\hat{r}$$

- *Interpretation and Implications:* Pressure decreases with height. The pressure gradient approximately two thirds of the way out of the Sun is more than ten thousand times greater than the atmospheric vertical pressure gradient at sea level on Earth.

Follow-on Exercise: Make a similar comparison for the pressure gradient at the solar surface.

Focus Box 3.2: Solar Coordinate Systems

We need a coordinate system for describing locations on the Sun. Figure 3-10 summarizes how we use two variables in this coordinate system to specify a point on the Sun's surface. Latitude and longitude underpin most solar coordinate systems.

Latitude (θ) is the angular distance from the solar equator, which is defined as the line equidistant from the two rotational poles. We measure solar latitude from $\theta = -90°$ (solar south pole) to $\theta = +90°$ (solar north pole). The solar equator is at $\theta = 0°$.

Central meridian (CM). The longitude on the disk directly facing Earth is the central meridian. To describe features on the visible disk the CM is given the value 0°. Distances from the CM are up to 90° E and 90° W on the Sun's visible hemisphere. On the Sun, west and east are reversed relative to terrestrial maps. When we view the Sun from Earth, we must look to the right to view the Sun's western hemisphere. Longitude increases toward the west limb and decreases toward the east limb.

Carrington longitude (L) is the east-west angular distance measured from a prime meridian that is a predetermined point of origin rotating with the Sun. Longitude ranges from 0° to 360°, increasing in the direction of the Sun's rotation. Because the Sun rotates at different rates according to latitude (differential rotation), we choose the average rotation rate: 13.2° per day, or one rotation every 27.3 days. Also, the Sun has no regular feature to help us identify the prime meridian by sight alone, so we simply track its position by timing. On November 9, 1853 at 00 UTC, Sir Richard Carrington arbitrarily assigned the north-south centerline on the visible solar disk as the meridian $L_0 = 0°$, and we've been tracking the prime meridian ever since. (This Sir Carrington is the one who observed the white light flare that we describe in Chap. 1).

The *Carrington rotation number* is the number of times the prime meridian has rotated around the Sun since 1853. At a rate of about 13.4 Carrington Rotations per year, the Carrington rotation number is now over 2050. We compute the Carrington rotation number with the Julian date (*JD*) using the formula provided by Duffet-Smith [1992]:

$$CR = 1690 + [(JD - 2444235.34) / (27.2753)]$$

The Julian date is the number of days and fractions of a day, since January 1, 4713 BC and can be calculated with simple applications on the internet. The Julian date for 00 UTC on July 4, 2009 was 2455016.50.

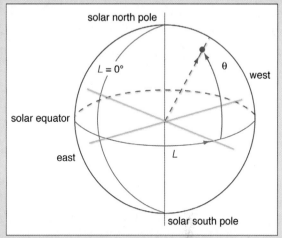

Fig. 3-10. A Heliographic Coordinate System. In most solar images north is at the top, west is at the right, and east is at the left. Because of the 7° tilt of the rotation axis from the ecliptic plane's perpendicular vector, we see the north pole for half of the year and the south pole for the other half. *(Courtesy of the AF Weather Agency)*

Pause for Inquiry 3-4

Calculate the current value of the Carrington rotation number and verify your value using the internet.

3.1.3 The Convective Sun

> **Objectives:** After reading this section, you should be able to...
>
> ♦ Relate convective instability to temperature and density variations and to buoyancy
> ♦ Contrast thermal convection with magnetic convection
> ♦ Compare and contrast granulation and supergranulation
> ♦ Describe the basic meridional circulation pattern in the convective zone

Convective Zone

Convection Basics. Between $0.7\ R_S$–$1.0\ R_S$, the matter in the Sun is neither dense enough nor hot enough to efficiently transfer interior heat energy outward via radiation. The Sun relies on columns of thermal convection to carry energy to the surface. Hydrostatic imbalance created by temperature differences drives *thermal convection* flow. Warmer parcels with their higher temperatures and lower densities experience buoyant acceleration, so they rise. This is the essence of *convective instability*. In a region called the solar *convective zone*, parcels with positive and negative buoyancy transport energy through the outer layer of the Sun (Figs. 3-11 and 3-12). Although convection is an inherently turbulent and seemingly disordered process, it promotes large-scale organization of the Sun's magnetic field and supports the solar dynamo and the sunspot cycle.

A rising, isolated parcel with excess thermal energy maintains a temperature greater than and a density less than its surroundings. Parcels less dense than their surroundings remain buoyant. As buoyant parcels flow upward they cool by doing work to expand into a lower pressure environment (The cooling process tends to reduce the buoyancy, but as long as the parcel remains less dense than its surroundings it experiences a net upward force (Fig. 3-13 in Focus Box 3.3). In natural environments, the temperature and density eventually adjust to be in equilibrium with the surroundings, whereby the buoyancy force disappears and the parcel no longer accelerates. The parcels may, however, remain in motion and overshoot their equilibrium position, similar to a skidding car that remains in motion even though the brakes have been applied.

Convective zone parcels continue to rise and cool until they reach a level where the temperature and density are sufficiently low that outward-emitted photons have a good chance of escaping into space. This photon escape level essentially defines the visible surface of the Sun (Fig. 3-14). After the matter cools by radiation at the surface, it plunges back to the base of the convective zone, to receive more energy from the top of the radiative zone. Scientists think the downward convective overshoot, caused by conservation of momentum, occurs at the base of the convective zone and carries turbulent downflows into the outer layers of the radiative zone.

A strong coupling exists between the charged particles in the solar plasma and solar magnetic fields. In fact, the magnetic fields are often "frozen" to the plasma (Chap. 6). A modified form of convection develops in magnetized plasma. Convection concentrates magnetic field lines, creating magnetic pressure and concentrated tubes of magnetic flux. Regions with high magnetic pressure tend to expel matter in an attempt to balance total pressure. The loss of mass renders magnetic flux tubes buoyant. Thus, regions of concentrated magnetic field are often buoyantly unstable, giving rise to *magneto-convection*. Magnetized parcel motion depends on the relative amount of energy in the kinetic motion of the plasma

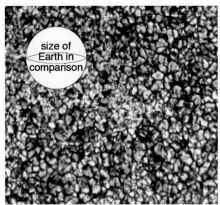

Fig. 3-11. Granular Convection Cells. Lighter regions are hotter and moving upward and outward. The narrow dark lanes show regions of cooler downward motion. Small dark blotches in the granular lanes are concentrations of the magnetic field called pores. The image resolution is about 100 km. The emission is from the 430 nm transition of the molecule CH, which exists in minute amounts in the cooler portions of the solar atmosphere. *(Courtesy of the Swedish Vacuum Telescope, operated by the Royal Swedish Academy of Sciences and Goran Scharmer)*

Fig. 3-12. The Outer 200 Mm of the Sun. The photosphere is at the top. The bottom edge stops just short of the tachocline. In this image, extracted from a simulation, granular structures are in gray. Dark finger-like structures show relatively fast downdrafts embedded within the slower upwellings. Scientists are trying to determine if the plunging downdrafts are associated with supergranular structure in the solar interior. *(Courtesy of Dewey Anderson at the University of Colorado and Robert Stein at Michigan State University)*

compared to that stored in the magnetic field configuration. The turbulent convection of the outer portion of the Sun creates a 'small-scale' dynamo that produces magnetic north and south poles all over the surface of the Sun. Convection aids in generating, concentrating, and expelling the solar magnetic field. In Chap. 9 we describe further the link between convective activity, sunspots, and active regions.

Focus Box 3.3: Convection

Above $0.85\ R_S$ the temperature increases sharply. As a result, the Sun's outermost region is subject to vigorous convection. Convection occurs if a rising fluid element becomes less dense than its surroundings. The Ideal Gas Law tells us that excess thermal energy is the source of the density perturbation. The slightly reduced mass in the parcel experiences a smaller gravitational force than similar parcels in the surrounding medium. The buoyant parcel accelerates upward as a result of the force imbalance.

Suppose a parcel of gas has a pressure equal to that in the surrounding gas but has a reduced density, $\rho^* < \rho_S$, where ρ^* is the mass density of the parcel and ρ_S is the mass density of the surrounding gas. For the surrounding, hydrostatically balanced environment

$$\left|\frac{dP_S}{dr}\right|_{out} = \left|-\frac{GM\rho_S}{r^2}\right|_{in} \quad (1a)$$

But for the parcel with the reduced density

$$\left|\frac{dP^*}{dr}\right|_{out} > \left|-\frac{GM\rho^*}{r^2}\right|_{in} \quad (1b)$$

This imbalance results in a buoyancy force (per unit volume) that we write as

$$\frac{\Sigma \mathbf{F}_{ext}}{Vol} = \rho \mathbf{a} = (\rho_S - \rho^*)\mathbf{g} \quad (2)$$

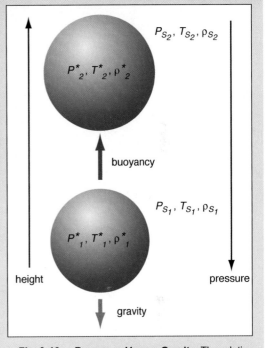

Fig. 3-13. **Buoyancy Versus Gravity.** The relative difference in density between the parcel (*-labeled parameters) and its surroundings (S-labeled parameters) creates a net upward force on the parcel.

Equation (2) says that low-density parcels accelerate upward, transporting momentum and energy as they do. The acceleration is proportional to the local gravitational acceleration ($\mathbf{g}$) and is positive upward.

Similarly, plasma parcels with embedded magnetic flux contribute to the flux's pressure. (Chapter 4 discusses this topic in greater detail.) To maintain pressure balance with the surrounding environment, the magnetized parcel typically expels plasma, thereby reducing the thermal pressure and the downward force of gravity on the parcel. Plasma expulsion reduces the density compared to the surrounding parcels and a situation described by Eq. (2) again prevails, with a resulting buoyancy. This process is the essence of magneto-convection.

Convection Scales. The convective thermal cells leave an imprint on the solar surface in the form of solar granulation and supergranulation. The outer solar envelope contains a hierarchy of convection cells (Figs. 3-11 and 3-14). Granular cells (*granulation*), with scale sizes of 1 Mm are evident in white light images. In granular cells, upward speeds are on the order of 1 km/s, and horizontal speeds are about 0.5 km/s. Down-welling speeds at the cell boundaries typically reach 2 km/s. These cells have lifetimes of ~10–15 minutes. They evolve rapidly through interactions with each other and bordering magnetic fields. The cell centers are generally about 300 K hotter than the surrounding gas, which makes granulation easily visible in white light.

Fig. 3-14. Granular and Super Granular Cells and Convective Zone Dynamics. Here we show various regimes of solar convection. Granular cells form with radii of 1 Mm, supergranular cells are 20 Mm–30 Mm in radius, and giant convection cells have a radius of 400 Mm. The downwelling plumes of concentrated magnetic field pierce the tachocline, where rotational motion further concentrates the magnetic field. *(Adapted from AF Weather Agency)*

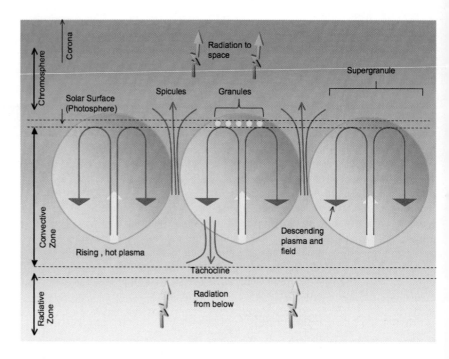

Pause for Inquiry 3-5

For a layered system, which of the following environmental conditions produces a setting most favorable to parcel instability?

a) Cold on top, cold on the bottom
b) Cold on top, warm on the bottom
c) Warm on top, warm on the bottom
d) Warm on top, cold on the bottom

Pause for Inquiry 3-6

Use the Ideal Gas Law to relate density to temperature in the acceleration equation in Focus Box 3.3. Show that a mass density reduction in a parcel is equivalent to a temperature increase.

Pause for Inquiry 3-7

Buoyant parcels that reach the top of the photosphere radiate their excess energy to space. How does their vertical motion change as a result of the energy loss?

Small-scale turbulence associated with granulation allows subsurface magnetic elements to rise. For the most part, these small dipolar elements diffuse into the background. Occasionally, the field becomes concentrated enough in the intergranular lanes to form tiny intense regions of magnetic field called *micropores* and *pores*. These structures are precursors to sunspots, but only the most concentrated and long-lived pores develop into sunspots (Fig. 3-11). *Spicules* are jets of plasma with lifetimes of 5–10 minutes. They also outline the supergranular cells in the chromosphere. Spicules originate in the photosphere and pass through the chromosphere into the corona (Fig. 3-14).

Scientists believe that granular motion creates acoustic noise that propagates upward, dissipating energy in the solar atmosphere via waves, and downward to create acoustic vibrations. These vibrations interfere with each other and refract off the denser regions below, creating concentrations of sound waves that cause patches of the solar surface to oscillate with a five-minute cadence (Fig. 3-15a).

The tops of larger scale convection cells are evident at and above the solar surface as *supergranulation*. Supergranular cells are about 20 Mm–30 Mm in size (~2 times Earth's diameter) and cover the entire quiet Sun (Fig. 3-15b). Each cell lasts about one day before being replaced by another. Material rises in each cell's center in relatively slow motion and then travels in the horizontal direction toward the edge of the cell with flow speeds of about 0.5 km/s. At the edges of the supergranular cells in the convective zone, the plasma downflows concentrate the magnetic field and draw it back to the deep convective zone. There the magnetic field participates in large-scale dynamo processes that produce the solar cycle. Simulations reveal that the downflow channels exert significant control of plasma and field motion. These high-speed, filamentary structures form merging downdrafts that penetrate into the tachocline and add recycled magnetic field to that created by rotational shear at the base of the convective zone.

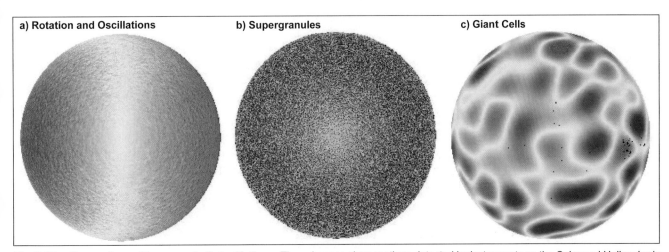

Fig. 3-15. Comparison of Large-scale Solar Motions. These images show motions detected by instruments on the Solar and Heliospheric Observatory (SOHO) satellite. **a) Rotation and Oscillations.** In the left image the rotational motion of the Sun is apparent. Material on the right edge moves away from the observer, while material on the left edge moves toward the observer, consistent with the Sun's counterclockwise rotation. The small-scale variations are associated with solar oscillations. **b) Supergranules.** the rising and falling motion of the supergranules appears after the Doppler shift associated with the left image has been removed. Again, blue indicates motion toward the observer (rising motion). **c) Giant Cells.** Here we show recently revealed large-scale motions. We can think of these giant cells as similar to the synoptic-scale "air mass" systems on Earth. *(Courtesy of David Hathaway at NASA)*

The pattern of super convection cells arises from deep within the convective zone. In locations where the plasma's kinetic energy density is much larger than its magnetic energy density, fluid flow concentrates magnetic fields in the cell boundaries. During the turbulent process, magnetic field lines sweep into the downflow lanes of the supergranules, where they stretch vertically (Fig. 3-14). Some

Chapter 3 The Quiescent Sun and Its Interaction with Earth's Atmosphere

of the field stretches upward into the solar atmosphere, forming a network outlining hundreds of granular cells. The upflow regions surrounding the supergranules may be the place where the solar wind takes root and flows outward in concentrated channels. In lower network regions, magnetic fields arch over the supergranules, forming a canopy of sorts. Field lines that stretch further into the upper solar atmosphere form loops in the corona. Simulations of the looping structures suggest that they form a carpet of interwoven magnetic field lines (Sec. 3.3).

Scientists long suspected that larger convective cells extend to the bottom of the Sun's convective zone and persist longer than a solar rotation. These *giant convection cells* (Fig. 3-15c) were discovered in the Doppler shift of photospheric emission lines [Hathaway et al., 2000]. The cells form small-amplitude climate-like patterns on which more vigorous, small-scale solar activities are superposed. Because the motions are small, scientists averaged many hours of data to reveal flows of about 1 m/s in the presence of material moving as fast as 2 km/s in both directions.

Convective zone dynamics are at the root of many space weather events. The strong coupling between plasma and magnetic fields, as well as the combination of convection and differential rotation, requires that convective processes be part of a three-dimensional circulation system. Data from space and the ground reveal *meridional* (along the meridians) and *zonal* (across longitudinal zones) *circulations* that are similar to a cross between atmospheric and oceanic circulations on Earth. Flow at the top of the convective zone moves gas and plasma from low latitudes to high latitudes at the rate of about 20 m/s. In addition, we see hints of upward motion at low latitudes and downward motion at high latitudes. To complete this circulation and preserve mass flow (mass continuity), there must be a return high-to-low latitude circulation deep in the convective zone. Because of the higher densities at the base of the convective zone, the flow needs a speed of about 2 m/s to close the circulation circuit.

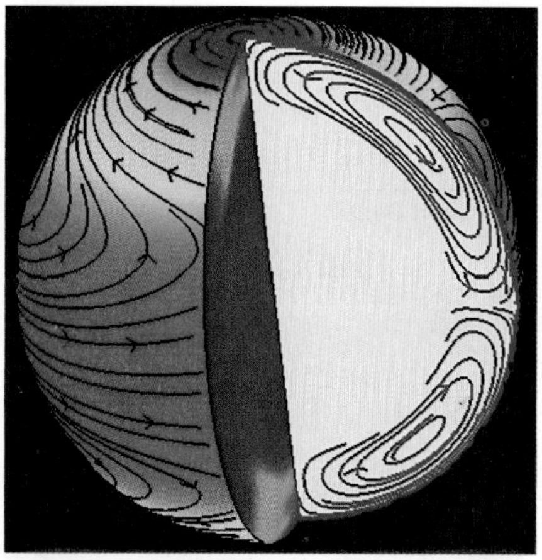

Pause for Inquiry 3-8

Are the speeds mentioned in the section above slower, faster, or about the same as normal highway speeds? How do they compare with typical aircraft speeds?

Pause for Inquiry 3-9

Consider a parcel near the top of the solar convective zone that moves from 20° to 80° latitude at the rate of 20 m/s. How long will the trip take? If the return flow is at the base of the convective zone where the material is roughly ten times denser, how long will the return trip take?

Fig. 3-16. Differential Rotation and Meridional Circulation. The colors on the left side of the image represent the difference in rotation speed between various areas on the Sun. Red-yellow is faster than average and blue is slower than average. Solar and Heliospheric Observatory (SOHO) observations indicate that these differences extend down approximately 20 Mm into the Sun. Sunspots, caused by disturbances in the solar magnetic field, tend to form at the edge of these bands. For simplicity we don't show the small-scale variations underlying these bands. The blue lines in the cutaway at the right represent the surface flow from the equator to the poles of the Sun, which extend to a depth of at least 26,000 km (4% of the solar radius). The return flow indicated at the bottom of the convective zone is from a simple model consistent with mass-flow continuity. *(Courtesy of ESA/NASA – the SOHO Program and the Stanford University Solar Oscillations Investigation Group)*

In summary, space- and ground-based observations and new simulation capabilities have given us the picture of solar material motions shown in Fig. 3-16. Below the tachocline the Sun rotates as a solid body. At the tachocline, sheared flows create alternating bands of slow and fast flow. Above the tachocline we observe rotational differences both radially and longitudinally. Convective motions added to meridional circulation drive a three-dimensional motion in the convective zone that twists and enhances magnetic field lines (not shown). These field lines may rise to the surface, forming sunspots and active regions. Some solar modelers have come to believe that variations in meridional flow speed may control the strength and period of the solar cycle. Their models now include flow-memory terms from previous solar cycles.

3.2 Regions of the Solar Atmosphere

3.2.1 The Sun's Lower Atmosphere

Objectives: After reading this section, you should be able to...

- Describe the features and emissions of the quiescent photosphere, chromosphere, and transition region
- Compare and contrast density, temperature, and magnetic properties with other regions
- Distinguish between sunspots, faculae, and plages
- Describe the use of Hα and Ca-II imagery in monitoring the Sun

The Photosphere

Above the convective zone, photons tend to have their last few interactions with solar matter. This thin boundary where visible photons have a good chance of flying free of the Sun is the "surface" of the Sun, the *photosphere* (Fig. 3-17a). The photosphere is less than 500 km thick (< 0.1 R_E), yet it is the source of 99% of the visible and infrared radiation reaching Earth. A close look at Fig. 3-17a reveals that the edge (limb) of the Sun is darker than the center. This effect is *limb darkening* and is a consequence of the photospheric temperature decreasing with height. At the limb, the photons come from the higher, cooler portion of the photosphere and thus produce a lower energy flux at the observer's position.

Although made up of a very tenuous gas ($\rho \sim 10^{-4}$ kg/m^3), the photosphere still offers some opacity to visible light. The opacity arises from the photosphere's combination of density and temperature that allows some hydrogen atoms to acquire an extra electron. The loosely attached electron is easily dislodged and set into motion by photons of infrared or visible light. The photosphere contains enough negative hydrogen ions to effectively absorb visible light, thus making the photosphere opaque and an effective blackbody. A blackbody spectrum is normally produced by a dense object, not a very thin gas like the photosphere. But because negative hydrogen ions make the photosphere as opaque as a dense object, the photosphere's visible and infrared spectrum is very similar to that of a blackbody. We illustrate the close resemblance in Fig 3-18.

Because the photosphere is relatively cool (~5800 K), much of the matter consists of neutral hydrogen and helium atoms (not ions). The fraction of ionization is similar to that in Earth's upper atmosphere (~1 ion per 10,000 neutral particles). But even this small fraction of ions interacts strongly with the photospheric magnetic field. The convective jostling in the photosphere shakes the roots of the magnetic field lines that extend into the upper atmosphere. This field line shaking probably accounts for part of the energy transport to the quiescent Sun's upper atmosphere.

Large magnetic features within the photosphere anchor active regions that extend higher into the solar atmosphere. Photospheric regions near magnetic concentrations and sunspots often appear brighter than the surrounding photosphere. These bright regions (*faculae*), are luminous patches of heated gas at the granular walls visible in G-band emissions (430.5 nm) in the upper

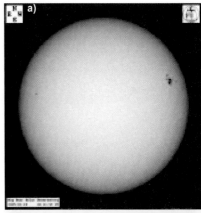
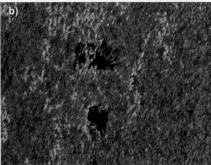

Fig. 3-17. The Sun in White Light. a) We observe the Sun as an edged sphere of photons—a photosphere. The edges of the disk appear darker because the photons are coming from a higher, cooler portion of the photosphere. *(Courtesy of the Big Bear Solar Observatory)* **b) Bright Faculae at 430.5 nm.** The small visible bright regions surrounding sunspots are the bright wall regions of hot granular cells. *(Courtesy of the Swedish Vacuum Telescope, operated by the Royal Swedish Academy of Sciences)*

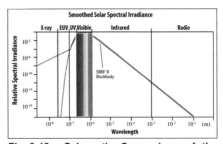

Fig. 3-18. Schematic Comparison of the Sun's Radiative Output with That of a Blackbody. At greater than X-ray wavelengths, most of the solar spectrum (red curve), emitted by the photosphere, approximates that of a blackbody (blue curve). The images above are in the visible part of the spectral region, shown by the color bar. *(Courtesy of the COMET Program at the University Corporation for Atmospheric Research. Adapted from NCAR/HAO)*

Pause for Inquiry 3-10

If you were to measure the energy flux across the center of Fig. 3-17a (moving left to right), the trace would look most like:

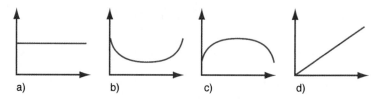

a)   b)   c)   d)

photosphere (Figs. 3-17b and 3-19). They exist in areas of enhanced magnetic fields, where the density of the gas is sharply reduced by the presence of strong magnetic fields. The low gas density allows transparency such that the deeper layers of the granule on the limb-side of the magnetic field concentration become visible. At these deeper layers, the gas is hotter and radiates more intensely, explaining the brightening. On the quiet Sun, faculae are most visible near the limb where limb darkening provides favorable contrast. We associate faculae lacking sunspots with developing or dissipating concentrations of magnetic field. While the sunspots tend to make the Sun look darker, the faculae make it look brighter. During the most active parts of the solar cycle the faculae actually overpower the sunspots and make the Sun appear slightly (about 0.1%) brighter at sunspot maximum than at sunspot minimum.

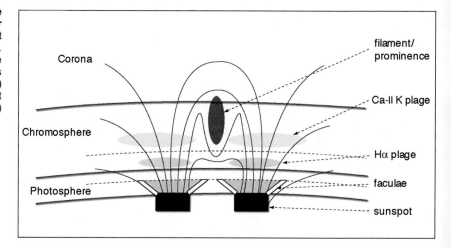

Fig. 3-19. Features of the Quiet and Active Sun That Emit at Various Levels in the Solar Atmosphere. The black curves represent magnetic field lines above the solar surface. The magnetic concentrations in sunspots give rise to many solar features we discuss in this section. Faculae and plages (defined below) may precede and supersede sunspot development. (The depiction is not to scale.) *(Adapted from the AF Weather Agency)*

The Chromosphere and Transition Region

The photospheric temperature decreases with height. The temperature minimum in Fig. 3-20 defines the base of the *chromosphere*. This region is about four times thicker than the photosphere (~2 Mm), but because of its low density, produces much less light. At the chromospheric base the temperature dips to 4300 K and then begins to rise, reaching about 20,000 K at the top of the layer. The chromosphere is transparent to visible wavelengths and is so tenuous that it contributes very little to the Sun's radiative output. It is, however, an effective

3.2 Regions of the Solar Atmosphere

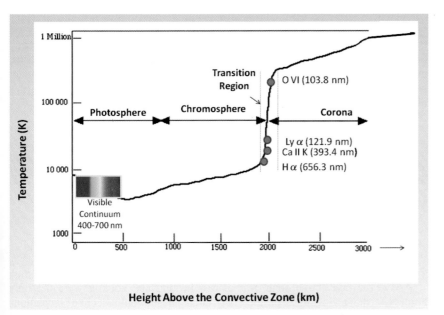

Fig. 3-20. Temperature Profile for the Solar Atmosphere. This log-linear plot shows the temperature profile for the solar atmosphere. Specific emissions for hydrogen, calcium, and oxygen from the transition region (TR) are identified by blue dots. *(Temperature data courtesy of Eugene Avrett at the Smithsonian Astrophysical Observatory)*

radiator of line emissions (Chap. 2). The high temperatures in the middle-to-upper chromosphere allow for an electronic transition in the resident neutral hydrogen atoms at 656.3 nm, known as the hydrogen-Balmer-α (H-α) line. This electronic n=3 to n=2 transition provides much of the visible chromospheric radiative emissions. If we screen out the broad-spectrum visible light of the photosphere with a narrow band filter or an occulting disk, we can image chromospheric features in the orange-red Hα spectral line. Figure 3-21a shows an Hα image of the Sun.

In the upper chromosphere, the particle density declines steeply, allowing magnetic field pressure (energy density) to approach and eventually exceed thermal pressure. Magnetism takes control. These magnetically dominated regions are a source of enhanced ultraviolet chromospheric emissions called *plages* (Chap. 9 further discusses faculae and plages). Radiation from singly ionized calcium (Ca-II) emanating from ~2 Mm above the solar surface, helps observers visualize the magnetic field. This spectral line absorbs about 98% of the light at its central wavelength (393.4 nm). The center of this spectral line is especially sensitive to the presence of magnetic fields in the material. If a magnetic field is present, then more light is transmitted (Fig. 3-22a). Bunched field lines, similar to those surrounding supergranules, allow more photon transmission from Ca-II ions. Observers therefore use Ca-II emissions to view the outline of the *supergranule network*, as well as regions of enhanced chromospheric magnetic field in plages and plage remnants, also known as *enhanced networks*.

Filaments are the relatively cool and dense thread-like features of chromospheric gas and plasma suspended in the hotter upper atmosphere by the generally horizontal magnetic field (Fig. 3-19). Their temperatures range between 7000 K and 15,000 K, while the surrounding material is much hotter—between 10^5 K and 10^6 K. The filaments absorb more photospheric continuum radiation in Hα than they emit, so they are darker than the background disk (Fig. 3-21a). However, when viewed near the Sun's limb, they appear bright against the dark space background. We often refer to limb filaments as *prominences*, because they are prominent Hα emitters glowing against a backdrop of cool space. These

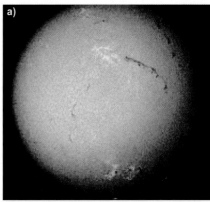

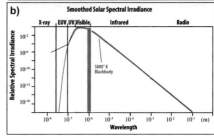

Fig. 3-21. a) The Hydrogen-Balmer-α (Hα) Spectral Line. Solar scientists use a narrow-band filter on a telescope to allow only the orange-red Hα-light to be transmitted, while all other light is rejected. *(Courtesy of Greg Piepol, Rockville, Maryland)* b) Relative Position of Hα Emission in the Solar Spectrum. The red vertical line on the inset plot indicates the Hα wavelength photons that are allowed to pass through the filter and into the telescope. *(Courtesy of the COMET Program at the University Corporation for Atmospheric Research. Adapted from NCAR/HAO)*

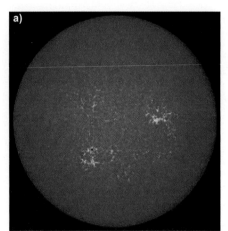

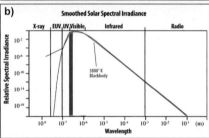

Fig. 3-22. **a) The Sun in the Ca-II K Spectral Line.** Here we show moderately magnetized regions of the Sun. The randomly distributed bright areas outline the supergranule network. The larger bright areas show the enhanced network (plages) associated with underlying sunspots The small sunspot in the lower part of the image has enough magnetic field strength to block CA-II emissions. *(Courtesy of Greg Piepol, Rockville, Maryland)* **b) Relative Position of Ca-II Emission in the Solar Spectrum.** The violet vertical line indicates the wavelength of photons associated with the image. *(Courtesy of the COMET Program at the University Corporation for Atmospheric Research. Adapted from NCAR/HAO)*

characteristics have prompted the sayings that prominences are "seen in emission," while filaments are "seen in absorption." Most filaments and prominences are stable and persist for weeks—hence the name "quiescent prominences." Sometimes the magnetic fields supporting the prominence or filament become unstable, allowing some of the plasma to fall to the photosphere and the remainder to accelerate into interplanetary space. Disintegrating filaments launched in Earth's direction are a source of space weather events at Earth.

> **Pause for Inquiry 3-11**
>
> Compare the origins of 656.3 nm and 393.4 nm radiation.

> **Pause for Inquiry 3-12**
>
> Contrast the uses of 656.3 nm and 393.4 nm spectral lines for solar monitoring.

Above the convective zone, if radiative heating and cooling were the only processes, the temperature of the solar atmosphere would decline as photons carried energy into space. However, in a thin layer (~100 km–200 km thick) just above the top of the chromosphere, the temperature rises sharply from ~10^4 K to over 10^6 K. This boundary zone is the *transition region* (Fig. 3-20).

We associate some of the temperature increase in the transition region with heating via the dissipation of compressive waves propagating upward from the photosphere. Mass density, and hence the electron density, decreases exponentially with increasing height (Sec. 3.4). Further, the lower portion of the transition region has relatively low thermal conductivity. The atmosphere's ability to cool is proportional to the square of the electron density and also increases with increasing temperature. With the rapid decrease in electron density, the solar atmosphere loses much of its cooling capacity. Because the heating rate from waves decreases with height more slowly than the cooling capacity, the temperature must increase with height.

Why does the wave heating rate decrease so slowly if at all? Turbulence in the convective zone generates mechanical energy in the form of low-frequency sound waves. Overshooting convective cells and jostling magnetic fields generate waves. These waves propagate upward where the rapid drop in the solar atmospheric density transforms the sound waves into shock waves. We recall that kinetic energy is proportional to ρv^2, so to conserve energy as particle density decreases, the particle velocities must increase. Certain types of wave motion are more favorable in magnetized media. The waves preferentially deposit their energy where densities are lowest. Magnetic regions in the upper chromosphere are therefore conspicuous in remotely sensed images.

Where the temperature is ~10^4 K, hydrogen atoms are readily excited to higher energy states and some are even ionized. De-excitation of these atoms produces photons that interact with and heat other atoms in the vicinity.

Some of these photons escape, producing solar emissions and acting as cooling agents for the region. Despite the cooling effect of escaping photons, the temperature continues to rise. As the temperature approaches 1.5×10^4 K, excited hydrogen atoms are ionized easily, and the number of neutral hydrogen atoms available to radiate away energy rapidly decreases. Thus, at temperatures above about 1.5×10^4 K, the radiative cooling rate (at constant pressure) begins to decline, and the unbalanced wave heating causes the temperature to increase until another cooling mechanism becomes effective in the solar corona.

Because ionization states of various minor constituents are also a strong function of altitude, imaging the Sun is also possible in the wavelengths associated with the emission lines from these ionization states (Figs. 3-20 and middle images in Fig. 3-23). What's more, because of the high level of ionization, the transition region is tightly constrained by the solar magnetic field. Magnetic flux tubes rising through the boundaries of the supergranule cells blossom into a magnetic carpet in the transition region. Scientists use these two facts to image the dynamics of the upper solar atmosphere and the state of the solar magnetic field (Fig. 3-23 and Focus Box 3.5).

Pause for Inquiry 3-13

In a few sentences explain why the solar atmosphere's ability to cool is related to electron density.

EXAMPLE 3.3

Wavelength of Solar Atmospheric Emissions

- *Problem Statement:* Determine the wavelength of emissions associated with the top image in Fig. 3-23.
- *Relevant Concept:* $E_{radiation} = hf$ and $E_{thermal} = k_B T$
- *Given:* Temperature is 3.5 MK
- *Assume:* Thermal equilibrium
- *Solution:*

$$E = hf = h\frac{c}{\lambda} = k_B T$$

$$\lambda = \frac{hc}{k_B T} = \frac{(6.63 \times 10^{-34}\,\text{J s})(2.998 \times 10^{8}\,\text{m/s})}{(1.38 \times 10^{-23}\,\text{J/K})(3.5 \times 10^{6}\,\text{K})} = \sim 4 \times 10^{-9}\,\text{m} = 4\,\text{nm}$$

- *Interpretation:* Emissions in the hot, outer corona correspond to X-ray wavelengths on the order of a few nanometers.

Follow-on Exercise 1: Make the same calculation for the other images in Fig. 3-23.
Follow-on Exercise 2: Plot the wavelengths on a curve similar to that in Fig. 3-18.

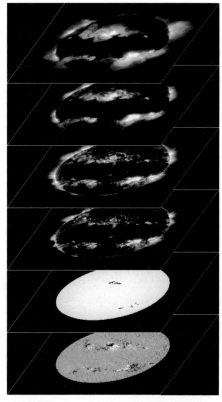

Fig. 3-23. **Images of the Sun at Different Altitudes in the Solar Atmosphere.** The bottom image in the stackplot shows a magnetic map (with magnetic polarities in different shades). Immediately above it is a white-light image from the SOHO satellite, taken on 2 August 1999. Above the chromosphere, the temperature increases rapidly to several million Kelvin. At these temperatures the few elements that still have electrons glow brightly at X-ray and extreme ultraviolet (EUV) wavelengths. We use different "colors" of the light to observe different temperatures. The four top images in the stack diagram show, from the top down, 3–4 million-K plasma observed by the Japanese soft X-ray telescope onboard the Yohkoh satellite, and the 2 million, 1.5 million, and 1 million-K plasma observed by the NASA small explorer satellite, Transition Region and Coronal Explorer (TRACE). These images are often displayed in false color. For simplicity we compare them all in gray tones. Different features are apparent in different wavelengths. *(Courtesy of Patricia Jibben at the Smithsonian Astrophysical Observatory) (TRACE is a mission of the Stanford-Lockheed Institute for Space Research, and the NASA Small Explorer Program)*

3.2.2 The Sun's Upper Atmosphere

Objectives: After reading this section, you should be able to...

- Describe the features and emissions of the quiescent corona
- Describe space environmental impacts of the quiescent corona
- Compare and contrast density, temperature, and magnetic properties of the corona with those of other regions
- Suggest heat sources for the corona
- Explain the sources of quiescent solar radio emission

The Corona

The Inner Corona in White Light, EUV, and X rays. The *corona* (Spanish for crown) is the hottest region of the outer solar atmosphere. Despite the high temperature, the coronal plasma is so rarified that the heat content is low (a visiting astronaut would feel cold). The corona produces virtually no light of its own in the visible part of the spectrum. However, electrons in the hot plasma scatter white photospheric light to produce an illumination, about half the brightness of the full Moon. Because Earth's blue sky is brighter than the corona, the corona historically was observed only during total solar eclipses. During solar eclipse, the Moon blocks the photospheric light that scatters in Earth's atmosphere. Blocking devices (occulting disks) of the same angular size as the lower solar atmosphere artificially create mini-eclipses for coronal imagers on balloon-borne, and ground- and space-based instruments. Spacecraft orbiting above Earth's atmosphere (and its scattered light) now provide near real-time views of the corona using an occulting coronagraph (Fig. 3-24).

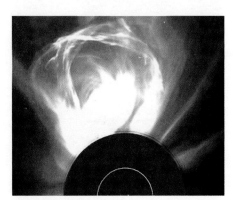

Fig. 3-24. View of the Solar Corona from the Solar and Heliospheric Observatory (SOHO) Spacecraft. In this image an occulting disk blocks the bright lower solar atmosphere and the innermost portions of the corona. The blocking disk covers two solar radii. The bright regions beyond the disk are the scattered white light of a coronal mass ejection (CME). The fainter regions are scattered white light of the quiescent corona. A few stars are visible in the background. *(Courtesy of ESA/NASA – the SOHO Program)*

Temperatures in the solar corona range from 1 MK–2 MK in magnetically quiet regions, to 2 MK–5 MK in magnetically active regions, and higher yet in flaring regions. Why is the corona so hot? By the Second Law of Thermodynamics, the corona cannot draw energy from below by heat transfer. The likely explanations are two-fold. The photosphere is an agitated mass of rising and falling columns of hot fluid. Convective overturning of the photosphere fills the atmosphere with sound waves. Some of this sound makes its way into the corona, where dissipative processes convert the audio energy into heat. A magnetic variant of the sound theory suggests that the presence of magnetic fields anchored in the photosphere enables the energy from the boiling motion to propagate upward as magnetized fluid waves. These magnetohydrodynamic waves are analogous to sound waves, but their properties depend on the magnetic field strength and direction. Some of these waves allow for particularly efficient transmission and deposition of energy.

An alternate explanation is that the heating arises from the interaction of the magnetic structures in the lower solar atmosphere with the convective motions. The near-surface layers of the Sun comprise countless small and large loops of magnetic flux, looking something like the field lines connecting the poles of a bar magnet. As the boiling motion twists and pushes the footpoints of the field lines around, strong electric currents are induced along the field lines. The surface layers become like a mass of twisted, braided, and tangled current-carrying wires. Eventually, by a process known as magnetic merging (similar to short-circuiting of electrical wires), the field lines rearrange themselves into a simpler pattern. In this process, large amounts of energy are released to heat the corona.

Because the corona is so thin and tenuous, only a tiny portion of all the energy in the photosphere needs to be absorbed to heat the corona to the observed temperatures. The relatively few particles occupying the corona receive copious amounts of energy and have limited means for cooling. Thus, they are energized to high temperatures and emit in EUV and X-ray wavelengths (Figs. 3-23 and 3-25). Short wavelength photons associated with coronal temperatures are responsible for completely ionizing coronal materials. Light ions lose all of their electrons; more massive ions such as iron may lose 10 or more of their outer electrons.

In the 1860s, observers recorded red and green line emissions from the corona during eclipses; emissions they could not associate with any known laboratory spectra. Later studies revealed that the observed spectral lines arise from what are known as forbidden emissions from very hot (10^6 K or more) atoms occurring under highly rarified conditions. That is, only in an ultra-low density, hot medium, such as the corona, are collisions between atoms so infrequent that atomic populations are maintained in the right kind of energy states to enable the observed emission.

The corona is not dense enough to qualify as a blackbody radiator. At best we treat the quiet corona as a quasi-blackbody with strong emission lines. The ionizing radiation from the corona has far-reaching effects, creating the ionospheres within Earth's upper atmosphere and in most other planetary atmospheres.

The temperature of the corona is so great that the Sun's gravity cannot prevent its expansion. Each second, a small portion of the Sun's coronal plasma escapes into space in an outflow called the solar wind. The solar wind extends tenuously for more than a hundred astronomical units. Earth and all of the planets are surrounded by this outflow. We live in the solar atmosphere and are subject to the disturbances in it.

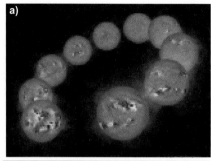

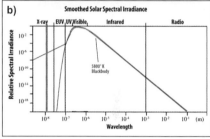

Fig. 3-25. a) **The Solar Corona during the Transition from Solar Cycle 23 to Solar Cycle 24.** These false color extreme ultraviolet (EUV) images show coronal emissions at 19.5 nm for solar maximum on the left, solar minimum in the middle and the subsequent solar maximum on the right. The outer corona is hotter during solar maximum and emits more radiation than during solar minimum. *(Courtesy of ESA/NASA – the Solar and Heliospheric Observatory (SOHO) Program)* b) **Relative Position of the 19.5 nm Emission in the Solar Spectrum.** The plot shows the emission band (vertical red line). The irradiance from the EUV emissions exceeds that expected from a 5800 K blackbody (shown by the blue curve). *(Courtesy of the COMET Program at the University Corporation for Atmospheric Research. Adapted from NCAR/HAO)*

Pause for Inquiry 3-14

Look for a solar imagery site on the internet. What are the wavelengths of the images you find?

Quiet Sun Radio Emissions above the Photosphere

All portions of the solar atmosphere emit radio waves, but the photospheric signal intensity is overwhelmed by radio emissions from plasma processes in the chromosphere and corona. The upper atmosphere emissions produce a background radio continuum. The amount of energy received at Earth in the narrow frequency bands observed by many solar radio telescopes is very small, so the value reported for such radiation is often in units of 10^{-22} W/(m^2·Hz), which we know as *solar flux units (SFUs)*. For historical reasons radio outputs from the Sun are often stated in terms of frequency rather than wavelength. Here we discuss the origin and observation of the background emissions.

Accelerating charged particles emit radiation (Sec. 2.3). In the plasmas above the photosphere, electrons and ions accelerate toward one another, but often cannot find the right balance of momentum and energy to recombine. The electrons, being less massive, experience more acceleration and emit the bulk of electromagnetic energy from these regions. At the temperatures and densities associated with the Sun's atmosphere, this energy, much of it in the form of bremsstrahlung, is emitted at radio wavelengths. And as the charged particles move around, they generate local concentrations of charge that cause electric and magnetic fields. These fields affect the motion of other distant charged particles.

Chapter 3 The Quiescent Sun and Its Interaction with Earth's Atmosphere

Entire regions of plasma oscillate with a frequency proportional to the free electron density. Thus, whenever a vast quantity of free and oppositely charged particles coexist in a relatively small space, the combination of their reactions causes intense, continuous, wideband, radio frequency radiation. These emissions are characterized as thermal because they depend on temperature and arise from the bulk behavior of the emitting particles.

The ambient solar radio emission from the chromosphere has two components: the background component and the slowly varying component. The *background component* is the contribution of a featureless solar disk and arises primarily from the chromosphere. During solar minimum, it is the dominant contributor to the radio Sun output. The intensity of the background component varies only slightly with frequency throughout the solar cycle. The *slowly varying component* also arises from the chromosphere but is more closely associated with active regions. During World War II (solar cycle 18), engineers developing radars for the Allied war effort noted that radars appeared to be jammed and feared that the enemy knew about the developmental efforts. In fact this "jamming" was the first direct observation of solar radio interference.

Since the 1940s, solar scientists have monitored the slowly varying component of the Sun at 2.8 GHz ($\lambda = 10.7$ cm). Daily observations of the 2.8 GHz irradiance integrated over the solar disk (Fig. 3-26) are made at the Penticton Observatory in British Columbia, Canada. This frequency penetrates Earth's atmosphere and is thus easy to observe from the ground, regardless of terrestrial weather conditions. As shown in Fig. 3-26, the sunspot number correlates strongly with the F10.7 cm solar flux. Sunspots are associated with magnetic disturbances in the upper reaches of the solar atmosphere that excite the plasma to emit at radio wavelengths. This plasma has a thermal behavior that also causes it to emit EUV radiation; hence the close correspondence between sunspot numbers, some solar radio emissions, and solar EUV emissions. In many applications, the F10.7 cm radio flux, rather than sunspot number, is used to describe the state of solar activity. For physics-based modeling, scientists use the EUV flux measured above Earth's atmosphere as a more appropriate index for solar activity and forcing.

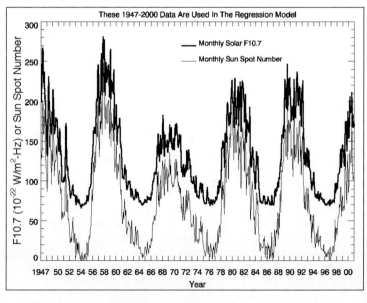

Fig. 3-26. The 50+-Year Record of the 10.7 cm Radio Output from the Sun (F10.7). Here we show the radio Sun flux at 10.7 cm, along with the sunspot record from the same interval. These two records are sometimes used interchangeably, but the agreement is not perfect. During some solar cycles, the Sun appears more active in the 10.7 cm portion of the spectrum than in its sunspot activity. *(Data courtesy of the National Geophysical Data Center)*

3.2 Regions of the Solar Atmosphere

As with shortwave emission, the Sun's appearance in radio emissions depends on the frequency (wavelength). The radio Sun at 3 GHz–30 GHz (λ = 10 cm–1 cm) is about the same size as the optical Sun, because this radiation comes from the chromosphere. At lower frequencies the size of the radio Sun appears to increase, because these lower frequencies are emitted at successively higher, less-dense layers in the Sun's corona. Figure 3-27a shows the Sun's extended atmosphere at 327 MHz. Figure 3-28 gives roughly the size of the radio Sun at various radio frequencies. Because the radio Sun is larger than the optical Sun, radio emissions from the Sun precede optical sunrise by several minutes. Conversely, solar radio emission continues for several minutes after the optical Sun has set. This detail is important for radar operators scanning the horizon during the minutes before dawn and after sunset.

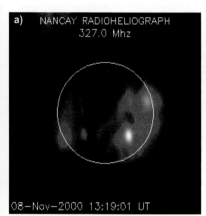
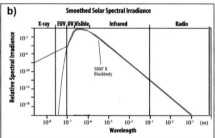

Fig. 3-27. a) The Sun Imaged in Radio Emission at 327 MHz (0.92 m). Here we show the radio Sun in an image obtained by the radio telescope at Nancay, France. The white circle in the image indicates the visible surface of the Sun (the photosphere). This image is less distinct than the 17 GHz image because of the coarser diffraction limit at the longer wavelengths of the lower frequency radio waves. *(Courtesy of the Nançay Radio Observatory)* **b) Relative Position of 0.92 m Emission in the Solar Spectrum.** The irradiance associated with one-meter emissions is much less than the irradiance in other portions of the solar spectrum. The red bar on the right that indicates the radio emission does not cross the emission curve for a solar-like blackbody. *(Courtesy of the COMET Program at the University Corporation for Atmospheric Research. Adapted from NCAR/HAO)*

The Sun emits across a wide spectrum; no single instrument or observatory can measure across the span of wavelengths. Figure 3-29 depicts several of the ground-based observatories that currently monitor the Sun. The observations from Big Bear Solar Observatory (Fig. 3-29a) provide a long record of the behavior of the lower solar atmosphere. This observatory is next to a high-altitude lake that helps maintain atmospheric stability and good viewing for the telescopes. The Dunn Solar Observatory (Fig. 3-29b) is also situated at a high altitude in a dry climate that helps prevent image blurring.

Pause for Inquiry 3-15

What formula relates radio frequency to wavelength?

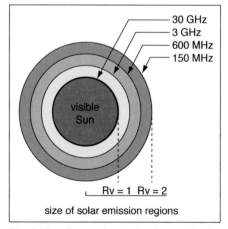

Fig. 3-28. Comparing the Optical and Radio Sun. The radio size of the Sun exceeds that of the visible Sun. One Rv is the radius of the visible Sun. *(Adapted from the AF Weather Agency)*

Chapter 3 The Quiescent Sun and Its Interaction with Earth's Atmosphere

Fig. 3-29. Ground-based Solar Telescopes. a) Big Bear Solar Observatory (BBSO). BBSO, noted for its long observational record of the lower solar atmosphere, is located on a lake peninsula at an altitude of 2070 m in the San Bernardino Mountains of California, USA. The mountain lake is characterized by sustained atmospheric stability, which is essential for clear viewing. *(Courtesy of BBSO)* **b) The Dunn Solar Observatory Vacuum Tower.** This telescope is nestled on a mountain top (2800 m) in New Mexico, USA. The structure contains an entrance window and two mirrors that guide sunlight down an evacuated, 100-m-long tube. The vacuum prevents light from interacting with and heating matter that would cause image blurring. *(Courtesy of the National Solar Observatory)* **c) Nobeyama Radioheliograph (NoRH) in Japan.** This radio imaging system measures two frequencies: 17 GHz and 34 GHz. The instrument is a radio interferometer consisting of 84 parabolic antennas, each 80 cm in diameter, sitting along lines 490 m long in the east-west direction and of 220 m long in the north-south direction. Scientists use the radio telescope to detect and identify sudden increases in radio energy emitted by the Sun. *(Courtesy of the Nobeyama Radio Observatory (NRO) and the National Astronomical Observatory of Japan (NAOJ))*

The challenge for radio telescopes is obtaining sufficient signal strength. Figure 3-29c shows an array of radio antennas used to gather the usually weak radio signals from the Sun to form a radio image. As we discuss in Focus Box 3.4, some radio antennas inadvertently gather solar radio output that collects in the receiver system. Weather radars are susceptible to radio noise, especially at sunrise and sunset. During strong solar radio activity, noise in cellular phone antennas also increases in Sunward facing dishes.

3.2 Regions of the Solar Atmosphere

EXAMPLE 3.4

Solar Emissions Compared to Blackbody Spectra

- *Problem Statement:* Plot the location of the solar emissions shown in Figs. 3-17 and 3-27 and the calculated emission from Example 3.3, at the appropriate wavelength on Fig. 3-30.

- *Given:* Chapter information and images, and the solar spectrum in Fig. 3-30.

- *Assume:* The diagram covers the full spectral range of the Sun.

- *Relevant Concept:* $c = \lambda f$

- *Solution:* Fig. 3-17 is a visible emission and should be located at the peak of the emission curve. Fig. 3-27 is an emission at 327 MHz, so the wavelength is at $(2.998 \times 10^8$ m/s$)/(327 \times 10^6$/s$) = \sim 0.92$ m ($\sim 10^9$ nm). The emission calculated in Example 3.3 is at ~4 nm.

- *Interpretation and Implications:* Emissions of interest for space weather are often in the non-blackbody regime of the Sun. The standard blackbody curves used by meteorologists and climatologists do not provide information about wavelengths of solar radiation important to space weather.

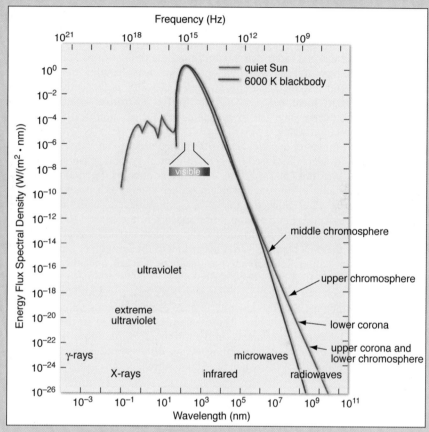

Fig. 3-30. **The Background Emission of the Quiet Sun.** Here we show the spectral range of the quiet Sun and compare it to a 6000 K blackbody radiation curve.

Follow-on Exercise 1: Plot and label the locations of all other emissions shown in the images of this section (3.2).

Follow-on Exercise 2: Labels on the diagram above show the wavelengths associated with chromospheric and coronal emissions. Label the wavelengths on the short side of visible that are associated with chromospheric and coronal emissions. Use the images in this chapter to guide your selections.

Chapter 3 The Quiescent Sun and Its Interaction with Earth's Atmosphere

Focus Box 3.4: Unintentional Observations of the Radio Sun

Not every radio observation of the quiescent Sun is intentional. Figure 3-31 illustrates a modern phenomenon familiar to all radar operators: detecting the Sun within the radar beam (*Sun echo*). Solar radio emissions interfere with the operations of civil and military radars. Operators need to be aware of the Sun's location in the radar's field of view and the fact that the radio Sun is larger than the optical Sun. The aviation community relies heavily on radio wavelengths for communication and also needs to be aware of the Sun in its daily progression through local time zones. Air traffic controllers often suggest radio frequency changes to maintain contact as aircraft fly through near-noon time zones.

The satellite community experiences a different quiescent Sun impact called *Sun-in-view*, which is caused by an alignment of ground sensors, geosynchronous satellites, and the Sun during near-equinox intervals (Fig. 3-32). This radio interference problem occurs when Earth-bound sensors aim near the Sun to track or communicate with a geosynchronous satellite. Excess radio noise appears in the sensor and degrades the ability of the sensor to receive the weaker satellite signal. The time of year that this happens varies according to the latitude of the observing ground station. Interference occurs about a week on either side of the maximum effect date. The interference lasts a few minutes to over an hour, depending on the receiving antenna's beam width. The time of day at which the Sun-in-view interference occurs depends upon the relative position of the satellite. Those satellites positioned in the western sky are subject to interference in the afternoon, while those in the eastern sky notice effects in the morning. As we discuss in Chap. 9, a more problematic situation arises when the Sun is active.

Fig. 3-31. Doppler Radars Fielded by the US National Weather Service. These radars operate at a wavelength of ~0.01 m (1 cm) and mostly detect return echoes from atmospheric phenomena. Near sunset on December 26 in the Midwest US, radar beams from seven Doppler radars intercepted radio emissions from the Sun as it dipped below the horizon. For each radar, these emissions appeared as narrow cone-like structures that we show as southwesterly vectors, numbered 1–7. In December, the Sun sets in the southwestern sky. *(Courtesy of the National Oceanic and Atmospheric Administration)*

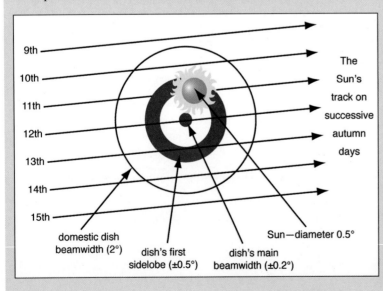

Fig. 3-32. Sensor Tracking Geosynchronous Satellite (Orbital Plane ~0° Latitude). The diagram shows the Sun in the field of view of a radar. Even if the Sun does not track through the primary beam, interference may occur from the radar beam side lobes. The dates on the left show the days in September that may have Sun-in-view radio interference. *(Adapted from Peter Vince)*

3.3 Characteristics of the Quiescent Solar Magnetic Field

3.3.1 Multi-scale Solar Magnetic Fields

Objectives: After reading this section, you should be able to...

- Describe and relate the features of the near-Sun magnetic field
- Explain when and why the Sun's magnetic field deviates from dipolar
- Compare the magnetic field strength at the solar surface with that in the solar wind near Earth
- Differentiate between open- and closed-field regions

In 1908, astronomer George Hale reported that the Sun exhibited an overall magnetic field that resembled a dipole. We now know the global-scale field is tilted, stretched, and in some cases, quite nondipolar. New imaging techniques reveal many smaller features that change as the Sun cycles its magnetic field. Magnetic fields on the Sun organize on several scales: global, synoptic, intermediate, small, and micro. Global-scale refers to the general tilted-dipole structure; synoptic-scale refers to polar open regions and equatorial closed regions. Intermediate-scale deals with active regions, prominences, networks, and large bipolar groups that we describe in Chap. 9. Small-scale is associated with sunspots and ephemeral regions. Micro-scale features form in inter-granular lanes.

The general global-scale magnetic field at the solar surface has an intensity of about 10^{-4} T (~1 gauss), which is 2–3 times stronger than Earth's magnetic field (3–6×10^{-5} T). Solar images, filtered in the "G-band" of the solar spectrum near 430 nm, reveal 100 km-scale magnetic features organized in the turbulent flows of the photosphere (the smallest resolved features in Fig. 3-11). The features appear as bright points of light, contrasted with larger, dark 1000 km-scale pore regions. These small-scale fields often appear during the decay of larger active regions. Scientists are uncertain how much of the small-scale flux is the result of the decay process and how much may be generated locally in a turbulent dynamo mechanism located in the upper convective zone of the Sun. Bright points, pores, and concentrations of magnetic field in the granular lanes (*intranetwork field*) are typical features of the photosphere. The mixed polarity of the intranetwork fields appears to originate continuously from localized source sites in granular cells. The intranetwork field concentrates into ribbon-like features at the supergranular boundaries.

Magnetic flux concentrated at the edges of the granulation cells further organizes into a larger *network* of magnetism in the chromosphere with field strengths on the order of 0.1 T. In the chromosphere, magnetic fields create much more heating than in the photosphere, resulting in the significantly brighter magnetic field pattern with dimensions on the order of 10 Mm. Images taken in the UV 396.8 nm (Ca-II H-band) spectral line show the magnetic structure of the chromospheric network (Fig. 3-22a).

Network fields are sometimes distinguished as quiet and active networks. The network field is the dynamic product of the merging and cancelling of granular or intranetwork fields, ephemeral regions, and the remnants of active regions. Network features have a characteristic lifetime of a few hours.

A significant portion of the total magnetic flux on or near the solar surface resides in *ephemeral active regions*. These small, shortlived, bipolar regions are 10 Mm–30 Mm across and have average lifetimes of 12 hours. Ephemeral regions are small, new bipoles that grow as a unit or a succession of bipolar units and whose poles move in opposite directions from their apparent site of origin. The decay of ephemeral regions allows remnant magnetic flux to migrate to the edges of supergranule regions. Although not usually associated with sunspots, ephemeral regions affect chromospheric structures and develop visible plages. Several hundred such regions may exist at any given time (white regions of Fig. 3-33). Similar to sunspots, their numbers vary with the 11-year cycle, but their magnetic orientation is more random than the sunspot polarity rules we describe in Sec. 3.3.2.

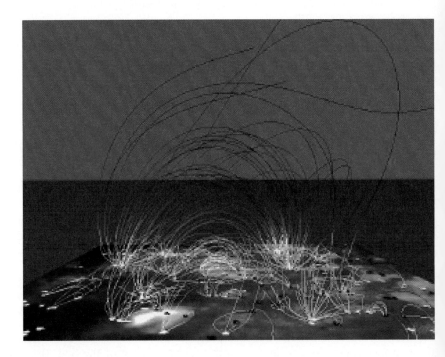

Fig. 3-33. Ephemeral Regions and Solar Magnetic Carpet. In this picture, white magnetic field lines emanate from the magnetic carpet and form arches from one magnetic polarity (white) to the other (black). The image underlying the arches is the heating observed by the Solar and Heliospheric Observatory (SOHO) Extreme Ultraviolet Imaging Telescope (EIT) in the iron line at 19.5 nm, with bright green corresponding to relatively hot regions and dark green corresponding to cool ones. *(Courtesy of Neal Hurlburt and Karel Schrijver at the Lockheed Martin Advanced Technology Center's Solar and Astrophysics Laboratory)*

A complex distribution and mixing of magnetic polarities from bipolar ephemeral regions form a *magnetic carpet* over the entire Sun. Magnetic loops, connecting regions of opposing magnetic polarity, rise far into the solar corona to form extensive loop structures (*arcades*) (Fig. 3-33). Small-scale current sheets are likely to form in the turbulent environment, and magnetic merging may promote coronal heating. Diffuse coronal heating appears to be associated with the ubiquitous magnetic carpet.

As Fig. 3-34 shows, on a global scale the solar equatorial regions contain fields with both ends attached to the Sun—*closed magnetic flux* regions. These regions trap hot plasma and glow in EUV and X-ray emissions. A large fraction of the closed regions stretch into space, forming structures called *helmet streamers* (they look like pointed helmets that soldiers wore in earlier centuries). Some streamers extend all the way to Earth. In contrast, the polar fields appear to open into interplanetary space and produce *open magnetic flux*. The open field lines connect to the other pole in distant regions beyond the solar system.

3.3 Characteristics of the Quiescent Solar Magnetic Field

Because these open field lines are conduits for the Sun's outward plasma flow, they are inefficient traps for plasma. In coronal X-ray images such regions appear as dark, relatively cool *coronal holes*. In Chap. 4 we develop some of the physics required to more fully describe the field variations. We give a more complete description of field variations in Chaps. 9 and 10.

During solar minimum, when the dipole pattern is most evident, the Sun's magnetic field aligns with the solar rotation axis (Fig. 3-34a). In solar maximum the field alignment with the rotation axis is more difficult to observe and may be absent. With increasing solar activity the dipolar field is often disrupted with the appearance of many localized *bipolar regions*, some of which appear as sunspots (Fig. 3-34b). And at all times, the solar field stretches outward into the heated and flowing solar atmosphere. Near the equator, at a distance of 2 R_S–3 R_S, the outward tug of the solar wind becomes stronger than the tension in the field lines, and the field lines are pulled outward, stretching into a thin sheet of oppositely directed field that ripples and warps as it projects into interplanetary space. Despite the extreme stretch, the field lines connect back to the Sun. Because of the Sun's rotation these waving field lines are transported outward in a spiral configuration, as we describe in Chap. 5. At Earth, the Sun's magnetic field decreases in strength to 5 nT–10 nT. The net effect of the battle between the expanding solar wind and the constraining magnetic field is the creation of two distinct regimes where open field regions are juxtaposed with regions of closed-loop field lines. Flow variations near the open-closed region boundaries create minor-to-moderate space weather disturbances at Earth.

3.3.2 Typical Solar Cycle Field Variations

> **Objectives: After reading this section, you should be able to...**
>
> ♦ Describe the features of 22-year and 11-year solar cycles
> ♦ Predict the polar field orientation for the next solar cycle
> ♦ Distinguish between Hale's Sunspot Polarity Law and Joy's Sunspot Tilt Law

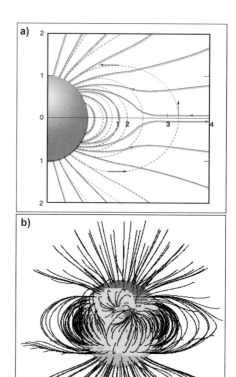

Fig. 3-34. a) A Comparison of a Dipole Field (dashed lines) with a More Realistic Field (solid curves). The rotation and dipole axes are along the left edge, and the solar equator is horizontal. The dashed curves indicate a dipole configuration. The solid curves give a view of the stretched interplanetary magnetic field. Equatorial regions near the Sun have approximately dipolar fields. The solar field reaching Earth has its footpoints rooted at middle latitudes. [After Pneumann and Kopp, 1971] b) Model of the Sun's Magnetic Field. Normally the polar field lines are nearly radial and splay outward into space. Though the field lines must close, the closure path is not obvious; hence the term "open flux." Red indicates an inward-directed field. Green locates the anchor of an outward-directed field. Near the equator the field lines emanating from the yellow areas are clearly closed. The small red-green region within the closed field lines hosts a bipolar region. During solar maximum, the yellow region is filled with such bipolar regions. The dynamics of the solar field allow a field reversal approximately every 11 years. *(Courtesy of Yan Li at the University of California at Berkeley)*

Here we describe the multi-year variation in solar features. The Sun is a variable magnetic star. The 11-year period generally garners the most attention because the cycling of the magnetic field creates disturbances in photons, fields, and particles that propagate to Earth. A more fundamental period of 22 years is the time required to completely cycle the global dipole orientation and return it to its original configuration.

At the start of an 11-year cycle, new sunspots form at 25°–30° latitude (north and south). Subsequently, active regions and sunspots form at progressively lower latitudes. George Hale, who first showed that sunspots were magnetic phenomena, observed a tendency of leading (in the sense of the Sun's rotation) polarity in active region fields to differ between the northern and southern hemispheres. *Hale's Law* has the following components:

- When sunspots come in pairs, one of them tends to have a magnetic field polarity that is opposite to the other—they are bipolar
- During a given 11-year cycle, the leading sunspots in each hemisphere tend to have opposite polarities

- During the subsequent sunspot cycle, the polarity of the leading spots in each hemisphere is opposite from what it was in the previous cycle

Additionally, Hale's student, Alfred H. Joy, found that sunspot pairs are likely to be tilted in latitude, with the leading spot closer to the equator than the following spot. This propensity is *Joy's Sunspot Tilt Law*. Recent studies reveal the non-leading (following) sunspots tend to diffuse and migrate poleward. This behavior contributes to the global field reversal shown in Fig. 3-35. When viewed on a large scale, the cycling of sunspot fields looks like a set of flapping butterfly wings, so the image in Fig. 3-35 is often called a magnetic butterfly diagram.

Figure 3-35 depicts the global- and small-scale field organization in time and space. For example, at the beginning of solar cycle 22 in the late 1980s, new bipolar sunspot groups formed at solar latitudes near 30°. The leading (equatorward) groups had opposite polarities in each hemisphere. In the northern hemisphere, the leading orientation was inward. As the cycle progressed, new sunspots formed at lower latitudes, while the remnant fields of the following spots migrated poleward. In the polar regions magnetic flux of one sign or the other accumulates. These regions are often called *unipolar*. The dark curves in the figure trace the path of surface plasma parcels that flowed poleward as sunspots formed at progressively lower latitudes.

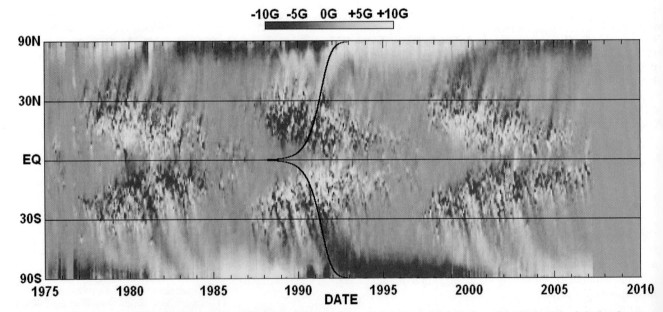

Fig. 3-35. Magnetic Butterfly Diagram of the Solar Surface. The vertical axis is heliographic latitude and the horizontal axis is time in years. The dark curves represent the expected trajectories of magnetic elements carried poleward by a meridional flow that has a peak velocity of 10 m/s. Yellow indicates the outward-directed field, and blue indicates the inward-directed field. Field strengths range from -10^{-3} T (blue) to $+10^{-3}$ T (yellow). An 11-year magnetic cycle begins when bipolar sunspots first appear at ~30° latitude. As time proceeds, the sunspots form closer to the equator. The magnetic field of the trailing sunspots drifts and diffuses poleward with paths indicated by the dark curves. At the time of maximum sunspot coverage, the polar field reversal occurs. Note the color change at 90° N and 90° S. (G is gauss, which = 10^{-4} T.) *(Courtesy of David Hathaway at NASA)*

3.3 Characteristics of the Quiescent Solar Magnetic Field

Although a polar field reversal occurred in 1991, the polarity of spots migrating equatorward was that of the polarity of the dipole field that existed at the beginning of the 11-year cycle. Toward the end of solar cycle 22, most sunspots were forming near (but not at) the equator. During the subsequent solar minimum, few sunspots formed, and magnetic flux concentrated at the poles. Solar cycle 23 began with the polar field and mid-latitude bipolar spot polarity opposite those of solar cycle 22.

Pause for Inquiry 3-16

What year did the polar field reversal occur in solar cycle 23? Did both poles reverse at the same time?

Pause for Inquiry 3-17

Will the southern hemisphere field be inward- or outward-directed for solar cycle 24?

The effects of field changes are not limited to the surface. In each cycle, the polar fields must eventually give way to the field migrating from the equator. Some of the polar field submerges and some releases into the solar wind at high altitudes via coronal holes. The multi-year interval between solar maximum and solar minimum is the declining phase of the solar cycle. In this time-frame, large, polar coronal holes expand equatorward, freeing unipolar magnetic fields into large sectors of the solar wind.

Chapter 3 The Quiescent Sun and Its Interaction with Earth's Atmosphere

Focus Box 3.5: Recent Solar Observatories

A much clearer understanding of the Sun and its interactions with Earth comes from the wealth of observations provided by new space- and ground-based solar observatories. We highlight a few of these here.

GONG: The Global Oscillation Network Group (GONG) is a world-wide ground-based program to study the solar internal structure and dynamics using helioseismology. Nearly 70 institutions from 20 countries support the GONG effort that began in 1995. GONG has developed a six-station network of extremely sensitive and stable plasma velocity imagers located around Earth to obtain nearly continuous observations of the Sun's "five-minute" oscillations. The combined instruments are capable of making imaged velocity measurements with a precision of significantly less than one m/s in plasma moving as rapidly as 10^7 m/s. Data from the GONG program are often used in concert with data acquired by spacecraft.

Fig. 3-36. **Coronal Loops above the Sun's Surface from TRACE.** Glowing brightly in extreme ultraviolet light (17.1 nm), the hot plasma attached to the arching magnetic field is cooling and raining back down on the solar surface. *(Courtesy of the Stanford-Lockheed Institute for Space Research and NASA)*

SOHO: Since its launch in December 1995, the Solar and Heliospheric Observatory (SOHO) has provided unprecedented views of the Sun (Figs. 3-14, 3-23, and 3-24). The spacecraft is in a halo orbit around the Sun-Earth L1 libration point, where the small difference between the Sun's and Earth's gravity provides exactly the centripetal force needed to match the spacecraft's orbit about the Sun with that of Earth. As a result, the spacecraft stays near the Sun-Earth line during its one-year orbit. It is about 1.5×10^9 m from Earth and 148.5×10^9 m from the Sun. The satellite orbits L1 once every six months, avoiding the exact L1 location so that radio communication with Earth is not contaminated by solar radio noise. SOHO supports a dozen experiments. Some of them provide quasi-operational data that are so important the spacecraft mission has been extended multiple times. One of the more significant long-term gains is a newly developed technique to recreate the conditions on the far side of the Sun (Fig. 3-4). SOHO is a joint effort between the European Space Agency (ESA) and NASA.

TRACE: NASA's Transition Region and Coronal Explorer (TRACE) satellite launched on April 2, 1998 into a low-Earth orbit on a Pegasus rocket. TRACE is exploring the three-dimensional magnetic structures in the solar atmosphere and helping to define the geometry and dynamics of the upper solar atmosphere (transition region and corona). The mission provides insight into the relation between the diffusion of the surface magnetic fields and the changes in heating and structure throughout the transition region and corona. Simultaneous images through the depth of the solar atmosphere help determine the rate of change of the magnetic topology and the nature of the local restructuring and merging processes (Figs. 3-23 and 3-36).

GOES SXI: The geosynchronous orbit of Geostationary Operational Environmental Satellite (GOES) allows it to have direct line-of-sight observations of the Sun, 24 hours per day, 7 days per week. The only exception to this is near the equinoxes, when GOES enters Earth's shadow for up to one hour each day. On board the most recent GOES satellites is a Solar X-Ray Imager (SXI) designed to collect a solar image once per minute. It uses exposure settings that follow a sequence optimized to observe three primary phenomenon of the solar atmosphere: coronal structures, active regions, and solar flares. The SXI telescope images the Sun in wavelengths ranging from soft X-rays to extreme ultraviolet. In January 2010 GOES-14 SXI began providing imagery for solar cycle 24.

HINODE: Hinode is a collaborative effort between Japan, the US, the United Kingdom, and Europe, launched in September 2006, to study the Sun and its magnetic field. The Hinode (Sunrise in Japanese) spacecraft circles Earth in a near-polar Sun-synchronous orbit. The mission includes a suite of three science instruments that cover optical, extreme ultraviolet, and X-ray wavelengths. Together, these instruments study, at high resolution, the generation, transport, and dissipation of magnetic energy from the photosphere to the corona. The mission provides data on how energy stored in the Sun's magnetic field is released, either gradually or violently, as the field rises into the Sun's outer atmosphere. The Institute of Space and Astronautical Science, Japan Aerospace Exploration Agency (ISAS/JAXA) operates Hinode.

Focus Box 3.5: Recent Solar Observatories *(Continued)*

STEREO: The Solar TErrestrial RElations Observatory (STEREO), launched in October 2006, employs two nearly identical space-based observatories to provide three-dimensional stereoscopic images to study the nature of coronal mass ejections (CMEs). One of the STEREO spacecraft is ahead of Earth in its orbit and the other trails behind. Similar to the slight offset between human eyes that provides depth perception, this spacecraft placement allows the STEREO observatories to obtain three dimensional images of the Sun. Figure 3-37a shows orbits of STEREO-A (Ahead) in red and STEREO-B (Behind) in blue relative to Earth's orbit in green. STEREO-A's elliptical orbit fits inside Earth's orbit and transits around the Sun faster than Earth; STEREO-B's orbit is larger than Earth's orbit and transits around the Sun more slowly. Figure 3-37b shows the spacecraft drift (at a rate of 22° per year) away from Earth relative to an Earth-Sun reference line (yellow). The "scalloped" curves are caused by the spacecrafts' elliptical orbits.

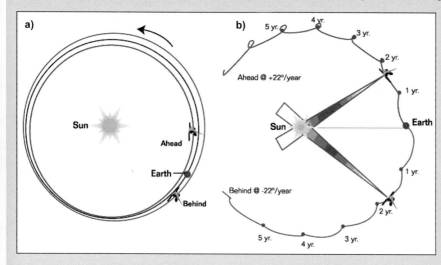

Fig. 3-37. STEREO's Orbits. This diagram illustrates how STEREO's spacecraft separate from each other to provide three-dimensional images of the Sun and its atmosphere. *(Courtesy of NASA)*

3.4 Basic Physics of Atmospheres

Now that we know about the structure of the Sun's atmosphere, we investigate some of the physics common to stellar and terrestrial atmospheres.

3.4.1 Law of Atmospheres

Objectives: After reading this section, you should be able to...

- Apply hydrostatic balance to explain the structure of a gaseous atmosphere
- Apply the barometric equation in the Law of Atmospheres
- Explain the concept of scale height

Here we investigate atmospheric density variation with height. In a typical atmosphere, density decreases with height at an exponential rate. This profile has important impacts on the atmosphere's ability to cool. In terrestrial applications it is a controlling element of satellite drag.

Next we explain why an atmosphere is dense at the bottom and tenuous at the top and why atmospheric atoms and molecules don't accumulate in a thick pool. The answers lie in a combination of factors described by the Ideal Gas Law and the hydrostatic equation, Eq. (3-3). Atmospheric constituents with sufficient energy to be a gas are free to respond to thermal agitation. Some of these molecules or atoms get bumped up to high altitudes by collisions with other energized molecules, but they remain constrained by gravity.

To quantify this process, we study the distribution of molecules in an atmosphere. We simplify our study by assuming that the atmosphere is at a constant temperature, T. We also draw on our knowledge of the Ideal Gas Law ($P = nk_BT$ form), which expresses the basic relationship between pressure P, number density of particles n, and temperature T for a gas in thermal equilibrium.

If we know the gas number density n and temperature T, we can find the pressure and vice versa. Clearly, the pressure must decrease as the altitude increases because a given layer of atmosphere has to support the weight of the atmosphere above it—the greater the altitude, the less the weight of the atmosphere above, and the lower the pressure. To determine how the pressure changes with height, we analyze an atmospheric layer of thickness dz and cross-sectional area A as in Fig. 3-38, where m is the mass of an individual particle in the gas. Summing the upward and downward forces gives the result in Eq. (3-3) that we use to develop the concept of hydrostatic equilibrium. As shown in Fig. 3-38, the pressure gradient force is balanced by weight. The weight of the gas in this infinitesimal layer of air is dW = mgdN = mgndV, or dW = $mgnA$dz. By summing the forces on the layer we find that $PA - (P + dP)A = mgnA$dz, which reduces to

$$dP = -mgn\,dz \qquad (3\text{-}4)$$

where

dP = pressure change [Pa]
m = mass of a single molecule or atom [kg]
g = local gravitational acceleration [m/s^2]
dz = change in height [m]

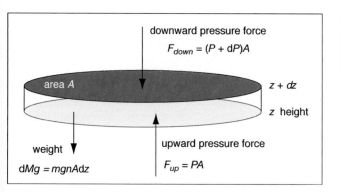

Fig. 3-38. An Atmospheric Layer in Hydrostatic and Thermal Equilibrium. Because the atmosphere is in hydrostatic equilibrium, the upward force on the bottom of this layer, $F_{up} = PA$, must exceed the downward force on the top of the layer, $F_{down} = (P + dP)A$, by an amount equal to the weight dW of the layer of gas.

Because $P = nk_BT$, and we assume T is constant, $dP = k_B T \, dn$. Combining this with Eq. (3-4) yields

$$\frac{dn}{n} = -\frac{mg \, dz}{k_B T} \qquad (3\text{-}5)$$

where

n = number density of particles [#/m³]

dn/n = relative change in number density compared to background [unitless]

k_B = Boltzmann's constant (1.38×10^{-23} J/K)

T = temperature [K]

We find the solution to Eq. (3-5) by integrating or by realizing that the function whose derivative is proportional to itself is an exponential function. The solution, known as the *Law of Atmospheres*, is

$$n(z) = n_0 \exp\left(-\frac{mg}{k_B T} \Delta z\right) \qquad (3\text{-}6a)$$

where

n_0 = constant density at a reference height $z = z_0$; that is, $n(z_0) = n_0$ [#/m³]

$\Delta z = z - z_0$ [km]

We write the density (n) as $n(z)$ to emphasize that it changes with height. Although the molecules of the atmosphere tend to fall under the force of gravity, collisions with other molecules cause some of them to move upward with enough energy to climb to great heights. This phenomenon is why atmospheres don't collapse into pools. The thermal agitation is constantly at work, even while gravity is pulling them down. The result is the equilibrium exponential distribution of Eq. (3-6a).

We restate this equation as

$$n(z) = n_0 \exp\left(-\frac{mg}{k_B T} \Delta z\right) = n_0 \exp\left(-\frac{\Delta z}{k_B T / mg}\right) = n_0 \exp\left(-\frac{\Delta z}{H}\right) \qquad (3\text{-}6b)$$

where we represent the terms, k_BT/mg (all assumed constant), as a new value H, the *scale height*. The scale height is a fundamental way of comparing the thermal energy (k_BT) to the gravitational potential energy ($mg\Delta z$) in a fluid. In convective fluids the scale height plays an important role in controlling the scale size of convection. Larger values of scale height are associated with thicker atmospheres and produce larger convective cells. As an interesting aside, the value of H is the depth an atmosphere would have if it did collapse into a pool of uniform density gas. In atmospheres that are stratified by mass (such as Earth's upper atmosphere) the scale height varies with altitude.

Equation (3-6) tells us that the density of an atmosphere in thermal equilibrium decreases exponentially with increasing height. The number density (n_0) at the base of the solar atmosphere is a few times 10^{23} particles/m³. At higher altitudes, the density is lower and so is the pressure.

Applying the Ideal Gas Law to Eq. (3-6a) we write the expression $P = n_0[\exp(-mg\Delta z/k_BT)]k_BT$. Therefore, the atmospheric pressure varies with height as

$$P(z) = P_0 e^{(-mg\Delta z/k_BT)} = P_0 e^{(-(\Delta z)/H)} \qquad (3\text{-}7)$$

where $P_0 = n_0 k_B T$. This equation is the *barometric equation*. Figures 3-39a and b show solar atmospheric pressure as a function of height. The nearly straight line on the semi-log plot of Fig. 3-39b is a clear indication of the exponential relationship between pressure and height and thus the applicability of our simple model.

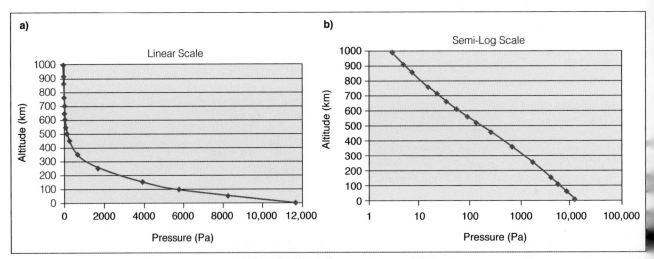

Fig. 3-39. Pressure Profile in the Lower Solar Atmosphere. **a) Linear Scale.** The horizontal scale begins with full atmospheric pressure (1 atmosphere) on the right. Pressure decreases with increasing altitude. **b) Semi-log Scale.** The horizontal scale is logarithmic. The plotting convention used here for atmospheric pressure is different from the norm, in that the independent variable (pressure) is on the horizontal axis, while the dependent variable (altitude) is on the vertical axis. We do this change to align with the human experience of "up" being vertical.

Pause for Inquiry 3-18

Do large scale heights imply small or large density changes with height?

EXAMPLE 3.5

Atmospheric Density and Optical Depth

- *Problem Statement:* What is the density of the solar atmosphere at an altitude of 500 km above the photosphere?
- *Concepts:* Apply the Law of Atmospheres
- *Given:* $\Delta z = 5 \times 10^5$ m
- *Assume:* The average molecular mass of the photosphere is 1.2 u where the unified mass unit, $u = 1.66 \times 10^{-27}$ kg. We use this value because the atmosphere is about 92% hydrogen atoms, with mass of $1u$ and 7.8% helium atoms, with mass of $4u$. We assume a temperature of 5800 K.
- *Solution:* Assuming a constant temperature atmosphere, we know that the density of the atmosphere decreases exponentially with height according to the Law of Atmospheres, Eq. (3-6b).

Substituting appropriate values gives

$$n(5 \times 10^5 \text{m}) =$$

$$n_0 \exp\left[-\frac{(1.2 \text{ u})(1.66 \times 10^{-27} \text{kg/u})(274 \text{ m/s}^2)(5 \times 10^5)}{(1.38 \times 10^{-23} \text{J/K})(5800 \text{K})}\right]$$

$$= n_0 \exp[-3.41]$$

$$= 0.033 \, n_0$$

- *Interpretation and Implications:* The atmospheric density at 500 km altitude in the photosphere is only 3.3% of the density at the base of the photosphere (for an atmosphere at $T = 5800$ K).

Follow-on Exercise: What is the value of the scale height (H) in the example above?

Follow-on Question: Suppose you wished to use the Law of Atmospheres in the corona. What value would you use for molecular mass? What value would you use for z and T? How would you modify g?

3.4.2 Absorption Interactions in an Atmosphere

Objectives: After reading this section, you should be able to...

- Explain the concept of optical depth and its relation to absorption in atmospheres
- Relate the production of terrestrial ozone, atomic oxygen, and ionization to solar radiation
- Understand the role of an exponential atmosphere in creating atmospheric layers and peaks of ionization
- Explain why the stratosphere and thermosphere are hot compared to surrounding layers
- Describe the impacts of long-term satellite exposure to atomic oxygen

In this section we combine our knowledge of atmospheric density with the basic physics of radiative absorption. When a beam of electromagnetic radiation emitted from elsewhere interacts with matter, the impinging photons are reflected, transmitted, or absorbed. We focus on absorption because absorbed photons add energy to the gas.

Absorption and Optical Depth

Absorption occurs when matter retains impinging radiant energy. It's an important step in converting electromagnetic radiation to heat. To learn more about absorption we model a column of atmosphere as an infinitely long tube filled with a gas, as shown in Fig. 3-40. A beam of photons enters the tube in the direction s. While in the tube, the photons leave the radiation beam by only two processes:

Chapter 3 The Quiescent Sun and Its Interaction with Earth's Atmosphere

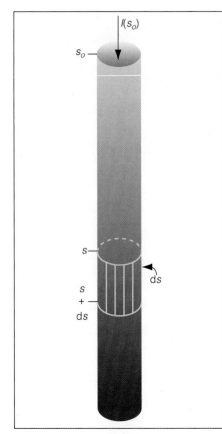

Fig. 3-40. Light Traveling through a Cylinder. Here we show a tube with a light beam flowing along it. The coordinate "s" is the "natural" coordinate that increases in the direction the radiation travels. We need to subtract the losses from the beam in the length, $s + ds$, to determine how much energy transfers from the beam to the gas in the tube.

absorption and scattering. Sources of radiation along the path also appear from emissions and scattering into the beam. However, we ignore such complications for now and see how well we predict changes in radiation based on absorption alone. The number of photons emerging at location $s + ds$ equals the source from the beam minus the losses (absorption) of photons to that point in the cylinder. The radiance (I) decreases as position (s) increases: $I(s_0) > I(s) > I(s + ds)$.

The change in radiance (dI) is proportional to the length (ds) and the initial radiance (I_s).

$$dI \propto I_s \, ds$$

Accounting for the particle density (n) and the *absorption cross section* (σ), which is the effective target size each particle presents to the beam, allows us to relate a decrement in radiance as

$$\frac{dI}{I_s} = -\sigma n ds \quad (3\text{-}8a)$$

where

dI/I_s = ratio of radiance change to the original intensity value [unitless]
σ = absorption cross section of the gas [m^2]
n = number density of gas particles [#/m^3]
ds = path length [m]

and the minus sign indicates that the radiance decreases with distance.

We recall that any quantity whose derivative depends on the original value of the quantity must change exponentially. To get the profile of radiation intensity as a function of distance we need to integrate both sides of Eq. (3-8a). Doing so gives the relative reduction in intensity from absorption. We write the integrated result in terms of a parameter called optical depth.

$$\int \frac{dI}{I_s} = \ln(I)\Big|_0^s = (-\sigma ns + C)\Big|_0^s$$

$$= (-\sigma ns + C) - (-\sigma n 0 + C) = -\sigma ns$$

Applying the anti-log to each side (exponentiating) yields:

$$\exp[\ln(I)]\Big|_0^s = \frac{I_s}{I_0} = \exp[-\sigma ns] \quad (3\text{-}8b)$$

If the path length is known or can be determined, then the exponential term converts into a parameter called *optical depth* (optical thickness), τ.

where

$$\tau = \int \sigma n ds = \text{optical depth of the gas [dimensionless]}$$

The result is a relation called *Beer's Law* that shows how the value of radiance coming through the tube is reduced with respect to the original value:

$$I_s = I_0 e^{-\sigma ns} = I_0 e^{-\tau} \quad (3\text{-}8c)$$

If σ and n are constant, their product is sometimes written as an absorption coefficient (k) that has units of 1/m. Optical depth tells us how effective the medium is at absorbing the radiation. It is a dimensionless measure of the virtual length of the tube. Mathematically, optical depth is a means for combining many factors that affect how far along the tube the radiation must travel until the photon count depletes to some value. For a given absorption coefficient, the optical depth increases as the tube length grows. The number of photons that reach the end of the tube decreases as the tube gets longer, hence the negative sign in the exponential term. To draw an analogy for Eq. (3-8c), we can think of irradiance as telling us about the number of photons on a radiation highway and optical depth telling us about the conditions on the radiation highway independent of the number of photons traveling on the highway.

The various wavelengths of radiation are likely to escape the solar atmosphere at and above a height where $\tau = 1$, specific to the wavelength. With respect to the Sun, layers with large optical depth are likely to trap and retain photons. Layers that are optically thin and likely to give up their photons have small optical depth and are considered exterior to the Sun. The small range of heights associated with $\tau \approx 1$ for visible wavelengths is the visible solar surface—the photosphere. At the top edge of the photosphere the photons are statistically unlikely to encounter additional absorbing material, and they are free to fly into space.

After some travel time, the escaping photons impinge on a detector or other matter, such as an atmosphere and are absorbed based on the characteristics of the new material they encounter. Figure 3-41a shows solar source regions for various wave bands of solar energy. Figure 3-41b shows terrestrial deposition regions for the photons in the wave bands of Fig. 3-41a. As photons penetrate into Earth's atmosphere, their likely absorption occurs at a height where τ approximately equals one.

Pause for Inquiry 3-19

Where $\tau = 1$, by how much is the intensity of a photon beam depleted?

Interactions with Earth's Atmosphere

Absorption and Heating in Earth's Atmosphere. As we calculate in Chap. 2, the Sun delivers approximately 1366 W/m^2 of *total solar irradiance (TSI)* to the top of Earth's dayside atmosphere. This value was formerly known as the solar constant. "Total" refers to energy across all wavelengths. Irradiance refers to the integration of radiance over the solid angle from which photons approach Earth's atmosphere—2π steradians. Absorption, reflection, and scattering in the atmosphere reduce the amount of radiation reaching Earth's surface to ~340 W/m^2. As Fig. 3-41b illustrates, Earth's atmosphere is transparent (optically thin) to some UV radiation, almost all visible radiation, and most IR radiation. Earth's atmosphere is opaque (optically thick) to X-ray, EUV, and some UV radiation.

Although constituting only ~1.5% of the TSI, wavelengths shorter than 300 nm are the primary heat source for the entire terrestrial atmosphere at altitudes from 15 km to more than 500 km. Atoms and molecules in Earth's atmosphere control the absorption and deposition of this energy. The shortwave radiation excites, dissociates, and ionizes atmospheric gases. Atoms subsequently de-excite or re-associate, and electrons and ions recombine. With each interaction some of the energy converts to heat. The gas kinetic energy is moved to lower levels of the atmosphere by heat conduction.

Photodissociation. By mass, oxygen is second to nitrogen throughout much of Earth's atmosphere. Nonetheless, it plays a singularly important role in absorption

Chapter 3 The Quiescent Sun and Its Interaction with Earth's Atmosphere

Fig. 3-41.   a) Solar Electromagnetic Source Regions. Visible wavelengths of solar radiation emanate from the photosphere and chromosphere. Shorter and longer wavelengths emanate for higher regions of the Sun's atmosphere. **b) Terrestrial Energy Deposition Regions.** At wavelengths shorter than visible, this energy-transfer roughly maps increasingly higher layers of the Sun's atmosphere to increasingly higher altitudes of Earth's atmosphere. At wavelengths longer than visible the atmosphere's interaction with solar radiation is more complex. Water vapor in the troposphere absorbs select wavelengths in the microwave band. In the radio band some wavelengths are absorbed, others transmitted and some are even reflected. (EUV is extreme ultraviolet; UV is ultraviolet; IR is infrared.)

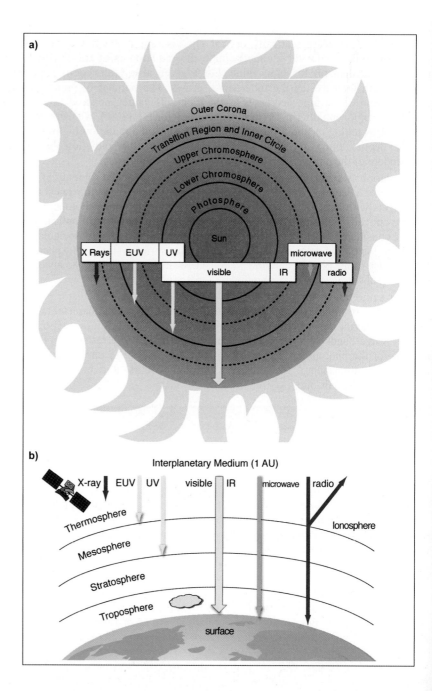

processes over a broad range of terrestrial altitudes. The double bonds in molecular oxygen atoms are more easily broken than the triple bonds of molecular nitrogen. Here we briefly review some of the solar radiation interactions important to Earth, using oxygen as the focus. Figure 3-42 shows the solar energy absorption profile for wavelengths shorter than 300 nm in the terrestrial atmosphere.

3.4 Basic Physics of Atmospheres

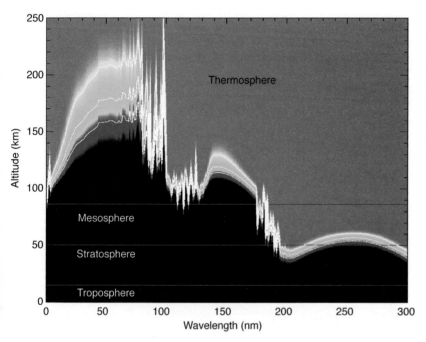

Fig. 3-42. **Absorption in Earth's Atmosphere as a Function of Altitude and Wavelength.** The red zone contains many photons, but little matter is available for interaction. The black zone has few photons, so production is zero or nearly so. The transition colors indicate where photons of a specific wavelength are actively absorbed. The topmost white curve corresponds to $\tau = 1$. The graph turns black at the height where all of the radiant energy has been absorbed by atmospheric gases. *(Courtesy of Dean Pesnell at the Goddard Space Flight Center, NASA)*

Pause for Inquiry 3-20

Starting at the top of Fig. 3-42, each significant color change represents a one unit change in optical depth, τ. Thus at the top of the black zone $\tau = 4$. What percentage of the original photons are still present in the solar beam at the top of the black zone?

Pause for Inquiry 3-21

Figure 3-42 does not cover the visible portion of the spectrum. If it did, how far down in altitude would the red zone extend?

Pause for Inquiry 3-22

What are the energies, in eV, of photons that interact with Earth's upper atmosphere?

Photons with wavelengths between 125 nm and 300 nm participate in photodissociation of molecular oxygen and other processes. We recall from Chap. 1 that the *stratosphere* extends from 15 km–50 km and envelops a layer of ozone with a peak density at an altitude of about 25 km. Solar photons with wavelengths less than 176 nm create O by dissociating O_2. Subsequently atomic oxygen recombines with O_2 to form tri-atomic ozone (O_3). Photons with wavelengths of 200 nm–300 nm dissociate ozone. Figure 3-43 shows the steady-state result of photon-oxygen interactions. Ozone appears to respond to subtle changes in solar UV radiation produced by the solar activity cycle. Because ozone absorption drives radiative and dynamical processes that couple the middle and lower atmospheres, this cycling is under intense investigation for space environment-climate connections in the stratosphere. Table 3-2 shows examples of the production and recombination of chemicals in the upper atmosphere.

Fig. 3-43. Ozone (O_3), Oxygen Ion (O^+), and Atomic Oxygen (O) Profile. The profile represents an equilibrium state between production and loss. These profiles are for the summer season over the equator. The density of individual constituents decreases exponentially above and below the altitude of peak density. The region below the lower red curve is the troposphere. The region with the highest concentration of ozone is the stratosphere. Atomic oxygen dominates the thermosphere. Oxygen ions (O^+) become more plentiful in regions where the Sun's shortwave radiation ionizes the atomic oxygen.

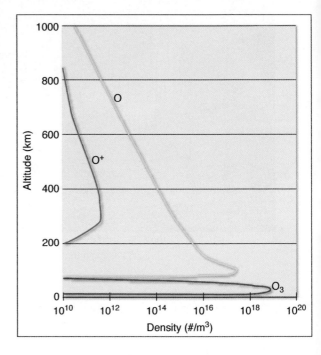

As we mention in Chap.1, the constituents of the *mesosphere*, the coldest region of Earth's atmosphere, are relatively poor absorbers of solar radiation. However, above 100 km is a region that contains little mass, but does have an abundance of high-energy solar photons available for interaction. This layer is the *thermosphere*, the region where low-Earth orbiting (LEO) satellites operate, and in which most human space activities occur. Because particle collisions are infrequent in the thermosphere, particles stratify based on their molecular mass. Oxygen atoms dominate at lower levels (120 km–600 km) and helium and hydrogen atoms dominate above 600 km. Solar activity changes the altitude at which particular species dominate. Solar radiation at wavelengths between 125 nm and 176 nm dissociates molecular oxygen to produce atomic oxygen. Figure 3-43 illustrates the distribution of atomic oxygen. Ultimately much of the excess kinetic energy of the dissociated and excited oxygen atoms becomes the thermal energy responsible for the high temperatures in the upper atmosphere. As Fig. 1-13 shows, the temperature at the base of the thermosphere increases rapidly with increasing altitude before achieving a relatively uniform value above 300 km. Table 3-2 gives examples of production and recombination of chemicals in the upper atmosphere.

Photoionization. Earth's uppermost atmosphere also absorbs photons at wavelengths shorter than 125 nm. The result is ionization. These photons create conducting layers of electrons and ions that are embedded in the neutral upper atmosphere from 70 km to 1000 km—the ionospheric layers. The process begins when an atom or molecule X intercepts an EUV or X-ray photon hf of sufficient energy to detach an electron. The result, as shown in Table 3-2, is a positive ion and a free electron ($X^+ + e^-$). Each interaction reduces the energy in the solar radiation beam by an amount hf. The rate of reduction per unit time and area is dI from Eq. (3-8a).

Table 3-2. Absorption and Emission Processes. Here we show production and loss processes for constituents of the terrestrial atmosphere. Oxygen reactions are primary examples of the absorption and emission processes.

Process	Production			Loss		
	Reactants	Results	*Examples*	Reactants	Results	*Examples*
Photo ionization						
Atomic	$X + hf$	$X^+ + e^-$	$O + hf$, $O^+ + e^-$	$X^+ + e^-$	$X + hf$	$O^+ + e^-$, $O + hf$
Photo dissociation						
Diatomic	$XY + hf$	$X^* + Y^*$	$O_2 + hf$, $O^* + O^*$	$X^* + Y^* + Z^*$	$XY + Z^* + hf$	$O^* + O^* + O^*$, $O_2 + O^* + hf$
Triatomic	$XYZ + hf$	$X^* + Y^* + Z^*$	$O_3 + hf$, $O_1^* + O_2^*$	$X^* + YZ^*$	$X + YZ + hf$	$O_1^* + O_2^*$, $O_3 + hf$

* Indicates electronically or thermally excited matter.

EXAMPLE 3.6

Ozone Example

- *Problem Statement:* Determine the effective attenuation of radiation by ozone (O_3) at the ozone peak in the stratosphere, given an ozone layer equivalent thickness and absorption coefficient.
- *Relevant Concept:* Optical depth
- *Given:* At 255 nm, where the O_3 absorption peaks, the product of absorption cross section and number density, called absorption coefficient k, is ~276 particles/cm

  The ozone layer has an integrated column thickness (equivalent thickness, $s = \Delta z$) of about 0.25 cm

- *Assume:* Photo dissociation is the dominant loss process and production rates equal loss rates
- *Solution:*

  $I = I_o e^{-\sigma n \Delta z}$

  $= I_o e^{-k \Delta z} = I_o \exp(-276 \text{ particles/cm})(0.25 \text{ cm})$

  $= I_o \times 10^{-30}$

- *Interpretation:* Although ozone represents only a few parts per million of the atmospheric density in the ozone layer, it is a very effective absorber of radiation at 255 nm

Follow-on Exercise 1: Use the data for the maximum ozone concentration in Fig. 3-43 along with the absorption coefficient to determine the ozone absorption cross section (σ).

Follow-on Exercise 2: In Fig. 3-42, each white curve represents a unit of optical depth. As photons pass down through the third white curve, how much of the original irradiance remains at a given wavelength?

Volume Production Rate. To quantify the photon-matter interaction leading to photoionization or photodissociation in the upper atmosphere we define a new quantity called the *production rate (R)*. This rate quantifies the relationship between the amount of radiation available to be absorbed, the amount of absorbing material and the likelihood the material will be involved in absorption. The latter quantity is characterized by the absorption cross section. The production rate (R) for ionization or dissociation is

$$R = -\sigma n I = \frac{dI}{dz} \qquad (3\text{-}9)$$

where

R = production rate of a new product [#/s]
σ = absorption cross section of the gas [m^2]
n = number density of gas particles [#/m^3]
I = solar irradiance [W/m^2]
dI/dz = change in solar irradiance with an altitude change [W/m^3]

The absorption cross section σ has values in the range of 10^{-22} m^2 – 10^{-27} m^2 for Earth's atmosphere at typical wavelengths of solar radiation.

Equation (3-9) indicates that production results in the decrement of electromagnetic radiation. The Law of Atmospheres tells us that the density n has an exponential character. The radiation intensity also decreases exponentially. Thus, the full form of Eq. (3-9) is a product of exponential terms that creates a production peak at some altitude that is species dependent.

> **Pause for Inquiry 3-23**
>
> Verify that the units of Eq. (3-9) are those of power density (W/m^3).

Volume Loss Rate. Production of ionization and dissociation continues only if there are mechanisms to restore or replenish the affected material. In steady-state situations recombination of materials balances production. The rate of recombination is proportional to the densities of the constituents involved. We formulate this recombination in terms of the *law of mass action*, in which the recombination rate is proportional to the densities of the interacting particles. For example, the recombination rate of atomic oxygen O to form O$_2$ via the process, O + O + M → O$_2$ + M, where M represents an atmospheric particle that carries away momentum and energy is

$$\text{recombination} \propto (n(O))^2 (n_{atmospheric\ particles})$$

Steady-state production and recombination processes create concentrations of absorption by-products at certain heights as shown in Fig. 3-43. This figure depicts three such profiles for ozone, atomic oxygen, and ionized atomic oxygen. The profiles have clear maxima and steep decreases on either side. The decreases are exponential and have differing slopes that depend on the specific characteristics of radiant interaction with different species. The profiles are characteristic of a combination of two natural features of absorption in atmospheres. High altitudes contain many photons that could be absorbed, but little matter to do the absorbing. At altitudes below the profile peak, there's plenty of matter, but the photons are lacking because they have been removed by absorption above. This process is the essence of the *Chapman mechanism*, named after Sydney Chapman, who first posed a simple explanation of electromagnetic radiation interactions with atmospheric matter in the 1930s [Chapman, 1931].

The matter is the neutral gas particles in Earth's atmosphere. Being gravitationally bound, the gas molecules follow Eqs. (3-6a and b), the Law of Atmospheres. The density $n(z)$ decreases with increasing altitude, as shown in Fig. 3-46, where we plot relative values of density and height rather than absolute values.

The rest of our discussion focuses on ionization and the ionosphere, although the general idea applies equally to dissociation in the neutral atmosphere. To

3.4 Basic Physics of Atmospheres

Focus Box 3.6: Neutral Atomic Oxygen Effects

Besides heating, production of atomic oxygen has significant space environment consequences in the upper atmosphere. Excited neutral oxygen produces an ultraviolet glow at 135.6 nm (Fig. 3-44). Scientists image this glow and its variations to trace the behavior and motions in the upper atmosphere. The airglow can be used as a proxy for neutral density upheaval associated with heating during solar flares and geomagnetic storms.

Atomic oxygen is also a highly reactive agent that corrodes satellite and component surfaces. LEO satellites fly in a bath of corrosive oxygen atoms. When atomic oxygen encounters other materials, it steals atoms of carbon, hydrogen, nitrogen, and other elements from the surface. Over time, this process erodes the material. It eventually weakens spacecraft components, changes their thermal characteristics, and degrades sensor performance. Corrosion is a major concern for spacecraft designers and operators. The space physics community has applied substantial resources to understanding the problem. The Long Duration Exposure Facility (LDEF) satellite was launched in 1984 to investigate this problem (Fig. 3-45a).

Because of photo-dissociation, oxygen in the thermosphere exists predominantly in its atomic form. Its density varies with altitude and solar activity and it's the predominant neutral species at altitudes of about 200 km–400 km during low solar activity.

At LEO altitudes atomic oxygen interacts with volatile species from silicones and hydrocarbons onboard spacecraft. These interactions deposit a contaminant that forms an optically absorbing film on the surfaces of Sun sensors, star trackers, and optical components, or increases the solar absorbance of thermal control surfaces. Simultaneous exposure to the solar ultraviolet radiation, micrometeoroid impact damage, sputtering, and contamination effects aggravates the atomic oxygen effects, leading to serious deterioration of mechanical, optical, and thermal properties of some material surfaces (Fig. 3-45b).

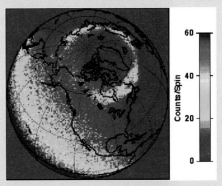

Fig. 3-44. Atomic Oxygen Emissions. Here we see the results of a simulation of far ultraviolet (FUV) emission from neutral oxygen (O_2) 135.6 nm airglow. The simulation illustrates the excited neutral oxygen airglow emissions (counts) in FUV that could be imaged by a spinning spacecraft equipped with an appropriate filter. The colors represent the number of photons emitted—from red (high) to blue (none). Earth's subsolar dayside is red because photo dissociation creates excited oxygen that emits FUV photons. The intensity of this radiation decreases at higher latitudes, because the Sun is less effective in dissociating molecular oxygen along slanted paths through the atmosphere. The northern auroral zone also glows in discrete patches, because oxygen in that region is excited by auroral particle bombardment. During quiet times this feature is faint or absent. *(Courtesy of James L. Green at NASA)*

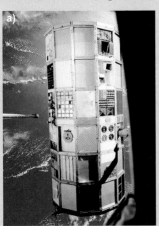 

Fig. 3-45. Atomic Oxygen Effects on The Long Duration Exposure Facility (LDEF). a) This is the LDEF in its early years of operation. **b)** Here we see corrosion and staining of some of the metals on LDEF after 5.7 years on orbit. *(Courtesy of NASA)*

Fig. 3-46. Plot of Relative Density versus Relative Altitude. The neutral atmosphere's density decreases exponentially with increasing altitude, according to the Law of Atmospheres. The horizontal axis is normalized with respect to the maximum density value at the surface. The vertical axis is altitude divided by scale height. At five scale heights above the surface the density is reduced by a factor of $e^{-5} = 0.0067$.

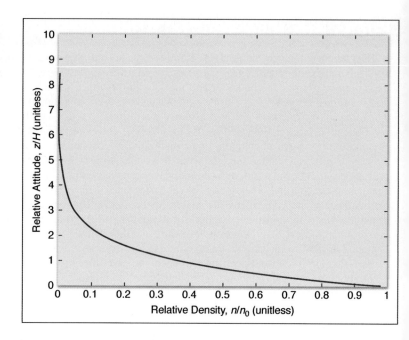

derive the shape of the profile for plasma density in the ionosphere, we need to determine the intensity of ionizing photons as a function of altitude. The variation with altitude is monotonic (that is, uniformly increasing with increasing altitude) but is not as simple as the expression for the Law of Atmospheres, because of density variations in the absorbing medium.

Typically we measure the atmosphere from the bottom up (increasing altitude), so as we move up in the atmosphere, the change in the light intensity should be positive. We simplify the equation by assuming that the Sun is directly overhead, so that ds from Eq. (3-8a) equals $-dz$, and $dI = \sigma n I dz$.

Because we already know from the Law of Atmospheres that the density n is a function of altitude z, we write the incremental intensity as

$$dI = \sigma n_0 I \exp\left(-\frac{z}{H}\right) dz \qquad (3\text{-}10a)$$

Rearranging Eq. (3-10a) so that intensity is on the left side of the equation and then integrating produces

$$I(z) = I_0 e^{(-\sigma n_0 H) e^{\left(-\frac{\Delta z}{H}\right)}} \qquad (3\text{-}10b)$$

Here we have an exponential raised to the power of another exponential. The shape of this curve is shown in Fig. 3-47. Until the sunlight penetrates to where the atmosphere is denser, its intensity is approximately constant, but as the atmosphere absorbs more photons, the intensity of the radiation decreases.

The rate at which photons are lost (with a negative sign) is the rate at which ions are produced (with a positive sign). Of course, ions and electrons want to recombine, because that is a lower potential energy state than being separated, so they continually produce ion-electron pairs to maintain the ionospheric plasma

3.4 Basic Physics of Atmospheres

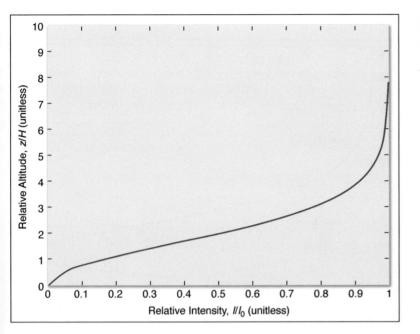

Fig. 3-47. Plot of Relative Radiation Intensity versus Relative Altitude. The intensity of sunlight is greatest at the top of the atmosphere and decreases as light is absorbed closer to Earth's surface.

density. In equilibrium the production rates and recombination rates must balance, so we say that the profile of the production rate $R(z)$ equals the profile of the ionosphere.

Because we know the variation of neutral density from the Law of Atmospheres (Eqs. 3-6a and b) and the variation with altitude of radiation intensity (Eqs. 3-10a and b), we substitute each of these into Eq. (3-9) with an appropriate sign change to get the ionization rate as a function of altitude

$$R(z) = -\frac{dI}{ds} = \sigma n(z)I(z) \tag{3-11a}$$

$$R(z) = \sigma n_0 I_0 e^{\left(-\frac{\Delta z}{H}\right)} \left(e^{-\sigma n_0 H e^{-\frac{\Delta z}{H}}} \right) \tag{3-11b}$$

So production is the product of two exponentially decaying functions. One accounts for the decrease of ionization above the peak level, and the other for the reduction in the ionization rate below the peak level. Together they produce the shape of the Chapman function. Figure 3-48 portrays this function in relative units for a particular set of parameters, $\sigma n_0 H = 5$. We also plot the relative intensity and density profiles on this graph for comparison. The curves in Fig. 3-43 are products of the two exponential functions in this plot, as well as the loss functions (not shown).

This model assumes a single wavelength (energy) of light ionizing a single species of atoms or molecules with a fixed cross section. It also assumes that each photon interaction produces a single ionization. In reality, the ionizing photons from the Sun cover a wide range of wavelengths and corresponding energies, and these interact with varying probabilities with the several species in the atmosphere. This simple model also assumes an overhead Sun (we relax this assumption in Chap 8).

Fig. 3-48. Chapman Ionization Production Rate Profile. The production rate for ions (proportional to the ion density) peaks at a certain altitude. In this graph, we also plot the relative intensity and density profiles for comparison.

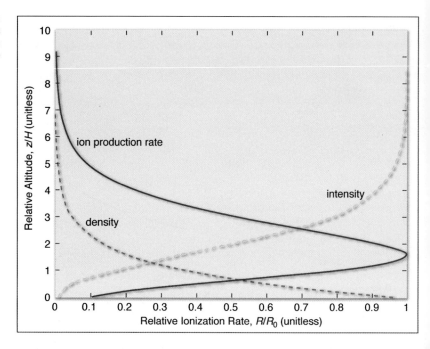

Moreover, we should bear in mind that as altitude increases and density decreases, the neutral atmosphere tends to stratify according to the relative masses of the constituents. So several local maxima form in the ionosphere. These maxima are the basis for the layers of the ionosphere: the D, E, F1, F2, and topside layers. This model also describes the equilibrium situation on Earth's dayside, while on the nightside, ion production ceases and these structures decay at different rates as ions and electrons recombine. Table 3-3 summarizes some important features of the various layers of Earth's ionosphere, including the major constituent ion species.

Table 3-3. Characteristics of Ionospheric Layers. Here we show some characteristics of the ionospheric layers.

Ionospheric Layer	Altitude Range (km)	Major Constituents	Notable Characteristics
D	70–90	NO^+ O_2^+ (molecular)	Disappears (recombines) very rapidly–minutes after sundown
E	90–140	O_2^+ (molecular) NO^+	Recombines rapidly–often disappears before midnight
F1	140–200	O^+ (atomic) NO^+	Mostly recombines after sundown, but pockets of ionization may remain
F2	200–400	O^+ (atomic)	Persistent because of low collision rates, but density decreases after sundown
Topside	> 400	O^+ (atomic) H^+	Merges into the plasmasphere, atomic oxygen dominates at lower altitudes, and hydrogen dominates at higher altitudes.

Pause for Inquiry 3-24

Verify that the product $\sigma n_0 H$ is dimensionless.

Summary

The Sun is a magnetically variably star whose output is the primary source of space weather. Remnant radiation from nuclear reactions in the solar core seeps outward to the convective zone, where the energy powers the solar dynamo and convective cells. Convective motions radiate electromagnetic energy to space and organize a magnetic field at the surface. The magnetic fields infiltrate the Sun's atmosphere. Differential rotation stretches and kinks the fields, producing knots of magnetic energy that are subject to destabilization, especially during the maximum of the ~11-year solar cycle. Although the solar magnetic field is often characterized as dipolar, the large-scale dipolar magnetic configuration is disrupted by numerous active regions.

The solar atmosphere consists of several layers. The lowest layer (photosphere) is opaque to radiation from below, but emits most of the Sun's light as visible and infrared radiation. Overlying the photosphere is the chromosphere, the coolest region of the Sun's atmosphere. From the base of the chromosphere temperatures rapidly increase. The steepest rise occurs in the transition region, a thin layer responsible for much of the Sun's extreme ultraviolet emissions. The transition region forms the base of the corona, the very hot, X-ray emitting portion of the Sun's outer envelope. The sources of upper atmospheric heating are not well understood, but at least some fraction of the heating is thought to come from merging of magnetic fields. Plasma motions in the Sun's upper atmosphere also produce radio emissions, sometimes used as solar activity proxy, that vary with the ~11-year activity cycle. Plasma (fluid pressure) dominates the lower portions of the Sun's atmosphere, while magnetic pressure dominates the outer layers.

At very large scales the Sun's magnetic field is dipolar, especially during solar minimum. During the cycle minimum, the solar magnetic field is poloidal (connected from pole to pole). Differential rotation allows the low-latitude portion of the field lines to outpace the polar portions and wrap around the Sun, forming a toroidal field. Velocity shears in the convective zone form flux ropes, or tubes, in which the field becomes strong and buoyant. When the tubes burst through the surface, they create sunspot pairs from which the field expands as small dipoles. The direction of the toroidal field determines the spot polarity.

At the outset of a new sunspot cycle the preceding spot has the same polarity as the polar field in the hemisphere where the spot originates. The latitude of the first appearance of sunspots depends on the differential rotation and magnetic field strength. When the first sunspots emerge at high latitudes, the magnetic pressure decreases. The process then moves to lower latitudes, with sunspots erupting closer to the equator. The preceding spots within each hemisphere merge. The trailing spots merge with the polar field. The polar fields reverse as the fields from the trailing spots dominate. Near minimum the field returns to a dipole-like configuration with the poles reversed.

We view the features of the solar magnetic field in the context of a layered solar atmosphere or by scale size. Sunspots are regions of intense magnetic field in the solar photosphere. They last from one day to several weeks and usually come in pairs of opposite polarity. Faculae are additional manifestations of enhanced magnetic field in the photosphere. They are bright regions outlining solar supergranular convection cells and are also present near some sunspots. These bright cells are most apparent where the density of the gas decreases abruptly by the presence of strong magnetic fields. The gas becomes nearly transparent, so that deeper layers of the granules on the limb-side of the magnetic field concentration glow more brightly. At these deeper layers, the gas is hotter and radiates more intensely, explaining the brightening. Faculae are likely to appear in a web-like pattern, sometimes called the network. The network outlines the supergranular cells and is caused by the presence of bundles of magnetic field lines that concentrate there because of the fluid motions within the supergranules.

Spicules are jets of plasma with lifetimes of 5–10 minutes. These jets, which also outline the supergranular cells in the chromosphere, originate in the photosphere and pass through the chromosphere into the corona. In the chromosphere, plages are bright regions surrounding and overlying sunspots. Plages also appear with concentrations of magnetic fields and form a part of the network of bright emissions that characterize the chromosphere. Filaments are dark, dense, somewhat cooler clouds of material that are suspended above the solar surface by loops of magnetic field. Prominences and filaments are different views of the same object. We see prominences projecting above the Sun's limb (edge). These structures can remain in a quiescent state for months. However, as the magnetic loops that support them change, filaments and prominences erupt and rise off the Sun over the course of a few minutes or hours.

In the corona, closed field lines that connect the emerging magnetic flux at the surface, and have associated faculae and plages, are called active regions. The concentrated plasma trapped on the closed field lines gives the eclipsed corona a weak glow in visible light. High intensity soft X-rays are emitted from broad diffuse regions above active regions. Coronal arcades connect regions with loops of magnetic field. Above the arcade the fields from the two adjacent regions become anti-parallel, separated by a current sheet. This structure is a helmet streamer. Evolution of multiple active regions produces large regions of unipolar magnetic field surrounding helmet streamers. The field from the interior of the unipolar region is nearly radial, creating a coronal hole that provides a source of high-speed solar wind. Open field regions do not trap plasma, and so appear dark in X-ray imagery. Coronal holes are usually present at the poles. Those that cross the equator are sources of high-speed solar wind that reaches Earth.

With respect to fluid behavior, the lower solar atmosphere follows the Law of Atmospheres. Pressure and density decrease exponentially with height. However, energy input to the upper solar atmosphere causes the temperature to increase with height. The extremely hot upper solar atmosphere emits shortwave radiation that interacts with Earth's upper atmosphere. This radiation ionizes, dissociates, and heats terrestrial atoms and molecules. It strongly affects oxygen molecules. It's also responsible for the heated layers in Earth's atmosphere (stratosphere and thermosphere) and for the ionized layer (ionosphere).

Key Words

absorption cross section, σ
arcades
background component
barometric equation
Beer's Law
bipolar regions
Carrington longitude (L)
Carrington rotation number
central meridian (CM)
Chapman mechanism
chromosphere
circulations
closed magnetic flux
convective instability
convective zone
core
corona
coronal holes
differential rotation
enhanced networks
ephemeral active regions
faculae
giant convection cells
granulation
Hale's Law
helmet streamers
hydrostatic equilibrium
intranetwork field
Joy's Sunspot Tilt Law
latitude (θ)
Law of Atmospheres
law of mass action
limb darkening
magnetic carpet
magneto-convection
meridional
mesosphere
micropores
network
open magnetic flux
optical depth
photosphere
plages
polodial magnetic fields
pores
production rate (R)
prominences
radiative diffusion
radiative zone
scale height
slowly varying component
solar dynamo process
solar flux units (SFUs)
spicules
stratosphere
Sun echo
Sun-in-view
supergranulation
supergranule network
tachocline
thermal convection
thermosphere
toroidal magnetic fields
total solar irradiance (TSI)
transition region
unipolar
zonal

Notation

$$\frac{d(m\boldsymbol{v})}{dt} = \text{Sum of external forces}$$

$$\rho\frac{d\boldsymbol{v}}{dt} = \rho\boldsymbol{a} = \Sigma \boldsymbol{F}_{ext}/volume$$

$$\rho\frac{d\boldsymbol{v}}{dt} = -\frac{GM\rho}{r^2}\hat{\mathbf{r}} - \frac{dP}{dr}\hat{\mathbf{r}} = \Sigma \boldsymbol{F}_{ext}/volume$$

Hydrostatic equilibrium $\frac{dP}{dr}\hat{\mathbf{r}} = -\frac{GM\rho}{r^2}\hat{\mathbf{r}} = -\rho g\,\hat{\mathbf{r}}$

Pressure change with height in a gravitational field
$dP = -mgn dz$

Law of Atmospheres - differential form $\frac{dn}{n} = -\frac{mg\,dz}{k_B T}$

Law of Atmospheres - integral form

$$n(z) = n_0 \exp\left(-\frac{mg}{k_B T}\Delta z\right) = n_0 \exp\left(-\frac{\Delta z}{k_B T/mg}\right)$$

$$= n_0 \exp\left(-\frac{\Delta z}{H}\right)$$

Scale height $H = k_B T/mg$

Barometric equation

$$P(z) = P_0 e^{(-mg\Delta z/k_B T)} = P_0 e^{(-(\Delta z)/H)}$$

Radiance as a function of path length - differential form

$$\frac{dI}{I_s} = -\sigma n ds$$

Radiance as a function of path length - integral form

$$\exp[\ln(I)]\Big|_0^s = \frac{I_s}{I_0} = \exp[-\sigma ns]$$

Optical depth $\tau = \int \sigma n ds$

Beer's Law $I_s = I_0 e^{-\sigma ns} = I_0 e^{-\tau}$

Production rate $R = -\sigma n I = \dfrac{dI}{dz}$

Intensity as a function of height - differential form

$$dI = \sigma n_0 I \exp\left(-\frac{z}{H}\right) dz$$

Intensity as a function of height - integral form

$$I(z) = I_0 e^{(-\sigma n_0 H) e^{\left(-\frac{\Delta z}{H}\right)}}$$

Ionization rate as a function of altitude - differential form

$$R(z) = -\frac{dI}{ds} = \sigma n(z) I(z)$$

Ionization rate as a function of altitude - integral form

$$R(z) = \sigma n_0 I_0 e^{\left(-\frac{\Delta z}{H}\right)} \left(e^{-\sigma n_0 H e^{\frac{\Delta z}{H}}}\right)$$

Answers to Pause for Inquiry Questions:

PFI 3-1: At 45° Fig. 3-1b shows the frequency to be 410×10^{-9} Hz.

$$Period = \frac{1}{f} = \frac{2\pi}{\omega} = \frac{2\pi}{(2\pi)(410 \times 10^{-9} \text{Hz})}$$

$$= 2.44 \times 10^6 \text{ s} = \frac{2.44 \times 10^6 \text{ s}}{86,400 \text{ s/day}} = 28.3 \text{ days}$$

PFI 3-2: $\left[\dfrac{Pa}{m}\right] = \left[\dfrac{N/m^2}{m}\right] = \left[\dfrac{\left(N\dfrac{m^2}{kg^2}\right)kg\left(\dfrac{kg}{m^3}\right)}{m^2}\right] = \left[\dfrac{N}{m^3}\right]$

PFI 3-3: The tachocline has a thickness of ~0.05 R_S, so $0.05 \times 7 \times 10^8$ m $= 3.5 \times 10^7 = 35$ Mm

The convective zone has a thickness of ~0.30 R_S, so $0.30 \times 7 \times 10^8$ m $= 2.1 \times 10^8 = 210$ Mm

PFI 3-4: For 14 September 2009, JD = 2455088.5, so Carrington Rotation = 1690 + (2455088.5 − 2444235.34)/27.2753 = 2087.912.

PFI 3-5: b) Cold on top, warm on the bottom

PFI 3-6: In Eq. (2) of Focus Box 3.3 the particle accelerates upward if the parcel density is less that the surrounding density. Substituting for density from the Ideal Gas Law:

$$\rho a = (\rho_S - \rho^*)g = \left(\frac{P_S}{RT_S} - \frac{P^*}{RT^*}\right)g$$

Assuming pressure equilibrium ($P_S = P^*$)

$$\rho a = \left(\frac{T^* - T_S}{T^* T^*}\right)g$$

Therefore, if the parcel density (n^*) is lower than the surrounding density (n_S) and the parcel temperature (T^*) is greater than the surrounding temperature (T_S), then the particle will accelerate upward.

PFI 3-7: After releasing their energy at the top of the photosphere, buoyant parcels have a lower temperature and pressure, so they begin to fall back to the tachocline.

PFI 3-8: The granular and supergranular speeds are on the order of 1 km/s. Normal highway speeds are on the order of 100 km/hr (0.028 km/s). So highway speeds are much slower than solar granular speeds. A typical passenger jet travels at ~220 m/s or 0.22 km/s, also much slower than granular motion.

PFI 3-9: The surface trip takes a little more than one year, and the return trip takes about 8 years. The entire cycle time is comparable to that of the 11-year sunspot number variation.

PFI 3-10: Fig c

PFI 3-11: The energy at 656.3 nm is from the hydrogen Balmer alpha (Hα) emission which is an n = 3 to n = 2 electronic transition in neutral hydrogen. The emission is in the red visible portion of the spectrum. Hα light comes from the photosphere and chromosphere where temperatures are cool enough to allow for a substantial density of neutral hydrogen. The emission at 393.4 nm is from singly ionized calcium. The outermost electron still attached to the calcium atom makes an electronic transition that creates higher energy photons (UV). This energy comes from the upper chromosphere, where higher temperatures create more violent collisions to ionize calcium and excite the remaining electrons.

PFI 3-12: Visible 656.3 nm emissions are used for monitoring structures in the lower solar atmosphere. This emission is enhanced by solar activity. The 393.4 nm emissions are modulated by the strength of the magnetic field. Moderate fields block the emissions. This shorter wavelength is used to monitor magnetic structures in the upper photosphere and chromosphere that also have a solar cycle dependence.

PFI 3-13: From Chap. 2: The cooling results from photon emission created by free electron interaction with the medium in which they are moving. The free electron impact knocks a bound electron to an excited state; this decays, emitting a photon. During collisions, a free electron may ionize a formerly bound electron, which takes energy from the free electron.

If a free electron recombines with an ion, the binding energy and the free electron's kinetic energy are radiated away. Additionally, free electrons, accelerated by an ion, emit bremsstrahlung radiation.

PFI 3-14: Typical images on the internet are in the visible to X-ray wavelengths.

PFI 3-15: $f = c/\lambda$

PFI 3-16: 2001. The southern reversal may have been slightly later.

PFI 3-17: The rise of solar cycle 24 will have an outward-directed field in the southern solar hemisphere. During the peak of solar cycle 24 the southern field will switch to inward-directed.

PFI 3-18: Large scale heights are associated with smaller density changes with height.

PFI 3-19: 1/e or ~0.37 times the original value

PFI 3-20: ~2%

PFI 3-21: To Earth's surface

PFI 3-22: The wavelength range is X ray to UV (a few nm to ~300 nm). These values correspond to ~300 eV to 4 eV.

PFI 3-23: σ has units of m^2, n has units of #/m^3. I has units of W/m^2; so R has units of (m^2)(#/m^3)(W/m^2) = W/m^3.

PFI 3-24: σ [m^2]. n [#/m^3]. H [m]. So, $\sigma n_0 H$ has units of (#), which we consider unitless.

References

Chapman, Sydney. 1931. The Absorption and Dissociative or Ionizing Effect of Monochromatic Radiation in an Atmosphere on a Rotating Earth. In *Proceedings of the Physical Society*. Vol. 43. London, UK.

Duffet-Smith, Peter. 1992. Carrington Rotation Numbers. In *Practical Astronomy with Your Calculator, 3rd Edition*. Cambridge, England: Cambridge University Press.

Snodgrass, Herschel B. and Roger K. Ulrich. 1990. Rotation of the Doppler features in the solar photosphere. *Astrophysical Journal*. Part 1, Vol. 351. American Astronomical Society. Washington, DC.

Figure References

Babcock, Horace W. 1961. The Topology of the Sun's Magnetic Field and the 22-YEAR Cycle. *The Astrophysical Journal*. Vol. 133. American Astronomical Society. Washington, DC.

Christensen-Dalsgaard, Jørgen. 2002. Helioseismology, *Review of Modern Physics*. Vol. 74. American Physical Society: College Park, MA.

Dikpati, Mausumi and Peter A. Gilman. 2006. Simulating And Predicting Solar Cycles Using A Flux-Transport Dynamo. *The Astrophysical Journal.* Vol. 649. American Astronomical Society. Washington, DC.

Hathaway, David H., John G. Beck, Richard S. Bogart, Kurt T. Bachmann, Gaurav Khatri, Joshua M. Petitto, Samuel Han, and John Raymond. 2000. The Photospheric Convection Spectrum. *Solar Physics.* Vol. 193, No. 1/2. Dordrecht, Netherlands: Springer Publishers.

Pneumann, Gerald W. and Roger A. Kopp. 1971. Gas-Magnetic Field Interactions in the Solar Corona. *Solar Physics.* Vol. 18, No. 2. Dordrecht, Netherlands: Springer Publishers.

Stein, Robert F. and Aake Nordlund. 2000. Realistic Solar Convection Simulations. *Solar Physics.* Vol. 192, No. 1/2. Dordrecht, Netherlands: Springer Publishers.

Thompson, Michael J., Jørgen Christensen-Dalsgaard, Mark S. Miesch, and Juri Toomre. 2003. The Internal Rotation of the Sun. *Annual Review of Astronomy and Astrophysics.* Vol. 41. Annual Reviews: Palo Alto, CA.

Further Reading

Lang, Kenneth. 2000. *The Sun from Space (Astronomy and Astrophysics Library).* Berlin: Springer-Verlag.

Living Reviews in Solar Physics. http://solarphysics.livingreviews.org/

Stix, Michael. 1991. *The Sun: An Introduction (Astronomy and Astrophysics Library).* Berlin: Springer-Verlag.

Space Is a Place...with Fields and Currents

4

UNIT 1. SPACE WEATHER AND ITS PHYSICS

Contributions by Devin Della-Rose

You should already know about...

- Definition of a vector
- Symbols and definitions of divergence and curl
- Line and area integrals
- Units of electric, magnetic, and gravitational fields
- Lorentz force on a charged particle
- Laws of Gauss, Faraday, and Ampere
- Field configurations from point charges and superposition of fields
- The basic form of Maxwell's Equations
- Ohm's Law
- How currents flow in continuous circuits
- Definition of plasma

In this chapter you will learn about...

- Use of the field concept in science
- Applications of the divergence and curl of a vector
- Understanding and interpreting Maxwell's Equations
- Induction
- Electric fields, magnetic fields, and current density
- Understanding and interpreting Ohm's Law
- Frozen-in magnetic flux
- Dynamos and magnetic merging in space
- Currents and fields in the near-Earth environment

Outline

4.1 Introduction to Fields
    4.1.1 The Field Concept
    4.1.2 Static Electric and Magnetic Fields
    4.1.3 Maxwell's Equations
    4.1.4 Electric, Magnetic, and Electromagnetic Wave Energy Densities (Pressure)

4.2 Electric Currents and Conductivity
    4.2.1 Charge Conservation, Current Continuity, and Current Density
    4.2.2 Ohm's Law and Electrical Conductivity

4.3 Magnetic Behavior in the Space Environment
    4.3.1 Frozen-in Magnetic Flux
    4.3.2 Electromagnetic Characteristics of Dynamos

4.4 Currents and Electric Fields in Near-space
    4.4.1 Electric Currents in the Near-Earth Environment
    4.4.2 Electric Fields in Space

4.1 Introduction to Fields

In space, electricity, magnetism, and gravity wage an unending battle for control of matter. Gravity concentrates matter. Various processes ionize the matter, sometimes creating charge imbalances. Electrical forces try to neutralize the charge imbalances. Flowing charges (currents) generate magnetic fields that intensify in the gathering materials. Sometimes the magnetic fields explosively adjust to a less-concentrated state. These powerful upheavals produce disruptive electromagnetic waves and accelerate individual charged particles to near light speed. Humans, hardware, and technology systems are at risk from such upheavals.

In Chap.1 we note that particles, photons, and fields mediate energy transfer. Chapter 2 deals with photons and particles. In this chapter, we briefly review the fundamentals of fields and currents and describe their basic roles in space weather. We first focus on fields as a general concept in physics, and then describe electric and magnetic field sources and their roles as reservoirs of energy. The discussion of electric currents in Sec. 4.2 provides the background for the topic of space environmental fields and currents in Secs. 4.3 and 4.4 and later in Chaps. 9–12. Anyone with a good grasp of electricity and magnetism should briefly review the examples in this section and Sec. 4.2 and then proceed to Secs. 4.3 and 4.4.

4.1.1 The Field Concept

> **Objectives:** After reading this section, you should be able to...
>
> ♦ Describe what the term "field" means in physics
> ♦ Depict fields associated with simple configurations of mass, charge, and moving charges
> ♦ Recognize and use the units of electromagnetic fields and forces
> ♦ Do order-of-magnitude comparisons of the forces acting in space

Fields. The term *field* describes the assignment of a quantity to every point in space. In physics the quantity is a physical entity, and is often a scalar such as temperature or pressure. Fields may also be vector quantities such as momentum or force. For the remainder of this chapter, we deal only with vector fields. With respect to forces, a field quantifies, at every point, the alteration of space caused by mass, electric charge, or moving electric charge. We can't directly touch or photograph such fields. We have to learn about their presence indirectly, perhaps by measuring the response of a charged particle to the presence of the field.

The originator of field theory as it applies to electromagnetism was Michael Faraday, who visualized the influences that charged particles exert on their surroundings, including empty space. Faraday's work eliminated the idea of action at a distance and replaced it with the concept that a field communicates the presence of a mass or charge. James Maxwell further developed the idea in a set of equations that describe how charges communicate at light speed using wavy fields.

In Chap. 1 we refer to three force equations (and combinations thereof) to describe fields: gravity: $F_g = mg$; electric: $F_E = qE$; and magnetic: $F_B = qv \times B$. Mechanical experiments usually measure forces directly, and from the forces we derive field characteristics. Knowing the field characteristics allows us to generalize the nature of the matter-field-force interactions. Mathematically specifying a field pattern allows us to predict the behavior of a mass, such as a charged particle, placed in that field without detailed knowledge of the mass, charge, or current distribution that was responsible for creating the field. The field concept helps describe how plasma behaves in the space environment.

In Fig. 4-1, we diagram a gravitational field as a vector field of force-per-unit-mass [N/kg] or equivalently, acceleration [m/s^2]. This allows us to draw the field independent of the mass placed in that field. If the mass is specified, then we easily calculate the force on it. Analogously, an electric field vector drawn in the vicinity of a point charge maps the force-per-unit charge [N/C].

All fields associated with forces have dimensions of newtons per unit source. We recall that the Lorentz force is $\boldsymbol{F} = q\boldsymbol{E} + q\boldsymbol{v} \times \boldsymbol{B}$, so the magnetic field has units of N/(C·m/s). These units are frequently rearranged into the form N/(A·m), also known as a tesla [T]. The tesla is the SI unit of magnetic field strength (magnitude of $\boldsymbol{B}$). However, this unit is quite large for most applications, so alternative units appear frequently.

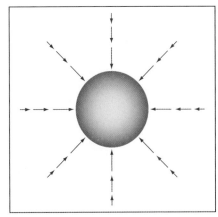

Fig. 4-1. A Gravitational Field in Two Dimensions. The blue arrows represent field vectors for a symmetrically distributed mass. Near the surface the field is stronger (closer to the center of mass) in accordance with Newton's Law of Universal Gravitation.

Pause for Inquiry 4-1

Verify, using the Lorentz force equation, that $\boldsymbol{B}$ has units of N/(A·m).

Pause for Inquiry 4-2

Suppose that Fig. 4-1 represents Earth. What magnitude would you assign to the vectors closest to Earth? How far would you have to travel from Earth's surface to get to a place where the vector has half the value of that at Earth's surface?

Pause for Inquiry 4-3

How would you modify Fig. 4-1 to represent:
a) The field of a positively charged particle?
b) The field of a negatively charged particle?
c) The field of a positive-negative charged particle pair?

Table 4-1 lists characteristics of static fields and currents of simple particle configurations. Static means constant in time. The information in Table 4-1 tells us that the gravitational field arising from a point mass m exerts a gravitational force on another massive object; the electric field of a charged mass q exerts an electric force on another charged mass; the magnetic field of a current $q\boldsymbol{v}$ (a moving charged mass) exerts a magnetic force on other moving charged masses.

Superposition of Fields. According to the superposition principle, simple charge and current elements add linearly to create more complex structures and field configurations. For example, individual charges can be arranged in strands to create sheets of charge. We show some relevant individual and superposed structures in Table 4-2, where we emphasize static dipole and sheet configurations. Static sheets of charge or current produce uniform electric and magnetic fields, respectively. Although the space environment is never truly static, the sheet approximation is particularly useful in describing large space plasma systems.

Chapter 4 Space Is a Place...with Fields and Currents

Table 4-1. **Mathematical Forms of Simple Fields.** Here we list the most basic static field configurations encountered in the space environment. In other applications we consider configurations that are more complicated.

Field: Law	Static Source	Form / Units	Field Units	Comment
Gravity: Newton's Law of Universal Gravitation	Point mass	$-\dfrac{Gm}{r^2}\hat{r}$	$\dfrac{N}{kg}$	May be simplified as $g\hat{r}$
Electric: Coulomb's Law	Stationary charged point mass	$\dfrac{kq}{r^2}\hat{r}$	$\dfrac{N}{C}$	May be written as $E\hat{r}$
Magnetic: Biot-Savart Law	Moving charged point mass with constant speed	$\dfrac{\mu_0 q}{r^2}(\mathbf{v}\times\hat{r})$	$\dfrac{N}{A\cdot m}$	Valid for non-relativistic motion

where

$k = \dfrac{1}{4\pi\varepsilon_0}$

G = universal gravitation constant = 6.67×10^{-11} m^3/(kg·s^2)

μ_0 = permeability of free space, $4\pi \times 10^{-7}$ N/A^2 = 1.26×10^{-6} N/A^2

ε_0 = permittivity of free space, 8.85×10^{-12} C^2/(N·m^2)

Figure 4-3 shows a segment of an infinite current sheet. Sheets of flowing electric charge are important structures in the space environment. Often current sheets are so thin in comparison to their length that they are treated as if they have zero thickness; so the charge flow occurs on a surface rather than in a volume. Typically electrons are more mobile than ions. The difference in mobility creates a current that must be accompanied by a sheared magnetic field. The oppositely directed magnetic fields in Fig. 4-3 are sheared fields. Any sheared magnetic field must have a current sheet (symbol J in Fig. 4-3); such sheets are present in the solar atmosphere, in the solar wind, at the magnetopause, in the magnetotail, and in the ionosphere. Current sheets are the physical structures that maintain separation of different plasma populations. Sheets with large current density tend to have strong magnetic fields and a large energy density associated with the fields. They are prone to instability, collapse, or a form disruption known as magnetic merging. We say much more about merging in the upcoming chapters.

Dipole configurations also play an important role in the space environment. To conveniently describe the field of dipoles, we use spherical coordinates in which r represents radial displacement, and θ is latitude about the origin, midway between the poles. Dipoles are symmetric in longitude (Fig. 4-2a) so there is no need to describe a longitudinal dependence. For a dipole, the field strength at large distances decreases as a function of $1/r^3$. Ideally steady currents flow in infinite loops, which at large distances also produce a dipole field configuration. Simple magnets may be represented by a current loop field.

Pause for Inquiry 4-4

In Fig. 4-3, if the upper magnetic field were absent, would a current still flow to the right? If the upper magnetic field increases in magnitude, what happens to the magnitude of the current?

4.1 Introduction to Fields

Table 4-2. **Fields for Source Distributions.** Here we list and show field configurations and dependencies for source distributions.

Distribution	Distance Dependence	Pictorial Form
Electric Dipole Dipole orientation is from positive to negative charge. The origin is midway between the charges. $$\mathbf{E} = \frac{kqd}{r^3}(2\cos\theta\,\hat{\mathbf{r}} + \sin\theta\,\hat{\boldsymbol{\theta}})$$ where d is the separation distance of the two poles, $\hat{\mathbf{r}}$ is the radial direction, $\hat{\boldsymbol{\theta}}$ is the latitude above or below the origin, and k is Coulomb's constant. $$k = \frac{1}{4\pi\varepsilon_0} = 8.9876 \times 10^9 \, \text{N} \cdot \frac{\text{m}^2}{\text{C}^2}$$	$\frac{1}{(r^3)}$ At large distances, the charges appear to cancel and the field strength falls off rapidly.	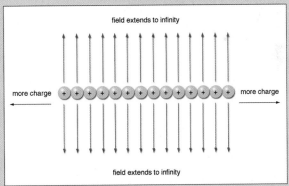 Fig. 4-2. a) Electric Dipole. The field is highly concentrated between the two charges. The dipole orientation is from the positive charge to the negative charge.
Infinite Uniform Charged Sheet $$|\mathbf{E}| = \left|\frac{\sigma}{2\varepsilon_0}\right|$$ where σ is surface charge density [C/m²]	The field strength is uniform throughout all space.	Fig. 4-2. b) Infinite Charged Sheet. The fields between charges cancel but beyond the charges they add to make a uniform electric field. The charge sheet extends infinitely to the right and left and also into and out of the page.

Chapter 4 Space Is a Place...with Fields and Currents

Table 4-2.   Fields for Source Distributions. (Continued) Here we list and show field configurations and dependencies for source distributions.

Distribution	Distance Dependence	Pictorial Form
Magnetic Dipole Dipole orientation is from magnetic south to magnetic north. The origin is midway between the poles. $$\mathbf{B} = \frac{\mu_0 IA}{4\pi r^3}(2\cos\theta\,\hat{\mathbf{r}} + \sin\theta\,\hat{\boldsymbol{\theta}})$$ Where I is the current [A] and A is the area [m²]. The product of IA is called the magnetic moment. In this geometry $\hat{\mathbf{r}}$ is the radial direction and $\hat{\boldsymbol{\theta}}$ is the polar coordinate.	$\dfrac{1}{(r^3)}$ At large distances, the field strength falls off rapidly.	**Fig. 4-2. c) Magnetic Dipole.** The field is highly concentrated between the two poles. The dipole orientation is from magnetic south to magnetic north.
Infinite Uniform Current Sheet $$|\mathbf{B}| = \left|\frac{\mu_0 J_S}{2}\right|$$ Where J_S is the sheet current density [A/m].	The field strength is uniform throughout all space.	**Fig. 4-2. d) Infinite Uniform Current Sheet.** The fields from the individual current elements cancel between the elements but add together outside of the elements to make a uniform magnetic field.

Fig. 4-3.   Current Sheet between Oppositely Directed Magnetic Fields. The curl (or shear) of **B** is associated with the sheet current **J**. Hence we have Ampere's Law for conduction currents: $\nabla \times \mathbf{B} = \mu_0 \mathbf{J}$. *(Courtesy of the COMET Program at the University Corporation for Atmospheric Research)*

At times, we simplify our discussion of the space environment by considering it as an ensemble of electric systems, filled with the structures in Table 4-2 as well as generators, resistors, capacitors, and other electrical devices, all electromagnetically linked by fields. Figure 4-4 is a schematic that shows how static electric fields, electric currents and magnetic fields are related. To make appropriate simplifications we need to know which fields are important in a given circumstance. In plasmas, gravitational fields are generally of minor importance in determining plasma motions. Electric and magnetic fields dominate. Example 4.1 provides perspective on the relative magnitudes of fields and forces for a charged particle in Earth's upper atmosphere.

4.1 Introduction to Fields

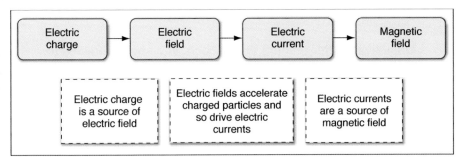

Fig. 4-4. Relation between Static Electric Charges, Electric Fields, Electric Currents and Magnetic Fields. The influence of an electric charge is characterized by the electric field. A net electric field puts charges into motion and thereby creates a current. An electric current produces a magnetic field.

Pause for Inquiry 4-5

Draw the electric field associated with an infinite line charge.

Draw the magnetic fields associated with a line current and ring current.

Pause for Inquiry 4-6

How do you configure multiple strands to create an infinite current sheet? What happens to the fields between the strands? (Consult a basic physics text if the answer isn't obvious.)

EXAMPLE 4.1

Force and Field Comparisons

- *Problem Statement:* Compare the magnitudes of the gravitational and electric forces on a proton in Earth's upper atmosphere. For this example, we use just the order-of-magnitude values for the quantities involved.

- *Relevant Concept or Equations:* Newton's Law of Universal Gravitation (simplified), Coulomb's Law, Lorentz's Law

- *Assume:* Constant field strength and each force acts in isolation

- *Given:* proton mass, $m_p \sim 10^{-27}$ kg, proton charge $q_p \sim 10^{-19}$ C, $|g| \sim 10$ m/s^2, $|E| \sim 10^{-3}$ V/m, $|B| \sim 10^{-5}$ T, $v_{p\perp} \sim 100$ m/s (the "$\perp$" subscript indicates the component of velocity perpendicular to the B-field)

- *Solutions:* Order of magnitude
  - Gravitational Force: $F = mg$
    $$F_g = m_p g \approx (10^{-27})(10) = 10^{-26} \text{ N}$$

  - Electrical Force: $F = qE$
    $$F_e = q_p E \approx (10^{-19})(10^{-3}) = 10^{-22} \text{ N}$$
    This force is a factor of 10^4 stronger than the gravitational force. For an electron, the factor is even 1000 times larger.

  - Magnetic Force: $F = q(v \times B)$
    $$F_B = q_p v_{p\perp} B \approx (10^{-19})(100)(10^{-5}) = 10^{-22} \text{ N}$$
    This value is the same order of magnitude as the electric force.

- *Interpretation:* In the upper atmosphere, the gravitational force on individual charged particles is weak compared to electric and magnetic forces.

  Though all the force magnitudes in this example may seem quite small, the plasma particle masses are tiny enough to allow electric and magnetic fields to be quite important in the space environment. We give further numerical examples to support this point as we develop the mathematical expressions for plasma motions in Chap. 6.

Follow-on Exercise: Verify that $(v \times B)$ has the same units as E.

4.1.2 Static Electric and Magnetic Fields

Objectives: After reading this section, you should be able to...

- Describe the sources of static fields
- Relate fields to currents
- Define magnetic flux

Electric and Magnetic Field Origins

Electric Field Sources and Currents. The second row in Table 4-1 indicates that charged particles are sources of electric fields. On Earth we rarely notice the electric field from individual charges, because we usually encounter matter as a matched set of negative and positive charges whose fields cancel on the large scale. Electrical equipment manufacturers fully shield users from the voltages and currents in systems that do carry a net charge. So, with the exception of static cling in some clothing and wintertime zaps when we reach for a metal door knob, nature appears free of electric fields. In our everyday ventures, we are generally unaware of the large forces that develop even when comparatively small numbers of charges are separated. Large charge accumulations that result in bolts of lightning really get our attention.

Electric fields act on relatively mobile charged particles to produce *conduction currents*. On Earth we are most familiar with the situations where electrons are pushed or pulled into motion while the heavy ions remain almost motionless. In space for the most part, electrons are also the dominant current carriers; however, where collisions are infrequent, ions may also carry current.

In space there is usually enough charge mobility to prevent significant charge accumulation. However, some processes in space plasmas continually separate charges that, in turn, produce currents that redistribute the charges. (For an example see Sec. 1.4 and Fig. 1-25.) In this manner nature tends to reign itself in. Sudden changes in electric field configuration, perhaps from a space storm, temporarily upset charge balance. Unlike currents in technology applications that tend to flow in well-shielded wires, space environment currents are free-flowing and unshielded. These currents generate magnetic fields that may be sensed over long distances. For example, storm-generated currents flowing a few hundred kilometers above Earth's surface may produce magnetic perturbations at Earth's surface of a few tens to hundreds of nanotesla. Table 4-3 provides relative values of magnetic field strength at Earth and the Sun.

Figure 4-4 shows a concept map for relations between static electric fields, currents, and magnetic fields. In this and upcoming sections we build on this chain and show how we modify it to describe time varying, interactive systems with feedback among the elements of Fig. 4-4. The space environment has multiple feedbacks.

Magnetic Field Sources, Configurations, and Strengths. As we describe in Sec. 1.4, the interaction between the magnetic fields in the solar wind and those confined in Earth's magnetosphere energizes charged particles and powers space weather storms. On the Sun, where magnetism is at the heart of space weather, currents flow in the convective zone and throughout the solar atmosphere. Where currents flow, magnetism exists. The third row in Table 4-1 indicates that currents are the source of magnetic fields and that the magnetic field at any point in space

Table 4-3. **Strengths of Typical Magnetic Fields.** Here we list some typical magnetic field strengths and alternate units.

Location	Tesla [T]	Gauss [G = 0.0001 T]	Nanotesla [nT]
Sunspots	1×10^{-1}	1000	1×10^8
Refrigerator magnet	1×10^{-2}	100	10^7
Quiet solar surface	1×10^{-3}	10	1×10^6
Earth's surface (Mid-latitude)	5×10^{-5}	0.5	50,000
Solar wind near Earth	5×10^{-9}	5×10^{-5}	5

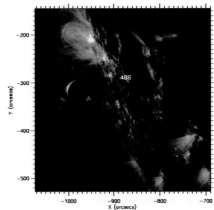

Fig. 4-5. Solar Magnetism in Active Region 10486. Here we see magnetic field lines encased in extreme ultraviolet (EUV) emitting particles. Plasma, forced to gyrate around the magnetic field lines, emits radiation at EUV wavelengths. This active region on October 24, 2003, imaged by the TRACE satellite, spawned several major solar flares and ensuing near-Earth space weather storms. The horizontal and vertical distance units are in arcseconds. *(Courtesy of the Lockheed Martin Solar and Astrophysics Laboratory)*

is perpendicular to the current that created the field. Currents may be controlled by the field, if the field strength is sufficiently high. Figure 4-5 shows electromagnetic emissions from trapped, flowing, charged particles (currents) that gyrate around strong solar magnetic fields. In solar active regions the magnetic field behaves as a virtual string around which the beads of plasma are forced to flow. Figure 4-6 shows the visible sunspot that anchors the active region and the magnetic loops.

Because field configurations are usually non-uniform, scientists sometimes describe field interactions in terms of the concentration of field lines. The number of field lines piercing a unit cross-sectional area has the units of field *flux* Φ. Electric flux has the symbol Φ_E. Magnetic field flux Φ_B has its own name, weber [$T \cdot m^2$ = Wb]. Solar scientists often use a related unit in the centimeter-gram-second (CGS) system called a maxwell (Mx). One weber of magnetic flux equals 10^8 Mx. Figure 4-7 shows the magnetic field lines associated with Earth's dipole. As described in Table 4-2, magnetic field lines are more concentrated near the poles. Thus, the flux values are relatively high in Earth's polar regions. During quiet conditions, within each of Earth's polar regions (above ~78° magnetic latitude), we find a magnetic flux of ~3×10^8 Wb. During disturbed times the flux in Earth's polar caps approaches 1 GWb.

As a word of caution, physicists use the term flux in two ways. One use refers to transfer or flow of a quantity through a unit area per unit time. Flux used in this way is a directed quantity, and hence, a vector. The solar wind mass flux is a mass of the solar atmosphere flowing across a unit area per unit time [kg/($m^2 \cdot s$)]. The total solar irradiance is the solar energy flux at Earth with units of W/m^2. In electromagnetism another use appears: the integral of a vector quantity over a surface area, where the area is treated as a vector. Electric and magnetic fluxes have dimensions of field strength times area and are scalar quantities. In the case of magnetic flux the units are $T \cdot m^2$. A quick unit check reveals which "flux" is being described.

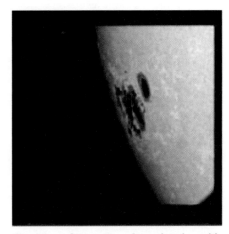

Fig. 4-6. Sunspots Associated with Active Region 10486. This visible image shows the underlying islands of magnetism in active region 10486 as dark sunspots. The magnetic loops are not visible in such images. However, remote sensing techniques that are sensitive to magnetic field orientation allow observers to determine the field directions. *(Courtesy of the Lockheed Martin Solar and Astrophysics Laboratory)*

Chapter 4 Space Is a Place...with Fields and Currents

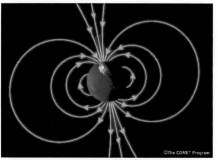

Fig. 4-7. Earth's External Magnetic Field in Two Dimensions. In this idealized drawing a dipole represents the field. Tangents to the field lines give the local direction of the magnetic field. *(Courtesy of the COMET Program at the University Corporation for Atmospheric Research)*

Pause for Inquiry 4-7

The magnetic dipole diagram (Fig. 4-2c) in Table 4-2 shows that the magnetic field enters the magnetic South Pole and exits the magnetic North Pole. Figure 4-7 shows magnetic field lines entering Earth's geographic North Pole region. Where is Earth's true magnetic South Pole located?

EXAMPLE 4.2

- *Problem Statement:* Show that the magnetic flux in either of Earth's polar caps is $\sim 3 \times 10^8$ Wb.

- *Relevant Concepts:* Area of a disk and magnetic flux.

- *Assume:* The polar cap subtends 12° of arc (90°–78°) and the magnetic field lines are vertical so that the polar area vector A is perpendicular out of the surface and antiparallel to B.

- *Given:* Polar magnetic field strength ~60,000 nT directed inward at the geographic north pole.

- *Solution:* First find the polar cap cross-sectional area, πr^2:
  Let $r = [\sin (12°) R_E] \sim [(0.21)(6378 \times 10^3 \text{ m})] = 1.326 \times 10^6$ m

  The polar cap area is $\sim 3.14 (1.76 \times 10^{12} \text{ m}^2) = 5.52 \times 10^{12}$ m^2

  Magnetic Flux $= B \bullet A = BA (\cos 180°) = (6 \times 10^{-5} \text{ T})(5.52 \times 10^{12} \text{ m}^2)(-1)$
  $= -3.31 \times 10^8$ Wb

- *Interpretation and Implications:* The magnetic flux into Earth's northern polar cap is -3.3×10^8 Wb. The minus sign indicates the flux is directed inward. For reasons we discuss in Chap. 7, this region of flux may connect to the solar wind, allowing for intermittent interplanetary-geomagnetic field interaction.

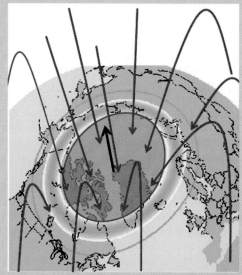

Fig. 4-8. The Northern Polar Cap. The northern polar cap is shaded in blue. Magnetic field lines entering the polar region are shown in dark gray. The black line represents the area vector for the polar cap; it's perpendicular to the surface and antiparallel to the local magnetic field lines. *(After the University of Alaska's Geophysical Institute)*

Follow-on Exercise: A typical sunspot covers about the same area as Earth's disk. Determine a reasonable value for the magnetic flux (Φ_B) in a sunspot.

4.1.3 Maxwell's Equations

Objectives: After reading this section, you should be able to...

- Distinguish between the divergence (div) and curl of a field
- Relate div and curl to the source terms in Maxwell's Equations
- Recognize that field sources may be non-material
- Explain the meaning of induction
- Apply Ampere's Law and Faraday's Law and relate these laws to induction

Curl and Divergence

Table 4-1 describes static fields created by matter whose behavior does not vary with time. Generalized forms of these equations were produced by Isaac Newton, Karl Friedrich Gauss, and Andre-Marie Ampere. Michael Faraday and James Maxwell investigated time-varying fields. Maxwell's studies revealed that electromagnetic field variations propagate at light speed and further, that sometimes fields, rather than being created by charged matter, are induced by changes in pre-existing fields. He thus determined that time-varying field configurations could act as field sources.

The non-matter sources were revolutionary concepts that required nonlinear feedback mechanisms between fields and temporal field variations. Electromagnetic fields have characteristics that exhibit this non-linear behavior. These characteristics come from the spatial configurations of the fields—in particular the tendencies to diverge (div) and curl. As an example, Maxwell found that *displacement current* arises when an electric field changes in a volume of space. In this situation, no charged particles move, but all of the effects of a current are present. The creation of such displacement currents requires large rates of flux variation that are usually not observed in space weather events, so we will continue to focus primarily on conduction currents.

Divergence ("$\nabla \cdot$" operator) describes how a vector quantity passes into or out of a volume of space. Divergence is a scalar value that indicates a source or sink of a quantity. Negative divergence is sometimes called convergence. Figure 4-9 depicts an electric field E diverging from its source, a positive charge.

Curl ("$\nabla \times$" operator) describes the "circulation" or shear of a vector field. To understand this operator, we can think of a vector field as the flow of some fluid, and then imagine placing a tiny paddle wheel in this flow, as shown in Fig. 4-10. If the flow circulates, then the paddle wheel will spin. The curl describes the vector properties of this circulation. Unlike the divergence, the curl is a vector with magnitude and direction. The magnitude of the curl vector describes the strength of the circulation (how fast the paddle wheel spins), and its direction follows the right-hand rule. In other words, if we curl the fingers of our right hand in the direction of the circulation (the direction of paddle wheel spin), then our extended right thumb will point in the direction of the curl vector. For example, in Fig. 4-10, the curl of B, which corresponds to a current in the illustration, points to the top of the page.

The operators "$\nabla \cdot$" and "$\nabla \times$" have the same dimensions: rate of change of a quantity per unit distance. James Maxwell noticed another similarity—both operators are related to forms of density. For example, in Fig. 4-9, the density of the diverging field lines relates to the charge density enclosed within the volume.

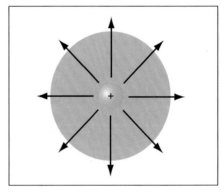

Fig. 4-9. Divergence of an Electric Field from a Point Charge. Here the electric field caused by a positive charge on a mass diverges away from the charge in all directions.

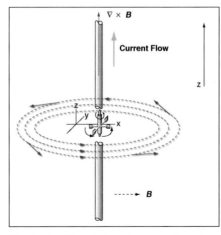

Fig. 4-10. Describing the Curl Operator. Here we show a tiny, massless paddle wheel spinning because it's immersed in a circulating vector field. Only a few of the B-field lines are drawn in a single plane. In reality the current is encased by concentric field lines. In this example, the curl vector and current point in the +z direction. [After Schey, 1973]

Chapter 4 Space Is a Place...with Fields and Currents

If we double the number of enclosed charges of the same sign, then we double the number of diverging field lines. One of Maxwell's Equations (Gauss's Law for electric charge) in Tables 4-4 and 4-5 describes this situation.

The density associated with curl is more subtle. It is the density of circulation. Referring to Fig. 4-10, the amount of current at the center relates to the density of magnetic field lines encircling the current. So, doubling the source, in this case the current, doubles the density of circulating field lines in a volume or space. The Maxwell Equation, called Ampere's Law for (conduction) current, describes this situation.

Pause for Inquiry 4-8

For a negative point charge, the field

a) Divergence is negative
b) Converges
c) Points inward everywhere
d) All of these are true

Pause for Inquiry 4-9

Suppose we were to line up an infinite number of individual current elements to the right and left of the one shown in Fig.4-10. Describe the configuration of B and the curl of B. (*Hint:* See the last entry in Table 4-2.)

Pause for Inquiry 4-10

The dipole configurations in Figs. 4-2a and 4-2c clearly have non-uniform fields. Which characteristic, curl or divergence, best describes the non-uniform two-dimensional fields in the diagrams? (*Hint:* If you placed a paddle wheel at any location other than the center (origin) of the diagrams, would it spin?)

Equations for Electromagnetism and Electrodynamics

Forms of Maxwell's Equations. Plasmas, like those we encounter in the solar and geospace environments, are described not only by conservation of mass, energy, and momentum equations, but by equations for vector electric and magnetic fields. A mathematical theorem for vectors says a vector field is uniquely defined if its divergence and curl are known. Maxwell related not only the divergence and curl, but also the time variations of electric and magnetic fields. His closed set of equations comes in two forms: differential and integral.

The differential form allows field calculations in the most general case and provides a more useful way of discussing the wave nature of light. In more advanced applications the differential forms describe electrically polarized (dielectric) or magnetized material. The integral form is useful for special cases where the fields may be discontinuous and for some special geometries of charges and currents. Usually these geometries are ones with a lot of symmetry. Solutions using either calculation must agree. For simplicity in Table 4-4, we show Maxwell's Equations only in their differential form and without their historical names. That table emphasizes the sources and variations of the electric and magnetic fields. Table 4-5 gives the historical names of the equations, provides prose descriptions, and compares their forms.

In the case of a local vacuum without distant accelerating charges, Maxwell's field equations have the form shown in the top section of Table 4-4. For such conditions nothing contributes to the curl or divergence of the fields, hence the zeros on the right sides of the equations. If we add local unbalanced charges to the system, the equations take the forms shown in the middle section of Table 4-4. Static charges create diverging local E fields, and local currents create curling B fields. Addition of local or distant time-varying sources produces the field equations shown in the bottom section of the table. The divergence equations don't change, but the curl equations each gain a new source term. A changing magnetic field induces curl in the electric field (*induction term*) and a changing electric field induces and adds to the curl in the magnetic field (*displacement current term*).

Table 4-4. Maxwell's Equations Describing Field Sources and Field Characteristics. In general, we apply these equations for vacuum or near-vacuum conditions. When we apply them to dense matter, we must specify the values of ε and μ appropriate to that form of matter.

Source	Divergence	Curl	Field
No local field sources	$\nabla \cdot E = 0$	$\nabla \times E = 0$	Electrostatic, E
Distant static field sources only	$\nabla \cdot B = 0$	$\nabla \times B = 0$	Magnetostatic, B
Local static field sources (and possibly distant sources as well)	$\nabla \cdot E = \rho_c/\varepsilon_0$	$\nabla \times E = 0$	Electrostatic, E
	$\nabla \cdot B = 0$	$\nabla \times B = \mu_0 J$	Magnetostatic, B
Local static and time-varying sources (and possibly distant sources as well)	$\nabla \cdot E = \rho_c/\varepsilon_0$	$\nabla \times E = \partial B/\partial t$	Electrodynamic, E
	$\nabla \cdot B = 0$	$\nabla \times B = \mu_0 J + \mu_0 \varepsilon_0 \partial E/\partial t$	Electrodynamic, B

where
E = electric field vector [V/m]
B = magnetic field vector [T]
ε_0 = permittivity of free space, 8.85×10^{-12} [$C^2/(N \cdot m^2)$]
μ_0 = permeability of free space, $4\pi \times 10^{-7}$ [N/A^2]
     = 1.26×10^{-6} [N/A^2]
ρ_c = charge density [C/m^3]
J = current density vector (described in Sec. 4.2.1) [A/m^2]

Pause for Inquiry 4-11

Do the zeros in Table 4-4 indicate an absence of fields?

Pause for Inquiry 4-12

Rewrite the bottom portion of Table 4-4 for vacuum conditions. (*Hint:* Which terms drop out of the bottom entries of Table 4-4 if local charges are removed?)

Chapter 4 Space Is a Place...with Fields and Currents

> **Pause for Inquiry 4-13**
>
> Can field sources exist in a vacuum?

Table 4-5. Maxwell's Equations in Differential and Integral Forms. Here we list the fundamental equations that describe the strengths and interactions of electric and magnetic fields. Space environment modelers use these equations to describe electric and magnetic fields (E and B), separately and as they interact. Individual equations carry the name of the historical developer. Maxwell's contribution was to put these equations together to describe time-varying fields. He also added the last term in Eq. (4-4), the displacement current term. The use of μ_0 and ϵ_0 indicates that the equations apply to vacuum or near vacuum conditions. (In the integral forms, dA and dl are differential area and differential path length respectively.)

Differential Form	Integral Form
Gauss's Law for Electric Fields	Charged particles create electric fields; unbalanced charges create divergent fields.
Div$E = \nabla \cdot E = \rho_c / \epsilon_0$ (4-1a)	$\oint E \cdot dA = Q/\epsilon_0$ (4-1b)
Faraday's Law for Electric Fields	Electric fields that curl or circulate are created by time-varying magnetic fields.
Curl$E = \nabla \times E = -\partial B/\partial t$ (4-2a)	$\oint E \cdot dl = -\dfrac{d\Phi_B}{dt}$ (4-2b)
Gauss's Law for Magnetic Fields	There are no diverging magnetic fields because magnetic monopoles don't exist.
Div$B = \nabla \cdot B = 0$ (4-3a)	$\oint B \cdot dA = 0$ (4-3b)
Ampere's Law for Magnetic Fields with Maxwell's Addition	Circulating or curling magnetic fields are created by flowing charge (conduction current) and time varying electric fields (displacement currents).
Curl$B = \nabla \times B = \mu_0 J + \mu_0 \epsilon_0 \dfrac{\partial E}{\partial t}$ (4-4a)	$\oint B \cdot dl = \mu_0 I + \mu_0 \epsilon_0 \dfrac{d\Phi_E}{dt}$ (4-4b)

where

$\oint$ = line (or surface) integral around a complete path (or a closed surface)
E = electric field vector [V/m]
B = magnetic field vector [T]
ϵ_0 = permittivity of free space (8.85×10^{-12} C^2/(N · m^2))
Φ_B = magnetic flux [T · m^2]
Φ_E = electric flux [V · m]
μ_0 = permeability of free space ($4\pi \times 10^{-7}$ N/A^2 = 1.26×10^{-6} N/A^2)
ρ_c = charge density [C/m^3]
Q = total charge (enclosed by the surface integral) [C]
J = volume current density vector (described in Sec. 4.2.1) [A/m^2]
I = total current (enclosed by the line integral) [A]

Interpreting Maxwell's Equations. The right sides of Gauss's Law for Electricity (Eq. (4-1)) and Faraday's Law (Eq. (4-2)) indicate that electric fields E are caused by net electric charges and time-varying magnetic fields. Similarly, the right side of Eq. (4-4) (Ampere's Law with an additional term) says that the magnetic field B comes from conduction currents and time-varying electric fields that generate displacement currents.

The zero on the right side of Eq. (4-3) means that magnetic fields never accumulate in a closed volume of space—what goes in comes out (Fig. 4-11). Magnetic poles always come in pairs; no magnetic monopoles exist. If, for example, we try to isolate one pole from the other by breaking a bar magnet in half, we find the broken end of the magnet always assumes a polarity opposite from the other end. Gauss's Law for magnetism is the mathematical expression of this. Magnetic field sources arise from moving charge (electric conduction current) and time varying electric fields, both of which create a sheared (circulating) magnetic field.

Maxwell's equations indicate that fields arise from charged matter and from changing fields. In some cases it is important to distinguish induction of fields from production of fields by static matter. The word *induction* indicates fields that arise not from matter but from time variations in pre-existing fields. Below we restate interpretations of Faraday's Law and Ampere's Law with Maxwell's Addition to denote how fields are induced. The concept map in Fig. 4-12 is the expanded version of Fig. 4-14 that includes electrodynamic interactions and suggests the sources of electromagnetic waves.

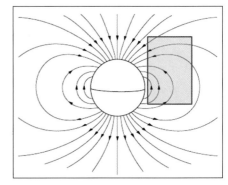

Fig. 4-11. Gauss's Law for Magnetism. Here idealized magnetic field dipole lines are associated with Earth's field configuration. All of the field lines entering a box leave the box. This is true even if the box encloses the dipole.

- *Faraday's Law:* Curl $E = \nabla \times E = -\partial B/\partial t$. An electric field is induced in any region of space in which a magnetic field is changing with time. The direction of the induced field is at right angles to the changing magnetic field. The magnitude of the induced electric field is proportional to the rate at which the magnetic field changes.

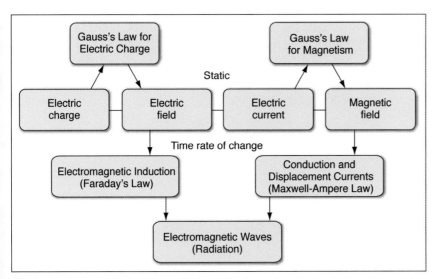

Fig. 4-12. Concept Map of Maxwell's Contribution to Describing Time-varying Fields. Time rates of change of electric and magnetic fields are at the root of electromagnetic waves and radiation.

- **Maxwell's Addition to Ampere's Law:** Curl $\boldsymbol{B} = \nabla \times \boldsymbol{B} = \mu_0 \boldsymbol{J} + \mu_0 \varepsilon_0 \partial \boldsymbol{E}/\partial t$. The second term on the right says that a magnetic field is induced in any region of space in which an electric field is changing with time. The direction of the induced magnetic field is at right angles to the changing electric field. The magnitude of the induced field is proportional to the rate at which the electric field changes.

Combining the induction equations reveals that oscillating electric charges produce wavy magnetic fields that propagate in a vacuum and in matter (Fig. 4-13). Similarly, changing magnetic fields produce wavy electric fields. One of Maxwell's greatest contributions was to recognize that the changes described in his set of equations feed on each other, ultimately producing electromagnetic waves that propagate at *light speed c*. In fact, the magnitudes of the $\boldsymbol{E}$ and $\boldsymbol{B}$ fields in a vacuum electromagnetic wave are related by light speed.

$$|\boldsymbol{E}| = c|\boldsymbol{B}| \tag{4-5}$$

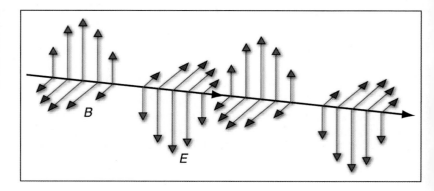

Fig. 4-13. An Instantaneous Snapshot of a Plane-polarized Wave. Here we show the vectors $\boldsymbol{E}$ and $\boldsymbol{B}$ along a particular ray. The wave is moving to the right (in accordance with the $\boldsymbol{E} \times \boldsymbol{B}$ direction) at light speed c. The plane containing the vibrating $\boldsymbol{E}$ vector and the direction of propagation is a plane of vibration. The direction of the ray is the direction in which energy propagates.

When the electric and magnetic fields are components of planar electromagnetic waves, they are perpendicular to each other and to the direction of wave propagation, forming a transverse wave, as shown in Fig. 4-13.

Applying Maxwell's Equations. Here we provide several opportunities for experience with Maxwell's Equations. Pause for Inquiry 4-4 and Example 4.5 illustrate static applications. Examples 4.4 and 4.6 and Pause for Inquiry 4-14 illustrate electrodynamic applications.

4.1 Introduction to Fields

Pause for Inquiry 4-14

Figure 3-34a illustrates how the outflowing solar wind extends the solar magnetic field into space. Figure 4-14 provides an oblique three-dimensional view. Oppositely directed magnetic field lines (lines with gray arrows) are separated by a shaded region of current. According to Ampere's Law, what is the direction of the current in the sheet? What direction would the current have 11 years later?

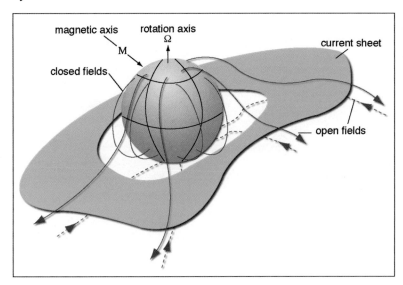

Fig. 4-14. A Three-dimensional Perspective of Interplanetary Magnetic Field (IMF) Lines. According to Ampere's Law, a sheet of current must separate IMF lines directed Sunward from those that flow away from the Sun. [Smith et al., 1978]

EXAMPLE 4.3

Electric Field Strength

- *Problem Statement:* During severe charging conditions at geosynchronous orbit, excess electrons may accumulate on or inside a spacecraft. Determine the approximate electric field associated with an excess charge buildup of 10^{11} electrons/cm^2.

- *Relevant Concepts:* Electric field and Gauss's Law of Electricity. Gaussian surface (a closed three-dimensional surface through which the flux of an electric field is calculated).

- *Assume:* The collecting surface is large compared to the size of an electron, so we use an infinite current sheet approximation.

- *Given:* Electron surface density = 10^{11} electrons/cm^2. The charge density is represented by σ. Total charge is represented by Q.

- *Solution:* Fig. 4-15 shows that for an infinite sheet of charge, the electric field is perpendicular to the surface. Therefore only the field piercing the ends of a cylindrical Gaussian surface contributes to the electric flux through an area A. Half of the field is directed upward from the sheet, and half is directed downward.

Gauss's Law for this situation is

$$\oint E \cdot dA = Q/\varepsilon_0 = E(2A) = \sigma A/\varepsilon_0$$

The resulting field is half that associated with the surface charge density

$$E = \frac{\sigma}{2\varepsilon_0}$$

$$E = \frac{10^{11} \text{electrons/cm}^2 (10^4 \text{cm}^2/\text{m}^2) \times (1.6 \times 10^{-19} \text{C/electron})}{2 \times 8.85 \times 10^{-12} \text{C}/(\text{V}\cdot\text{m})}$$

$$\approx 9.04 \times 10^6 \text{V/m}$$

- *Interpretation and Implications:* Excessive electric fields cause breakdown in some materials, resulting in discharge, shorting, arcing, and damage to circuitry in satellites. Breakdown and discharge occur when dielectric materials are not present or have been compromised (Fig. 4-16). To assure survivability in harsh radiation-belt environments, dielectric materials must withstand electric fields on the order of tens of megavolts per meter.

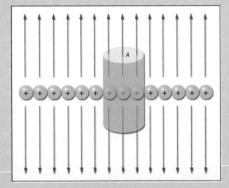

Fig. 4-15. A Thin Infinite Charged Sheet. For illustration some of the charge is surrounded by an imaginary Gaussian cylinder. The electric field **E** from the charges pierces the ends (area **A**) of the cylinder, but not the sides.

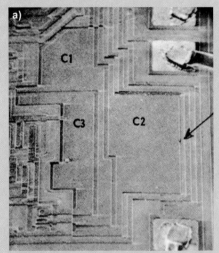
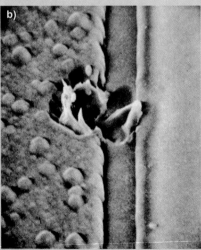

Fig. 4-16. Circuit Damage Caused by Electric Discharge. A C2 MOS capacitor damaged by an electrostatic discharge. The upper image is magnified 175 times. The lower image is magnified 4300 times. The black arrow in the upper image points to the enlarged region at bottom. The capacitor, originally intended to be part of a satellite instrument, was rendered inoperable. *(Courtesy of NASA/Jet Propulsion Laboratory)*

EXAMPLE 4.4

Faraday's and Ampere's Laws

- *Problem Statement:* Determine the response in a loop of wire to a change of magnetic flux threading the loop.

- *Relevant Concept or Equations:* The curl paddle wheel concept and Eq. (4-2) (Faraday's Law); Curl $E = \nabla \times E = -\partial B/\partial t$

- *Solution:* Consider a rectangular loop of wire, as in Fig. 4-17. Consider a uniform, vertically-directed magnetic field B. If the magnitude of this field is constant, Faraday's Law predicts that no current will be induced in the wire. However, if the magnitude of B changes, a current will be induced, but only while the field is changing. When the field stops changing, the current disappears. Further, an increase in field strength $|B|$ results in a current in one direction, whereas a decrease in $|B|$ causes the current to flow in the opposite direction. In either case, the direction of the induced electric field E_{ind} is the same as the direction of the current. To show how this works, we use Eq. (4-2).

In Fig. 4-17a, we suppose that the magnitude of B is increasing. Thus, $\partial B/\partial t$ points upward, as shown by the heavy gray arrow. Faraday's Law states that the curl of the induced current $\nabla \times E_{ind}$ is then downward. The induced current E_{ind} generates a magnetic field to oppose the change in the original field. This effect is called *Lenz's Law* and is a great example of nature's tendency to "reign itself in" toward the lowest possible energy state. The figure shows that the electric field is associated with an induced current I_{ind} that circulates in a clockwise direction. The magnetic field perturbation (dashed lines) from this induced current points in the direction opposite to the original $\partial B/\partial t$.

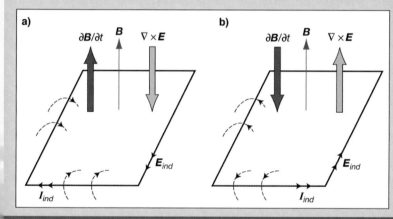

Fig. 4-17. Analysis of Faraday's Law. Using a simple rectangular circuit, in a time-varying magnetic field, we measure the induced current and induced magnetic field. The field B points in the vertical direction, perpendicular to the plane of the circuit. In diagram **a)** $|B|$ is increasing. In diagram **b)** $|B|$ is decreasing. The induced current direction, I_{ind}, is shown in each diagram, along with the resulting magnetic field (dashed lines with arrows). [Wangsness, 1986]

Follow-on Exercise: In a manner similar to the discussion above, analyze and discuss the application of Faraday's Law to Fig. 4-17b. In this case, $|B|$ decreases with time.

EXAMPLE 4.5

Auroral Zone Electric Field Strength Variation

- *Problem Statement:* Under quiet conditions, the magnetic flux in either of Earth's polar caps is $\sim 3 \times 10^8$ Wb. Suppose that during a magnetic storm the inner edge of the auroral zone expands from 78° to 50° latitude over the course of an hour. What happens to the electric field in the auroral zone?

- *Relevant Concepts:* Magnetic flux variation and Faraday's Law (Eq. (4-2))

- *Assume:* The polar cap subtends 40° of arc (90°–50°) and the magnetic field lines are vertical so that the area vector A is perpendicular to the surface and aligned with B. The magnitude of B is uniform. We treat the area of the polar cap as a disk and determine the average of the circumference of the polar cap.

- *Given:* Expanded polar magnetic field strength ~60,000 nT and the geometry in Fig. 4-18.

- *Solution:* First find the approximate new polar cap area, πr^2. We let $r = [\sin(40°)R_E] \sim [(0.64)(6378 \times 10^3 \text{ m})] = 4.1 \times 10^6$ m.

  The polar cap cross-sectional area A is $\sim 3.14 (1.68 \times 10^{13} \text{ m}^2) = 5.28 \times 10^{13}$ m^2

  Magnetic Flux = $(B \cdot A) = (6.0 \times 10^{-5} \text{ T})(5.28 \times 10^{13} \text{ m}^2) \times (\cos 180°) = -3.2 \times 10^9$ Wb

  Now we determine the rate of change of magnetic flux and relate it to $\oint E \cdot dl$

  $\partial B/\partial t = (-3.2 \times 10^9 \text{ Wb} - (-3 \times 10^8 \text{ Wb}))/(3600 \text{ s}) = 8.0 \times 10^5$ Wb/s

  Now we determine the average circumference C of the polar cap during the interval.

  The line length,
  $l = C = 2\pi r_{avg} = 2\pi[(1.3 \times 10^6 + 4.1 \times 10^6)/2] \text{ m} \approx 1.7 \times 10^7$ m

  Now we solve for $|E| = |-\partial B/\partial t|/l = (8.0 \times 10^5 \text{ Wb/s})/(1.7 \times 10^7 \text{ m}) = 0.049$ V/m or 49 mV/m.

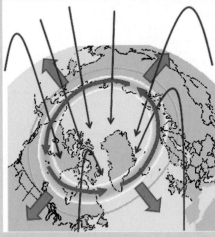

Fig. 4-18. An Expanding Polar Cap with Induced Electric Field. Here we show the geometry of a polar cap expanding during a geomagnetic storm. The thick blue arrows indicate the equatorward expansion of the polar cap. The expansion captures more magnetic flux, shown by the gray curves, in the polar cap. Consistent with Faraday's Law, an induced electric field flows in the conducting auroral zone. The electric field, shown by the red arrows, produces a current that creates a magnetic field that opposes the increased magnetic flux in the polar cap. (*After the University of Alaska's Geophysical Institute*)

- *Interpretation:* This electric field is very large compared to typical values of a few mV/m in the polar cap region. However, a large magnetic storm may increase electric field strengths by an order of magnitude.

Follow-on Exercise 1: What is the significance of the minus sign in Faraday's Law?
Follow-on Exercise 2: What happens to the electric field as the expansion ceases?
Follow-on Exercise 3: What happens to the electric field as the polar magnetic flux returns to quiescent values?

4.1.4 Electric, Magnetic, and Electromagnetic Wave Energy Densities (Pressure)

Objectives: After reading this section, you should be able to...

- Relate the Poynting flux vector to electric and magnetic field variations that propagate energy
- Calculate the pressure and energy density associated with electricity and magnetism

In Sec. 2.2.1 we briefly introduce the idea of energy density and its relation to pressure. The entries in Table 2-5 show that as electric and magnetic field intensity increases, the pressure from these fields grows. As pressure grows, so does its potential to do work. Thus energy density of fields is a convenient measure of potential energy that may become available to accelerate or heat particles or reconfigure systems.

We learned in the previous section that changing electric and magnetic fields produce electromagnetic waves. These waves propagate energy (Chap. 2). The rate and direction (S) at which this wave energy flows through a unit surface area perpendicular to the flow is the *Poynting flux vector*, and is given by

$$S = \frac{E \times B}{\mu_0} \quad (4\text{-}6)$$

where

S = Poynting flux vector (energy flux of the electromagnetic wave) [W/m^2]

The wavy fields carry momentum that allows electromagnetic waves to exert pressure.

Electromagnetic waves travel at light speed c. Although from the standpoint of classical mechanics, a photon has zero mass, and thus carries no momentum, special relativity shows that a photon carries energy and momentum. When electromagnetic momentum changes in time, it exerts a force. This electromagnetic force, acting over an area of space, creates a pressure (radiation). This pressure is equivalent to an energy density u that we express in terms of the component electric and magnetic field magnitudes as follows:

$$u_E = \frac{1}{2}\varepsilon_0 E^2 \quad (4\text{-}7)$$

$$u_B = \frac{B^2}{2\mu_0} \quad (4\text{-}8)$$

where

u_E = electric energy density (pressure) [J/m^3] or [Pa]
ε_0 = permittivity of free space (8.85 × 10^{-12} C^2/(N · m^2))
E = electric field magnitude [V/m]
u_B = magnetic energy density (pressure) [J/m^3] or [Pa]
B = magnetic field magnitude [T]
μ_0 = permeability of free space (4π × 10^{-7} N/A^2 = 1.26 × 10^{-6} N/A^2)

For the total electromagnetic energy density in the wave, u_{EB}, we simply add the component energy densities. Equations (4-5), (4-7), and (4-8) combine to produce a further simplification that we use to describe the energy density of light waves in a vacuum.

$$\frac{E}{B} = c = \frac{1}{\sqrt{\mu_0 \varepsilon_0}}$$

$$\varepsilon_0 E^2 = \frac{B^2}{\mu_0}$$

then, the total electromagnetic energy density is given by

$$u_{EB} = \frac{1}{2}\varepsilon_0 E^2 + \frac{B^2}{2\mu_0} = \varepsilon_0 E^2 = \frac{B^2}{\mu_0} = \frac{EB}{\mu_0 c} \quad (4\text{-}9)$$

The expression u_{EB} [J/m^3] or [Pa] represents the total electromagnetic energy density (the pressure) associated with the wave.

Because the E-field and B-field vary over each wave cycle, the average energy density is

$$\langle u_{EB} \rangle = \frac{1}{2}\varepsilon_0 E_{max}^2 = \frac{B_{max}^2}{2\mu_0} = \frac{E_{max} B_{max}}{2\mu_0 c} \quad (4\text{-}10)$$

The factor of two in the denominator comes from averaging the amplitude of E and B over the wave cycle. Dividing Eq. (4-10) by light speed gives the momentum density carried by the wave.

These expressions have important applications in space science. We use them to describe the amount of energy per volume in electromagnetic fields, i.e., the ability of these fields to transfer energy or do work. In Chap. 12 we evaluate the energy delivered to the auroral zone by field-aligned currents that perturb Earth's geomagnetic field.

Pause for Inquiry 4-15

Verify that energy density can also be written as average Poynting flux divided by light speed. Verify that dividing Eq. (4-10) by light speed produces units of momentum density.

EXAMPLE 4.6

Energy Density I – Sunlight at Earth

- *Problem Statement:* Determine the electromagnetic energy density in sunlight at the top of Earth's atmosphere.
- *Relevant Concept:* Poynting flux S
- *Given:* Total solar irradiance (TSI) = 1366 W/m^2, light speed, $c = 2.998 \times 10^8$ m/s
- *Assume:* No variations in TSI
- *Solution:* The Poynting flux S is the energy per unit area per unit time or the power per unit area. We have power per unit area and we want energy density.

  The energy incident on a unit area A in time Δt is $E = SA\Delta t$
  We need to divide by volume to get energy density.
  Light travels a distance d in a unit of time (Δt), so $d = c \Delta t$
  The volume V = distance × area = $(c \Delta t) A$
  Energy per volume = $[(S A \Delta t)/(c \Delta t A)] = S/c$
  The sunlight energy in a cubic meter at the top of the atmosphere is
  $E = (1366 \text{ W/m}^2)/(2.998 \times 10^8 \text{ m/s}) = 4.56 \times 10^{-6}$ J/m^3

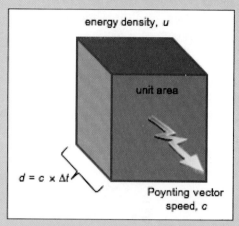

Fig. 4-19. Volume through Which Poynting Flux Flows. We use this volume in determining electromagnetic energy density.

Follow-on Exercise 1: What is the energy density of sunlight at the solar corona (~3 R_S)?

Follow-on Exercise 2: The energy flux at wavelengths that affect the space environment (X ray and EUV) is ~1.5 mW/m^2. Determine the energy density at the top of the atmosphere for these wavebands.

EXAMPLE 4.7

Energy Density II – Sunspot

- *Problem Statement*: Determine the magnetic energy stored in an active (sunspot) region.
- *Relevant Concept*: Definition of magnetic energy density
- *Assume*: Energy density is constant across the volume of the active region in the Sun's atmosphere
- *Given*: In the solar atmosphere above a strong sunspot region, the magnitude of the Sun's magnetic field can be about 0.1 tesla (Table 4-1). A reasonable vertical depth for the active region is about 2000 km (near the top of the chromosphere); a reasonable horizontal length is about five Earth radii (~30,000 km) on a side.
- *Solution*: The magnetic energy density is:

$$u_B = \frac{(0.1 \text{ T})^2}{(2)(4\pi \times 10^{-7} \text{ N/A}^2)} \approx 4 \times 10^3 \text{ J/m}^3$$

If we estimate the volume, then we can compute the total magnetic energy stored in this active region. This stored energy represents the maximum energy available for release in a solar flare that would include particle acceleration, thermal heating, and electromagnetic wave generation.

Active region volume = $(2000 \times 10^3 \text{ m})(3 \times 10^7 \text{ m})^2 \approx 10^{21} \text{ m}^3$

Thus the total stored magnetic energy is

Magnetic energy = $(4000 \text{ J/m}^3)(10^{21} \text{ m}^3) = 4 \times 10^{24}$ J

- *Interpretation*: Only the largest solar flares release this much energy, and the energy output is spread over a period of tens of minutes to hours as we calculate at the end of Chap. 1. While energy released from a large flare equals only about one tenth of the energy the Sun emits each second, the energy release is very localized. This focused energy release can disrupt solar magnetic field configurations and launch mass and magnetic field from the solar corona.

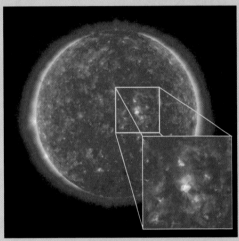

Fig. 4-20. Solar Active Region. The bright spot in the solar atmosphere overlies a photospheric sunspot with high energy density. *(Courtesy of ESA/NASA – the Solar and Heliospheric Observatory (SOHO) Program)*

EXAMPLE 4.8

Energy Density III – Electric Energy Stored in a Battery

When NASA's Mars Exploration Rover "Spirit" landed on the red planet in January 2004, it was partially powered by high-energy-density rechargeable lithium-ion batteries. Solar power recharged the batteries to provide energy during time spent shaded from the Sun and during intense system operations. Spirit's battery, developed as part of the US Air Force's Energy Storage and Thermal Science Program, was rated at 300 W-hr/liter. The battery is designed for 2000 charge-discharge cycles.

- *Problem Statement*: Convert the W-hr/liter rating to SI units of energy density.
- *Relevant Concept*: Units of energy and power
- *Solution*: $300 \text{ W-hr/liter} = 300 \frac{\text{W-hr}}{\text{liter}} \left[\frac{\text{J/s}}{\text{W}}\right]\left[\frac{3600 \text{s}}{\text{hr}}\right]\left[\frac{\text{liter}}{10^{-3} \text{m}^3}\right] = 1080 \text{ MJ/m}^3$

- *Interpretation*: Comparing this result to the energy density in a sunspot discussed in Example 4.7, shows that the battery energy density is much higher than the energy density in a sunspot. However, the shear volume of a sunspot creates a much larger reservoir of energy that can be tapped for explosive energy release.

Fig. 4-21. The Spirit Mars Rover. The lithium-ion batteries on Spirit are rated at 300 W·hr/liter, and they're good for 2000 charge-discharge cycles. *(Courtesy of NASA)*

Follow-on Exercise: Determine the energy density of sunlight at Mars.

4.2 Electric Currents and Conductivity

4.2.1 Charge Conservation, Current Continuity, and Current Density

Objectives: After reading this section, you should be able to...
- Describe the basic implications of current continuity
- Apply the divergence concept to current
- Use different forms of current density
- Explain how differential velocity of charged particles gives rise to currents

Currents are one of the communication links between regions of the space environment. Electric currents flow whenever relative motion occurs between positive and negative charges. In space these currents depend on the conductivity of the plasma, which may in turn depend on the strength and orientation of the magnetic field. Maxwell's Equations, together with Ohm's Law, describe the self-consistency between space charges, electric fields, conductivity, currents, and magnetic fields. In this section, we describe different forms of currents, be they in a volume, on a surface, or in a line. In the next section, we look at the detailed connection between currents, electric fields, and the conductivity of the medium.

Charge Conservation and Current Continuity

We already know that energy and momentum are conserved quantities. Now we consider charge conservation. We capture this idea with an equation for charge conservation (*current continuity*)

$$\nabla \bullet \bm{J} + \partial \rho_c / \partial t = 0 \qquad (4\text{-}11a)$$

$$\nabla \bullet \bm{J} = -\partial \rho_c / \partial t \qquad (4\text{-}11b)$$

where
$\bm{J}$ = current density vector [A/m^2]
ρ_c = charge density [C/m^3]

Equations (4-11a and b) say that if the time rate of change of charge density equals zero, then the current density has no divergence: $\nabla \bullet \bm{J} = 0$. A further implication of Eq. (4-11) is that current flow must be continuous. Currents do not abruptly begin or end. Focus Box 4.2 discusses how the concept of current continuity relates to Maxwell's Equations. In all of the figures in this section, current systems must complete their circuits beyond the image boundary.

Chapter 4 Space Is a Place...with Fields and Currents

Focus Box 4.1: Deriving the Current Continuity Equation

We arrive at the charge conservation law by first applying Ampere's Law and then combining it with Gauss's Law. Ampere's Law relates a curl in the magnetic field to current.

$$\nabla \times \boldsymbol{B} = \mu_0 \boldsymbol{J} + \mu_0 \varepsilon_0 \partial \boldsymbol{E}/\partial t$$

The divergence of this equation is zero because pure circulations do not diverge; that is div (curl (any field)) ≡ 0.

$$\nabla \bullet \nabla \times \boldsymbol{B} = \mu_0 \nabla \bullet \boldsymbol{J} + \mu_0 \varepsilon_0 \partial (\nabla \bullet \boldsymbol{E})/\partial t = 0$$

$$\nabla \bullet \boldsymbol{J} + \varepsilon_0 \partial (\nabla \bullet \boldsymbol{E})/\partial t = 0$$

Gauss's Law for electric fields is $\nabla \bullet \boldsymbol{E} = \rho_c/\varepsilon_0$. We substitute this quantity into the equation above to produce the three-dimensional current continuity equation

$$\nabla \bullet \boldsymbol{J} + \partial \rho_c/\partial t = 0$$

This equation shows that if a charge is moving out of (into) a differential volume, the amount of charge within that volume is going to decrease (increase), so the rate of change of charge density is negative (positive). The current continuity equation amounts to a conservation of charge law. Figure 4-30 illustrates current continuity in the near-Earth system.

Current Density

Unlike the currents near Earth's surface that are usually confined to wires, cables, and channels of lightning, currents in space flow in ill-defined structures. To characterize the currents, we describe the integrated current I and the volume current density $\boldsymbol{J}_V$. If current flows in a volume, its magnitude can be given in terms of the total charge ΔQ divided by the time interval Δt:

$$I = \frac{\Delta Q}{\Delta t} = \frac{n_{ci} q_i A l}{l/v} = n_{ci} q_i A v = \rho_{ci} A v \quad (4\text{-}12a)$$

where

I	= integrated current [A]
$\Delta Q/\Delta t$	= change of charge over time interval [C/s]
n_{ci}	= number of charges of species "i" per unit volume [#/m³]
q_i	= magnitude of charge of species "i" [C]
v_i	= drift speed of species "i" [m/s]
A	= area through which the charge flows [m²]
l	= length of the volume in which the charge flows [m]
ρ_{ci}	= volume charge density of species "i" [C/m³]

Because charge motion has speed and direction, we use a vector equation to relate the motion to the vector quantity $\boldsymbol{J}_V$, the volume current density.

$$\boldsymbol{J}_V = \frac{I}{A}\hat{\boldsymbol{v}} = n_{ci} q_i \boldsymbol{v} = \rho_{ci} \boldsymbol{v} \quad (4\text{-}12b)$$

We combine Eqs. (4-12a) and (4-12b) to get the current magnitude $\int J_V \cdot dA$. The dot product of the two vector quantities signifies that electric current is a scalar. Volume current density has units of $C/(m^2 \cdot s)$ rather than $C/(m^3 \cdot s)$. Multiplying the C/m^3 units in charge density by units of flow velocity (m/s) produces a result of either charge flow density $[(C/m^3)(m/s)]$ or volume current density $[A/m^2]$. The magnitude of volume current density (J_V, also written simply as J where the context is clear) indicates how much current passes a unit cross-sectional area and how much charge moves in a volume of space at a given speed. Figure 4-22 shows the relationship between I and J for uniform current flow in a volume. The current in Fig. 4-22 could result from positive ions moving right, negative charges moving left, or both.

In space, charge may flow in a volume, in a sheet, or in a line. To accommodate these configurations, we need to modify the meaning of charge density to fit the geometry. If the geometry is that of a sheet (or line), the dimensions of the current density change to A/m and A, respectively. Having multiple meanings (and units) for current density can be confusing. We provide some schematics and examples to help clarify the usage. Figure 4-23 shows the relationship between I and J for uniform current flow on a surface or a sheet. The length and width of the sheet are usually large compared to its thickness, similar to a sheet of paper. For example, currents in the low- and mid-latitude ionosphere have a vertical dimension of about 20 km, whereas their horizontal extent is usually thousands of kilometers. We often treat ionospheric currents as a surface or sheet current. Example 4.11 provides some insight into the current densities in the dayside sheet currents associated with low latitude current systems. Example 4.13 gives some perspective on nightside current densities.

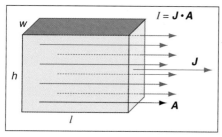

Fig. 4-22. Volume Current Density for a Uniform Current. A uniform volume current flows past a cross-sectional area of magnitude A [m^2], as defined by the width w and height h of the volume. The I is the total integrated current that flows perpendicular to the area given by the product of w and h. J is the volume current density vector $[C/(m^2 \cdot s)]$ or $[A/m^2]$. We show only a segment of the continuous current circuit.

EXAMPLE 4.9

Current Density

- *Problem Statement:* Write out the formula for the current density J_V for the case of two charge species with differing velocities.
- *Relevant concept:* Definition of volume current density
- *Given:*
  1) Ions of charge $+e$, density n, and velocity v_i
  2) Electrons of charge $-e$, density n, and velocity v_e
- *Solution:* $J_V = n \, e \, (v_i - v_e)$

- *Interpretation:*
  - We obtain the value for current density J_V by positive charges moving in a given direction and an equal number of negative charges moving in the opposite direction.
  - Current is defined by the direction that positive charges move (if they are actual charge carriers). A current can be pictured as positive charges moving in one direction or equivalently as negative charges moving in the opposite direction.
  - If all species of charge move with the same velocity, the total (net) current density is zero.

Follow-on Exercise: Assume that the charge density at ionospheric heights is 10^5 particles/cm^3. If the field-aligned current density is 1×10^{-6} A/m^2, what is the relative speed of the electrons with respect to the ions?

Pause for Inquiry 4-16

Verify that the solution to Example 4.9 has units of volume current density $[A/m^2]$.

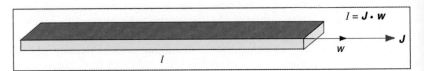

Fig. 4-23. Geometry of the Surface Current Density. Here we have collapsed the volume to a sheet, thus eliminating the height h. A uniform surface current flows along the sheet, spanning a width w. The surface current density is in A/m. Surface current flows along the length of the current sheet. I is the total integrated current [A] across the width.

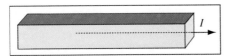

Fig. 4-24. A Line Current. A filamentary line current is similar to current in an electrical wire. Here J and the current I have the same units [A].

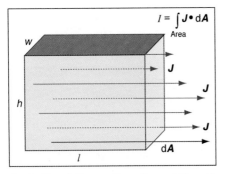

Fig. 4-25. A Non-uniform Current Flow. To quantify this flow, we integrate the current density of small areas, dA. We calculate the current I by integrating $J \cdot dA$ over the cross-section.

We can take the simplification of a volume current one step further by reducing the dimension w to an infinitesimally small value. In doing so we create a filament or line of current, as in Fig. 4-24. In such a current the current density J has the magnitude I, and the direction of the net velocity of positive charge.

Current flow in the space environment often varies with distance (Fig. 4-25). For some applications, we may need to consider the volume current as slices of sheet current and integrate the current density for each sheet. In other applications, we add the current density in individual filaments (Fig. 4-25) and multiply each filament by a unit cross-sectional area to create a bundle of line currents in a volume. Full scale modeling of the currents in the space environment requires such an approach.

A question about how auroras and currents are related often arises. Most quiet-time visible auroras correspond to regions of upward current and have a generally sheet-like structure. But the auroras are not the sheet currents *per se*. The invisible currents that consist of downward-directed electrons excite electrons in mostly neutral atoms, which then glow as auroras. The auroras share a common space with currents but in general are not charged particles capable of carrying current.

4.2.2 Ohm's Law and Electrical Conductivity

Objectives: After reading this section, you should be able to...

- Use Ohm's Law in its electric field form
- Distinguish between conductance and conductivity
- Relate voltage, resistance, electric field, and conductivity to current

Applying an electric field to charged particles in a conducting medium produces charge motion—a conduction current. The electric field-current relationship is specified by *Ohm's Law*. A simple version of this law is

$$J_V = \sigma E \qquad (4\text{-}13)$$

where

J_V = volume current density [A/m^2]
σ = conductivity [1/($\Omega \cdot$m) = mho/m]
E = electric field [V/m]

The equation says that in a steady-state environment, the volume current density is directly proportional to the electric field strength and is in the same direction. Variations in conductivity or the introduction of a magnetic field alters this version of Ohm's Law, which assumes that the cause of charge motion is limited to an applied electric field. Space plasma contains other sources of charge motion, which we explore later.

The electrical *conductivity* σ expresses how freely the charges in the conductor move. The previous section states that we can analyze many current flows in the space environment as thin sheets—their thickness is tiny compared to the area of the sheet. In many locations, including the polar latitudes, these current sheets extend to great heights. If we ignore conductivity changes over the conductor thickness and simply add all the conductivity contributions over the length of the sheet, then we have the height-integrated conductivity, which is the *conductance* Σ. The units of conductivity and conductance are the source of much confusion. The name for the conductivity unit is mho/m ≡ 1/(ohm·m) and symbolized by 1/($\Omega \cdot$m). The name for the conductance unit is siemens = mho and symbolized by S or 1/Ω.

Pause for Inquiry 4-17

Show that the units of conductivity are A·C/(N·m^2) and the units of conductance are A·C/(N·m).

Focus Box 4.2: Ohm's Law

The macroscopic circuit version of Ohm's Law relates voltage difference ΔV, total current I, and resistance R: $\Delta V = IR$. If we multiply both sides of Eq. (4-13) by area A, which has dimensions of length times length, we arrive at

$$JA = \sigma l E l$$

$$I = \Sigma \Delta V$$

$$IR = \Delta V$$

where the conductance Σ is the inverse of resistance, R. By doing this manipulation we have converted Eq. (4-13), which is a microscopic form of Ohm's Law, into the familiar circuit relationship. The microscopic form has more general applicability in the space environment and allows for current-driving terms in addition to an electric field. We often apply the steady-state Ohm's Law to space physics problems even though changes do occur. We make the application by assuming that for the space environment, after any changes in conductivity or electric field, steady state conditions are usually restored rapidly. The assumption is valid most of the time.

Equation (4-13) describes a linear relationship between the electric field and the resulting current density, and it implies that the two vectors point in the same direction. If the direction of E changes, then J assumes the new direction of E. In physics we know this situation as an *isotropic* process. In space physics, however, isotropy does not apply in general, because Earth's magnetic field establishes a preferred direction in space for currents to flow, which removes the isotropy. We emphasize this situation in the next section. The currents and electric fields in the space environment generally do not point in the same direction. Therein lies some of the complexity and intrigue of space physics and space weather. The mathematics of this complexity involves linear algebra and tensors, and so is beyond the scope of this book. Here we simply consider one of the results: the ionospheric dynamo.

The presence of Earth's magnetic field results in three distinct types of conductivity. One is conductivity along (parallel to) the direction of the magnetic field (called the direct, field-aligned, or longitudinal conductivity). The other two conductivities lie in the plane perpendicular to B. The longitudinal conductivity grows exponentially with height in Earth's ionosphere and is a result of the neutral density decaying exponentially with height. Longitudinal conductivity is always much higher than the perpendicular conductivities because charged particles move more easily along the magnetic field than across it (a Lorentz force effect). The perpendicular conductivities, on the other hand, are significant only in a small altitude range (about 90 km to 120 km), where ion and electron motions are very different.

As Earth rotates beneath the Sun, differential heating of the atmosphere causes atmospheric expansion on the dayside and contraction on the nightside. Tides, driven substantially by the rotation of the Earth under the gravitational field of the Moon, add to this disturbance. This combination of periodic forcing creates ionospheric winds, which in turn produce fluid motion across magnetic-field lines. Because electrons are more mobile than the ions, electric currents are induced (Fig. 4-27). The result is an *ionospheric dynamo*—a region where mechanical energy converts to electromagnetic energy. We discuss dynamos further in Sec. 4.3 and in Chaps. 8 and 12.

Examples 4.10 and 4.11 focus on one of the current systems related to this dynamo.

EXAMPLE 4.10

Ohm's Law Applied to the Dayside Ionosphere

- *Problem Statement:* Determine the current density in the dayside quiet dynamo current system
- *Relevant Concepts and Equations:* Ohm's Law $\boldsymbol{J} = \sigma \boldsymbol{E}$
- *Assume:* Steady state conditions and that the currents flow in sheets
- *Given:* In the daytime ionospheric dynamo region, the typical electrical conductivity (in the plane perpendicular to $\boldsymbol{B}$) is about 5×10^{-4} mho/m. Electric field strengths are about 1 mV/m.
- *Solution:*

$$\boldsymbol{J} = \sigma \boldsymbol{E}$$
$$\boldsymbol{J} = (5 \times 10^{-4} \text{ mho/m})(1 \text{ mV/m}) = 0.5 \ (\mu A)/m^2$$

- *Interpretation:* While this value may seem very small, when the current density is integrated over a low and mid-latitude fraction (the circulation contour area in Fig. 4-26) of Earth's dayside, the total current is on the order of 10 MA.

EXAMPLE 4.11

Equivalent Ionospheric Surface Current Density

- *Problem Statement:* Determine the equivalent surface current density in the dayside quiet system current (Fig. 4-26).
- *Relevant Concepts and Equations:* $\boldsymbol{J}_V = \sigma \boldsymbol{E}$, and the concepts of conductivity, conductance, and surface current density $\boldsymbol{J}_S$. To obtain the surface (sheet) current density, we integrate the conductivity contributions over the height range of the dynamo region. This process converts the conductivity σ [mho/m] into the height-integrated conductance Σ [mho].
- *Assume:* Steady-state conditions and uniform conductivity in the dynamo region
- *Given:* Depth is 2×10^4 m (20 km)

$$\boldsymbol{J}_S = \Sigma \boldsymbol{E}$$

- *Solution:*

We obtain the conductance by integrating σ over the depth of the dynamo region in meters. When σ is uniform, we can simply multiply it by the depth Δz.

$$\Sigma = \sigma \Delta z$$
$$\Sigma = (5 \times 10^{-4} \text{ mho/m})(2 \times 10^4 \text{ m}) = 10 \text{ mho}$$

We multiply the conductance by the electric field to calculate the sheet current density

$$\Sigma E = (10 \text{ mho})(10^{-3} \text{ V/m}) = 10^{-2} \text{ A/m} = 10 \text{ mA/m}$$

- *Interpretation:* Physically, this value is the equivalent current we'd get if the dynamo region were flattened into a sheet of current, as in Fig. 4-26. The interpretation of this current magnitude is that each horizontal meter of the dynamo region carries about 0.01 A. Just such a current sheet flows in the mid-latitude ionosphere.

Fig. 4-26. **Height-integrated Current.** This current is flattened to its equivalent sheet current flow (counterclockwise circulation) over the northern dayside region of Earth. The southern circulation is clockwise. *(Courtesy of US Geological Survey)*

Follow-on Exercise: Strong vertical auroral zone currents have values of 100 mA/m and electric fields of 5 mV/m. Find the associated conductance. What is the conductivity if the region is 100 km thick?

4.3 Magnetic Behavior in the Space Environment

Sections 4.1 and 4.2 describe how electric fields, currents, and magnetic fields interact. Here we describe some of the far-reaching implications of magnetic field interactions with conducting plasma. These implications include frozen-in magnetic flux, dynamos, and magnetic merging. We treat all of these topics in more detail later, but an overview is warranted to describe plasma behavior.

4.3.1 Frozen-in Magnetic Flux

Objectives: After reading this section, you should be able to...

- Describe frozen-in magnetic flux
- Compare the conditions of frozen-in magnetic flux to those that allow magnetic merging

In highly conducting plasma, particles that circulate about the magnetic field lines separate the lines of magnetic force. The Lorentz force requires that the unbound charged particles spiral about magnetic field lines. Any attempt by a magnetic field line to change its configuration within a plasma induces an electric field that drives particle motion, which in turn further encapsulates and isolates the field line. Hannes Alfvén captured the concept of this interaction in the term *frozen-in magnetic flux*. The field and plasma are locked (frozen) to each other.

A slightly more advanced explanation is this: the magnetic flux through a surface moving with the plasma is conserved. Thus, plasma elements that are initially located on the same magnetic field line will remain there. Plasma moves freely along field lines, similar to beads on a string. However, in motion perpendicular to the field, the plasma and field must move together. Figure 4-27 portrays this situation. Although adequate for many applications, the frozen-in flux model is not a perfect representation of reality. Over long enough intervals some magnetic field lines diffuse through their accompanying plasma. The degree to which this happens depends on the plasma conductivity.

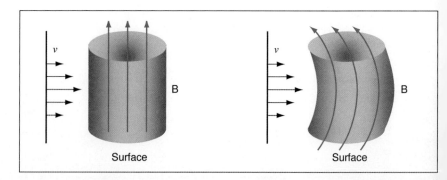

Fig. 4-27. Frozen-in Magnetic Flux. If a magnetic field is frozen to a plasma, then the field may distort as the plasma moves or vice versa; however, they remain interconnected. Were it not for the frozen-in condition, magnetic fields and plasma would quickly decouple and our universe would have a different structure.

Does the plasma or field control the motion in a frozen-in situation? The answer depends on the energy densities of the plasma and magnetic field. If the thermal energy density in the plasma dominates, then the plasma controls the

4.3 Magnetic Behavior in the Space Environment

field. If the magnetic field energy density is higher than that of the plasma, then the field controls the plasma motion. This competition is reflected in the *plasma beta*, defined as the ratio of thermal and magnetic energy density (pressure).

$$\beta = \frac{nk_B T}{\left[\dfrac{B^2}{2\mu_0}\right]} \quad (4\text{-}14)$$

where

- β = plasma beta [dimensionless]
- n = plasma number density [#/m^3]
- k_B = Boltzmann's constant (1.38×10^{-23} J/K)
- T = temperature [K]
- B = magnetic field strength [T]
- μ_0 = permeability of free space ($4\pi \times 10^{-7}$ N/A^2 = 1.26×10^{-6} N/A^2)

Pause for Inquiry 4-18

Under which of the conditions would the magnetic field clearly dominate the behavior of the plasma?

a) $\beta \gg 1$
b) $\beta \ll 1$
c) $\beta \approx 1$

EXAMPLE 4.12

Plasma Beta in the Corona

- *Problem Statement:* Calculate the plasma beta parameter in the corona.
- *Assume:* Coronal magnetic field strength of 10^{-3} T and a coronal mass density of 5×10^{-14} kg/m^3. The Sun is composed of hydrogen. Coronal temperature is approximately 1×10^6 K.
- *Solution:*

$$\beta = \frac{nk_B T}{B^2/2\mu_0}$$

convert mass density to number density

$$\beta = \frac{[(5 \times 10^{-14}\text{kg/m}^3)/(1.67 \times 10^{-27}\text{kg})](1.38 \times 10^{-23}\text{J/K})(1 \times 10^6 \text{K})}{(10^{-3}\text{T})^2/(2 \times 4\pi \times 10^{-7}\text{N/A}^2)}$$

$$\beta = \frac{0.0004 \text{ J/m}^3}{0.3981 \text{ N/m}^2} \approx 0.001 \text{ [dimensionless]}$$

- *Interpretation:* The solar coronal plasma beta value is very small, so the magnetic field energy density dominates the plasma energy density. Since the magnetic field strength in the corona is high, variable, and height dependent, values of beta range from 0.01 to 0.0001.

Follow-on Exercise: Calculate the plasma beta in the photosphere. The average mass density in the corona is ~1×10^{-4} kg/m^3. The quiet photospheric magnetic field strength is ~10^{-2} T. Verify that the plasma energy density dominates the magnetic energy density.

4.3.2 Electromagnetic Characteristics of Dynamos

Objectives: After reading this section, you should be able to...

♦ Describe the cycle of electromagnetic interactions that support a dynamo
♦ Explain the basic features of magnetic merging

Dynamos are nature's means of converting mechanical energy to an electromagnetic form. Magnetic fields store energy in a potential form. The lowest energy magnetic structure is a dipole. An external force must do work to reshape a dipole field into other configurations such as a magnetosphere (Fig. 1-11) or the sheared fields associated with the heliospheric current sheet (Fig. 4-14). The result may take the form of cellular and rope-like magnetic structures that confine plasma. The heliosphere and magnetosphere are large-scale examples. Coronal mass ejections (CMEs) are smaller, transient forms of magnetic cells that transport energy and magnetic flux.

Energy transforms when motions of conducting materials—rotating wires, spinning disks, plasma in space, falling charged cloud droplets in thunderstorms, conducting liquids in planetary cores—generate magnetic fields. Complex motions of the moving conductor cause the magnetic field to develop a very distorted shape, and consequently the field stores more energy. This magnetic energy comes at a mechanical-energy cost to the medium responsible for the motion. For example, energy of deep solar convection is transformed to magnetic energy in solar active regions.

Figure 4-28 presents a concept map for the electromagnetic energy of dynamo action. The process requires a seed (background) magnetic field in a plasma that can be acted on by external forces. The external forces and natural thermal motion put the magnetized plasma into motion. The charged particles produce their own inherent electric fields. When moving in the background magnetic field, they generate an additional electric field. These fields put charges into differential motion, creating new currents that change the background magnetic field and create a bulk (body) Lorentz force ($J \times B$). This in turn creates new plasma motion. This process is the essence of a magnetic dynamo. The Lorentz electromotive force associated with the flow has to overcome the magnetic dissipation in the fluid for the dynamo to maintain itself. A small-amplitude seed magnetic field is sustained and amplified by the flow. The field strength increases until the resultant electromagnetic forces are sufficient to feed back on the flow field.

Nature ultimately reigns in this process via the electromagnetic field's reverse reaction on the fluid motion and by dissipating the magnetic energy (the opposite effect of frozen-in flux). The dissipation may be slow, as in viscous interactions, or it may be explosive, providing bursts of energy to heat plasma and accelerate flows and particles. *Magnetic merging*, in which oppositely directed magnetic field components rapidly join to release their stored energy, is a dominant mechanism for dissipating magnetic energy. In magnetized plasmas, magnetic energy often explosively converts into photons (flares) and particle kinetic energy (CMEs and energetic particles) via the merging process.

4.3 Magnetic Behavior in the Space Environment

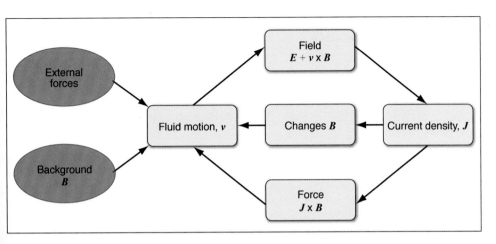

Fig. 4-28. The Dynamo Cycle. The feedback in the cycle often amplifies magnetic fields. Dynamos must have external forces that supply the energy to keep the dynamo going in the face of resistive forces (not shown).

Pause for Inquiry 4-19

Show that the $\mathbf{J} \times \mathbf{B}$ force is the force density equivalent of the magnetic force, $q\mathbf{v} \times \mathbf{B}$.

Magnetic Merging

Magnetic merging is an intrinsic part of the dynamo mechanism, but in a sense it is an anti-dynamo process. In accord with Ampere's Law, current sheets form wherever magnetic fields shear; this typically occurs at magnetic cell boundaries or where dipole structures are severely deformed. Merging alters magnetic field topology by disrupting current sheets and breaking and rearranging field lines as suggested by Fig. 4-29. Merging transforms magnetic energy into other forms (photons and kinetic energy of plasma) as the magnetic field topology changes from the original configuration.

Magnetic merging requires unique conditions not met in typical plasmas. For one thing, the frozen-in flux condition disappears. The currents that sustain the magnetic field's shear dissipate. When this happens, the fields approach and cancel each other (Fig. 4-29). Plasma jets away from the merging region, creating a reduced pressure zone that pulls in more field lines for merging. In this way, the field lines reconfigure and energize their surroundings. The magnetic field configuration of Fig. 4-29 resembles an "X". If the merging location is spatially extended, the region is called a *merging "X-line"*.

Magnetic merging is associated with most solar flare-CME processes. Near-Earth magnetic merging allows the interplanetary magnetic field to link with the geomagnetic field and extract energy, as we discuss in Sec. 4.4.2. In the magnetotail, intermittent merging is the source of energy for particles that ultimately create the beautiful auroras associated with magnetic storms.

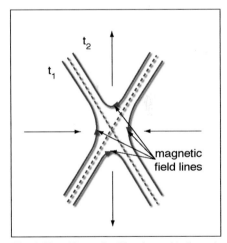

Fig. 4-29. Magnetic Merging. At time t_1 oppositely directed magnetic field lines flow into a small region with unstable currents. The approaching field lines splice into highly kinked structures that accelerate away from the merging region at time t_2. The separated field lines form an "X", indicated by the dashed lines in the figure.

Chapter 4 Space Is a Place...with Fields and Currents

EXAMPLE 4.13

Energy Release by Magnetic Merging in Earth's Magnetic Tail

- *Problem Statement:* Determine the amount of energy released to other forms by magnetic merging in a magnetized slab of plasma that is 2 R_E high, 7 R_E long, and 10 m deep. The magnetic field strength is 5 nT, typical of the interplanetary magnetic field.
- *Relevant Concept:* Energy density in a magnetic field
- *Given:* Slab volume = 10 m deep × 2 R_E high × 7 R_E long, $|B|$ = 5 nT
- *Assume:* 100% energy conversion

- *Solution:*

$$\text{Magnetic Energy Density} = \frac{B^2}{2\mu_0} = \frac{(5 \times 10^{-9}\,\text{T})^2}{2(1.26 \times 10^{-6}\,\text{N}^2/\text{A})} \approx 10^{-11}\,\text{J/m}^3$$

Total Energy = $(10^{-11}\,\text{J/m}^3)(10\,\text{m})(2)(6{,}378{,}000\,\text{m})(7)(6{,}378{,}000\,\text{m})$
= 5.7×10^4 J

- *Interpretation and Implications:* Magnetic merging in a slab that is 2 R_E high × 7 R_E long × 10 m deep produces tens of kilojoules of energy. Even if the efficiency of energy conversion is much lower, magnetic merging provides an impressive energy re-allocation. Over time this energy excites the plasma in the near-Earth environment or stores it in other forms for further redistribution. Explosive release of this energy energizes plasma at the merging line. The plasma accelerates along the newly reconfigured magnetic field lines and slams into Earth's upper atmosphere, creating an aurora.

4.4 Currents and Electric Fields in Near-space

Now we describe the nature of some of the fields and currents in the space environment. The zones of influence range from relatively small volumes of space in auroral structures to the scale of the entire magnetosphere and larger. The lifetime of these structures extends from short (minutes to hours) to permanent.

4.4.1 Electric Currents in the Near-Earth Environment

Objectives: After reading this section, you should be able to...

- Describe the magnetopause boundary currents
- Describe the cross magnetotail currents
- Describe the electrojet currents in the ionosphere

Currents flow at all spatial gradients in magnetized plasmas and in areas of conductivity gradients—these gradients act as boundaries. Some of the current systems are permanent residents, whereas others are active only during space weather storms.

Magnetotail Currents

The solar wind flows past Earth's magnetic field, creating a volume—the magnetosphere—shaped like a bullet. Because the magnetopause boundary separates the relatively strong magnetic field in the magnetosphere from the relatively weak magnetic field in the magnetosheath, a surface current must exist on the boundary in accordance with Ampere's Law. This current is nature's way of separating two regions with differing magnetic characteristics. We show the *magnetopause current* system in Fig. 4-30. On the dayside (locations "D" in the figure), eastward-flowing (dawn-to-dusk) surface currents enhance the magnetic field Earthward of the sheet, and nulls it on the Sunward side. This current largely closes on the magnetopause, forming a complete loop of current. Part of the dayside magnetopause current also closes using currents aligned with Earth's polar magnetic field lines, currents that we call the field-aligned current (FAC) system (Figs. 4-30 and 4-31). Similar current loops flow farther back on the tail of the magnetopause (at locations A connecting to location E), and also connect to currents flowing in the interior of the magnetotail, as we describe next.

Inside the extended tail of the magnetosphere are two regions of oppositely directed geomagnetic field. These too must be separated by a current sheet. We show two strands of that current system at the point labeled "E" in Fig. 4-30. Current continuity requires that this current system be part of a larger current structure. In fact, the relatively flat magnetotail current is part of the complete current circuit that forms semicircular charge flows in Fig. 4-30. The tail current is sometimes called neutral sheet current, because it flows in the magnetically neutral region of Earth's geomagnetic field. Example 4.14 describes the density of current in the geomagnetic tail.

Fig. 4-30. The Primary Space Weather Current Systems. The diagram depicts the bullet-shaped magnetosphere, and associated magnetopause currents (locations A and D). The magnetopause currents close through the ionosphere via field-aligned currents, which we detail further in Fig. 4-31. Currents associated with locations A flow on the tailward magnetopause and close through the central magnetotail (locations E). Currents associated with location B form a circular current in the inner magnetosphere. Some of that current diverges into the ionosphere via field-aligned currents and closes in the auroral zone or across the polar caps. Currents associated with locations D close largely on the dayside magnetopause, although some current diverges as field-aligned current into the dayside ionosphere and closes in the auroral zone and polar cap. *(Courtesy of the COMET Program at the University Corporation for Atmospheric Research)*

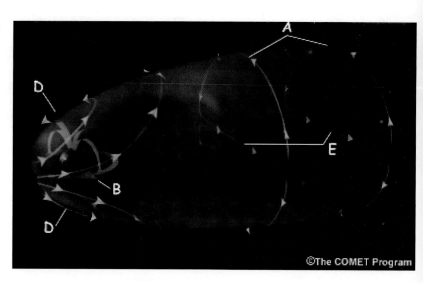

EXAMPLE 4.14

Magnetotail Current Density

- *Problem Statement:* Determine the current density necessary to support the central magnetotail structure. The geotail field near the central current sheet has a strength of ~10 nT.
- *Relevant Concept:* Ampere's Current Law
- *Given:* B = 10 nT on either side of the current sheet
- *Assume:* A uniform current flows in a volume 10 km thick and 50 R_E long.
- *Solution:* The volume current density can be derived from:

$$\nabla \times \boldsymbol{B} = \mu_0 \boldsymbol{J}_V$$

$$\frac{\partial B}{\partial z} = \mu_0 J_V$$

The magnitude of the volume current density is

$$J_V = \frac{[10^{-8}\text{T} - (-10^{-8})\text{T}]}{10^4 \text{m} \times 1.26 \times 10^{-6} \text{N/A}^2}$$

$$J_V = 1.6 \times 10^{-6} \text{ A/m}^2 \approx 2 \times 10^{-6} \text{ A/m}^2$$

From the perspective of the magnetospheric system, this current is flowing in a thin volume that can be viewed as a sheet of current. The equivalent sheet current density is calculated by multiplying the volume current density by the thickness of the sheet, 10 km.

$$J_S = 2 \times 10^{-6} \text{ A/m}^2 \times 10^4 \text{ m} = 20 \text{ mA/m}$$

The total current I flowing in this region is the sheet current density multiplied by the length of the sheet, 50 R_E

$$I = 0.02 \text{ A/m} \times 50 \times 6.38 \times 10^6 \text{ m} \approx 6.4 \text{ MA}$$

- *Interpretation and Implications:* The volume current density supporting the magnetotail is very small. However, when integrated over the volume in which the current flows the total current is substantial.

Follow-on Exercise: Determine the direction of this current density.

Currents in the Magnetospheric Cavity

Ring Current. Close to Earth is a variable current system that encircles Earth in and near the equatorial plane (location "B" in Fig. 4-30). Charged particles, energized by intense electric fields created during intervals of strong solar wind-magnetosphere interactions, drift around Earth in a ring at a distance of about 4 R_E. Some of the particles with critical energies (20 keV to 300 keV) become trapped in the near-Earth dipole field, forming a ring of current. The ring is really more of a toroidal volume, rather than a sheet, and is usually described in terms of total current instead of current density. Ring current particle fluxes increase dramatically during geomagnetic storms. The charged particles make multiple trips around Earth. The quiet-time ring current consists predominantly of electrons and ionized hydrogen, while the storm-time ring current also contains a significant component of ionospheric ionized oxygen. The magnetic field inside the ring current has a direction that opposes Earth's magnetic field at low latitudes. During storm time, superposition of the ring current magnetic field and Earth's magnetic field may reduce Earth's surface magnetic field by a few tenths of a percent. During the strongest storms the field reduction is approximately one percent. Chapters 7 and 11 discuss ring current further.

Field-aligned Current. Field-aligned (Birkeland) currents (named after Kristian Birkeland, a Norwegian auroral scientist), extend from the magnetosphere to the ionosphere. These currents flow along Earth's high-latitude field lines in the two primary regions shown in Figs. 4-30 and 4-31. Region 1 field-aligned currents are at high latitudes and flow into the ionosphere on the dawn side and out on the dusk side. These currents flow along 1) magnetic field lines that connect the dawn and dusk magnetopause charge-separation regions to Earth's high latitudes; 2) magnetic fields connected to the solar wind in Earth's polar regions; and 3) Earth's magnetic fields connected to the inner edge of the plasma sheet on the nightside. The Region 1 currents tend to outline the high-latitude edge of the auroral oval. Region 2 field-aligned currents flow into the ionosphere on the dusk side and out on the dawn side at lower latitudes. They tend to outline the low-latitude edge of the auroral oval. They originate from charge separation caused by plasma flows in the magnetotail and from irregularities in the ring current. These currents also intensify during intervals of strong solar wind interaction with Earth's magnetic field. During the strongest forcing, currents from both regions contribute to currents that flow in the auroral zone.

Ionospheric Currents and Fields

The ionosphere supports several current systems. Some are continuously active. Others are more obvious during space weather storms, when the dynamo electric field is strong.

Solar Quiet (Sq) Current System. As the name implies, this current system is evident during relatively calm space weather conditions. It doesn't disappear during storm times, but ionospheric space weather effects, propagating equatorward from the auroral zone, modify its structure somewhat. The *Sq current system* is created by a dynamo process that converts the kinetic energy of the thermospheric (neutral) particles into electromagnetic energy.

Solar heating of Earth's atmosphere is strongest at low latitudes. In the daytime hours, a high-pressure zone develops near the sub-solar point (intersection of the Sun-Earth line with Earth's surface). This zone creates a daytime thermospheric wind blowing outward from the zone. The outward flowing wind is controlled by the Coriolis acceleration, creating a clockwise circulation of air. The neutral particles drag the conducting lower ionosphere

Fig. 4-31. Field-aligned and Polar Cap Current Circuit. Noon is at 12 local time (LT), dawn is at 06 LT, and dusk is at 18 LT. Distant current closure for the Region-1 currents is at the magnetopause. Distant current closure for the Region-2 currents is in the ring current. The high latitude circuit is three-dimensional, with near-vertical currents along high-latitude field lines and near horizontal currents closing across the polar cap. Only a longitudinal slice of the system is shown. Much of the current is carried by electrons, since the ions are too massive to be accelerated significantly by typical electric field strengths. Storm time electric fields are much stronger and do accelerate ions. *(Courtesy of the COMET Program at the University Corporation for Atmospheric Research)*

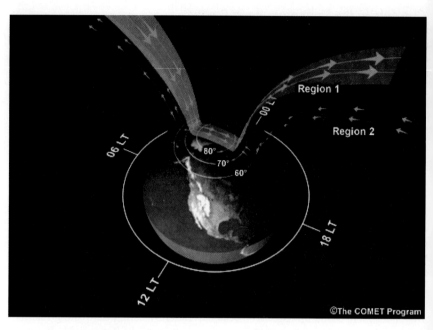

(~90 km–120 km altitude) along. The ions move with the neutral particle circulation but suffer many collisions. Electrons, on the other hand experience fewer collisions and bypass the ions. This relative motion of electrons creates a counterclockwise current. Thus, solar heating indirectly induces charged-particle motion across Earth's magnetic field, resulting in dynamo action. These are the same current circulations we show in Fig. 4-26.

Equatorial Electrojet. The strong current system in the lower ionosphere, flowing near the magnetic equator, is the equatorial electrojet. This "jet" of current is part of the Sq current system and is especially strong because of enhanced electrical conductivity σ at Earth's magnetic equator. Conductivity is high because the strong sunshine there promotes larger plasma densities.

The electrojet exists within about ±2° of the magnetic equator. At higher latitudes, the increasing vertical tilt of Earth's magnetic field eliminates the electrojet effect. The magnetic field of the equatorial electrojet current is about 150 nT–200 nT, as measured on the ground at the magnetic equator. The field is 3–4 times stronger than those generated by other portions of the Sq current system. Observers must reckon with the large magnetic signature of the electrojet current when measuring the strength of other variable storm-time current systems that are hazardous to technology.

Auroral Electrojet. The auroral electrojet is an ensemble of large horizontal currents flowing in the lower regions of the auroral ionosphere (90 km–120 km altitude). The current jets occur because the conductivity and horizontal electric field strength are larger in the auroral ionosphere than at lower latitudes. Although horizontal ionospheric currents flow at any latitude where horizontal ionospheric electric fields are present and conductivity is significant, the auroral electrojet currents are remarkable for their strength and persistence.

During magnetically quiet periods, the electrojet is relatively weak and confined to high latitudes. However, during disturbed periods, it increases in strength and expands to higher and lower latitudes (Fig. 4-32). This expansion results from two factors: enhanced particle precipitation and stronger ionospheric electric fields. The strongest currents tend to flow in the nighttime auroral zones. An eastward-flowing (dusk to midnight) jet occurs in the pre-midnight hours, and a westward (dawn to midnight) jet flows in the post-midnight hours. The auroral electrojets connect to magnetospheric current systems using the field-aligned currents, and the connection is typically stronger with the westward jet. Chapter 11 discusses this topic further.

> **Pause for Inquiry 4-20**
>
> The ribbon-like structures in Fig. 4-31 represent a pair of field-aligned currents in the high-latitude regions. Assume the ribbon tips indicate the direction of current flow. They must have a magnetic field surrounding them. Between the right-most set of ribbons the magnetic field should point
>
> a) down   c) into the page
> b) up   d) out of the page

4.4.2 Electric Fields in Space

> **Objectives: After reading this section, you should be able to...**
> - Describe sources of electric fields in the space environment
> - Define polarization electric field
> - Understand how scale analysis produces estimates for field magnitudes
> - Calculate the magnitude and direction of fields produced by magnetized plasma moving with respect to a stationary frame of reference

Electrostatic Fields from Conductivity Gradients

Electric fields have two sources: electric charges and time-varying magnetic fields. Fields from charges are often called *polarization electric fields*, because positive and negative charges (poles) must separate to obtain a net charge density at a given location in space. This raises two questions: what causes this charge separation, and where does it occur in the space environment?

Currents in the space environment are rarely steady; one reason is conductivity variations. A prime example of this effect is the nighttime auroral zone boundary. Inside the auroral zone, energetic particles from the magnetotail collide with neutral particles in the lower ionosphere, enhancing the charge density and electrical conductivity (Figs. 4-31 and 4-32). Outside the auroral region, however, plasma density and conductivity are much smaller. Thus, current flow across this boundary causes polarizing charges to accrue because of the conductivity variations. As another example, the day-night terminator is a location where conductivity changes substantially, resulting in upper atmospheric polarization electric fields. Just how much polarizing charge is required to produce a significant electric field in space? We tackle this problem in Focus Box 4.3.

Figure 1-25 in Sec. 1.4 provides another example of currents resulting from conductivity changes. Solar wind charged particles flowing past the magnetosphere experience differing mobility based on charge as they approach the magnetospheric flanks. Protons are trapped in the dawnside magnetosphere, electrons in the duskside. Polarization charge builds an electric field that drives a cross magnetotail current, as shown in Fig. 1-25 and Fig. 4-30. Table 4-6 gives rough magnitudes of fields and currents in the space environment near Earth.

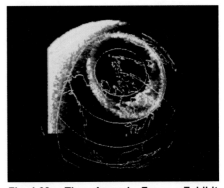

Fig. 4-32. The Auroral Zones Exhibit Variable Conductivity. In the auroral oval, particle impacts create enhanced ionization that increases conductivity compared to surrounding areas. Currents flow more freely in the auroral zones. *(Image from the POLAR Mission, Courtesy of NASA)*

Chapter 4 Space Is a Place...with Fields and Currents

> **Pause for Inquiry 4-21**
>
> Where are the charge accumulations associated with the current in locations "E" of Fig. 4-30?

Table 4-6. Typical Field and Current Magnitudes in the Space Environment. Here we list typical field strengths for the space environment.

Quantity	Magnitude
Electric Fields	
ionospheric	1 mV/m – 10 mV/m
magnetospheric	0.1 mV/m
solar wind	0.1 mV/m
Magnetic Fields	
ionospheric:	
Earth's surface field	3 nT × 10^4 nT
space currents	10 nT – 10^3 nT
magnetospheric:	
inner magnetosphere	10^2 nT – 10^3 nT
distant tail	1 nT – 10 nT
solar wind	few nT
Electric Current Densities	
ionospheric	1 μA/m^2
magnetospheric	1 μA/m^2 – 10^3 μA/m^2
solar wind	10^{-4} μA/m^2 – 10^{-6} μA/m^2

EXAMPLE 4.15

Field Strength and Charge Density at the Magnetopause

- *Problem Statement:* Determine the static charge density associated with an electric field of 1 mV/m at the magnetopause flanks (Fig. 1-25). The boundary layer containing the space charge is ~ $0.01 R_E$ thick. On one side of the boundary is solar wind flow. On the other side charges are trapped inside the magnetosphere.
- *Relevant Concepts:* Gauss's Law for electricity, charge density
- *Given:* $dl = (0.01)(6378 \times 10^3 \text{ m})$; proton and electron charge magnitude $q = 1.6 \times 10^{-19}$ C
- *Assume:* Electrostatic conditions apply in one dimension so that $\nabla \cdot E$ is written as $\dfrac{dE}{dl}$
- *Solution:*

$$\nabla \cdot E = \rho/\varepsilon_0 = qn/\varepsilon_0$$

$$\frac{dE}{dl} = \frac{qn}{\varepsilon_0}$$

$$n \approx \frac{\varepsilon_0 \Delta E}{q \Delta l}$$

$$n \approx \frac{(8.85 \times 10^{-12} \text{C}^2/\text{Nm}^2)(1 \times 10^{-3} \text{V/m})}{(1.6 \times 10^{-19} \text{C})(6.37 \times 10^4 \text{m})}$$

$$n \approx 0.87 \text{ particles/m}^3$$

- *Interpretation and Implications:* An excess charge density of only ~1 electron/m^3 creates a polarizing electric field of 1 mV/m. This field is sufficient to drive currents as shown in Fig. 4-30 that help reduce the mid-magnetotail electric field strength to ~0.1 mV/m.

Follow-on Exercise: Verify the claim in Focus Box 4.3 that ionospheric charge densities of a few tens of ions or electrons per cubic meter create ionospheric electric fields of 1 mV/m.

4.4 Currents and Electric Fields in Near-space

Electrodynamic Fields from Plasma Moving Relative to a Magnetic Field

The previous section describes fields and currents associated with charge accumulation. Now we turn to a different perspective on electric fields—one associated with changes in reference frame. While the laws of physics are the same in all inertial (non-accelerating) reference frames—and thus the net force measured by all inertial observers is equal—the fields associated with the force may differ between any two reference frames. For instance, an observer at rest with respect to a certain experiment may detect only an electric field, whereas an observer moving with respect to the same experiment may see both electric and magnetic fields.

To explain this difference in observations, we construct an experiment, illustrated in Fig 4-33, to observe the force on a particle of charge $+q$ as seen by two observers. One observer is at rest in a reference frame S, and the other is at rest in an inertial frame S', moving with velocity v in the y direction relative to the observer in S. A magnetic field B exists in frame S, and the charged particle is initially at rest in frame S'. Without loss of generality, we align the z-axis with the magnetic field direction. The coordinate axes of both frames are parallel. Assuming that v is much less than light speed, we know that the magnetic fields in S and S' are equal, and we analyze the Lorentz force on the charged particle (as measured in both frames) as follows:

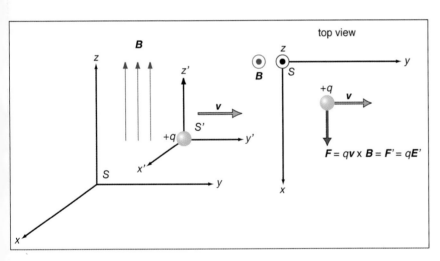

Fig. 4-33. Charged-particle Acceleration as Viewed by Two Inertial Observers. The left panel shows reference frame S' moving at velocity v relative to the frame S. The magnetic field, present in both frames, is aligned with the z-axis (and the z'-axis). The right panel shows a top-down view of the situation, along with a charge ($+q$) held at rest in the S' frame (the charge moves at velocity v relative to S). At the moment of release from its restraint, the net force on the charged particle, identical in both frames, is in the x-direction. *Note:* The unprimed quantities often represent Earth's reference frame.

Suppose the particle is initially restrained by the observer in S', and then released. Because both observers are in inertial frames, they must agree on the magnitude and direction of the force vector acting on the charged particle at the moment it is released. However, the two observers will disagree on the cause of this force. The observer in S sees the charged particle accelerate in the +x direction and describes the Lorentz force to be $F = qv \times B$, because the particle had velocity v (relative to S) upon release. On the other hand, the observer in S' sees the charge accelerate in the +x' direction from rest, and so describes the Lorentz force as purely an electrical force because a magnetic force cannot accelerate a particle from rest. In frame S', the Lorentz force is $F' = qE'$. We know that $F = F'$, so mathematically we write:

$$E' = v \times B$$

Focus Box 4.3: Upper Atmospheric Electric Fields and Scale Analysis [Kelley, 1989]

Here we estimate how much net polarization electric charge is required to generate a sizable electric field in space. What is a sizable field strength? According to Table 4-6, typical ionospheric field strengths are about 1 mV/m, and this amount is enough to greatly influence plasma motions in the upper atmosphere. We estimate from Gauss's Law (Eq. (4-1b)) the amount of charge density needed to create the field.

We use a technique called scale analysis of differential equations. This tool is very powerful for estimating the rough size of unknown quantities in a differential equation. It works by replacing a derivative with an equivalent algebraic expression; for instance, the instantaneous speed of an object can be re-cast in the following way:

$$\frac{dx}{dt} \Rightarrow \frac{x}{t}$$

where x is the distance traveled in the time interval t.

This method is similar to the inverse of the limiting process that converts an average slope into a derivative. We apply the transformation to Gauss's Law as an example. We recall that the ∇ operator has dimensions of 1/length.

$$\nabla \bullet E \approx \frac{E}{L} = \frac{\rho_c}{\varepsilon_o}$$

where L is the scale length over which the electric field varies between the values of zero (no field) and E (the field value produced by the net charge density, ρ_c).

Given a suitable estimate for L, we compute the net charge density required to produce an ionospheric electric field of 1.0 mV/m. For example, in an auroral arc with an L of 1.0 km

$$\rho_c = \frac{\varepsilon_0 E}{L} = \frac{(8.85 \times 10^{-12} \text{C}/(\text{V} \cdot \text{m})) \times 10^{-3} \text{V/m}}{10^3 \text{m} \times 1.6 \times 10^{-19} \text{C/particle}} \approx 55 \text{ particles}/\text{m}^3$$

The required excess of ions or electrons is a few tens per cubic meter. This amount is only a tiny fraction of the minimum ionospheric plasma density of 10^{10} particles per cubic meter, which means that significant electric fields in space are generated from very small charge imbalances (Example 4.15).

Further, if an electric field E exists in frame S, this equation becomes

$$E' = E + v \times B$$

We interpret this equation to mean that the electric field in the two inertial frames differs by $v \times B$.

This result has extremely important applications in space physics. We attribute most of the electric fields measured in the magnetosphere to the relative motion v_{SW} of the solar wind (primed frame) past Earth's unprimed, motionless frame. This electric field is a prominent player in dynamo conversion of the solar wind's kinetic energy into electrical energy in the magnetosphere.

Next we use the electric field relation given above to mathematically describe this dynamo electric field. Assume that $v = v_{SW}$, so that the charged solar wind particles are stationary in this frame. Now, because the solar wind is a very good conductor, we conclude that $E' = 0$; otherwise, the charges wouldn't be stationary. In other words, the solar wind has no embedded electric field in its own rest frame. Therefore, the previous equation becomes:

$$E = -v_{SW} \times B_{SW} \tag{4-15}$$

where

E = interplanetary electric field measured in Earth's reference frame [N/C or V/m]

The motion of the magnetized, highly conducting solar wind past Earth generates an electric field in Earth's reference frame. This electric field is perpendicular to v_{SW} and B_{SW} and accelerates (energizes) particles in Earth's reference frame (the magnetosphere). The net result is that the solar wind does work on and in Earth's near-space environment.

Figure 4-34 brings together the ideas of Secs. 4.3 and 4.4. If the magnetic field in the solar wind has a southward (negative B_z, opposed to Earth's dipole orientation) component, then merging at the dayside magnetopause occurs between the solar wind field and Earth's magnetic field. During such events, the frozen flux conditions that we describe in Sec. 4.3.1 break down. If energy transfers repeatedly, but aperiodically, to the magnetosphere this way, then over the course of several hours or days, magnetospheric currents continually amplify and decay at the magnetopause and in the magnetosphere. The current variations create waves in Earth's internal magnetic field that in turn resonate with charged particles and energize them. Electrons, in particular, with their smaller masses are very susceptible to these processes. Energized electrons in the vicinity of the geosynchronous orbital altitude collide with satellites and embed their charges on satellite surfaces or in components, sometimes causing severe charging and arcing events. (Example 4.3)

Our results are illustrated in Fig. 4-34 in the example below. The solar wind velocity v_{SW} points to the right in the figure. The IMF B_{SW} points downward, which leads to a dynamo connection to Earth's magnetic field. In Earth's reference frame, this dynamo produces an electric field E, pointing out of the page, given by Eq. (4-15).

Pause for Inquiry 4-22

If the IMF B_{SW} points northward, in what direction will the electric field imposed on the magnetosphere point? If B_{SW} points southward, in what direction will the electric field imposed on the magnetosphere point?

EXAMPLE 4.16

Electric Field Strength in the Solar Wind

- *Problem Statement:* Calculate the electric field strength in the solar wind dynamo pictured in Fig. 4-34.
- *Relevant Concepts:* Electric fields in different reference frames
- *Assume:* The magnetized solar wind flows past Earth and carries a magnetic field perpendicular to the flow.
- *Given:* Solar wind speed $|v_{SW}|$: ~ 500 km/s

  IMF field strength $|B_{SW}|$: ~ 5×10^{-9} T
- *Solution:* Using Eq. (4-15), we find

$$|E| \approx (5 \times 10^5 \text{ m/s})(5 \times 10^{-9} \text{ T}) = 2.5 \times 10^{-3} \text{ V/m} = 2.5 \text{ mV/m}$$

E is in the +y direction (out of the page)

- *Interpretation:* An electric field is observed in Earth's rest frame that is not present in the moving frame. The field strength may seem small; however, when the field is summed over the entire dayside contact region, the potential drop (voltage) across the width of the magnetosphere may be well over 100 kV. This electromotive force powers the entire magnetospheric electric circuit, generating a current in the +y direction (out of the page). When the potential difference grows rapidly and remains so, because of sudden increases in v_{SW} or B_{SW}, widespread effects occur in the space environment.

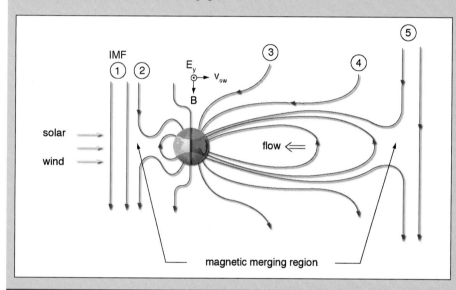

Fig. 4-34. A Side View of the Solar Wind-Magnetosphere Dynamo from Nose to Tail. The numbers on the field lines indicate the sequence of field line motion. The solar wind impinges on the magnetosphere from the left. Its flow energy converts to electrical energy in the magnetosphere's frame of reference at the dayside magnetic merging region. Subsequently, the electrical energy converts back to flow energy in the nightside magnetosphere in the magnetic merging region. In this example the interplanetary magnetic field B_{SW} points downward (–z) and the resulting electric field imposed by the solar wind points out of the page (+y). The dark regions at the poles indicate where the magnetic field has one foot attached to Earth and the other foot attached to the solar wind. [After Kelley, 1989]

Summary

The term *field* describes the assignment of a quantity to every point in space. This chapter describes electric and magnetic fields that modify the surrounding space to exert force on charged masses. The field units are force per unit source. Accumulations of electric and magnetic fields produce a pressure or energy density. We describe the fields for some common static charge and current distributions. An important concept used often in space studies is that of magnetic flux, the product of magnetic field strength and the area it pierces.

In static situations, electric fields energize charged particles, producing currents that lead to magnetic fields. In dynamic situations we need Maxwell's equations to describe the field interactions and configurations, and ultimately how field variations become a field source. The static version of individual Maxwell equations allows us to describe the current sheet configurations that separate strongly sheared magnetic fields. Gauss, Ampere, and Faraday studied the behavior of fields and led Maxwell to conclude that changing magnetic flux induces changing electric flux and vice versa. Together these changes produce electrodynamic waves that travel at light speed. The Poynting flux vector quantifies the amount of energy that electrodynamic waves transfer from one location to another.

We also describe how differential velocity of charged particles creates current in continuous circuits. Ohm's Law expresses the direct relation between current and electric field. Conductivity of the current-carrying medium modulates the amount of current driven by an electric field.

Under typical conditions in plasma, a magnetic field and the plasma elements are tied to each other by the Lorentz force. Plasma and field move together for all motion perpendicular to the field lines, and we call this a frozen-in-flux condition. Motion parallel to the field line is constrained not by the magnetic field, but by electric fields and collisions.

Dynamos are processes that tap the mechanical energy of a system and convert it to electromagnetic energy. When magnetized plasma is subject to motion from external forcing, the magnetic field stretches, shears, and concentrates to form energetic configurations. Nature keeps this process in check through dissipative forces and through an anti-dynamo process known as magnetic merging. Merging, which occurs during the breakdown of a frozen-in flux condition, changes the topology of field lines with opposing directions to splice together to form simpler structures. The energy of the complex field structure is released back to the surrounding plasma.

When plasmas with different magnetic characteristics meet, they form cell-like structures separated by sheets of current. In the near-Earth system, the magnetopause current system and the cross-tail current system connect to Earth's ionosphere via sheets of magnetically field-aligned currents. Within the ionosphere, currents are driven in the resistive medium by mechanical heating on the dayside and by electrical forcing in the auroral zone. Many of the electric fields controlling auroral zone processes have their origin in solar wind–magnetosphere interactions. Motion of the magnetized solar wind past the relatively stationary geomagnetic field energizes Earth's field and plasma. Earth observers detect this interaction as an electric field whose source is the mechanical motion of the solar wind. Hence the solar wind acts as a dynamo to energize Earth's magnetic domain.

Chapter 4 Space Is a Place...with Fields and Currents

Key Words

Ampere's Law for Magnetic Fields with Maxwell's Addition
conductance Σ
conductivity σ
curl
current continuity
displacement current term
divergence
Faraday's Law for Electric Fields
field
flux Φ
frozen-in magnetic flux
Gauss's Law for Electric Fields
Gauss's Law for Magnetic Fields
induction
induction term
ionospheric dynamo
isotropic
Lenz's Law
light speed c
magnetic merging
magnetopause current
merging "X-line"
Ohm's Law
plasma beta
polarization electric fields
Poynting flux vector
Sq current system
static

Notations

Gauss's Law for Electric Fields
$$\text{Div}\mathbf{E} = \nabla \cdot \mathbf{E} = \rho_c/\varepsilon_0$$
$$\oint \mathbf{E} \cdot d\mathbf{A} = Q/\varepsilon_0$$

Faraday's Law for Electric Fields
$$\text{Curl}\mathbf{E} = \nabla \times \mathbf{E} = -\partial \mathbf{B}/\partial t$$
$$\oint \mathbf{E} \cdot d\mathbf{l} = -\frac{d\Phi_B}{dt}$$

Gauss's Law for Magnetic Fields
$$\text{Div}\mathbf{B} = \nabla \cdot \mathbf{B} = 0$$
$$\oint \mathbf{B} \cdot d\mathbf{A} = 0$$

Ampere's Law for Magnetic Fields with Maxwell's Addition
$$\text{Curl}\mathbf{B} = \nabla \times \mathbf{B} = \mu_0 \mathbf{J} + \mu_0 \varepsilon_0 \frac{\partial \mathbf{E}}{\partial t}$$
$$\oint \mathbf{B} \cdot d\mathbf{l} = \mu_0 I + \mu_0 \varepsilon_0 \frac{d\Phi_E}{dt}$$

Relation for electric and magnetic field strength in an electromagnetic wave $|\mathbf{E}| = c|\mathbf{B}|$

Energy flux of an electromagnetic wave (Poynting flux vector) $\mathbf{S} = \dfrac{\mathbf{E} \times \mathbf{B}}{\mu_0}$

Energy density in an electric field $u_E = \dfrac{1}{2}\varepsilon_0 E^2$

Energy density in a magnetic field $u_B = \dfrac{B^2}{2\mu_0}$

Energy density in an electromagnetic wave
$$u_{EB} = \frac{1}{2}\varepsilon_0 E^2 + \frac{B^2}{2\mu_0} = \varepsilon_0 E^2 = \frac{B^2}{\mu_0} = \frac{EB}{\mu_0 c}$$

Average energy density in an electromagnetic wave
$$\langle u_{EB}\rangle = \frac{1}{2}\varepsilon_0 E_{max}^2 = \frac{B_{max}^2}{2\mu_0} = \frac{E_{max}B_{max}}{2\mu_0 c}$$

Conservation of charge
$$\nabla \cdot \mathbf{J} + \partial \rho_c/\partial t = 0; \; \nabla \cdot \mathbf{J} = -\partial \rho_c/\partial t$$
$$I = \frac{\Delta Q}{\Delta t} = \frac{n_{ci}q_i A l}{l/v} = n_{ci}q_i A v = \rho_{ci} A v$$

Volume current density $\mathbf{J}_V = \dfrac{I}{A}\hat{\mathbf{v}} = n_{ci}q_i \mathbf{v} = \rho_{ci}\mathbf{v}$

Ohm's Law $\mathbf{J}_V = \sigma \mathbf{E}$

Plasma beta $\beta = \dfrac{nk_B T}{\left[\dfrac{B^2}{2\mu_0}\right]}$

Relationship of $\mathbf{E}$ in Earth's reference frame to $\mathbf{B}$ in the reference frame of the solar wind $\mathbf{E} = -\mathbf{v}_{SW} \times \mathbf{B}_{SW}$

Answers to Pause for Inquiry Questions

PFI 4-1: $[N] = [C] \,[(m/s)] \times$ units of B. Therefore, B must have units of $[N/(C \cdot m \cdot s)] = [N/(A \cdot m)]$

PFI 4-2: At the surface, $a_g = 9.81$ m/s$^2 = GM/r^2$ so to get $(1/2)a_g$ we need
$$r = \sqrt{2R_E} \approx 1.4 R_E \approx 9020 \text{ km}$$
from the center of the Earth.

PFI 4-3: Positive charge—leave it the same. Negative charge—reverse the arrow directions. Positive-negative pair—draw field lines from the positive charge curving to the negative charge.

PFI 4-4: Since a current sheet necessarily produces a B-field, the absence of the upper B-field would indicate an absence of current (and the absence of the lower B-field as well). If the upper field increases in magnitude, then so does the current (and the lower B-field).

PFI 4-5: a) The electric field associated with an infinite line charge is uniform and outward directed.

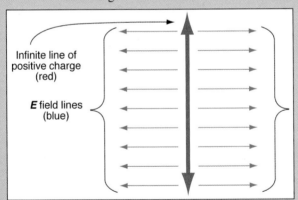

b) The magnetic field associated with a line current encircles the line of current and curls in the same direction the fingers of the right hand curl if the right thumb is aligned in the direction of current flow.

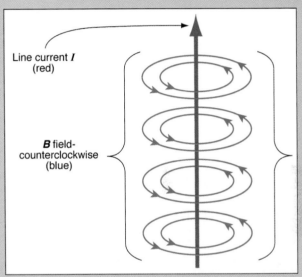

c) The magnetic field for a ring of current encircles the current and points upward on the inside and downward on the outside.

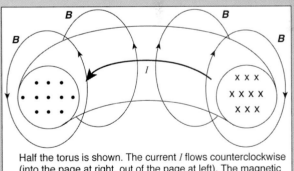

Half the torus is shown. The current I flows counterclockwise (into the page at right, out of the page at left). The magnetic field B inside the torus points upward; outside the torus, it points downward.

PFI 4-6: To form an infinite current sheet, align the strands parallel to each other. The fields between the strands cancel each other.

PFI 4-7: Earth's magnetic south pole is near the geographic north pole, north of Canada. So, the north end of magnetic compasses points to the south magnetic pole because opposite ends of magnets attract each other.

PFI 4-8: d) all of these are true.

PFI 4-9: The fields of the individual current elements (the single line in Fig 4-10) would combine to create a uniform B-field pointing in opposite directions on either side of the current bundles. The fields between the bundle elements cancel each other.

PFI 4-10: Curl

PFI 4-11: The zeros don't necessarily indicate an absence of fields. Where the derivatives are 0, the fields are not changing.

PFI 4-12: When charges are removed, the charge density ρ_c becomes 0 and so, $\nabla \cdot E = 0$. The current density also becomes 0, so dJ = 0. Since J goes to 0, $\nabla \times B$ reduces to $\mu_0 \varepsilon_0 \partial E / \partial t$.

PFI 4-13: Yes, changing electric and magnetic fields create new fields.

PFI 4-14: In Fig. 4-3, the current along the thin sheet flows counterclockwise around the Sun's equator. Eleven years later, the flow will be in the opposite direction, because the inbound and outbound field lines reverse directions.

PFI 4-15: Poynting flux magnitude $S = EB/\mu_0$ (E and B perpendicular). Energy density $u_{EB} = EB/\mu_0 c$. So, dividing S by c makes them equal. Equation (4-10) has units of J/m^3, so dividing by light speed produces $J \cdot s/m^4 = N \cdot m \cdot s /m^4$ = $N \cdot s/m^3$ = (kg $\cdot$ m/s)/m^3 which is momentum density.

PFI 4-16: (#/m^3) C(m/s) = A/m^2.

PFI 4-17: $\sigma = |J|/|E| = [A/m^2]/[N/C] = [A \cdot C/(N \cdot m^2)]$. Also, conductance is conductivity integrated over distance, so Σ has units of $(A \cdot C/(N \cdot m^2))(m) = (A \cdot C/(N \cdot m))$.

PFI 4-18: b) $\beta \ll 1$

PFI 4-19: The force $q\mathbf{v} \times \mathbf{B}$ has units of C(m/s) T = A $\cdot$ m $\cdot$ T. To obtain the force density, we divide by m^3, and get A $\cdot$ T/m^2. $\mathbf{J} \times \mathbf{B}$ has units of (A/m^2) T, the same as force density.

PFI 4-20: c) Into the page

PFI 4-21: At the magnetospheric flanks as shown in Fig. 1-25.

PFI 4-22: For a northward IMF, the E-field points from dusk to dawn. For a southward IMF, the E-field points from dawn to dusk.

References

Kelley, Michael C. 1989. The Earth's ionosphere: Plasma physics and electrodynamics. *International Geophysics Series*, Vol. 43. San Diego, CA: Academic Press.

Figure References

Schey, Harry M. 1973. *Div, Grad, Curl, and all That: an Informal Text on Vector Calculus*. New York, NY: W.W. Norton and Company.

Smith, Edward J., Bruce T. Tsurutani, and Ronald L. Rosenberg. 1978. Observations of the Interplanetary Sector Structure up to Heliographic Latitudes of 16°: Pioneer 11. *Journal of Geophysical Research* Vol. 83. American Geophysical Union: Washington, DC.

Wangsness, Roald. 1986. *Electromagnetic Fields*. 2nd Edition. Hoboken, NJ: Wiley Publishing Company.

Further Reading

Hargreaves, John K. 1992. *The solar-terrestrial environment*. Cambridge, England: Cambridge University Press.

Serway, Raymond A. 1997. *Principles of Physics*. Fort Worth, TX: Saunders College Publishing.

Wolfson, Richard and Jay M. Pasachoff. 1999. *Physics with modern physics for scientists and engineers*. 3rd Edition. Reading, MA: Addison-Wesley.

The Quiescent Solar Wind: Conduit for Space Weather

5

UNIT 1. SPACE WEATHER AND ITS PHYSICS

You should already know about...

- Sound waves and Mach number
- Electric potential energy and thermal energy (Chap. 2)
- The Lorentz force (Chap. 3)
- The structure of the Sun (Chap. 3)
- Fields and their interactions with matter (Chap. 4)
- Definitions of thermal, dynamic, and magnetic pressure (Chaps. 2 and 4)
- Plasma β (Chap. 4)
- Frozen-in magnetic flux and magnetic merging (Chap. 4)

In this chapter you will learn about...

- The expanding solar atmosphere beyond three solar radii—the heliosphere
- Characteristics and acceleration of the solar wind
- The quasi-stationary solar wind
- Mean free path and collision cross section
- Supersonic and Alfvénic speeds
- Interplanetary magnetic field (IMF)
- Solar wind plasma and IMF plots
- The interplanetary current sheet and sector structures
- Additional aspects of coronal holes
- Earth's bow shock
- The edge of the Sun's influence

Outline

5.1 **The Quiescent Solar Wind**
    5.1.1 Solar Wind Plasma Characteristics
    5.1.2 The Supersonic and Super-Alfvénic Solar Wind
    5.1.3 Interplanetary Magnetic Field Characteristics
    5.1.4 The Current Sheet and Magnetic Sectors

5.2 **Solar Wind Interactions in the Heliosphere**
    5.2.1 Solar Wind Deceleration at a Magnetized Barrier
    5.2.2 Solar Wind Beyond the Inner Planets

Chapter 5 The Quiescent Solar Wind: Conduit for Space Weather

5.1 The Quiescent Solar Wind

The *solar wind* is the supersonic outflow of plasma from the upper solar atmosphere. The plasma carries with it strands and bundles of the solar magnetic field called the *interplanetary magnetic field (IMF)*. Together these flows constitute the *interplanetary medium*. Extending far beyond the orbits of the outermost planets, the interplanetary medium terminates in a discontinuity called the *heliopause*, where it interfaces with the weakly ionized interstellar medium. In this chapter, we explore the physics of the quiescent interplanetary medium and some of the space weather it causes.

5.1.1 Solar Wind Plasma Characteristics

Objectives: After reading this section, you should be able to...

- Describe pre-space age evidence for the existence of the solar wind
- Describe basic characteristics of the near-Earth quasi-stationary solar wind
- Apply the concept of mean free path to collisions
- Explain why the solar wind is considered collisionless

Observations and Statistics of the Bulk Solar Wind near Earth

Because the solar wind is so tenuous, its minimal radiation is little use in determining its presence or distinguishing its important features. Nonetheless, scientists have long suspected its existence. Before the Russian spacecraft LUNIK made the first observations of the interplanetary medium in 1959, the most readily available intimation of the solar wind was from its interaction with comets. Sir Isaac Newton noted that comet tails preferentially pointed away from the Sun. Others noticed that comet tails often kinked or split. The split tails (Fig. 5-1) are traces of ions, in addition to photons and dust particles, flowing outward from the Sun. The ion flow aligns with the variable direction of the solar wind. We now know that kinks develop when solar wind structures interact with the comet tail. In 1896 Norwegian scientist Kristian Birkeland suggested that electrically charged particles, hurled from the Sun, somehow linked into Earth's polar regions, causing the aurora borealis.

In 1962, proof of persistent solar particle flow came as NASA's Mariner II spacecraft probed interplanetary space on its way to Venus. Soon a series of experiments yielded detailed, but not always continuous, measurements of the solar atmosphere flowing past Earth. During five Apollo missions to the Moon, astronauts exposed metal foils to the solar wind and returned them for laboratory analysis. The Helios and Voyager spacecraft sampled near and far: Helios to within 0.3 AU of the Sun, and Voyager to 100 AU and beyond. The Ulysses spacecraft sampled out of the ecliptic plane, exploring the three-dimensional heliosphere. Since 1996, the Advanced Composition Explorer (ACE) satellite has orbited upwind of Earth, providing nearly full-time monitoring of the solar wind material headed for Earth.

Table 5-1 provides statistics about the solar wind plasma gleaned from over 40 years of near-Earth measurements. These data reveal that the solar wind is hot, tenuous, and magnetized. It's also highly conducting, virtually collisionless, and

Fig. 5-1. Evidence of the Solar Wind: The Two Tails of Comet Hale-Bopp. The whitish-turquoise dust tail consists of larger particles of cometary debris in orbit about the Sun. These particles are of a size that preferentially reflects white and turquoise light. They lag their parent body, because solar photons (radiation) exert a force directed away from the Sun. They act as a dusty tracer of where the comet has been. The blue plasma tail consists of tiny ions that flow away from the Sun in the direction of the solar wind. The ions preferentially reflect shorter wavelength blue light. *(Courtesy of NASA)*

supersonic. The solar wind has three states—slow, fast, and transient—that are linked to origins near streamers, in coronal holes, and with coronal mass ejections (CMEs), respectively. Each of these states creates unique modes of interaction with the near-space environment. We consider the slow and fast quasi-stationary flow in this chapter and discuss the transient and highly variable solar wind in Chap. 10.

Table 5-1. Properties of the Non-transient Solar Wind Near Earth's Orbit. This table lists the properties for the quasi-stationary fast and slow solar wind. For convenience, scientists usually report solar wind speed in kilometers per second and density in the number of ions per cubic centimeter. The density value refers to the number of positive ions. A given volume may contain slightly more electrons because some ions lose more than one electron. [Schwenn, 1990 and Gosling, 2007]

Quiescent Flow Type Property	Fast Wind	Slow Wind	Average Wind
Proton temperature	1.5×10^5 K	5×10^4 K	1.2×10^5 K
Electron temperature	7.5×10^4 K	2×10^5 K	1.4×10^5 K
Bulk speed	450 km/s–800 km/s	250 km/s–450 km/s	468 km/s
Velocity variance	5%	25%–50%	~15%
Sound Mach number	11–20	6–11	11
Alfvén Mach number	5–10	3–5	5
Ion # density	~3 ions/cm^3	~10 ions/cm^3	8.7 ions/cm^3
Composition	~95% H, 4% He, ~1% other ions	~94% H, ~5% He, ~1% other ions,	~95% H, 4% He, ~1% other ions
Magnetic field	$|B|$ ~5 nT	$|B|$ < 5 nT	$|B|$ = ~6.2 nT[*]
Origin at Sun	Coronal holes	Streamer boundaries	--
% Occurrence	44%[*]	34%[*]	--

[*] Transients in the solar wind account for ~23% of the flow and the high value of magnetic field strength in the average solar wind.

Electrons, protons, alpha particles (twice-ionized helium), and heavy ions flow from the Sun at speeds of hundreds of kilometers per second. Approximately 10^9 kg of material leaves the Sun each second (Example 5.1). Data from the Solar and Heliospheric Observatory (SOHO) satellite and the Hinode mission suggest that the solar wind originates in funnel-shaped structures rooted in the magnetic network (Fig. 5-2). Some of the outflow may come from edges of active regions. The fast solar wind appears to originate about 20 Mm above the photosphere. Typical speeds at that location are ~10 m/s. The fast solar wind accelerates to speeds of ~800 km/s within 10 R_S. The slow solar wind, which originates in the streamer belt, accelerates within the first 20 R_S and reaches speeds near 300 km/s.

As part of the extended hot corona, solar wind ions and electrons carry the imprint of their energetic origin out into the heliosphere. To some degree their energy dissipates with solar wind expansion. Without additional heat sources (adiabatic expansion) the temperature should decrease roughly as $T(r) \propto 1/r^{2/3}$. If the random motion of escaping charged particles conducts heat away from the Sun,

Chapter 5 The Quiescent Solar Wind: Conduit for Space Weather

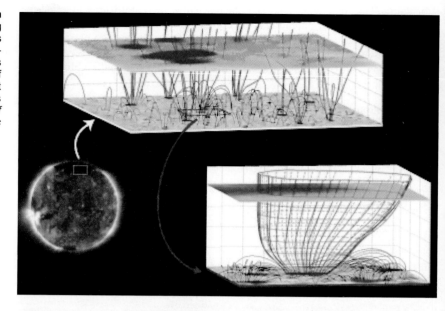

Fig. 5-2. The Source of the Solar Wind in Regions of Open Magnetic Fields. By tracing the flow of highly ionized neon, scientists observed the solar wind moving out of funnel-shaped magnetic fields anchored in the lanes of the magnetic network above the surface of the Sun. In the image at lower right we see that crowding of neighboring magnetic loops helps to create the funnel constriction. *(Courtesy of ESA/NASA – the Solar and Heliospheric Observatory (SOHO), After Tu et al., 2005)*

EXAMPLE 5.1

Mass Loss from the Sun

- *Problem Statement:* Verify that the mass loss rate from the Sun due to the solar wind is on the order of 10^9 kg/s, given that the number density observed in the solar wind near Earth is ~5 proton-electron pairs per cubic centimeter.

- *Relevant Concept:* Mass conservation and flow continuity—the time rate of change of mass in a volume must equal the mass flowing through the volume's surface.

- *Given:* Solar wind density near Earth is $n_{sw} = 5$ ions/cm^3

- *Assume:* Steady state solar wind speed of 400 km/s; spherical symmetry. Electron mass is negligible compared to ion mass.

- *Solution:* By invoking flow continuity we write the rate of change of mass in a volume in terms of mass flux through the surrounding surface. That is, if mass leaves a volume of space, it must flow through the surrounding surface. Without loss of generality, we assume that the surrounding surface is a sphere.

$$\frac{\partial m}{\partial t} = \int_V \frac{\partial \rho}{\partial t} dV = -\int_S \rho v \, \mathrm{d}s = -\rho v (4\pi r^2)$$

At one AU the mass flux is dominated by the proton mass.

$$-\rho v (4\pi r^2) = -\left[\left(\frac{5 \times 10^6 \text{ protons}}{\text{m}^3}\right)\left(\frac{1.67 \times 10^{-27} \text{ kg}}{\text{proton}}\right) \right.$$
$$\left. (400 \times 10^3 \text{ m/s})(4\pi(1.5 \times 10^{11} \text{ m})^2)\right]$$

$$-\rho v (4\pi r^2) = -9.44 \times 10^8 \text{ kg/s} \approx -10^9 \text{ kg/s} = \frac{\partial m}{\partial t}$$

- *Implication and Interpretation:* The Sun sends a billion kilograms of material (roughly the mass of 20 battleships) via the solar wind into interplanetary space each second. In Earthly terms this amount is significant, but for the Sun, even over thousands of years this is insignificant compared to the solar mass.

Follow-on Exercise 1: Compare the mass loss rate into the solar wind with the mass loss rate from coronal mass ejections (CMEs). Assume one CME per day. Use the values from Sec. 1.3.

Follow-on Exercise 2: Determine the mass sent into interplanetary space each year by the solar wind. Compare this to the mass in Mar's larger moon, Phobos, and to that of Earth's Moon.

5.1 The Quiescent Solar Wind

then the solar wind should cool more slowly with distance ($T(r) \propto 1/r^{2/7}$). The actual cooling rate is between the two. On average, between the Sun and Earth, electrons appear to cool slowly, while ions seem to cool at roughly the adiabatic rate. Table 5-1 shows that the particles arrive at Earth with a temperature in the range of 10,000 K–100,000 K. Although these temperatures seem to imply a high thermal pressure, the particle number density falls off as $\sim 1/r^2$, so near Earth the solar wind exerts minimal thermal pressure.

As a fluid, the solar wind exerts a dynamic (or ram) pressure on heliospheric objects, including Earth's magnetosphere. The pressure has two components: speed and density. Figure 5-3 shows the distributions of speed and density for the near-Earth solar wind. Rather than being bell-shaped, the speed distributions are asymmetric, with a rough edge at low values and a long tail at high values. The density distribution has a skew toward lower values but never reaches zero. The solar wind always has some matter in it. Additional processing of the data reveals that the solar wind has two modes: fast and slow. Solar wind density correlates inversely with the speed variations so that fast flow tends to appear with low density and slow flow with high density. In a rather interesting balancing act, the Sun manages to keep the energy flux density needed for each flow type roughly equal. The energy flux in slow flow goes predominantly to lifting the solar wind mass out of the Sun's gravitational field. The energy flux in the fast (and less dense) solar wind mostly accelerates the particles to high speeds.

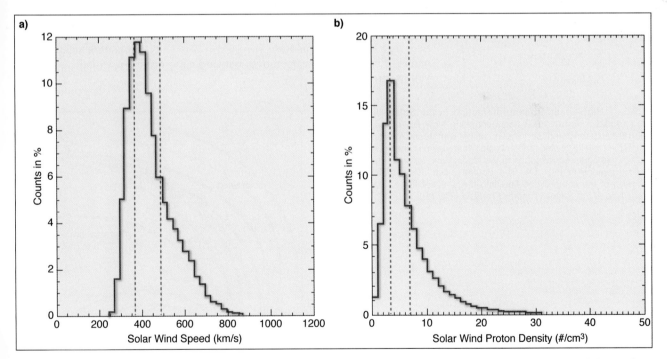

Fig. 5-3. Solar Plasma Distributions for the Interval 1990-2007. a) Solar Wind Speed. This plot gives solar wind speed observations in bins of 25 km/s at Earth (1 AU). **b) Solar Wind Number Density.** This plot shows solar wind ion density observations at Earth in bins of 1 proton per cubic centimeter. The dashed lines indicate the limits for each 33% of the distributions. Data from several solar wind monitors were merged to create these plots. All three types of solar wind (slow, fast, and transient) contribute to the distributions. *(Data from NASA OMNIWeb)*

Chapter 5 The Quiescent Solar Wind: Conduit for Space Weather

Fig. 5-4.   Flow Regimes in the Solar Wind. Model output shows high-speed flow in the dark blue and red regions. Moderate and slow flows emanate from the green areas. The high-speed flow regions in each hemisphere have different colors to indicate that they follow magnetic fields with opposite directions. *(Courtesy of Munetoshi Tokumaru at The Solar-Terrestrial Environment Laboratory (STELab) of Nagoya University in Japan)*

> **Pause for Inquiry 5-1**
>
> Figure 5-3 displays solar wind number density. Assuming the solar wind is made up of hydrogen atoms, what is the range of mass density in the solar wind?

Flow Variations

Figure 5-4 depicts a cut-away volume view of solar wind model output. As we describe in Chap. 3, the out-flowing plasma comes from closed and open magnetic regions of the solar corona. In general, the solar regions above 60° latitude, where the magnetic field is open, generate fast (high-speed, low density) solar wind from coronal holes. The blue and red regions in Fig. 5-4 (as well as the green and red regions in Fig. 3-34b) represent high-speed flow from coronal holes. At streamer belt locations in regions of closed magnetic field near the solar equator, the solar wind escapes at a lower speed with higher density.

During the declining phase of the solar cycle, the solar magnetic field restructures. The high-speed wind from high-latitude coronal holes expands into and overtakes the slow flow regions near the ecliptic plane, thus allowing the slow and fast wind to interact. During such times the solar wind contains a succession of *high-speed streams (HSS)* with intervening low-speed solar wind. Figure 5-5 illustrates an ecliptic cross section, in which fast-moving plasma overtakes slower-moving steamer belt plasma ahead and behind. As a result, a series of compression-rarefaction zones develops. The pattern co-rotates with the Sun, as a *co-rotating interaction region (CIR)*. Only the pattern rotates—each parcel of solar wind plasma moves outward almost radially, as indicated by the yellow arrows.

Fig. 5-5.   High-speed Stream Interaction in the Solar Wind. This artist's rendition portrays fast flow overtaking slow flow ahead of and behind the stream interface. The resulting compression region plows through the radially flowing ambient solar wind plasma. The compression region in this figure corresponds to the leading and trailing edges of the red flow regions in Fig. 5-4. The double black arrow indicates the expansion of the compressed flow into the ambient flow. *(Courtesy of Victor Pizzo at the Space Weather Prediction Center)*

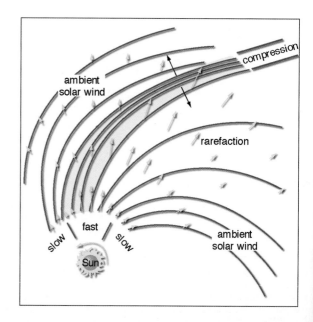

5.1 The Quiescent Solar Wind

Pause for Inquiry 5-2

Using Fig. 5-5, draw speed and density profiles for flow along an imaginary line that extends from the top "ambient solar wind" label to the bottom "ambient solar wind" label.

Particle Collisions in the Solar Wind

The data in Table 5-1 reveal that the solar wind is very dilute plasma, suggesting that particle collisions beyond the corona are rare. In Example 5.2 we estimate that the time between collisions is roughly comparable to the time for the solar wind to expand from the Sun to one AU. This is the basis for saying that the solar wind is nearly collisionless plasma. With few collisions, the solar wind easily conducts heat and charged particles away from the Sun. In fact, collisions are so rare that electrons, protons, and other species in the solar wind develop their own temperature, energy, and momentum balances, as summarized in Table 5-1. Modelers must account for this decoupling in their solar wind simulations.

Pause for Inquiry 5-3

Why should the interaction radius (effective length) decrease as temperature increases?

Despite the limited number of collisions, the solar wind generally behaves as a fluid, because electric and magnetic effects provide it with the characteristics of a collisional fluid. Rather than billiard-ball-like collisions, the charged particles engage in path deflections caused by plasma instabilities, gyro-motions of particles tied to magnetic field lines, and occasional interactions with other charged particles.

We quantify this collisionless condition by defining the mean free path of a particle in a non-ionized gas and then extending the idea to plasma. The mean free path is the average distance a particle travels between collisions. For hard-body collisions where the cross section of the individual particles is σ, we calculate the mean free path by recognizing it as the distance a test particle travels into the gas before it has roughly a 100% chance of colliding. Figure 5-6 illustrates the concept.

We consider a tube with a cross-sectional area A and length l filled with individual spheres, each with a radius r_s. The probability of the test particle colliding with material in the sphere is the ratio of the cross-sectional area of N particles to the endcap's surface area

$$\frac{\text{cumulative cross-sectional area of the } N \text{ particles in the volume}}{\text{surface area of the endcap}}$$

$$= \frac{n(\pi r_s^2)(A \times l)}{A} = nl(\pi r_s^2) = nl\sigma$$

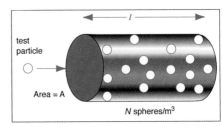

Fig. 5-6. Mean Free Path. In a volume of N individual spheres, each with a cross section of πr_s^2, the cumulative cross section is $N\pi r_s^2$. When a test particle enters the volume, it travels straight until it hits another particle. We define the mean free path as the average distance between collisions.

where

n = particle number density [#/m³]

l = mean free path (distance) the test particle travels to have effectively a "100%" probability of collision [m]

σ = each sphere's cross-sectional area [m²]

In the event of a collision, the probability of collision is one

$$nl\sigma = 1$$

Rearranging the equation for a collision event gives

$$l = \frac{1}{n\sigma} \tag{5-1}$$

Applying this idea to plasma: if a path deflection comes from a charged particle q_1, passing close to another (q_2) (for simplicity, we assume q_2 is so massive it doesn't move), we call the interaction a Coulomb collision. The deflecting particle has an effective cross-sectional area given by πr_e^2, where the effective radius r_e is an interaction radius. In Example 5.3 we estimate this radius by assuming it is the distance at which the Coulomb (electric) potential energy roughly equals the kinetic energy (thermal energy). Table 5-2 lists typical mean free paths for space environment plasmas and non-ionized gases.

Pause for Inquiry 5-4

Is the Coulomb mean free path in the high-speed solar wind longer or shorter than in low-speed flow?

Table 5-2. **Characteristics of Plasmas and Non-ionized Gases.** Here we list examples of densities, temperatures, mean free paths for charged and neutral particles for neutral particles.

Plasma	Density [# ions/m³]	Temperature [K]	Mean Free Path Length: Coulomb [m]
Upper photosphere	10^{18}	5×10^3	10^{-2}
Chromosphere	10^{16}	10^4	10^3
Corona	10^{13}	$1\text{-}2 \times 10^6$	10^6
Solar wind	10^1	10^5	10^{10}
Non-ionized Gases	Density [# particles/m³]	Temperature [K]	Mean Free Path Length: Collisions [m]
Atmosphere at sea level	2×10^{25}	300	10^{-8}
Base of thermosphere	10^{19}	200	10^{-1}
Mid-thermosphere	10^{13}	1000	10^4

5.1 The Quiescent Solar Wind

EXAMPLE 5.2

Mean Free Path Length in the Solar Wind

- *Problem Statement:* Estimate the mean free path length in the solar wind due to Coulomb collisions.
- *Relevant Concept:* Electric potential energy between two point charges; mean free path length.
- *Given:* The charge is 1.6×10^{-19} C, solar wind temperature $T = 1 \times 10^5$ K, solar wind number density $n_{sw} = 7$ ions /cm^3 and results from Focus Box 5.2.
- *Assume:* The distance scale is associated with an effective length r_e where the Coulomb (electric) potential energy roughly equals the kinetic (thermal) energy.

potential energy $\approx$ *thermal energy*

$$PE = \frac{q_1 q_2}{4\pi\varepsilon_0 r_e} = \frac{q^2}{4\pi\varepsilon_0 r_e} = k_B T$$

- *Solution:*

The effective length is a function of temperature T.

$$r_e = \frac{q^2}{4\pi\varepsilon_0 k_B T} = \frac{(1.6 \times 10^{-19} \text{C})^2}{4\pi(8.85 \times 10^{-12} \text{C}^2/(\text{N} \cdot \text{m}^2))(1.38 \times 10^{-23} \text{J/K})T}$$

$$= \frac{1.67 \times 10^{-5} \text{m} \cdot \text{K}}{T(\text{K})} = \frac{1.67 \times 10^{-5} \text{m} \cdot \text{K}}{10^5 \text{K}} = 1.67 \times 10^{-10} \text{m}$$

The Coulomb mean free path is

$$l = \frac{1}{n\sigma} = \frac{1}{n\pi r_e^2} = \frac{1}{7 \times 10^6 \text{ions/m}^3 \times \pi \times (1.67 \times 10^{-10} \text{m})^2} \approx 1.63 \times 10^{12} \text{m}$$

where $\sigma [\text{m}^2]$ is the collision cross section.

- *Implication and Interpretation:* With the assumptions used here, the electric mean free path in the solar wind, sometimes called the *Landau length (L)*, is about 10 AU. Coulomb collisions between particles in the solar wind from the Sun to 1 AU are unlikely. In reality magnetic deflections and particle interactions with solar wind structures tend to reduce the mean free path, so that a few collisions occur in the solar wind between the Sun and Earth. Hence the mean free path value of 10^{10} m shown in Table 5-2.

Follow on Exercise: Shock fronts in advance of some coronal mass ejections (CMEs) may have densities of $n = 50$ ion-electron pairs/cm^3. Determine the Landau length for these structures.

5.1.2 The Supersonic and Super-Alfvénic Solar Wind

> **Objectives:** After reading this section, you should be able to...
> - Describe the acceleration profile of the solar wind and the origin of the supersonic solar wind
> - Explain the difference between solar wind speed and solar wind temperature
> - Calculate sound speed and Alfvén speed
> - Locate the L1 libration point between the Sun and Earth
> - Interpret an ACE solar wind plasma chart

Solar Wind Acceleration

In Sec. 3.1.2, we consider the hydrostatic one-dimensional motion of solar fluid parcels. For isothermal, unaccelerated motion, momentum conservation alone allows an adequate description of particle motion. To describe accelerated motion, modelers need information from other conservation equations. During the 1950s, Sydney Chapman applied momentum conservation (the barometric equation, Eq. (3-7)) along with energy conservation to model the solar atmosphere. His conclusions implied that the solar atmosphere is gravitationally bound to the Sun, with only a tiny fraction of particles having sufficient energy to escape (Ex. 5.3).

However, in the mid-1950s, Eugene Parker, now widely recognized for formulating modern solar wind theory, noticed a mismatch between the idea of a gravitationally bound solar atmosphere and dynamic comet tails. He included conservation of mass and a simple form of conservation of energy to describe the accelerated solar wind. His results suggest that stars similar to the Sun (ones with the appropriate temperature and gravitational variations in their atmosphere) accelerate gases outward from their atmosphere. The acceleration is similar to the gas flow through a rocket nozzle that produces supersonic flow under the proper conditions.

The proper conditions for the upper solar atmosphere are associated with high temperatures in the much-extended corona. Conservation of mass requires that the density fall off as $1/r^2$. If the density decreases faster than $1/r^2$ (which it does when T is high and g falls off rapidly with distance), then the radial velocity $v(r)$ must increase to maintain constant mass flow. The result is outward acceleration between 5 R_S and 20 R_S.

The physical mechanism responsible for the accelerated solar wind is the pressure gradient between the corona (high pressure) and points in distant space (low pressure). Figure 5-7 illustrates the difference between Earth's atmosphere (or a non-accelerating stellar atmosphere) that is in hydrostatic equilibrium and the accelerated atmosphere of the Sun. Since gravity varies as $1/r^2$, its influence decreases rapidly with altitude. In an extended, hot stellar atmosphere, the pressure at the base of a parcel is much greater than the pressure at the top. The difference is so large that gravity cannot restrain the parcels.

Rather than deriving the Parker solution here, we present a flow chart for the solution. Figure 5-8 illustrates the information needed to solve for flow speed and acceleration as a function of distance from the Sun. Details of the solution require spherical coordinates and the simultaneous solution of several equations. The

EXAMPLE 5.3

Particle Escape from the Sun

- *Problem Statement:* Estimate the random thermal speed of particles in the outer corona ($2R_S$) and determine if this speed is sufficient for escape from the Sun's gravitational field.

- *Relevant Concepts:* Thermal energy and gravitational potential energy

- *Assume:* Coronal temperature $T = 2 \times 10^6$ K and charged particles are constrained by the magnetic field to have only two degrees of freedom in their motion, so thermal energy $= k_B T$

- *Solution:* 1) Relate thermal energy to random motion of individual particles (protons in this case).

  2) Equate gravitational potential energy to random motion to determine the escape speed.

- *Interpretation:* Thermal escape of hot particles is not the source of the solar wind. Escape speed at the outer corona is ~4.4×10^5 m/s. The average thermal speed of a proton (1.82×10^5 m/s) from the solar corona is not sufficient to allow average-energy protons to escape. Of course, random collisions could boost the energy of individual protons, so some could escape and fill the heliosphere with a tenuous plasma. Even so, no mechanism associated with purely thermal processes exists that would give the protons an average speed of 440 km/s (4.4×10^5 m/s).

- $\frac{1}{2}mv_{th}^2 = k_B T$

  $v_{th} = \sqrt{\frac{2k_B T}{m}} = \sqrt{\frac{2(1.38 \times 10^{-23} \text{J/K})(2 \times 10^6 \text{K})}{1.67 \times 10^{-27} \text{kg}}} = 1.82 \times 10^5 \text{m/s}$

- $\frac{1}{2}mv_{esc}^2 = \frac{GMm}{r}$

  $v_{esc} = \sqrt{\frac{2GM_S}{2R_S}} = \sqrt{\frac{2(6.67 \times 10^{-11} \text{N} \cdot \text{m}^2/\text{kg}^2)(2 \times 10^{30} \text{kg})}{1.4 \times 10^9 \text{m}}} = 4.4 \times 10^5 \text{m/s}$

---

result is the multi-class solution set shown in Fig. 5-9. The figure depicts the solutions in terms of ratios. The vertical axis is the ratio of the solar wind speed to sound speed. The horizontal axis is the ratio of distance from the base of the corona to the *critical distance*, where the solar wind becomes supersonic.

Not all of the mathematical solutions are physical, and of those that are physical, only one class of solutions is consistent with observations. Class I and II solutions produce non-physical results. They give two values of the solar wind at the same location, or no solutions at all in some locations. The class III curves suggest that the solar wind is supersonic at all locations, even close to the solar surface, which is inconsistent with observations. The class IV results indicate the solar wind is subsonic in all locations, which is also inconsistent with observations. It's appropriate for very massive stars but not for the Sun. The class VI solution is supersonic at the Sun and subsonic at Earth—again inconsistent with observation.

We are left with class V solutions, which show the solar wind as subsonic near the Sun and supersonic beyond a critical radius. A class V solution yields a family of curves for monotonically increasing solar wind speed. Figure 5-10 illustrates examples of a class V solution for various initial coronal temperatures. Most of the acceleration occurs within approximately 20 R_S.

Chapter 5 The Quiescent Solar Wind: Conduit for Space Weather

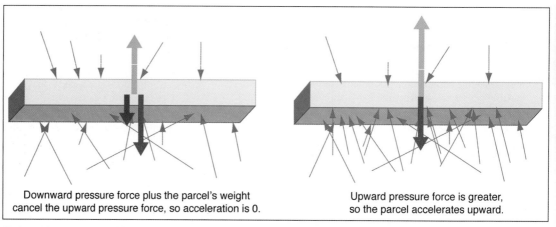

Downward pressure force plus the parcel's weight cancel the upward pressure force, so acceleration is 0.

Upward pressure force is greater, so the parcel accelerates upward.

Hydrostatic Atmosphere (Earth)
- Gravity is assumed to vary slowly with altitude
- More collisions transfer momentum to the bottom side than to the topside
- The parcel would accelerate upward if not for the downward tug of gravity
- In a hydrostatic atmosphere, gravitational force helps balance the pressure gradient force

Non-Hydrostatic Atmosphere (Corona of a "Cool" Star such as the Sun)
- Gravity is not constant; it decreases rapidly over the depth of the parcel, allowing a large density gradient
- More collisions transfer momentum to the bottomside than the topside
- Pressure gradient (thermal push) is greater than gravitational pull
- The result is a net outward acceleration
- Because of the acceleration, the speed ultimately exceeds local sound speed, resulting in a supersonic solar wind

Fig. 5-7. Comparison of a Slab of Gas in Earth's Atmosphere with a Slab in the Solar Atmosphere. Pressure gradients in the hot, extended solar atmosphere accelerate the gas to supersonic speeds. The variation of gravity is very important in determining the nature of the atmosphere. A rapid decrease with height supports the possibility of atmospheric escape.

Fig. 5-8. Concept Map for Solving for Solar Wind Speed. This flow chart traces the solution for the solar wind speed as a function of distance from the corona. (Here r is distance from the Sun's center; ρ is density of the Sun's constituents; v is particle velocity; F_{ext} is the external forces; m is mass; a is acceleration; F_g is the force of gravity; and F_{pg} is the pressure gradient force.)

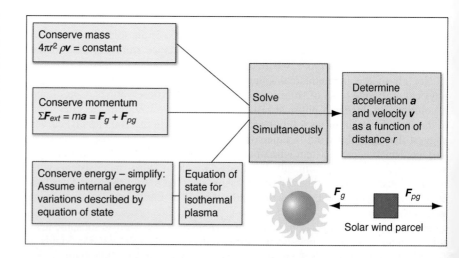

5.1 The Quiescent Solar Wind

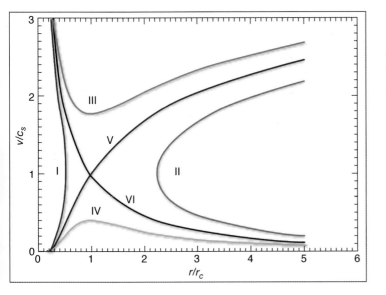

Fig. 5-9. Possible Solutions to Parker's Simplified, Isothermal Solar Wind Problem for Relative Speed and Distance. This plot gives multiple solutions in terms of the flow speed divided by the sound speed (Mach number) on the vertical axis. The horizontal axis indicates the relative location at which supersonic speed is achieved, in terms of the critical distance. Six different solution classes exist. Only the Class V solution is consistent with observations and produces the needed pressure to stand off the interstellar wind. Figure 5-10 presents more details of the Class V solution. (Here v = flow speed; c_s = sound speed; r is where v becomes supersonic; r_c is the critical distance.) [Parker, 1958]

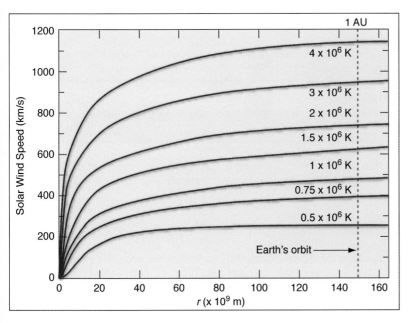

Fig. 5-10. Class V Solution of Solar Wind Speed for Different Coronal Temperatures. This plot shows solar wind speed solutions for several initial temperatures in the solar corona. The horizontal axis is distance from the base of the corona. Each temperature is for an isothermal atmosphere. *(Courtesy of Christopher Russell at UCLA and Eugene Parker [1963])*

Although the solar wind reaches an equilibrium velocity, between 400 and 500 km/s for typical coronal temperatures, the exact solution is a function of the initial temperature conditions at the Sun. Depending on the flow source (coronal hole or coronal streamer) any of the curves in Fig. 5-10 may be appropriate for the solar wind at Earth. Focus Box 5.1 describes solar wind measurements in the vicinity of Earth.

Example 5.4 combines the mass loss rate from Example 5.1 with the solar wind speed to show that the kinetic power of the solar wind at one AU is roughly 10^{20} W. The disk of Earth's magnetosphere intercepts about ten-millionth of this energy. However, kinetic power is only part of the story; radiative and electromagnetic power also flows to Earth in the interplanetary medium.

Pause for Inquiry 5-5

What do the data in Fig. 5-10 indicate about the pressure difference between 1 AU and 5 AU?

Pause for Inquiry 5-6

What do the data in Fig. 5-10 suggest about the variation of flow speed over the course of a solar cycle?

EXAMPLE 5.4

Solar Wind Kinetic Power

- **Problem Statement:** Determine the Sun's rate of kinetic energy loss to the solar wind
- **Relevant Concepts:** Kinetic energy and power
- **Given:** Mass loss rate is 10^9 kg/s and flow rate is 425 km/s
- **Assume:** The flow speed is constant

- **Solution:** We compute the kinetic energy loss rate as $\frac{d}{dt}\left(\frac{1}{2}mv^2\right)$

$$\text{Loss rate} = \frac{d}{dt}\left(\frac{1}{2}mv^2\right) = \frac{1}{2}v^2\frac{dm}{dt}$$

$$\text{Loss rate} = 0.5\ (4.25 \times 10^5 \text{ m/s})^2\ (10^9 \text{ kg/s}) = 9.0 \times 10^{19}\text{ W} \approx 10^{20}\text{ W}$$

- **Interpretation:** This process is very powerful compared to most Earth processes; however, the kinetic power is a small fraction of the power dissipated by a solar flare or a coronal mass ejection. (See Chap. 1 examples.)

Follow-on Exercise 1: Compare the kinetic energy loss rate to the solar radiation loss rate.
Follow-on Exercise 2: Assume the magnetosphere has a radius of 15 R_E. Show that it intercepts about 10^{13} W of solar wind kinetic energy.

Now that we know the solar wind is very fast in the vicinity of Earth, we want to relate its speed to other natural speed characteristics: the sound speed and the magnetic wave speed.

Sound Speed

The transmission of sound depends on collisions, with the average distance between collisions and the average frequency of collisions being important parameters. We already know that the solar wind lacks the mechanical collisions and Coulomb collisions needed to support information transfer, but another possibility exists—magnetic "collisions." For ideal (thermalized) plasma, the average kinetic speed $v_\perp$ of particles gyrating in a direction perpendicular to the magnetic field equates roughly to the sound speed (c_S). Equating gyrational kinetic energy to the approximate thermal energy ($k_B T$) gives

$$\frac{1}{2}mv_\perp^2 = k_B T$$

$$v_\perp \approx \sqrt{\frac{2k_B T}{m}} \approx c_s \qquad (5\text{-}2)$$

We know that the solar wind is very fast, but how do we determine if it is supersonic? We form the ratio of flow speed to sound speed (or the ratio of ordered motion to random motion, as shown in Fig. 5-11). Example 5.5 shows that the speed of the solar wind bulk flow at Earth is about 10 times the sound speed ($v/c_s = M = 10$), that is, the solar wind is supersonic at Mach 10. Space weather modelers and forecasters consider bulk flow speed, thermal speed, and density to be fundamental parameters needed to describe the state of the solar wind.

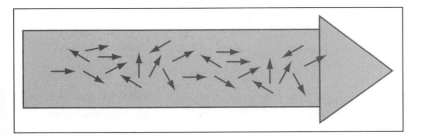

Fig. 5-11. Comparison of Directed and Random Velocity for the Solar Wind Parcel. The directed flow represented by the large orange vector is typically ~450 km/s. We usually refer to this flow rate as the solar wind speed. The random velocity, represented by the short blue vectors, is usually ~40 km/s. This is the motion associated with solar wind temperature.

Pause for Inquiry 5-7

The solar wind thermal speed is a form of random kinetic energy. Relate the random speed value (Fig. 5-11) to energy in eV for a solar wind ion.

Alfvén Wave Speed

We already know that the solar wind can't support sound waves mediated by collisions. However, other types of waves do propagate in the magnetized medium. One type, called an Alfvén wave, after Nobel Laureate Hannes Alfvén, is particularly important in the space environment. Alfvén waves are three-dimensional magnetic field oscillations. As we describe in Chap. 6, these transverse magnetic waves travel along the field lines. The motion of the ions and the perturbation of the magnetic field are in the same direction, transverse to the direction of propagation. The speed of the magnetic perturbation is a function of magnetic field strength and plasma density. The Alfvén speed compares the magnetic field tension to the plasma density, much the way wave speed in a taut string is determined by the ratio of tension to string mass density. The Alfvén magnetic wave speed is

$$v_A = \sqrt{\frac{B^2}{\rho \mu_0}} \qquad (5\text{-}3)$$

where

v_A = Alfvén wave speed [m/s]
B = magnetic field strength [T]
ρ = solar wind mass density [kg/m³]
μ_0 = permeability of free space ($4\pi \times 10^{-7}$ N/A²)

Comparing the Alfvén speed in the solar wind to the solar wind bulk flow speed yields an important measure of solar wind behavior (Ex. 5.5). In doing so, we discover that the solar wind is super-Alfvénic as well as supersonic.

Chapter 5 The Quiescent Solar Wind: Conduit for Space Weather

EXAMPLE 5.5

Sound and Alfvén Speed in the Solar Wind

- *Problem Statement:* Determine the sound and Alfvén speeds in the solar wind
- *Relevant Concepts:* Alfvén speed and Mach number
- *Given:* Solar wind number density = 7 ions/cm^3; solar wind speed v_{sw} = 400 km/s; solar wind temperature $T = 1.0 \times 10^5$ K; IMF strength $|B|$ = 10 nT
- *Assume:* Steady-state conditions
- *Solution:*

Sound speed from Eq. (5-2)

$$c_s = \sqrt{\frac{2k_B T}{m}} = \sqrt{\frac{2(1.38 \times 10^{-23} \text{J/K})(1 \times 10^5 \text{K})}{1.67 \times 10^{-27} \text{kg}}}$$

$$\approx 4 \times 10^4 \text{ m/s}$$

Alfvén speed = $v_A = \sqrt{\frac{B^2}{\mu_0 \rho}}$

$$v_A = \sqrt{\frac{(1 \times 10^{-8} \text{T})^2 / (4\pi \times 10^{-7} \text{C}^2/(\text{N} \cdot \text{m}^2))}{(7 \times 10^6 \text{ions/m}^3)(1.67 \times 10^{-27} \text{kg/ion})}}$$

$$= 8.25 \times 10^4 \text{ m/s}$$

Sound Mach = $v_{sw}/c_s = (400 \times 10^3 \text{ m/s})/(4.0 \times 10^4 \text{ m/s}) \approx 10$

Alfvénic Mach = $v_{sw}/v_A = (400 \times 10^3 \text{ m/s})/(8.25 \times 10^4 \text{ m/s}) \approx 5$

- *Implication and Interpretation:* The solar wind is supersonic and super-Alfvénic.

Follow-on Exercise: On rare occasions the solar wind becomes sub-Alfvénic. What combination of field and plasma variations would lead to such a condition?

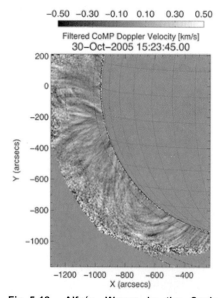

Fig. 5-12. Alfvén Waves in the Sun's Corona. In this image, made by a ground-based instrument at the National Solar Observatory in New Mexico, USA, the oscillations of the plasma velocity are made clearer by filtering the velocity data to show only oscillations that recur with a five-minute period. *(Courtesy of Steve Tomczyk and Scott McIntosh at the National Center for Atmospheric Research)*

Pause for Inquiry 5-8

Which curve in Fig. 5-14 corresponds to the orange arrow (bulk speed) in Fig. 5-11?

Which curve in Fig. 5-14 corresponds to the blue arrows (random speed) in Fig. 5-11?

Pause for Inquiry 5-9

Determine the highest and lowest sound Mach numbers for the interval shown in Fig. 5-14.

The high-speed solar wind emanating from large coronal holes requires an energy enhancement in the supersonic region of the flow. In the corona, Alfvén waves carry a significant energy flux even for a small wave energy density. These waves propagate through the corona without increasing the solar wind mass flux substantially, and deposit their energy flux to the supersonic flow. The solar wind measured at Earth is filled with fluctuations whose largest amplitudes are in the high-speed wind. Many of these have coupled changes in flow velocity and magnetic field vectors. The Alfvénic fluctuations are probably remnants of waves and turbulence that heat the corona and accelerate the solar wind (Fig. 5-12). Fluctuation amplitudes decrease with increasing distance from the center of the Sun; their dissipation heats the wind far from the Sun.

Focus Box 5.1: Reading the Solar Wind Plasma Record from the Advanced Composition Explorer (ACE) Satellite

The Advanced Composition Explorer (ACE) spacecraft is stationed in the free-flowing supersonic solar wind. The satellite orbits the Sun in tandem with Earth from a location called Lagrange Point (L1), about 1.5×10^6 km from Earth and 148.5×10^8 km from the Sun (Fig. 5-13). The Italian-French mathematician Joseph Lagrange, who studied the dynamics of orbiting bodies, discovered five special locations near two orbiting masses, where a third, smaller mass orbits at a fixed distance from the larger masses. These *Lagrange points* mark positions where the gravitational pull of the two large masses precisely cancels the radial acceleration required to rotate with them. Of the five Lagrange points, three are unstable (L1, L2, and L3), and they lie along the line connecting the centers of the two large masses. Each of the stable Lagrange points (L4 and L5) is at a vertex of an equilateral triangle, with the large masses at the other vertices. The L1 point affords an uninterrupted view of the Sun; however, it's unstable on a time scale of approximately 23 days. Satellites near this position must correct their course and attitude regularly. They actually orbit L1. The ACE spacecraft has enough propellant to maintain its orbit about L1 until about 2019. In the future, the L4 and L5 points are likely locations for satellites specifically designed to view the Sun from multiple vantage points.

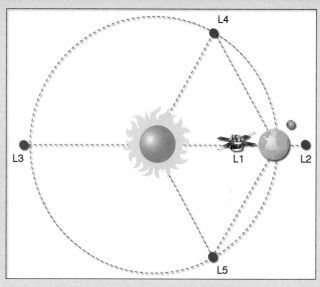

Fig. 5-13. Lagrange Points. This figure portrays a not-to-scale drawing of the Lagrange points for the Earth-Sun system. *(Courtesy of AF Institute of Technology)*

Figure 5-14 shows a seven-day record of solar wind magnetic field and plasma flow at the ACE satellite. During the interval two large magnetic disturbances passed Earth. The first traveled with the speed of the solar wind, but drove a significant density enhancement. The second arrived with slightly high speed, but less of a density increase.

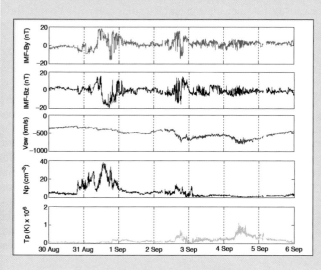

Fig. 5-14. Advanced Composition Explorer (ACE) Magnetic Field and Plasma Measurements for 30 August–5 September 2005. The top curve shows the east-west (B_y) component of the interplanetary magnetic field. The second curve shows the north-south (B_z component). The middle curve is the solar wind speed [km/s]. Both magnetic field components and the solar wind flow are shown in a coordinate system centered at Earth. The negative sign indicates flow directed outward from the Sun past Earth. The fourth curve displays proton density [#/cm³], and the lower curve shows particle temperature [K]. The large rotations in the magnetic field on 31 August and 2 September are associated with coronal mass ejections. *(Data from NASA OMNIweb)*

Conductivity in the Solar Wind

The *conductivity* of a substance is its ability to conduct or transmit electricity or heat. The corona and solar wind are highly ionized, so their electrical conductivity is effectively infinite. It is dominated by the highly mobile electrons, and depends strongly on temperature ($\propto T^{5/2}$). For typical coronal conditions the conductivity is about twenty times higher than the thermal conductivity of copper at room temperature. The high conductivity allows heat to exit the corona and move outward into the heliosphere. As a result, the solar wind exhibits a cooling rate with distance that is slower than expansional (adiabatic) cooling alone, as mentioned on pg. 203.

5.1.3 Interplanetary Magnetic Field Characteristics

> **Objectives: After reading this section, you should be able to...**
>
> ♦ Explain the basic idea of frozen-in magnetic flux as it applies to the solar wind
>
> ♦ Calculate the garden hose angle of the solar wind at any distance from the Sun
>
> ♦ Distinguish between radial plasma flow and the spiral interplanetary field configuration

The combination of solar wind expansion and frozen-in magnetic field stretches the solar magnetic field away from its approximate dipole configuration. By the time the field reaches Earth, it is distorted and rippled.

Magnetic Flux Frozen in the Solar Wind

The Lorentz force causes the solar plasma and the solar magnetic field to form a closely coupled system. Where the magnetic energy is large compared to thermal energy, the magnetic field confines the plasma. Where the field is weak, it tends to follow the plasma motions. When approximated as a dipole in a vacuum, the Sun's magnetic field strength decreases as $1/r^3$. The decrease is sufficiently rapid with height that at altitudes greater than ~1 R_S the magnetic field is less capable of containing the coronal expansion. The *inner corona* is that region of the Sun where the magnetic field is strong enough to direct the motion of coronal plasma. In the inner corona the magnetic field energy density dominates the kinetic energy density (low β, as we describe in Sec. 4.3.1). Because magnetic fields transfer angular momentum to the plasma, the inner corona co-rotates with the Sun with semi-rigid-body motion. The surface encompassing the inner corona (often taken to be $r \approx 2\ R_S$) is the *source surface*, which is the apparent location where the quiescent magnetic field exiting the Sun becomes radial. Beyond 3 R_S, the magnetic energy density decreases faster than the kinetic energy density of the plasma. The *outer corona* has much weaker gravitational and magnetic ties to the Sun. From this region the outflowing plasma stretches solar magnetic field lines into interplanetary space.

The solar wind plasma is an excellent electrical conductor, so the solar magnetic field is encapsulated in the solar wind as it expands away from the Sun. The condition in which the magnetic field flows along with the solar wind is the *frozen-in flux* condition, which we describe in Chap. 4 and illustrate in Fig. 4-26. Frozen-in flux develops because of electromagnetic interactions. We recall from Chap. 4

that the motion of a conductor through a magnetic field causes an electric field E. In turn, the electric field produces a current that keeps the magnetic field B constant in the conductor. When a perfect conductor containing a magnetic field moves away from its field source, induced currents keep the conductor's internal flux at a constant value, which means the magnetic field must ride along with the conductor.

This frozen-in condition means a magnetic flux tube enclosing a single set of field lines remains intact as long as

- The magnetic flux in the volume remains constant as it moves, even if the volume changes shape
- All plasma initially lying along a magnetic field line continues to do so

Radial Polar Field

Plasma parcels departing the Sun stretch magnetic field lines into interplanetary space. At high latitudes, where the magnetic field is primarily radial, conservation of magnetic flux requires that the field strength diminish with distance at the rate of $1/r^2$. We can therefore describe the high altitude radial field strength at an arbitrary distance from the original position:

$$B_r A = B_0 A_0$$

$$B_r \left(\frac{4\pi r}{4\pi r_0}\right)^2 A_0 = B_0 A_0$$

$$B_r = B_0 \left(\frac{r_0}{r}\right)^2 \qquad (5\text{-}4)$$

The relationship is modified at lower latitudes where, although plasma parcels travel radially outward, the Sun's rotation causes low-latitude field lines to bend into spirals, as we describe next.

Lower-latitude Spiral Field

Within two to three solar radii, the Sun's magnetic field tends to rotate with the Sun, but at greater distances the field stretches into the interplanetary medium with the supersonic plasma motion. The imaginary surface where this happens, between 2 R_S and 3 R_S, becomes the *source surface* of the radially expanding magnetic field. Figure 5-15 illustrates these two regimes.

While the magnetic field maintains one foot at the Sun, the other foot stretches along the plasma path. Thus, the field acquires a longitudinal (east-west) component described by the ratio of the angular speed to the radial speed:

$$B_\psi = -B_r \left(\frac{\omega_S r}{v_{SW}}\right) \qquad (5\text{-}5a)$$

The subscript ψ indicates a longitudinal, or east-west, component.

We can recast this equation in terms of the original value of the magnetic field, B_0, at the starting radius r_0, using Eq. (5-4):

Fig. 5-15. The Interplanetary Magnetic Field Spiral. This depicts the ecliptic magnetic field viewed from above the Sun's north pole. Although the solar wind plasma motion is radial, the Sun's rotation combined with frozen-in magnetic flux creates a spiral configuration for the interplanetary magnetic field. The solid and dashed curves illustrate the field above and below the ecliptic plane. Above the plane the field points outward (+); below the plane the dashed curves indicate an inward (−) directed field. The solar surface is 1 R_S. Between 1 R_S and 2 R_S the solar magnetic field rotates with the Sun and has various directions associated with active regions. Beyond 3 R_S the field no longer rotates with the Sun. [Schatten, 1969]

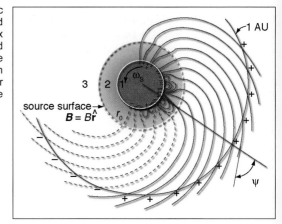

$$B_\psi = -B_0 \left(\frac{\omega_S r}{v_{SW}}\right)\left(\frac{r_0^2}{r^2}\right) = -B_0 \left(\frac{\omega_S r_0^2}{v_{SW} r}\right) \tag{5-5b}$$

The interpretation of Eq. (5-5) is that near the ecliptic plane the solar magnetic field is stretched and flattened away from a dipole configuration. The longitudinal component depends on solar rotation rate, solar wind speed, and the distance from the source surface. The extreme stretching of the field has a further result: in one hemisphere the field points toward the Sun and in the other hemisphere it points away from the Sun. Earth sits at a location that samples a variety of field orientations.

Magnetic field lines, with one footpoint at the magnetic source surface and the other frozen to the expanding plasma, must twist. The result is a spiral structure of magnetic field lines near the equatorial plane. An advanced version of Eugene Parker's solar wind theory predicted this feature. Hence we sometimes call the spiral configuration the *Parker spiral*. We also call it the *garden hose spiral*, because the pattern resembles that of water flowing out of rotating garden sprinkler. Each water droplet moves radially away from the head, but an arc forms in the pattern of the successive droplets.

The angle that develops between the direction of the magnetic field and the radial vector is the *garden hose angle* ψ. The size of the angle depends on the distance from the rotating source surface and on the solar wind speed, and is given by

$$\psi = \tan^{-1}\left(-\frac{B_\psi}{B_r}\right) = \tan^{-1}\left(\frac{\omega_S r}{v_{SW}}\right) \tag{5-6}$$

At Earth's orbit, the garden hose angle has two likely values: 45° or 225°.

Figure 5-16 provides a heliospheric view of the spiraling solar magnetic field at all latitudes. Since the rotation at the solar poles is slower than at the equator, the field at the highest latitudes has very little twist, and appears nearly radial. The higher speed flow from the poles also contributes to a loosening of the polar spiral. Near the equatorial plane, on the other hand, the spiral of the magnetic field is more tightly wound.

5.1 The Quiescent Solar Wind

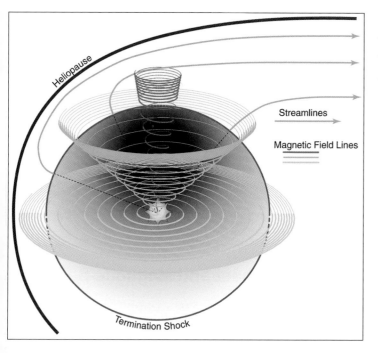

Fig. 5-16. The Solar Wind Spiral at All Latitudes. This is an oblique view from above the solar north pole. The magnetic field lines have been stretched into a spiral configuration. At 1.0 AU, the average field line spiral in the equatorial plane is inclined ~45° to the radial direction from the Sun. The field lines above and below the ecliptic have looser twists. *(Courtesy of Steve Suess at NASA)*

Pause for Inquiry 5-10

Out of the ecliptic plane the solar rotation rate is slower and the solar wind speed is higher. If all other parameters are held constant, what is the effect of these variations on the garden hose angle?

Pause for Inquiry 5-11

Why does a higher solar wind speed loosen the spiral configuration?

Data from the Voyager 2 spacecraft indicate that the interplanetary magnetic field (IMF) in the ecliptic plane decreases in intensity as a function of ~$1/r^2$ for a few tens of AU beyond Earth. At larger distances irregularities in the solar wind and solar wind interaction with the interstellar medium cause bunching of the field (Sec. 5.2).

Our discussion idealizes the smooth field configuration in the ecliptic plane. In reality these field lines attach to moving and oscillating sources on the Sun, so the field lines are in constant motion in all directions. Field variability, rooted in the photosphere, imprints on the solar wind and has been observed in interplanetary space by all solar wind monitors. Embedded within the stretched dipole and spiral motion are a multitude of perturbations that change the orientation of the IMF on short time and small space scales. Analysis of IMF data reveals that the magnetic perturbations out of the ecliptic plane (**Z** direction) are large relative to perturbations in the east-west (**X** and **Y**) directions ($\Delta B_z/B_z > \Delta B_y/B_y$ or $\Delta B_x/B_x$). As we describe at the end of Chap. 4, out-of-ecliptic IMF efficiently couples to Earth's geomagnetic field and delivers energy and momentum to the geospace.

EXAMPLE 5.6

Interplanetary Magnetic Field Spiral Angle

- *Problem Statement:* Determine the two most likely IMF spiral (garden hose) angles at Jupiter.
- *Relevant Concept:* Parker or garden hose angle in Eq. (5-7b)
- *Given:* Solar wind speed, v_{sw} = 400 km/s, solar rotation rate = 2.7×10^{-6} rad/s, distance to Jupiter is 7.78×10^{11} m
- *Assume:* Steady-state conditions

- *Solution:*

$$\tan(\psi) = \omega_s r / v_{SW}$$

$$\psi = \tan^{-1}(\omega_s r / v_{SW})$$

$$\psi = \tan^{-1}[(2.7 \times 10^{-6}\,\text{rad/s})(7.78 \times 10^{11}\,\text{m})/(4 \times 10^{5}\,\text{m/s})]$$

$$\psi = 79.2° \text{ or } 259.2° \text{ for the opposite sector}$$

- *Implication and Interpretation:* With greater distance from the Sun the garden hose angle increases.

Follow-on Exercise: Determine the typical garden hose angles at Saturn during a high-speed flow event.

5.1.4 The Current Sheet and Magnetic Sectors

Objectives: After reading this section, you should be able to...

- Explain why the heliospheric current sheet exists and calculate its current density
- Describe the solar wind's sector structure
- Distinguish between toward (–) and away (+) sectors

The Heliospheric Current Sheet as a Hemispheric Boundary

We know that the Sun's magnetic field is not very dipole-like far from the source surface. The *heliospheric current sheet (HCS)* is the near-equatorial boundary at which the polarity of the Sun's magnetic field changes sign. We illustrate the geometry of the magnetic field lines (without the spiral feature) in Fig. 4-13. Figure 5-17a is an idealized rendering of the Sun's dipole-like magnetic field stretching into the heliosphere, minus the spiral configuration for simplicity. The disk in the figure represents the current sheet that separates oppositely directed magnetic fields. Because of the tilt of the solar dipole axis, Earth intercepts the IMF from either hemisphere as the Sun rotates. From Earth's perspective, the current sheet tilts upward and downward. The opposing fields on either side of the solar equator produce the two peaks in the spiral IMF angles shown in Fig. 5-17b. The observations show the expected spiral angle at $\psi = 45°$ and an additional peak at 225°. The former is associated with the magnetic field pointing outward from the Sun, and the latter occurs when the inward-pointing field in the opposite hemisphere arrives at Earth.

Pause for Inquiry 5-12

What is the likely Parker spiral angle at large distances from the Sun?

5.1 The Quiescent Solar Wind

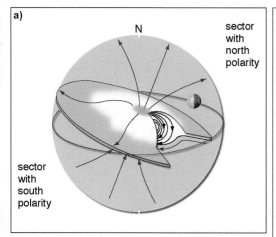
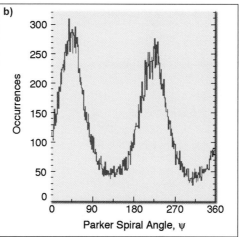

Fig. 5-17. a) The Interplanetary Current Sheet. This is an artist's rendering of a tilted, but smooth, heliospheric current sheet without the spiral structure. A few of the field lines diverge from or converge to the poles. The outward field lines shown on the light green surface have been severely stretched from their dipole configuration by the solar wind outflow. Underneath the surface are similar, but oppositely directed, field lines. The sheared magnetic fields must be separated by a current sheet. The red and blue circle representing Earth's orbit intersects the current sheet at two points. Each time Earth crosses the current sheet a different solar magnetic polarity interacts with Earth's magnetosphere. The changing polarity causes the tangent angle to change by 180 degrees. This change accounts for the double peak in the distribution of spiral angles in Fig. 5-17b. [After Hundhausen, 1997] **b) Interplanetary Spiral Angle Data.** These measurements of the Parker spiral angle were observed at 1.0 AU from 1981–1990 and binned in one-hour increments. The peaks in the angle are separated by 180°. During this time, the field from northern latitudes pointed outward. Values slightly away from the peaks are associated with variations in the solar wind speed. Values significantly different from the peak values are associated with Earth being immersed in the current sheet or with coronal mass ejections. The data are plotted with respect to the Sun's radial direction. [After Luhmann et al., 1993]

Figure 5-17 shows the heliospheric current sheet as a smooth surface. More often than not the dynamic boundary is irregular, similar to the ruffled ballerina skirt in Fig. 5-18. Ulysses spacecraft data reveal a tendency for the current sheet to tilt slightly or cone toward the southern hemisphere (–**Z** direction). Other spacecraft confirm the tilt has persisted for at least two solar cycles. As an amusing analogy, scientists suggest that the Sun with its tilted current sheet is similar to a bashful ballerina who repeatedly pushes her flaring skirt downward.

Pause for Inquiry 5-13

Figure 5-18 shows a ballerina skirt depiction of the heliospheric current sheet. Can you think of some ways the sheet could become more distorted?

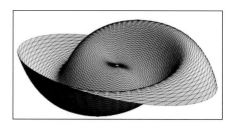

Fig. 5-18. A Computed Heliospheric Current Sheet Out to 6 AU. This model shows the 22.5 degree tilt of the rotating axis, in a 400 km/s wind. The image shows the warping and undulating current sheet consistent with the idea of a spinning ballerina skirt. The magnetic field lines are omitted. *(Courtesy of Steve Suess at NASA)*

EXAMPLE 5.7

Current Sheet Density Example

- *Problem Statement:* Determine the total current flowing in the heliospheric current sheet.
- *Relevant Concepts:* Ampere's Law
- *Assume:* A static current sheet. The extent of the current sheet is about 100 AU.
- *Given:* The magnetic field strength equals 0.05 nT at 70 AU. We take this as the average value over the region of interest, and so treat B as a constant in the following integrals. The thickness of the current sheet is $\sim 3.0 \times 10^4$ km
- *Solution:*

$$\oint \boldsymbol{B} \cdot d\boldsymbol{l} = \boldsymbol{B} \cdot \boldsymbol{l}_{up} + \boldsymbol{B} \cdot \boldsymbol{l}_{out} + \boldsymbol{B} \cdot \boldsymbol{l}_{in} + \boldsymbol{B} \cdot \boldsymbol{l}_{down}$$

In a static current sheet, the magnetic field is perpendicular to the dl_{up} and dl_{down} legs, so we have no contribution to the dot product

$$\oint \boldsymbol{B} \cdot d\boldsymbol{l} = 0 + \boldsymbol{B} \cdot \boldsymbol{l}_{out} + \boldsymbol{B} \cdot \boldsymbol{l}_{in} + 0 = 2(B)(dl) = \mu_0 I$$

$$I = \frac{2(B)(dl)}{\mu_0} = \frac{2(0.05 \times 10^{-9} \text{T})(100 \times 1.5 \times 10^{11} \text{m})}{4\pi \times 10^{-7} \text{N/A}^2}$$

$$= 1.1 \times 10^9 \text{A}$$

- *Interpretation:* More than a billion amperes of current flows in the heliospheric current sheet. Of course, the current sheet covers a huge area, so the current density is low.

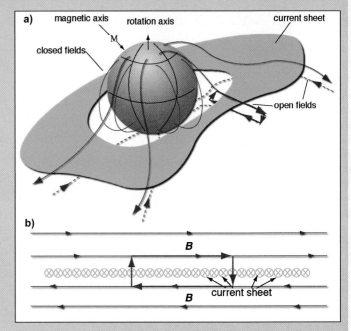

Fig. 5-19. a) **Heliospheric Current Sheet.** This diagram depicts an oblique view of the stretched magnetic field orientation in both hemispheres. b) **Edge-on Illustration.** This image shows the cross section of an idealized current sheet with appropriately directed current flow separating the oppositely directed magnetic fields. The red arrows are the integration path for the problem. [Smith et al., 1978]

Follow-on Exercise: Determine the current density in the heliospheric current sheet.

Sector Structure

In addition to being tilted, the current sheet is often rippled. The sheet dips and undulates in response to variations in the solar outflow. During quiescent times, it evolves slowly, over months, responding to changes in the solar magnetic field. But transient events deform the sheet significantly in a matter of hours. The warps create structure in the IMF passing Earth. Early IMF measurements revealed the sector structure of the field as illustrated in the equatorial cut through the ballerina skirt in Fig. 5-20. Each *sector* encompasses a region in interplanetary space within which the magnetic field has the same polarity, i.e., either toward the Sun (– radial direction) or away from the Sun (+ radial direction). The polarity switches from one sector to the next. An entire sector typically passes Earth every few days, but unipolar events lasting up to two weeks have been observed. The Sun rotates once every 27 days, so similar sector structures sweep past Earth periodically. To an observer in a stationary frame of reference, these magnetic sectors appear to rotate with the Sun. Near the solar equatorial plane, the IMF typically organizes into two to four sectors (Fig. 5-20).

Earth's speed in its orbit is not fast enough to keep pace with the spinning current sheet, so sharp sector boundaries sweep past Earth in a matter of hours. The *sector boundaries* are relatively thin. Spacecraft data show them to be several R_E thick near Earth and Jupiter. Space weather analysts are interested in tracing the sector boundaries back to their roots at the solar source surface as a first step in forecasting solar wind structures likely to intersect Earth.

Figures 5-21a and b are model output of the open and closed field regions that produce the sector structures. The format is the flat map analog of Fig. 3-34b. The red and green areas are sources of high-speed flow with a preferred magnetic field orientation. Each blue curve starts in one hemisphere, arches over the equator, and closes in the opposite hemisphere. In Fig. 5-21a, open fields and associated high-speed flows are present only at high latitudes, so sector structures do not intercept Earth. Figure 5-21b is from a later interval in which the dipole field has flipped, and the current sheet is more warped. When Earth's magnetosphere intercepts flow from the region near 180 degrees, the IMF will be directed inward (toward the Sun). As the Sun rotates, Earth will encounter IMF with an outward (from the Sun) component indicated by the red coloring. The warped current sheet produces sector structures at Earth. Figures 5-22a and b show X-ray snapshots of the Sun consistent with images in Fig. 5-21b.

Pause for Inquiry 5-14

Use Fig. 5-21b to characterize the sequence of solar wind flow past Earth during the Carrington Rotation in 1996–1997. (Focus Box 3.2 discusses Carrington Rotations.) When will the flow be slow? When will it be fast?

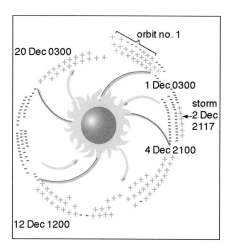

Fig. 5-20. Solar Wind's Spiral Structure. Here we view the spiral and sector structures from above the solar north pole. Dates and UT times are annotated. Illustrated are measurements made by the Interplanetary Monitoring Platform 1 (IMP-1) spacecraft during three 27-day solar rotational periods beginning in late 1973. The beginning of the first 27 day interval is marked as orbit no. 1. Wedge-shaped sectors between the curved black lines show regions with a consistent magnetic field direction, either away from (+) or toward (−) the Sun. Each + or − sign denotes the direction of the field according to three hours of IMP-1 data. The outermost circle of + and − signs represents the first 27-day period; the two circles within represent the next two periods in succession. The sector structure persisted over most of the long interval of observation. *(Courtesy of NASA and Wilcox and Ness [1965])*

Chapter 5 The Quiescent Solar Wind: Conduit for Space Weather

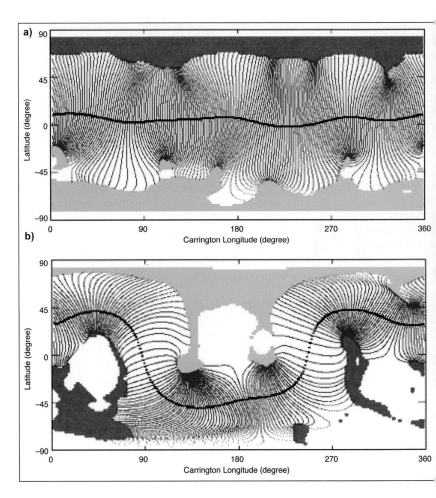

Fig. 5-21. Solar Magnetic Orientation. a) This plot is a flattened view of the magnetic sectors (toward and away) in the solar corona and solar wind for Carrington Rotation 1928–1929. The red area depicts the magnetic field oriented away from the Sun (+), and the green area depicts the magnetic field oriented toward the Sun (–). The blue area shows the last closed magnetic field lines with both feet on the Sun. Other closed field lines (not shown) are nested beneath the blue lines. The black curve indicates the current sheet at its approximate source at 2.5 R_S. The current sheet originates near the apex of the blue magnetic field lines. **b)** This plot portrays the magnetic sectors (toward and away) in the solar corona and solar wind for the Carrington Rotation in 1996–1997. *(Courtesy of NASA's Community Coordinated Modeling Center Potential Field Source Surface Model)*

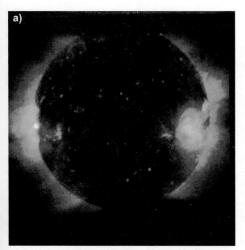 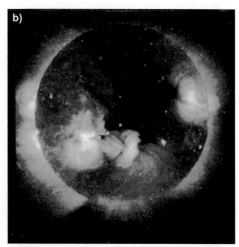

Fig. 5-22. X-ray Images of the Sun from the Yohkoh Spacecraft. a) This image (taken in 1995) is near solar minimum, with closed field loops near the equator and a polar coronal hole in the southern hemisphere. The darkness of the image indicates minimal X-ray emissions. **b)** This image is from the declining phase (taken in 1993) with a large coronal hole dipping toward the solar equator. Earth intercepted high-speed flow from this region. *(Courtesy of the Montana State University's Solar Science and Engineering Laboratory and the Yohkoh Legacy Data Archive)*

Pause for Inquiry 5-15

The solar wind arriving at Earth from the dark region in Fig. 5-22b would be best characterized as

a) nil or stagnant

b) slow and high density

c) high speed and low density

d) hot

e) c and d

We summarize this section with the images in Fig. 5-23. They provide a visual perspective on solar wind climatology for approximately two solar cycles. The solar wind data are from the Ulysses spacecraft that was operated jointly by NASA and ESA for 18 years, and made three trips around the Sun, of which we show two. The spacecraft collected the interplanetary data as it swung from Jupiter, over both solar poles and back to Jupiter's orbit. The data in the left image show that during solar minimum the solar wind flow is bimodal (fast in the high latitudes and slow in the low latitudes). At low latitudes, the solar wind speeds are often less than 400 km/s, but are occasionally punctuated by high-speed flows. At higher latitudes, the solar wind speeds often exceed 700 km/s. The dipole field orientation is consistent with that of solar cycle 22, with the north magnetic pole at the north heliographic pole. During solar maximum, Ulysses's second orbit encountered a mix of flow types. A much larger portion of the Sun exhibited closed field structures, and the coronal holes were smaller.

Chapter 5 The Quiescent Solar Wind: Conduit for Space Weather

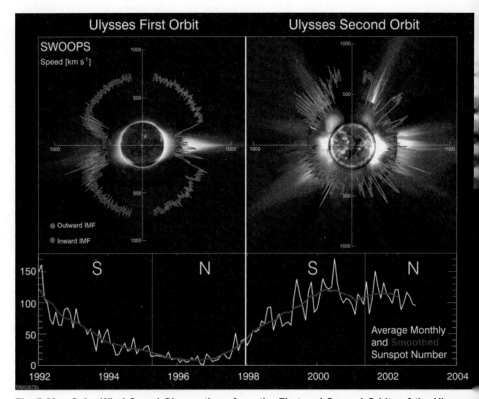

Fig. 5-23. **Solar Wind Speed Observations from the First and Second Orbits of the Ulysses Spacecraft.** The data are from the Solar Wind Observations Over the Poles of the Sun (SWOOPS) instrument package. The top panels show solar wind speed as a function of latitude; with the numbers on the vertical and horizontal axes denoting 500 and 1000 km/s. The spacecraft sampled the Sun at high latitudes, but did not pass directly over the pole, hence the gap in displayed values near the pole. Images of the Sun from the Solar and Heliospheric Observatory (SOHO) are superposed. The color coding of the flow curve indicates the orientation of the interplanetary magnetic field (IMF). Red signifies an outward-directed magnetic field, while blue means inward-directed. The bottom panel shows the sunspot number through the two orbits. The first orbit occurred through the solar cycle declining phase and minimum, while the second orbit spanned solar maximum. The solar wind is overplotted on solar images, characteristic of solar minimum (17 August 1996) and maximum (7 December 2000). The solar images are composites of (from the center out): the SOHO's Extreme Ultraviolet Imaging Telescope (Fe XII at 195 Å), the Mauna Loa K-coronameter (700 nm–950 nm), and the SOHO's C2 Large Angle Spectrometric Coronagraph (white light). Each Ulysses orbit begins in the east (to the left of the Sun), in 1992 and 1998, and the orbit progressed counterclockwise. The letters 'S' and 'N' refer to the heliographic hemisphere in which the data were obtained. [McComas et al., 2003]

5.2 Solar Wind Interactions in the Heliosphere

In this section we describe how plasma flow and magnetic structures in the interplanetary medium adjust as they encounter objects in the solar system. We follow the effects of these on Earth. We also describe Earth's bow shock and its interaction with the solar wind.

5.2.1 Solar Wind Deceleration at a Magnetized Barrier

Objectives: After reading this section, you should be able to...

- Explain the existence of a bow shock in the solar wind upstream from Earth
- Describe how fluid parameters vary across a shock
- Describe how magnetic parameters vary across a shock
- Distinguish between the free-flowing solar wind and the magnetosheath

Previous sections describe the quiescent solar wind that passes Earth on its way through the heliosphere. Here we investigate how the interplanetary medium reacts to obstacles such as Earth's magnetosphere.

Now that we know the solar wind accelerates, we ask what happens when it decelerates as it interacts with solar system bodies and barriers. The flow must deflect around the obstacle, (e.g., Earth's magnetosphere), but the fact that the obstacle is "in the way" comes with little notice to the supersonic solar wind. A *shock* (a sudden deceleration of the inbound supersonic flow) develops when an object moves at supersonic speeds through a fluid or equivalently, a supersonic stream flows past a barrier. Solar wind shocks develop near magnetized planets, at flow boundaries, and near the edge of the heliosphere.

In an ordinary fluid, individual particles interact so often via collisions that the fluid acts as a single entity rather than as separate particles. In a supersonic flow the sound waves cannot travel away from the obstacle in the flow path fast enough to carry useful information for orderly flow communication and response. Thus, in supersonic flow the orderly conveyance of information to the moving fluid breaks down. Instead a shock (pile-up of sound waves) develops. Particle mean-free-path is an organizing parameter for this behavior and controls the sound speed.

From afar, a moving fluid parcel approaches a stationary object. Initially the fluid is unaffected by the obstacle in its path. Within the flow, a fluid boundary layer develops next to the obstacle. Information leaves the boundary via pressure perturbations at sonic speeds. (The pressure perturbations of the sound waves are one means a fluid interacts with itself and its surroundings.) The obstacle's presence begins to affect the fluid by the way it perturbs the flow. Sound waves cause the flow to deflect. The parameters of the flow (density, speed, etc.) determine how far upstream this influence propagates. For an airplane flying in a dense atmosphere, the shock layer is very small compared to the size of the plane. The solar wind fluid is quite tenuous (low density), so that the distance between the obstacle (for example, Earth wrapped in its magnetic field) and the shock is relatively large.

The propagation speed of sound waves, shocks, and many other wave modes depends on the wave amplitude. If the propagation speed increases rapidly enough with wave amplitude, then the waves steepen nonlinearly (because of the intense parts of a wave packet piling up) and form a shock. At a shock, flow energy changes from a high-quality, directed-form in the upstream flow to lower-quality heat in the downstream flow. Downstream of the shock, it becomes high-density turbulent flow.

Earth's Bow Shock and Magnetosheath

The diversion created by a planetary *bow shock*, a non-linear wave standing in the solar wind flow upstream of the planetary magnetosphere, marks the sudden transition from supersonic, directed flow to subsonic, turbulent flow in the downstream. We call the change "sudden" because it occurs in a few tens of kilometers compared to tens of thousands of kilometers in interplanetary space. Based on shock crossings determined from a large set of spacecraft measurements, the bow shock is curved, with a blunt nose at the sub-solar point (the point closest to the Sun), as shown in Fig. 5-24. The shock is strongest at the nose and weakest on the flanks.

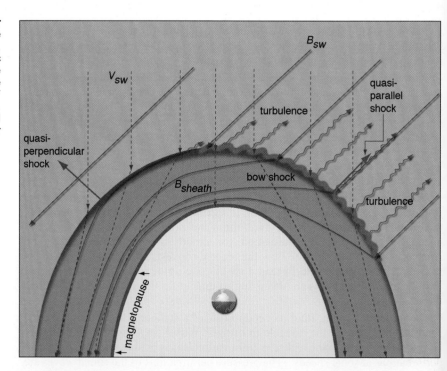

Fig. 5-24. North Pole View of the Solar Wind-Earth Hydromagnetic Cavity. At the bow shock, the solar wind plasma decelerates, heats, and diverts. The solar wind's magnetic field strength, density, and temperature increase as the plasma transits the shock. The bow shock reflects charged particles. The IMF guides the particles as they return upstream. Reflected particles are often energized and flow back to interact with the approaching solar wind. *(Adapted from Bruce Tsurutani)*

We can compare Earth's bow shock to the bow wave of a boat. The bow wave in front of a boat is a high-density, subsonic example of a fluid wave. In a bow wave, information about a disturbance in the flow moves at sound speed, which is much too slow to convey information in supersonic flow. A higher-amplitude, faster wave is required. In a sense, Earth's bow shock is just a nonlinearly steepened bow wave that moves fast enough to deflect the supersonic fluid approaching it. As a point of interest, the Moon does not have a classic bow shock. When the Moon is between Earth and the Sun, the solar wind collides with the lunar surface. When humans first visited the Moon, one of their first activities was to set up sampling stations to collect particles from the solar wind.

Earth's shock region has several features. Upstream from the bow shock is a foreshock created by energetic particles that travel upstream from the bow shock and interact with the incoming plasma. This region is usually filled with plasma waves. Earthward of the bow shock is a magnetized sheath (*magnetosheath*) of heated plasma a few Earth radii thick that separates the shock from the magnetopause. At the magnetopause the subsonic solar wind and the IMF encounter Earth's magnetic barrier. Inside the magnetopause, the fields and flows are those of Earth's magnetosphere. The magnetosheath has properties distinct from the solar wind and the magnetosphere, though the plasma and magnetic field observed in the magnetosheath area are clearly of solar origin. In the flanks of the magnetosheath, downstream from Earth, the shocked solar wind is eventually re-accelerated to supersonic speed, to rejoin the free-flowing solar wind.

Figure 5-24 depicts a free-flowing solar wind zone, a turbulent zone associated with the bow shock, and a laminar zone associated with the magnetosphere. Upstream of the shock, the fluid's properties, such as density, temperature, flow speed, etc., are different from downstream of the shock. Pressure, temperature, and total magnetic field strength increase on the downwind side of the shock. The lines in Fig. 5-24 labeled IMF represent the magnetic field arriving at Earth with a typical Parker spiral angle. When the IMF crosses the bow shock, it compresses and slightly deflects. The figure also shows regions where the IMF vector is roughly parallel to the shock normal wave (*quasi-parallel shock*). These regions tend to be the thickest, most turbulent portion of the shock. The shock thickens in these regions because ions reflected from the shock travel upstream to interact with the IMF. In the region where the IMF is roughly perpendicular to the shock normal (*quasi-perpendicular or tangential*) the flow is more laminar upstream.

Observations of the bow shock often occur when the shock moves over a spacecraft as solar wind changes cause the shock position to fluctuate. At other times the motion of the spacecraft with respect to the shock results in a shock crossing. A spacecraft's bow-shock crossing is usually of short duration (seconds to minutes). In that time, spacecraft instruments record at high resolution the electric and magnetic fields and particle distribution functions in the shock. From these measurements we learn about plasma and shock processes and how variations in the solar wind parameters affect these processes, and hence the shock structure.

Because the solar wind is a fully ionized plasma with an embedded magnetic field, large-scale penetration of another magnetic domain is not possible under usual conditions. A planetary magnetosphere presents a virtually impermeable obstacle. Upon close approach, the solar wind must divert, and both magnetic fields must modify their form. This interaction occurs at all planets that have an intrinsic magnetic field, but we concentrate on Earth. The shock diversion has three important effects for the space environment:

- A shock develops in the solar wind at a relatively constant distance upstream
- Energy dissipation leads to a range of energetic particles and plasma waves
- The planetary magnetic field is reshaped to a streamlined magnetosphere

The magnetized plasma features magnetic wave modes. Scientists think that some of the shock energy transformation that would normally happen via particle collisions in a gas shock, instead takes place through plasma wave dissipation. The shock also reflects solar wind ions. Some of the structure in the quasi-parallel shock regions comes from these reflected ions. Wave activity also depends on thermal and magnetic energy density. In Chap. 4 we define the plasma beta (β) parameter as the ratio of thermal energy density to magnetic energy density. At low β values (relatively high magnetic energy density), wave activity is low. As β increases, the amplitude of waves at and downstream from the shock becomes very large.

Table 5-3 provides information about Earth's bow shock. In space physics, the norm is to measure shock distances in terms of the radius of the object creating the disturbance. The bow shock is typically about 15 R_E upstream from Earth. The position and width of the shock is highly variable and depends on the solar wind parameters, most importantly the solar wind's density and velocity. The shock position changes on a time scale of minutes in response to changes in the solar wind.

Table 5-3. **Bow Shock Characteristics.** Here we list the characteristics of Earth's bow shock.

Characteristic	Range	Typical
Position	12 R_E to 20 R_E upstream from Earth	15 R_E
Motion	10 km/s–100 km/s	50 km/s
Thickness	100 km–1000 km	200 km
Shape	Round to elongated	Roughly hyperbolic

Heating and compression typify the magnetosheath. In crossing Earth's bow shock, the solar wind plasma slows to ~200 km/s and kinetic energy converts to thermal energy, increasing the temperature to 5×10^6 K (about 5–10 times hotter than the free-flowing solar wind). The magnetic field strength generally rises by a factor of 4. The increased temperature in the magnetosheath means that the sound speed c_s is higher, which makes the solar wind speed subsonic. At the nose of the magnetospheric obstacle, the flow stagnates. Nearby it diverges to flow around the flanks. After arriving at the flanks, the modified solar wind parcels accelerate to maintain flow continuity with the rest of the solar wind.

Pause for Inquiry 5-16

Is the magnetosheath a high- or low-entropy region?

In an ordinary gas shock, sudden changes in the mean free path of the gas molecules disrupt ordered flow. This flow breakdown scatters the molecules and increases the width of the shock as the collisional mean free path increases. In space plasmas, the mean free path is often so large that the system is nearly collisionless. However, the idea of mean free path is still useful. Plasma particles interact instead by way of electric and magnetic fields. Flowing charged particles create electric and magnetic fields that then organize the particles' collective motion. These electromagnetic interactions allow the charged particles in the solar wind to act collectively. Hence, what appears to be an oxymoron, *collisionless shock*, is an important interaction mechanism in a plasma environment.

For magnetic interactions the mean free path is the average distance a particle moves before changing direction. In Chap. 1, we see that the Lorentz force causes charged particles of charge q and mass m, circulating in a magnetic field $\mathbf{B}$, to have an effective interaction distance equal to the gyroradius r and an effective collision frequency equal to the gyrofrequency ω. These quantities are defined by the Lorentz force interaction, $\mathbf{F} = q\mathbf{v} \times \mathbf{B}$ (Sec. 1.4), from which we derive the gyroradius r and gyrofrequency ω

$$|F| = \left|\frac{mv_\perp^2}{r}\right| = |qv_\perp B|$$

$$r = \frac{mv_\perp}{qB}, \qquad \omega = \frac{v_\perp}{r} = \frac{qB}{m}$$

EXAMPLE 5.8

Shock Thickness

- *Problem Statement:* Compare a hydrogen ion gyroradius downstream of Earth's bow shock to the thickness of the bow shock.
- *Relevant Concept:* Gyroradius
- *Given:* Solar wind speed in the magnetosheath, v_{SH} = 200 km/s; amplified magnetosheath field strength $|\mathbf{B}|$ = 5 nT × 4 = 20 nT
- *Assume:* No other forces act on the ion

- *Solution:*

$$\text{Gyroradius} = r = \frac{mv_\perp}{qB} = \frac{(1.67 \times 10^{-27}\,\text{kg})(2 \times 10^5\,\text{m/s})}{(1.6 \times 10^{-19}\,\text{C})(20 \times 10^{-9}\,\text{T})}$$

$$r \approx 104 \text{ km}$$

- *Implication and Interpretation:* The gyroradius of a hydrogen ion in the magnetosheath is comparable to the thickness of the bow shock shown in Table 5-3. As the magnetosheath compresses the IMF and solar plasma, the gyroradius changes and the shock width adjusts to the new conditions.

Follow-on Exercise: Determine the gyroradius and gyrofrequency of a hydrogen ion in the free solar wind.

Shock Jump Conditions

We can make some general statements about the behavior of magnetized fluids that flow through shocks such as Earth's bow shock, if we put ourselves in the shock reference frame. By applying conservation laws to the one-dimensional fluid flow, we relate the downstream fluid properties to those upstream from the shock. We ignore the fluid properties within a very short distance of the shock. We also assume there are no source terms. The fluid *jump conditions*, described by the Rankine-Hugoniot Equations (Focus Box 5.2) come from conservation of magnetic flux frozen in the plasma and conservation of mass, momentum, and energy. Figure 5-25 illustrates how the plasma and magnetic field change as the fluid transits (jumps) the shock. The magnetic field is described in terms of its orientation with respect to a vector that is perpendicular to the shock. The component of the magnetic field aligned or anti-aligned with the shock normal vector is $B_{tangent}$. The component of the magnetic field perpendicular to the shock normal vector is B_{normal}. Table 5-4 contrasts the upstream and downstream parameters for a shock. Upstream of the shock the plasma is supersonic, relatively cool, and in an uncompressed equilibrium state. Downstream the

Chapter 5 The Quiescent Solar Wind: Conduit for Space Weather

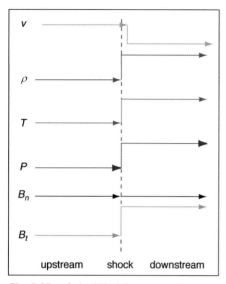

Fig. 5-25. **Solar Wind Parameter Changes across a Bow Shock.** This plot shows how the interplanetary medium parameter values jump as they cross the bow shock at the edge of Earth's magnetosphere. (Here v is speed; ρ is mass density of particles; T is temperature; P is pressure; B_n is normal magnetic field strength; B_t is tangential field strength.)

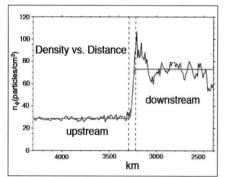

Fig. 5-26. **Solar Wind Density Changes Measured across Earth's Bow Shock.** Density [#/cm³] is on the vertical axis. Distance [km] is on the horizontal axis. In this figure, the Earth is on the right and Sun is on the left. The upstream flow density is low compared to the compressed fluid downstream behind the shock. For this transit across the bow shock the density increases by a factor of 2.5, on average. The shock thickness is slightly less than 500 km (red dashed lines). The green curve is the best fit to the data. The upstream condition is on the right. [Bale et al., 2003]

plasma is subsonic, hot, and in a compressed equilibrium state. Focus Box 5.2 provides a brief summary of the relevant equations for shock conditions. These equations describe the flux of conserved quantities such as mass, energy, and momentum across the shock.

Table 5-4. **Shock Jump Conditions for Quasi-ideal Magnetized Plasma.** Here we list the conditions upstream and downstream of a plasma shock wave.

Upstream		Downstream
Bulk speed: High Free flowing, v > sound speed		**Bulk speed:** Low Shocked, v < sound speed
Density: Low Free flowing		**Density:** High Pile-up
Temperature: Low Little internal motion	Shock	**Temperature:** High Much internal motion
Pressure: Low Consistent with T and ρ		**Pressure:** High Consistent with T and ρ
B tangent (B_t): Low Flows with plasma		**B tangent (B_t):** High Piles-up with plasma
B normal (B_n): Constant Conductivity high along **B**		**B normal (B_n):** Constant Conductivity high along **B**

The conceptual solution for the shock jump equations follows that of Parker's supersonic solar wind calculation, with magnetic flux added to the set of conserved quantities. For an ideal gas, applying the conservation conditions in Focus Box 5.2 results in a ratio between upstream (u) and downstream (d) speed and density given by $\frac{v_d}{v_u} = \frac{1}{4} = \frac{\rho_u}{\rho_d}$. The maximum changes in either direction are limited to a factor of four.

Although the solar wind is not an ideal gas, the magnetic interactions that control particles allow it to roughly mimic the properties of one, so the jump conditions for speed and density at a tangential shock in the solar wind are nearly the same as for an ideal gas. In most shocks, only the magnetic field component aligned with the vector normal to the shock is maintained in the transition. The component perpendicular to the shock normal may increase in magnitude by up to a factor of 4. Other aspects of the interplanetary medium must adjust accordingly.

Figure 5-26 shows the adjustment of the solar wind density across the bow shock. Immediately downstream of the shock the density rises by a factor of ~3.5. Further downstream the density settles to an enhancement of ~2.5. This measurement was made by the CLUSTER satellites on December 25, 2000.

Focus Box 5.2: Rankine-Hugoniot Equations for a Magnetized Shock

Several flow constraints exist for a quasi-ideal magnetized shock. The constraints are called the Rankine-Hugoniot Equations and help us determine the downstream state of the plasma from the upstream state. The main plasma quantities conserved across a quasi-ideal magnetized shockwave are written with a special notation. The notation [X] indicates the difference of a quantity X in the upstream and downstream media: $[X] = X_u - X_d$. For conserved quantities $[X] = 0$. These are vector equations. The notation below assumes that all cross and dot product operations have been performed. Thus we use the vector components normal n and tangential t. For more details, see Kivelson and Russell [1995].

Conserved Quantity

Mass flow:

$[\rho v_n] = 0$

Interpretation

→ If the plasma slows, then the plasma density increases.

Momentum flow along the shock normal:

$[\rho v_n^2 + p + \mu_0 B_t^2] = 0$

→ Dynamic pressure, gas pressure, and magnetic pressure adjust across the shock to maintain momentum balance in the directions across and along the shock.

Momentum flow transverse to the shock normal:

$\left[\rho v_n v_t - \dfrac{B_n}{\mu_0} B_t\right] = 0$

Energy flow:

$\left[\dfrac{1}{2}\rho v_n^2 + \left(p + \dfrac{B^2}{2\mu_0}\right)v_n + \left[\left(\dfrac{\gamma}{\gamma-1}\right)p + \dfrac{B^2}{2\mu_0}\right]v_n\right] = 0$

→ Flow energy, internal energy, and electromagnetic energy adjust across the shock to maintain energy balance.

Magnetic flux: $[B_n] = 0$ → Gauss's Law (no monopole sources for magnetic field)

Electric field: $[v \times B]_t = [v_n B_t - v_t B_n] = 0$ → Faraday's Law of Induction

where
ρ = mass density [kg/m^3]
v_n = solar wind speed normal to shock [m/s]
v_t = solar wind speed tangential to shock [m/s]
p = thermal pressure [N/m^2]
B = magnetic field strength [T]
γ = ratio of specific heats, usually equal to 5/3 [unitless]
B_n = magnetic field parallel to shock normal
B_t = magnetic field tangent to shock normal

The conservation laws consist of 6 equations. If we want to find the downstream quantities from upstream ones, we have 6 unknowns: $(\rho, v_n, v_t, p, B_n, B_t)$.

Pause for Inquiry 5-17

How do the values of solar wind density, speed, temperature, and magnetic field in the magnetosheath compare to those in the free solar wind?

Chapter 5 The Quiescent Solar Wind: Conduit for Space Weather

5.2.2 Solar Wind Beyond the Inner Planets

Objectives: After reading this section, you should be able to...

- Describe the variations in solar wind plasma beyond 1 AU
- Calculate solar wind density in the heliosphere as a function of distance from the Sun
- Distinguish between the termination shock and the heliopause
- Give an estimate of the heliopause and termination shock locations
- Relate the shape of the heliopause to the magnetopause

Solar Wind Speed, Density, and Temperature in the Heliosphere

The interstellar medium confines the heliosphere and bounds the scale of the Sun's influence. Beyond the planets the streaming interplanetary medium expands into the interstellar medium, until it reaches a dynamic pressure balance (Example 5.9). When the supersonic solar wind reaches that boundary it must decelerate, so a termination shock forms. In fact, three discontinuities form because of the interaction between the supersonic solar wind flow and the interstellar wind: the termination shock, the heliopause and, if the interstellar wind is supersonic, a bow shock. At the *termination shock*, the supersonic solar wind passes to a subsonic regime. The *heliopause* is a discontinuity that separates the magnetic field components of the two media. Probably on the interstellar side of the heliopause an interstellar bow shock exists. Figure 5-27 gives the locations of the Voyager spacecraft as they leave the heliosphere.

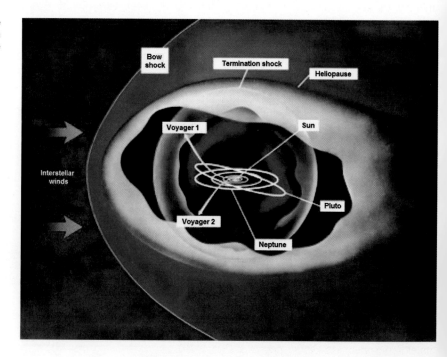

Fig. 5-27. Idealized Drawing of the Heliosphere and its Boundaries. Both Voyager spacecraft are now beyond the termination shock. *(Courtesy of NASA)*

5.2 Solar Wind Interactions in the Heliosphere

Voyager 1 crossed the termination shock at ~95 AU in late 2004. Figure 5-28 shows the sharp increase in magnetic field encountered at the termination shock. In 2007, Voyager 2 encountered the termination shock closer to the Sun, as the boundary moved inward in response to the lower pressure of solar minimum.

Data for the Voyager 2 spacecraft (Fig. 5-29) reveal that the heliospheric solar wind is variable even at 80 AU. Beyond 20 AU are speed increases with superposed fluctuations separated by roughly a solar rotation. The average solar wind speed is close to 440 km/s, consistent with the discussion in Section 5.1.2. At approximately 84 AU, Voyager 2 encountered the subsonic solar wind at the termination shock. Beyond the termination shock, in the heliosheath, the solar wind velocity decreases with radial distance until it reaches the heliopause.

We recall from Sec. 2.2.2 that the density of any quantity flowing outward through concentric spherical shells must diminish as $1/r^2$. So the solar wind number density (n_{sw}) varies as

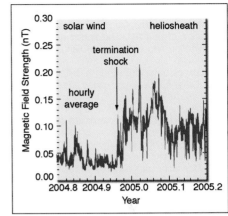

Fig. 5-28. Hourly Averages of the Magnetic Field Strength (B) for Voyager 1. The horizontal axis is decimal year. In the latter part of 2004, Voyager 1 was approaching the termination shock. On or about December 16, 2004, it crossed the shock (~95 AU) and entered into a region of stronger magnetic field thought to be the heliosheath. The ratio of $B_{solar\ wind}$ to $B_{heliosheath}$ is ~1/3. *(Courtesy of Leonard Burlaga at NASA)*

$$n_{sw}(r) = n_o \left(\frac{a}{r}\right)^2 \qquad (5-7)$$

where

n_o = initial or reference number density [#/m^3]

a = reference distance, often one solar radius [km or AU]

r = distance to location of interest from solar center [km or AU]

According to the Voyager 2 density data (shown on a logarithmic scale in Fig. 5-29), this expected relationship holds to beyond 80 AU. Accordingly, the number density near Earth is 10 particles/cm^3, and decreases to approximately 0.001 particles/cm^3 at roughly 90 AU.

On a smaller scale, not well displayed in a logarithmic plot, are density structures associated with merged interaction regions in the solar wind. The merged regions are the results of high-speed streams that eventually catch each other. Their superposition creates compression of the solar wind, resulting in enhanced density in the compression region, and presumably an enhanced magnetic field as well. We describe these in more detail in Chap. 10.

Magnetic merging and deposition of wave energy from below the corona tend to heat and ionize coronal particles, creating high electrical and thermal conductivity in the plasma. The heated plasma pushes outward, opposing and overcoming the Sun's gravitational and magnetic restraints. In essence the solar wind is the expansion-cooled out-gassing of the solar corona. The temperature seems to fluctuate throughout the whole distance. Figure 5-29 shows the Voyager 2 ion temperature measurements on a linear scale between 1 AU and ~80 AU. Stream interactions, magnetic merging, and cosmic particle interactions are possible heating mechanisms to counteract cooling, especially in the distant heliosphere. At the termination shock the plasma temperature rises rapidly, consistent with a flow regime change.

The Sun and the interstellar medium are in motion relative to each other. Current estimates suggest that the interstellar ions and neutral atoms flow at ~25 km/s relative to the Sun. The supersonic solar wind diverts the interstellar plasma flow to create a heliospheric structure similar to Earth's magnetosphere. Interstellar neutral atoms penetrate the heliopause, but interstellar ions tend to divert around it. The exception to the diversion is a highly energetic particle population called cosmic rays. The heliopause, as a grand extension of the coronal magnetic field, deflects roughly 90% of energetic particles arriving from interstellar space. It acts as a leaky umbrella, allowing the other 10% of high-energy cosmic particles access to the inner heliosphere and Earth's geospace (Fig. 5-30).

Chapter 5 The Quiescent Solar Wind: Conduit for Space Weather

Fig. 5-29. Voyager 2 Plasma Data from 1 AU – 90 AU. The solar wind speed remains high to a distance of ~84 AU. The speed decreases at the termination shock. The solar wind plasma density decreases as $1/r^2$. The solar wind temperature decreases beyond Earth to a distance of about 25 AU. Other heat sources cause a minor temperature increase beyond 30 AU. At the termination shock the temperature rises dramatically. The dark smooth line in each of the plots indicates the trend line for the data in the absence of the inward-moving termination shock. Voyager 2 entered the termination shock in August 2007. *(Data courtesy of the Massachusetts Institute of Technology's Space Plasma Group)*

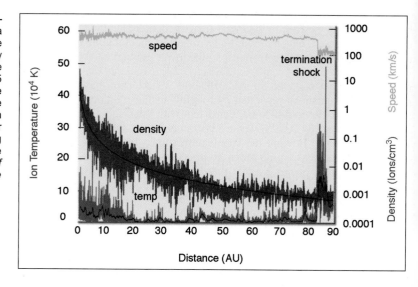

Pause for Inquiry 5-18

Use Eq. (5-7) to estimate the solar wind density at Saturn. Assume the solar wind density is ~3×10^{10} particles/m³ at 4 R_S.

Fig. 5-30. The Heliosphere. This artist's rendering shows the solar wind gray streamlines slowing to subsonic speeds at the termination shock. Beyond this, the solar wind turns toward the heliotail, carrying with it the interplanetary magnetic field. The fields form a tail and a nose similar to Earth's magnetosphere. Most interstellar ions divert at the heliopause. However, interstellar neutral atoms do penetrate into the solar system. Some of these ionize as they approach the Sun and then the heliospheric field redirects them. The small orange area inside the termination shock represents the Sun's entire planetary system. *(Courtesy of Steven Suess at NASA)*

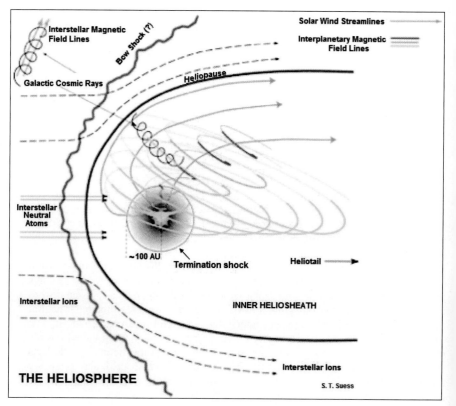

Both Voyager spacecraft should reach the heliopause, about 130 AU – 150 AU, by the year 2017. The heliopause is a dynamic boundary that expands and contracts, with some time delay, as a function of solar activity. In Example 5.9 we estimate the heliopause distance based on pressure balance arguments. Rough verification of the heliopause location comes from the radio noise observations of the interaction of shocks generated by large interplanetary disturbances that race to the heliopause.

Pause for Inquiry 5-19

After traveling for ~13 months, a large structure of tangled magnetic field and mass (merged CME and shock) reached the heliopause, creating radio hiss sensed by the Voyager spacecraft. Assume the structure traveled slightly faster than the typical solar wind, say 450 km/s. Determine the distance between the Sun and the heliopause.

EXAMPLE 5.9

Location of the Heliopause

- *Problem Statement:* Determine the approximate location of the heliopause based on pressure balance.
- *Relevant Concept:* Dynamic pressure and momentum balance
- *Given:* Pressure of the interstellar medium is $\sim 1.3 \times 10^{-13}$ Pa. Use a proton mass for that of the particle, and a typical solar wind density of 5×10^6 particles/m^3.
- *Assume:* Dynamic pressure decreases as a function of $1/r^2$. This pressure balances that of the interstellar medium at the heliopause stagnation point.
- *Solution:* The dynamic pressure in the solar wind at Earth is

$$m_p n_p v_{sw}^2 = (1.67 \times 10^{-27} \text{kg})(5 \times 10^6 \text{particles/m}^3)(4 \times 10^5 \text{m/s})^2$$

$$\approx 1.34 \times 10^{-9} \text{Pa}$$

If the pressure decreases as $1/r^2$, then the dynamic pressure at the heliopause is

$$P_{HP} = P_{Earth}\left(\frac{1 \text{AU}}{R_{HP}}\right)^2$$

The heliopause location is

$$R_{HP} = \sqrt{\frac{P_{Earth}}{P_{HP}}}(1 \text{AU}) = \sqrt{\frac{1.34 \times 10^{-9} \text{Pa}}{1.3 \times 10^{-13} \text{Pa}}}(1 \text{AU}) \approx 100 \text{AU}$$

- *Implication and Interpretation:* By comparing the solar wind pressure at Earth with the solar wind pressure at the heliopause (equal to the interstellar medium pressure), we see that the Sun's influence extends to at least 100 AU.

Follow-on Exercise 1: Determine the magnetic pressure on the interstellar side of the heliopause. Assume that $B_{ISM} = 0.5$ nT (ISM is interstellar medium).

Follow-on Exercise 2: Determine the dynamic and thermal pressure on the interstellar side of the heliopause assuming: a proton density of 0.1 particles/cm^3, a flow speed of 25 km/s, and a temperature of 10^4 K.

Chapter 5 The Quiescent Solar Wind: Conduit for Space Weather

Summary

Comet tails provided the first tantalizing evidence of an interplanetary medium. Space-based observations reveal that a hot, highly conducting, supersonic plasma moves past Earth. The source of the plasma is the solar corona, where pressure gradients accelerate the solar atmosphere to speeds of 300 km/s–800 km/s. As the solar wind exits the Sun, it drags along thin strands of solar magnetic field. These field lines curve and stretch into space, producing a cocoon-like structure called the heliosphere that encases the planets and other distant objects to a distance of ~100 AU.

The solar wind has two primary quiet states: slow and fast. The former emanates from mostly equatorial regions of the Sun. The latter comes from open magnetic regions that dominate at polar latitudes, but may extend to equatorial latitudes during the declining phase of the solar cycle. The solar wind is created in an ionizing environment and remains fully ionized as it expands into space. As the solar wind expands spherically, the Sun's magnetic field stretches into a spiral structure by solar rotation. Stretched field lines from both hemispheres approach a neutral region called the heliospheric current sheet that hovers about the solar magnetic equator. The sheet has the appearance of a twirling ballerina skirt. The field, anchored in the Sun, rotates faster than Earth moves in its orbit, so the sectors of unipolar field with ruffles and disturbances sweep by Earth. The disturbances are often generated by high-speed streams of solar wind that rush into slower flow regions, creating compression regions that may interact with Earth's magnetosphere. Earth regularly experiences sector boundary passages. Near the sector boundaries, field irregularities may contribute to moderate geomagnetic storms.

Before interacting with Earth, the solar wind must decelerate to subsonic speeds. The bow shock in front of Earth's magnetosphere decelerates the flow. The shock is somewhat unusual in that it occurs in a collisionless environment. The changes in solar wind particle speeds and trajectory are brought about by magnetic interactions rather than true particle collisions. The ACE satellite, stationed in the free solar wind upstream of the bow shock, measures the characteristics of the solar wind before it arrives at Earth. The conducting magnetized plasma does work on Earth's geospace, so space weather forecasters want to know how much work and the form of the work, for purposes of predicting space weather events.

Beyond Earth the solar wind continues until it encounters the local interstellar medium at the heliopause. Another shock, the termination shock, slows the solar wind to subsonic speeds on the approach. The termination shock is a location of heated plasma subject to thermal and magnetic turbulence. It's where low-energy ions coming from inside the heliosphere may be accelerated to medium energies.

John (Jack) Gosling provided useful discussions and insight related to the materials in this chapter.

Keywords

bow shock
collisionless shock
conductivity
co-rotating interaction region (CIR)
critical distance
frozen-in flux
garden hose angle ψ
garden hose spiral
heliopause
heliospheric current sheet (HCS)
high-speed streams (HSS)
inner corona
interplanetary magnetic field (IMF)
interplanetary medium
jump conditions
Lagrange points
Landau length (L)
magnetosheath
outer corona
Parker spiral
quasi-parallel shock
quasi-perpendicular or tangential
sector
sector boundaries
shock
solar wind
source surface
termination shock

Notations

Mean free path $l = \dfrac{1}{n\sigma}$

Relation between sound speed and particle velocity in a magnetic field $v_\perp \approx \sqrt{\dfrac{2k_B T}{m}} \approx c_s$

Alfvén wave speed $v_A = \sqrt{\dfrac{B^2}{\rho\mu_0}}$

High-latitude magnetic field strength of the solar wind at distance r from the Sun $B_r = B_0 \left(\dfrac{r_0}{r}\right)^2$

Longitudinal component of the solar wind magnetic field at distance r from the Sun $B_\psi = -B_r\left(\dfrac{\omega_s r}{v_{SW}}\right)$,

$$B_\psi = -B_0 \left(\dfrac{\omega_s r}{v_{SW}}\right)\left(\dfrac{r_0^2}{r^2}\right) = -B_0\left(\dfrac{\omega_s r_0^2}{v_{SW} r}\right)$$

Garden hose angle $\psi = \tan^{-1}\left(-\dfrac{B_\psi}{B_r}\right) = \tan^{-1}\left(\dfrac{\omega_s r}{v_{SW}}\right)$

Gyroradius $r = \dfrac{mv_\perp}{qB}$

Gyrofrequency $\omega = \dfrac{v_\perp}{r} = \dfrac{qB}{m}$

Solar wind number density at a distance r from the Sun
$$n_{sw}(r) = n_o\left(\dfrac{a}{r}\right)^2$$

Answers to Pause for Inquiry Questions

PFI 5-1: The ion count ranges from 1 ion/cm^3 to ~30 ion/cm^3. So, mass density ranges from 1.67 × 10^{-27} kg/cm^3 to ~5 × 10^{-26} kg/cm^3.

PFI 5-2: Speed and density profiles. Starting at the top, the speed would be moderate (~450 km/s). The speed rises rapidly just after the edge of the compression region (~600 km/s–800 km/s) and then slowly decreases to perhaps below average speed (~350 km/s) near the bottom. In a similar manner, the density would be moderate to start (~7 ion/cm^3). The density value rises sharply at the compression region to (~20 ion/cm^3, or more). Beyond the compression regions the density values drop to below normal (2–3 ion/cm^3) before returning to normal values at the second marker.

PFI 5-3: As temperature increases, the particles' velocities increase, making interactions more frequent, so the interaction radius is smaller.

PFI 5-4: Because the high-speed flow is less dense, its mean free path is longer.

Chapter 5   The Quiescent Solar Wind: Conduit for Space Weather

PFI 5-5: Because the speed plots for all temperatures are nearly level at 1 AU and probably remain level to 5 AU, we conclude that no pressure differential exists between 1 AU and 5 AU.

PFI 5-6: The flow speed depends on the coronal temperature, which in turn is tied to the solar cycle. So the flow speed is indirectly a function of the solar cycle.

PFI 5-7: With the random speed equal to 40 km/s, we find the temperature (eV) with: $T = 1/2\ (mv^2)/k_B = 1/2\ (1.6 \times 10^{-27}\ \text{kg})(1.6 \times 10^9\ \text{m}^2/\text{s}^2)/(1.38 \times 10^{-23}\ \text{J/K}) = 9.3 \times 10^4$ K, which corresponds to an ion energy of 8 eV.

PFI 5-8: The Speed curve corresponds to the orange arrow (bulk speed). The Temperature curve corresponds to the blue arrows (random speed).

PFI 5-9: From Example 5.4 and the temperature plot in Fig. 5-14, we find the sound speed at the lowest and highest temperatures, using $c_s = \sqrt{2k_B T/m}$. The lowest temperature is 1×10^4 K and the highest is 3×10^5 K. So, the lower $c_s = 129\ \sqrt{1 \times 10^4\ \text{m}^2/\text{s}^2} = 12,900$ m/s. And the higher $c_s = 129\ \sqrt{3 \times 10^5\ \text{m}^2/\text{s}^2} = 70,700$ m/s. For the Mach number at each temperature, we use the solar wind speed at the same times. So $v_{SW} = 380$ km/s at the lowest temperature and $v_{SW} = 530$ km/s at the highest temperature. The highest Mach number is 29.5 and the lowest is 7.5.

PFI 5-10: The slower rotation speed and the higher solar wind speed reduce the garden hose angle.

PFI 5-11: A higher solar wind speed stretches the frozen-in magnetic field lines, so they have a smaller spiral angle. In the spiral angle equation, the dr term is larger while the $d\psi$ stays the same, yielding a smaller spiral angle.

PFI 5-12: As the distance becomes large, the Parker angle approaches 90 degrees.

PFI 5-13: Coronal mass ejections and high-speed streams that create compression regions would disrupt the smooth nature of the current sheet.

PFI 5-14: The flow speed will be high when the solid colors, indicating open fields, are near the equator. The speed will be slow when the blue stripes, indicating closed fields, are near the equator.

PFI 5-15: c. high speed and low density.

PFI 5-16: Because of the heating and compression within the magnetosheath, it is a volume of high entropy.

PFI 5-17: In the magnetosheath, speed decreases, while density and temperature increase. The magnetic field component perpendicular to the shock (parallel to the shock normal) is unchanged. The magnetic field component parallel to the shock (perpendicular to the shock normal) increases.

PFI 5-18: Using Eq. (5-7), we substitute the reference distance (4 R_S) in the numerator and the distance to Saturn (9.582 AU) in the denominator and square the value, $= 3.8 \times 10^{-6}$. We multiply that ratio times the reference density (3×10^{10} ion/m^3). Thus, $n_{SW} = 1.2 \times 10^5$ ion/m^3 = 0.12 ion/cm^3.

$$n_{SW}(r_{saturn}) = n_0 \left(\frac{a}{r_{saturn}}\right)^2$$

$$= (3 \times 10^{10}\ \text{ion/m}^3)\left(\frac{2.8 \times 10^9\ \text{m}}{1.4 \times 10^{12}\ \text{m}}\right)^2$$

$$= 1.2 \times 10^5\ \text{ion/m}^3$$

PFI 5-19: $R_{heliopause}$ = rate × time = 450 km/s × 13 months. $R_{heliopause} = 1.5 \times 10^{10}$ km = 100 AU.

References

Kivelson, Margaret G. and Christopher T. Russell. 1995. *Introduction to Space Physics.* Cambridge, UK: Cambridge University Press.

Figure References

Bale, Stuart D. et al. 2003. Density-Transition Scale at Quasiperpendicular Collisionless Shocks. *Physical Review Letters.* Vol. 91. American Physical Society. College Park, MD.

Gosling, John T. 2007. The Solar Wind in *Encyclopedia of the Solar System*, Second Edition, edited by Lucy-Ann McFadden, Paul R. Weissman, and Torrence V. Johnson. San Diego, CA: Academic Press.

Hundhausen, Arthur. 1977. "An interplanetary review of coronal holes." *Coronal Holes and High Speed Wind Streams.* (ed. Jack Zirker) Boulder, Colorado. University of Colorado Press.

Luhmann, Janet G., Tie-Long Zhang, Steven M. Petrinec, Christopher T. Russell, Paul Gazis, and Aaron Barnes. 1993. Solar Cycle 21 Effects on the Interplanetary Magnetic Field and Related Parameters at 0.7 and 1.0 AU. *Journal of Geophysical Research.* Vol. 98, No. A4. American Geophysical Union. Washington, DC.

McComas, David J., Heather A. Elliott, Nathan A. Schwadron, John T. Gosling, Ruth M. Skoug, and Bruce E. Goldstein. 2003. The three-dimensional solar wind around solar maximum. *Geophysical Research Letters.* No. 30. American Geophysical Union. Washington, DC.

Parker, Eugene N. 1958. Dynamics of the Interplanetary Gas and Magnetic Field. *Astrophysical Journal.* Vol. 128. American Astronautical Society. Springfield, VA.

Parker, Eugene N. 1963. *Interplanetary Dynamical Processes.* New York, NY: Interscience Publishers.

Schatten, Kenneth H., John M. Wilcox, and Norman F. Ness, 1969. A Model of Interplanetary and Coronal Magnetic Fields, *Solar Physics.* Vol. 6. Springer. Dordrecht, Netherlands.

Schwenn, Rainer. 1990. Large Scale Structure of the Interplanetary Medium in Schwenn, Rainer and Eckert Marsch (eds): *Physics of the Inner Heliosphere I: Large Scale Phenomena.* Berlin: Springer-Verlag.

Smith, Edward J., Bruce T. Tsurutani, and Ronald L. Rosenberg. 1978. Observations of the Interplanetary Sector Structure up to Heliographic Latitudes of 16°: Pioneer 11. *Journal of Geophysical Research.* Vol. 83. American Geophysical Union. Washington, DC.

Tu, Chuan-Yi, Cheng Zhou, Eckart Marsch, Li-Dong Xia, Liang Zhao, Jing-Xiu Wang, and Klaus Wilhelm. 2005. Solar Wind Origin in Coronal Funnels. *Science Magazine.* Vol. 308. American Association for the Advancement of Science. Washington, DC.

Wilcox, John M. and Norman F. Ness. 1965. Quasi-stationary Co-rotating Structure in the Interplanetary Medium. *Journal of Geophysical Research.* Vol. 70. American Geophysical Union. Washington, DC.

Further Reading

Bingham, Robert. 1993. *Space Plasma Physics: Microprocesses in Plasma Physics.* (Edited by Richard Dendy.) Cambridge, UK: Cambridge University Press.

Cranmer, Steven R. 2002. Coronal holes and high speed solar wind. *Space Science Review*, Vol. 101. Springer. Dordrecht, Netherlands.

Kallenrode, May-Britt. 2004. *Space Physics: An Introduction to Plasmas and Particles in the Heliosphere and Magnetospheres,* 3rd Edition. Berlin: Springer-Verlag.

Kamide, Yohsuke and Abraham C-L. Chian. 2007. *Handbook of the Solar Terrestrial Environment.* Berlin: Springer.

Chapter 5 The Quiescent Solar Wind: Conduit for Space Weather

Space is a Place...with Plasma

6

UNIT 1. SPACE WEATHER AND ITS PHYSICS

Contributions by Linda Krause and Michael Dearborn

You should already know about...

- Newton's laws of motion
- Vector identities for cross products
- Conservation laws for mass, energy, and momentum
- Charged particle response to E and B fields (Chaps. 1 and 3)
- Energy and energy density formulations for neutral particles (Chap. 2)
- Coulomb collisions (Chap. 3)
- Electric potential (Chap. 4)
- Energy and energy density formulation for magnetic field (Chap. 4)
- Alfvén speed (Chap. 5)

In this chapter you will learn about...

- The range of energies, temperatures, and ionization ratios in plasmas
- Models describing plasma behavior in single-particle, thermal, and kinetic modes
- Quantifying ionization ratios in a thermal plasma using the Saha equation
- How plasma particles respond to balanced and unbalanced forces
- Drift velocity for electric fields, gravitational fields, and non-uniformities in magnetic fields
- Guiding center drift
- Energy density in plasma
- Magnetohydrodynamic (MHD) approximations
- Magnetic induction
- Magnetohydrodynamic applications in the space environment
- Plasma parameters: Knudsen Number, magnetic Reynolds Number, plasma beta
- Basic plasma waves and oscillations

Outline

6.1 Plasma Characteristics and Behaviors
    6.1.1 Qualitative Plasma Behavior
    6.1.2 Defining Characteristics of a Plasma

6.2 Single-particle Dynamics—I
    6.2.1 Single-particle Accelerated Motion
    6.2.2 Single-particle Unaccelerated Motion

6.3 Thermal Plasma Fluid Dynamics
    6.3.1 Maxwell-Boltzmann Distribution and the Saha Equation
    6.3.2 Energy Density in Plasma

6.4 Magnetized Plasma Fluid Dynamics
    6.4.1 The Basics of Magnetohydrodynamics (MHD)
    6.4.2 Applying Magnetohydrodynamics (MHD)

6.5 Elementary Plasma Waves
    6.5.1 Wave Propagation
    6.5.2 Plasma Oscillations and Waves

6.1 Plasma Characteristics and Behaviors

The Sun and the space environment create and support long-lived plasmas. Unlike terrestrial weather, space weather occurs in a plasma environment. As society places more reliance on technology, our activities spill into regions where plasma is common. Plasma is a key element of over-the-horizon radio communication. All satellites operate in a plasma environment and modern satellite precision-navigation signals traverse it. In this chapter we describe plasma behavior and how it makes space weather unique.

6.1.1 Qualitative Plasma Behavior

> **Objectives:** After reading this section, you should be able to...
> - State basic plasma characteristics
> - Describe sources of ionization energy
> - Distinguish between energetic particles and plasma
> - Explain why plasma is called a fourth state of matter

Characteristics and Sources

Plasma is an electrically neutral, ionized fluid (usually gas, but sometimes mixed with dust grains) composed of ions, electrons, and neutral particles. It is a type of matter distinct from solids, liquids, and fully neutral gases (Chap. 1). We live in the 1% of the cool, condensed universe where plasma is rare. Natural, thermal plasmas develop in high-temperature regimes (typically $T > 10^4$ K). Life as we know it finds a more hospitable environment at much lower temperatures (a few 100 K), where organic molecules maintain their structure. Figure 6-1 shows the range of energies and ionized density for many of the space components we present in this text. Table 6-1 gives several examples of plasma.

In an ideal neutral gas, particles interact with each other via collisions and have speed distributions similar to those shown in Fig. 2-8. The same fluid behavior dominates weakly ionized plasma—sometimes called cold or thermal plasma. In strongly ionized (hot) plasma, particles also interact with each other via electric and magnetic force fields. These fields in turn undergo changes from the collective behavior of the plasma particles. Such plasmas are called non-thermal, because so many forces besides collisions control their behavior. In the space environment different plasma populations are distinguishable by temperature and the degree to which electromagnetic fields and collisions influence their behavior.

Characteristics. Below are the basic characteristics of plasmas:

- Plasmas tend to be neutral. The fluid, as a whole, does not have a net charge even though it is composed of charged particles.

- Plasmas conduct electric current. Some plasmas are better conductors than others.

- Plasmas respond to electric and magnetic fields in a collective manner. Because the charged particles are subject to the Lorentz force, electric and magnetic fields may affect the plasma flows.

- Plasmas often exhibit wave motion. Plasma perturbations create electric and magnetic fields that cause restoring forces. In turn, the forces create oscillations that may propagate.

- Plasmas are prone to instabilities (uncontrolled wave growth). Unstable configurations result in plasma turbulence that attempts to restore a more stable configuration. Fluid (thermal) and kinetic (non-thermal) instabilities are prevalent in plasmas.

Note: A partially ionized gas may be plasma, depending on its temperature and density. We explain how to quantitatively distinguish plasma from gas in Sec. 6.1.2.

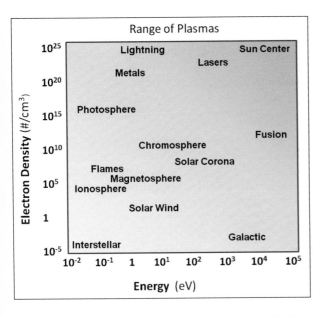

Fig. 6-1. Temperature and Density Ranges in Plasmas. This figure displays the locations of various natural and artificial plasmas on a density-temperature chart. Color gradations give a qualitative picture of how hot or cold a plasma is. [After Peratt, 1966]

Pause for Inquiry 6-1

In which corner of Fig. 6-1 would you find Earth's neutral atmosphere?

Pause for Inquiry 6-2

On the right side of Fig. 6-1 place a scale of density in electrons/m^3. On the top place a temperature scale in Kelvin.

Table 6-1. Natural and Artificial Plasmas. This table lists natural and artificial plasmas roughly by scale size and by energy. Both columns are in descending order.

Natural Plasma Listed Roughly by Scale Size	Artificial Plasma Listed Roughly by Energy
Interstellar wind	Nuclear fusion plasma
Solar wind	Spacecraft pulsed plasma thrusters
The Sun and other stars	Plasma TV displays
Ionosphere	Neon lights
Lightning	Fluorescent lights

Although plasma behavior tends to be collective, nature differentiates some plasma activities: often, the more massive ions carry momentum and kinetic energy, while the less massive, more mobile electrons carry current and thermal energy. Any charge separation in the plasma produces polarization electric fields that in turn influence the plasma and keep it locally neutral. If a magnetic field B permeates the plasma, then the plasma properties are different in directions parallel and perpendicular to B.

Ionizing a substance to produce plasma takes a fair amount of energy (per atom or molecule). We instinctively know that, when we see a lightning-like discharge and glow in a laboratory demonstration, we shouldn't stick our hand in it. Luckily for us, most natural Earthly plasma discharges are short-lived and far away. And in our neutral world, electrons and ions recombine so rapidly that plasma easily returns to a non-ionized state. Rapid recombination is why we won't (normally) find plasma in our coat pocket—we don't have an energy source nearby large enough to maintain it. On Earth some energetic mechanisms must artificially produce and constantly replenish the plasmas in fluorescent light bulbs and plasma TVs. On the other hand, in space, natural plasmas flourish and have extended lifetimes. Stellar sources often create long-lived plasmas, if the ions and electrons don't find the right combination of energy and momentum to recombine.

Pause for Inquiry 6-3

Figure 6-1 shows that metals can be plasmas. Consider the characteristics of plasmas. Which of these characteristics are most applicable to metals? Which might be most applicable to ionized interstellar dust grains?

Plasma Sources. Processes that produce ionization are many. Strong electric fields rip electrons from parent atoms. Shortwave photons from the Sun free electrons from neutral matter in a process called *photoionization*. Extreme ultraviolet waves are the primary source of ionization in Earth's upper atmosphere. Energetic particles in beams or in individual units, as in cosmic rays, knock electrons from neutral matter (*impact ionization*). Collisions among individual particles of a hot thermal gas produce a similar result. Electrons in some types of matter may develop a resonance with radio waves and are excited to a level that allows them to escape their parent matter.

Plasma Model Classification

The description of plasma behavior tends to be situational. In space physics, for example, if a plasma is immersed in a magnetic field and the collision frequency for the particles is very small (compared to the frequency of particle orbits about the field line), we represent its behavior with single-particle dynamics (Sec. 6.2). If the particles experience enough collisions to achieve equilibrium, then the plasma will become thermalized. A thermal plasma (in the absence of external forces) is usually described by modification to fluid (hydrodynamic) theory. That is, we represent a thermal plasma as a fluid with a single temperature, average thermal speed, bulk velocity, etc. One of the modifications usually involves the addition of magnetic characteristics (magnetohydrodynamics). In thermal plasmas so many collisions occur that the individual plasma particles do not have time to make an uninterrupted orbit about a magnetic field line. If the plasma behavior is such that the ionized particles are clearly modifying the ambient electric and magnetic fields, then we look at the collective effects via kinetic theory, which involves more math. Figure 6-2 shows the general relationships between treatments of plasma.

Focus Box 6.1: The Plasma "State" and Energetic Particles

The argument for calling plasma a "fourth state" is based on the fact that the presence of electric and magnetic fields produces behavior that is very unlike any behavior observed in neutral gases. But "fourth state" in the technical sense may not be appropriate. For solids, liquids, and gases the equilibrium between particle thermal energy (i.e., random kinetic energy) and binding energy of the particles determines the state. Some latent heat is released or absorbed as materials transition from one state to another. When latent processes occur, their temperature is typically constant. Conversely, transition from gas to plasma occurs gradually and continuously, even while temperature changes. If we adopt a natural model that requires a latent energy transition to define phase states, then plasma doesn't quite fit the model.

An example of the transition from gas to plasma is a molecular gas that is heated until it dissociates into an atomic gas. If the atomic gas heats further, it becomes progressively more ionized as the number of inter-particle collisions increases and more collisional energy goes into ionizing the gas atoms. The resulting plasma may consist of a mixture of neutral particles, positive ions (atoms or molecules that have lost one or more electrons), and negatively charged electrons. As the gas heats further, the fraction of ionized particles increases.

People often want to know if the aurora is a plasma. The auroral light is not plasma. However, the displays are stimulated when electron currents with voltages in excess of 6 kV impact the atmosphere. These beams of energetic particles excite bound electrons to higher energy states within their parent atoms or molecules. When the bound electrons return to lower states they emit photons in the visible and UV portions of the spectrum. The energetic particle beams also ionize some atmospheric constituents, so some plasma interactions occur in the auroral zones, but the visible light of the aurora is just that—light.

In many cases the thermal environment allows a mix of plasma and high-energy neutral particles. The neutral particles may have just as much thermal energy, but be in a neutral state as a result of a recent recombination. In addition to thermal ions, electrons, and neutrals, plasma often contains a hot *(suprathermal)* species called *energetic particles*. The energetic particles are typically much hotter and less dense than their thermal counterparts. Sometimes, though, the tradeoff between temperature and density allows energetic particles to exert a pressure comparable to that of thermal ions. Energetic particles are more prone to individual than to collective action. When they do act collectively, they may appear as beams of particles in which the quasi-neutral condition breaks down. Interaction between energetic particles and plasmas often produces waves and instabilities. Table 6-2 provides rough guidance for distinguishing the energies associated with neutrals, plasmas, and energetic particles.

Table 6-2. **Energetic Particles.** Here we list general categories of particles and plasmas. The energy limits for the categories are approximate.

Particle/Plasma Type	Energy	Example
Energetic particles	~0.5 MeV to GeV and beyond	Cosmic rays, solar energetic particles, radiation belt particles
Energetic plasma	~100 eV to 0.5 MeV	Radiation belt particles, solar and stellar flare constituents
Plasma	~0.1 eV to 100 eV	Ionosphere, plasmasphere, solar wind, candle flame
Neutral particles	< 0.1 eV	Gas, liquid, dust, etc.

Fig. 6-2. Descriptions of Plasma Behavior. Kinetic behavior in a plasma corresponds to a low frequency of particle collisions in a non-thermal environment and the tendency for the particles to modify the ambient electric and magnetic fields. A kinetic description of plasma behavior is the most general but also the most mathematically complex. We use a fluid approximation when a plasma reaches equilibrium, where its particles undergo enough collisions to make its thermal properties dominant. A single-particle approximation applies to a stable plasma in the presence of fields that are not significantly modified by the particle behavior.

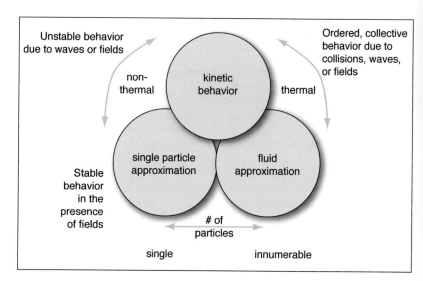

Non-thermalized, collisional plasmas are unstable and may undergo rapid changes, called kinetic instabilities, to achieve thermal equilibrium. Even thermal plasmas may become unstable if external forces act on them. Such instabilities often result in turbulent mixing that brings the plasma out of thermal equilibrium. Fluid instabilities may arise when two different particle or plasma populations, e.g., thermal plasma and energetic particles, occupy the same volume. Plasma dynamics becomes complicated quickly. In this text, we focus on plasmas represented by 1) non-collisional, single-particle dynamics and 2) single-temperature, collisonally thermalized plasmas represented by fluid dynamics.

6.1.2 Defining Characteristics of a Plasma

Objectives: After reading this section, you should be able to...

- Calculate the Debye length in a plasma
- Calculate the natural oscillation frequency of a plasma
- Determine if a fluid is a plasma using the Debye criteria

Debye Length: The Scale of Influence

In exploring the fluid aspects of plasma behavior we consider

- Debye length—the distance over which a particle influences its surroundings
- Plasma frequency—the oscillation frequency of the electrons about the relatively massive and immobile ions
- The formal definition of a plasma—using Debye length and plasma frequency

We recall from Chap. 4 that when a mass or charge influences the space around it, we call the influence a field. Coulomb's force law describes electrical influence by $F = qE$. Here we consider the electric field E and the plasma particles that it influences. From Table 4.1, the electrostatic field generated by a charge in a vacuum is

$$E(r) = \frac{kq}{r^2}\hat{r}$$

where
- E = electrostatic field vector [N/C]
- k = Coulomb's constant (9×10^9 N·m²/C²)
- q = point charge [C]
- $\hat{r}$ = radial unit vector from the charged particle to the point of interest [direction only]
- r = distance from the charged particle to the point of interest [m]

The electric potential difference relates to the electric field as

$$\Delta V = -\int E(r) \cdot dr = -\int \frac{kq}{r^2}\hat{r} \cdot dr$$

Because we are interested in potential difference, we customarily set the lower potential to zero for convenience. The integration is performed over an arbitrary path (usually the shortest one), connecting the zero-potential point with the point under consideration. The electrical potential created by a point charge q at a distance r from the charge (relative to the potential at infinity, which we assume is zero) is

$$V = \frac{kq}{r} \qquad (6\text{-}1)$$

How far does the influence of a point charge extend? To infinity. However, if the charge is immersed in a fluid of mobile electrons and ions, then a phenomenon called *Debye shielding (electrostatic screening)* occurs. The presence and mobility of other charges reduces the distance over which the charge influences its surroundings. If a point charge is positive, it will exert a repulsive force on other ions and an attractive force on the electrons. The surrounding plasma fluid adjusts itself so that more electrons and fewer ions are close to the charge. (The Debye description is named after Peter Debye, a 1936 Nobel Laureate in Chemistry.) We characterize the length of any point charge's influence in a plasma by the *Debye length* (λ_D),

$$\lambda_D = \sqrt{\frac{k_B T_e \varepsilon_0}{n_e e^2}} \qquad (6\text{-}2)$$

where
- λ_D = Debye length [m]
- k_B = Boltzmann's constant (1.38×10^{-23} J/K)
- T_e = temperature of the background electrons in the plasma [K]
- ε_0 = permittivity of free space (8.85×10^{-12} C²/(N·m²))
- n_e = electron density = ion density [#/m³]
- e = electronic charge (1.6×10^{-19} C)

What do we mean by "close?" The Debye length represents the characteristic length scale of influence. The interaction of the charge and the plasma fluid causes the electric field strength to decrease rapidly with distance from q. At a distance of one Debye length, the electric field is reduced to 1/e of its original value. We modify the above expressions as follows:

$$\boldsymbol{E}(r) = \frac{kq}{r^2} \exp\left(\frac{-r}{\lambda_D}\right) \hat{\mathbf{r}} \qquad (6\text{-}3)$$

$$V(r) = \frac{kq}{r} \exp\left(\frac{-r}{\lambda_D}\right) \qquad (6\text{-}4)$$

Figure 6-3 shows the electric potential for a charge in a vacuum and the same charge in a plasma, shielded by surrounding charges.

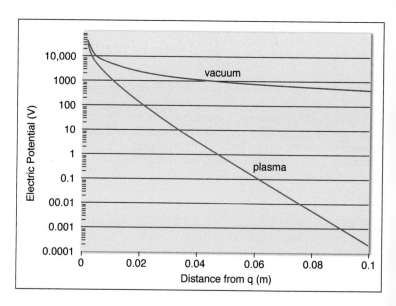

Fig. 6-3. Debye Shielding. This diagram compares the electric potential from a point charge (q = 5 nC) in a vacuum (red trace) to that in a plasma (blue trace). The plasma, with a temperature of 1000 K, particle density of 10^{11} particles/m³, and Debye length of 0.0069 m, shields the rest of the plasma from the point charge. The values are associated with the calculations in Examples 6.1 and 6.2. In a vacuum, the potential difference falls off with distance rapidly at first, but a significant potential difference still exists up to a meter away and beyond—an entire universe to an electron! This distance is what we mean by "long range interactions" —an electron in a vacuum experiences enough of an electrical force in this region to "collide" with the point charge. However, when the point charge is immersed in plasma, other charged particles in the plasma shield the point charge with a characteristic length scale of the Debye length. In fact, at 0.1 m from the point charge, the plasma potential is less than 10^{-5} times that of a vacuum potential!

Pause for Inquiry 6-4

Does a larger charged particle density reduce the shielding distance?

Pause for Inquiry 6-5

Suppose an instrument on a spacecraft tends to become charged. When it does so, it causes interference with other instruments. To reduce this interference we should operate it in which of the following environments?

a) A high plasma density environment

b) A low plasma density environment

c) A high temperature environment

d) A low temperature environment

If more than one answer applies, state why.

EXAMPLE 6.1

Debye Shielding I

- *Problem Statement:* Determine the shielding length for a specific plasma.
- *Given:* A 5 nC charge q placed in a 1000 K plasma with an electron density of 10^{11} particles/m^3.
- *Relevant Concept:* Debye length
- *Solution:* The Debye length is

$$\lambda_D = \sqrt{\frac{k_B T_e \varepsilon_o}{n_e e^2}} = \sqrt{\frac{(1.38 \times 10^{-23}\,\text{J/K})(1000\,\text{K})(8.85 \times 10^{-12}\,\text{C}^2/(\text{N}\cdot\text{m}^2))}{(10^{11}/\text{m}^3)(1.6 \times 10^{-19}\,\text{C})^2}}$$

$$= 0.0069\,\text{m} = 6.9\,\text{mm}$$

- *Interpretation:* Beyond ~0.007 m the charge is effectively hidden from the rest of the plasma. The Debye length does not depend on the magnitude of the point charge q placed in the plasma.

Follow-on Exercise: Determine the Debye length for a region in the solar corona where the plasma density is 10^7 particles/cm^3 and the temperature is 10^6 K.

EXAMPLE 6.2

Debye Shielding II

- *Problem Statement:* Determine the electrostatic potential associated with the point charge of 5 nC at a distance of 0.1 m from the point charge.
- *Assume:* A 5 nC charge placed in a 1000 K plasma with an electron density of 10^{11} particles/m^3
- *Given:* The Debye length for the plasma is 6.9 mm, as calculated in Example 6.1.
- *Relevant Concept:* Shielded electrostatic potential (Eq. (6-4))

- *Solution:* The electrostatic potential from the point charge in the presence of the plasma is

$$V(r) = ((kq/r)\exp(-r/\lambda_D))$$

$$V(r) = (9 \times 10^9\,\text{N/C}^2)(5 \times 10^{-9}\,\text{C})(1/0.1\,\text{m})[\exp(-0.1\,\text{m}/0.0069\,\text{m})]$$

$$V(0.1\,\text{m}) = 0.229\,\text{mV}$$

- *Interpretation:* The electric potential value depends on the distance from the point charge as well as on the value of the Debye length. We plot the result and compare it to the potential in a non-plasma environment in Fig. 6-3.

Follow-on Exercise: Determine the vacuum potential and plasma potential 1 m from the 5 nC point charge.

The shielding effectiveness depends on the temperature of the most mobile charges (usually the electrons) and on how many oppositely-charged particles are near a given charged particle. We express the number N of particles in a *Debye sphere*, whose volume is

$$\frac{4}{3}\pi \lambda_D^3$$

The larger N is, the more effective the shielding. Figure 6-4 shows a Debye sphere of radius λ_D in a plasma. The number of particles inside the sphere is the volume of the sphere multiplied by the plasma density. Because the plasma has a finite temperature, the charges are constantly moving across the imaginary boundary. Statistically the shielding charges stay a bit closer to the charge they are shielding. Charges with the same sign as the shielded charge stay further away.

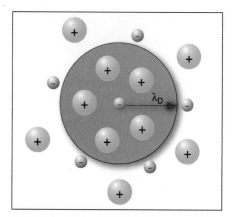

Fig. 6-4. Debye Sphere. Charges of either sign are constantly moving in and out of the sphere. Beyond the imaginary sphere's edge other plasma particles are only minimally perturbed by the negative point charge at the sphere's center.

Although Fig. 6-4 suggests that the Debye sphere is a volume within a definite surface, this is not really the case. In a plasma, shielding of a point charge is only partially effective. The Debye boundary is blurred. Some distance from the partially shielded charge, charges of opposite sign will have enough kinetic energy to escape the charge they are shielding. The result is that the weakened electric field of the partially shielded charge leaks into the plasma. This leakage allows the electric field to influence other charges, and produces the collective behavior that is essential to a plasma. In any plasma, collective effects only occur on scale sizes larger than λ_D. On shorter scales, charged particles demonstrate individual behavior (still influenced by the fields of all surrounding particles).

Plasma Frequency

Next we consider a plasma that is cold (little thermal motion). Suppose that a temporary electric field is applied to the plasma. The electrons will be uniformly displaced in the opposite direction, while the much more massive ions will hardly move at all. The displacement generates a small electric field between each ion-electron pair, as shown in Fig. 6-5. (The figure depicts plasma as a well-ordered crystalline structure, which real plasma is not.) Each electron experiences a restoring force in a direction opposite to the new electric field, which causes it to move back toward its starting point. However, once it's in motion, its momentum makes it overshoot its equilibrium point. Another electric field is generated, producing another restoring force, and we have a setup for simple harmonic motion. The result is a plasma oscillation.

Fig. 6-5. Electrons Separated from Ions. Because of a temporary external force (electric field **E**) dragging the electrons (small blue dots) to the right, they separate from the ions (large dots). When this field disappears, the charge separation creates a situation for simple harmonic motion because of the restoring force of the inter-charge electric fields.

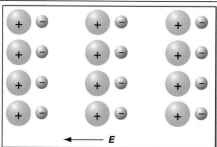
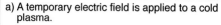

a) A temporary electric field is applied to a cold plasma.

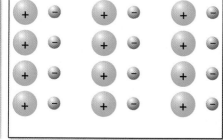

b) The electrons are displaced from their equilibrium positions. The heavy ions remain relatively stationary.

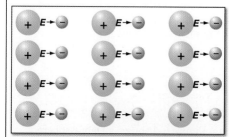

c) The displacement sets up a small electric field between each ion-electron pair. The electrons are accelerated to the left.

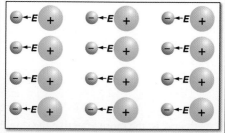

d) Momentum carries the electrons past their equilibrium positions, setting up an opposite electric field between each pair. The result is plasma oscillation.

The oscillations of the electrons about the (relatively massive and immobile) ions are characterized by the plasma's natural frequency, appropriately called the *plasma frequency* (ω_p)

$$\omega_p = \sqrt{\frac{n_e e^2}{\varepsilon_0 m_e}} = \frac{1}{\lambda_D}\sqrt{\frac{1}{k_B T_e m_e}} \qquad (6\text{-}5)$$

where

ω_p = angular plasma frequency [rad/s]
n_e = electron density [#/m^3]
e = electric charge (1.6×10^{-19} C)
ε_0 = permittivity of free space (8.85×10^{-12} C^2/(N·m^2))
m_e = electron mass (9.11×10^{-31} kg)
λ_D = Debye length [m]
k_B = Boltzmann's constant (1.38×10^{-23} J/K)

The plasma frequency describes how rapidly the charges have to move to maintain the Debye shielding. Combining the constants and recognizing that the angular frequency ω and the linear frequency f are related by $\omega = 2\pi f$, we rewrite the expression in a convenient form

$$f_p \approx 9\sqrt{n_e} \qquad (6\text{-}6)$$

where

f_p = linear plasma frequency [Hz]
n_e = electron density [#/m^3]

This example of a plasma oscillation involves no externally propagating energy, so all of the energy stays in the plasma.

Formal Definition of Plasma

Now that we've defined some quantitative characteristics of plasma, we can give the formal definition of plasma. In plasma the following criteria are met:

1. The Debye length must be much smaller than the characteristic scale L of the medium.

$$\lambda_D \ll L$$

What values would we use for L? For a laboratory plasma it is the approximate length of the plasma. In the ionosphere, we typically use the scale height H. In the magnetosphere we use the length of the field line to which the plasma is attached.

2. The plasma frequency must be much greater than the collision frequency.

$$\omega_p \gg v_c$$

3. The number of charged particles N_D inside a *Debye sphere* must be much greater than 1.0.

$$N_D = \left(\frac{4}{3}\pi \lambda_D^3\right) n_e \gg 1$$

Chapter 6 Space is a Place...with Plasma

EXAMPLE 6.3

Plasma Determination

- *Problem Statement:* Determine if the constituents of the ionosphere qualify as plasma according to the criteria above.

- *Given:* Temperature of the ionized gas is about 1000 K and the charged particle density is 10^{11} particles/m^3, collision frequency is $\sim 2 \times 10^4$ Hz.

- *Assume:* The given values are representative of the overall ionosphere.

- *Relevant Concepts:*

  Criterion 1: $\lambda_D \ll L$

  Criterion 2: $\omega_p \gg v_c$

  Criterion 3: $N_D = \left(\frac{4}{3}\pi \lambda_D^3\right) n_e \gg 1.0$

- *Solution:*

  1. Length scale

  $$\lambda_D = \sqrt{\frac{k_B T_e \varepsilon_o}{n_e e^2}} = \sqrt{\frac{(1.38 \times 10^{-23} \text{J/K})(1000 \text{ K})(8.85 \times 10^{-12} \text{C}^2/(\text{N} \cdot \text{m}^2))}{(10^{11} \text{particles/m}^3)(1.6 \times 10^{-19} \text{C})^2}} = 6.9 \text{ mm}$$

  The length scale L is hundreds of km (the thickness of the ionosphere). This value is far larger than 6.9 mm. The length criterion is met.

  2. Plasma frequency compared to collision frequency:

  $$\omega_p = \sqrt{\frac{(10^{11} \text{particles/m}^{-8}) \times (1.69 \times 10^{-19} \text{C})^2}{8.85 \times 10^{-12} \text{C}^2/(\text{N} \cdot \text{m}^2) \times 9.11 \times 10^{-31} \text{kg}}} = 1.78 \times 10^7 \text{Hz}$$

  $$\omega_p = 1.78 \times 10^7 \text{Hz} \gg 2 \times 10^4 \text{Hz}$$

  The frequency criterion is met.

  3. Number density within a radius of a Debye length is

  $$N_D = \left(\frac{4}{3}(3.14)(6.9 \times 10^{-3} \text{m})^3\right)(10^{11} \text{particles/m}^3) = 1.4 \times 10^5 \text{particles} \gg 1 \text{ particle}$$

- *Interpretation:* According to all of the criteria the ionosphere is plasma. Amazingly, on average, the fraction of ionization in the ionosphere ranges from 10^{-4} to 10^{-6}. The plasma state exists with only a tiny fraction of ionized particles within a background neutral gas.

Follow-on Conceptual Question: During strong solar flares, Earth's mesosphere can have a population of ionized particles. But the mesosphere constituents are not considered a plasma. Why do you think they fail to meet the plasma criteria even during flare events?

6.2 Single-particle Dynamics—I

6.2.1 Single-particle Accelerated Motion

Objectives: After reading this section, you should be able to...

- Describe and calculate the force and acceleration on a charged particle in a uniform electric field
- Describe and calculate the force and acceleration on a charged particle in a uniform magnetic field
- Calculate the gyroradius and gyrofrequency of particles influenced by a magnetic field
- Describe and calculate the force and acceleration on a charged particle under the influence of uniform but unbalanced magnetic and electric fields
- Explain how the motion in a combined electric and magnetic field differs from the influence of either field alone

We ultimately want to understand the collective behavior of plasma. To do so we need to know which of the individual plasma particle behaviors we can smooth away in statistical treatments of plasma fluids. This section treats briefly of single-particle dynamics. We begin with Newton's Second Law of Motion and incorporate the electric and magnetic portions of the Lorentz Force and other forces as needed. The other forces may include pressure gradient forces, gravity, non-uniformity in the magnetic field, and collisions. The result is the rate of change of momentum for a particle of charge q and mass m.

$$m\frac{d\mathbf{v}}{dt} = \mathbf{F}_{total} = q\mathbf{E} + q(\mathbf{v} \times \mathbf{B}) + \mathbf{F}_{other} \qquad (6\text{-}7)$$

where

$\mathbf{E}$ = electric field vector [N/C or V/m]

$\mathbf{B}$ = magnetic field vector [N/(A·m)]

$\mathbf{v}$ = charged particle's velocity vector [m/s]

In combination these forces produce linear and centripetal accelerations that result in gyrating motion. We separate the dynamics of Eq. (6-7) into categories: accelerated (linear and gyro) motion and across-field drift motion. Table 6-3 summarizes the results of Newton's Second Law applied to individual plasma particles upon which unbalanced forces act. We fill in the details in the subsections below.

Uniform Electric Field and No Magnetic Field: Constant Linear Acceleration

We start with the simple situation of a charged mass in an electric field (first entry in Table 6-3). The electric field exerts a force on a charge, accelerating it along the direction of the field. By assumption, no other forces are acting, so the charge continues to accelerate as long as the field is present. Newton's Second Law of Motion yields an equation of motion

Chapter 6 Space is a Place...with Plasma

$$\Sigma F = m\frac{dv}{dt} = qE$$

$$a = \frac{dv}{dt} = \frac{qE}{m} \quad (6\text{-}8)$$

Table 6-3. Accelerated Motion for an Individual Charged Particle. This table lists various ways that uniform electric and magnetic fields accelerate charged particles.

Accelerated Motion	$F = \frac{d(mv)}{dt}$ $= qE + q(v \times B) + F_{Other}$	$a = \frac{dv}{dt} = \frac{F}{m}$	Velocity	Comments
Uniform E	$F = qE$	$a_{\parallel} = \frac{qE}{m}$	$v_{\parallel} = v_{\parallel 0} + a_{\parallel}t$ $v_{\perp}$ is constant	Linear motion caused by the electric portion of the Lorentz force
Uniform B	$F = q(v \times B)$	$|a_{\perp}| = \frac{qBv_{\perp}}{m}$	$|v_{\perp}| = \frac{rqB}{m}$ $|v_{\perp}| = r\Omega$ $v_{\parallel}$ is constant	Circular motion caused by the magnetic portion of the Lorentz force $\Omega = \frac{qB}{m}$
Uniform E and B	$F = qE + q(v \times B)$	$a_{\parallel} + a_{\perp}$	$v_{\parallel} + v_{\perp}$	Helical motion

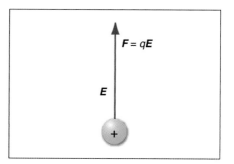

Fig. 6-6. Electric Force and a Positive Charge. The force depends on the total charge and electric field strength and direction.

The equation of motion tells us that the charge experiences a force given by qE. The resulting acceleration is the momentum change divided by the mass. We can integrate the equation of motion to find the particle's speed and trajectory. If the electric field is constant in space and time, and the charge has no transverse velocity component, the trajectory will be a straight line (Fig. 6-6).

Unbounded electric fields do not exist in space. However, local electric fields (such as those in the auroral acceleration region) produce significant accelerations over short distances. In these regions, the field transfers energy to the charged particles and changes their momentum.

Uniform Magnetic Field: Constant Circular Acceleration and Uniform Speed

The second entry in Table 6-3 is the magnetic component of the Lorentz force. In the absence of an initial motion relative to the field, this force is zero. However, thermal motions invariably cause particles to have a random component of v perpendicular to B. An ion with a velocity component perpendicular to a uniform external magnetic field experiences a force equal to

$$F = q(v \times B)$$

$$m\frac{dv}{dt} = m\frac{v^2}{r}\hat{r} = q(v \times B)$$

Using Newton's Second Law to equate the centripetal force to the Lorentz Force produces Eq. (6-9). The acceleration is perpendicular to both v and B and results in circular motion.

6.2 Single-particle Dynamics—I

$$a_\perp = \frac{dv}{dt} = \frac{qBv_\perp}{m} = \Omega v_\perp \tag{6-9}$$

where

$v_\perp$ = component of the particle's velocity vector perpendicular to the magnetic field [m/s]

Ω = gyrofrequency [Hz]

Recall from Example 1-5 that manipulation of Newton's Second Law reveals an associated radius of gyration (also called the *gyroradius* or *Larmor radius*)

$$r = mv_\perp/qB = v_\perp/\Omega \tag{6-10}$$

where

$$\Omega = qB/m \tag{6-11}$$

is the *gyrofrequency*.

For this situation, there is no force in the direction the charge is moving, so the particle's speed does not change. However, the direction of motion must continuously change, producing a circular (cyclotron) trajectory in the plane of motion (Example 6.4). Figure 6-7 shows that for a particle orbiting in a uniform magnetic field, the center of the orbit remains fixed in space in the absence of other external forces. The orbiting particle produces a tiny current and associated magnetic field that opposes weakly the direction of the external, ambient magnetic field. This property of a magnetized plasma is *diamagnetic*. We find a few situations in the space environment where the collective diamagnetism of plasma is important.

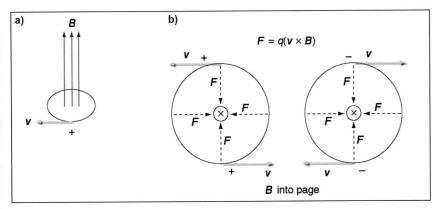

Fig. 6-7. Particle Motion in a Magnetic Field. a) We show a positive ion gyrating in a uniform magnetic field. If we lay our right hand along the instantaneous particle velocity vector **v** and curl our fingers in the direction of **B**, our extended thumb points in the direction of the centripetal force **F** that points radially inward. **b)** This diagram is a view of the motion of positive and negative charges from below the plane of circulation in part a. The gyroradius is not to scale.

Chapter 6 Space is a Place...with Plasma

EXAMPLE 6.4

Ion Mass Spectrometer

A mass spectrometer is an instrument that measures relative concentrations of ion constituents in the space environment. It generates a magnetic field to separate components of singly ionized atoms and molecules by mass. Figure 6-8 shows the ions entering the device through an opening and being exposed to the magnetic field. The field bends the trajectories of the ions with a radius r that depends on the mass. We consider one of the two major species prevalent in Earth's upper atmosphere—atomic oxygen.

- *Problem Statement:* What magnetic field strength is required so the atomic oxygen ions entering the instrument strike the collector of an ion spectrometer?

- *Given:* Ram velocity of the spacecraft is 7.5 km/s

  Atomic oxygen collector is placed 0.02 m from the entrance slit of the instrument

- *Assumptions:* We assume that the entrance speed of the ions is independent of mass. This assumption is not bad for measuring constituents in Earth's upper atmosphere: the ram speed of the spacecraft carrying the instrument is typically much greater than the ion speeds.

  We also assume that the large-scale electric field near the spacecraft is negligible and that the velocity vector is perpendicular to the magnetic field.

- *Solution:*

$$F = q(v \times B)$$

which is the force responsible for a centripetal acceleration

$$\frac{v^2}{r}(\hat{r}) = \frac{q(v \times B)}{m}$$

and considering only the magnitude of the acceleration

$$\frac{v^2}{r} = \frac{qvB}{m}$$

Solving for B yields:

$$B = \frac{mv}{qr}$$

The mass of O^+ is 16 amu = 16 (1.67 × 10^{-27} kg) = 2.67 × 10^{-26} kg. Since the radius equals half the diameter, r = 0.01m

Substituting gives

$$B = (2.67 \times 10^{-26} \text{kg})(7500 \text{ m/s})/((1.6 \times 10^{-19}\text{C})(0.01 \text{ m}))$$

$$\approx 0.13 \text{ T}$$

- *Interpretation:* This magnetic field is strong by Earth standards. We use the one-to-one relationship between mass and radius to separate ions by mass.

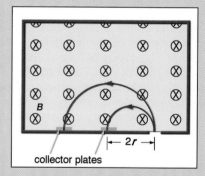

Fig. 6-8. A Mass Spectrometer. Particles enter the slit and react to the magnetic field. Depending on their mass and charge, they accumulate at different positions on the collector plates. *B* points into the page.

Pause for Inquiry 6-6

Is an electron cyclotron frequency larger or smaller than an ion cyclotron frequency? By what order of magnitude?

Pause for Inquiry 6-7

Write a formula for the work done by an electric field in accelerating a charged particle.

Pause for Inquiry 6-8

Verify the units of gyroradius and gyrofrequency.

Pause for Inquiry 6-9

How much work does the magnetic field do in creating this gyromotion?

Uniform *B* Field and Parallel *E* Field: Stretched Helical Motion

The presence of an *E* field parallel to *B* causes the particle trajectory to stretch into a helix along the axis of a magnetic field line. Equation (6-12) describes the components of acceleration. Figure 6-9a shows helical motion that is common to charged particles in a magnetized medium.

$$a_\parallel = \frac{dv_\parallel}{dt} = \frac{qE}{m}, \quad a_\perp = \frac{dv_\perp}{dt} = \frac{qBv_\perp}{m} = \Omega v_\perp \qquad (6\text{-}12)$$

Where the $\parallel$ and $\perp$ symbols indicate components parallel and perpendicular to the magnetic field, respectively.

Figure 6-9b is an image of electrons spiralling about Earth's magnetic field lines above Fairbanks, Alaska, USA. The light trail is created as electrons collide with neutral particles in Earth's atmosphere.

Pause for Inquiry 6-10

Why do the coils in Fig. 6-9b appear to have an opposite sense twist compared to those in Fig. 6-9a?

6.2.2 Single-particle Unaccelerated Motion

Objectives: After reading this section, you should be able to...

- Describe and calculate the force and acceleration on a charged particle under the influence of a uniform magnetic field and other fields perpendicular to that field
- Describe why charges in a uniform field perpendicular to *B* exhibit drift motion
- Calculate charged particle drift associated with electrical and gravitational fields
- Calculate charged particle drifts, associated gradients, and curvatures in the magnetic field
- Define "guiding center motion"

Now we investigate how plasma particles respond to balanced forces, that is, to forces that sum to zero. Uniform drift motion is the particle's only motion. At the micro scale the particles accelerate, while at the macro scale they appear to

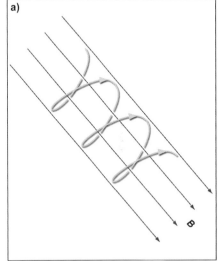

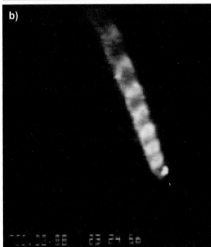

Fig. 6-9. a) **Helical Motion of Positively Charged Particles about Magnetic Field Lines.** The effect of the along-the-field acceleration is minimized. b) **Electrons Spiraling around a Magnetic Field Line in Earth's Northern High Latitudes.** This image comes from the ECHO 7 sounding rocket test conducted in February 1988 from the Poker Flats Research Range near Fairbanks, Alaska. The two-stage sounding rocket carried a large electron gun, operated here at an altitude of ~100 km. The 36-keV, 180 mA electron beam spirals upward along the magnetic field line. The image was made from a deployed subsatellite with a low-light-level TV. [Winckler, 1989]

drift along a guiding center. In a twist unique to magnetized plasma, a combination of forces may cause the charged particles to drift perpendicular to the driving fields. Tables 6-4 and 6-5 summarize the results of Newton's Second Law applied to individual plasma particles. We describe the details in the subsections below.

To analyze the net drift motion of a single plasma particle acted on by multiple forces, we recognize that the magnetic portion of the Lorentz force must be balanced by the perpendicular component of F_{other}

$$F_{other} = -q(v \times B) \tag{6-13}$$

where F_{other} may be gravity, an electric field, a pressure gradient, or other forces. The cross product operation guarantees that F_{other} will be perpendicular to both v and B.

We determine the drift velocity v_d by taking the cross product of both sides of Eq. (6-13) with B:

$$F_{other} \times B = -q(v_d \times B) \times B = qB^2 v_d$$

Where the magnitude of the double cross product is $qB^2 v$. Solving for v gives

$$v_d = \frac{F \times B}{qB^2} \tag{6-14a}$$

where

v_d = instantaneous guiding-center drift velocity vector [m/s]
F = external force vector acting on the particle [N]

The drift is always perpendicular to the external force and to the magnetic field. The subscript d is often replaced by a subscript indicating the source of the drift, such as E for electric field and g for gravity.

Table 6-4. Drift Motion for an Individual Charged Particle. We assume the fields are uniform. On the macro scale the acceleration is zero, but the individual particles can be accelerated.

Drift Motion	$\Sigma F = \frac{d(mv)}{dt} = 0$	$v_d = \frac{F \times B}{qB^2}$	Comments
Uniform E and B	$qE = -q(v \times B)$	$v_E = \frac{E \times B}{B^2}$	Drift perpendicular to E and B, independent of charge
Uniform g and B	$mg = -q(v \times B)$	$v_g = \frac{mg \times B}{qB^2}$	Drift perpendicular to mg and B Charge-dependent
Other uniform net forces, collisions, pressure gradient	$F_{other} = -q(v \times B)$	$v_{other} = \frac{F_{other} \times B}{qB^2}$	Drift motion determined by the nature of the force and the charge of the particle

Uniform Fields Perpendicular to a Uniform *B* Field: Cycloidal Drift Motion

E × ***B*** **Drift.** Consider a vacuum permeated by two uniform fields: a magnetic field ***B*** and an electric field ***E*** perpendicular to ***B***. The fields are generated by source charges and currents that are not part of our problem, so they are independent of the plasma charges that we analyze here. We wish to determine the plasma particle motion.

$$v_E = \frac{q\mathbf{E} \times \mathbf{B}}{qB^2}$$

With the charges canceling, we are left with the general expression for the ***E*** × ***B*** drift

$$v_E = \frac{\mathbf{E} \times \mathbf{B}}{B^2} \qquad (6\text{-}14b)$$

Figure 6-10 shows ions and electrons drifting under the influence of the ***E*** and ***B*** fields. The particles' orbits consist of narrow and broad loops. The motion is a combination of cross-field drift and gyration. In the broader portions of the curves, ***E*** and ***v*** × ***B*** are in opposite directions, weakening the net force on either particle. This reduced force allows the gyroradius to grow. In the tighter portions of the curves, ***E*** and ***v*** × ***B*** are in the same direction, effectively creating a larger force, thereby making the radius of gyration smaller. The combination of alternating small and large gyroradii deforms the motion from circular to cycloidal. The drift is independent of the properties of the drifting particles (q, m, v), and hence, is equal for both electrons and ions. Relativistic effects have been ignored. They would be important if v_d were a significant fraction of c.

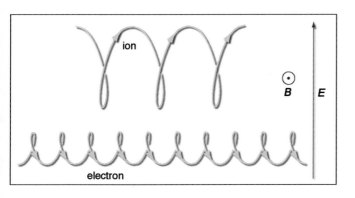

Fig. 6-10. Ion and Electron Drifts in Perpendicular *E* and *B* Fields. This diagram shows electrons and ions drifting in the ***E*** × ***B*** direction (***B*** out of the page and ***E*** upward). Where the forces are in the same direction, a large force with smaller gyroradius develops. Where forces oppose each other, a larger gyroradius develops. Over the full gyration the particles' trajectories look like a cycloid.

The motions are greatly simplified to emphasize the difference in radii at the top and bottom only. The net motion is perpendicular to both fields. Although we show the motions separately for clarity, in fact the particle drifts are superposed along a common guiding center. Examples 6.5 and 6.6 illustrate particle behavior for ***E*** × ***B*** drifts.

In Focus Box 6.2, we slightly relax one of the assumptions associated with single-particle motion and allow the charged particles to experience collisions. The collisional influence is an example of F_{other}. Doing so allows us to study how plasmas behave in the lower thermosphere, where plasma mixes with the relatively dense neutral atmosphere.

Chapter 6 Space is a Place...with Plasma

> **Pause for Inquiry 6-11**
>
> Suppose the electric field in Fig. 6-10 had a component directed into the page. How would the motions of the ion and electron change as a result? Would there be a current?

EXAMPLE 6.5

Relative Influence of E and $v \times B$

- *Problem statement:* Determine the relative influence of the E and $v \times B$ forces on the ion in Fig. 6-10.
- *Relevant Concepts:* Lorentz force and Newton's Second Law of Motion.
- *Given:* Figure 6-10 and information therein.
- *Assume:* The ratio of radii at the top and bottom of the ion loops is 5:1.

- *Solution:* $r_{top} = 5 r_{bottom}$ so
  $F_{bottom} = 5 F_{top}$
  $|q(E - v \times B)| = |5q(E - v \times B)|$
  $E + vB = 5E - 5vB$
  $6vB = 4E$
  $E = (3/2)vB$

- *Interpretation and Implications:* For the situation illustrated in Fig. 6-10 the ratio of E to $v \times B$ is 1.5:1.

EXAMPLE 6.6

$E \times B$ Drift Motion

- *Problem statement:* Determine the drift speed and direction of a magnetospheric ion and electron moving under the influence of typical electric and magnetic field values.
- *Relevant Concepts:* $E \times B$ drift motion
- *Given:* Magnetospheric dawn-to-dusk electric field value of 0.1 mV/m and northward magnetic field value of 20 nT
- *Assume:* Uniform field strength and perpendicular fields

- *Solution:*
  $$v_E = \frac{E \times B}{B^2}$$
  $$v_E = \frac{(1 \times 10^{-4}\ \text{V/m})(20 \times 10^{-9}\ \text{T})}{(20 \times 10^{-9}\ \text{T})^2}$$
  $$v_E = \frac{(1 \times 10^{-4}\ \text{V/m})}{(20 \times 10^{-9}\ \text{T})} = 5 \times 10^4\ \text{m/s} = 50\ \text{km/s}$$

  Direction is toward Earth

- *Interpretation and Implications:* For typical magnetospheric field conditions, particles drift Earthward at the rate of 50 km/s.

Follow-on Exercise: Occasionally flow bursts are observed in the magnetotail with speeds of 300 km/s. What electric field value would be needed to create such flows from $E \times B$ drifts? Use the value of magnetic field given in the example.

Gravitational Drift. We now consider the impact of the gravitational field g on charged particles immersed in a magnetic field. The "other" force on the particle is $F_{other} = mg$, where m is the mass of the particle. Substituting this expression into the guiding-center drift equation (Eq. (6-14a)), gives

$$v_g = \frac{m g \times B}{qB^2} \qquad (6\text{-}14c)$$

Focus Box 6.2: Charged Particles in a Magnetic Field and a Collisional Environment

Previously we considered a non-collisional environment. Now we briefly consider the collisions to see how they modify the non-collisional results. Collisions interrupt the gyro-motion of charged particles. In the case of a pre-existing external force that causes drift motion, any collisions further distort the motion. This situation is complicated, but we show particle motion with a few simple diagrams. We use an electric field as the source of the external force, but any force has a similar impact.

Suppose we begin with an electron in the E and B fields as shown. If the E field is particularly strong with respect to the B field, then the cycloid motion of the electron will tend toward an extreme that appears as semicircular motion.

As a result of collisions, the electron is partially freed from its drifting cycloid motion and takes on a component of motion associated with the external electric field. In the extreme case, when collisions dominate the particle motion, the cycloidal motion is destroyed and the particles behave as if the magnetic field is not present.

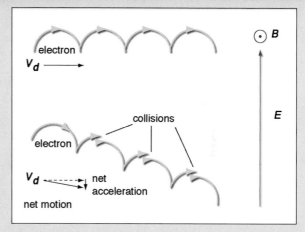

Fig. 6-11. **The Influence of Collisions on Single-particle Motion.** In the top portion of the diagram an electron drifts under the influence of a strong E perpendicular to B. Without collisions the particle drifts in the $E \times B$ direction. In the bottom portion the drift is interrupted by collisions that allow the electric field to have slightly more influence. The electron accelerates opposite the direction of E, in addition to drifting in the $E \times B$ direction.

Figure 6-12 shows ions and electrons drifting under the influence of perpendicular g and B fields. The force is larger at the bottom of the orbits than at the top.

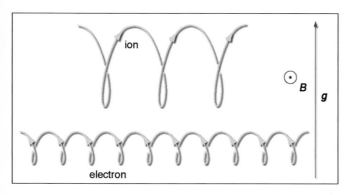

Fig. 6-12. **Particles Moving under the Combined Influence of Magnetic and Gravitational Fields.** Ions drift to the right in a looping path. Electrons drift to the left in a smaller, faster looping motion. The resulting current is directed toward the right, following the ions.

This drift depends on the charge. In plasma, the oppositely charged ions and electrons experience charge separation from the gravitational drift. Furthermore, the drift is proportional to the mass of the particle. Because the ions are so much more massive than the electrons, they have much higher drift speeds. The flow of charged particles in opposite directions creates a current, and this example is one of the rare cases where the ions are the primary current carriers. In the circuits of most electrical devices the electrons tend to be the current carriers. Because of their low mass, electrons are more easily accelerated by electric fields.

> **Pause for Inquiry 6-12**
>
> In Fig. 6-12, the charged particles encounter no external electric field, yet a current develops. Explain why this is so.

Charged-particle Motion in a Non-uniform *B* Field: Cycloidal Drift Motion

Non-uniformities in the magnetic field strength such as curvature and gradient also produce drift motions.

Gradient B Drifts. If magnetic field lines are straight but the magnitude of ***B*** varies in space, the particles' orbits look like those in Fig. 6-13. The increased field strength at the top of the loop produces a larger force that causes higher speeds and a tighter radius of curvature. This situation is called the *gradient drift* and is produced, not by ***B***, but rather by the gradient in ***B***. Because the particles are gyrating in opposite directions, the longer arcs in the loops result in opposite drift directions. The current is in the direction of motion of the ions.

Fig. 6-13. Particles Moving under the Combined Influence of a Magnetic Field and a Gradient in that Field. Ions flow to the left in a looping path. Electrons move to the right in a looping path. The resulting current is directed toward the left, following the ions.

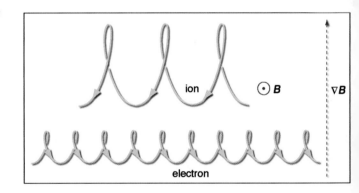

The nature of this force is beyond the scope of this text. Here we state the result, assuming that the force created by a gradient in the field is

$$F = -\frac{1}{2}mv_\perp^2 \frac{\nabla B}{B}$$

We recall from Table 6-4 that the drift velocity is given by $v_d = \dfrac{F \times B}{qB^2}$

Combining these equations we get

$$v_{\nabla B} = -\frac{1}{2}mv_\perp^2 \frac{\nabla B \times B}{qB^3} \qquad (6\text{-}15)$$

6.2 Single-particle Dynamics—I

Table 6-5. Drift Motion for an Individual Charged Particle in a Non-Uniform Magnetic Field. Magnetic fields are rarely uniform; non-uniform fields produce particle drifts. On the macro scale, the acceleration is zero, but the individual particles can be accelerated.

Drift Motion	$\Sigma F = \dfrac{d(m\mathbf{v})}{dt} = 0$	$\mathbf{v}_d = \dfrac{\mathbf{F}_\perp \times \mathbf{B}}{qB^2}$	Comments
Gradient in $\mathbf{B}$	$\mathbf{F}_{\nabla B} = -\dfrac{1}{2}mv_\perp^2 \dfrac{\nabla \mathbf{B}}{B}$ $= -q(\mathbf{v} \times \mathbf{B})$	$\mathbf{v}_{\nabla B} = \dfrac{-mv_\perp^2}{2}\dfrac{\nabla \mathbf{B} \times \mathbf{B}}{qB^3}$	Drift perpendicular to $\nabla \mathbf{B}$ and $\mathbf{B}$ Charge-dependent
Curvature in $\mathbf{B}$	$\mathbf{F}_{\nabla \times B} = \dfrac{mv_\parallel^2}{R_C^2}\mathbf{R}_C$ $= -q(\mathbf{v} \times \mathbf{B})$	$\mathbf{v}_C = \dfrac{mv_\parallel^2 \mathbf{R}_C \times \mathbf{B}}{R_C^2\, qB^3}$	Drift perpendicular to $\mathbf{F}_C$ and $\mathbf{B}$ Charge-dependent R_C is radius of curvature

Curvature B Drifts. Another variation in the spatial structure of a dipolar magnetic field comes from its curvature, which is a natural feature of all dipole fields. As a gyrating particle moves with speed $v_\parallel$ along a $\mathbf{B}$-field line that curves, some additional force must act on the particle and make it follow the field line. This force is inherent to the curved field line. Figure 6-14 shows the curved particle trajectory.

The force associated with the curvature is

$$ \mathbf{F} = \dfrac{mv_\parallel^2}{R_C}\hat{\mathbf{R}}_C = \dfrac{mv_\parallel^2}{R_C^2}\mathbf{R}_C $$

where

- F = force vector causing the curved motion [N]
- m = particle mass [kg]
- $v_\parallel$ = particle instantaneous velocity vector parallel to the curving magnetic field [m/s]
- $\hat{\mathbf{R}}$ = unit vector along the radius of curvature at an instant [unitless]
- $\mathbf{R}_C$ = vector along the radius of curvature [m]

The curvature drift is given by

$$ \mathbf{v}_C = \dfrac{mv_\parallel^2 \mathbf{R}_C \times \mathbf{B}}{R_C^2\, qB^3} \tag{6-15e} $$

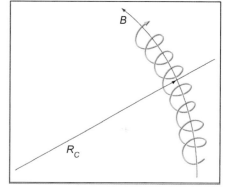

Fig. 6-14. Particle Moving under the Influence of a Curved Magnetic Field. This drawing shows a charged particle following a magnetic field line, as it curves with radius R_C.

A magnetic dipole field has both curvature and gradient, so in dipole applications these forces are often combined into a single force that contributes to a drift called the *gradient-curvature drift*.

$$ \mathbf{v}_{\nabla BC} = \left(mv_\parallel^2 \dfrac{\mathbf{R}_C}{R_C^2} - \dfrac{1}{2}mv_\perp^2 \dfrac{\nabla \mathbf{B}}{B} \right) \times \dfrac{\mathbf{B}}{qB^2} \tag{6-15f} $$

In Eq. (6-15f), the vectors $\mathbf{R}_C$ and $\nabla \mathbf{B}$ both form cross products with $\mathbf{B}$.

The strength of the force has elements of kinetic energy dependence (mv^2). The kinetic energy due to the parallel component of velocity controls the strength of the curvature force, while the kinetic energy due to the perpendicular component controls the strength of the gradient force. Equation (6-15f) looks intimidating. However, we can learn quite a lot about the magnetosphere by considering the directions of the drifts without doing any math calculations. We examine this in Example 6.7.

> **Pause for Inquiry 6-13**
>
> Are the contributions of curvature drift and gradient drift for a given particle in the same or opposite directions?

EXAMPLE 6.7

Curvature-gradient Drift

- *Problem statement:* Determine the direction of the curvature-gradient drift of a positive ion in Earth's nighttime equatorial magnetosphere.
- *Relevant Concepts:* Curvature and gradient drift of charged particles
- *Given:* Dipolar shape of Earth's magnetic field in the inner and middle magnetosphere (Fig. 6-15) and Eqs. (6-15d) and (6-15e).
- *Assume:* Static field configuration
- *Solution:*

1) The curvature drift is:

$$v_c = \frac{mv_\parallel^2 R_C \times B}{R_C^2 q B^2}$$

Therefore the drift direction follows $R_C \times B$. The R_C vector (shown in red in the figure) points away from Earth. Near the equatorial plane the magnetic field points northward. Therefore, $R_C \times B$ points out of the page at midnight. At dusk the curvature drift is directed toward the dayside.

2) Gradient drift is

$$v_{\nabla B} = -\frac{1}{2} m v_\perp^2 \frac{\nabla B \times B}{qB^2}$$

The drift direction is $B \times \nabla B$ (we eliminate the minus sign by changing the order of the cross product operation). The ∇B vector (shown in green in Fig. 6-15) points toward Earth in the equatorial plane, meaning that B becomes larger closer to Earth. Near the equatorial plane the magnetic field points northward. $B \times \nabla B$ or $(-\nabla B \times B)$ is directed out of the page at midnight. At dusk the gradient drift is directed toward the dayside.

- *Interpretation and Implications:* The combined effect of gradient and curvature drift on a positively charged particle in the equatorial plane of Earth's inner magnetosphere is to guide the particle around Earth from midnight to dusk to noon and from there around to dawn and back to midnight. A negatively charged particle moves in the opposite direction. The opposing motion of the charged particles creates a current, called a ring current, that has measurable effects at Earth's surface. This is the physical origin of the ring current we discuss at the end of Chap. 4.

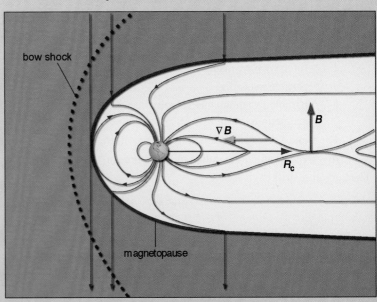

Fig. 6-15. Earth's Magnetosphere. This drawing depicts an edge-on equatorial view of Earth's dipolar magnetic field and the surrounding space that it affects.

A Macroscopic Perspective: Guiding-center Drift Motion

Most space plasma particles are influenced by a combination of forces. We get a "big picture" view of the single-particle motion by combining the information presented above into a concept called "guiding-center drift," which we illustrate in Fig. 6-16.

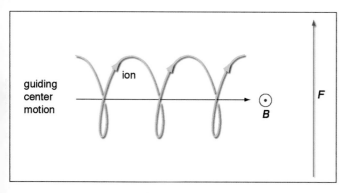

Fig. 6-16. Cycloidal and Guiding-center Motion. The horizontal arrow represents the drift motion (v_d). The magnetic field **B** points out of the page, and the perpendicular force $F_\perp$ points to the top of the page.

If a perpendicular external field acts on a gyrating charged particle on a time scale that is long compared to the period of one orbit, then the motion may be separated into two forms: uniform circular motion and a relative displacement (drift) of the center of the particle's orbit. To put this in a more familiar context, we consider the motion of a point on the edge of a car tire. As the car moves, the point executes a cycloidal trajectory that we separate into two simpler motions: a circular motion whose angular rate is the same as that of the tire, and a linear motion at the speed of the moving car. If we keep pace with the car, all we see is the circular motion of the point on the tire. Alternatively, we can track the tire's linear motion component by standing at the roadside and observing the movement of the tire's center as the car moves past. In other words, if we ignore the particle's rotational motion, we see a uniform-drift motion (drift velocity), which we describe in the discussions above. We also call this motion the *guiding-center drift* and show it for a plasma particle in Fig. 6-16. In many macroscopic applications in plasma, the guiding center drift is more important than the cycloidal motion, so the latter is often averaged out, but the drift wouldn't exist without the underlying cycloids.

Pause for Inquiry 6-14

Is the guiding-center motion in Fig. 6-16 valid for both signs of charge? What happens to the direction of $v_\perp$ if the sign of the charge is reversed?

For a given net force $F_\perp$, the drift velocity given by Eq. (6-14a) is constant in space and time, and so is consistent with the notion of guiding-center drift motion. But what about the uniform circular (accelerated) motion similar to that of the point on the tire's edge in the example above? This component of motion obviously does not disappear, but it does average out. For a fixed force $F_\perp$, if we take measurements of the particle's location over a time interval that is much longer than the rotational period of the circular motion, then Eq. (6-14a) will adequately describe the particle's motion. In other words, actual observations of the particle's position versus time match well against the graph of the function $x(t) = v_d t$, where t is the elapsed time, and v_d is the guiding-center drift speed.

6.3 Thermal Plasma Fluid Dynamics

In space plasmas, each cubic meter contains 10^6–10^{12} charged particles. Modeling the behavior of even a few million of these particles would tax even the fastest computers. Fluid dynamics provides an alternative means of describing the bulk plasma behavior.

6.3.1 Maxwell-Boltzmann Distribution and the Saha Equation

> **Objective:** After reading this section, you should be able to...
>
> ♦ Describe the terms in a Maxwell-Boltzmann distribution function
> ♦ Describe, according to the Saha equation, situations in which the ionization fraction in a plasma will increase

Here we ignore magnetic fields for the moment and concentrate on plasma particles that collide frequently enough to be in thermal equilibrium (thermalized). We estimate how much ionization is likely to occur in plasma of a given density and temperature and we estimate statistically the velocity of a plasma particle.

Distribution Functions. Space plasma consists of vast numbers of particles. Instead of calculating the motion and position of every particle and then averaging the results, we use equations for the statistical quantities. We recall from Chap. 2 that the mean kinetic energy of a molecule in a gas relates to the temperature of the gas by the relation:

$$\frac{1}{2}mv_{rms}^2 = \frac{3}{2}k_B T$$

where $v_{rms} = \sqrt{\langle v^2 \rangle}$

This is the average kinetic energy of an atom or molecule in the ensemble of particles. These particles are colliding with each other continually, and these collisions can accelerate a few particles to very high speeds. If a collision is sufficiently energetic, it will knock an electron free, ionizing the particle.

In equilibrium distributions of gas molecules, in which concepts such as temperature make sense, we typically find (and often assume) that the distribution of atomic or molecular speeds follows the *Maxwell-Boltzmann distribution*. The speed distribution function, $\chi(v)$ describes the probability of a particle having a speed between $[v, v + dv]$. For a gas in thermodynamic equilibrium, and in which the particles are free to move between short duration collisions, this probability is given by

$$\chi(v)dv = 4\pi \left(\frac{m}{2\pi k_B T}\right)^{\frac{3}{2}} v^2 \exp\left(-\frac{mv^2}{2k_B T}\right) dv \qquad (6\text{-}16)$$

We show this function for several different speeds in Fig. 2-8. The first term in parentheses is a normalizing factor, which forces the probability distributed over all speeds to equal one. The exponential term accounts for the decreasing probability of particles with very high kinetic energy in a thermalized gas. The function describes the probability of a particle's velocity being near a given value, as a function of the mass m of the particle, the speed v, and the system temperature T.

From this we obtain the particle distribution, which describes how many gas particles populate a given speed bin:

$$N(v)dv = 4\pi N_T \left(\frac{m}{2\pi k_B T}\right)^{\frac{3}{2}} v^2 dv \qquad (6\text{-}17a)$$

where

dv = the speed increment [m/s]
N_T = the total number of particles

We can relate this speed function to other forms of energy. For example, the speed and potential energy in an atmospheric distribution, where altitude is a factor, includes a potential energy term (mgz). For an atmosphere in thermodynamic equilibrium the particle distribution function is:

$$N(v)dv = (4\pi N_T)\left(\frac{m}{2\pi k_B T}\right)^{\frac{3}{2}} v^2 \exp\left(-\frac{(1/2)mv^2 + mgz}{k_B T}\right) dv \qquad (6\text{-}17b)$$

Integrating over speed and dividing by volume produce $n = n_o \exp(-mgz/k_B T)$, which is the Law of Atmospheres (Chap. 3). In Chap. 2 we call mgh (or mgz) the potential energy of position. The exponential term means that the probability of finding particles at great heights is small.

For a plasma in thermodynamic equilibrium, the more relevant potential energy for the system often takes the form qV, where the potential at large distances from the charge is zero.

$$N(v)dv = (4\pi N_T)\left(\frac{m}{2\pi k_B T}\right)^{\frac{3}{2}} v^2 \exp\left(-\frac{(1/2)mv^2 + qV}{k_B T}\right) dv \qquad (6\text{-}17c)$$

Integrating simplifies this equation to $n = n_o \exp(-qV/k_B T)$. The exponential term indicates that the likelihood of finding particles with large electric potential energy is low.

For a given volume, the Boltzmann Equations (6-17a–c) predict more particles with low kinetic or potential energy, and fewer high-energy particles. Although space plasmas are often far from equilibrium, and non-Maxwellian distributions are common, the Maxwell-Boltzmann distribution is often a good approximation.

Saha Equation. How much energy does a gas need to be called plasma? The *Saha equation*, named after Indian astrophysicist, Meghnad Saha, answers this question. The Saha equation quantifies the amount of ionization in a thermalized gas energized to a given temperature. The ratio of ionized to neutral particles is given by

$$\frac{n_i}{n_n} = \frac{n_e}{n_n} = \frac{(2\pi m_e k_B T)^{3/2}}{n_i h^3} \exp\left(\frac{-U_i}{k_B T}\right) \quad (6\text{-}18a)$$

where

n_e = number density of electrons [particles/m^3] (equals the number density of ions, n_i)

n_n = number density of neutral particles [particles/m^3]

m_e = electron mass (9.11×10^{-31} kg)

U_i = ionization energy of the gas [J]

k_B = Boltzmann's constant (1.38×10^{-23} J/K)

h = Planck's constant (6.63×10^{-34} J·s)

We combine the constants and simplify the expression:

$$n_i = [n_n (2.4 \times 10^{21} T^{3/2}) \exp(-U_i/k_B T)]^{1/2} \quad (6\text{-}18b)$$

The Saha equation quantifies one of nature's ongoing battles: maintaining ionization in a thermal plasma. As the relative fraction of ionization increases, the number of ions n_i increases and the number of neutral particles n_n decreases. Thus the raw material for ionization is depleted. The greater the number of ionized particles in the gas, the more free electrons are available for recombination (the process by which an electron and an ion combine to form a neutral particle). If more free electrons are available for recombination, then the relative fraction of ionization decreases. This situation allows interplanetary plasma to exist: so few electrons in a given volume are available for recombination that the particles remain as a plasma long after whatever created the plasma ceases to affect it.

In a thermal plasma there are three ways to maintain or increase the ionization ratio:

- Vigorous thermal agitation ($k_B T \approx U_i$). The high kinetic energy of the particles causes ionization by collision. This occurs in the solar core, solar corona, and Earth's upper atmosphere.

- Low density. This condition makes the probability of recombination low: after an atom is ionized, finding an electron to recombine with so that energy and momentum are conserved (e.g., in the solar wind) is unlikely.

- Departure from thermal equilibrium. In this case the Saha equation loses its validity. For space plasmas, collision mean free paths are usually long and collision frequencies low. This means that the plasma takes a very long time to reach thermal equilibrium.

Pause for Inquiry 6-15

Assuming constant temperature, what happens to the relative amount of ionization as ionization potential energy increases?

Pause for Inquiry 6-16

If $n_i = n_e$, show that the electron density in a thermal plasma is proportional to $(n_n)^{1/2}$.

EXAMPLE 6.8

Kinetic Motion and Ionization Potential

- *Problem statement:* Estimate the temperature required to ionize a ground-state hydrogen atom.
- *Relevant Concepts:* Kinetic motion and ionization potential
- *Given:* Ionization energy for a ground-state hydrogen atom is 13.6 eV
- *Assume:* Thermal equilibrium and no magnetic field
- *Solution:*

  $k_B T_{ionization} = 13.6$ eV $(1.6 \times 10^{-19}$ J/eV$)$

  $T_{ionization} = (13.6$ eV$)(1.6 \times 10^{-19}$ J/eV$)/1.38 \times 10^{-23}$ J/K
  $= 1.6 \times 10^5$ K

- *Interpretation and Implications:* The temperature required for "thermal" ionization of hydrogen atoms is very high, such as that found in the solar corona. Yet hydrogen ions are present at lower temperatures. The lower and cooler solar atmosphere contains hydrogen atoms that are excited above the ground state. Ionizing the already excited particles requires only a fraction of the full ionization energy.

6.3.2 Energy Density in Plasma

Objectives: After reading this section, you should be able to...

- Describe and calculate the energy density of ions and electrons in plasma
- Recognize when to simplify analysis to treat plasma as a single species

Plasmas are usually quite amorphous, so we often describe the energy content in terms of energy density u and then integrate over the appropriate volume of space to determine the total energy content. The forms of energy density we list below are very similar to those we describe in Chaps. 2 and 4. The most obvious difference is the need to separately quantify the energy for electrons and ions (and sometimes for specific species of ions).

In a plasma, the energy of random motion is

$$u_{thermal} = \frac{3}{2} k_B (n_i T_i + n_e T_e) \quad (6\text{-}19a)$$

Generally the number of ions and electrons is the same, but their temperatures are different. Under these circumstances Eq. (6-19a) becomes

$$u_{thermal} = \frac{3}{2} n_i k_B (T_i + T_e) \quad (6\text{-}19b)$$

If the magnetic field threading a plasma is relatively strong, the number of degrees of freedom of motion decreases from three to two, which makes the leading constant in Eqs. (6-19a and b) equal to 1 rather than 3/2. If the ion and electron temperatures are the same, the energy density is roughly double that of an ideal gas

$$u_{thermal} = 2nk_BT \qquad (6\text{-}19c)$$

Ions and electrons both contribute to the pressure. Plasma in a steady state flow also has kinetic energy of directed motion

$$u_{KE} = \frac{1}{2}n_i m_i v_i^2 + \frac{1}{2}n_e m_e v_e^2 \qquad (6\text{-}20a)$$

If the ions and electrons do not have the same speed, the only simplification is to convert to mass density

$$u_{KE} = (1/2)(\rho_i v_i^2 + \rho_e v_e^2) \qquad (6\text{-}20b)$$

If the ions and electrons do have the same speed, the kinetic energy density is

$$u_{KE} = (\rho_i v_i^2) \qquad (6\text{-}20c)$$

The electromagnetic energy density in plasma is field-dependent rather than particle-dependent.

$$u_{B+E} = \frac{1}{2}\left(\frac{B^2}{\mu_0} + \varepsilon_0 E^2\right) \qquad (6\text{-}21a)$$

Usually the electric field term is small compared to the magnetic field term, so the result is

$$u_{B+E} \approx u_B = \frac{1}{2}\left(\frac{B^2}{\mu_0}\right) \qquad (6\text{-}21b)$$

In later chapters, we often compare the relative contributions of each of these terms to determine the processes that dominate in a given plasma regime. For example, in the solar wind the kinetic flow energy usually dominates, whereas in Earth's magnetosphere the temperature and field terms usually dominate.

Pause for Inquiry 6-17

Given the solar wind speed values from Chap. 5, which energy density in Eqs. (6-19c), (6-20c), or (6-21b) has the highest value? Which has the lowest?

Pause for Inquiry 6-18

Verify that Eq. (6-20a) simplifies to Eq. (6-20c), if the ions and electrons have the same speed.

EXAMPLE 6.9

Thermal and Kinetic Energy Density

- *Problem Statement:* Compare the ratio of thermal to kinetic energy density for electrons in the solar wind.
- *Relevant Concepts:* Thermal and kinetic energy density (Eqs. (6-19b) and (6-20b)).
- *Given:* $T_e = 1.4 \times 10^5$ K (slightly hotter than solar wind ions).
- *Assume:* Constant temperature, average solar wind speed, solar wind density of 5 ion-electron pairs/cm^3 and two degrees of freedom of motion.
- *Solution:*

  1. Electron thermal energy density

  $$\frac{2}{2} k_B n_e T_e = [(1.38 \times 10^{-23} \text{ J/K})(5 \times 10^6 \text{ electrons/m}^3)(1.4 \times 10^5 \text{ K})] = 9.66 \times 10^{-12} \text{ J/m}^3 \approx 10 \text{ pPa}$$

  2. Electron kinetic flow energy density

  $$u_{KE} = (1/2) m v_e^2 = (1/2)(5 \times 10^6 \text{ electrons/m}^3)(9.11 \times 10^{-31} \text{ kg/electron})(400 \times 10^3 \text{ m/s})^2 \approx 3.6 \times 10^{-13} \text{ J/m}^3 = 0.36 \text{ pPa}$$

  Ratio of electron thermal energy to kinetic energy = 10/0.36 = 26.8

- *Interpretation and Implications:* Electrons are more effective at moving energy via thermal motion than via flow motion.

Follow-on Exercise 1: Repeat the calculation for ions.

Follow-on Exercise 2: Compare the ratio of ion-to-electron thermal energy density and ion-to-electron flow energy density. Are ions and electrons equally good at both forms of energy transfer?

6.4 Magnetized Plasma Fluid Dynamics

When a magnetic field exerts a strong influence on a plasma, the fluids behave collectively as a single fluid that carries current. This behavior happens if the magnetized fluid is sufficiently "collisional." Model designers use magnetized fluid dynamics (magnetohydrodynamics) in regimes where the electrical conductivity of the fluid is high, changes are relatively slow, and the magnetic force acts as a unifying quasi-collisional force on charged particles.

6.4.1 The Basics of Magnetohydrodynamics (MHD)

Objectives: After reading this section, you should be able to...

- State the limitations for applying MHD
- Describe what additions and simplifications are needed to convert fluid dynamics to ideal MHD
- Describe the key parameters for determining if MHD approximations apply to plasma

Magnetohydrodynamics (MHD) Equations and Approximations

In Chaps. 4 and 5 we characterize non-magnetized fluids with hydrodynamics—a combination of conservation laws and the equation of state (which relates the macroscopic properties of a substance to one another). As shown in Fig. 5-8, hydrodynamics describe how different forces cause fluid motion. When describing fluids that are additionally influenced by magnetic fields, we keep the same basic hydrodynamic equations and add Maxwell's equations and Ohm's Law to accommodate the magnetic and electric field behaviors. *Magnetohydrodynamics* is a special limit, or approximation, of magnetized plasma physics that is characterized by the fluid-like behavior of a charged particle gas. Figure 4-28 shows the additional field and force interactions that occur with MHD). In Fig. 6-19 we blend the ideas of Fig. 5-8 and Fig. 4-28 to provide a flow diagram for MHD.

When modelers use MHD, they hope to specify (and ultimately forecast) the magnetized fluid behavior as a function of time and space. For example, they use MHD when they want to know how mass, energy, and momentum transfer from the solar wind to the magnetosphere, and how the energy deposits in and influences the coupled magnetosphere-ionosphere system.

In the MHD approximation we assume:

- We can ignore gyro-motions of individual particles in favor of guiding-center motions
- Electrical conductivity in the plasma is high, so there is no charge buildup and the magnetic field and the fluid are frozen together (Sec. 4.3)
- The characteristic length scale of the dynamics is much greater than the ion gyroradius and the mean free distance between collisions

6.4 Magnetized Plasma Fluid Dynamics

- The characteristic time of the dynamics is much longer than an ion gyro-period and the plasma oscillation period. Thus, time variations of the fields are slow—meaning that we can drop the displacement current term in the Maxwell-Ampere equation without penalty.
- Plasma velocities are not relativistic

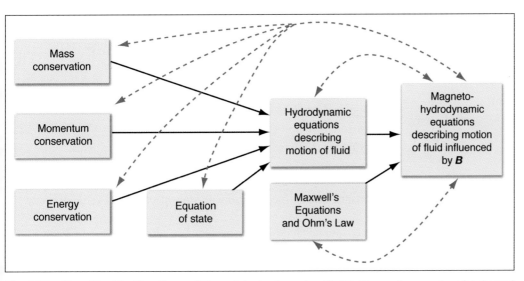

Fig. 6-17. Flow Chart for Equations of Magnetohydrodynamics (MHD). We use the equations for charged particles and electromagnetic fields to help describe the motion of a fluid influenced by magnetic fields.

Table 6-6 summarizes the remainder of the material in this subsection.

Table 6-6. The Highly Conducting Plasma (Ideal) Form of Magnetohydrodynamic (MHD) Equations. Here we list the simple form of the MHD equations in their per-unit-volume form. These equations include the MHD approximations we discuss in this section.

Fluid Equations	Maxwell's Equations	Constitutive Equations
Vector Equation	**Vector Equations**	**Vector Equation**
$\rho \dfrac{d}{dt}\mathbf{v} = -\nabla P + \mathbf{J} \times \mathbf{B}$ (1)	$\nabla \times \mathbf{E} = -\dfrac{\partial \mathbf{B}}{\partial t}$ (2)	Ohm's Law
	$\nabla \times \mathbf{B} = \mu_0 \mathbf{J}$ (3)	$\mathbf{J} = \sigma(\mathbf{E} + \mathbf{v} \times \mathbf{B})$ or (4) if $\sigma \to \infty$ then $\mathbf{E} = -\mathbf{v} \times \mathbf{B}$
Scalar Equation	**Scalar Equation**	**Scalar Equation**
$\dfrac{d\rho}{dt} = \dfrac{\partial \rho}{\partial t} + \nabla \cdot (\rho \mathbf{v}) = 0$ (5)	$\nabla \cdot \mathbf{B} = 0$ (6)	--
$\dfrac{d}{dt}(P \rho_m^{-\gamma}) = 0$ (7) where γ is the ratio of the heat capacity at constant pressure to the heat capacity at constant volume	$\nabla \cdot \mathbf{E} = 0$ (8)	$P = 2(\rho/m_i)k_B T$ (9)

Pause for Inquiry 6-19

The basic MHD equations in Table 6-6 capture several aspects of nature in an MHD system. Match the following statements and descriptions to the appropriate equation number in Table 6-6. Equation numbers may be used more that once.

- Mass is transported but not created or destroyed. Eq. _____
- Energy is conserved and, in its simplest form, it is thermal energy in an adiabatic system. Eq. _____
- Momentum changes come from pressure gradient forces and from currents that interact with the magnetic field. Eq. _____
- Charge buildup does not occur. Eq. _____
- Magnetic monopoles do not exist. Eq. _____
- Varying magnetic fields are the only source of electric fields. Eq. _____
- Currents are the only source of magnetic fields. Eq. _____
- The equation modified to deal with slow electric field variations only. Eq. _____
- The equation that describes sources of acceleration. Eq. _____
- The equation of state. Eq. _____
- This equation is the Maxwell equation modified for high conductivity. Eq. _____
- This equation relates magnetized plasma motion to the electric field. Eq. _____
- This equation includes the Lorentz term. Eq. _____

Fluid Equations. (Table 6-6 Eqs. (1), (5), and (7)). In the Earthly domain where MHD applies, mass is conserved. We further assume that gravity and viscous forces can be ignored and that large-scale electric fields are absent in highly conducting systems. Thus, we ignore gravity, electric, and viscous forces in the momentum equation, leaving only pressure gradients and the body forces created by currents and magnetic fields ($\mathbf{J} \times \mathbf{B}$). We revisit the pressure gradient term at the end of this section. Further, we often treat energy conservation with an adiabatic assumption, so the energy equation contains only pressure and density in an adiabatic relation. Equation (7) in Table 6-6 implies no external energy sources or losses. So we can ignore viscous heating, Joule heating, radiative cooling, and conductive cooling.

Constitutive Equations. (Table 6-6 Eqs. (4) and (9)). These equations describe observable properties of the MHD fluid. We show the simplest form of Ohm's Law (Eq. (4)). We assume that only two fluids, ions and electrons, create gas pressure (Eq. (9)).

Simplification of Gauss's Law for Electric Fields and Ampere's Law for Currents. (Table 6-6 Eqs. (8) and (3)). In assuming that the plasma is highly conducting, we're saying that charges will move freely and adjust themselves so that there is no local charge buildup. This means an absence of local particle sources of electric fields.

$$\nabla \bullet \mathbf{E} = \frac{\rho_c}{\varepsilon_0} = 0$$

This equation does not mean that electric fields are absent, but rather that Faraday's Law and the Magnetic Induction Equation describe them.

We recall from Chap. 4 that the vacuum values for the permeability (μ_0) and permittivity (ε_0) are related to the light speed as $\varepsilon_0 \mu_0 = 1/c^2$. (In space physics we usually assume a vacuum environment.) If we are dealing with slow variations, which must be true for MHD to apply, then the last term of Ampere's Law, when divided by c^2, is small. This assumption means the conduction current term is large compared to the displacement current term; the result is in Table 6-6 Eq. (3).

$$\nabla \times \mathbf{B} = \mu_0 \mathbf{J} + \frac{1}{c^2} \frac{\partial \mathbf{E}}{\partial t} \approx \mu_0 \mathbf{J}$$

With the slow-change assumption we recover the original version of Ampere's Law.

Faraday's Law and the Magnetic Induction Equation. (Table 6-6 Eq. (2)). Next we eliminate $\mathbf{E}$ (in favor of $\mathbf{J}$ and $\mathbf{B}$) from Faraday's Law. This step involves a bit of manipulation because the electric field does cause some of the plasma motion. We recall from Chap. 4 that when a conducting plasma is moving with respect to an external magnetic field, the electric field experienced by the plasma is the sum of whatever part of $\mathbf{E}$ exists in the rest frame plus a form of $\mathbf{E}$ generated by the bulk motion $\mathbf{v}$ of the plasma in the magnetic field

$$\mathbf{E}' = \mathbf{E} + (\mathbf{v} \times \mathbf{B}) \qquad (6\text{-}22)$$

where

$\mathbf{E}'$ = combined electric field vector [N/C]
$\mathbf{E}$ = rest frame electric field vector [N/C]
$\mathbf{v}$ = plasma velocity [m/s]
$\mathbf{B}$ = magnetic field strength vector [T]

Rearranging in Ohm's Law (Table 6-6 Eq. (4)) gives

$$\frac{\mathbf{J}}{\sigma} = \mathbf{E} + (\mathbf{v} \times \mathbf{B})$$

which we then substitute into Faraday's Law to produce the *magnetic induction equation*:

$$\frac{\partial \mathbf{B}}{\partial t} = \nabla \times (\mathbf{v} \times \mathbf{B}) - \left(\nabla \times \frac{\mathbf{J}}{\sigma}\right) \qquad (6\text{-}23\text{a})$$

Faraday's Law specifies two components of $\mathbf{E}$; one describes the relative ease of current flow ($\mathbf{J}/\sigma$) and the other describes the motion of the plasma with the respect to a stationary frame of reference ($\mathbf{v} \times \mathbf{B}$).

Frozen-in Magnetic Flux. If the plasma is highly conducting, then the electric fields caused by charge build-up disappear, leading to one further simplification:

$$\frac{\partial \mathbf{B}}{\partial t} \approx \nabla \times (\mathbf{v} \times \mathbf{B}) \qquad (6\text{-}23\text{b})$$

This equation says that the evolution of the magnetic flux field in a highly conducting plasma is controlled primarily by convective motion of the plasma. This control is the gist of "frozen-in" magnetic flux (Sec. 4.3). The magnetic flux threading a loop of plasma does not change over time, even if the area bounding the field deforms. Tubes of magnetic flux cannot pass through each other. If the conductivity is high enough, the induced currents and magnetic fields become large enough to prevent any change in the external magnetic field. Under these conditions the field lines are "frozen" into the plasma. Thus B cannot escape the plasma if the conductivity is large.

This magnetic flux trapping has extremely important effects in the space environment: space plasmas of different origins and characteristics do not mix; they tend to exclude each other, except under special circumstances. Thus, the solar wind generally does not enter the magnetosphere except when the ideal MHD assumption breaks down and magnetic merging can take place.

Diffusive Magnetic Flux. In some locations in the space environment, we must retain the $\nabla \times J/\sigma$ term of Eq. (6-23a), because spatial changes (curl) in currents do contribute to magnetic field variations. If the plasma is resistive (σ is small), then ideal MHD gives way to resistive MHD. By substituting $J = (\nabla \times B)/\mu_0$ from Ampere's Law into the magnetic induction equation, we create a modified version of the equation. It describes contributions from convecting magnetic fields and from magnetic fields that diffuse through the plasma rather than being completely frozen to it.

$$\frac{\partial B}{\partial t} = \nabla \times (v \times B) - \nabla \times \frac{(\nabla \times B)}{\mu_0 \sigma} \qquad (6\text{-}23c)$$

Convection Term + Diffusion Term

This equation says that under conditions of low conductivity the evolution of the magnetic field arises from two sources: a convection term involving v and a term that describes the ease of magnetic field diffusion through the plasma (B/σ). With some resistivity in the plasma, the magnetic field lines move (or diffuse) through the plasma, rather than being frozen to it. For high resistivity (low conductivity) the convection term ($v \times B$) becomes unimportant, and we return to a simple Ohm's Law and the original version of Faraday's Law. In locations where magnetic merging occurs, the diffusion term is important. The magnetic field is subject to significant amplification in regions where the convective term dominates the diffusion term. We observe the results of convective amplification in features such as sunspots and active regions that bubble out of the solar convective zone when the magnetic field becomes too intense.

The MHD fluid equations consist of three vector field equations (nine component equations in three dimensions) and four scalar equations. Within these equations are 14 unknowns—two scalar unknowns, P and ρ, and 12 vector component unknowns: v, B, E, J. With 14 unknowns and only 13 equations, the system is not mathematically closed. (We use the state equation in the energy conservation equation, so the state equation does not provide independent information.) Ohm's Law supplies the additional equation needed. A system of 14 equations with 14 unknowns is complicated and best solved by computer program. The equations must be solved for each of hundreds if not thousands of grid points to produce results that approximate even simple conditions in the space environment. In the next subsection, we select some problems that are solvable by subsets of the equations in Table 6-6.

The equations in this section are applied, sometimes in more general forms, to describe very dynamic situations. Figure 6-18 shows an image from an MHD simulation of the coronal mass ejections moving into the solar wind. This output would not be possible without the combined sets of these equations we discuss above.

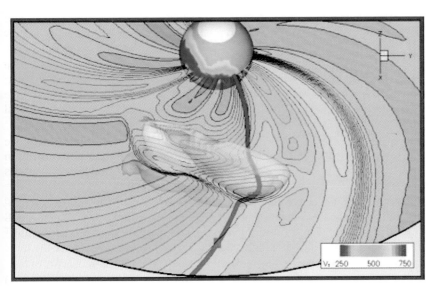

Fig. 6-18. Magnetohydrodynamic (MHD) Applications. At a very basic level, MHD modelers want to describe how plasma and magnetic fields circulate (convect) in the heliosphere. This image depicts a coronal mass ejection (CME) moving radially outward from the Sun through the magnetized solar wind. The translucent gray shape is the CME. Only one magnetic field line is shown. The color contours correspond to the solar wind speed. The plasma is hidden except in the region of enhanced density near the CME. *(Courtesy of Dusan Odstrcil at the University of Colorado and the Center for Integrated Space Weather Modeling [CISM] at Boston University)*

Plasma Parameters

We have three parameters useful for discussing magnetic plasmas and deciding which simplifications to the equations of Table 6-6 to use:

- The Knudsen Number reveals if we can reasonably treat a plasma as an MHD fluid
- The magnetic Reynolds Number reveals when the "frozen-in" conditions apply
- The plasma beta parameter tells if the magnetic pressure or the thermal pressure dominates a plasma

Knudsen Number. The *Knudsen Number* K_n is the ratio of the molecular mean free-path length to a representative physical length scale of the plasma:

$$K_n = \frac{l}{L} \qquad (6\text{-}24)$$

where

l = the mean free path [m]
L = a characteristic length of the plasma [m]

If the Knudsen Number is much greater than 1.0, then the flow is mostly collisionless and the plasma may be described with MHD equations, as long as time variations of **B** are slow. In the solar wind $K_n \gg 1$.

Magnetic Reynolds Number. Elements of the induction equation along with scale analysis tell us when "frozen-in magnetic field" conditions apply. We define a quantity called the *magnetic Reynolds Number* R_M by

$$R_M = \frac{\text{convection term}}{\text{diffusion term}} = \frac{|\nabla \times (\mathbf{v} \times \mathbf{B})|}{\left[\frac{|\nabla^2 \mathbf{B}|}{\mu_0 \sigma}\right]} \approx \frac{vB/L}{\left[\frac{B/L^2}{\mu_0 \sigma}\right]} = \mu_0 \sigma v L \quad (6\text{-}25)$$

where

R_M = magnetic Reynolds Number [dimensionless]
$\mathbf{v}$ = fluid velocity vector [m/s]
σ = fluid electrical conductivity [siemens/m = 1/ohm]
μ_0 = permeability of free space ($4\pi \times 10^{-7}$ N/A^2 = 1.26×10^{-6} N/A^2)
L = approximate size of the plasma system [m]

If $R_M \gg 1$, then the convection term dominates and the $\mathbf{B}$ field will be frozen to the fluid, MHD analysis applies, and the field moves with the flow. On the other hand, if $R_M \ll 1$, then diffusion dominates and the fluid does not appreciably affect $\mathbf{B}$. The solar wind has a high magnetic Reynolds Number.

Plasma Beta Parameter. The ideal MHD momentum equation consists of two terms: pressure gradient force and the $\mathbf{J} \times \mathbf{B}$ force. The magnetic field may contribute to the pressure gradient term, thus producing two components in the pressure gradient term: a thermal term and a magnetic term. For certain applications we are interested in knowing if one of the terms dominates the other. From Chap. 4 we know that the *plasma beta parameter* β (Eq. 4-14) is the ratio of the thermal energy density ($nk_B T$) to the magnetic energy density ($B^2/2\mu_0$). Assuming that the magnetic field is frozen to the plasma, the value of the thermal pressure relative to the magnetic pressure allows us to determine whether the particles go where they want, dragging the field along, or if the field determines where the plasma goes. The plasma beta parameter equation is

$$\beta = \frac{P_T}{P_B} = \frac{n_e k_B (T_e + T_i)}{(B^2/2\mu_0)} \quad (6\text{-}26)$$

where

β = plasma beta parameter [dimensionless]
P_T = thermal pressure [J/m^3]
P_B = magnetic pressure [J/m^3])
B = magnetic field strength [T]
T_e and T_i = electron and ion temperatures, respectively [K]
μ_0 = permeability of free space ($4\pi \times 10^{-7}$ N/A^2 = 1.26×10^{-6} N/A^2)
k_B = Boltzmann's constant (1.38×10^{-23} J/K)

EXAMPLE 6.10

Plasma Beta in the Solar Wind

- *Problem Statement:* Determine the plasma beta β value for the solar wind.
- *Given:* Solar wind density = 5 protons /cm^3, solar temperature ~10^5 K, IMF strength = 5×10^{-9} T.
- *Assumptions:* Solar wind ion and electrons temperatures are the same.

- *Solution:*

$$\beta = \frac{P_T}{P_B} = \frac{n_e k_B (T_e + T_i)}{(B^2/2\mu_0)}$$

$$\beta = \frac{(5 \times 10^6 \text{ protons/m}^3)(1.38 \times 10^{-23} \text{ J/K})(2 \times 10^5 \text{ K})}{\left[\frac{(5 \times 10^{-9} \text{ T})^2}{2(4\pi \times 10^{-7} \text{ N/A}^2)}\right]} = 1.39 \text{ dimensionless}$$

- *Interpretation:* In the solar wind, the magnetic field pressure (energy density) and the plasma thermal pressure (energy density) are nearly equal.

Follow-on Exercise: Some plasma beta calculations include the dynamic pressure term in the top of the equation. How does the calculation change if dynamic pressure is considered?

6.4.2 Applying Magnetohydrodynamics (MHD)

Objectives: After reading this section, you should be able to...

- Describe the key parameters for determining if MHD approximations apply to plasma
- List the most important equations that describe magnetic dynamos
- Calculate the magnetic pressure in a static sunspot

Simple MHD Applications to Dynamos

Momentum and Magnetic Induction: The Making of a Dynamo. Curly and wavy magnetic fields fill the cosmos. A magnetic field "factory" must exist somewhere. Indeed, the universe has many such "factories": the interiors of stars, perhaps the centers of galaxies, and the molten cores of planets. In all of these situations, magnetic fields and plasmas are in motion with respect to each other. As a result, the convective motion of the magnetized fluid converts mechanical energy to electromagnetic energy. The simplified momentum equation (Table 6-6, Eq. (1)) and Faraday's Law (induction equation in Eq. (6-23c)), assuming frozen-in magnetic flux, are the governing equations for dynamos. They define how v and B evolve in time and space. If we combine them with the appropriate boundary conditions for density and pressure, they give us two vector equations with two vector unknowns, v and B. (In reality, we have six scalar equations with six unknowns, but at least they are a closed set.)

For magnetic dynamos we are particularly interested in situations where the Knudsen Number, magnetic Reynolds Number, and plasma beta values are large. When these conditions are met, the feedback loops shown in Fig. 6-17 produce dynamo action. The Sun's convective zone is a case in point. There the mechanical motion of the bubbling plasma puts frozen-into-the-plasma magnetic fields into motion, which in turn creates electric fields and currents whose curls further amplify the magnetic field. The currents interact with the amplified magnetic field through the $\boldsymbol{J} \times \boldsymbol{B}$ force (momentum equation) to accelerate the plasma. In such cases the magnetic field is subject to significant amplification.

The results are features such as sunspots and their associated active regions that billow into the corona when the magnetic field becomes too intense. The 11- and 22-year solar cycles are the result of deep, long-term dynamos.

Magnetic Pressure Balance. The previous discussion suggests that MHD applications tend to be quite complicated. Here we pursue one situation that, with a minimal amount of simplification, sheds some light on sunspot characteristics. We will use the momentum equation (Eq. (1) from Table 6-6).

First we use the right-most term of the momentum equation: $\boldsymbol{J} \times \boldsymbol{B}$. Using Ampere's Law, we convert this term to one involving only the magnetic field: $\boldsymbol{J} = (\nabla \times \boldsymbol{B})/\mu_0$. The momentum equation becomes:

$$\rho \frac{d\boldsymbol{v}}{dt} = -\nabla P + \frac{1}{\mu_0}(\nabla \times \boldsymbol{B}) \times \boldsymbol{B}$$

The double cross product on the right can be reformulated using a vector identity (see Vector Identity #11 of the Naval Research Laboratory Plasma Formulary, 2009):

$$(\nabla \times \boldsymbol{B}) \times \boldsymbol{B} = -\nabla\left(\frac{1}{2}\boldsymbol{B} \bullet \boldsymbol{B}\right) + \boldsymbol{B}(\boldsymbol{B} \bullet \nabla)$$

Dividing by μ_0 gives

$$\frac{1}{\mu_0}(\nabla \times \boldsymbol{B}) \times \boldsymbol{B} = \left[-\frac{1}{2\mu_0}\nabla(B^2) + \frac{1}{\mu_0}\boldsymbol{B}(\boldsymbol{B} \bullet \nabla)\right]$$

$$= \text{magnetic pressure} \quad + \text{magnetic tension}$$

The left side of the equation is equal to $\boldsymbol{J} \times \boldsymbol{B}$, which is also the right-most term of the momentum equation. It has two parts: 1) a magnetic pressure term that acts to oppose any bunching in the field (or tries to make the field uniform) and 2) a magnetic tension term that attempts to straighten any curvature in the magnetic field.

We put these terms back into the momentum equation and gather the pressure terms with the result:

$$\rho \frac{d\boldsymbol{v}}{dt} = -\nabla\left[P + \frac{(B^2)}{2\mu_0}\right] + \frac{1}{\mu_0}(\boldsymbol{B} \bullet \nabla \boldsymbol{B})$$

The momentum equation has two pressure terms, thermal and magnetic, as well as a magnetic tension term. For a magnetostatic situation with relatively straight field lines the rate of momentum change (left hand side) equals zero and the magnetic tension is approximately zero. Thus, the magnetic and thermal pressures must balance. We discuss such a case for a simple sunspot in Example 6.11.

EXAMPLE 6.11

Sunspots in Magnetostatic Equilibrium

- *Problem Statement:* Sunspots are cool, rather long-lived features of the solar surface. Show that thermal pressure in sunspots is less than that in surrounding regions. We model the sunspot as a set of time-steady, vertical magnetic field lines.

- *Assumptions:* 1) Electrical conductivity in the sunspot and in the surrounding environment is high, so the current density $J \approx 0$, except at a thin region at the flux tube boundaries. 2) The magnetic field B_S is vertical and only exists inside the sunspot. 3) The thermal pressure of the spot is P_S which equals the thermal pressure of the environment P_E. 4) Static equilibrium. 5) The field lines are straight, so we ignore the magnetic tension term.

- *Solution:* Conserve momentum in the spot and its surrounding environment.

Under static conditions the MHD equations reduce to

$$\rho \frac{\partial v}{\partial t} = 0 = -\nabla \left[P_{thermal} + \left(\frac{1}{2\mu_0} B^2 \right) \right]$$

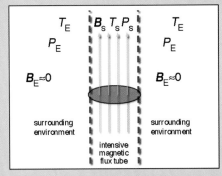

Fig. 6-19. Magnetic Flux Tube. This diagram represents a flux tube containing a steady-state sunspot.

This condition sets the equilibrium for the total pressure gradient, meaning that the total pressure must be equal inside and outside the sunspot. Outside of the sunspot the magnetic field is negligible. Therefore

$$\nabla \left[P_{thermal\ environment} + \frac{B_{environment}^2}{2\mu_0} \right] = \nabla \left[P_{thermal\ spot} + \frac{B_{spot}^2}{2\mu_0} \right]$$

$$P_{thermal\ environment} + 0 = P_{thermal\ spot} + \frac{B_{spot}^2}{2\mu_0}$$

$$P_{thermal\ spot} = P_{thermal\ environment} - \frac{B_{spot}^2}{2\mu_0}$$

- *Interpretation and Implications:* The plasma thermal pressure inside the sunspot must be less than the plasma pressure outside. This pressure difference happens in one of several ways: either the particle density in the spot is low or the temperature is low, or a combination of these. We know that sunspots are cooler than their surroundings. In general, their density is also low. Magnetic pressure pushes the particles out along the vertical field lines, partially evacuating the spot region.

6.5 Elementary Plasma Waves

In many instances the free ions and electrons in a plasma respond to forces with a wave or wave-like behavior that falls into two general categories: 1) oscillations in which individual plasma components move to and fro, but the energy remains localized (no propagation) and 2) waves that rely on the collective behavior of the plasma, or its components, to propagate energy. We provide qualitative descriptions of these behaviors here.

6.5.1 Wave Propagation

Objectives: After reading this section, you should be able to...

- Distinguish between electrostatic and electromagnetic plasma waves
- Distinguish between parallel, perpendicular, and oblique plasma waves

All wave motion results from restoring forces. Various aspects of wave propagation are strongly linked to the wave frequency ω and the *wave number k* (magnitude and direction of the electromagnetic wave). Plasma is a dispersive propagation medium, meaning that the speed of the energy propagation *(group speed v_{gr})* and the wave pattern speed (or *phase speed v_{ph}*) depend on the wave's frequency.

$$v_{gr} = \partial\omega/\partial k \qquad (6\text{-}27)$$

$$v_{ph} = \omega/k \qquad (6\text{-}28)$$

This behavior is similar to visible light passing through a prism that disperses light into colors, because the propagation speed within the prism material depends on the frequency of the light. In magnetized plasma, the direction of wave propagation with respect to the direction of the magnetic field is particularly important. Motions that are merely oscillatory and localized do not propagate energy. For these the wave vector is zero.

We describe waves in plasma with respect to the direction of the electric field vector E and the wavefront. According to Maxwell's Equations, all longitudinal waves (k aligns with E) are electrostatic waves, and all transverse waves (k is perpendicular to E) are electromagnetic and feature an oscillating magnetic field. If the wave vector is perpendicular to E then an oscillating magnetic field exists. By definition, this is an *electromagnetic wave*.

We also describe waves with respect to the direction of the external magnetic field threading a plasma. Waves propagating through a magnetized plasma (that is, in the presence of a significant external magnetic field), are either *parallel waves* (k is parallel to B), *perpendicular waves* (k is perpendicular to B), or *oblique waves* (k has parallel and perpendicular components relative to B).

We further classify waves by the type of collisional medium, if any, required for propagation. Acoustic (sonic) waves require matter for propagation. Electromagnetic waves can propagate without a medium. In plasmas with an ambient magnetic field, we find a hybrid type of wave that has characteristics of both acoustic and electromagnetic waves. This latter type of wave is an *Alfvén wave*, named for Hannes Alfvén, who developed the theory of wave propagation in magnetized plasma. Magnetic tension is the restoring force in an Alfvén wave.

6.5.2 Plasma Oscillations and Waves

> **Objectives:** After reading this section, you should be able to...
> - Distinguish between plasma oscillations and plasma waves
> - State the directions of wave propagation for different types of plasma waves
> - Calculate Alfvén wave speeds in magnetized plasma

Oscillations are non-propagating, wave-like behaviors of ions and electrons. The wave vector **k** equals zero, and the energy remains with the oscillating particles. Wave propagation requires a restoring force. The restoring force for a sound wave in a gas, for example, is the local pressure gradient force. Below we consider other forces in plasma and provide a qualitative description of the associated plasma motion. Later in the text we provide wave formulas as needed.

Particle Gyromotion—Magnetized Plasma

An example of an oscillation from the single-particle approximation is the two-dimensional gyromotion of a charged particle about a magnetic field line. From Sec. 6.2, the frequency of oscillation is $\Omega = qB/m$. It's proportional to the field strength and is called the *cyclotron frequency*.

Plasma Oscillation—Cool, Unmagnetized Plasma

Plasma oscillations (electron oscillations) with little thermal motion result from electric field disturbances. We discuss this behavior in Sec. 6.1.2 where we define the plasma frequency: Eq. (6-5) $\omega_p = \sqrt{(n_e e^2)/(\varepsilon_0 m_e)}$. This oscillatory motion (also called *Langmuir oscillation*) is proportional to the square root of electron density. For purposes of remote sensing, if we measure the plasma frequency, then we can determine the plasma electron density.

Without energy propagation the electron plasma frequency might seem uninteresting. Just the opposite is true. If an oscillating wave, such as a radio wave, has a frequency lower than the electron plasma frequency and attempts to enter a plasma, the electrons in the plasma will respond and vibrate at that frequency. They absorb the energy of the wave, thus blocking the wave penetration. This phenomenon is very important for radio waves in Earth's ionosphere. Low-frequency waves are absorbed. But the electrons are unable to respond to the higher-frequency waves that penetrate the ionosphere and carry information. The plasma electron frequency is sometimes called the *critical frequency,* and is crucial to radio communication.

A related but much slower plasma oscillation involves ion motion with a similar formula: $\omega_{pi} = \sqrt{(n_i e^2)/(\varepsilon_0 m_i)}$.

Plasma Waves—Warm Plasma

Electron waves in warm plasma where thermal collisions occur also result from electric field disturbances. Most of the above statements regarding plasma oscillations still apply. However, thermal motions allow the plasma particles to interact in such a way that energy propagates from the disturbance. The Coulomb force acts as the restoring force to generate an electrostatic, longitudinal wave ($k \parallel E$). This wave is 1-, 2-, or 3-dimensional, and we call it a *Langmuir wave*.

Ion Acoustic Waves—Warm Plasma

These waves propagate as the result of small pressure disturbances in the plasma generated by an electric field perturbation. The ions have a slight motion in response to the pressure change, so the wave is similar to a sound wave, requiring a compressible fluid. Pressure pulses propagate away from their source at a speed characteristic of a compressible fluid. The pressure gradient acts as the restoring force. As with a pressure or sound wave in a neutral gas, this wave is dispersive. And like the plasma wave we discuss in the previous paragraph, this wave is electrostatic and longitudinal ($k \parallel E$).

Alfvén Waves—Magnetized Plasma

Alfvén waves are a hybrid, combining properties of material waves and electromagnetic waves (Sec. 5.1.2). They arise from a slight disturbance B_1 to the background magnetic field B_0. They need a medium in which to propagate, but it need not be compressible. Alfvén waves exist because the fluid resists compression and the magnetic field lines have a tension that tends to restore them to the lowest energy configuration. The restoring force comes from a magnetic tension rather than the thermal pressure gradient. Alfvén waves propagate as transverse electromagnetic waves with k perpendicular to B_1 and E_1. However, because $k \parallel B_0$, it falls into the class of parallel waves. In effect these waves are a transverse shaking of the field lines. They propagate as shown in Fig. 6-20 with a speed given by $v_A = \sqrt{B^2/(2\mu_0 \rho)}$. Alfvén waves are very common in plasmas and modelers have to account for their behavior.

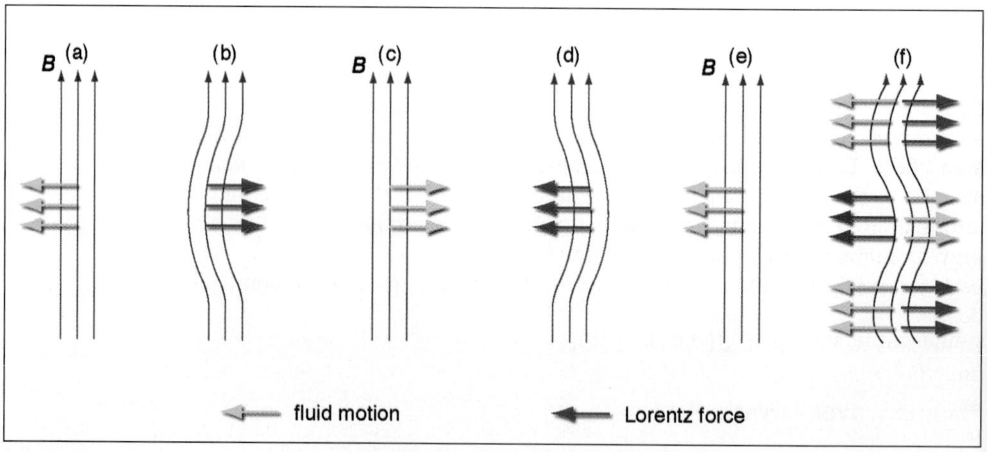

Fig. 6-20. Alfvén Waves. These waves propagate along the magnetic field lines. Here we show a sequence of how the fluid velocity induces a Lorentz force that causes an Alfvén wave to move along the magnetic field lines. The wave vector points to the top of the page.

Magnetosonic Waves—Warm, Magnetized Plasma

Even though plasma resists being compressed, pressure waves still develop, as they do in a neutral gas. So sonic waves are usually present in plasma. The disturbance propagates across (perpendicular to) the field lines. Magnetosonic waves are similar to ion acoustic waves in that both have some restoring force from thermal pressure gradients. However, magnetosonic waves also have a magnetic pressure gradient force. Compressibility, and hence the wave speed, is controlled by the plasma and the magnetic field. The frozen-in magnetic field rarifies and compresses with the wave motion. This wave is classified as a transverse electromagnetic wave ($k \perp B_1$ and $k \perp E_1$), as well as a perpendicular wave ($k \perp B_0$).

Summary

Plasma consists of a collection of free-moving electrons and ions. In a weakly ionized plasma, only a tiny fraction of electrons are detached from their atoms. In a strongly ionized plasma, most atoms have lost at least one electron. Electric fields keep the ions and electrons near each other, so plasmas are neutral in general. Because of the good mobility of electrons, plasma is a good conductor of heat and electricity. In a plasma, mutual electromagnetic interactions substantially affect the dynamics of the system and cause the system to behave collectively. Plasmas are the most common form of matter, constituting more than 99% of the visible universe. They permeate the interplanetary, interstellar, and intergalactic environments. Ionization energy comes in several forms: thermal, electrical, or light (e.g., ultraviolet light or intense visible light from a laser). Plasma creation does not involve a phase change; rather plasma production can be a gradual process depending on the ionization source.

Three factors determine if a collection of charged particles is a plasma:

1. The Debye (shielding) length must be much smaller than the characteristic scale of the system
2. The plasma frequency must be much greater than the collision frequency
3. The number of charged particles inside a Debye sphere must be much greater than one

The degree of ionization is described by the Saha equation. If so many neutral particles exist that the dynamics of charged particles is essentially affected by their collisions with the neutrals, the plasma is said to be weakly ionized. Earth's ionosphere is an example. In contrast, if collisions with the neutrals have no significant effect on the dynamics of plasma particles, the plasma is said to be strongly ionized. Much of Earth's magnetosphere is a strongly ionized plasma.

The motion of individual charged particles under the influence of various fields provides insight into many important physical properties of plasmas. Single particle motion and guiding center drift allow us to look at the plasma particle behavior, assuming the individual particles are not changing the ambient environment. Plasma particles are accelerated by electric and magnetic fields and undergo linear acceleration and gyromotion. A low-collision plasma with an embedded magnetic field usually has properties that differ along and perpendicular to the magnetic field. Combinations of fields perpendicular to the magnetic field produce drift motion described by the guiding-center of motion concept. Only charge-independent drifts produce currents.

In a moving conductive fluid, magnetic fields induce currents that exert forces on the fluid, and also change the magnetic field. The equations that describe these magnetohydrodynamic (MHD) interactions are a combination of the fluid dynamic equations and Maxwell's Equations. MHD is a simplified model of a magnetized plasma, in which the plasma is treated as a single fluid that can carry an electric current. Three parameters provide insight about which simplifications we apply to plasmas:

The Knudsen Number is the ratio of mean free path to the characteristic scale length of the plasma. A plasma with a large Knudsen Number can be modeled with MHD. The magnetic Reynolds Number describes the degree to which the magnetic field is frozen to the plasma. If convective motions dominate, this number is large and the frozen-in assumption applies. The plasma beta is the ratio of thermal energy density to magnetic energy density. If thermal energy dominates, the field moves with the particles; if the magnetic energy dominates, the particles are controlled by the magnetic field.

In many instances the free ions and electrons in a plasma respond to forces with a wave or wave-like behavior that falls into two general categories: 1) oscillations in which individual plasma components move to and fro, but the energy remains localized (no propagation) and 2) waves that rely on the collective behavior of the plasma, or its components, to propagate energy. Plasma can sustain several different wave phenomena that may be transverse or longitudinal, have low or high frequency, and propagate along or perpendicular to the magnetic field. Alfvén waves combine aspects of material and electromagnetic waves. These hybrid waves are a response to a slight disturbance in the background magnetic field.

Waves and particles may interact with each other. Wave energy may also be lost by particle collisions. Some plasma particles may resonate with the waves and absorb energy from them. The result is particle acceleration and wave damping. The opposite may also occur, with energy transferred from particles to growing waves.

Key Words

Alfvén wave
critical frequency
cyclotron frequency
Debye length (λ_D)
Debye shielding (electrostatic screening)
Debye sphere
diamagnetic
electromagnetic wave
energetic particles
gradient drift
gradient-curvature drift
group speed v_{gr}
guiding-center drift
gyrofrequency
gyroradius
impact ionization
Knudsen Number K_n
Langmuir oscillation
Langmuir wave
Larmor radius
magnetic induction equation
magnetic Reynolds Number R_M
magnetohydrodynamics
Maxwell-Boltzmann distribution
oblique waves
parallel waves
perpendicular waves
phase speed v_{ph}
photoionization
plasma
plasma beta parameter (β)
plasma frequency (ω_p)
Saha equation
suprathermal
wave number (k)

Notation

Unshielded electric potential $V = \dfrac{kq}{r}$

Debye length $\lambda_D = \sqrt{\dfrac{k_B T_e \varepsilon_0}{n_e e^2}}$

Shielded electric field vector $\boldsymbol{E}(r) = \dfrac{kq}{r^2}\exp\!\left(\dfrac{-r}{\lambda_D}\right)\hat{\boldsymbol{r}}$

Shielded electric potential $V(r) = \dfrac{kq}{r}\exp\!\left(\dfrac{-r}{\lambda_D}\right)$

Plasma angular frequency

$$\omega_p = \sqrt{\dfrac{n_e e^2}{\varepsilon_0 m_e}} = \dfrac{1}{\lambda_D}\sqrt{\dfrac{1}{k_B T_e m_e}}$$

Plasma linear frequency $f_p \approx 9\sqrt{n_e}$

Momentum change rate $m\dfrac{d\boldsymbol{v}}{dt} = \boldsymbol{F}_{total}$

$$= q\boldsymbol{E} + q(\boldsymbol{v}\times\boldsymbol{B}) + \boldsymbol{F}_{other}$$

Charge acceleration from an $\boldsymbol{E}$ field $\boldsymbol{a} = \dfrac{d\boldsymbol{v}}{dt} = \dfrac{q\boldsymbol{E}}{m}$

Charge acceleration perpendicular to a $\boldsymbol{B}$ field

$$a_\perp = \dfrac{dv}{dt} = \dfrac{qBv_\perp}{m} = \Omega v_\perp$$

Gyroradius $r = mv_\perp/qB = v_\perp/\Omega$

Gyrofrequency $\Omega = qB/m$

Charge acceleration parallel to an $\boldsymbol{E}$ field

$$a_\| = \dfrac{dv_\|}{dt} = \dfrac{qE}{m}, \qquad a_\perp = \dfrac{dv_\perp}{dt} = \dfrac{qBv_\perp}{m} = \Omega v_\perp$$

$\boldsymbol{F}_{other} = -q(\boldsymbol{v}\times\boldsymbol{B})$

General guiding-center drift $\boldsymbol{v}_d = \dfrac{\boldsymbol{F}\times\boldsymbol{B}}{qB^2}$

$\boldsymbol{E}\times\boldsymbol{B}$ drift $\boldsymbol{v}_E = \dfrac{\boldsymbol{E}\times\boldsymbol{B}}{B^2}$

Gravitational drift $\boldsymbol{v}_g = \dfrac{m\boldsymbol{g}\times\boldsymbol{B}}{qB^2}$

Gradient $\boldsymbol{B}$ drift $\boldsymbol{v}_{\nabla B} = -\dfrac{1}{2}mv_\perp^2\dfrac{\nabla B\times\boldsymbol{B}}{qB^3}$

Curvature drift $\boldsymbol{v}_C = \dfrac{mv_\|^2 \boldsymbol{R}_C\times\boldsymbol{B}}{R_C^2 qB^3}$

Gradient-curvature drift

$$\boldsymbol{v}_{\nabla BC} = \left(mv_\|^2\dfrac{\boldsymbol{R}_C}{R_C^2} - \dfrac{1}{2}mv_\perp^2\dfrac{\nabla B}{B}\right)\times\dfrac{\boldsymbol{B}}{qB^2}$$

Chapter 6 Space is a Place...with Plasma

Maxwell-Boltzmann distribution

$$\chi(v)dv = 4\pi\left(\frac{m}{2\pi k_B T}\right)^{\frac{3}{2}} v^2 \exp\left(-\frac{mv^2}{2k_B T}\right) dv$$

$$N(v)dv = 4\pi N_T \left(\frac{m}{2\pi k_B T}\right)^{\frac{3}{2}} v^2 dv$$

$$N(v)dv = (4\pi N_T)\left(\frac{m}{2\pi k_B T}\right)^{\frac{3}{2}} v^2 \exp\left(-\frac{(1/2)mv^2 + mgz}{k_B T}\right) dv$$

$$N(v)dv = (4\pi N_T)\left(\frac{m}{2\pi k_B T}\right)^{\frac{3}{2}} v^2 \exp\left(-\frac{(1/2)mv^2 + qV}{k_B T}\right) dv$$

Saha equation $\frac{n_i}{n_n} = \frac{n_e}{n_n} = \frac{(2\pi m_e k_B T)^{3/2}}{n_i h^3} \exp\left(\frac{-U_i}{k_B T}\right)$

$$n_i = [n_n(2.4 \times 10^{21} T^{3/2}) \exp(-U_i/k_B T)]^{1/2}$$

Thermal energy density in a plasma

$$u_{thermal} = \frac{3}{2} k_B (n_i T_i + n_e T_e)$$

$$u_{thermal} = \frac{3}{2} n_i k_B (T_i + T_e)$$

$$u_{thermal} = 2n k_B T$$

Kinetic energy density in a plasma

$$u_{KE} = \frac{1}{2} n_i m_i v_i^2 + \frac{1}{2} n_e m_e v_e^2$$

$$u_{KE} = (1/2)(\rho_i v_i^2 + \rho_e v_e^2)$$

$$u_{KE} = (\rho_i v_i^2)$$

Electromagnetic energy density in a plasma

$$u_{B+E} = \frac{1}{2}\left(\frac{B^2}{\mu_0} + \varepsilon_0 E^2\right)$$

$$u_{B+E} \approx u_B = \frac{1}{2}\left(\frac{B^2}{\mu_0}\right)$$

Electric field in a moving magnetized plasma

$$E' = E + (v \times B)$$

Magnetic induction $\frac{\partial \mathbf{B}}{\partial t} = \nabla \times (\mathbf{v} \times \mathbf{B}) - \left(\nabla \times \frac{\mathbf{J}}{\sigma}\right)$

Magnetic induction in a highly conducting plasma

$$\frac{\partial \mathbf{B}}{\partial t} \approx \nabla \times (\mathbf{v} \times \mathbf{B})$$

Magnetic pressure balance

$$\frac{\partial \mathbf{B}}{\partial t} = \nabla \times (\mathbf{v} \times \mathbf{B}) - \nabla \times \frac{(\nabla \times \mathbf{B})}{\mu_0 \sigma}$$

Knudsen Number $K_n = \frac{l}{L}$

Magnetic Reynolds Number

$$R_M = \frac{\text{convection term}}{\text{diffusion term}} = \frac{|\nabla \times (\mathbf{v} \times \mathbf{B})|}{\left[\frac{|\nabla^2 \mathbf{B}|}{\mu_0 \sigma}\right]} \approx \frac{vB/L}{\left[\frac{B/L^2}{\mu_0 \sigma}\right]}$$

$$= \mu_0 \sigma v L$$

Plasma beta parameter $\beta = \frac{P_T}{P_B} = \frac{n_e k_B (T_e + T_i)}{(B^2/2\mu_0)}$

$v_{gr} = \partial \omega / \partial k$

$v_{ph} = \omega / k$

Answers to Pause for Inquiry Questions

PFI 6-1: In the lower left corner, where electron density and temperature are low.

PFI 6-2: Electron density range is 10^1 electrons/m^3 to 10^{31} electrons/m^3 (from the bottom to the top). The temperatures range from 10^2 K to 10^9 K (left to right).

PFI 6-3: Metals are neutral in bulk. Some are good charge-carriers by virtue of their valence electrons. The mobile charges in current-carrying metals respond to electric and magnetic fields; however the responses may be tempered by collisions. Waves and instabilities may develop under strongly forced conditions. Ionized dust grains have detached charges somewhere in the local vicinity. Only on very large scales are the current carriers subject to the electric and magnetic forces that would give rise to waves

Answers to Pause for Inquiry Questions

and instabilities. On very large scales these dust grains can act collectively. They are then called "dusty plasma."

PFI 6-4: The Debye length is the scale over which mobile charge carriers (usually electrons) screen electric fields in plasmas.

Materials with higher charge density have more mobile electrons to do the screening. So the effective length for shielding is reduced because more electrons surround a positive charge to null its electric field.

PFI 6-5: To reduce the effective screening distance, reduce the temperature and increase the plasma density.

PFI 6-6: Larger by a factor of $m_i/m_e \approx 1836$, so approximately three orders of magnitude.

PFI 6-7: $W = q \int \mathbf{E} \cdot d\mathbf{r}$

PFI 6-8: Gyroradius: $r = mv/qB$. The units are $((kg \cdot m/s))/(A \cdot s \ (kg/(A \cdot s^2)))$. This reduces to m.

Gyrofrequency: $\Omega = qB/m$. The units are $(A \cdot s)(kg/A \cdot s^2)/kg$, which reduces to 1/s.

PFI 6-9: $W = 0$ because the magnetic field is perpendicular to the motion.

PFI 6-10: The moving particle in Fig 6-11a is an ion. The moving particle in Fig 6-11b is an electron, which spins in the opposite direction around a magnetic field line.

PFI 6-11: The ion will accelerate in the direction of the electric field (into the page) and the electron will accelerate opposite the direction of the electric field (out of the page). Yes, there will be a current, directed into the page.

PFI 6-12: The electrons are drifting in a direction opposite that of the ion drift. This constitutes a current.

PFI 6-13: Same direction.

PFI 6-14: Guiding-center motion depends on the nature of the force causing the drift. Forces that are charge-independent produce drifts in the same direction for ions and electrons. Forces that are charge-dependent cause oppositely directed guiding center motions for opposite charges. If the sign of the charge is reversed, the sign of $v_\perp$ is reversed.

PFI 6-15: For constant temperature, the relative amount of ionization decreases as the ionization potential energy increases.

PFI 6-16: In Eq. (6-18b) multiply both sides by $n_i n_n$. Assume $n_i = n_e$. The left side of the equation equals n_e^2. The right side is proportional to n_n. Taking the square root of both sides leads to $n_e \propto \sqrt{n_n}$.

PFI 6-17: $u_T = 2nk_B T = 2(5 \times 10^6 \text{ ions/m}^3)(1.38 \times 10^{-23} \text{ J/K}) 10^5 K = 1.38 \times 10^{-11} \text{ J/m}^3$

$u_{KE} = (\rho_i v_i^2) = (5 \times 10^6 \text{ ions/m}^3)(1.67 \times 10^{-27} \text{ kg/ion}) (4 \times 10^5 \text{ m/s})^2 = 1.3 \times 10^{-9} \text{ J/m}^3$

$u_B = B^2/2\mu_0 = (5 \times 10^{-9} \text{ N}/(A \cdot m))^2/(2 \ (1.26 \times 10^{-6} \text{ N} \cdot A^2)) \approx 10^{-11} \text{ J/m}^3$

$u_{KE} \gg u_T \approx u_B$

PFI 6-18: $u_{KE} = (1/2)(n_i m_i v_i^2 + n_e m_e v_e^2)$
Assume $n_i = n_e = n$
$u_{KE} = (1/2)n(m_i v_i^2 + m_e v_e^2)$
Assume $v_i = v_e = v$
$u_{KE} = (1/2)(2)nv^2(m_i + m_e) = nv^2(m_i + m_e)$
$m_i \sim 1000 \ m_e$
$u_{KE} = nv^2 m_i \approx \rho_i v^2$

PFI 6-19:
- Mass is transported but not created or destroyed (Eq. (5)).
- Energy is conserved and, in its simplest form, it is thermal energy in an adiabatic system (Eq. (7)).
- Momentum changes come from pressure gradient forces and from currents that interact with the magnetic field (Eq. (1)).
- Charge buildup does not occur (Eq. (8)).
- Magnetic monopoles do not exist (Eq. (6)).
- Varying magnetic fields are the source of electric fields (Eq. (2)).
- Currents are the only source of magnetic fields (Eq. (3)).
- The equation modified to deal with only slow electric field variations (Eq. (3)).
- The equation that describes sources of acceleration (Eq. (1)).

- The equation of state (Eq. (9)).
- This equation is the Maxwell equation modified for high conductivity (Eq. (8)).
- This equation relates magnetized plasma motion to the electric field (Eq. (4)).
- This equation includes the Lorentz term (Eq. (1)).

Figure References

Peratt, Anthony L. 1966. Advances in Numerical Modeling of Astrophysical and Space Plasmas. *Astrophysics and Space Science.* Vol. 242. Springer. Dordrecht, Netherlands.

Winckler, John R. and Robert J. Nemzek. 1989. Observation and Interpretation of Fast Sub-visual Light Pulses from the Night Sky. *Geophysical Research Letters.* No. 16. American Geophysical Union. Washington, DC.

Further Reading

Baumjohann, Wolfgang and Rudolf A. Truemann. 1997. *Basic Space Plasma Physics.* London: Imperial College Press.

Chen, Francis F. 1984. *Introduction to Plasma Physics and Controlled Fusion.* Vol. 1, 2nd Ed. New York, NY: Plenum Press.

Cravens, Thomas E. 1997. *Physics of Solar System Plasmas.* Cambridge, UK: Cambridge University Press.

Kivelson, Margaret G. and Christopher T. Russell. 1995. *Introduction to Space Physics.* Cambridge, UK: Cambridge University Press.

Naval Research Laboratory. 2009. *NRL Plasma Formulary.* Washington, DC: Beam Physics Branch. NRL/PU/6790-09-523 (available online).

Earth's Quiescent Magnetosphere: Its Role in the Space Environment and Weather

7

UNIT 1. SPACE WEATHER AND ITS PHYSICS

Contributions by Jeffrey Love

You should already know about...

- Spherical and polar coordinates
- The general nature of Earth's dipolar magnetic field
- Definitions of thermal, dynamic, and magnetic pressure (Chaps. 2 and 4)
- Magnetic merging and dynamos (Chap. 4)
- How magnetized plasmas induce electric fields (Chaps. 4 and 6)
- The solar wind (Chap. 5)
- How charged particles move in the presence of a magnetic field (Chap. 6)
- How charged particles drift in the presence of a non-uniform magnetic field (Chap. 6)
- How plasma drifts in uniform magnetic and electric fields (Chap. 6)

In this chapter you will learn about...

- Earth's geomagnetic field
- Coordinate systems for dipolar magnetic field lines and stretched field lines
- Structures in the magnetosphere
- Particles in the magnetosphere
- Trapping mechanisms in the magnetosphere
- The role of electric fields in the magnetosphere
- Quiescent solar wind-magnetosphere interactions
- Substorms in the quiescent magnetosphere

Outline

7.1 **Earth's Geomagnetic Field**
  7.1.1 Geomagnetic Field Basics
  7.1.2 Coordinate Systems for Earth's Magnetic Domain

7.2 **Geomagnetic Structures and Charged Particle Populations**
  7.2.1 Earth's Magnetized Geospace
  7.2.2 Inner Magnetosphere
  7.2.3 Connections to Distant Regions

7.3 **Single Particle Dynamics—II**
  7.3.1 Gyration and Bounce
  7.3.2 Longitudinal Drift

7.4 **Quiescent Magnetospheric Processes**
  7.4.1 Forming the Dayside Magnetopause
  7.4.2 Mass, Energy, and Momentum Transfer to the Magnetosphere
  7.4.3 Trapping Processes in the Plasmasphere

Chapter 7 Earth's Quiescent Magnetosphere: Its Role in the Space Environment and Weather

7.1 Earth's Geomagnetic Field

Earth's magnetic field acts as an umbrella that protects Earth and its space assets. Electric currents flowing deep within Earth generate about 90% of the field. Additional contributions come from the ionosphere, the magnetosphere, and even the movement of ocean water. Space professionals need to understand the structure of Earth's magnetic umbrella and its evolution in time. We must learn about "inner space" to understand "outer space."

7.1.1 Geomagnetic Field Basics

Objectives: After reading this section, you should be able to...

- Describe the general configuration of the geomagnetic field
- Identify locations and sources of strength and weakness in Earth's field
- Describe the time scale of geomagnetic field changes

Global Features of the Near-surface Field

Earth's near-surface geomagnetic field is approximately a dipole, offset from the rotational axis by ~11°. Earth's core and crustal fields dominate at the surface. At present the field lines emanate from near the south geographic pole and converge near the north geographic pole (Figs. 1-11 and 4-7). Non-dipolar elements cause compass needles to deviate from a north-south alignment. For example, large iron ore deposits in Earth's crust create a slight bulge in the northern Pacific Ocean and a notable weakness in the south Atlantic Basin. Temporary field deviations occur with geomagnetic storms. The amount of the deviation, the declination, varies as a function of geographic location and level of space weather activity. These deviations have been a nuisance historically for navigators, and an early source of space weather intrigue.

Above Earth's surface, at distances greater than 20,000 km, other departures from a dipole configuration are created by the magnetospheric current systems illustrated in Figs. 4-30 and 4-31. Figure 7-1 depicts a nearly dipolar set of field lines associated with the inner magnetosphere and a stretched set of field lines associated with the mid-magnetotail current system and with solar wind distortion of the distant magnetosphere. The diagram also shows in bright colors magnetically trapped particle regions inside the dipole field lines. We describe these in Sec. 7.3.

Field intensity maps reveal that the lowest field intensity is in the vicinity of the geomagnetic equator and the strongest is in the polar regions. Figure 7-2 is a map of the total field strength of much of Earth's surface, excluding the extreme polar regions.

Figure 7-3 gives details from a computer simulation of the quasi-dipolar field in Earth's interior. Regions where the colors mingle represent non-dipolar elements in the field. These elements may add or subtract from the main field. The concentration of blue field lines in the southern hemisphere is associated with a weakness in Earth's main field intensity called the *South Atlantic Anomaly (SAA)*. This feature is evident in field intensity maps (Fig. 7-2) and in spacecraft anomaly maps (Fig. 7-4).

7.1 Earth's Geomagnetic Field

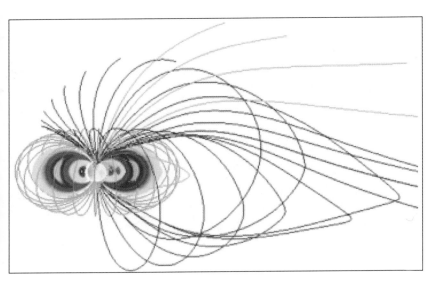

Fig. 7-1. An Idealized Rendering of Earth's External Magnetic Field and Some of the Particles Trapped within It. The field points into Earth in the geographic northern hemisphere. The aqua-colored lines crossing the equator near 6 R_E are approximately dipolar. Field lines that cross the equator at more distant locations (purple color) are distorted. A few field lines (yellow) connecting to high latitudes are attached to the solar wind and do not cross the equator. *(Courtesy of the AF Research Laboratory)*

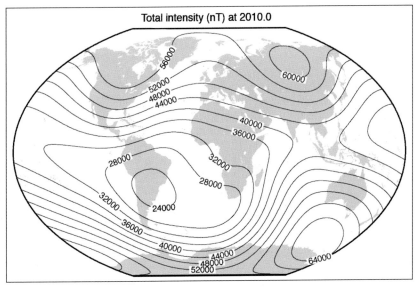

Fig. 7-2. Computed Contours of Constant Surface Magnetic Field Strength in Units of Nanoteslas (nT). The contours represent points of intersection of constant magnetic field with Earth's surface. In general, the field is strongest near the magnetic poles and weakest near the magnetic equator, as expected for a dipole field. The contours on this map represent the footpoints of the families of field lines shown in Fig. 7-1. *(Courtesy of NOAA's National Geophysical Data Center)*

A simple measure of the field's geometry is the position of the magnetic poles. At the north geomagnetic pole, a freely moving magnetic needle points down (toward Earth's center), while at the south geomagnetic pole, the needle points up (away from Earth's center). For this reason, the geomagnetic poles are sometimes referred to as 'dip poles.' One geomagnetic pole is currently located in the Arctic Ocean just beyond the Canadian border at about 84° N latitude and 245° E longitude (geographic coordinates). The other pole is in the Antarctic Ocean south of Australia at about 65° S latitude and 138° E longitude. This antipodal asymmetry is one indicator of the geomagnetic field's complexity.

To be mathematically consistent we should call the pole in the southern geographic hemisphere the north geomagnetic pole—this location has outward directed magnetic field lines. However, the common reference to the north geomagnetic pole is to the pole in the northern geographic hemisphere. In our discussions we stay with this convention.

Chapter 7 Earth's Quiescent Magnetosphere: Its Role in the Space Environment and Weather

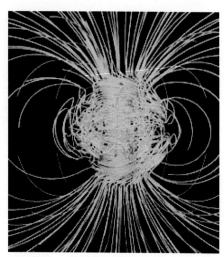

Fig. 7-3. Earth's Interior Magnetic Field Generated by the Geodynamo. Orange field lines are outward directed and blue field lines are inward directed. *(Courtesy of Gary Glatzmaier at the Pittsburgh Supercomputing Center)*

Fig. 7-4. The South Atlantic Anomaly (SAA). A sensitive camera aboard NASA's Terra spacecraft recorded energetic proton hits over and just east of South America between 3 and 16 February 2000. The camera cover had not yet been removed. Energetic protons penetrated the instrument and registered as light. Individual Terra orbital tracks are shown by the colored stripes in between the regions of black that indicate lack of coverage during the interval. Maps of spacecraft anomalies show a similar pattern. *(Courtesy of NASA/GSFC/ JPL, Multi-angle Imaging Spectro Radiometer Science Team)*

The Internal Field and Its Variability

Geodynamo. Earth's interior consists of a solid iron inner core, an iron-filled liquid outer core, and an overlying rocky mantle. Thermal and chemical energy drives convection in the outer core. This energy release, combined with Earth's rotation, powers the *geodynamo*. Electric current sustained by the geodynamo generates the main part of Earth's magnetic field. Paleomagnetic (magnetic history) measurements of rocks suggest that Earth has possessed a magnetic field for at least 3.5 billion years. Like the Sun, Earth has a regenerative process that offsets the inevitable ohmic (electrical resistance) dissipation of electric currents; otherwise the geomagnetic field would vanish after about 15,000 years.

This regenerative process relies on the principles of magnetic induction (Chap. 6). In effect, the core is a natural electric generator, where convection kinetic energy, driven by chemical energy release and the heat of internal radioactivity, changes into electromagnetic energy. Electrically conducting fluid flowing across magnetic field lines induces an electric current, and this generated current supports its own magnetic field. With the proper geometrical relationship between the fluid flow and the magnetic field, the generated magnetic field reinforces the pre-existing magnetic field, so the dynamo is self-sustaining.

Variability. Earth's main field is much more variable than most people realize. Field maps, provided by ground- and space-based observatories, provide a global view of field changes. Over the past 150 years, the main component of Earth's magnetic field has decayed by nearly 10%. Much of the reduction in the global magnetic field comes from field changes in the vicinity of the South Atlantic Anomaly (Figs. 7-2 and 7-4). In this region the surface magnetic field is now about 35% weaker than for an ideal dipole. This field weakness has implications for satellites in low-Earth orbit, because it increases their radiation exposure, as we see in Fig. 7-4. The present field decay rate is about ten times faster than if the dynamo were simply switched off. To that extent, the dynamo today is operating more as an anti-dynamo that destroys part of the field. This decay rate is characteristic of magnetic *polar reversals* that probably occur on average, though with great variability, about once every quarter- to half-million years. Paleomagnetic records suggest that Earth's field has reversed its polarity many hundreds of times during our planet's history. By contrast, the Sun has a rather regular polarity reversal every 11 years or so. The average time between polarity flips is about 250,000 years, making us overdue for the next polarity reversal by about 500,000 years.

Field variability clearly exists in the polar regions as well. Over the last century, the geomagnetic poles have been moving at the rate of ~10 km/yr. During that time the magnetic pole in the northern geographic hemisphere has translated a remarkable 1100 km. Since about 1970, the pole movement has accelerated and is now more than 40 km/yr. These polar wanderings are part of the larger scale variability. The magnetic poles also wander daily around their average positions, and on days when the magnetic field is disturbed, may wobble by 80 km or more. Although the poles' motion on any given day is irregular, the average path forms an ellipse-like figure on Earth's surface.

Focus Box 7.1: The Geodynamo, Field Weaknesses, and Polarity Reversal

Comparing high-fidelity field maps reveals an increasing number of reversed-polarity patches in both magnetic hemispheres, many of which are migrating poleward. The process starts with an east-west directed region of magnetic field, often called a flux tube, submerged in the outer core. Turbulence and eddies in the core's convective motions lift the flux tube outward. As the tube rises, Earth's rotation twists it (via the Coriolis force) into the north-south direction, producing a feature that emerges through the mantle to form a region whose polarity opposes the magnetic pole in that hemisphere (Fig. 7-5). Reversed patches begin to erode Earth's main magnetic field. This process is similar to the one we think is responsible for the Sun's polarity flips, though Earth's cycle is more irregular and much slower.

Planetary rotation, through the Coriolis force, deflects the convective outer core material into helical trajectories (Fig. 7-5). Scientists think this mechanism continually regenerates a new magnetic field to replace the lines that diffuse away. The convective process creates turbulence. Core turbulence appears to be a key element of polarity reversals.

Researchers are developing three-dimensional computer models to simulate the process. It may be 2020 before computing speeds enable detailed computations. But even now, models are yielding promising results, as we see in Fig. 7-6. As time elapses, the number and extent of patches grow, and over a transition period of about nine millennia, the main field polarity flips. The next field reversal, which may start within 2000 years, will surely have implications for space weather events and ecosystems on Earth. We note the similarities with the Sun's dynamo that we describe in Chap. 5.

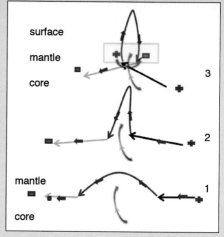

Fig. 7-5. Formation of a Reversed Polarity Patch. At time 1, turbulent flow in the outer core lifts an east-west directed magnetic flux tube toward Earth's mantle. Thousands of years later (time 2) rotation, combined with upwelling, kinks field lines and provides a perpendicular field component in the mantle. Further twisting (time 3) produces small patches of field reversal in the mantle that ultimately extend to the surface and space.

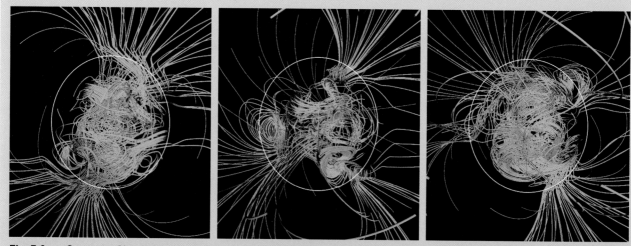

Fig. 7-6. Computer Simulation of a Spontaneous Main Field Polarity Reversal from the Glatzmaier-Roberts Geodynamo Model. The white circle represents Earth's surface. Magnetic field lines are blue where the field is directed inward and yellow where directed outward. Over a transition of thousands of years, reversed flux patches, driven by core turbulence and Earth's rotation, flip the main field polarity. The beginning of this process is shown in Fig. 7-3. *(Courtesy of Gary Glatzmaier, University of California, Santa Cruz and Los Alamos National Laboratory)*

7.1.2 Coordinate Systems for Earth's Magnetic Domain

> **Objectives:** After reading this section, you should be able to...
> ♦ Determine the field strength and orientation within 6 R_E of Earth's surface
> ♦ Interpret magnetic field intensity maps
> ♦ Use L-shell coordinates to locate magnetic field lines near Earth
> ♦ Distinguish between geocentric coordinate systems

Quantifying Earth's Main Field as a Simple Dipole

Here we provide a short discussion of important coordinate systems used for magnetospheric descriptions. Perhaps the most familiar demonstration of the reality of Earth's magnetic field is the north-seeking tendency of compass needles. The magnetic needle in a compass is free to rotate in the horizontal plane. But if we permit it to have full directional freedom, such as suspending it from a thread so that it's free to orient itself horizontally and vertically, then it will vary continuously from one point in space to another in the vertical, as well as the horizontal. Furthermore, if we measure the force on the magnetic needle, we find that the strength of this force, proportional to the intensity of the magnetic field, also varies continuously with its location. We use these properties to map a continuous family of magnetic force vectors.

To describe Earth's field quantitatively, we use the magnetic dipole configuration developed in most introductory physics texts and described in Table 4-2. Spherical coordinates of distance r, latitude λ, and longitude ϕ are most appropriate. Assuming that the dipole field is symmetric about the rotational axis allows us to ignore the longitudinal component. We also ignore non-dipole effects. This arrangement is equivalent to having all the curves in Fig. 7-2 parallel to the magnetic equator. The field variation in latitude and distance from Earth's center for this simple dipole approximation consists of a grouping of constants that includes information about the electrical current strength required to produce the magnetic field ($M\mu_0/4\pi$), and information about geometry in terms of r and λ.

$$B_r = \left(\frac{M\mu_0}{4\pi}\right)\left(\frac{-2\sin\lambda}{r^3}\right) \tag{7-1}$$

$$B_\lambda = \left(\frac{M\mu_0}{4\pi}\right)\left(\frac{\cos\lambda}{r^3}\right) \tag{7-2}$$

where
B_r = magnitude of radial field component [T]
B_λ = magnitude of latitudinal field component [T]
μ_0 = permeability of free space ($4\pi \times 10^{-7}$ N/A^2 ≈ 1.26×10^{-6} N/A^2)
M = Earth's magnetic dipole moment (8.1×10^{22} A·m^2)
λ = magnetic latitude [deg or rad], $\lambda = 0°$ at the equator
r = distance from Earth's center [m]

Figure 7-7 illustrates the geometry of Earth's idealized dipole field in a two-dimensional cross section. Equations (7-1) and (7-2) provide the magnitude of the field. A magnetic field line is everywhere tangent to a line of magnetic force. The magnetic field strength and direction vary along a line of force.

We can determine $|\boldsymbol{B}|$ for any location using the trigonometric identity $\sin^2(\lambda) + \cos^2(\lambda) = 1$, and the results of Eqs. (7-1) and (7-2). The idealized field strength B_S anywhere on Earth's surface is

$$B_S = \sqrt{B_r^2 + B_\lambda^2} = \frac{M\mu_0}{4\pi R_E^3}\sqrt{1 + 3\sin^2(\lambda)} \qquad (7\text{-}3)$$

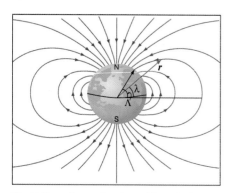

Fig. 7-7. Cross Section of Earth's Simple Dipole Field. Together the position vector r and polar latitude angle λ map a magnetic field line. An individual field line intersects Earth's surface at a geomagnetic latitude angle Λ. The hemispheric labels N and S refer to geographic coordinates. In dipolar coordinates the outward directed (positive) field lines are associated with magnetic north.

The field intensity increases by a factor of two from the equator to the pole. Earth's magnetic field component values at the equatorial surface are

$$B_r = 0 \qquad B_\lambda = \left(\frac{M\mu_0}{4\pi R_E^3}\right)$$

So the total surface field magnitude, B_{Seq}, at the equator is:

$$B_{Seq} = \sqrt{B_r^2 + B_{\lambda=0}^2} = \left(\frac{M\mu_0}{4\pi R_E^3}\right) \approx 31{,}000 \text{ nT}$$

Pause for Inquiry 7-1

Values of constant magnetic field strength B can be found
a) In symmetrical rings in either hemisphere.
b) Along the field lines connecting hemispheres
c) Only at the vertical lines in the polar regions

Pause for Inquiry 7-2

Earth's dipole field at the poles is purely latitudinal/radial (choose one) and has only a vertical/horizontal (choose one) component.

Pause for Inquiry 7-3

According to the simple dipole approximation, at the equator Earth's field
a) Is horizontal
b) Has no radial component
c) Has no longitudinal component
d) Has a magnitude solely dependent on distance from Earth's center
e) All of the above

7.1 Earth's Geomagnetic Field

EXAMPLE 7.1

Geomagnetic Field Example

- *Problem Statement:* Determine the strength and orientation of the idealized magnetic field at Earth's surface at 45° N latitude.
- *Relevant Concepts and Equations:* Equations for Earth's dipole field
- *Given:* $B_{Seq} = 3.1 \times 10^{-5}$ T
- *Assume:* The field is a perfect dipole.
- *Solution:* For 45°

$$B_r = B_{Seq} R_E^3 \left(\frac{-2\sin(\pi/4)}{r^3}\right) = (-2)(3.1 \times 10^{-5} \text{T})(0.707)$$

$$= -4.38 \times 10^{-5} \text{ T}$$

$$B_\lambda = B_{Seq} R_E^3 \left(\frac{\cos(\pi/4)}{r^3}\right) = (3.1 \times 10^{-5} \text{T})(0.707)$$

$$= 2.19 \times 10^{-5} \text{ T}$$

$$B_{S45} = \sqrt{(-4.38 \times 10^{-5} \text{T})^2 + (2.19 \times 10^{-5} \text{T})^2}$$

$$= -4.9 \times 10^{-5} \text{ T}$$

- *Interpretation:* At 45°, the radial portion of the field is twice as strong as the latitudinal portion of the field. The minus sign indicates the radial portion of the field points inward, as it does in the Northern Hemisphere.

Follow-on Exercise 1: Verify that the results above are consistent with using $B_{S45} = B_{Seq} \frac{R_E^3}{r^3} \sqrt{1 + 3\sin^2(\lambda)}$

Follow-on Exercise 2: Determine the strength and orientation of the surface magnetic field at 60° latitude in the Northern Hemisphere and at an altitude of 1 R_E above the surface at 70° latitude in the Southern Hemisphere.

Surface Field Orientation Referenced to Geographic Coordinates

Many magnetic field measurements are made at Earth's surface, so we often need to describe Earth's field with reference to geographic coordinates (Fig. 7-8). The Geographic Coordinate System (GCS) is defined so that its **X**-axis is in Earth's equatorial plane but is fixed with Earth's rotation, passing through the Greenwich meridian (0° longitude). Its **Z**-axis is parallel to Earth's rotation axis, and its **Y**-axis completes a right-handed orthogonal set (**Y** = **Z** × **X**).

At any point on Earth's surface, we can describe the orientation of the geomagnetic vector relative to a geographic coordinate system using two angles: *declination D*, the angle between the horizontal component of the magnetic-field vector and true north (Earth's spin axis), and *inclination I*, the angle between the local horizontal plane and the total field vector. These angles help reckon the *horizontal magnetic intensity vector* **H** and the *total magnetic field vector* **F**. The length of **F** represents the intensity of the magnetic field, which is independent of the orientation of the reference coordinate system. The magnetic-field components, F_X, F_Y, and F_Z (usually written as **X**, **Y**, **Z**) are Cartesian components (north, east, down). The following geometric relations apply to **H** and *D*:

$$X = H \cos D \quad \text{(7-4a)}$$

$$Y = H \sin D \quad \text{(7-4b)}$$

$$Y/X = \tan D \quad \text{(7-4c)}$$

Chapter 7 Earth's Quiescent Magnetosphere: Its Role in the Space Environment and Weather

The following geometric relations apply to Z, H, and I:

$$\tan I = Z/H \qquad (7\text{-}4d)$$

$$|F|^2 = |B|^2 = |Z|^2 + |H|^2 \qquad (7\text{-}4e)$$

> **Pause for Inquiry 7-4**
>
> At a particular location the intensity of the magnetic field is 48,500 nT, the declination angle is $-1°$ and the inclination angle is $54°$. What is the magnitude of the horizontal field component? What is the eastward component of the field? How far from geographic north is the field deviated? Is the location polar, equatorial, or mid-latitude?

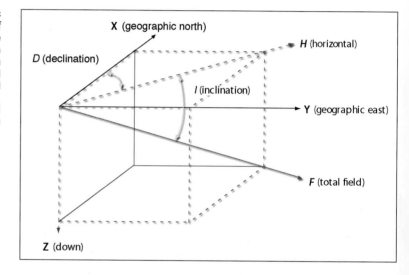

Fig. 7-8. Surface Geographic and Geomagnetic Coordinates. The total magnetic vector, F, consists of horizontal, H, and vertical, Z, components, which are themselves vectors. The horizontal vector is often expressed in a geographic reference frame of X, north and Y, east. The magnitudes of X, Y, and Z are defined in terms of the declination angle D, between horizontal component and north and the inclination angle I, between the horizontal component and the total field vector F. The angles are measured in degrees positive east for D and positive down for I.

Magnetic Field and *L*-shell Coordinates for the Magnetosphere

Early space-age data revealed that energetic particles were organized by Earth's magnetic field in shells around Earth. Space physicist Carl McIlwain developed a coordinate system that takes advantage of some of the symmetry of Earth's quasi-dipolar field. His system applies to low and middle latitudes close to Earth (typically 6 R_E or less).

Because Earth's magnetic field is approximately symmetric in longitude, a given magnetic line of force may be revolved about the dipole, to yield a hollow toroidal shell (as shown by the aqua colored lines in Fig. 7-1). The distance from the dipole center to the location where the shell intersects the magnetic equator is a constant at any longitude. The shell is associated with a single parameter L, which is the distance in Earth radii between the center of the dipole and where the outer part of the shell intersects the magnetic equator. Figure 7-9 shows a cross section of three L shells. The latitude where a particular field line crosses Earth's surface ($r = R_E$) is defined as Λ (Fig. 7-9). A particular position on an L shell at any longitude is determined by using the value of the magnetic field B. L and B combined produce the L-B coordinate system.

A magnetic field line is everywhere tangent to a line of magnetic force. In a dipole system without longitudinal dependence, the field line equation is:

$$r\, d\lambda/B_\lambda = dr/B_r \qquad (7\text{-}5)$$

With slight manipulation and substituting values from Eqs. (7-1) and (7-2) we get

$$dr/r = (-2\sin\lambda\, d\lambda)/\cos\lambda$$

Integrating produces the equation of a field line

$$r = r_{eq}\cos^2(\lambda) \qquad (7\text{-}6a)$$
$$= L\cos^2(\lambda) \qquad (7\text{-}6b)$$

where
- r = distance from the dipole center [km or R_E]
- $r_{eq} = L$ = the distance to the equatorial crossing of the field line for $\lambda = 0$ [km or R_E]
- λ = geomagnetic latitude angle [° or rad]

(Note that at Earth's surface $\lambda = \Lambda$)

Pause for Inquiry 7-5

Verify that $\Lambda = \cos^{-1}(1/L)^{1/2}$ at Earth's surface.

EXAMPLE 7.2

Geomagnetic Latitude Angle

- **Problem Statement:** Determine the geomagnetic latitude angle Λ for a magnetic field line crossing a geosynchronous orbit.
- **Relevant Concepts and Equations:** $L\text{-}\mathbf{B}$ coordinate system
- **Given:** Result from Pause for Inquiry 7-5 and a location at geosynchronous orbit, 6.6 R_E
- **Assume:** The field is a perfect dipole and the geosynchronous orbital plane is in the magnetic equatorial plane
- **Solution:**

$L = r/\cos^2(\lambda)$,
$L = 6.6\ R_E\, /\cos^2(0) = 6.6\ R_E$
$\Lambda = \cos^{-1}(1/L)^{1/2} = \cos^{-1}([1/6.6]^{1/2}) = 67°$

- **Interpretation:** Geosynchronous orbit magnetically maps to 67° at Earth's surface, which is often the lower edge of the auroral zone.

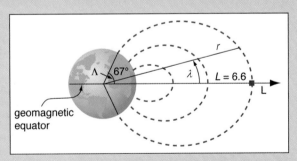

Fig. 7-9. Dipole Field Diagram. The geomagnetic latitude angle at a field line at radius r is λ. The magnetic latitude on Earth's surface where a field line pierces the surface is the geomagnetic latitude angle Λ. The value of L is the distance in Earth radii between the center of the dipole and where the outer part of the shell intersects the magnetic equator.

Follow-on Exercise: Determine which L shell intersects Earth's surface at a latitude of 60°.

Magnetospheric Coordinates Referenced to the Sun

Data from the 1960s and 1970s reveal that the magnetosphere organizes particles and fields in ways that preferentially align in Sun-solar wind-magnetosphere geometry. We illustrate several Earth-centered (geocentric) coordinate systems in Fig. 7-10.

Geocentric Solar Magnetospheric Coordinates (GSM). The **X**-axis of the GSM coordinate system connects Earth's center to the Sun's center. The **Y**-axis is defined to be perpendicular to the Earth's magnetic dipole so that the **X-Z** plane contains the dipole axis. The **Y**- and **Z**-axes rotate around the **X**-axis every day, because they depend on the direction of Earth's magnetic dipole, which moves on the edge of an imaginary cone as the Earth rotates. Data from satellites located in the magnetosphere are often displayed in GSM coordinates. Data from solar wind monitoring satellites are often transformed into this coordinate system to link solar wind effects to geomagnetic activity.

Solar Magnetospheric Coordinates (SM). In the SM coordinate system the **Z**-axis is parallel to the north magnetic pole and the **Y**-axis is perpendicular to the Earth-Sun line, positive toward dusk. The **X**- and **Z**- axes rotate around the **Y**-axis every day. Thus, the **X**-axis is generally not pointed directly at the Sun. The amount of rotation is the dipole tilt angle. As with the GSM system, the SM system rotates with a yearly and daily period with respect to inertial coordinates.

Geocentric Solar Ecliptic Coordinates (GSE). The GSE system has its **X**-axis pointing from Earth toward the Sun and its **Y**-axis in the ecliptic plane pointing toward dusk (opposite to planetary motion). Its **Z**-axis is parallel to the ecliptic pole. This system is often used to show satellite trajectories, the solar wind, and the interplanetary magnetic field.

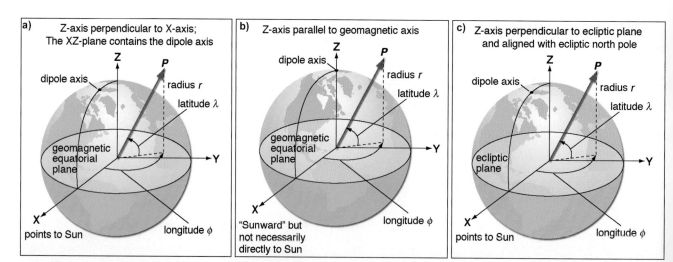

Fig. 7-10 Magnetospheric Coordinate Systems. a) The Geocentric Solar Magnetospheric Coordinates (GSM). XY-plane tilts toward and away from the Sun by an angle equal to the dipole tilt, so the **Y** and **Z** axes are in constant motion with respect to inertial coordinates. b) The Solar Magnetospheric Coordinate System. The **X**-axis is in constant motion with respect to the Sun. c) The Geocentric Solar Ecliptic Coordinates. The magnetosphere wobbles to and fro about the axes of the system. The ***P*** vector shows how to locate a point in each coordinate system. *(Based on Russell [1971] and Fraenz and Harper [2002])*

7.2 Geomagnetic Structures and Charged Particle Populations

7.2.1 Earth's Magnetized Geospace

Objectives: After reading this section, you should be able to...

- Categorize magnetospheric regions and boundaries
- Describe the scale of the magnetosphere
- Identify the magnetosphere's inner, middle, and outer regions

Earth's magnetic field extends into space, but the solar wind confines it to the volume called the *magnetosphere*. The solar wind shapes the system into a blunt bullet shape (magnetonose) on the dayside and a long cylinder shape (magnetotail) on the nightside. The magnetosphere is defined as the region of space above Earth's ionosphere in which charged-particle motion is dominated by the geomagnetic field. Figure 7-11 shows a scaled noon-to-midnight diagram of the region that includes exterior features such as the bow shock and the magnetosheath. Surrounding the magnetosphere are currents that form a boundary called the *magnetopause*. The magnetopause current system effectively excludes the solar wind from the magnetosphere. Threading the magnetosphere are current systems that assist in regional "communication" within the magnetosphere and with ionosphere-magnetosphere coupling. Some current systems are permanent, but highly variable. Others develop and decay while transients in the solar wind interact with the magnetosphere. Particles, quasi-static magnetic fields, and low-frequency electromagnetic fields play a large role in energy transfer in the magnetosphere (the role of photons is generally rather small).

Geomagnetic field lines fit into one of three regimes. Figure 7-11 shows dipolar magnetic field lines that have both feet (ends of field lines) attached to Earth. These are usually associated with the inner "closed field" magnetosphere. We associate regions that have open-ended field lines with the outer or distant magnetosphere. In the middle is a vast and variable region with closed but very stretched field lines. The middle magnetosphere is the location of much of the interesting dynamics associated with space weather.

Pause for Inquiry 7-6

Show the location of the lunar orbit on Fig. 7-11.

In addition to the field line configuration, we describe the magnetosphere in terms of particles that populate the field lines. Interestingly, the magnetosphere does not have a native particle population. The particles come from cosmic rays, solar outflows, and Earth's ionosphere. Collisions are so rare in most of the magnetosphere that its electrical conductivity is nearly infinite. Thus, magnetic fields and very-low-energy particles are frozen together. As the plasma moves, so do the magnetic field lines that thread it. Particles in the magnetosphere exhibit different behaviors. They drift on gradients of the magnetic field. These drifts create electric currents, because they are energy- and charge-dependent. In turn

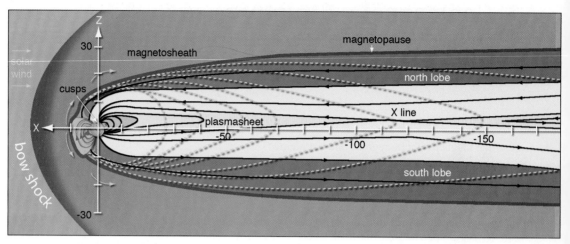

Fig. 7-11. **A Noon-to-midnight View of the Magnetosphere.** The Sun is to the left. Earth is at the origin. The axes are labeled in Earth radii. Positive **X** is toward the Sun. Positive **Z** is to the top of the page. Positive **Y** is out of the page. The solar wind, streaming radially into space, compresses the Sunward side of the magnetosphere to a distance of 6 R_E–12 R_E, and drags it out the nightside (into a magnetotail) to 1000 R_E or more. For clarity in subsequent discussions, we show diagrams in which the length scale is compressed. The shading in the different regions indicates different particle populations and temperatures, which we describe in Tables 7-1 and 7-2. [After Pilipp and Morfill, 1978]

these currents create magnetic fields that add to the geomagnetic field originating within Earth. Because of these currents, the total field in the middle and outer magnetosphere (beyond several Earth radii) departs radically from what is calculated by simply extrapolating Earth's internal field.

We provide more context for magnetospheric features in the not-to-scale perspective of Fig. 7-12. The labeled regions represent a spectrum of field lines and particle populations. The cusps are open to the solar wind-magnetosheath particles. The inner magnetosphere comprises the plasmasphere, the radiation belts, and portions of the plasmasheet. The extended plasmasheet becomes approximately flat in the outer magnetotail. The entire system is wrapped in a cocoon of current (the magnetopause current). Beyond this current lies the disturbed solar wind in the magnetosheath. In the subsections below we describe the magnetic structures in the magnetosphere. Tables 7-1 and 7-2 provide average values for particle populations associated with these structures.

7.2 Geomagnetic Structures and Charged Particle Populations

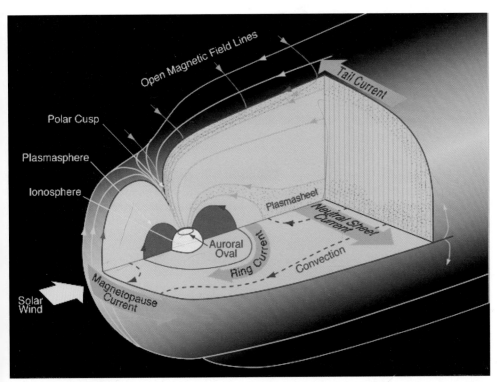

Fig. 7-12. An Oblique Cross-sectional Cut of the Magnetosphere. This view is from approximately 1400 local time (two hours after noon) and illustrates many of the regions and boundaries identified in the text. For simplicity the radiation belts that overlap the plasmasphere are not shown. Currents are depicted as orange arrows. The distant magnetotail is off the page to the right. *(Courtesy of NASA's Goddard Space Flight Center)*

EXAMPLE 7.3

Earth's Magnetotail Particle Mass

- *Problem Statement:* Estimate the mass of particles contained in Earth's magnetosphere to a distance of 100 R_E downtail and compare the value to the mass of Earth's atmosphere—5×10^{18} kg.

- *Relevant Concepts and Equations:* Volume and mass density

- *Given:* Fig. 7-11 and average particle density in the magnetosphere is 10 protons/cm^3

- *Assume:* That the volume of interest is modeled as a hemisphere that extends from 10 R_E upstream of Earth attached to a cylinder that extends a distance 100 R_E downtail.

- *Solution:* We add the volume of the Sunward magnetosphere to the volume of the long magnetotail cylinder. Then we multiply the total volume by the average density to get the total mass in the magnetosphere.

Volume of hemisphere = $(2/3)\pi r^3$

Volume of cylinder = $\pi r^2 l$

Volume of hemisphere = $(2/3)\pi (10\ R_E)^3 = (2/3)\pi (10 \times 6378 \times 10^3\ m)^3 = 5.4 \times 10^{23}\ m^3$

Volume of cylinder = $\pi r^2 l = \pi (10 \times 6378 \times 10^3\ m)^2 (100 \times 6378 \times 10^3\ m) = 8.1 \times 10^{24}\ m^3$

Total volume = $5.4 \times 10^{23}\ m^3 + 8.1 \times 10^{24}\ m^3 = 8.6 \times 10^{24}\ m^3$

Mass in magnetosphere = $(10 \times 10^6\ \text{protons/m}^3)(1.67 \times 10^{-27}\ \text{kg/proton})(8.6 \times 10^{24}\ m^3) \approx 1.4 \times 10^5\ kg$

- *Interpretation:* The mass in the magnetosphere is about 2.8×10^{-14} of the mass in Earth's atmosphere.

Follow-on Exercise: Model Earth's plasmasphere as a shell that extends from 1 to 6 R_E. Use the density value in Table 7.1 to compare the mass in the plasmasphere to that in the entire magnetosphere.

7.2.2 Inner Magnetosphere

> **Objectives: After reading this section, you should be able to...**
>
> ♦ Compare and contrast the field configurations and particle populations of the inner magnetosphere
> ♦ Know the relative locations of the plasmasphere and the radiation belts

The inner magnetosphere is home to closed field lines and two distinctly different types of plasma. The plasmasphere is cold (thermal) plasma that we characterize by temperature and density. In the same region of space are two very sparse population of hot particles that we characterize by energy flux. These are radiation belt particles. The two types of plasma have little interaction during quiescent intervals.

Inner Magnetosphere

Plasmasphere. Our magnetospheric tour begins in a region of space that has multiple identities: the plasmasphere and the radiation belts. The plasma populations occupying the region come from different places and behave quite differently. Some of the region's atoms are neutral and thus form part of the geocorona we describe in Chap. 8.

The *plasmasphere* is the upward extension of the ionosphere (Fig. 7-13a). Its outer edge is often near geosynchronous orbit. This torus-shaped volume of closed field lines contains dense, relatively cold, low-energy plasma that has evaporated from Earth's ionosphere. We identify the plasmasphere as a separate entity where the composition becomes rich in plasma. The ionosphere-plasmasphere boundary is nominally at an altitude of 1000 km, although the transition is not sharp.

The plasmasphere ions and electrons have characteristic energies of 0.5 eV–1.0 eV. The ions are primarily H+ and He+, with a smaller percentage (usually) of ionized nitrogen and oxygen. Geosynchronous satellites are typically in the plasmasphere from mid-afternoon to early evening (~1500 to ~2000 Local Time), and in the adjacent plasmasheet (which we describe shortly) from early evening to just past dawn (~2000 to ~0600 Local Time). The satellites measure an ambiguous region of mixed plasmas at other local times.

While the Earthward, inner boundary is quite stable, the plasmasphere as a whole shifts in size and density in response to disturbances elsewhere in the magnetosphere. The upper plasmaspheric boundary begins at 4 R_E–6 R_E at the equator. At high latitudes the boundary is much closer to Earth, because the plasma follows the magnetic field lines to a high-latitude intercept in the ionosphere, at magnetic latitudes from 60°–70°. During storm conditions, plasma may detach from the plasmasphere and become part of the larger plasma convection process in the magnetosphere. We describe this process in more depth in Chap. 11.

The *plasmapause* separates the plasmasphere proper and the plasmasheet, as observed on Earth's nightside. The cold plasma is attached to Earth's magnetic field and must rotate with it, meaning the plasmasphere co-rotates or drifts with Earth in near rigid-body motion (Sec. 7.4). Corresponding to this motion is an electric field, just sufficient to produce the observed $E \times B$ drift (Sec. 7.4). This effect decreases dramatically beyond the plasmasphere. The plasmapause is the boundary at which co-rotation ceases.

Radiation Belts. The radiation belts occupy much of the plasmasphere (Fig. 7-13b). So why do we use a different name? Because the radiation belt particles are so energetic that they have very different effects on detectors and on the local environment. We learned this in 1958 when a Geiger counter mounted onboard the United States' first satellite, Explorer 1, detected the belts of particles with energies in the MeV range. Subsequent missions and experiments collected data on these particles and found that two donut-shaped zones of trapped electrons and protons encircle Earth. These belts lie approximately within the plasmasphere and bear the name of James Van Allen, who first explained to the public the significance of these high-energy particles. Particles in the Van Allen radiation belts are confined to their regions by a combination of processes we describe in Secs. 7.3 and 7.4.

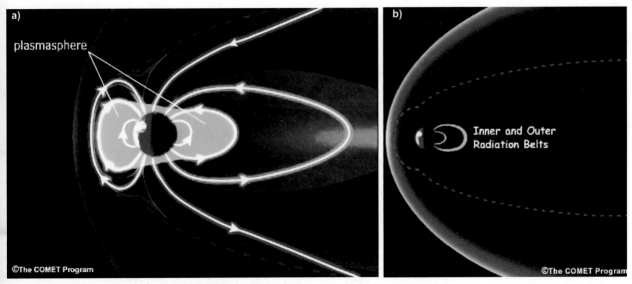

Fig. 7-13. Artist's Rendition of Regions of the Magnetosphere with Closed Field Lines. The regions shown contain Earth's magnetic field lines with both feet (field-line ends) anchored to Earth. **a)** The plasmasphere, a region of cold plasma of ionospheric origin, is shown in green. The red dashed line is the magnetopause. **b)** Here we show in aqua the regions of hot trapped plasma and energetic particles of multiple origins in blue. The red dashed line is the magnetopause. *(Courtesy of the COMET Program at the University Corporation for Atmospheric Research)*

Trapped electrons and protons exist throughout the magnetosphere, but they are mostly concentrated in the Van Allen belts. The inner belt is dominated by very high-energy protons produced when cosmic rays interact with molecules in the upper atmosphere. Inner-belt protons remain trapped for days to years. Hence this belt is considered a stable feature of the magnetosphere. Many ions have energies on the order of tens of MeV or greater. The inner belt extends from an altitude of about 400 km to about 10,000 km and occasionally to 12,000 km. The peak proton concentrations are near 3200 km (~0.5 R_E above the surface), but the altitude varies depending on the protons' energies. Figure 7-14 displays output from a NASA radiation belt model. The small red arc close to the equator represents the peak of the inner belt.

Variability is a key characteristic of the outer radiation belt. The solar wind and interplanetary magnetic field affect this weaker field region of the magnetosphere more than the inner zone, leading to shorter lifetimes of trapped

Chapter 7 Earth's Quiescent Magnetosphere: Its Role in the Space Environment and Weather

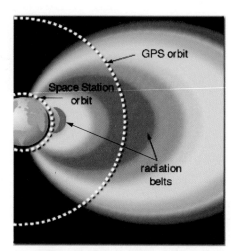

Fig. 7-14. A Cross-sectional View of NASA's Radiation Belt Model. The peak of the inner belt is shown in red close to Earth's equator. The peak of the outer belt is darker red. Global Positioning System (GPS) orbits traverse the heart of the outer radiation belt. The lighter colors indicate lower fluxes of energetic particles. *(Courtesy of NASA)*

particles and more dynamics. The acceleration mechanisms for these particles are an active area of research.

The outer belt is populated primarily with high-energy electrons produced by cosmic rays, solar particles, and magnetospheric acceleration processes. The outer belt extends from the top of the inner belt, an altitude of 10,000 km to about 12,000 km, up to about 60,000 km and sometimes higher, depending on the electrons' energies and the level of solar activity. During quiescent intervals, the heart of the outer radiation belt coincides with Global Positioning System (GPS) orbits. The peak electron concentrations are near 16,000 km (~2.5 R_E above the surface). The relatively low particle density region between the inner and outer belts, shown in yellow in Fig. 7-14, is the *slot* (*slot region*).

During steady-state conditions in the magnetosphere, particles don't escape these trapping regions. During magnetospheric disturbances (Chap. 11), however, accelerated particles often enter and leave the radiation belts. As fluxes and energies of these particles rise, the opportunity for undesirable particle-spacecraft interaction also rises. Occasionally, new radiation belts form between the inner and outer zones when interplanetary disturbances originating at the Sun impact the magnetosphere. We devote more discussion to this in the next section and in Chap. 11.

Trapped energetic charged particles in the radiation belts are a persistent threat to satellites passing through them. Some incidents are merely nuisances—background noise in sensors increase while they pass through the belts. Other impacts are truly damaging—high-energy particles penetrate a spacecraft's skin and embed themselves in the microchip electronics within, causing physical damage and phantom commands (Example 4-3). As the particles enter the spacecraft material they deposit kinetic energy. This causes atomic displacement and produces streams of charged atoms in the incident particles' wake. Solar cells that power nearly all Earth-orbiting satellites deteriorate slowly as the satellites repeatedly traverse the belts. Problems are particularly acute for satellites in low orbits that pass through the South Atlantic Anomaly, where Earth's magnetic field is weak (Sec. 7.1).

Table 7-1. **Thermal Plasma and Field Values in Earth's Magnetosphere.** Here we list characteristics of the thermal plasmas and fields in Earth's magnetosphere.

	Density (n)	Electron Temperature (T_e)	Proton Temperature (T_i)	Magnetic Field	Comments
Plasmasphere	>10^2 particles/cm^3	10^4 K	0.1 eV – 1 eV	5000 nT	Cold plasma of terrestrial origin
Plasmasheet	0.1–1.0 particles/cm^3	2×10^6 K to 2×10^7 K	Always higher by a factor of 3–5 than the electron temperature	9 nT in deep tail	Thickness of 4–6 R_E, forms into the ring current at 5–6 R_E from Earth
Plasmasheet boundary layer	0.1–1.0 particles/cm^3	2×10^6 K to 1×10^7 K	1×10^7 K to 5×10^7 K	20 nT–50 nT at 20 R_E	Region that maps to auroral zone, producing discrete arcs
Lobe	10^{-3}–10^{-2} particles/cm^3	< 10^6 K	< 10^7 K	Few nT	Lowest densities found in the magnetospheric cavity
Mantle	0.5–50 particles/cm^3	10^5 K–10^6 K	5×10^5 K to 8×10^6 K	10 nT–30 nT	Entry layer into the magnetosphere of magnetosheath plasma

7.2 Geomagnetic Structures and Charged Particle Populations

Table 7-2. **Non-thermal Energetic Particles and Field Values in Earth's Magnetosphere.** Here we list characteristics of the non-thermal energetic particles and fields in Earth's magnetosphere.

	Flux	Electron Energy	Proton (Ion) Energy	Magnetic Field	Comments
Inner belt and South Atlantic Anomaly	10^5 ions/(cm$^2 \cdot$s) >10 MeV at 1.5R_E	--	1 MeV– 500 MeV Typical: 10 MeV	20,000 nT	Created by cosmic ray interaction with terrestrial atmosphere
Outer belt	10^6 e$^-$/(cm$^2 \cdot$s) >1 MeV at 4R_E	1 keV to 10 MeV Typical: 0.1 MeV	--	100 nT	May be filled with terrestrial and solar wind electrons
Slot region	Very low 10^4 e$^-$/(cm$^2 \cdot$s) >1 MeV at 2R_E	0.01 MeV to 20 MeV	0.01 MeV to 50 MeV	1000 nT– 5000 nT	Energized during unusual solar wind conditions
Ring current ions	Very low	--	10 keV to 300 keV	5000 nT	Energized during magnetic storms

7.2.3 Connections to Distant Regions

Objectives: After reading this section, you should be able to...

- Compare and contrast the field configurations and particle populations of the middle magnetosphere
- Know the relative locations of the plasmasheet and the cusps

Dayside Solar Wind-Magnetosphere-Ionosphere Connections

Polar Cusps. As viewed in the noon-midnight cross section shown in Fig. 7-15, the *polar cusps* separate the magnetic field lines that extend on Earth's Sunward side from those that stretch into the tail of the magnetosphere. These high-latitude features are located where geomagnetic field lines extend out to meet the magnetopause at nearly right angles. Earth's field is normally very weak in these regions and acts more as a funnel for solar wind particles than a barrier. When solar wind particles enter the cusp, they follow the magnetic field lines toward Earth. Through the polar cusps, high-speed charged particles bombard the upper atmosphere in rather narrow slits centered near ~75° geomagnetic latitude. The particles also interact with the field lines to create electromagnetic waves that dissipate as heat.

The cusp may act as a local particle accelerator that provides energized particles to very distant locations. Many of the particles are energized in the narrowing magnetic channel. Some then escape along magnetic field lines to become part of the reservoir of particles in the magnetotail.

Inner Magnetosphere Connections to the Magnetotail

Plasmasheet. Beyond the plasmasphere and radiation belt area is a region of relatively hot, tenuous plasma called the *plasmasheet*. It is a region of variable thickness and density in which solar wind plasma coexists with Earth's plasma. The plasmasheet contains a mixture of particles originating in the solar wind (H+) and ionosphere (O+). As we describe in Chap. 4 and Sec. 7.4, the magnetic merging process is a likely suspect in allowing solar plasma to enter the magnetotail region. The motions of magnetic field lines associated with nightside merging (also called reconnection) largely guide the behavior of plasmasheet particles.

Chapter 7 Earth's Quiescent Magnetosphere: Its Role in the Space Environment and Weather

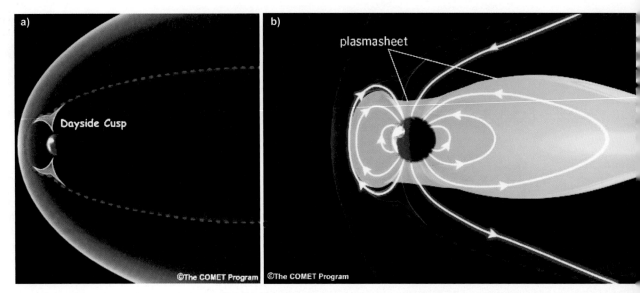

Fig. 7-15. Regions That Connect Earth with the Solar Wind and Magnetotail. a) Cusp Regions. Nulls in the magnetic field allow the cusp regions to admit particles from the shocked solar wind. The red dashed line is the magnetopause. b) Plasmasheet. Here is a region of energized (warm to hot) plasma originating in the solar wind and Earth's ionosphere. Plasmasheet particles are found in the compressed dayside region as well as in the tail region; they have access to the auroral zones. The red dashed line is the magnetopause. *(Courtesy of the COMET Program at the University Corporation for Atmospheric Research)*

The plasmasheet connects the inner and outer magnetospheric regions. This connection is an important way for energy to redistribute in the magnetosphere, as we describe in Sec. 7.4 and in Chap. 11. A very thin boundary between the plasmasheet and the tail lobes is the *plasmasheet boundary layer*, which is the source of most auroral events. Geosynchronous satellites spend a portion of their orbital lifetime in this region, and thus are exposed often to hot plasma particles that may attach or embed themselves on or in the satellites.

Figure 7-16 shows how the plasmasheet conforms to the shape of the stretched outer magnetosphere. Some distance downtail, the plasmasheet becomes relatively flat. In the flattened region a thin neutral current sheet divides the plasmasheet. The name comes from the fact that magnetic fields from Earth's northern and southern hemispheres nearly cancel, making the region magnetically "neutral."

Outer Boundary Regions

Plasma Mantle. Tailward of the cusps is the magnetotail, where Earth's magnetic field is stronger and acts as a partial barrier to the shocked solar wind in the magnetosheath. Some solar wind particles become trapped within the first few kilometers of this enhanced field region. This partially trapped plasma is the *plasma mantle*.

Lobes. Inside of the plasma mantles, in both hemispheres, are *lobes* of magnetic field. We show these lobes as the dark blue chambers in Fig. 7-16. The small amount of plasma on these field lines originates as cool plasma evaporating from the ionosphere. These regions host magnetic field lines that emanate from the north and south polar regions. Figure 7-11 shows two regions where the field lines are oppositely directed. In these regions the field is of sufficient strength to create the bulk of the pressure that keeps the magnetosphere from collapsing on itself. In the upper (northern) half of the magnetolobe, the magnetic field is directed toward

7.2 Geomagnetic Structures and Charged Particle Populations

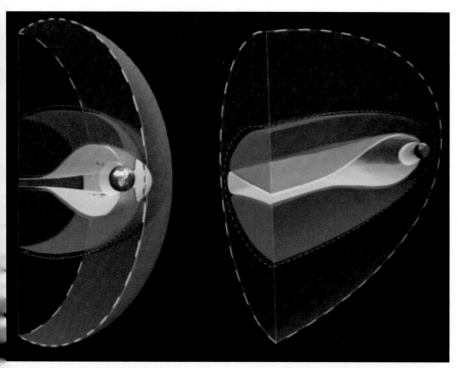

Fig. 7-16. The Configuration of the Plasmasheet and the Tail lobes Within the yellow regions, particles are subject to various forces. The dark blue lobe regions tend to have cooler particles and low density. Red indicates the magnetopause. The plasma mantle occupies the thin regions between the magnetopause and the lobes. The bowshock is shown in semi-transparent blue. *(Courtesy of the COMET Program at the University Corporation for Atmospheric Research)*

Earth, and in the lower (southern) half, it's directed away from Earth. As long as the solar wind remains fairly steady, the lobes exist in equilibrium.

Distant or Deep Magnetotail

The magnetosphere streams into an elongated magnetotail that stretches hundreds of Earth radii downstream from Earth away from the Sun, further than most of our diagrams show. What is the distant magnetotail like? Only a few spacecraft in highly elliptical equatorial orbits have visited the deep magnetotail, beyond 100 R_E. The spacecraft data, when merged with models, suggest that the far magnetotail becomes wispy and subject to the turbulence in the solar wind. If it were visible, it would most likely look like a flag flapping in the (solar) wind.

The magnetotail is also where the solar wind field lines that merged with Earth's field lines must "let go" and return to being full-fledged solar wind. This occurs between 30 R_E–150 R_E and only happens if a magnetic disconnection is present similar to that associated with coronal mass ejections in the solar atmosphere. In fact, strong evidence shows small versions of mass ejections, called *plasmoids*, exploding out of the magnetotail region during active times. The process creates a cloud of detached Earth plasma heading away from Earth, and recoiling bursts of plasma flows in the magnetotail that stream Earthward. These join other accelerated particles in a process called convection. We describe this further in Sec. 7.4 and in Chap. 11.

Chapter 7 Earth's Quiescent Magnetosphere: Its Role in the Space Environment and Weather

Pause for Inquiry 7-7

Match the descriptions and names of the various magnetopause regions and boundaries below.

Description	Names
• _____ Dayside regions that divide the magnetic field lines that intersect Earth's surface on the sunlit side from those that stretch into the nightside magnetotail. These regions have either a null field value or low magnetic field strength. • _____ Outer boundary region in which solar wind plasma may be trapped. • _____ Outer nightside, high-latitude regions nearly devoid of plasma. • _____ A predominantly nightside region of moderate particle energy and density that extends into the magnetotail. It is magnetically connected to the auroral ovals and extends into the mid-tail region. • _____ A region close to Earth populated by relatively cold plasma (energies < 1 eV) which co-rotates with Earth because of magneto-coupling. • _____ A region occupying much the same volume as the plasmasphere in which high-energy electrons and protons (energies 100 keV–100 MeV) gyrate around closed magnetic field lines. • _____ A region far from Earth with open or extremely stretched field lines. • _____ The outer boundary of the magnetosphere that divides the magnetosphere and solar wind plasma. • _____ Extends behind Earth in the anti-solar direction. It divides the geomagnetic tail into lobes with magnetic field lines pointing toward and away from Earth. • _____ A current system that develops near the edge of the radiation belts in response to geomagnetic activity. • _____ The boundary between the plasmasphere and the plasmasheet	*Plasmasheet* *Plasma Mantle* *Cusp* *Plasmasphere* *Distant Magnetotail* *Trapped Radiation Belts* *Lobes* *Plasmapause* *Ring Current* *Neutral Current Sheet* *Magnetopause*

Pause for Inquiry 7-8

What direction does the current flow in the neutral sheet? *Hint:* Apply Ampere's Law.

a) Dayside to nightside

b) Nightside to dayside

c) Dusk to dawn

d) Dawn to dusk

Focus Box 7.2: Sensing and Operating in Regions of the Magnetosphere

Some of what we know about the magnetosphere comes from ground observations, but the bulk of our knowledge is from satellite measurements. Figure 7-17 depicts satellite orbits used for magnetospheric sensing. Geosynchronous satellites orbit at the edges of the particle-trapping region in Earth's magnetic field and Table 7-3 lists orbit characteristics. Satellites in medium-Earth orbits (MEO [semi-synchronous and others]) travel through the heart of the radiation belts, as do some satellites in highly elliptical orbits (HEO). Satellites in highly inclined low-Earth orbits may sample particles that fall out of the lower edges of the radiation belts and particles moving along field lines open to the solar wind in the polar regions. Satellites in low-Earth orbits (LEO) often sample particles from nearly every magnetospheric region because geomagnetic field lines must have at least one foot anchored at Earth. Some LEO satellites are instrumented to sample the drizzle of energetic particles to low altitudes. Others become unintended particle detectors during strong geomagnetic disturbances or when they pass through the South Atlantic Anomaly.

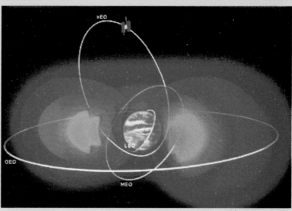

Fig. 7-17. Four Satellite Orbits in the Magnetosphere. The color gradations indicate particles of different energies. Blue regions contain particles with lower energies. *(Courtesy of B. Jones, Peter Fuqua, and James Barrie, The Aerospace Corporation)*

Satellites, especially those operating near the radiation belts, are subject to radiation stress that may damage electronic components. The dynamics of this hostile environment are characterized by time scales ranging from a few minutes to one solar cycle (11 years). Satellite designers and operators want to model the distributions of ionizing particles in space at any given time. This modeling assists them in designing future satellites and provides diagnostic tools for anomalies in flight.

Table 7-3. Some Common Orbits for Spacecraft Exploring or Operating in Geospace. Here we show altitudes for common orbits.

Definition	Designator	Period (τ) or Altitude Range (h)
Low-Earth orbit	LEO	$\tau < 225$ min $h < 2000$ km
Middle-Earth orbit	MEO	h: 18,000 km–25,000 km
Semi-synchronous orbit	SSO	h: 20,200
Geosynchronous orbit	GEO	h: 35,486 km–36,086 km
Highly elliptical orbit	HEO	Perigee < 5000 km, Apogee > 5000 km
Beyond Earth influence	BEI	Heliosphere transit or planetary orbits

7.3 Single Particle Dynamics—II

The magnetosphere consists of a sea of charged particles threaded by non-uniform magnetic field lines. Some of the particles are so strongly influenced by the geomagnetic field that they are effectively trapped. The particle's guiding-center behavior contributes to trapping via three superposed motions—gyration, bounce, and longitudinal drift. Gyration, combined with any parallel motion, creates a spiral trajectory. We briefly describe particle drift in non-uniform fields in Chap. 6. Here we describe in more depth what types of particle motions are conserved in slowly varying magnetic fields. We illustrate how these motions apply to particles trapped in the Van Allen radiation belts.

7.3.1 Gyration and Bounce

Objectives: After reading this section, you should be able to...

- Describe and calculate the magnetic moment of a current loop
- Determine the pitch angle of a gyrating particle
- Calculate the field strength at a magnetic mirror point
- Relate magnetic mirroring to pitch angle variations and velocity loss cone
- Explain how particles are lost to the atmosphere from trapping regions
- Describe characteristic gyration and bounce periods

Gyromotion

From Chaps. 1 and 6 we know that a particle with charge q and mass m in a magnetic field of strength $|\mathbf{B}|$ gyrates about its guiding center with gyrofrequency Ω and gyroradius r given by

$$\Omega = qB/m$$

$$r = mv_\perp / qB$$

This motion is consistent with that of a miniature loop of current I, created by the charged particle gyration

$$I = qv_\perp / (2\pi r) \tag{7-7}$$

where

- I = current from charged particle motion [A]
- q = individual particle electric charge [C]
- $v_\perp$ = particle speed perpendicular to the magnetic field [m/s]
- r = gyroradius [m]

Current loops have a type of stability, a magnetic moment that is analogous to the stability of a spinning wheel. The vector *magnetic moment* μ of the loop is the product of the current I and the area A of the loop. (Sometimes a specific magnetic moment, such as that of Earth, may be called M, as in Eq. (7-1)).

Substituting from the formulas for cyclotron frequency and cyclotron radius produces the magnitude of the magnetic moment for a gyrating particle, and its relationship to the kinetic energy associated with the perpendicular component of the velocity. (For convenience, we refer to this energy as the "perpendicular kinetic energy," even though energy is a scalar.)

$$|\mu| = IA = \left(\frac{q\Omega}{2\pi}\right)\pi r^2 = \left(\frac{q^2 B}{2m}\right)\left(\frac{mv_\perp}{qB}\right)^2 = \frac{(1/2)mv_\perp^2}{B} \quad (7\text{-}8)$$

where

$|\mu|$ = magnetic moment magnitude [A·m²]
A = area of the current loop, πr^2 [m²]
$(1/2)mv_\perp^2$ = perpendicular kinetic energy [J]

Pause for Inquiry 7-9

Verify that the substitutions made in Eq. (7-8) give the stated results.

The magnetic moment follows the direction an extended right hand thumb points when the fingers align (curl) with the current. Nature tends to adjust the motion of gyrating charged particles to conserve the particles' magnetic moment. Thus, by Eq. (7-8) loops of current created by gyrating particles tend to maintain a constant ratio of perpendicular kinetic energy to magnetic field strength.

EXAMPLE 7.4

Magnetic Moment of an Energetic Particle

- **Problem Statement:** Determine the magnetic moment of a typical energetic ion in the inner radiation belt.
- **Given:** Data in Table 7-2
- **Assume:** The particle's velocity is perpendicular to the magnetic field line.

$$\mu = \frac{KE}{B}$$

- **Solution:**

$= 10 \text{ MeV}/(2 \times 10^4 \text{ nT}) = (10^7 \text{ eV} \times 1.67 \times 10^{-19} \text{ J/eV})/(2 \times 10^{-5} \text{ T})$
$\approx 8.0 \times 10^{-8} \text{ J/(N/(A·m))} = 8.0 \times 10^{-8} \text{ A·m}^2$

- **Solution:** Although this value seems small, it is a large magnetic moment for an individual particle. Significant rapid perturbations of the magnetic field would be needed to change the magnetic moment for this particle. On rare occasions, such perturbations are caused by fast coronal mass ejections that impinge on the magnetosphere and produce violent field changes.

Follow-on Exercise: Calculate the magnetic moment for a typical energetic electron in the outer radiation belt.

To understand what this means in terms of a particle gyrating around a magnetic field line, we introduce the idea of the pitch angle (Fig. 7-18). The *pitch angle* α is the angle the particle velocity makes with the magnetic field.

$$\alpha = \tan^{-1}\left(\frac{v_\perp}{v_\parallel}\right) \quad (7\text{-}9)$$

Chapter 7   Earth's Quiescent Magnetosphere: Its Role in the Space Environment and Weather

Fig. 7-18.   A Particle Executing Spiral Motion around a Magnetic Field Line B. The yellow spiral represents the motion of a charge q about the dark gray **B**-field line. This particle has a component of motion along **B**. It thus has an pitch angle α with respect to the line. A pitch angle of zero corresponds to motion parallel to the field line. A pitch angle of 90° indicates that the particle circles the field line, and doesn't move along it.

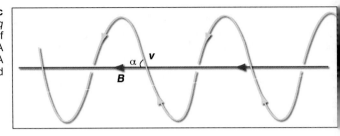

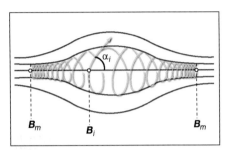

Fig. 7-19.   The Configuration of a Magnetic Bottle Associated with Mirroring. Particles spiraling around field lines are reflected as they approach the mirror points (B_m). At those points the pitch flattens into circular motion as the pitch angle approaches 90°. This motion is magnetic mirroring. Although the particle experiences considerable acceleration, all of the motion and changes in motion conserve energy. That is, $(1/2)mv_\perp^2 + (1/2)mv_\parallel^2$ is constant. *(Courtesy of the AF Research Laboratory)*

Particles that gyrate about dipole field lines (or other non-uniform or non-static field lines) must adjust their perpendicular kinetic energy, if they are to maintain a constant magnetic moment as they move in the field. Such adjustments produce changes in the pitch of the particle about the field lines (Fig. 7-19).

In regions of stronger **B**-field the gyromotion becomes less spiral-like and more circular (and vice versa for a weaker **B**-field).

Pause for Inquiry 7-10

Which of the following is true about $v_\perp$?

a) $v_\perp = v\cot(\alpha)$

b) $v_\perp = v\sin(\alpha)$

c) $v_\perp = v\cos(\alpha)$

d) $v_\perp = v(\alpha)$

Pause for Inquiry 7-11

Assume an effective current loop inside of Earth that creates the dipolar field. In what direction must the current be flowing to be consistent with the dipole orientation?

If the magnetic field varies only on time scales much longer than the gyroperiod and distance scales much larger than the gyroradius, then the magnetic moment μ is invariant. We call it an adiabatic invariant, because μ does not vary unless energy is added to the system at a rapid rate. Conservation of the magnetic moment largely controls the distribution of particle energy parallel and perpendicular to the magnetic field. In the absence of external forces the total kinetic energy must be constant; meaning the perpendicular velocity $v\sin(\alpha)$, only increases at the expense of the parallel velocity, $v\cos(\alpha)$. So the pitch angle adjusts as **B** varies.

$$\mu = \frac{(1/2)m[v\sin(\alpha)]^2}{B} = \frac{KE_\perp}{B} \approx \text{constant} \qquad (7\text{-}10)$$

Bounce Motion in a Converging or Diverging Magnetic Field

In dipolar magnetic fields and other non-uniform field configurations, charged particles undergo a bouncing motion. As a particle moves into a region of increasing B, the gyroradius shrinks, but μ remains constant. Because B is increasing, $v_\perp$ must increase. And since the total energy of the particle must also remain constant, $v_\parallel$ must decrease. In effect, the particle bounces away from the convergent field region, as if in a magnetic mirror. The place where the pitch angle becomes 90° and all parallel motion ceases is the *mirror point*.

The converging magnetic field exerts a repulsive force on the particle, which then gains a component of motion back toward regions of weaker field. Figure 7-19 shows a magnetic field mirror configuration. In planetary dipole-like fields, where the magnetic field strength grows with increasing magnetic latitude, such mirror points exist in each hemisphere. At these points, particles traveling along a field line reflect, leading to a bouncing (oscillating) motion. The periodic bounce motion is a second invariant motion that is conserved if the magnetic field changes slowly compared to the bounce period. The superposition of bounce and gyration motion is a significant factor in trapping particles in the radiation belts (Fig. 7-20).

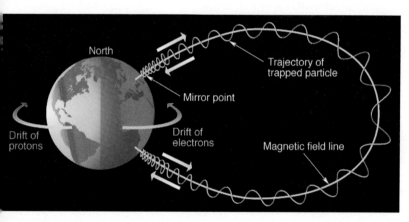

Fig. 7-20. Periodic Bounce and Gyration Motions. Charged particles moving along dipole field lines experience periodic motion in two forms: gyration and bounce. The first and second adiabatic invariants quantify the motion about and along the field line. The helical trajectory, shown along the single dipole field line, combines gyration and bounce. Particles whose mirror points lie above the atmosphere are trapped in ongoing bounce motions unless outside forces act on them. Particles with small pitch angles have mirror points in the atmosphere. They are likely to collide with atmospheric particles, and so their bounce motion is interrupted. They then form a population of 'loss-cone' particles that precipitate into the upper atmosphere. *(Courtesy of Joseph Mazur at the Aerospace Corporation)*

The distance a trapped particle moves from the magnetic equator before mirroring is determined by its equatorial pitch angle, which is the angle at the equator between the velocity of the particle and the magnetic field. Because the converging field exerts a repulsive force on charged particles, the pitch angle changes to accommodate this circumstance. A particle with a pitch angle nearer to 90° at the equator mirrors at a location closer to the equator than does a particle with an equatorial pitch angle of 70°.

Although the bounce period cannot be determined analytically, numerical integration yields good approximate results. For protons and electrons the bounce periods are, respectively

$$\tau_{bounce\ e} = 0.15 \frac{L}{\sqrt{E}} (3.7 - 1.6\sin(\alpha_{eq})) \qquad (7\text{-}11a)$$

$$\tau_{bounce\ p} = 0.65 \frac{L}{\sqrt{E}} (3.7 - 1.6\sin(\alpha_{eq})) \qquad (7\text{-}11b)$$

Chapter 7 Earth's Quiescent Magnetosphere: Its Role in the Space Environment and Weather

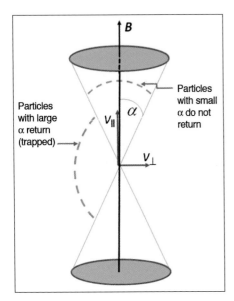

Fig. 7-21. Loss Cone Diagram. Particles with velocities within the loss cone do not mirror and are lost from the general population of particles. Particles with velocities outside of the loss cone remain trapped in the gyro-bounce motion characteristic of the magnetic bottles.

where

- τ = bounce period [s]
- L = value of equatorial crossing point of magnetic field line [unitless]
- E = particle energy [MeV]
- α = equatorial pitch angle [rad or °]

While most particles on closed field lines are trapped in dipolar magnetic "bottles," a few have equatorial pitch angles so close to 0° or 180° that their mirror points are below the altitude where the atmosphere thickens. The external collisions with atmospheric particles interrupt the bounce motion, causing some particles to be lost to the atmosphere. We categorize particles as subject to loss or subject to trapping by introducing the idea of a velocity loss cone. The particles subject to collision and loss have velocity vectors that fill double-ended cones from 0° to some critical pitch angle α and from 180° to 180°–α. We represent this *loss cone* by a cone of velocity vectors relative to the magnetic-field vector as in Fig. 7-21. At geostationary orbit (6.6 R_E), the loss cone is less than 3° wide for particles with typical energies.

EXAMPLE 7.5

Charged Particle Mirroring

- **Problem Statement:** Consider a particle at the magnetic equator, at $L = 2$ where the magnetic field strength $|B| = 3.75 \times 10^{-6}$ T. Determine the pitch angle α_{eq} that the particle must have to mirror at an altitude where B has a magnitude of 4.5×10^{-5} T.

- **Conceptual Relations:** Magnetic moment invariance (Eq. (7-10)) mirroring constraint ($\alpha = 90°$) and mid-latitude surface magnetic field strength

- **Given:** Equatorial field strength at one Earth radius above the surface = 3.75×10^{-6} T
  Mirror field strength $|B| = 4.5 \times 10^{-5}$ T

- **Assume:** A static magnetic field configuration where B increases with decreasing altitude. The particles do not interact with the atmosphere.

- **Solution:** Mirroring occurs if α_{eq} meets the condition

$$\frac{\sin^2(\alpha_{eq})}{3.75 \times 10^{-6} \text{ T}} = \frac{\sin^2(90°)}{4.5 \times 10^{-5} \text{ T}}$$

$$\sin^2(\alpha_{eq}) = \frac{3.75 \times 10^{-6} \text{ T}}{4.5 \times 10^{-5} \text{ T}} = 0.083$$

$$\alpha_{eq} = 16.8°$$

- **Interpretation:** Particles with an equatorial pitch angle $\alpha >$ 16.8° mirror and are trapped. Particles with equatorial pitch angles less than that loss cone value do not mirror, and hence hit Earth's surface. In the real world, most particles in this situation collide with atmospheric particles before they reach Earth's surface.

Follow-on Exercise: Determine the pitch angle needed to retain particles for $L = 3$.

7.3.2 Longitudinal Drift

Objectives: After reading this section, you should be able to...

◆ Determine drift directions associated with non-uniformities in the terrestrial magnetic field

◆ Compare and contrast longitudinal drift periods with gyration and bounce periods

In Chap. 6 we show that if the magnetic field varies in space, charged particles may drift across field lines. This result comes directly from the higher field strength closer to Earth causing the particles' orbital radii of curvature to be reduced. In dipolar configurations the field gradient produces a drift that results in a third type of periodic motion. Figures 7-22 and 7-23 illustrate the results of a combination of curvature and gradient drift motion. These particles precess around Earth. Charged particles deviate from straight-line motion as they approach the bunched field lines closer to Earth. Some particles experience only slight path deviations as they travel Sunward (Fig. 7-24). More energetic particles follow more deviated paths. Particles with sufficient energy form full circular drift paths around Earth. They exhibit an invariant motion as they circle Earth with periods of a few seconds to a few hours. Oppositely charged particles naturally gyrate in opposite directions. The result is a longitudinal current that flows westward (from midnight to dusk to noon to dawn and back to midnight). This current is the *ring current* that we describe in Sec. 4.4.

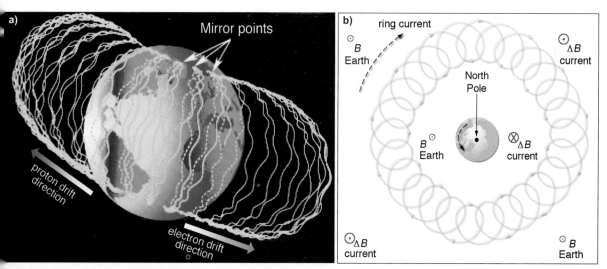

Fig. 7-22. Charged Particles in Periodic Motion. a) Superposed Gyration, Bounce, and Drift Motion of a Charged Particle in a Dipole Magnetic Field. In the simulation that produced this graphic a positive charge (pink) and a negative charge (green) were arbitrarily introduced into the geomagnetic field at the locations shown by the small triangles in the diagram. The yellow curves show the stretched gyration, bounce, and drift motions of the particles. For clarity some of the curves on which the particles bounce from the northern hemisphere to the southern hemisphere are shown as dashed. The particles only drift across magnetic field lines of the same strength. More energy would be required to move the particles into regions of stronger field closer to Earth. *(Courtesy of Joseph Mazur at the Aerospace Corporation)* **b) An Equatorial Cross Section of the Motion for a Positive Ion.** The dashed curves indicate the directions of the current associated with the looped motion. The net current flows counterclockwise in this view, creating a downward magnetic perturbation inside the ring of current and an upward magnetic perturbation on the outside of the ring. The downward magnetic perturbation reduces the magnetic field measured at Earth's surface. [After Shulz and Lanzerotti, 1974]

Fig. 7-23. Above-the-Pole View of Longitudinal Drift Motion. The drift motion occurs at values of L < 8 and is across the field lines. The most energetic particles drift in full circles around Earth. *(Courtesy of the US Air Force Academy)*

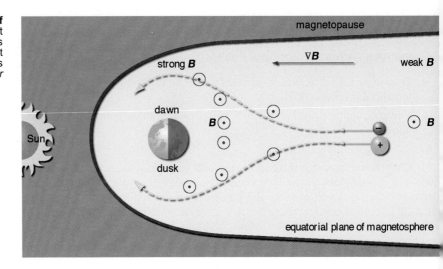

The longitudinal drift period for protons and electrons of the same energy is similar and is approximated by

$$\tau_{drift\ e\ or\ p} = \frac{0.367}{E \cdot L(0.35 + 0.15\sin(\alpha_{eq}))} \quad (7\text{-}12)$$

where

τ = longitudinal drift period [hr]

E = particle energy [MeV]

L = L-value of equatorial crossing point of magnetic field line [unitless]

α = equatorial pitch angle [rad or deg]

Table 7-4 gives values of periods for two levels of energetic particles.

Table 7-4. Periods of Invariant Motion. This table lists the periods associated with adiabatic (no energy added) motion for a 50-MeV proton and a 0.5 MeV electron at two L-shell values. We assume an equatorial pitch angle of 45°.

	Approximate Period in Seconds			
	50-MeV Proton, $\alpha = 45°$		0.5-MeV Electron, $\alpha = 45°$	
Location	L = 1.5	L = 4.5	L = 1.5	L = 4.5
Gyration	0.007	0.190	4×10^{-6}	1×10^{-4}
Bounce	0.35	1.1	0.08	0.25
Drift	39	13	3860	1290

The westward current from the longitudinal drift creates a magnetic field opposed to the direction of the ambient field inside of the ring of current. When many particles participate in this ring current, the effect can be sensed at Earth. This is the storm-time disturbance in Earth's magnetic field, measured by network of magnetometer stations near the equator spread in longitude across the globe. During geomagnetic storms, moderate-energy (~100 keV) ions produce a current that points in the opposite direction from Earth's internal field, and so the disturbance reduces the strength of the magnetic field measured at these magnetometers. The reduction in the surface field is usually less than 100 nT. A really strong ring current can produce a field depression of –300 nT or even more.

Pause for Inquiry 7-12

Compare and contrast the period of the drift motion with that of gyromotion and bounce motion. How does the dependence on L control the motion?

Pause for Inquiry 7-13

Which type of invariant motion is most likely to be interrupted because of magnetic field fluctuations in a quiescent magnetosphere? Explain your answer.

Pause for Inquiry 7-14

Calculate the drift period for a 1 MeV particle with a 90° pitch angle at $L = 4.5$.

7.4 Quiescent Magnetospheric Processes

The magnetosphere is a highly responsive buffer against the fluctuating solar wind. Magnetospheric boundaries expand, contract, and vibrate even under quiescent conditions. Here we describe some of the processes that exchange mass and momentum, and convert energy from one form to another during quiescent intervals. We start at the outer boundaries and work our way inward.

7.4.1 Forming the Dayside Magnetopause

Objectives: After reading this section, you should be able to...

- Use pressure balance to locate the dayside, sub-solar magnetopause
- Explain the presence and direction of the dayside, sub-solar magnetopause current

The dayside (toward the Sun) magnetopause is about 10 R_E upstream from Earth. The boundary location is controlled by momentum balance between the solar wind and Earth's magnetosphere. We calculate a typical, quiescent location for the *sub-solar magnetopause* (noon-time, equatorial magnetopause) by determining which physical pressures (energy densities) are in balance. The situation requires

$$pressure_{solar\ wind} = pressure_{magnetosphere}$$

In principle thermal, dynamic, and magnetic pressures could be contributors to momentum balance. We calculate some of these pressures in Sec. 6.3.

Thermal pressure in the solar wind and the magnetosphere is usually small, because the plasma is so tenuous. Thus, we ignore thermal pressure in the first order approximation of the magnetopause location. Dynamic pressure in the magnetosphere is also a small contributor because most particles are moving slowly compared to solar wind bulk flow. We are left with dynamic and magnetic pressure in the solar wind and magnetic pressure in the magnetosphere to include in a balance equation

$$n_{SW} m_{ion} v_{SW}^2 + \frac{B_{SW}^2}{2\mu_0} = \frac{B_{MG}^2}{2\mu_0} \qquad (7\text{-}13a)$$

where
- *SW*   indicates solar wind
- *MG*   indicates magnetosphere
- n_{SW} = number density of solar wind ions [#/m^3]
- m_{ion} = mass of solar wind ion [kg]
- v_{SW} = speed of solar wind ion [m/s]
- B = magnetic field strength at a specified location [T]
- μ_0 = permeability of free space (1.26 × 10^{-6} N/A^2)

To learn more about the magnetic pressure terms, we need to consider small-scale, charged-particle behavior in the vicinity of the magnetosphere. Solar wind particles in the magnetosheath that encounter Earth's strong magnetic field at the magnetopause boundary gyrate around the magnetopause field lines. Many of the particles gyrate so vigorously in the strong field that they are effectively reflected. As we show in Fig. 7-24, these particles perform a half gyration as they reverse course. Electrons behave similar to ions, but move in smaller orbits. The half gyration of charged particles produces a current sheet in the boundary region. The direction of the dayside magnetopause current creates a magnetic field in the solar wind near the magnetopause that often opposes or nulls that of the solar wind. Therefore we set $B_{SW} = 0$. On the inside of the magnetopause the magnetic field from the magnetopause current doubles the geomagnetic field. This current is the origin of the magnetopause field that is a necessary part of the momentum balance at the magnetopause.

$$n_{SW} m_{ion} v_{SW}^2 + 0 = \frac{(2B_{MG})^2}{2\mu_0} \qquad (7\text{-}13\text{b})$$

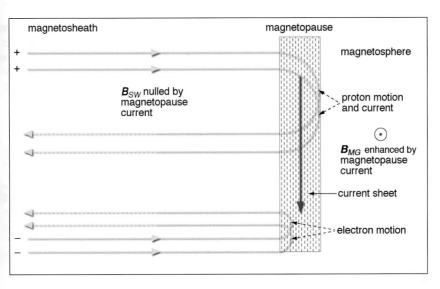

Fig. 7-24. **Charged Particles from the Solar Wind Encounter the Strong Field of the Magnetosphere over a Small Region.** Particles travel toward the magnetosphere from the left. They complete a half gyration and return to the left. The effect on the right side is to create a current sheet that doubles the magnetic field. On the left side the current sheet's field nulls the solar wind field. (B_{SW} is the magnetic field in the solar wind; B_{MG} is the magnetic field in the magnetosphere.) [After Willis, 1971]

Now we apply our knowledge of dipole magnetic field structure to find a distance relationship. In the equatorial plane, where the cosine of the latitude equals one, Earth's magnetic field strength is

$$B_{MG} = B_{Seq} \frac{(1) R_E^3}{r^3}$$

Substituting this expression into Eq. (7-13b) gives

$$n_{SW} m_{ion} v_{SW}^2 = \frac{2 B_{Seq}^2}{\mu_0} \frac{R_E^6}{r^6} \qquad (7\text{-}13\text{c})$$

From which we determine the location of the sub-solar magnetopause.

$$r = R_E \left(\frac{2B_{Seq}^2}{\mu_0 n_{SW} m_{ion} v_{SW}^2} \right)^{\frac{1}{6}} \approx 10 \, R_E \qquad (7\text{-}13d)$$

Pause for Inquiry 7-15

Verify, using typical solar wind values, that the magnetopause distance is about 10 R_E.

EXAMPLE 7.6

Width of the Magnetopause

- *Problem Statement:* Estimate the width of the magnetopause
- *Relevant Concepts and Equations:* Particle gyration radius in a magnetic field
- *Given:* $B_S = 3.1 \times 10^{-5}$ T, the shocked solar wind speed is 1×10^5 m/s
- *Assume:* Typical solar wind conditions and the magnetopause boundary is the width of one half of a proton gyroradius. The magnetopause distance is 10 R_E.
- *Solution:* Determine the magnetic field strength at the magnetopause

$$B = 2B_S \frac{R_E^3}{r^3}$$

$$B = (2)(31{,}000 \times 10^{-9} \, \text{T}) \frac{R_E^3}{(10 R_E)^3}$$

$$= 62 \, \text{nT}$$

Determine the proton gyroradius

$$r_C = \frac{mv}{qB}$$

$$r_C = \frac{(1.67 \times 10^{-27} \, \text{kg})(1 \times 10^5 \, \text{m/s})}{(1.6 \times 10^{-19} \, \text{C})(62 \times 10^{-9} \, \text{T})}$$

$$r_C \approx 16{,}800 \, \text{m}$$

One half of the gyroradius is ~8 km

- *Interpretation:* The magnetopause is a very thin boundary, typically a few to tens of kilometers thick.

Follow-on Exercise: Use a typical thermal particle speed instead of the shocked solar wind speed to determine the magnetopause thickness.

7.4.2 Mass, Energy, and Momentum Transfer to the Magnetosphere

Objectives: After reading this section, you should be able to...

- Explain why solar wind flow interactions produce electric fields in the magnetosphere
- Calculate the direction and magnitude of the viscous electric field
- Explain the significance of the interplanetary magnetic field (IMF) orientation to dayside merging
- Describe the sequence of events leading from dayside merging to nightside reconnection
- Determine the magnitude and direction of the magnetospheric electric field associated with merging and convection

During quiet times, small amounts of solar wind mass, energy, and momentum enter the magnetosphere through the slightly permeable magnetopause boundary. We describe two paths by which this solar wind injection occurs.

Solar Wind Injection by Viscous Interaction

During quiet times the magnetosphere is not completely isolated from the solar wind. Particle scattering and wave interaction allow the solar wind to influence plasma inside the magnetopause. On the flanks of the magnetosphere, thin boundary layers of plasma move anti-Sunward. The frozen-in flux approximation suggests that field lines around which the plasma gyrates should also move anti-Sunward. Mass and magnetic flux conservation requires a return Sunward transport elsewhere inside the magnetosphere (Fig. 7-25).

We show how the Lorentz force supports such a motion in Sec. 1.4 and Fig. 1-25. Positively charged particles in the equatorial plane gyrate into the magnetosphere on the dawn side, while negative charges gyrate inward on the dusk side. Small waves on the boundary enhance the process. A charge separation develops between the magnetopause flanks and creates an electric field from dawn to dusk. In effect the captured, charged particles produce an electric field that energizes magnetospheric processes at a low level. Inside the magnetosphere the drift motion associated with the electric field creates a return flow that roughly completes the circulation, as suggested in Fig. 7-25. The process that leads to this circulation is called *viscous interaction*.

The viscous interaction is a rather weak, but ongoing, process in the quiescent magnetosphere. As we show in Example 7.7, it provides a few kilovolts of electric potential. This potential difference sets charged particles in motion, thus transferring energy from the solar wind to the magnetosphere. This transfer extends into the ionosphere. Viscous interaction accounts for ~10%–20% of solar wind energy transfer to the magnetosphere. A more efficient, but episodic, transfer comes with magnetic merging, which we describe next.

Chapter 7 Earth's Quiescent Magnetosphere: Its Role in the Space Environment and Weather

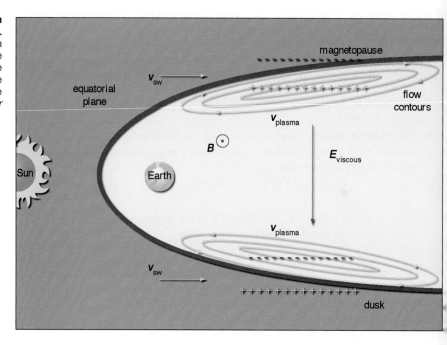

Fig. 7-25. The Solar Wind Interacts with Earth's Magnetic Field along the Flanks. Here we show the viscous interaction between the solar wind (from the left) and the magnetosphere along the flanks of the magnetopause. The view is from above the magnetosphere looking down onto the equatorial plane. *(Courtesy of the US Air Force Academy)*

Pause for Inquiry 7-16

Use your knowledge of the relation between an electric field and an electric potential difference to explain how an electric field imposed from the solar wind could "energize" particles in the magnetosphere.

Pause for Inquiry 7-17

Use your knowledge of electric and magnetic field interactions to explain why charged particles drift Sunward in the central magnetosphere under the influence of an electric field oriented as shown in Fig. 7-25.

Pause for Inquiry 7-18

Describe how the Lorentz Force law produces the charge separation depicted in Fig. 7-25.

Solar Wind Injection by Magnetic Merging and the Convection Cycle

Dayside Merging. Magnetic merging is the primary means by which mass, energy, and momentum transfer from the solar wind to the magnetosphere. This process was first described by James Dungey in 1961. As we note in Chap. 4, magnetic merging allows magnetic field lines to change their topology (layout). This change means the frozen-in flux assumption breaks down in localized resistive regions of high current density. We label these special sites as current layers. During merging, magnetic flux of solar-wind origin transfers across the

Pause for Inquiry 7-19

Which solar wind flow type would create the largest viscous electric field?

a) Low speed, low density flow

b) Low speed, medium density flow

c) High speed, low density flow

d) High speed, high density flow

EXAMPLE 7.7

Electric Field in the Magnetotail Caused by Viscous Interaction

- *Problem Statement:* Estimate the magnitude of the electric field caused by viscous merging at 20 R_E downtail in the magnetosphere.

- *Conceptual Relations:* $|E| = \left|-\dfrac{dV}{dl}\right|$

- *Given:* Voltage from merging is a few tens of kilovolts: ~30 kV

- *Assume:* We use Fig. 7-12 as a good indicator of the scale for the problem. At 20 R_E on Earth's nightside, the tail is about 50 R_E wide (= $50 \times 6378 \times 10^3$ m).

- *Solution:*

$$|E| = \left|\frac{dV}{dl}\right| = \frac{30 \times 10^3 \text{ V}}{50 \times 6378 \times 10^3 \text{ m}} = 6.27 \times 10^{-5} \text{ V/m}$$

~0.06 mV/m

- *Interpretation:* A weak electric field forms in the magnetotail as a result of energy entering the magnetosphere from the solar wind. The field drives a dawn-to-dusk current, consistent with the direction of the current in the neutral current sheet. This process is a dynamo interaction that converts kinetic energy of the solar wind into electrical energy in the magnetosphere. It occurs no matter what the orientation of the solar wind magnetic field.

Follow-on Exercise: Assume that the magnetic field lines that thread the magnetospheric flanks flow into the high-latitude ionosphere and that the excess charges are free to flow along the field lines, making these field lines equipotential. Will the electric potential in the polar ionosphere be larger or smaller than the potential in the magnetosphere? Will the electric field in the polar ionosphere be smaller or larger than in the magnetosphere?

magnetopause in *flux transfer events*. At the same time, magnetic field energy converts into plasma kinetic energy. Flux transfer events occur when oppositely oriented magnetic fields closely approach, forming an X-line (Figs. 4-29 and 7-29).

The X-line formation allows gyrating particles to move from one field line to another, or for field lines to take on new identities and orientations. Merging produces an instant disconnect from the old topology and re-formation of different magnetic field configuration. When solar wind field lines merge with Earth's closed dipolar field lines, the result is a hybrid field line with one foot in the solar wind and one foot in the magnetosphere. The newly merged magnetic field lines are highly kinked and under tension. Nature tries to undo the kink by redirecting the field lines and the flowing plasma out of the merging region. The plasma heats up and accelerates. The field straightens.

Magnetic merging ultimately affects the entire magnetosphere on large and small scales. On the micro-scale, the plasma particles gain random kinetic energy—they heat up. On the large scale, frozen-in plasma and field lines move together, producing an extensive stirring of the magnetosphere. As we note in Chaps. 4 and 6, field line motion is equivalent to producing an electric field imposed by the solar wind. During intervals of southward IMF, the electric field E_{SW} nominally points from dawn to dusk, and is on the order of a few millivolts per

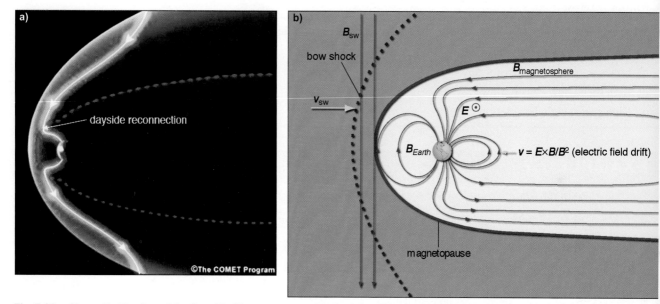

Fig. 7-26. **Magnetic Merging at the Dayside Magnetopause with the Interplanetary Magnetic Field (IMF) Pointing South. a)** The view includes the magnetopause. The view is the noon-midnight meridian plane. *(Courtesy of the COMET Program at the University Corporation for Atmospheric Research)* **b)** The view is of field lines inside the magnetopause. During periods of southward-oriented IMF, the solar wind IMF merges with Earth's magnetic field. Plasma is redirected out of the merging region, to a tailward direction. *(Courtesy of the US Air Force Academy)*

meter. High conductivity along magnetospheric field lines allows the solar wind electric field to project into Earth's polar regions, as we describe shortly.

A south-pointing IMF is sometimes regarded as an energy switch because southward IMF ($-B_z$) easily transfers energy in the form of an electric field from the solar wind to the magnetosphere. During prolonged merging events, energy loads into the magnetotail via the electric field. In the magnetotail the energy dissipates as it drives various current systems, puts plasma into bulk motion, and adds to the random motion of individual particles. Field-aligned currents help maintain a connection between the field lines and the plasma anchored to the feet of the field lines in the ionosphere.

During southward IMF conditions, the magnetosphere as a whole is magnetically better-connected to the solar wind. Quiescent periods of southward IMF come from small ripples in the field created by magnetic field motions at the Sun. Other events associated with slightly more disturbed conditions arise from out-of-ecliptic IMF orientations, from co-rotating interaction flows, and coronal mass ejections whose edges intercept Earth's magnetosphere. On average, the out-of-ecliptic IMF has a southward orientation half of the time.

Other IMF orientations support merging, but they transfer energy less efficiently. For example, merging may occur at very high latitudes when the IMF has a northward component. There, the northward IMF and Earth's magnetic field have opposite polarity, with Earth's field pointing southward in the vicinity of the North Pole (Fig. 7-27). The approaching solar wind field lines connect to lobe field lines that already have an open end. The newly connected open end flows away with the solar wind. The Earth-attached end of the lobe field line merely stirs around, maintaining the lobe configuration. The limited energy transfer is highly localized and is deposited near the footpoints of the magnetic cusp.

7.4 Quiescent Magnetospheric Processes

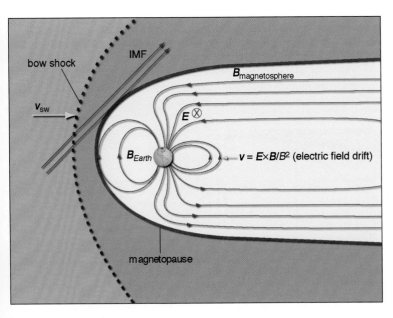

Fig. 7-27. Magnetic Merging at the Dayside Magnetopause with the Interplanetary Magnetic Field (IMF) Pointing Northward. The view is of the noon-midnight meridian plane. During periods of a northward-oriented IMF, the solar wind IMF merges with Earth's magnetic lobe field. This form of merging creates minimal interaction with Earth's closed field lines. *(Courtesy of the US Air Force Academy)*

Imposed Electric Field and Convection. A consequence of magnetic merging with closed geomagnetic field lines is the imposition of the solar wind electric field onto the magnetosphere. This imposed field has several names, depending on where it is measured: the solar wind electric merging field, the dynamo electric field, the magnetosphere convection field, and the cross-polar-cap field.

We recall from Eq. (6-14b) that in a magnetoplasma, an applied E-field results in a plasma drift

$$v_{plasma} = \frac{E \times B}{B^2}$$

Thus, E and v_{plasma} are tightly coupled in conducting plasmas such as those in the magnetosphere and solar wind. Motion of magnetic field lines is such that an observer moving with the magnetoplasma detects zero electric field. However, if a magnetoplasma moves at a velocity v_{plasma} with respect to a stationary observer, the observer will measure an electric field

$$E = -v_{plasma} \times B$$

From a stationary, Earth-fixed observer's viewpoint, the dynamics of the magnetosphere appears as a dawn-to-dusk electric field instead of as moving field lines.

Down tail, the stretched magnetic field lines eventually reconnect with each other in the mid-magnetotail region, causing a tension force on the plasma. Together with the pressure gradient and the potential difference applied across the magnetosphere by the flowing solar wind, these forces produce motion of the magnetospheric plasma on closed field lines toward the Sun, and an associated dawn-to-dusk magnetospheric electric field in the tail.

Earth's magnetic field lines (open or closed) are often treated as equipotential, because the conductance along the field lines is high. To first order, the E_{SW}-field imposed by merging maps along Earth's magnetic field to the footpoints in the ionosphere. Because the reconnected magnetic field lines are further apart at the magnetopause than in the polar cap, the electric field must be stronger in the polar cap. There, the dawn-to-dusk E_{SW}-field maps along the field lines, creating an anti-Sunward $E \times B$ drift of plasma in the polar cap ionosphere. The mapping of this electric field into the ionosphere produces a dusk-to-dawn electric field and a Sunward plasma flow at latitudes below the polar cap. An imposed, two-cell pattern of ionospheric motion called *ionospheric convection* forms (Fig. 7-28).

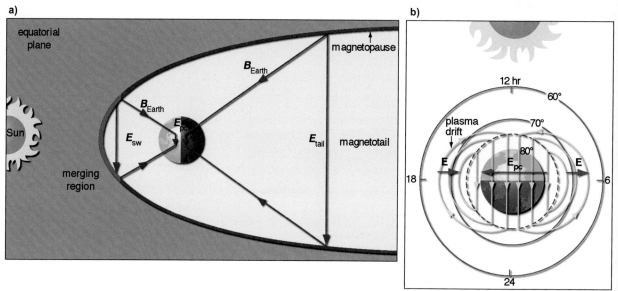

Fig. 7-28. **Mapping the Solar Wind's Electric Field into Earth's Polar Cap. a) Solar Wind and Electric Field Mapping into the Ionosphere.** The view is from above the North Pole and depicts the merging and convecting electric fields projected into the equatorial plane of the magnetosphere. The diagram shows how the solar wind's electric field maps into Earth's polar cap ionosphere and magnetosphere. E_{SW} and E_{PC} are the electric fields in the solar wind and polar cap, respectively. This mapping is for a southward IMF. *(Courtesy of the US Air Force Academy)* **b) Polar Convection.** Here we see a close-up of the two-cell ionospheric plasma convection pattern driven by solar wind-magnetosphere-ionosphere interaction during intervals of southward IMF.

From the imposed solar wind electric field we can determine a characteristic parameter of the magnetosphere: the electric potential difference across the magnetosphere. We recall that the product of the electric field acting over a distance relates the field to the electric potential

$$\Delta V = |V_f - V_i| = \left| \int -E \bullet dl \right|$$

Because of high parallel conductivity everywhere along two separate magnetic field lines, the electric potential difference ΔV is roughly the same no matter if the field lines are on the dayside, over the polar cap, or on the nightside. Typical electric potential differences in the polar cap ionosphere imposed by merging are between 50 kV–100 kV. The electric field is one of the few quantities that instruments measure locally that provides a large-scale view of the coupled magnetosphere-ionosphere system. Figure 7-28a illustrates the concept of mapping the E_{SW}-field from the magnetopause to the magnetotail and the ionosphere. The mapping allows scientists to make ground- or space-based measurements of the electric potential in the polar cap and use it to determine how energy is moving through the entire system.

Pause for Inquiry 7-20

The merging region in Fig. 7-28a is about 5 R_E long. Use the solar wind electric field value from Example 4.16 to show that the merging voltage is ~80 kV.

Convection and Nightside Merging during Intervals of Southward IMF. Because of momentum conservation, merged field lines with one foot in the solar wind must move in the direction of the bulk solar wind flow. Thus, the merged field lines flow tailward over the poles and stretch into the tail lobe. In the magnetotail, layers of open field lines are sandwiched on top of each other, creating a magnetic pressure. Storing these reconfigured magnetic field lines in the magnetotail leads to a stockpile of magnetic energy that may become unstable and explosively release energy in events known as *magnetospheric substorms*. Focus Box 7.3 and Fig. 7-29 (noon-to-midnight cross-section view) highlight the sequences of magnetic merging, convection, and reconnection, where the numbers represent the time sequence of field line motion. We show the motion of the terrestrial end of the field lines in the polar cap inset.

When magnetic pressure in the tail becomes high enough, the quasi-open field lines merge or reconnect with partners from the opposite hemisphere at a distant X-line in the magnetotail (location 6 in Fig. 7-29). The physical process is the same as that of dayside merging, but it occurs at a different location—deep in the magnetotail. During reconnection, magnetic tension shortens the field lines and returns magnetic flux to the dayside, where the process begins anew, as long as the IMF remains southward.

After a field line has reconnected to reform as a quasi-dipolar field line, it has a much kinked configuration (similar to the dayside kinks) that represents an energy state well above that of a dipole field line. Nature seeks to reduce the energy by several means. One means is *dipolarization*, in which field lines accelerate away from the merging region, as they attempt to achieve something closer to a dipole configuration. Those field lines Earthward of the merging line race toward Earth, building a field gradient. Plasma particles attempt to stay with their field lines as the lines begin their Earthward trek; however, magnetic field curvature and gradients, along with the addition of electrical energy from the reconnection, make it harder for particles to stay attached. The particles become involved in many kinds of drifts that we mention in Secs. 6.2 and 7.3 and describe further in Chap. 11.

The motion of the field lines and their flowing plasma accounts for some of the lost potential energy in the magnetic field configuration. Downtail from the reconnection point, mass is ejected in the form of a plasmoid. This flowing mass is also a portion of the energy budget—energy lost from the magnetosphere.

In addition to energy release during substorms, an overall change occurs in the way energy is partitioned in the magnetosphere when merging processes are active. Rather than being stored as mostly potential energy in the field lines, the energy is redistributed to motion of the field lines and plasma. Therefore, kinetic energy in the system rises as a result of external energy input from the solar wind. The merging process increases large-scale potential and kinetic energy (convection) and creates opportunities for explosive energy release (substorms).

Fluctuations in the IMF affect the rate of dayside merging, and cause corresponding fluctuations in the convection electric field. Nightside merging in the center of the current sheet occurs, but its onset is delayed relative to the dayside. When merging begins on the nightside, it explosively releases the energy stored in the tail's magnetic field. Multiple substorms driven by a long interval of fluctuating southward IMF cause many particles to be energized and trapped in the radiation belts.

EXAMPLE 7.8

Electric Field in the Ionosphere Caused by Merging

- *Problem Statement:* If the polar cap ionosphere has a diameter of 2.0×10^6 m, what is the electric field strength in the polar cap? What is the field direction?

- *Conceptual Relations:* $|E| = \left|-\dfrac{dV}{dl}\right|$

- *Given:* $|dl| = 2 \times 10^6$ m, $\Delta V = 63$ kV

- *Assume:* Earth's magnetic field lines are highly conducting, and thus equipotentials

- *Solution:*

$$|E| = \left|-\frac{dV}{dl}\right| = \frac{6.3 \times 10^4 \text{ V}}{2 \times 10^6 \text{ m}} = 3.15 \times 10^{-2} \text{ V/m}$$

$E = -v_{plasma} \times B$ and points in the dawn-to-dusk direction

- *Interpretation:* The merging electric field maps to the polar cap. The potential difference between field lines is roughly constant, but the electric field strength increases in regions of converging magnetic field lines.

Follow-on Exercise: We know the magnetic field lines converge in the polar cap. Assume that the magnetic field strength in the polar cap (~60,000 nT) is twice as strong as the surface strength at the equator. Use your knowledge of the $E \times B$ drift to calculate the drift speed and direction of the plasma in the polar cap ionosphere.

Focus Box 7.3: Merging and Convection in the Magnetosphere during Southward IMF

The magnetospheric elements of merging, open-field convection, reconnection, and closed-field convection are as follows:

1–2 and 1′–2′. A southward IMF line approaches a northward terrestrial field line and they merge. The kinked field lines accelerate away from the merging point, straightening as they go.

2–5 and 2′–5′. The connected hybrid field lines sweep back because of the momentum of the solar wind and accumulate in the magnetotail, where they pack together.

6 and 6′. Intense packing creates a region of oppositely directed field lines in the magnetotail. The terrestrial field lines reconnect.

7′. Disconnected field lines, those with feet in the solar wind, continue their motion outward into the heliosphere, probably carrying some terrestrial plasma with them.

7–9. Kinked field lines move away from the merging point. Reconnected terrestrial field lines with feet connected to Earth return to the dayside (a process that maintains the flux balance).

7.4 Quiescent Magnetospheric Processes

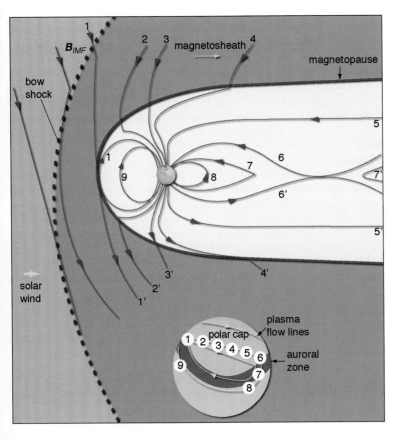

Fig. 7-29. **Magnetic Merging, Convection, and Reconnection.** Merging reconfigures Earth's outermost magnetic field lines, opens the magnetosphere to solar wind plasma, and drives large-scale organized plasma flow in the magnetosphere and in the coupled ionosphere. [After Cowley, 1996]

Substorms and Aurora. Merging, convection, and reconnection usually don't proceed in a steady manner. Substorms driven by the quiescent, but unsteady, solar wind occur, on average, three to four times a day because of small fluctuations in the IMF direction. After the neutral merging line forms, energy stored in the magnetotail releases in a spasm. Magnetic field lines accelerate toward Earth. Some particles, with small pitch angles, accelerate along magnetic field lines and crash into the upper atmosphere. This activity leads to visible phenomena such as *auroral substorms* (Fig. 7-30). When the plasmasheet is disturbed in this manner, accelerated particles move along Earth's magnetic field and bombard the upper atmosphere around the poles in regions known as *auroral ovals*. These ovals normally remain at high latitudes. From Earth's perspective, they are "windows" into the machinery of the magnetosphere. During a typical two-hour substorm, energy comparable to that of a very strong earthquake (about 10^{17} J) is released into the upper atmosphere.

Outside of solar minimum, the auroral lights appear nightly. A small fraction of auroral light also comes from particles that have entered the magnetosphere through the polar cusps on the dayside. Auroras occur on Earth's dayside as well, but are masked by sunlight. They are produced when energetic charged particles accelerate into the auroral oval, then collide with and excite neutral atoms and molecules in the upper atmosphere. Auroras appear in a variety of forms and colors, according to the degree of turbulence and the energy level in the magnetosphere. We say more about auroral colors and forms in Chaps. 8 and 12.

Fig. 7-30. **Aurora over Bear Lake, Alaska, USA.** This image of the Aurora Borealis shows how bright these space weather phenomena are. *(Senior Airman Joshua Strang, USAF, Eielson Air Force Base, Alaska, USA)*

When conditions are relatively quiet, the aurora drifts across the sky as green or white curtains. Inhabitants of far northern and southern latitudes witness the motion of beautiful, multicolored forms across the night sky. In the northern hemisphere, the brilliant spectacles are the *northern lights (Aurora Borealis)*. In the southern hemisphere, the displays are the *southern lights (Aurora Australis)*. During a moonless night, the aurora may be so bright that reading a book by its natural glow is possible. People living at relatively low latitudes are seldom aware of the beautiful and dynamic substorms. Only rarely, during violent magnetic storms, does the aurora flow to low latitudes. We say more about this visible manifestation of the solar wind-magnetosphere-ionosphere-thermosphere connection in Chap. 12.

Steady Magnetospheric Merging, Convection, and Reconnection. On rare occasions the magnetosphere processes the energy delivered by magnetic merging in a steady manner. During steady southward IMF conditions lasting several hours, the entire magnetosphere sometimes remains free of instabilities. These periods are *steady magnetospheric convection (SMC)* events. Such events are intervals of enhanced convection without classic substorm signatures. When the magnetosphere is driven to this relatively steady state, a balance exists between the creation of merged magnetic flux on the dayside and the closure of flux by reconnection on the nightside.

The ionospheric plasma moves in a stable, two-cell, convection pattern, with the dayside auroral oval at unexpectedly low latitudes. The nightside oval is quite active with bulging and surging auroral formations. We say more about steady magnetospheric convection in Chap. 11.

7.4.3 Trapping Processes in the Plasmasphere

Objectives: After reading this section, you should be able to...

- Explain how particles in the plasmasphere are trapped in co-rotating motion
- Calculate the direction and magnitude of the co-rotating electric field
- Describe the source of plasmaspheric material

Earth's plasmasphere is a grand extension of the ionosphere and an inner part of the magnetosphere. Located inside $L = 5$, it is generally protected from much of the turbulence and activity in the magnetosphere. Nonetheless, the plasmasphere is not dormant.

Filling the Plasmasphere

As we describe in Chap. 3, solar shortwave energy ionizes Earth's upper atmosphere. The low mass electrons gain substantial kinetic energy; many of them have sufficient energy to escape Earth's gravity, but remain bound to Earth's magnetic field. The electric field created by the separating space charge draws light ions from the ionosphere. Over intervals of hours to days this escaping plasma accumulates until it reaches an equilibrium where as much plasma flows into the ionosphere as flows out. This donut-shaped region of cold (~1 eV) plasma encircling Earth is called the plasmasphere.

Although not bound by gravity, the cold plasma connects with Earth through the geomagnetic field. As the inner geomagnetic field rotates with Earth, so does the plasma attached to the field, although the plasma tends to lag behind. The plasmasphere shrinks with increased space weather activity and expands or refills during times of inactivity. During times of shrinkage some of the plasmasphere is drawn away from Earth in a process called plasmaspheric erosion, toward the dayside magnetopause.

The plasmasphere has several features. It has a distinct edge (plasmapause) at which the plasma density drops sharply. When viewed in ultraviolet light from space, the plasmasphere often shows a plume associated with the tendency to rotate slightly slower than Earth (Fig. 7-31). Space weather disturbances create density bite-outs (notches) and density enhancements (fingers).

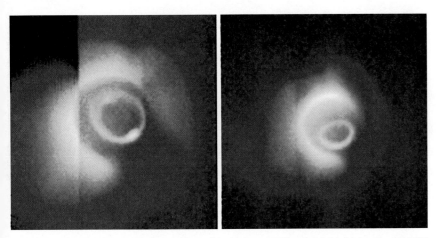

Fig. 7-31. Earth's Plasmasphere Viewed in Extreme Ultraviolet Light by the IMAGE Spacecraft. Earth is at the center of the images. The Sun is to the upper left. The view looks down on Earth's North Pole and shows the northern aurora. The 30.4-nm emission from the plasmaspheric helium ions appears in false color as a pale blue cloud surrounding Earth. The emission is brightest near Earth on the dayside. The dark "bite out" seen in the lower right portion of the emission is Earth's shadow. A faint tail or plume of plasmaspheric material is entrained in the Sunward flow of plasma from the magnetotail and is being carried toward the dayside magnetopause in the right hand image. *(Courtesy of Bill Sandel and Terry Forrester at the University of Arizona and NASA)*

Co-rotation in the Plasmasphere

We recall from Chaps. 4 and 6 and the discussion of the previous section, that the motion of a conductor across magnetic field lines creates an electric field $E = -v \times B$ that keeps the plasma and B frozen to each other. For the plasmasphere this process means that the closed dipolar field lines trap the relatively cold plasma welling up from the ionosphere in a co-rotation type of motion as suggested in Fig. 7-32.

To describe this situation we view the region using the spherical magnetic coordinates we introduce in Sec. 7.2. The plasma and field in the plasmasphere are strongly attached to Earth, rotating with a ~24-hour period. Therefore, the plasma is in an Earth-co-rotating motion U with respect to the Sun and the rest of the magnetosphere.

Chapter 7 Earth's Quiescent Magnetosphere: Its Role in the Space Environment and Weather

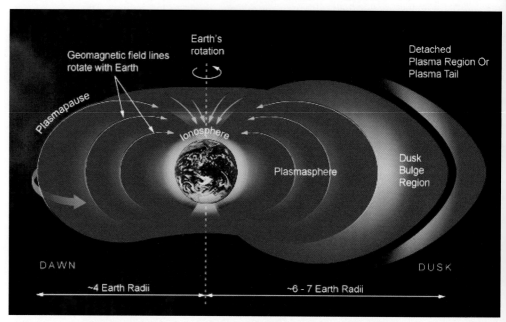

Fig. 7-32. Cross Section of the Plasmasphere Viewed from Earth's Dayside. Here we look at Earth's plasmasphere as if we were standing on the Sun. We see a slight asymmetry in that the duskside extends farther into space. The translucent arrow at left indicates the direction of motion of the co-rotating field lines and plasma. *(Courtesy of the University Corporation for Atmospheric Research and Windows to the Universe)*

Near Earth we have

$$\boldsymbol{U} = r\omega\hat{\varphi} \tag{7-14}$$

where

$\boldsymbol{U}$ = linear velocity vector associated with Earth's rotation [m/s]

r = distance from Earth's rotation axis [m]

$\omega\hat{\varphi}$ = angular rotation velocity vector in Earth's spin direction [rad/s]

$$\boldsymbol{B} = B_{Seq}\left(\frac{R_E}{r}\right)^3 \hat{\boldsymbol{\lambda}} \tag{7-15}$$

where

$\boldsymbol{B}$ = latitudinal component of the magnetic field vector [nT]

B_{Seq} = surface value of Earth's magnetic field at the equator [nT]

R_E = Earth's equatorial radius (6378 × 10³ m)

$\hat{\boldsymbol{\lambda}}$ = latitudinal unit vector [unitless]

The frozen-in field condition gives

$$\boldsymbol{E}_{\text{co-rotation}} = -\boldsymbol{U} \times \boldsymbol{B} = -r\omega\hat{\varphi} \times B_{Seq}\left(\frac{R_E}{r}\right)^3 \hat{\boldsymbol{\lambda}} \tag{7-16a}$$

$$E_{\text{co-rotation}} = \omega \left(\frac{B_{Seq} R_E^3}{r^2} \right)(-\hat{r}) \quad (7\text{-}16b)$$

Referring to Fig. 7-32, $U_{co\text{-}rotation}$ is the direction of the thick arrow while $E_{co\text{-}rotation}$ is inward toward Earth. The magnetic field is tangent to the direction of the thin, curved arrows. The plasmasphere is controlled by the co-rotating electric field and associated velocity. The cold plasma is trapped on closed drift paths inside the plasmapause. The magnetosphere is controlled by the solar wind merging, the convection electric field, and the associated velocity. Because electric fields are additive, we superpose the field and the motion associated with the field to show plasma particle trajectories in the equatorial plane of the magnetosphere (Fig. 7-33).

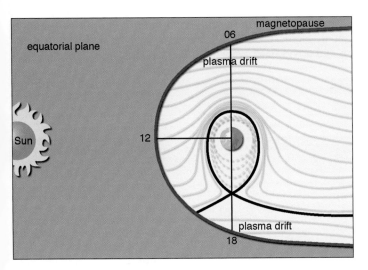

Fig. 7-33. Plasma Convection in the Equatorial Plane of the Magnetosphere. Particles generally drift Sunward along the yellow curves in the outer magnetosphere. In the plasmasphere, cold particles are trapped on co-rotating drift paths. These particles move in roughly circular motion along the yellow dashed curves. The solid black inner curve is a boundary that separates open drift paths (solid yellow) from closed drift paths (dashed yellow). Within this boundary, particle motions are counterclockwise in the direction of Earth's rotation. Consistent with Fig. 7-32, the edge of the plasmapause is closer to Earth on the dawnside, where the flow paths are close together. The bulge shown in Fig. 7-32 is shown in this figure as an elongation of the plasmapause (solid black line) on the duskside. [After Kavanagh et al., 1968]

Pause for Inquiry 7-21

Fig. 7-33 shows the trajectories of plasma particles in the equatorial plane of the magnetosphere. In what direction is the electric field in the magnetotail? In the tear-drop shaped region, what is the direction of the electric field? Where is the strongest electric field in the diagram? Where is the weakest electric field in the diagram?

As we describe in Sec. 7.2.2, Earth is surrounded by two intense belts of energetic particles with sufficient energy to ionize other matter. The radiation belt region contains electrons, protons, helium, carbon, oxygen, and other ions with energies from less than 1 keV to hundreds of MeV. The two zones of trapped energetic electrons and protons encircle Earth like donuts. The belts are charge-neutral and overlap the plasmasphere, as shown in Fig. 7-13. In Sec. 7.2 we describe the behavior of single plasma particles in non-uniform magnetic fields—they are trapped. During steady-state magnetospheric conditions, particles do not escape these trapped orbits. A few quasi-steady-state processes allow particles to enter the trapping regions.

The quiescent inner radiation belt is generated by cosmic rays whose journeys begin with the births and deaths of distant stars. These energetic particles collide with atoms in Earth's atmosphere and produce showers of secondary products. Some of these products are neutrons that subsequently decay into energetic protons. The neutron half-life is short enough (~10 minutes) that many of the protons are born in Earth's lowest L shells. These are trapped in Earth's strong magnetic field. The telltale clue for the decay source is the dominance of protons over other types of ions. The inner zone's stability results from a combination of long particle lifetimes caused by limited collisions in this part of the magnetic field and the slowly varying cosmic ray inputs.

The quiescent outer radiation belt consists of energetic electrons created by cosmic rays similar to energetic electrons from the solar wind. The ionosphere may contribute cold particles that are energized by magnetospheric processes. Recurrent reconnection leads to higher energy electrons that form the outer radiation belt. Electrons are also likely to receive energy from electromagnetic waves that are excited by long periods of merging or by periods when merging turns off and on in rapid succession. Thus, solar wind control of magnetic merging controls and drives outer radiation belt behavior.

EXAMPLE 7.9

Electric Field in the Plasmasphere from Co-rotation

- **Problem Statement:** Compare the co-rotation electric field in the plane of Earth's equator at (1) the base and (2) the top of the plasmasphere.

- **Conceptual Relations:** Electric field caused by co-rotation and Earth's rotation rate, Eq. (7-16)

- **Given:** B_{Seq} = 30,000 nT, the base of the plasmasphere is 1000 km altitude, the top of the plasmasphere is r = 5 R_E

- **Assume:** Earth rotates about its axis every 24 hours (2π rad/86,400 s)

- **Interpretation:** The electric field from co-rotation points toward Earth and increases in strength by about two orders of magnitude from the top to the base of the plasmasphere. Where the electric field is strongest, the plasma is most tightly controlled by the co-rotation. Conversely, at the outer reaches of the plasmasphere the electric field is weaker. Other perturbing forces that increase in strength during storm time often strip the plasma from the weaker field regions. During disturbed intervals, the smooth shells of co-rotating plasma, illustrated in Fig. 7-32, become dimpled, corrugated surfaces that support dynamic wave structures. We follow up on this idea in Chap. 11.

- **Solution:**

$$\omega = \frac{2\pi}{86,400 \text{ s}} = 7.27 \times 10^{-5} \text{ rad/s}$$

$$E_{top} = \frac{\omega B_{Seq} R_E^3}{r^2}(-\hat{r}) = \frac{(7.27 \times 10^{-5} \text{ rad/s})(3 \times 10^{-5} \text{ T})(6378 \times 10^3 \text{ m})^3}{(5 \times 6378 \times 10^3 \text{ m})^2}(-\hat{r}) = 5.56 \times 10^{-4} \text{ V/m}(-\hat{r})$$

$$E_{base} = \frac{\omega B_{Seq} R_E^3}{r^2}(-\hat{r}) = \frac{(7.27 \times 10^{-5} \text{ rad/s})(3 \times 10^{-5} \text{ T})(6378 \times 10^3 \text{ m})^3}{(1000 \times 10^3 + 6378 \times 10^3 \text{ m})^2}(-\hat{r}) = 1.03 \times 10^{-2} \text{ V/m}(-\hat{r})$$

Follow-on Exercise: Describe the condition that occurs where the directions of $E_{co\text{-}rotation}$ and B_{Earth} produce a plasma motion that allows the plasma to move with the co-rotating field.

Neither belt consists exclusively of one type of particle. In fact, each belt is charge-neutral. The name "proton belt" means that the protons carry the bulk of the energy and momentum in the volume. But in steady state it has an equal number of electrons. However, their small mass means they contribute less energy than do the protons. The outer electron belt is also charge-neutral, but the protons there are much slower than the electrons and carry significantly less energy.

A strong electric field and convection also cause positive ions and electrons to drift Earthward from the reconnection region. These drifting particles encounter a gradient in the B-field that produces a deviation of the plasma trajectory, sometimes a ring of current. The strength of the ring current serves as a measure of the energy density of low-to-medium energy particles (10s of keV) that sweep toward Earth. During quiet times the particles may only deviate around Earth and drift toward the dayside. In that case, only a partial ring current exists. During more disturbed times, which we describe in Chap. 11, the particles sometimes encircle Earth for hours and even days. The particles are of sufficient density at geosynchronous altitude that they create a magnetic field detectable at the ground.

Summary

The geomagnetic field fills a cavity around Earth called the magnetosphere. Earth's field arises in interior movements of iron- and nickel-rich fluids. The geomagnetic field varies in time because of inner fluid motion and electric currents in the magnetosphere and upper atmosphere. The outer boundary of the magnetosphere is well-defined at a distance of about 10 R_E in the Sunward direction. The boundary is produced by a sheet of current that cancels most of interplanetary magnetic field outside the boundary while doubling the field inside. This current layer is the magnetopause, and it separates the geomagnetic field from the field and plasma of the solar wind. This layer is not impermeable. During quiet times, mass, momentum, and energy cross the boundary in small amounts.

Combinations of field line orientation and particle populations produce distinct regions in the magnetosphere. Close to Earth in the plasmasphere the cold plasma is locked to the rotation of the nearly dipolar field. In the same region but with much less mass density are the inner and outer radiation belts. The energy of the radiation belt particles is sufficient to ionize other matter. Energy density, rather than mass density, characterizes the trapped particles in the radiation belts.

The radiation belt region contains electrons, protons, and helium, carbon, oxygen, and other ions with energies from less than 1keV to hundreds of MeV. The two zones of trapped energetic electrons and protons encircle Earth like donuts. The belts are charge-neutral and overlap the plasmasphere. A few quasi-steady-state processes allow particles to enter the trapping regions.

Cosmic rays generate the quiescent inner radiation belt. These energetic particles collide with atoms in Earth's atmosphere and produce showers of secondary products. Some of these products are neutrons that subsequently decay into energetic protons. The neutron half-life is short enough that many of the protons are born in Earth's lowest L shells. These are trapped in Earth's strong magnetic field. The telltale clue for the decay source is the dominance of protons over other types of ions. The inner zone's stability results from a combination of long particle lifetimes caused by limited collisions in this part of the magnetic field and the slowly varying cosmic ray inputs.

Chapter 7 Earth's Quiescent Magnetosphere: Its Role in the Space Environment and Weather

The quiescent outer radiation belt consists of energetic electrons created by cosmic rays, energetic electrons from the solar wind, and ionospheric particles energized by magnetospheric processes. Solar wind control of magnetic merging drives outer radiation belt behavior. Electrons are also likely to be energized by electromagnetic waves that are excited by long periods of merging or by periods when merging cycles off and on in rapid succession.

Neither belt consists exclusively of one type of particle. In fact each belt is charge-neutral. The belts are designated "proton" or "electron" to specify the particles that carry most of the momentum and energy.

Convection electric fields cause positive ions and electrons to undergo an $E \times B$ drift Earthward from the reconnection region. These drifting particles encounter a gradient in the B-field that produces a deviation of the plasma trajectory in the form of a ring of current. The strength of the ring current serves as an energy density measure of low-to-medium energy particles (10s of keV) that sweep toward Earth. During quiet times the particles may only deviate around Earth and drift toward the dayside, producing only a partial ring current. During more disturbed times the particles encircle Earth for hours and even days. The particles are of sufficient density at geosynchronous altitude to create a magnetic field that's detectable at the ground as a storm-time disturbance magnetic field.

More distant regions of the magnetosphere are influenced by the solar wind through two major processes: viscous interaction and magnetic merging. Viscous interaction occurs along the flanks of the magnetosphere, transferring solar wind momentum to closed field lines inside the magnetopause. These field lines flow tailward and create boundary layers with flows internal to the magnetosphere. Magnetic merging allows the interplanetary magnetic field (IMF) to connect with the geomagnetic field. These merged field lines are transported by the solar wind over the poles and deposited behind Earth as a long, comet-like tail. Magnetic flux conservation requires that these field lines eventually reconnect and return by internal flows to their origin. Flows produced by the viscous and the dominant reconnection process are called magnetospheric convection. The electric field produced in the magnetosphere by convection is projected onto the ionosphere along magnetic field lines and causes a stirring of ionospheric plasma.

Earth's rotation is a second source of electric field for low-energy particles. The plasma polarizes so that the resulting electric field is everywhere perpendicular to the magnetic field. In the absence of other effects, the combination of electric and magnetic fields causes charged particles to co-rotate eastward with Earth. When added to the dawn-dusk magnetospheric electric field, this second field source imposes a fundamental asymmetry on the magnetosphere. As a result, electrons drift closer to Earth on the dusk side and ions closer on the dawn side.

The magnetic merging rate depends on the angle between the IMF and Earth's field. When the IMF has a southward component (anti-parallel) to Earth's field, then strong interaction is possible. Because this angle is constantly changing, the level of magnetospheric convection and associated electric field also changes. These fluctuations allow the magnetosphere to energize and trap particles in its inner portions, creating the radiation belts. The fluctuations are also responsible for bursts of energy from the distant magnetotail. These substorms are the source of auroras, the most visible manifestation of solar wind-magnetosphere interactions.

Key Terms

auroral ovals
auroral substorms
declination D
dipolarization
flux transfer events
horizontal magnetic intensity vector H
inclination I
ionospheric convection
lobes
loss cone
magnetopause
magnetosphere
magnetospheric substorms
mirror point
northern lights (Aurora Borealis)
pitch angle α
plasma mantle
plasmasheet
plasmasheet boundary layer
plasmapause
plasmasphere
plasmoid
polar cusps
polar reversals
ring current
slot
slot region
South Atlantic Anomaly (SAA)
southern lights (Aurora Australis)
steady magnetospheric convection (SMC)
sub-solar magnetopause
total magnetic field vector F
viscous interaction

Notations

Magnitude of radial component of Earth's dipolar magnetic field $B_r = \left(\dfrac{M\mu_0}{4\pi}\right)\left(\dfrac{-2\sin\lambda}{r^3}\right)$

Magnitude of latitudinal component of Earth's dipolar magnetic field $B_\lambda = \left(\dfrac{M\mu_0}{4\pi}\right)\left(\dfrac{\cos\lambda}{r^3}\right)$

Magnitude of Earth's dipolar magnetic field
$$B_S = \sqrt{B_r^2 + B_\lambda^2} = \dfrac{M\mu_0}{4\pi R_E^3}\sqrt{1 + 3\sin^2(\lambda)}$$

x- and y-components of Earth's B-field as a function of declination
$X = H\cos D$
$Y = H\sin D$
$Y/X = \tan D$

z-component of Earth's B-field as a function of inclination $\tan I = Z/H$

Total magnitude of Earth's B-field
$$|F|^2 = |B|^2 = |Z|^2 + |H|^2$$

Equation of a B-field line $r = r_{eq}\cos^2(\lambda) = L\cos^2(\lambda)$

Current from charged particle gyration $I = qv_\perp/(2\pi r)$

Magnetic moment
$$|\mu| = IA = \left(\dfrac{q\Omega}{2\pi}\right)\pi r^2 = \left(\dfrac{q^2 B}{2m}\right)\left(\dfrac{mv_\perp}{qB}\right)^2 = \dfrac{(1/2)mv_\perp^2}{B}$$

Pitch angle $\alpha = \tan^{-1}\left(\dfrac{v_\perp}{v_\parallel}\right)$

Magnetic moment when B varies only over large time and distance scales
$$\mu = \dfrac{(1/2)m[v\sin(\alpha)]^2}{B} = \dfrac{KE_\perp}{B} \approx \text{constant}$$

Proton bounce period
$$\tau_{bounce\,e} = 015\dfrac{L}{\sqrt{E}}(3.7 - 1.6\sin(\alpha_{eq}))$$

Electron bounce period
$$\tau_{bounce\,p} = 0.65\dfrac{L}{\sqrt{E}}(3.7 - 1.6\sin(\alpha_{eq}))$$

Longitudinal drift period in hours
$$\tau_{drift\,e\,or\,p} = \dfrac{0.367}{EL(0.35 + 0.15\sin(\alpha_{eq}))}$$

Pressure balance at the sub-solar magnetopause
$$n_{SW}m_{ion}v_{SW}^2 = \dfrac{2B_{Seq}^2}{\mu_0}\dfrac{R_E^6}{r^6}$$

Location of the sub-solar magnetopause

$$r = R_E \left(\frac{2B_{Seq}^2}{\mu_0 n_{SW} m_{ion} v_{SW}^2} \right)^{\frac{1}{6}} \approx 10\, R_E$$

Near-Earth plasma co-rotation motion with respect to the Sun $U = r\omega\hat{\varphi}$

Latitudinal component of the magnetic field vector

$$B = B_{Seq}\left(\frac{R_E}{r}\right)^3 \hat{\lambda}$$

Co-rotating electric field in the plasmasphere

$$E_{\text{co-rotation}} = -U \times B = -r\omega\hat{\varphi} \times B_{Seq}\left(\frac{R_E}{r}\right)^3 \hat{\lambda}$$

$$E_{\text{co-rotation}} = \omega\left(\frac{B_{Seq} R_E^3}{r^2}\right)(-\hat{r})$$

Answers to Pause for Inquiry Questions

PFI 7-1: a) In symmetrical rings in either hemisphere.

PFI 7-2: Earth's dipole field at the poles is purely radial and has only a vertical component.

PFI 7-3: e) All of the above.

PFI 7-4: Refer to Eq. (7-4) and Fig. 7-8.
$H^2 = F^2 - Z^2$
$Z = H \tan I$, so
$H^2 + (H \tan I)^2 = F^2$
$H^2 + (1.38H)^2 = 2.89\, H^2 = (48{,}500\text{ nT})^2$
$H = 2.85 \times 10^{-5}$ T $= 28{,}500$ nT
$Y = H \sin D = (28{,}500\text{ nT}) \sin(-1°) = -498$ nT
The field deviates 1 degree west of north.

Based on the large declination angle and value of the horizontal component, the location is mid-latitude.

PFI 7-5: L = equatorial crossing of a field line in R_E; From Eq. (7-6b) $r = L\cos^2(\Lambda)$; $\Lambda = \cos^{-1}[(r/L)^{1/2}]$

At Earth's surface $r = 1$ R_E; $\Lambda = \cos^{-1}[(1/L)^{1/2}]$

PFI 7-6: The moon orbits Earth at a distance of 60.4 R_E. The lunar orbit passes through the nightside magnetotail at $r = -60.4$ R_E. The lunar orbit is 60.4 R_E upstream from Earth on the dayside.

PFI 7-7: Cusp; Plasma Mantle; Lobes; Plasmasheet; Plasmasphere; Trapped Radiation Belts; Distant Magnetotail; Magnetopause; Neutral Sheet Current; Ring Current; Plasmapause

PFI 7-8: d) Dawn to dusk

PFI 7-9: A current loop consists of a charge traveling in a circle of radius, r

The area of the circle is πr^2

The current has dimensions of charge/time which is equivalent to (charge × distance)/(time × distance) = charge × velocity/distance.

The distance of interest is the circumference of the current loop $2\pi r$

The velocity can be rewritten as $r\Omega$

Thus the magnetic moment $= IA = (q\Omega r^2/2)$

Substituting for gyrofrequency and gyroradius gives $(mv_\perp^2)/(2B)$, which is the ratio of perpendicular kinetic energy to the magnetic field strength.

PFI 7-10: b) $v_\perp = v\sin(\alpha)$

PFI 7-11: The current must flow from east to west to create a loop of current with a magnetic moment that is outward-directed in the southern geographic hemisphere.

PFI 7-12: The drift interval is much longer than the bounce or gyration interval. The drift period is inversely proportional to energy and L. The bounce period is directly proportional to L and inversely proportional to the square root of energy.

PFI 7-13: The gradient drift motion is most likely to be interrupted because it requires the longest interval of unchanging B-field. The IMF is rarely steady for more than a few tens of minutes.

PFI 7-14:

$$\tau_{\text{longitudinal drift}} \approx \frac{2.53}{1(1 + 0.43\sin 90°)} \approx 1.77 \text{ hours}$$

PFI 7-15:

$$r = R_E \left(\frac{2(31{,}000 \times 10^{-9}\,T)^2}{(1.26 \times 10^{-6}\,N/A^2)(5 \times 10^6\,ion/m^3)} \right)^{\frac{1}{6}} \times$$

$$\left(\frac{1}{(1.67 \times 10^{-27}\,kg/ion)(450{,}000\,m/s)^2} \right)^{\frac{1}{6}} = 9.8\,R_E$$

PFI 7-16: Voltage is the integrated product of electric field and distance. Charged particles derive energy from the voltage difference based on the product of charge and voltage (qV). Thus, energizing charged particles is the product of the charge, the electric field, and the distance over which the field acts, $q \int (-E \bullet dl)$.

PFI 7-17: A dawn-to-dusk electric field creates an $E \times B$ drift in the Sunward direction, as determined by the right hand rule. Since the drift is charge-independent, both ions and electrons have velocity $v_E = E \times B/B^2$ (Eq. (6-14b)).

PFI 7-18: As particles drift Sunward and enter the region where Earth's field is more dipolar, they experience a gradient in the magnetic field. Because the gradient drift is charge-dependent, it separates charges and sends positively charged particles to Earth's duskside and negatively charged particles to the dawnside.

PFI 7-19: d) High speed, high density flow

PFI 7-20: The imposed solar wind $|E|$ = 2.5 mV/m,

$dl = 5\,R_E = \sim 3.2 \times 10^7$ m, $V = \int (-E \bullet dl)$;

$V = 79{,}600$ V ≈ 80 kV.

PFI 7-21:
- Dawn to dusk
- Toward Earth
- The strongest electric field is near the dawnside plasmapause where the flow lines are close together
- The weakest electric field is near the duskside plasmapause where the flow lines spread apart

References

Dungey, James W. 1961. Interplanetary Magnetic Field and the Auroral Zones. *Physical Review Letters*, Vol. 6. American Physical Society. College Park, MD.

McIlwain, Carl E. 1961. Coordinates for Mapping the Distribution of Magnetically Trapped Particles. *Journal of Geophysical Research*, Vol. 66. American Geophysical Union. Washington, DC.

Figure References

Cowley, Stanley W. H. 1996. The Earth's Magnetosphere. *Earth in Space*. Vol. 8, No. 7. American Geophysical Union. Washington, DC.

Fraenz, M. and D. Harper. 2002. Heliospheric Coordinate Systems. *Planetary and Space Science*. Vol. 50. Elsevier Science. Amsterdam, Netherlands.

Kavanagh, Lawrence D., Jr., John W. Freeman Jr., and A. J. Chen. 1968. Plasma Flow in the Magnetosphere. *Journal of Geophysical Research*. Vol. 73. No. 17. American Geophysical Union. Washington, DC.

Pilipp, Werner G. and Gregor Morfill. 1978. The Formation of the Plasma Sheet Resulting from Plasma Mantle Dynamics. *Journal of Geophysical Research*. Vol. 83(A12). American Geophysical Union. Washington, DC.

Russell, Christopher T. 1971. Geophysical Coordinate Transformations. *Cosmic Electrodynamics*, Vol. 2. Dordrecht-Holland: Reidel Publishing Company.

Schulz, Michael and Louis J. Lanzerotti. 1974. *Particle Diffusion in the Radiation Belts*. Physics and Chemistry in Space. New York, NY: Springer-Varlet.

Willis, D. M. 1971. Structure of the magnetopause. *Reviews in Geophysical Space Physics*. Vol. 9. American Geophysical Union. Washington, DC.

Further Reading

Jursa, Adolph S., Ed. 1985. *Handbook of Geophysics and the Space Environment*. Air Force Geophysics Laboratory.

Cravens, Thomas E. 1997. *Physics of Solar System Plasmas*. Cambridge, UK: Cambridge University Press.

Glatzmaier, Gary A. and Peter Olson. 2005. Probing the Geodynamo; Our Ever Changing Earth. *Scientific American*. Nature America. New York, NY.

Hargreaves, John K. 1992. *The Solar Terrestrial Environment*. Cambridge, England: Cambridge University Press.

Lemaire, Joseph F. and Konstantin I. Gringauz. 1998. *The Earth's Plasmasphere*. Cambridge, UK: Cambridge University Press.

Stern, David P. 1989. A brief history of magnetospheric physics before the spaceflight era, Reviews of Geophysics and Space Physics. Vol. 27(1). American Geophysical Union. Washington, DC.

Stern, David P. 1996. A brief history of magnetospheric physics during the space age. Reviews of Geophysics and Space Physics. Vol. 34(1). American Geophysical Union. Washington, DC.

Walt, Martin. 1994. *Introduction to Geomagnetically Trapped Radiation*. Cambridge, UK: Cambridge University Press.

Earth's Quiescent Atmosphere and Its Role in the Space Environment and Weather

8

UNIT 1. SPACE WEATHER AND ITS PHYSICS

Contributions by Stanley C. Solomon, Richard C. Olsen, and M. Geoff McHarg

You should already know about...

- How to multiply, integrate, and differentiate exponential functions
- The general structure of Earth's atmosphere (Chap. 1)
- Scale height (Chap. 3)
- Solar radiation and formation of atmospheric layers (Chap. 3)
- Hydrostatic equilibrium (Chap. 3)
- The location of field-aligned currents in the high-latitude regions (Chap. 4)

In this chapter you will learn about...

- Earth's mesosphere, thermosphere, and geocorona
- The composition and structure of Earth's neutral upper atmosphere
- Earth's ionosphere as a function of latitude and altitude
- More details on ionospheric layers
- Electron density profiles and total electron content
- Airglow and aurora
- Radio propagation in the ionosphere
- Current and particle linkages between the ionosphere and geospace
- Heating of the upper atmosphere

Outline

8.1 **Earth's Neutral Atmosphere**

8.2 **Earth's Ionized Atmosphere**
    8.2.1 Transient Luminous Events
    8.2.2 Ionization as a Function of Height
    8.2.3 The Ionosphere Characterized by Latitude and Layers

8.3 **Radio Wave Propagation in the Ionosphere**
    8.3.1 Operational Aspects of High-frequency (HF) Radio Propagation
    8.3.2 The Physics of Ionospheric Radio Wave Propagation

8.4 **Ionospheric Interactions with Other Regimes**
    8.4.1 Quiescent High-latitude Ionospheric Connections to the Magnetosphere
    8.4.2 Ionospheric Mass Outflows

Chapter 8 Earth's Quiescent Atmosphere and Its Role in the Space Environment and Weather

The coldest and hottest portions of Earth's neutral atmosphere are the background for the weakly ionized ionosphere. The ionosphere joins with the bottommost region of the plasmasphere. The outward extension of Earth's quiescent neutral atmosphere extends to geosynchronous altitude. In this chapter we characterize the quiescent mid-latitude ionosphere's layered structure and the linkages between the ionosphere and the magnetosphere.

8.1 Earth's Neutral Atmosphere

Objectives: After reading this section, you should be able to...

- Explain why the mesosphere is cold
- Describe where and why noctilucent clouds develop
- Describe the location and origin of the sodium airglow layer
- Explain why the thermosphere is hot and how it redistributes its energy
- Relate enhanced thermospheric heating to satellite drag
- Explain why thermosphere constituents are differentiated in height
- Describe the exosphere and relate it to Earth's neutral atmosphere
- Explain why hydrogen dominates the geocorona

Middle Atmosphere

At an altitude range of 50 km–90 km lies the mesosphere—the coldest region of Earth's atmosphere. The mesosphere is notoriously difficult to investigate. At mesospheric altitudes the atmosphere is too thin to support standard balloon flights and too dense to allow satellite flight. Therefore, rocket soundings and remote sensing are the primary options for exploring the dynamics of this region. Few of its constituents absorb sunlight, so most of the energy in the region comes from thermal conduction of extreme ultraviolet (EUV) energy from the thermosphere above. The mesosphere is home to trace amounts of hydroxyl (OH) and carbon dioxide (CO_2), which are very effective infrared radiators. Radiation is an important means of cooling. Collisions occur less often than at lower altitudes. The reduced collision frequency allows atmospheric constituents to sort themselves by mass, with the heavier molecules diffusing downward. Collisions are sufficiently frequent to allow some momentum transport via small air vortices.

The global circulation in the mesosphere is driven by momentum deposited by upward-propagating atmospheric tides, internal atmospheric gravity waves, and planetary waves. Many of these waves originate in the troposphere and stratosphere. As the waves propagate into the low density mesosphere, their amplitudes become so large that the waves destabilize, depositing energy and momentum. Recent studies have shown that some of the wave energy and momentum transcend the middle atmosphere and arrive in the thermosphere.

The mesosphere is home to a sodium airglow layer based between 80 km and 105 km. Excited neutral sodium atoms radiate weakly at 589.3 nm, thus contributing to a faint glow that is present day and night. Most of the sodium comes from meteors, although sea salt may contribute a small amount. Sodium airglow intensity exhibits a solar cycle variation. The glow provides a remote signal that allows scientists to investigate this difficult-to-reach region.

In late summer, when temperatures drop below 180 K in the mesosphere, constituents may sublimate to form *polar mesospheric clouds (PMCs)*. These vapor-ice mixes are called *noctilucent ("night-shining") clouds*. They are visible from the ground in the evening, when they are illuminated from below by the setting Sun (Fig. 8-1). Combined ground-based and space-based observations reveal that noctilucent clouds consist of crystals of water ice 20 nm–100 nm across. Ice crystals attach to and grow on water molecules and nuclei provided by dust and pollutants. Upwelling winds in the summertime carry water vapor from the moist lower atmosphere toward the mesosphere. The source of the nuclei is less clear. Ordinary tropospheric clouds get their dust and sea salts from Earth's surface. Lifting dust to the mesosphere is more difficult, but volcanoes sometimes lift dusty material that high. Space launches and interplanetary space dust may provide some material as well. Every day, Earth receives tons of meteoroids—tiny bits of debris from comets and asteroids. Many are just the right size to seed noctilucent clouds.

Fig. 8-1. Noctilucent Clouds. The billows and waves in these mesospheric clouds show that the upper atmosphere is a highly dynamic region. The electric-blue color is typical of these clouds. *(Courtesy of Richard Keen and Gary Thomas at the Laboratory for Atmospheric and Space Physics, University of Colorado)*

In recent years, PMCs have appeared more frequently and at lower latitudes than in the past. Studies are underway to determine whether their occurrence relates to global climate change. Greenhouse gases in the upper atmosphere may be effective radiators that enhance cooling in and above the mesosphere. In mid-2007, NASA's Aeronomy of Ice in the Mesosphere (AIM) spacecraft began the first dedicated space-based measurements of PMCs. The AIM data reveal an amazing global extent of these mysterious clouds.

Thermosphere

The region of Earth's atmosphere containing neutral atmospheric constituents and located between about 90 km and 1000 km is the *thermosphere*. This region is heated by absorption of short wavelength solar energy, as we mention in Sec. 3.4 and discuss further in Sec. 8.2. Most human space activities take place and many satellites orbit in the thermosphere. The satellites experience a drag that eventually reduces their altitude and causes some of them to de-orbit. Collisions in the thermosphere are infrequent, so thermospheric particles tend to stratify based on their molecular weight. More massive gases settle to lower altitudes, while the lighter ones migrate upward in a process called *diffusion*. Densities of Ar, O_2, and N_2 decrease the fastest with height, because they are the heaviest gases. Each gas behaves according to its own barometric equation (Eq. 3-6) and develops its own scale height. From 180 km to 600 km, atomic oxygen is the dominant species. Helium and hydrogen dominate above 600 km. (Fig. 8-2). Some thermospheric atoms and molecules may attain sufficient energy to escape Earth's gravity. They become part of the exosphere.

Fig. 8-2. Diffusive Equilibrium Profiles of Constituent Density for Solar Minimum and Solar Maximum. Here we plot hydrogen, helium, argon, molecular oxygen, molecular nitrogen, and atomic oxygen profiles for altitudes of 100 km–1000 km. At altitudes above 200 km, atmospheric density varies substantially with the solar cycle. *(Courtesy of AF Weather Agency)*

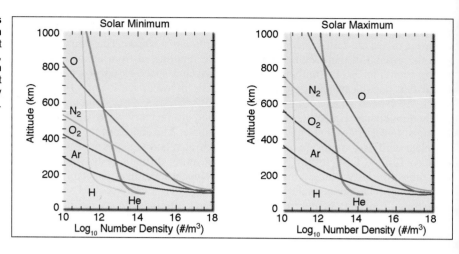

Scientists study thermospheric motion and composition with satellite-borne mass spectrometers (Example 6.4), accelerometers, rockets, radars, and models that assimilate observations from the thermosphere and ionosphere. Additionally, tracking meteor ionization trails has proven effective in studying lower thermospheric motions.

As we see in Fig. 8-3, the temperature in the lower thermosphere increases rapidly with increasing altitude from a minimum near 90 km. Eventually, temperature becomes altitude independent, approaching an asymptotic value known as the *exospheric temperature* at heights greater than 200 km. Figure 8-3 shows exospheric temperatures for solar minimum and solar maximum. Solar radiation interacts with the rarified atmosphere in this region, creating the extreme temperatures. The low density at great heights means individual particles gain tremendous energy from the continuous stream of incoming shortwave photons (Chap. 3). Each thermospheric particle shares in the abundance of radiative energy that changes to random thermal motion—heat—whose measure is temperature. Solar wavelengths on the short side of 160 nm have enough energy to free electrons from atoms and dissociate molecules in the thermosphere. The hot particles exchange some of their kinetic energy with other matter, with a thermospheric heating efficiency of about 33%. The remainder of the energy is reradiated to space or to the lower atmosphere. Thermospheric temperature, density, and composition are sensitive to the solar cycle because of heating associated with the solar EUV radiation. The average mid-thermospheric temperature varies by ~1000 K over the solar cycle. Over long intervals, thermospheric temperature achieves a balance between heating and cooling. Short-term variations produced by internal and external energy easily throw the system out of thermodynamic balance.

The thermosphere also receives energy from tidal and gravity waves generated in the lower atmosphere. Gravity waves, primarily from long-lived tropospheric thunderstorm complexes and frontal systems, propagate through the lower atmosphere and into the thermosphere. Scientists are trying to determine the magnitude and preferred locations of this wave activity.

At high latitudes, the aurora, driven by the interaction of the highly variable solar wind with Earth's magnetosphere, is another thermospheric energy source. Energetic auroral electrons and protons from the magnetosphere collide with thermospheric neutrals, causing ionization, dissociation, and heating. Auroras

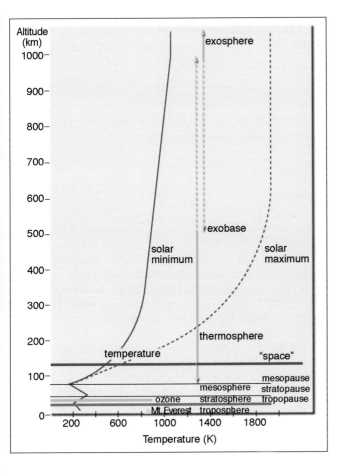

Fig. 8-3. **Temperature Profile of Earth's Atmosphere to 1000 km for Solar Minimum and Solar Maximum.** Temperature increases associated with absorbed radiation by ozone and other constituents of the upper atmosphere counter the natural tendency of the atmosphere to cool with height. *(Courtesy of the AF Weather Agency)*

and airglow result from these interactions. In most instances, auroral particles are accompanied by enhancements in atmosphere-magnetosphere coupling driven by a million-ampere field-aligned current system (Sec. 4.4). Dissipation of current also strongly heats the high-latitude thermosphere. When the heating is impulsive, energy moves out of the auroral zones in waves called traveling atmospheric disturbances.

Pause for Inquiry 8-1

What is the dominant species at 700 km during solar maximum? During solar minimum?

Pause for Inquiry 8-2

How does the amount of upper atmospheric mass at 500 km vary over the solar cycle? Why does it vary in this manner?

> **Pause for Inquiry 8-3**
>
> Which phase, solar maximum or solar minimum, has the larger scale heights?

> **Pause for Inquiry 8-4**
>
> Referring to Fig. 8-2, at 500 km altitude, the change in atomic oxygen density between solar minimum and solar maximum is about
>
> a) 1%
>
> b) 10%
>
> c) An order of magnitude
>
> d) Several orders of magnitude

Conduction and Eddy Transport. Above 120 km, much of the thermospheric cooling comes from downward heat conduction by thermospheric molecules. Below 120 km, where the atmosphere thickens enough to allow frequent collisions, energy is transported via mass motion in eddies or small vortices of motion. Eddy mixing in the lower thermosphere moves excess thermospheric energy downward to the cold mesosphere.

Radiation. Another mechanism for thermospheric energy loss is radiation, but generally at wavelengths much longer than that of the incoming photons. This radiation is one form of *airglow emission*. Airglow is caused by the conversion of chemical energy into light. During daytime, short-wavelength solar radiation photodissociates molecular oxygen into individual atoms. Because it does not efficiently recombine, atomic oxygen has a long lifetime. It is a reservoir of chemical energy that powers nighttime airglow. At night, recombination of atomic oxygen to form molecular oxygen produces faint visible airglow at 557.7 nm wavelength. The upper atmosphere also glows in several infrared bands because of vibration and rotation transitions in its constituents: nitric oxide (5.3 μm) and carbon dioxide (15 μm). Carbon dioxide is also a strong thermospheric radiator at 63 μm. Numerous airglow detectors operate in the nighttime to observe this emission for the study of thermospheric dynamics. Radiative cooling is most effective in the denser parts of the thermosphere.

Upper-atmospheric Tides and Winds. Earth's rotation constantly exposes different longitudes to solar heating and nighttime cooling. Solar heating of the thermosphere causes atmospheric expansion and an increase in potential energy of thermospheric parcels in the sunlit sector. The nightside cools and contracts. Associated pressure gradients instigate horizontal winds, flowing from the dayside to the nightside. At altitudes near 300 km, observations show a close correspondence between the solar heating pattern and the thermal wind pattern during quiescent conditions. The maximum temperature occurs in the early afternoon sector. Along with this temperature is a pressure increase that drives winds toward the cooler nightside. Figure 8-4 depicts the flow pattern associated with dayside thermospheric heating during quiescent conditions. Pressure gradient-driven horizontal winds may exceed 450 m/s at altitudes near 300 km. Because Earth rotates underneath, the heating pattern for these winds is technically "tidal," but we refer to them as *thermospheric winds*, since the vertical winds propagating upward from below are already called "tides."

At lower altitudes (~120 km) the pattern in Fig. 8-4 is influenced by ion-neutral particle collisions, as well as pressure gradients. Lorentz forces inhibit partially ionized air from crossing Earth's magnetic field lines. Where ions are numerous, they collide with and slow the neutrals. Wind speeds at 120 km altitude are typically on the order of 100 m/s. Tides and waves propagating upward from the lower atmosphere also influence the lower thermosphere. The diurnal pattern of Fig. 8-4 gives way to a more structured semi-diurnal pattern at higher altitudes. A semi-diurnal pattern (two maxima and two minima in a 24-hour period) arises naturally in Earth's fluid atmosphere.

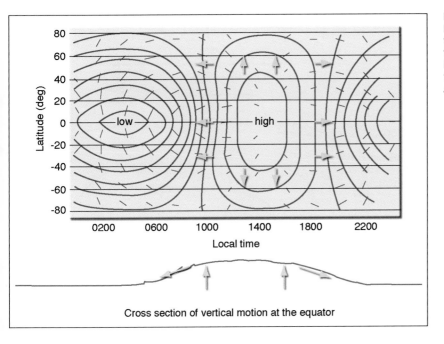

Fig. 8-4. The Upper Atmospheric Zonal Neutral Wind Flow Pattern. Here we see how the winds in the upper atmosphere move mass from regions of high temperature, and hence high pressure. Mass flows to regions of low temperature. *(Courtesy of the AF Weather Agency)*

In addition to the broad zonal (across time zones) circulation patterns caused by vertical motions from daily solar heating and cooling cycles, seasonal variations cause meridional circulations. During the equinoxes (September and March), solar heating concentrates along the equator. The rising motion in the equatorial regions produces a circulation cell in each hemisphere, as shown in Fig. 8-5a. These circulation patterns change according to geomagnetic activity as heated air flows out of the auroral zones. During the solstices (December and June), solar heating concentrates near one of the poles. The rising motion at high latitudes results in a single large circulation pattern that covers both hemispheres (Fig. 8-5b). Either circulation system can be altered by geomagnetic activity. When geomagnetic activity is low, little thermal input occurs at high latitudes, resulting in only a small circulation pattern, as suggested in Fig. 8-5a. When geomagnetic activity increases, auroral thermal input is larger, causing the polar circulation pattern to dominate and expand to mid-latitudes.

Fig. 8-5. Meridional Thermospheric Circulation Patterns. Altitude is on the vertical axis, and latitude is on the horizontal scale. The solid lines represent circulation caused by solar heating. The dotted lines represent circulation caused by geomagnetic storms. **a)** Quiescent equinox circulations are hemispherically symmetric. Mass rises at the equator and sinks near the poles. The return flow is not shown. **b)** The solstice circulation consists of one dominant cell, with mass moving from the summer hemisphere to the winter hemisphere. [Roble, 1977]

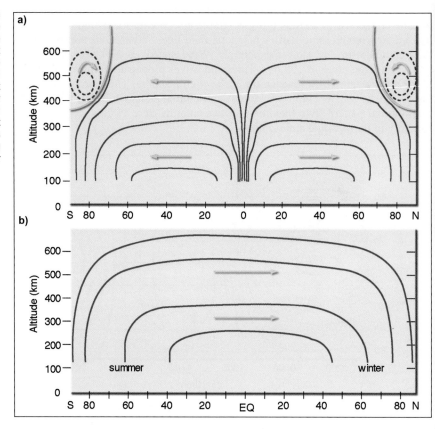

Thermosphere Coupling to the Ionosphere

Earth's upper atmosphere is in constant motion. It is subject to accelerations by 1) tides from solar heating, 2) low-frequency planetary waves, 3) high-frequency gravity waves, and 4) traveling wave disturbances, pulsing out of the auroral zone during periods of intense geomagnetic activity. The ionosphere is also forced from below. Recent studies suggest that between 15% and 25% percent of observed ionospheric electron density variations are attributable to waves of tropospheric origin. In the thermosphere, *in situ* tides associated with solar heating reach velocities approaching 200 m/s. Mid- and low-latitude neutral-atmosphere waves with 2-, 5-, 10-, and 16-day periods modulate the tropopause and the atmosphere above it. Absorption of solar photons by stratospheric ozone and tropospheric water vapor is the source of the dominant, solar-driven diurnal and semidiurnal tides.

Other tides, some from lunar effects, add to the complex motions. Were it not for electromagnetic braking provided by ionized particles that are tied by gyration to magnetic field lines, the tides' speeds would be even higher. The neutral wind tries to force ionized particles across the magnetic field, but the gyrating particles must be struck again and again by the neutral gas to be transported. This slows the neutral wind. The neutral wind induces a current when it tries to carry the plasma across the field lines. In effect, this process is an electric generator.

EXAMPLE 8.1

Thermospheric Mass

- *Problem Statement:* Determine the percentage of mass in the thermosphere relative to the rest of the atmosphere.
- *Given:* The standard pressure at Earth's surface is ~101,300 Pa and at 90 km is 0.18 Pa. Earth's radius is 6378 km. The base of the thermosphere is at 90 km.
- *Assumptions:* Hydrostatic equilibrium and uniform mass distribution. Earth's gravitational acceleration varies with height.
- *Solution:* Balance the vertical pressure gradient force F_p with the force of gravity to obtain the mass of a column of atmosphere resting on one square meter of Earth's surface. Multiply by the total surface area.

$$F_p = mg = G\frac{mM_E}{R_E^2}$$

For one square meter, this becomes

$$P = G\frac{mM_E}{(1\,m^2)R_E^2}$$

So the mass per square meter of a column of atmosphere is

$$mass/(unit\ area) = \frac{PR_E^2}{GM_E} = \frac{(101{,}300\ N/m^2)(6{,}378{,}000\ m)^2}{(6.67\times10^{-11}\,N\cdot m^2/kg^2)(5.97\times10^{24}\,kg)}$$

$$= 10{,}300\ kg/m^2$$

The total mass of the atmosphere is:

$$(10{,}300\ kg/m^2)(4\pi)(6{,}378{,}000\ m)^2 = 5.27\times10^{18}\ kg$$

At $R_E + 90$ km

$$mass/(unit\ area) = \frac{PR_E^2}{GM_E} = \frac{(0.18\ N/m^2)(6{,}468{,}000\ m)^2}{(6.67\times10^{-11}\,N\cdot m^2/kg^2)(5.97\times10^{24}\,kg)}$$

$$= 0.0189\ kg/m^2$$

The total mass above 90 km altitude is

$$(0.0189\ kg/m^2)(4\pi)(6{,}468{,}000\ m)^2 = 9.94\times10^{12}\ kg$$

The percentage of mass above 90 km is

$$(9.94\times10^{12}\ kg)/(5.27\times10^{18}\ kg)\times100\% = 0.00018\%$$

- *Interpretation:* Earth's thermospheric mass is insignificant compared to the lower atmosphere's mass. However, the mass that is present is heated to high temperatures by the constant input of solar shortwave photons. Low-Earth orbiting satellites constantly collide with these hot particles. Over time, these collisions cause the satellites to decelerate and drop in altitude. Satellites operating below 500 km need re-boosting from time to time to maintain an operational altitude.

Follow-on Exercise: Data show that the average speed of the rising thermospheric air at 300 km altitude on the dayside is 1 m/s. How far does the air rise during 12 hours? How much work is done to lift the air mass this distance? Assume that the pressure at 300 km is $\sim 1\times 10^{-5}$ Pa. Can solar heating provide this energy?

Additional Exercise: Use Figs. 8-2 and 8-3 to determine the scale height of atomic oxygen at 400 km during solar maximum and solar minimum.

Focus Box 8.1: Satellite Drag

We know from the discussion of the Law of Atmospheres in Chap. 3 that atmospheric density decreases exponentially with height. Figure 8-2 shows the density profiles of several individual atmospheric constituents. Variations in total neutral atmospheric density have profound effects on the time that some satellites stay on orbit. Below about 500 km, the atmosphere is thick enough to impart a substantial drag force on orbiting spacecraft. Unless a satellite's propulsion system compensates for this drag force, its orbit will decay until it re-enters. Density effects also contribute to the torques on the spacecraft due to the aerodynamic interaction between the spacecraft and the atmosphere, and must be considered in the design of the attitude control system. In addition, density variations influence the spacecraft's precision pointing and our ability to track the location of satellites and debris in low-Earth orbit.

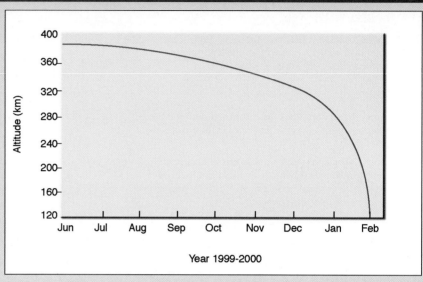

Fig. 8-6. Altitude Versus Time Profile of the Starshine 1 Satellite. The satellite remained in orbit for about eight months. It lost mechanical energy and exchanged some of its potential energy for kinetic energy as it sped to a fiery disintegration in the atmosphere.

For satellites orbiting below 500 km, orbital lifetimes decrease when Earth's upper atmosphere becomes hotter than normal. The energized molecules collide more frequently and the atmosphere expands upward. Denser material, upwelling from below, causes more satellite-particle collisions that create drag and thus shorten on-orbit lifetimes. The International Space Station needs frequent and costly re-boosts to stay on orbit, particularly when the Sun emits excess energy. Figure 8-6 shows the orbital decay of the Starshine 1 satellite from 1999 to 2000. Because the satellite continuously collided with atmospheric molecules and atoms, it lost the mechanical energy it needed to remain in orbit.

Mission planners closely monitor long-term forecasts of solar cycle variations. Pre-launch difficulties sometimes lead to a satellite reaching orbit during a part of the solar cycle that significantly shortens the mission lifetime because of enhanced drag.

Gravity waves, some generated by thunderstorms and mountain ranges, grow with altitude, causing mass displacements in the region collocated with the ionospheric plasma. Assuming these gravity waves induce electric fields in a manner similar to that of tides, they should affect the coupled ionosphere-magnetosphere significantly. The highly conducting magnetic field lines allow the electric fields generated by the atmosphere to map throughout the magnetosphere. For low levels of magnetic activity, Earth's wind field (and the associated electric field) modulates the electrical structure of the inner magnetosphere.

Geocorona

Moon-based images from the Apollo 16 mission reveal that Earth is shrouded in a low-density envelope of neutral hydrogen that extends from ~500 km to beyond geosynchronous orbit. Figure 8-2 shows that hydrogen is a secondary constituent in the upper atmosphere. Reduced collisions in the upper neutral atmosphere allow hydrogen atoms to float to the top. Some of these atoms, after being energized by solar radiation, escape, forming the exosphere. Exospheric particles may have ballistic trajectories or trajectories that allow them to make many Earth orbits. The thin veil of neutral hydrogen, which scatters solar far ultraviolet (FUV) hydrogen Lyman-α radiation, produces a *geocorona*. This faintly glowing region is the grand extension of Earth's atmosphere to space.

At geosynchronous altitude, neutral hydrogen has roughly the same density as plasma (about 10 particles/cm^3), depending upon the local time. Earth's hydrogen crown has been observed to distances of 100,000 km (~16 R_E). Figure 8-7 shows a Dynamics Explorer satellite image of the glowing geocorona as well as lower altitude features in the ionosphere. In the geocorona, particle behavior is influenced by radiation pressure from solar far-ultraviolet photons, charge exchange with the magnetosphere, photoionization, impact ionization, and atmospheric variations below the exobase. Solar radiation pressure pushes the exospheric hydrogen away from Earth to form a tail of neutral hydrogen.

Earth's geocorona acts as an imaging screen for magnetospheric and ionospheric ions. These ions exchange charge with cold neutral particles in the geocorona, creating cold ions and fast neutrals. Space-based detectors view and map the escaping neutrals, allowing scientists to gain information about inner magnetosphere-ionosphere-exosphere interactions. The exosphere and geocorona are the sources of most magnetospheric plasma, the bulk of which resides in the plasmasphere.

Pause for Inquiry 8-5

What is the relationship between the plasmasphere, the exosphere, and the geocorona?

Fig. 8-7. View of Earth, the Aurora, the Equatorial Airglow Bands, and the Geocorona from the Dynamics Explorer Spacecraft. The Sun is behind Earth. Features of Earth's disk, including dayglow from the sunlit atmosphere, auroral oval, and equatorial airglow appear primarily in the emissions of atomic oxygen at 130.4 nm and 135.6 nm and in bands of molecular nitrogen. Beyond the limb, the glow is from the solar Lyman-α radiation, scattered by Earth's extended hydrogen atmosphere (the geocorona). The image is from an altitude of 16,500 km and 67° N latitude. The northern auroral oval forms a halo of light above Earth's limb, while the equatorial airglow bands in the pre-midnight sector straddle the magnetic equator. Isolated points of light are bright stars emitting in UV. *(Courtesy of Louis Frank at the University of Iowa and NASA)*

8.2 Earth's Ionized Atmosphere

8.2.1 Transient Luminous Events

Objectives: After reading this section, you should be able to...

- Define transient luminous events (TLEs)
- Distinguish between blue jets and red sprites
- Relate gamma ray bursts to lightning

For decades, aircraft pilots reported fleeting upward-directed lightning from the tops of thunderstorms. In the late 1980s, scientists began imaging vertical luminous events called red sprites and blue jets (Fig. 8-8). Space Shuttle video footage from 1989 confirmed the transient events. Most of these accompanied intense thunderstorms. Growing evidence exists for a lower atmosphere-upper atmosphere linkage in the form of *transient luminous events (TLEs)* associated with lightning. Lightning is the most visible ionization process in the lower atmosphere. Although ionization events are short-lived in the dense regions of the lower atmosphere, they have a longer lifetime in regions above thunderstorms. Transient luminous events are the result of electrostatic fields arising from unbalanced charge distributions in clouds. *Blue jets* are blue cone-shaped discharges that extend upward at speeds on the order of 10^5 m/s from the electrical cores of thunderstorms. The blue color is thought to be related to the 427.8 nm emission from ionized molecular nitrogen in the atmosphere. The jets usually terminate in the stratosphere around an altitude of 50 km, but some may extend to 70 km–80 km.

Sprites begin in the mesosphere near the base of the ionosphere and develop very rapidly downward at speeds in excess of one tenth light speed and typically last a few milliseconds. *Red sprites* occur when positive charge is brought from a cloud to the ground by a potent stroke of lightning. This positive cloud-to-ground stroke creates an intense electrostatic field above the cloud from which it emanates. The quasi-static electric field is large enough to initiate atmospheric breakdown at altitudes between 70 km–80 km. The sprites' red color is from visible emissions by neutral nitrogen molecules. Electrified streamers move downward, then upward from this initial altitude, thus electrically connecting the lightning in the stratosphere with the ionosphere. Sprites are very bright, brighter than Venus, but because of their brief duration, seeing them requires specialized high-speed cameras with low-light imaging capability. Sprites propagate currents of a few thousand amps and deposit 1 MJ–10 MJ of energy in the mesosphere and lower ionosphere. The disturbed volume may exceed 10^4 km^3. Electromagnetic disturbances from the lightning that cause the sprites may spread laterally as a transient electromagnetic disturbance called an *elf*.

Jets develop out of moderate thunderstorms. Sprites seem to need the more energetic situations arising in severe thunderstorms that move copious amounts of positively charged ice crystals to great heights. These thunderstorms may be so strongly energized that they discharge some of their positive charge from the cloud tops to the ground or to another cloud, but still retain residual charges whose field is strong enough to cause an upward-directed discharge. While much remains unknown about the various forms of TLEs, scientists are reasonably certain that they represent an important part of the global electrical circuit linking Earth's surface to space.

In the 1990s, the Compton Gamma Ray Observatory spacecraft, searching for energetic events in distant parts of the galaxy, found evidence of extremely energetic gamma-ray bursts in Earth's atmosphere. These bursts, called *terrestrial gamma ray flashes (TGFs)*, are associated with intense thunderstorms and last about 1 ms. Because they radiate in the gamma-ray spectrum, the discharges are not visible to the human eye.

Above a thunderstorm, powerful electric fields stretch upward into the upper atmosphere. These electric fields accelerate free electrons, driving them to near light speed. When these electrons collide with molecules in the air, the collisions release high-energy bremsstrahlung in the form of gamma rays and more electrons, setting up a cascade of collisions and perhaps more TGFs. Individual particles in a TGF acquire energy sometimes exceeding 20 MeV, roughly a thousand times more energetic than the particles that produce the aurora. The electric field strength required for the particle acceleration is most likely to be found at 15 km–20 km, the altitude at which blue jets have been observed (Fig. 8-8). Investigators are not sure if TGFs and blue jets are related.

Scientists don't know how many TGFs occur or even very much about their global distribution. Limited observations suggest about 50 TGFs each day, many of them near the equator. However, many more such bursts may develop at sufficiently low altitudes that they are not sensed from space. Scientists are actively investigating the link between the gamma ray flashes and lightning. It is not clear if the TGFs are the result of lightning or the initiators of particularly strong lightning events.

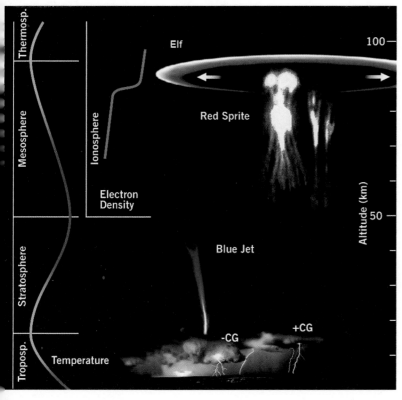

Fig. 8-8. Transient Luminous Events (TLEs). This figure gives an artist's rendition of the associations between various forms of TLEs and tropospheric thunderstorms. On the left are notional temperature and ionospheric density profiles. In the region of the atmosphere that supports TLEs, the warmest temperatures are at Earth's surface (~288 K) and at the stratopause (~255 K). The electron density rises sharply at the upper reaches of the mesopause (Sec. 8.2.2). In the troposphere, lightning causes negative and positive cloud-to-ground (CG) strikes. Above some thunderstorms, blue jets travel to the stratosphere. Very strong thunderstorms may be accompanied by red sprites. Sprites start at 70 km–80 km and produce downward streamers. The disturbed volume above the sprite may lead to a signal called an elf. [Pasko, 2003]

8.2.2 Ionization as a Function of Height

Objectives: After reading this section, you should be able to...

- Compare and contrast electron density with neutral density in the ionosphere
- Compare electron density profiles on and off the solar zenith
- Distinguish between electron density profiles (EDPs) and total electron content (TEC)
- State the number of electrons per unit area that correspond to 1 TEC unit

In Sec. 3.4.2 we explore solar photoionization in Earth's atmosphere. Here we provide a more quantitative description of ionospheric layers, electron density profiles, and total electron content. We also extend the concept of electron density profiles to off-solar zenith angle geometries. Ionospheric radio wave propagation depends strongly on electron density variation with height. Observers determine the electron density at various altitudes by examining how radio waves behave at these altitudes (Sec. 8.3).

Electron Density Profiles. Above 90 km, a fraction of the atmospheric constituents are in a continual state of ionization. This electrically charged portion of the atmosphere is the *ionosphere*, and the ionized gas within this layer is the *ionospheric plasma*. Figure 8-9 reveals that neutral particles far outnumber charged particles below 800 km. Nonetheless, the small fraction of charged particles strongly influences the upper atmosphere's behavior. The ion-electron pairs are very important because of their effect on the passage of radio waves, and because they act as a source of plasma for Earth's magnetosphere. The charged particles make the upper atmosphere an electrical conductor supporting electric currents that dissipate energy during geomagnetic storms.

Electron density corresponds to the degree of ionization, or "strength" of the ionosphere. "Electron density" is frequently used in place of "ion density." The substitution is appropriate if we assume that each ionization event creates only one ion-electron pair. Surface-based ionosonde stations measure the electron densities of the lower ionosphere using specially tuned radio antennas. These data reveal vertical *electron density profiles (EDPs)* below the peak ionization at specific locations. Satellite observations yield information about EDPs above the ionization peak.

Figure 8-9a depicts mid-latitude electron density and neutral density profiles for solar maximum and solar minimum conditions. Figure 8-9b shows day-night variations of the same quantities. At mid-latitudes, electron density decreases at all altitudes at nighttime. The profiles from solar maximum years reveal enhanced ionization, particularly in the upper portion of the ionosphere. Actual electron and ion densities vary dramatically with altitude, latitude, geomagnetic activity, and solar activity.

Steady-state Electron Density. In Figs. 8-9 and 8-10 we see the steady-state electron density profiles for solar minimum and maximum, night and day conditions at mid-latitudes. At each point the ion continuity equation, Eq. (8-1), applies. This equation says that local changes in ionization density are caused by transport and local production (R), and loss (L) of ionization.

8.2 Earth's Ionized Atmosphere

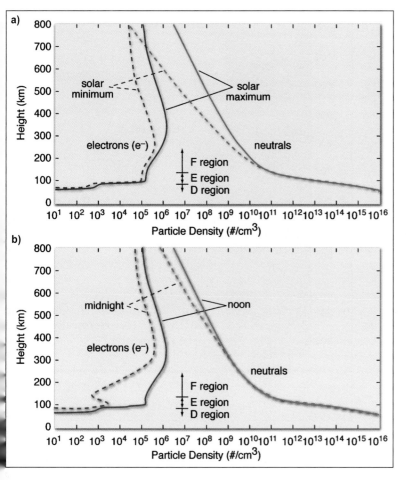

Fig. 8-9. a) Electron and Neutral Density Profiles for Solar Maximum and Solar Minimum. The solid (dashed) red curve shows globally averaged neutral density for solar maximum (minimum); the solid (dashed) blue curve shows globally averaged electron density for solar maximum (minimum). Approximate heights of the ionospheric layers are indicated by letters D, E, and F. Additional shortwave solar radiation impinges on the upper atmosphere during solar maximum, causing increased ionization and electron density at all layers in the ionosphere. The profiles and total electron content (TEC) vary significantly from place to place and time to time. b) Day-Night Variation of Electron and Neutral Densities. The solid (dashed) red curve shows globally averaged neutral density for noon (midnight); the solid (dashed) blue curve shows globally averaged electron density for noon (midnight). *(Data from the Mass Spectrometer Incoherent Scatter model, NASA)*

$$\frac{\partial n_e}{\partial t} = -\nabla \bullet (n_e \mathbf{v}) + R - L \qquad (8\text{-}1)$$

where

$\partial n_e / \partial t$ = local rate of change of electron density [#/(m³·s)]
n_e = number density of free electrons [#/m³]
$\mathbf{v}$ = velocity of free electrons [m/s]
R = volume production rate (source) of ionization [#/(m³·s)]
L = volume loss rate of ionization [#/(m³·s)]

If we assume the electron density is in steady state, so transport can be ignored, then production and loss of ionization must balance. First we explore production (in more detail than in Chap. 3).

Chapter 8 Earth's Quiescent Atmosphere and Its Role in the Space Environment and Weather

Fig. 8-10. Electron Density Profiles (EDPs) for Solar Minimum and Maximum, Night and Day Conditions at Mid-latitudes. Here we see EDPs for solar maximum and solar minimum from ~70 km to 1200 km. The ledges (bumps) in the electron density profile show where local maxima in electron density produce the ionospheric layers D-F, whose characteristics are described in the next section. The sources of the bulk of the electrons for the layers are on the left. *(Courtesy of the Air Force Weather Agency)*

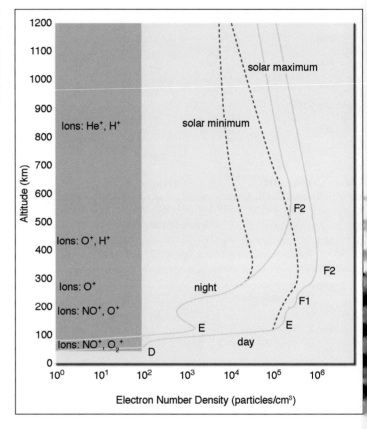

Solar photons with λ = 10 nm–125 nm produce the dayside ionospheres of most planets. As we note in Chap. 3, the formation of an ionosphere requires two basic conditions: (1) a neutral atmosphere, and (2) a source of ionization for the atmospheric gases. Two processes create ionospheric plasma: photo-ionization and impact by energetic particles. To ionize, the energy of the photon or particle must exceed the ionization potential of the atom or molecule. Table 8-1 summarizes the ionization energy for several ionospheric constituents.

Table 8-1. Various Ionization Energies. The energies on the right are for singly ionizing the species on the left.

Neutrals	Ions	Ionization Energy
N_2	N_2^+ + e	(15.5 eV)
O	O^+ + e	(13.6 eV)
O_2	O_2^+ + e	(12.1 eV)
NO	NO^+ + e	(9.3 eV)

Pause for Inquiry 8-6

What is the longest wavelength of light that will ionize molecular nitrogen?

Impact Ionization by Energetic Particles. Energetic charged particles created in the magnetosphere flow into the high-latitude upper atmosphere along dipole-like magnetic field lines. Some of them strike particles in the neutral atmosphere. Above about 100 km, oxygen and nitrogen atoms are the primary targets. In a collision, kinetic energy lost by the impinging particle causes dissociation, ionization, heating, bremsstrahlung, and electronic excitation. Ionization energy for upper atmospheric particles is in the range of a few eV. The energetic particles are mostly electrons, but protons also deposit energy.

Figure 8-11 gives the ionization profiles for several electron energies. The profile shape is similar to that of the Chapman profile we discuss in Chap. 3. The shapes are similar because the same limiting factors control the production rate: availability of ionizing particles and availability of material to be ionized. Clearly, higher energy particles penetrate deeper into the atmosphere. Ionizing a single particle takes only a few eV of energy, but the particles in Fig. 8-11 have keV energy levels. A particle with only the minimum energy needed for one ionization is stopped high in the atmosphere. To effectively ionize, an impinging particle needs lots of energy to distribute to many particles.

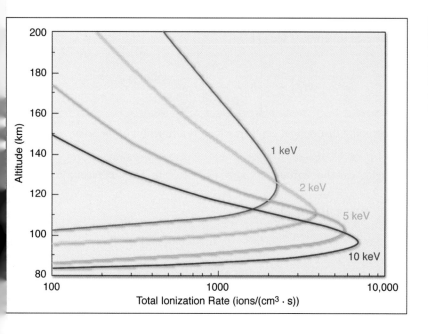

Fig. 8-11. Model Ionization Profiles for Energetic Electrons. Electrons with relatively low energy (~1 keV) produce maximum ionization near 120 km, while 10 keV electrons cause maximum ionization below 100 km. Most ionization events create secondary ionizations. *(Courtesy of Stan Solomon at the National Center for Atmospheric Research)*

Off-Zenith Photoionization. In Sec. 3.4, we use the ideas of cross section and optical depth to describe how the light from the overhead Sun forms a layer of ionization. Equation (3-11a) indicates that for a single-species, plane-parallel atmosphere, at any particular wavelength and altitude z, the ionization rate $R(z)$ is a product of radiation intensity I_z, ionization cross section σ, and neutral density $n(z)$.

Reproduced below is the ion production rate equation, Eq. (3-11b). It accounts for the effects of the exponential variation of the neutral atmosphere and for the exponential reduction of incoming radiant intensity from an overhead Sun. The production, or photoionization rate is a function of three terms: cross section, density, and radiant flux:

$$R(z) = \sigma n_0 I_0 = \sigma \left[n_0 e^{\left(-\frac{\Delta z}{H}\right)} \right] \left[I_0 e^{(-\sigma n_0 H) e^{\left(-\frac{\Delta z}{H}\right)}} \right]$$

where

$R(z)$ = ionization rate as a function of height [ion/s]
Δz = distance [m]
σ = absorption cross section [m^2]
n_0 = neutral particle density at a reference height [particles/m^3]
I_0 = radiation intensity at reference height, usually the top of the atmosphere [W/m^2]
H = scale height [m]
$\sigma n_0 H$ = optical depth [unitless]

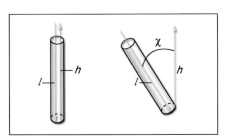

Fig. 8-12. Geometry for Zenith and Off-axis Line-of-site Views. The upward vertical arrow represents the zenith. The downward arrow is a measurement path. Here $l = h\cos(\chi)$. Corrections must be made to models and observations that deal with off-zenith axis geometries.

We include the effects of the longer travel path for the off-zenith sunlight (Fig. 8-12) by multiplying the optical depth term in the radiation intensity exponent by sec (χ). This modification gives a electron production equation that can be written in terms of sec (χ).

$$R(z) = \sigma \left[n_0 e^{\left(-\frac{\Delta z}{H}\right)} \right] \left[I_0 e^{(-\sigma H \sec(\chi) n_0) e^{-\frac{\Delta z}{H}}} \right] \quad (8\text{-}2)$$

Recognizing that optical depth τ, which is a function of the solar zenith angle and height, can be written as $\tau_z = (\sigma H \sec(\chi) n_z) = \left(\sigma H \sec(\chi) n_0 e^{\left(-\frac{\Delta z}{H}\right)} \right)$,

allows us to simplify Eq. (8-2) to:

$$R(z) = \sigma I_0 n_0 e^{\left(-\frac{\Delta z}{H} - \tau_z\right)} \quad (8\text{-}3)$$

Equation (8-3) describes the variation of ion (or electron) production with height in a manner that accounts for solar zenith angle, exponentially decreasing atmospheric density with height, and the exponential reduction of radiant intensity with path length.

We determine where the production is greatest by differentiating Eq. (8-3) and setting the result equal to zero.

Differentiating results in

$$\frac{dR(z)}{dz} = \sigma I_0 n_0 \left(-1 + (\sigma H \sec(\chi) n_0) e^{\left(-\frac{\Delta z}{H}\right)} \right) e^{\left(-\frac{\Delta z}{H} + (\sigma H \sec(\chi) n_0) e^{\left(-\frac{\Delta z}{H}\right)} \right)} = 0 \quad (8\text{-}4a)$$

The intensity term on the near right hand side of the equation and the exponential term on the far right side of the equation are positive terms. Only the middle term can equal zero. Therefore, to satisfy Eq. (8-4a):

$$(\sigma H \sec(\chi) n_0) e^{\left(-\frac{\Delta z}{H}\right)} = 1 \qquad (8\text{-}4b)$$

Substituting for the neutral density variation with height $n_z = n_0 e^{-\frac{\Delta z}{H}}$ gives:

$$\sigma H \sec(\chi) n_z = 1 \qquad (8\text{-}4c)$$

The left side is the optical depth τ_z.

The production function for ions (or electrons) is a maximum when optical depth $\tau_z = 1$. On either side of the maximum the production rate decreases exponentially (Figs. 3-48 and 8-12). The local maxima in the curve in Fig. 8-10 correspond to local ionization maxima created by ionization of the species listed on the left of the figure. Because of the low recombination rate at high altitudes, the equilibrium ionization values are rather high above the F2 peak.

Pause for Inquiry 8-7

Suppose that the Sun strikes the atmosphere at an angle of 30° off zenith. The peak of the ionization layer occurs

a) At lower density and greater heights
b) At lower density and lower heights
c) At higher density and greater heights
d) At higher density and lower heights

Figure 8-13 shows the height of peak radiant energy absorption (optical depth becomes unity) as a function of wavelength. Much of the EUV radiation is absorbed in the thermosphere, where it ionizes oxygen and nitrogen. One notable exception is the radiation associated with hydrogen Lyman-α. Some of this energy at 121 nm passes through the thermosphere and is deposited in the mesosphere to form the lowest region of the ionosphere. Longer wavelengths penetrate to the stratosphere, where they are absorbed by ozone.

Pause for Inquiry 8-8

Figure 8-13 gives the height of the $\tau_z = 1$ surface in the atmosphere for an overhead Sun. What happens to the locations of the $\tau_z = 1$ surface for off-zenith sunlight?

Ionization Loss. Equations (8-2) and (8-3) provide information about the production of ionization, but we also need to know about the loss of ionization. The ionosphere has two principal ways to remove ionization: recombination and attachment. The recombination rate $L_{recombination}$ varies with the presence of electrons and ions available for recombination and with the recombination coefficient α

Fig. 8-13.   Plot of Unitary Optical Depth as a Function of Wavelength. As photons transit the ionosphere, they are absorbed by ionospheric constituents. At the height of the curve, the radiation intensity has been reduced by 1/e. Photons at longer wavelengths dissociate rather than ionize atmospheric constituents. [Chamberlain, 1978]

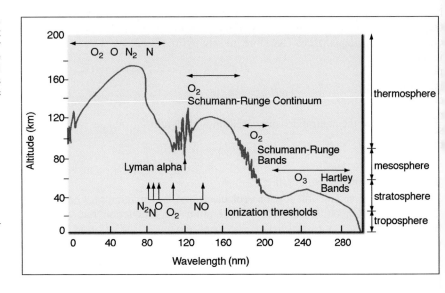

$$L_{recombination} = \alpha n_e n_i \approx \alpha n_e^2 \quad (8\text{-}5a)$$

where

α = recombination coefficient [m^3/s]
n_e = electron density [#/m^3]
n_i = ion density [#/m^3]

Near 100 km, the recombination coefficient is ~10^{-13} m^3/s

The dominant loss process at high altitudes is recombination. Attachment allows electrons to form a bond with a heavy neutral particle, forming a negative ion. The attachment rate L_{attach} varies with the presence of electrons and with the attachment coefficient β as:

$$L_{attach} = \beta n_e n \quad (8\text{-}5b)$$

where

β = attachment coefficient [m^3/s]
n_e = electron density
n = neutral density [#/m^3]

Near 85 km, the attachment coefficient for molecular oxygen is ~10^{-12} m^3/s

In general, attachment dominates at low altitudes, where ion densities are small compared to neutrals. When we account for loss processes, we arrive at the average behavior of the ionosphere's electron density distribution in Fig. 8-10.

Figure 8-14 provides a notional overview of the processes that produce interactions with upper atmospheric constituents. Ionization and dissociation in the high- and mid-latitudes result primarily from photon interaction or energetic electron impact (protons may also impact). The resulting atmospheric products may be excited and emit auroral light or be ionized to have further chemical interactions. The products may also recombine to create new products such as

NO or NO_x. These may further interact with ionizing photons or particles, or if they remain in an excited state, may later decay via less energetic emissions—airglow. Airglow from various forms of NO is one of the most important thermospheric cooling mechanisms. The glow is mostly in the infrared portion of the spectrum. After large storms, the upper atmosphere glows for hours and in some cases days. The figure provides only a conceptual view of the complex upper atmospheric chemistry behind aurora and airglow. In particular, it does not capture the cascade of energy from energetic photons and electrons to secondary (and even tertiary) photons and electrons. In fact the secondaries are the real agents in balancing ionization production and loss.

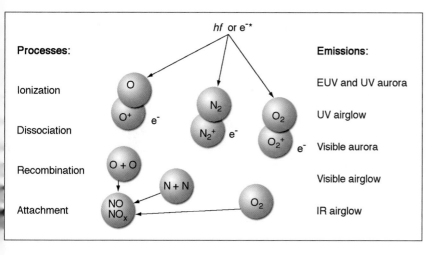

Fig. 8-14. **A Notional Diagram of Ionosphere-Thermosphere Energy Flow from Photons and Energetic Particles.** The flux of energetic photons and energetic electrons (e^{-*}) initiates a cascade of energy and subsequent interactions that lead to non-linear behavior. Ionization occurs throughout the thermosphere but is most likely at high altitudes. Ionization converts photon energy to chemical potential energy. Dissociation also results from a radiant photon or energetic electron impact. Dissociative recombination converts ionization energy to dissociation products and kinetic energy. Recombination is most likely at lower altitudes. *(Courtesy of Stan Solomon at the National Center for Atmospheric Research)*

Total Electron Content. A concept closely related to electron density profile is column total electron content. The *total electron content (TEC)* is the sum (or integral) of ionospheric electrons over a square meter at Earth's surface. If we slice the profiles in Fig. 8-9 into small slabs and add the contents, we have the TEC over a location for which the profiles are representative. The TEC gives a column-integrated level of overall ionospheric density (Eq. (8-6)). Operators use a summary version of TEC called a Total Electron Content Unit (TECU). One TECU equals 10^{16} electrons/m^2. The TEC ranges from 5 TECU–120 TECU. Figure 8-15 depicts the global range of TEC during a geomagnetic storm on April 17, 2002. As expected, the dayside ionosphere has the highest TEC values. We discuss the elongated peaks of TEC near the geomagnetic equator in Sec. 8.2.3.

$$N(z_0) = \int_{z_0}^{\infty} n(z)dz \qquad (8\text{-}6)$$

where

N = TEC [electrons/m^2]

z_0 = base of the ionosphere [m], usually taken to be 90,000 m

$n(z)$ = electron density as a function of altitude [electrons/m^3]

The upper limit of the integral is taken to be where the electron density becomes negligible.

Fig. 8-15. Total Electron Content (TEC) Maps for April 17, 2002. TEC values in the sunlit sector are higher than those in the dark sector. Two crests of ionization bound the dayside equatorial regions. *(Courtesy of NASA/ Jet Propulsion Laboratory)*

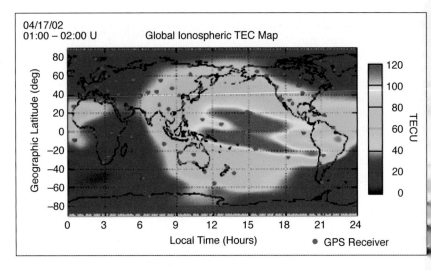

Focus Box 8.2: Total Electron Content and the GPS System

A new sub-discipline of radio science has developed to estimate TEC from satellite observations. The Global Positioning System (GPS) broadcasts timing codes and data on two frequencies, L1 (1575.42 MHz) and L2 (1227.60 MHz). Radio signal delays in the GPS system are related to TEC and are used to estimate it.

Many TEC estimates made by satellites come from a slant path observation over an extended line of site rather than a direct vertical path. So the satellite senses electron content between the ground and its location over a longer path than one associated with a zenith position. These values must be reckoned to vertical values. The GPS system doesn't measure the TEC directly. Rather, it measures the delay in the signal between the transmitter and receiver. Ionospheric effects on GPS measurements depend on the signal frequency and the level of ionospheric activity, with the L1 frequency subject to more variability. More electrons equate to more delay. Increased signal delay corresponds to larger GPS position errors. The maximum vertical delay on GPS L1 measurements is about 15 m, but low elevation angles, which cause the signal to traverse more ionospheric material, increase this by a factor of three. The ionosphere may also cause intermittent signal fading that in severe cases causes loss of availability.

For the typical single-frequency user, each satellite transmits information that represents the global ionosphere in terms of the amplitude of the vertical delay and the period of the model. The information is updated about once a week and provides at least a 50% reduction in the single-frequency user's error. Information about sudden changes in regional ionospheric behavior is not incorporated in these transmissions.

Use of the L2 frequency, and its associated expensive equipment, can provide additional reference information about the ionosphere that helps users eliminate much of the ionospheric error. Some regional systems provide L1 corrections for specific grid points in the service area. The density of the grid points roughly matches the expected spatial variations in the ionospheric vertical delay during periods of high solar activity. The parameters are updated at least every few minutes, and the residual grid point ionospheric vertical delays are expected to be less than 0.5 m.

EXAMPLE 8.2

Electron Density

- *Problem Statement:* An average quiescent value of total electron content (TEC) over the mid-latitudes is 10 TECU. Show that this value is consistent with the electron density profiles in Fig. 8-9.
- *Given:* TEC over the mid-latitudes is 10 TECU.
- *Assumptions:* Quiescent conditions. The ionosphere extends from 90 km–1000 km.
- *Solution:* TEC is related to electron density as

$$N(z_0) = \int_{z_0}^{\infty} n(z)\,dz = n_{avg}\Delta z$$

We find the average electron number density by dividing the column density by the ionospheric depth.

$$n = \frac{N}{\Delta z} = \frac{10 \times 10^{16}\,\text{electrons/m}^2}{(1000 \times 10^3\,\text{m} - 90 \times 10^3\,\text{m})} = 1.1 \times 10^{11}\,\text{electrons/m}^3$$

- *Interpretation:* After converting the values in Fig. 8-9 from electrons/cm^3 to electrons/m^3, we show that an average electron density of ~1×10^{11} electrons/m^3 is a reasonable electron density estimate for an average ionospheric column.

Follow-on Thought Question: Measurements of electron density from the ground are possible only from the bottom to the maximum of the density profile. Some satellite measurements are possible only from the satellite position down to the maximum of the density profile. If you had to make a choice, which measurement would be preferable? Why?

EXAMPLE 8.3

Scale Height in an Ionized Atmosphere

- *Background:* Fig. 8-9 shows that the electron density profiles in the ionized atmosphere look very different from the neutral density profiles at low altitudes. However, above 400 km, the plasma and neutral profiles become more aligned. These observations suggest that the high-altitude plasma density may be characterized by a scale height with a few adjustments.
- *Problem Statement:* Determine a characteristic plasma scale height in the F2 layer.
- *Relevant Concepts:* scale height = $H = \frac{k_B T}{mg}$; in a plasma, ion and electron temperatures, T_i and T_e, often differ.
- *Given:* Altitude = 400 km, T_i = 1200 K, T_e = 3500 K
- *Assume:* The controlling ionospheric species at 400 km is O$^+$ ($m = 2.68 \times 10^{-26}$ kg).

- *Solution:* First determine g at 400 km

$$\frac{6.67 \times 10^{-11}\,\text{Nm}^2/\text{kg}^2 (5.97 \times 10^{24}\,\text{kg})}{(6780 \times 10^3\,\text{m})^2} = 8.66\,\text{m/s}^2$$

Now calculate the plasma scale height

$$H_p = \frac{k_B T_p}{mg} = \frac{k_B (T_i + T_e)}{mg}$$

$$H_p = \frac{\left(\frac{1.38 \times 10^{-23}\,\text{J}}{\text{K}}\right)(1200 + 3500)\,\text{K}}{2.68 \times 10^{-26}\,\text{kg}\left(\frac{8.66\,\text{m}}{\text{s}^2}\right)} \approx 280\,\text{km}$$

- *Interpretation:* The plasma scale height is larger than the neutral scale height at the same location. The light electrons are hotter and more mobile. Even if the ions and electrons have the same temperature, the plasma scale height is twice that of neutrals with the same temperature. Because electrons add to the pressure, they increase the scale height.

Follow-on Exercise: Compare the scale heights for molecular and atomic neutral and ionized oxygen and nitrogen during solar minimum and solar maximum. Make the simplifying assumption that the electrons, ions, and neutrals all have the same temperature.

8.2.3 The Ionosphere Characterized by Latitude and Layers

Objectives: After reading this section, you should be able to...

- Distinguish between features of the low- and high-latitude ionosphere
- Know which regions of the ionosphere are connected most directly to the magnetosphere
- Compare and contrast levels of ionization in the ionospheric layers and give the approximate locations of the ionospheric layer maxima
- Know which layers are the most persistent
- Explain the mid-latitude trough and equatorial anomaly

Altitude

We identify ionospheric regions by latitude and by altitude. In this section we overview the polar-, auroral-, mid-, and low-latitude ionosphere. We refer to these regions as layers, but their structure is far from simple layers.

Ionospheric Layers in the Mid-latitudes. In Fig. 8-10, we identify ionospheric layers by letter. The *D layer* is the lowest layer of the ionosphere. Its base is between 70 km and 90 km, depending on solar activity, but fluxes of solar energetic protons may create ionization as low as 50 km. X-ray radiation ($\lambda < 1.0$ nm) is responsible for ionizing all atmospheric gases, while the solar ultraviolet line at 121.6 nm (hydrogen Lyman-α emission) ionizes oxygen and nitrogen. Because the D layer is mostly solar-controlled, its influence is strongest near local noon and during the summer months when the Sun angle is high. When the Sun sets, the D layer vanishes rapidly as electrons and ions quickly recombine. The D layer is home to most of the molecular oxygen ions as well as most of the negative ions in the ionosphere. Negative ions form when extra electrons temporarily attach to atoms or molecules. The D layer is often slightly stronger during solar maximum years.

In the lower *E layer*, between 90 km–140 km, X-ray and extreme-ultraviolet (EUV) (0.8 nm $< \lambda <$ 10.4 nm) radiation ionizes O and O_2. The E layer is solar-controlled and is strongest during the local noon hours and during the summer months. Although it often disappears after midnight, it dissipates more slowly than the D layer because the lower collision rate delays the recombination. The E layer densities often double during solar maximum years.

In the *F layer* (140 km–500 km), solar X-ray and EUV radiation ionize oxygen atoms and nitrogen molecules. Electrons reach their highest concentrations in the F layer. At night, there is usually just a single F layer. During the day, a low altitude (140 km–200 km) enhancement of ionization often appears, called the *F1 layer*. It contains a relatively high concentration of nitric oxide ions (NO+). The F1 layer dissipates after sundown, though pockets of increased electron densities persist. These are, at best, a nuisance to radio communication. At worst, radio signals are severely deviated or even absorbed.

The E and F1 layers dissipate because of higher collision rates and because they contain molecular species that are more likely to recombine with electrons than are atomic species. Atoms and electrons have difficulty reaching the proper balance of energy and momentum that allows for recombination. In the ionosphere, the probability for electron-molecule recombination is about 10^5 greater than the probability for electron-atom recombination. This difference partially explains the longevity of the F2 layer, which we discuss next.

The *F2 layer* extends above 200 km, where X-ray and EUV radiation with wavelengths less than 80 nm readily ionize atomic oxygen (O), the dominant constituent. The relatively low collision rate and low recombination rate for electrons with atomic species in the F2 layer mean that this layer persists at night. Only the long dark winters near the geographic poles provide sufficient time for recombination of some of the electrons and ions in the F2 layer. The F2 layer is slightly stronger during the winter months between the geomagnetic latitudes of 45° and 55°, especially during high geomagnetic activity. The transport of E and F layer plasma from the summer hemisphere causes this anomaly, where the Sun strengthens the ionosphere in these layers.

The *topside ionosphere* starts at the height of the maximum density of the F2 layer and continues upward to a transition height where O+ ions become less numerous than H+ and He+. The transition height varies, but seldom drops below 500 km at night or 800 km in the daytime, although it may reach as high as 1100 km. The topside ionosphere co-exists with the geocorona.

Solar radiation is not the only mid-latitude ionization source. Cosmic rays, solar energetic particles, and radiation belt particles are responsible for ionization that may go to very low levels, even less than 50 km. Between 85 km and 105 km, meteors provide ionization and ionization trails. Meteors also introduce material to the ionosphere that is subsequently ionized. Section 14.2 discusses meteor effects in detail. Figure 8-16 illustrates the depth of penetration of various types of energetic particles.

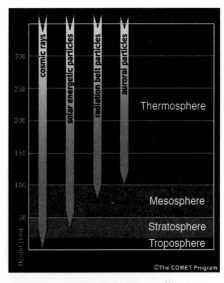

Fig. 8-16. Penetration Depth for Various Forms of Energetic Particles. The upper atmosphere interacts with all forms of energetic particles. The most energetic ones (cosmic rays) penetrate deeper into the atmosphere than the others. *(Courtesy of the COMET Program at the University Corporation for Atmospheric Research)*

Latitude

Polar Ionosphere. The polar ionosphere is extremely variable, from a state of full illumination to total darkness, depending on the season. The polar cap ionosphere (~75°–90° N and S) often connects directly to the solar wind, as we describe in Chap. 7. The dark region surrounded by the auroral oval in Fig. 8-17 corresponds to this region. Solar wind particles impact neutral atoms and molecules and ionize them. In the polar ionosphere, plasma density changes because of variations in the solar EUV flux, precipitating particle ionization, and plasma motion from lower latitudes.

The region inside the auroral oval, near the magnetic poles, experiences seasonal extremes in electron density. Ionospheric layers are fairly strong and regular in the summer months; in the winter, the lack of solar radiation in the lower thermosphere creates the *polar hole*, a region of very low electron density. The uppermost ionospheric layer may persist because it is exposed longer to solar radiation at high altitudes and recombination is reduced. Occasionally particle streams entering through the polar cusps are energetic enough to enhance the polar ionosphere even when it is in darkness. On rare occasions, solar wind field-aligned beams of energetic particles have direct access to the polar cap. These produce excess ionization and cause the usually dark polar caps to glow in X-ray emissions.

Under ordinary conditions, an outflow of plasma, called the *polar wind*, comes from the polar regions. Some satellites interact with this wind when crossing the polar cap at altitudes from 1000 km–10,000 km. The flux varies with time by more than an order of magnitude. In addition, localized regions of higher polar wind density are sometimes present in the polar cap. These enhancements suggest a relationship between the polar wind and disturbed conditions in the ionosphere below or in the magnetosphere above.

Auroral and Mid-latitudes. The most prominent feature of the high-latitude ionosphere is the aurora, illustrated in Figs. 8-17a and b. The oval-shaped region expands and contracts as the magnetosphere's energy state varies, as we describe in Example 4.5. The *auroras* result from magnetosphere-ionosphere interactions. Particles energized in magnetic merging events in the magnetotail are further energized within a few Earth radii by voltage variations along the magnetic field lines. When they arrive at the upper reaches of the ionosphere, the particles have energies in excess of a few keV. The auroras appear at high latitudes—typically near 60°–65° (N and S) magnetic latitude—in an oval-shaped distribution, centered on the magnetic poles. The particles mostly responsible for auroras are electrons, but during particularly energetic events, protons may provide a large fraction of local auroral energy.

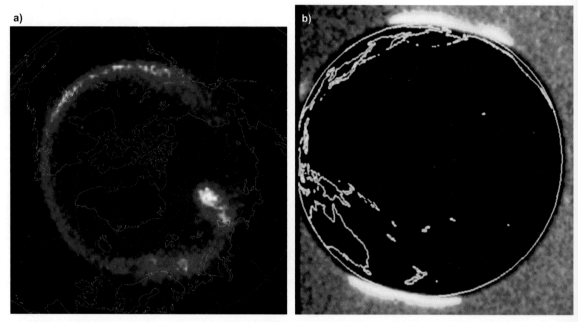

Fig. 8-17. **a) The Auroral Oval Created by Precipitating Protons.** The false-color image displays the northern auroral oval observed with the IMAGE Far Ultraviolet Spectrographic Imager. These auroral emissions are from precipitating energetic protons. The bright spot near the geographic pole is the footpoint of the magnetospheric cusp, which is located on Earth's day side near local noon. *(Courtesy of NASA's IMAGE Program and the Southwest Research Institute)* **b) The Aurora in Both Hemispheres.** This false color image at ultraviolet wavelengths is overlaid with a coastline map to show the auroral ovals in the polar regions. These auroral emissions arise predominantly from atomic oxygen at about 130.4 nm and 135.6 nm and from the Lyman-Birge-Hopfield bands of molecular nitrogen. The Dynamics Explorer-1 satellite took the image over the Pacific Ocean. *(Courtesy of Louis Frank at the University of Iowa and NASA)*

Most auroral light comes from impact excitation and most auroral conductivity comes from impact ionization. Bright, visible auroral displays happen most often near local midnight. Because the *auroral zones* are characterized by excess ionization, strong jets of currents (electrojets), found in the equatorial and high-latitude ionosphere, lead to plasma instabilities that radars sense as variations in Doppler spectra. At high latitudes, auroral electrojets produce "radar aurora" that clutter and may confuse the interpretation of radar returns from over-the-horizon scanning radars. During undisturbed periods, a steady stream of low-energy electrons precipitates into the oval, creating what is called the *diffuse aurora*. This influx keeps the ionospheric layers from decaying as quickly as they otherwise would. In fact, the auroral ionosphere layers often persist through nighttime.

In nature, boundary regions are often the most interesting, and the boundary between the auroral and mid-latitude ionosphere is a case in point. On the low-latitude edge of the nighttime auroral oval, a trough (depletion) in electron densities often occurs in the E and F layers. The *mid-latitude trough* is more pronounced at night time and winter. Ionospheric plasma likely "leaks" upward into the magnetosphere and is not replenished by surrounding ionospheric plasma. This comes from net ion loss caused by the field-aligned currents that increase during geomagnetic substorms.

Observations reveal a December gain in equatorial and mid-latitude ionospheric electron densities related to Earth's perihelion. Midday electron densities between the geographic latitudes of 35° S and 50° N increase because Earth is a little closer to the Sun during this time. The Northern Hemisphere tilts away from the Sun during December. The fact that enhancements are present all the way to 50° N underscores the strong influence of transport from the summer hemisphere to the winter hemisphere, as illustrated in Fig. 8-5.

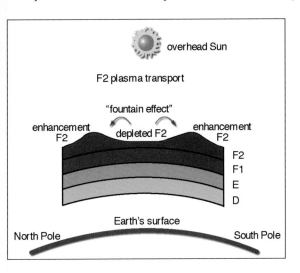

Fig. 8-18. Equatorial Anomaly. Here we see a view of the rising motion ($E \times B$) of the F2 layer in the equatorial ionosphere. The easterly direction is into the page. Dayside solar heating initiates the motion. The charged particles in the F2 layer generate an electric field as they move through the horizontally aligned magnetic field at the equator. But the field also constrains the particle motion and builds a pressure gradient that forces the particles to spread horizontally and sink at latitudes below ±15° magnetic. Gravity also contributes to the sinking motion. The sinking motion increases the electron density at these latitudes and creates strong latitudinal gradients throughout the equatorial zone. *(Courtesy of the AF Weather Agency)*

The Equatorial Ionosphere and Its Crests. The *equatorial ionosphere* (~0°–30° N and S) is often a frothy fountain of plasma that generates strong electron density gradients. These gradients significantly disturb some types of communication and navigation signals. When the Sun passes directly overhead, the electron density is a maximum because of the intense solar radiation. Large-scale rising motion of the neutral atmosphere lifts the ionosphere (Sec. 8.1). Near the equator, the magnetic field B is nearly parallel to Earth's surface. Ions and electrons embedded in the neutral atmosphere are forced to move across the magnetic field. An electric field ($E = -v \times B$) results from the rising motion of the charged particles. During daytime, E is directed eastward. (Example 8.5). The B field constrains the uplifted plasma, which is forced to move north and south along B in response to gravity and pressure-gradient forces. The result is a fountain effect. This equatorial plasma fountain depletes plasma along the equator and enhances it just a few degrees north and south. Figure 8-18 depicts this phenomenon. Ionization maxima form on either side of the equator—the so-called equatorial anomaly (equatorial ionization crests). The ionization crests are at approximately ±15° from the magnetic equator. They stretch in longitude, as shown in Figs. 8-15 and 8-19.

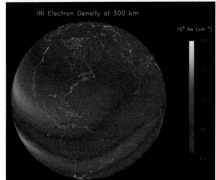

Fig. 8-19. Electron Density over the Globe. In this image, created from the International Reference Ionosphere Model, the dayside is on the left. High electron density appears in the subsolar region. The fountain effect lifts ionization to higher altitudes on either side of the equator. The longer recombination times at higher altitudes allow the crests of ionization to persist into the nighttime. GPS signals passing through the crests or through electron density gradients on either side are often scattered or scintillated by the uneven distribution of electron density. The magnetic equator dips south of the geographic equator in the south-Atlantic region. This dip explains the slight southerly shift of the crest in the central part of the image. *(Courtesy of Stan Solomon at the National Center for Atmospheric Research)*

Focus Box 8.3: Auroral Colors

The aurora often displays a striking green color, but may also be pink, blue, and purple. When the energetic particles from the magnetosphere impact atmospheric constituents, they excite atoms and molecules, producing light. Different materials emit different colors when they are excited. Atomic oxygen at high altitudes (about 300 km) produces red light at 630 nm. The emission occurs only at high altitudes, because these atoms are de-excited by collisions with nitrogen molecules below about 180 km. Excited nitrogen interacts with the atomic oxygen, causing a yellow-green emission at a wavelength of 557.7 nm between 180 km and 120 km. The auroral green line disappears below 100 km, where little atomic oxygen exists. Very energetic particles piercing to low altitudes trigger the magenta and crimson emissions of molecular nitrogen at the lower edge of rippled auroral curtains. Sometimes waves of green light appear to chase the magenta light, an effect caused by the time lag between excited oxygen and nitrogen. Oxygen atoms persist in their unstable state for about one second before decaying, while the nitrogen molecule emissions are nearly instantaneous. Figure 8-20 is an artist's rendition of these emissions. Figure 8-21 is a visible image of aurora curtains viewed from the side. The astronaut who snapped the picture also captured glow emissions from the Space Shuttle.

Occasionally a statement appears claiming that the auroral particles come directly from the Sun. In general this is not true. A few solar wind particles do access the narrow cusp ionosphere directly, or rarely the open polar cap. However, emissions from these particles are not in the visible range. The auroral emissions that we can see originate in the magnetosphere.

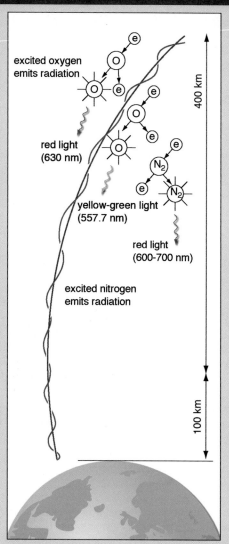

Fig. 8-20. Auroral Emissions Created by Energetic Electrons from the Magnetosphere. In this artist's rendition, electrons spiral along the magnetic field lines before colliding with atmospheric neutrals. In the figure, the starburst pattern indicates an excited state. *(Courtesy of NOAA's Space Weather Prediction Center)*

Fig. 8-21. The Space Shuttle and the Aurora. Atomic oxygen is responsible for the red light above 180 km and the blue-green light below 180 km. Auroral curtains outline the paths of incoming electrons. *(Courtesy of NASA)*

8.3 Radio Wave Propagation in the Ionosphere

8.3.1 Operational Aspects of High-frequency (HF) Radio Propagation

Objectives: After reading this section, you should be able to...

- Explain the interaction between radio waves and electrons in the ionosphere
- Distinguish between the maximum and lowest usable HF frequencies
- Explain the HF radio propagation window
- Understand why the HF propagation window varies with solar illumination

Now we turn to an important ionospheric effect on technology. Actually, the ionosphere enables a technology we all take for granted—radio propagation. Shortly after radio became a communication medium, anomalies began to appear in the effective ranges of radio equipment. Occasionally, transmissions were received at distances that far exceeded what appeared to be the maximum equipment range. These peculiar effects were especially noticeable at night. Researchers finally concluded that a layer of ionization existed at high altitude in the atmosphere, and that radio waves reflected back to Earth at points far from their origin. The development of radio science stirred an interest in the ionosphere, long before humans or spacecraft passed through the region.

When an electromagnetic wave propagates through the ionosphere, the wave's oscillating electric field causes the electrons to oscillate with respect to the much heavier, and therefore relatively stationary, ions. The interaction changes the wave direction. As a result, radio waves refract (depending on their frequency) from the ionosphere, as we describe in more detail in the sections below. Because of its reflective properties at sufficiently low radio frequencies, the ionosphere behaves like a "mirror in the sky" and is used routinely for short-wave communications, over-the-horizon radar (OTHR), and radio navigational systems. In general, the influence of the ionosphere on wave propagation decreases with increasing frequency. Waves with frequencies greater than the plasma frequency pass through the ionosphere to reach satellites. High-frequency (3 MHz–30 MHz) radio waves reflect off of the F layer or the daytime E layer, while very-low-frequency (less than 30 kHz) radio waves reflect off of the D layer.

When a radio wave travels through the ionosphere, the electrons oscillate at the wave's frequency. Some of the oscillation energy leaves the wave when electron collisions with neutrals convert this energy to random kinetic energy (heat). Consequently, the radio wave attenuates. The higher the electron collision rate with neutral molecules, the greater the attenuation. Hence, most attenuation takes place in the D layer, where the neutral molecules' density and the electrons' collisional frequency is greater than in the upper regions.

Sky wave radio propagation uses the ionosphere to provide a signal path between two antennas. Radio waves refract away from regions of higher electron density, as Fig. 8-22 illustrates. Vertical gradients in electron density refract radio waves toward and away from Earth. Depending on their frequency and angle, radio waves return to Earth. We call these conditions a *hop* or a *skip transmission*. The radio waves reflect from the surface and repeat the process in multi-hopping. Waves may also become trapped between ionospheric layers—a phenomenon called *ionospheric ducting*. If a radio wave doesn't bend enough as it travels upward, so that it penetrates through the ionosphere, it becomes a *trans-ionospheric wave*.

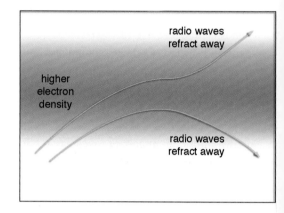

Fig. 8-22. Ionospheric Refraction of Radio Waves. The geometry of skywave propagation depends on the radio wave frequency and the ionospheric electron density profile.

At D layer altitudes a large number of neutral air atoms and molecules coexist with the ionized particles. As a passing radio wave causes the ions and free electrons to oscillate, they collide with the neutral air particles, and the oscillatory motion dampens and converts to heat. Thus the D layer absorbs passing radio wave signals. The lower the frequency, the greater the signal absorption. The *lowest usable frequency (LUF)* is that frequency below which the ionosphere absorbs too much of the radio signal to permit it to pass through the D layer. Normally the LUF lies in the lower portion of the HF band. The part of the ionosphere with the most ionization is the F layer (200 km–400 km altitude). Free electrons in the F layer cause radio waves to refract (bend), but the higher the frequency, the less the bending. The *maximum usable frequency (MUF)* is that frequency above which radio signals encounter too little ionospheric refraction (for a given take-off angle) to bend back toward Earth's surface (i.e., they become trans-ionospheric). Normally the MUF lies in the upper portion of the HF band. The MUF and LUF allow surface-to-surface radio communication at medium-to-high frequencies (300 kHz to 30 MHz). Satellite communication (SATCOM) operators use very high to extremely high frequencies (30 MHz to 300 GHz) to penetrate the ionosphere.

The *HF radio propagation window* is the range of frequencies between a LUF (complete D-layer signal absorption) and a MUF (insufficient F-layer refraction to bend back the signal). This window varies by location, time of day, season, and the level of solar and geomagnetic activity. HF operators choose propagation frequencies within this window so their signals pass through the ionosphere's D layer and subsequently refract from the F layer. As seen in Fig. 8-23, typical LUF-MUF curves exhibit a normal, daily variation. During early afternoon, photo-ionizing solar radiation (some X rays and EUV, but mostly FUV) is at a maximum, so the D and F layers are strong and the LUF and MUF are elevated.

During the night, the loss of ionizing sunlight reduces the electron density in all ionospheric layers (some layers disappear altogether), and the LUF and MUF become depressed. Figure 8-23 shows that communication at 6 MHz, for example, should only succeed from sunset to a couple of hours past sunrise. Conversely, communication at 10 MHz should succeed during the day between 0800 and 1700 local solar time. No one frequency works well during both night and day. Consequently, many HF broadcasting systems publish at least two transmission frequencies—the lower frequencies for night transmissions and higher frequencies for day transmissions.

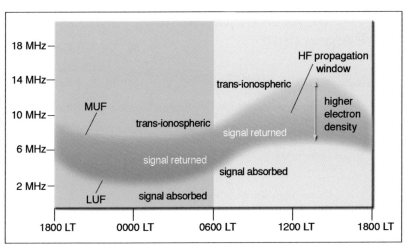

Fig. 8-23. High-frequency Radio Propagation Window. The propagation band varies with time of day, mostly because of the solar radiation absorbed in the ionosphere. Radio propagation is possible at frequencies between the LUF and MUF. Signals with frequencies above the MUF exit the ionosphere. Signals with frequencies below the LUF are absorbed and do not propagate. (MUF is maximum usable frequency; LUF is lowest usable frequency.) *(Courtesy of the Air Force Weather Agency)*

Pause for Inquiry 8-9

Below which ionospheric level does most surface radio communication take place?

Pause for Inquiry 8-10

Assuming that Fig. 8-23 represents solar minimum years, what happens to the curves and the HF window during solar maximum?

Pause for Inquiry 8-11

Consider Fig. 8-23. Imagine that you are operating communication equipment on the dayshift. You've been notified of a large solar flare. Should you shift your frequency up or down? What may happen to the propagation window?

> **Pause for Inquiry 8-12**
>
> Why does the MUF rise after noon?
>
> a) Because the D-layer height rises
> b) Because the F-layer height rises
> c) Because the electron density in the D layer rises
> d) Because the electron density in the F layer rises

8.3.2 The Physics of Ionospheric Radio Wave Propagation

> **Objectives: After reading this section, you should be able to...**
>
> ♦ Calculate the frequency at which a vertical radio wave will refract so sharply that it reflects
> ♦ Explain how a continuously varying refractive index allows reflection of an oblique radio wave

To understand how radio waves refract in the ionosphere, we need to develop a few concepts: refractive index, Snell's refraction law, characteristic frequency, and critical frequency.

As we note in Chap. 6, waves progress in two ways: phase speed or group speed. *Group speed* is the speed with which the overall shape or envelope of the wave's amplitude moves through space. Group speed describes how information or energy propagates through the medium. *Phase speed* is the speed at which the crest or trough, or other marker, propagates. The phase speed of electromagnetic radiation can be greater than light speed in a vacuum, but this doesn't mean that energy transfers at greater than light speed. Our interest here is in phase speed.

Refractive Index. We define a medium's *refractive index n* as the ratio of the speed c of an electromagnetic wave in a vacuum to the phase speed v_n of the wave in the medium

$$n = c/v_n \tag{8-7}$$

where

n = refractive index of a medium, such as air [unitless]
c = speed of electromagnetic waves in a vacuum (2.998×10^8 m/s)
v_n = phase speed of electromagnetic waves in a medium [m/s]

In most substances, $n > 1$, and varies with frequency. It's larger for violet light (high frequency) and smaller for red light (low frequency). In a plasma, the refractive index (neglecting the geomagnetic field) is less than one, and is given by

$$n = \sqrt{1 - \frac{N_e e^2}{\varepsilon_0 m_e \omega^2}} \approx \sqrt{1 - \frac{80.6 N_e}{f^2}} \tag{8-8}$$

8.3 Radio Wave Propagation in the Ionosphere

where

N_e = electron density [electrons/m^3]
m_e = electron mass (9.11 × 10^{-31} kg)
e = electron charge (1.6 × 10^{-19} C)
ε_0 = permittivity of free space (8.85 × 10^{-12} C^2/(N·m^2))
ω = angular frequency [rad/s]
f = wave frequency [Hz]

(In this section, we deviate from the usual practice of using the lower case "n" for electron density, because "n" here designates the refractive index.) Equation (8-8) gives a characteristic frequency that determines the boundary between a real and an imaginary refractive index. In Chap. 6 we define this characteristic frequency, also known as the plasma frequency, by the condition of $n = 0$ or equivalently, the equation

$$f_p = \sqrt{80.6 N_e} \quad \text{or} \quad \omega_p = \sqrt{\frac{N_e e^2}{\varepsilon_0 m_e}} \qquad (8\text{-}9)$$

where

f_p = plasma frequency [Hz]
ω_p = plasma angular frequency [rad/s]

Equation (8-9) presents the natural frequency of oscillation of the electrons around the stationary ions.

Pause for Inquiry 8-13

A vertically transmitted wave originating at the ground encounters

a) An increasing concentration of electrons and a decreasing refractive index
b) An increasing concentration of electrons and an increasing refractive index
c) A decreasing concentration of electrons and a decreasing refractive index
d) A decreasing concentration of electrons and an increasing refractive index

Snell's Law

Snell's Law describes wave behavior during refractive index changes. It tells us how an incident wave refracts (bends) at the interface between two media (Fig. 8-24).

$$n_i \sin \theta_i = n_r \sin \theta_r \qquad (8\text{-}10)$$

where

θ_i = incident angle [rad or deg]
θ_r = refracted angle [rad or deg]

Consider a layered medium in which each layer has a lower refractive index than the one below it, as shown in Fig. 8-25.

Chapter 8 Earth's Quiescent Atmosphere and Its Role in the Space Environment and Weather

Fig. 8-24. Snell's Law for a Single Boundary. The incident wave approaches the boundary between the media at an angle θ_i and departs the boundary at an angle θ_r.

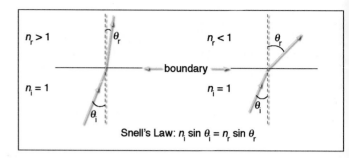

Fig. 8-25. Snell's Law for a Continuously Varying Refractive Index. Here we show how incident radiation arrives at each boundary at some incident angle θ_i, and departs at a refracted angle θ_r. To find the final angle, we apply Snell's Law at each boundary in sequence.

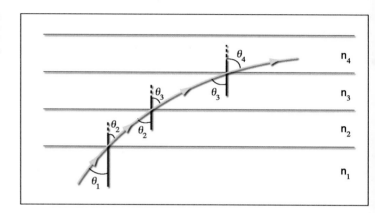

Applying Snell's Law at each boundary we obtain

$$n_1 \sin\theta_1 = n_2 \sin\theta_2 = n_3 \sin\theta_3 \ldots = n_m \sin\theta_m$$

or in general,

$$n \sin\theta = \text{constant} \tag{8-11}$$

We apply this process to the ionosphere, as depicted in Fig. 8-26.

Fig. 8-26. Snell's Law in the Ionosphere. The refractive index varies continuously in the ionosphere and ultimately causes a total reflection of the incident wave.

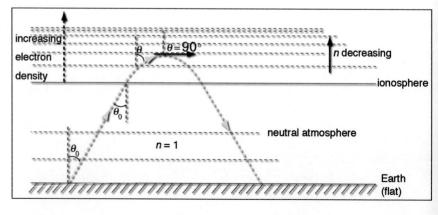

380

Equation (8-11) must hold for all points of the ray's path. Before the wave enters the ionosphere, θ equals θ_0 and $n \approx 1$. As the wave enters the ionosphere, the electron density increases and the refractive index begins to decrease. To satisfy Snell's Law, the angle θ must increase and the ray must bend further and further away from the vertical. For reflection, θ must reach 90° at the top of the trajectory, after refractions:

$$n_m = \sin\theta_0 \quad (8\text{-}12)$$

We calculate what the electron density must be for this to happen, using Eqs. (8-8) and (8-12):

$$n = \sqrt{1 - \frac{N_e e^2}{\varepsilon_0 m_e \omega^2}} = \sqrt{1 - \frac{80.6 N_e}{f^2}}$$

The minimum electron density $N_{e\ min}$ needed to reflect a ray that left the ground at an angle θ_0 is given by

$$\sin\theta_0 = n_m = \sqrt{1 - \frac{80.6 N_{e\ min}}{f^2}}$$

Solving for $N_{e\ min}$, we obtain our answer, using the identity $1 - \sin^2\theta_0 = \cos^2\theta_0$

$$N_{e\ min} = \frac{f^2}{80.6}\cos^2\theta_0 \quad (8\text{-}13)$$

where $N_{e\ min}$ has units of electrons/m^3

The theory associated with Eqs. (8-7) through (8-13) is valid for boundaries and waves of infinite extent and indicates that at the reflection height no wave energy penetrates the medium. In the real world, where waves and boundaries are finite, a small amount of energy does penetrate the boundary. The waves generated on the far side of the boundary, called *evanescent waves*, decay exponentially with distance from the boundary.

The phenomenon we describe above is a form of total internal reflection, which occurs when an electromagnetic wave propagates from a region of higher refractive index to one with a lower refractive index. According to Eq. (8-8), the ionospheric refractive index decreases as the electron density increases, as we illustrate in Fig. 8-26. Figure 8-27 is a visualization of ionospheric refraction of two radio waves in a model atmosphere. The two waves in the image have different frequencies, and bend to return to Earth at different altitudes.

Critical Frequency

As the refractive index approaches zero, the wave normal becomes horizontal, and the wave is reflected and begins its return trip to the ground. At the height of reflection no wave energy penetrates the medium.

Fig. 8-27. Radio Wave Propagation. Here we see an example of high-frequency radio beams at 7.5 MHz and 15 MHz propagating in the polar ionosphere. The cross section shows the ionospheric structure and how low-frequency waves bend back toward Earth as they are refracted by electron density gradients. Different frequencies penetrate to different altitudes in the ionosphere before being reflected. The reflection height, and therefore the transmission distance, depends on the electron density. *(Courtesy of the Arctic Region Supercomputing Center)*

EXAMPLE 8.4

Radio Wave Propagation

- *Problem Statement:* Suppose a radio wave of frequency 4.032 MHz is transmitted from the ground at an elevation angle of 30° and reflects from a height of 220 km. Determine the electron density at the reflection point and estimate how far away the signal will be received.

- *Given:* $f = 4.032$ MHz, height of reflection layer = 220 km, elevation angle = 30°

- *Assume:* A narrow beam, a single reflection, and neglect Earth's curvature

- *Solution:* First determine the electron density associated with the critical frequency:

$$\text{Oblique: } N_{e\ min} = \frac{f^2}{80.6}\cos^2\theta_0$$

If the elevation angle is 30°, then the angle with respect to the vertical, θ_0, = 60°

$\cos(60°) = 0.5$

The electron density is

$N_e = [(0.5)(4.032 \times 10^6 \text{ Hz})]^2/80.6 = 5.04 \times 10^{10}$ electrons/m³

Next determine the wave hop distance. One half of the hop distance equals the length of the side adjacent to the 30° angle.

$\tan(30°) = 220$ km/(half hop distance)

(half hop distance) = 220 km/$\tan(30°)$ = 381 km

The hop distance = 2 × 381 km = 762 km

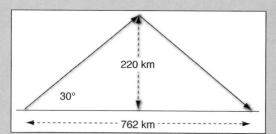

Fig. 8-28. Geometry for Simple Radio Wave Propagation. This simplified, straight-line representation of the problem permits easy calculation and yields reasonable results.

- *Interpretation:* This rather low-frequency wave is reflecting at a location somewhat below the typical height of the F2 layer peak. If it travels this distance along a north-south path it will traverse approximately 7° of latitude (each degree of latitude ~110 km).

In the special case of vertical incidence ($\theta_0 = 0°$), the refractive index n vanishes:

$$n = \sqrt{1 - \frac{80.6 N_{e\,min}}{f^2}} = \sqrt{1 - \frac{f_p^2}{f^2}} = 0 \qquad (8\text{-}14)$$

where
- f = radio frequency
- f_p = plasma frequency

Thus, for vertical incidence, $n = 0$ when the incident wave frequency f equals the plasma frequency f_p. Reflection occurs at a level in the ionosphere where
- $v_n = \infty$, phase velocity becomes infinite
- $v_g = 0$, group velocity = velocity of energy transport is zero

The minimum electron density required for reflection in the vertical is provided by Eq. (8-13), with $\theta_0 = 0°$:

$$\text{Vertical: } N_{e\,min} = \frac{f^2}{80.6}$$

where $N_{e\,min}$ has units of electrons/m^3

A smaller electron density is required for oblique incidence. For a given minimum electron density, Eq. (8-13) allows us to relate the oblique wave frequency f_θ to the vertical reflected frequency f_p by

$$f_\theta = \frac{f_p}{\cos\theta} = f_p \sec\theta \qquad (8\text{-}15)$$

so in general $f_\theta \geq f_p$.

If $N_{e\,max}$ is the maximum (peak) electron density in an ionospheric layer, then the highest frequency that reflects at vertical incidence is the *critical frequency* of the layer and is given by

$$f_c = \sqrt{80.6 N_{e\,max}} \approx 9\sqrt{N_{e\,max}} \qquad (8\text{-}16)$$

where $N_{e\,max}$ is the maximum electron density.

For oblique incidence, the critical frequency for a flat Earth is

$$f_c(\theta) = f_c/\cos\theta = f_c \sec\theta \qquad (8\text{-}17)$$

For large values of θ (approaching 90°) this critical frequency leads to a substantial increase in the number of frequencies available for communications. This formula does not account for Earth's curvature, and further modifications are necessary for large θ.

All of this analysis assumes that the ionosphere has only one critical frequency. The real ionosphere has several critical frequencies, one associated with each layer. At and below each layer, and at the critical frequency, electrons re-radiate the incident wave energy and the radio waves propagate, thus permitting radio communications by reflecting waves. If the operating frequency is above the critical frequency (cut-off frequency), then the wave is not reflected and it penetrates the layer and travels upward. If the operating frequency is above the maximum critical frequency for the F2 layer, the wave escapes into space (perhaps to a satellite).

Figure 8-29 shows information about two types of waves. After a transmitted radio signal begins to penetrate the ionosphere, Earth's magnetic field splits the wave into two characteristic components called the *ordinary wave (o-wave)* and the *extraordinary wave (x-wave)*. For the ordinary wave, the E field accelerates electrons parallel to the magnetic field. The magnetic field has no influence, because a magnetic field only imposes a force on charged particles moving perpendicular to it. For the extraordinary wave, the E field of the incident radiation accelerates the free electrons normal to the magnetic field. The magnetic field exerts a force on the electrons and therefore modifies the motion, causing the refractive index of the extraordinary wave to be different from that of the ordinary wave. Wave components with different velocities have different refractive indices.

Each of these waves travels an independent path through the ionosphere. Further, each propagating signal component contributes to a different mixture of power levels, which together total the power of the signal before it entered the ionosphere. The x-wave is usually the weaker of the two. At higher frequencies, the ordinary and the extraordinary waves often follow very similar paths. At lower frequencies they diverge.

The existence of the ordinary waves and extraordinary waves is evident in soundings of the ionosphere called *ionograms* (Fig. 8-29). The waves have different critical frequencies for the different ionospheric layers. This is most evident for the F2 layer. Both waves also have their own maximum usable frequency (MUF) values. The extraordinary wave MUF is always higher. Nearly all propagation prediction programs account for only the ordinary wave critical frequency when computing the MUF and other output parameters. The actual MUF might be higher than predicted with these programs due to the extraordinary wave refraction properties.

8.3 Radio Wave Propagation in the Ionosphere

Focus Box 8.4: Sounding the Ionosphere

Knowing how radio waves reflect in the ionosphere, we next explore how we might use this information to "sound" the ionosphere. A device called an ionosonde generates a radio pulse that travels upward and reflects back to the device. The travel time is recorded and divided by two to determine the height of reflection. By smoothly varying the radio frequency of subsequent pulses and recording their travel times as a function of frequency, we generate the record of frequency versus height. As the frequency increases, each wave is refracted less by the ionization in the layer, and so each penetrates further before it is reflected. Eventually, a frequency is reached that enables the wave to penetrate the layer without being reflected. For ordinary mode waves, this occurs when the transmitted frequency just exceeds the peak plasma (critical) frequency of the layer. This ionogram (frequency versus height plot) provides information about the altitude of various reflecting layers and their associated electron density. Figure 8-29 shows an ionogram in which we identify the height and critical frequency of the E, F1, and F2 layers. The critical frequencies are identified by the cusps in the ionogram. From these we characterize the ionosphere's bottomside electron density profiles (EDPs).

A satellite with a radio transmitter can make a similar measurement of the ionosphere's topside. A topside sounder is best for determining the ionosphere's structure above the F2 layer. Emissions from topside sounders may produce resonances or disturbances that are useful in locating plasma irregularities in the ionosphere.

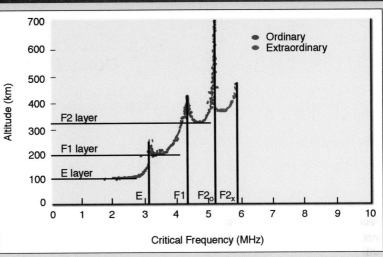

Fig. 8-29. An Ionogram Made Near Local Noon from a Northern European Location. The flattened base of each parabola is the location of the layer peak. So, the base of the E layer is 100 km, and the base of the F2 layer is about 320 km.

EXAMPLE 8.5

Electron Density at Ionosphere Layer Maxima

- *Problem Statement:* Determine the electron density at the F1 and F2 layer peaks in Fig. 8-29.
- *Given:* Figure 8-29 is an ionogram from a vertical incidence sounder.
- *Assumptions:* A horizontally stratified ionosphere
- *Solution:* The critical plasma frequency is f_p (in Hz) = $\sqrt{80.6 N_e}$

$N_e \approx f_p^2 / 80.6$

$N_{eF1} \approx (3.3 \times 10^6)^2 / 80.6 = 1.4 \times 10^{11} /m^3 = 1.4 \times 10^5$ electrons/cm^3

$N_{eF2} \approx (4.7 \times 10^6)^2 / 80.6 = 2.7 \times 10^{11} /m^3 = 2.7 \times 10^5$ electrons/cm^3

- *Interpretation:* These electron densities are slightly larger than those shown in Fig. 8-10, but certainly at the right order of magnitude. However, the data is from a high-latitude station. We expect the layer peak values to be less than those at middle and low latitudes.

Follow-on Exercise 1: Use the E- and F-layer densities to determine the critical frequencies for Fig. 8-10.
Follow-on Exercise 2: What is the wavelength associated with the critical frequency of each layer in Fig. 8-29?

8.4 Ionospheric Interactions with Other Regimes

8.4.1 Quiescent High-latitude Ionospheric Connections to the Magnetosphere

Objectives: After reading this section, you should be able to...

- Describe the current linkages between the high-latitude ionosphere and the magnetosphere
- Distinguish between Hall and Pederson currents
- Distinguish between Poynting flux and particle flux
- Describe how the high-latitude ionosphere reacts to energy input from the magnetosphere

Magnetosphere-Ionosphere Coupling. Roughly 75% of ionospheric energy comes from solar irradiance. The bulk of the remainder comes from processes involved in magnetosphere-ionosphere-atmosphere coupling. *Magnetosphere-ionosphere coupling* refers to the processes that connect the ionospheric plasma with the energized plasmas and mechanisms of the magnetosphere. At the magnetopause and inside the magnetosphere, magnetic field-aligned currents connect to the high-latitude ionosphere. Figure 8-30 illustrates a few of the magnetic field lines that map to the high-latitude ionosphere from distant magnetosphere regions. The fields that drive these currents arise from viscous charge separation (Fig. 1-25 and Sec. 7.4.2) and from solar-wind-induced electric fields (Secs. 4.4.2 and 7.4.2 and Figs. 4-34 and 7-28). The currents pass momentum and energy between the ionosphere and magnetosphere.

The two major mechanisms of energy transfer at the ionosphere are energy delivered by the electromagnetic fields that accompany the field-aligned currents (Poynting flux) and energy delivered by particle precipitation. The energy is dissipated in the ionosphere 1) as Joule (Ohmic) heating produced when field-aligned currents close across equipotential surfaces; 2) by momentum transfer that accelerates neutrals; and 3) by heating and ionization caused by kinetic energy loss from energetic particles.

Field-aligned Currents (FACs). In Chaps. 4 and 7, we describe how the magnetosphere imposes electric fields on the ionosphere. Magnetospheric plasma pressure gradients and magnetic field stresses create local electric fields. The electric fields and their sheet-like, field-aligned currents project along magnetic field lines into the ionosphere. These currents are carried mostly by electrons. Upward currents are associated with precipitating electrons from the magnetosphere. Downward currents are supported by upwelling ionospheric electrons. The vertically flowing currents diverge to horizontal currents as they enter the ionosphere (Figs. 8-30 and 8-31).

The magnetosphere acts as a current generator that separates charged particles, making them available to flow as organized currents, and as a voltage generator that drives the particles through the resistive ionosphere. Four regions participate in field-aligned current formation:

8.4 Ionospheric Interactions with Other Regimes

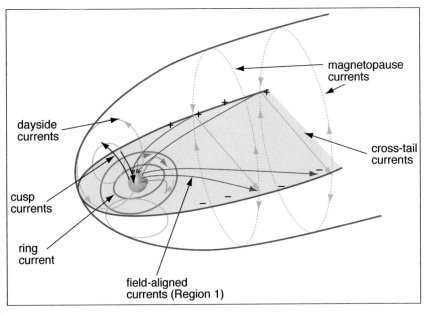

Fig. 8-30. **Ionosphere-Magnetosphere Current Systems.** This is a view of the currents from the distant and inner magnetosphere that connect to and close in the high-latitude ionosphere. The equatorial plane is in yellow. Green lines in the equatorial plane show the cross-tail currents flowing from dawn to dusk. The magnetopause currents are indicated by the dotted green curves. The currents shown in red connect currents and charge at the magnetospheric flanks with the ionosphere. Currents shown in black connect the dayside magnetosphere and cusps to the ionosphere. The ring current connection to the ionosphere (blue) can be intermittent and highly localized. Viscous interactions and induced electric fields drive currents from the dawnside of the magnetosphere down to and across the polar and auroral ionosphere and back out to the duskside.

- The dayside magnetopause at high latitudes produces *cusp currents*. These are the signature of the dayside interaction of the interplanetary magnetic field with the geomagnetic field.

- The equatorial edge of the magnetopause is a region of polarized electric fields. The resulting currents are called *Region-1 currents*. They arise from solar wind plasma and interplanetary magnetic field interaction with the geomagnetic field.

- The plasma sheet supports the divergence of the ring current. The associated currents are called *Region-2 currents*. These highly variable currents result from partial ring current gradients caused by interplanetary magnetic field interaction with the geomagnetic field. Region-2 currents become prominent during magnetic storms.

- The reconnection region in the equatorial plasmasheet intermittently produces surges of currents that flow to the nightside auroral oval. These surges are called *substorm current wedges* and result from substorms that divert some portion of the neutral sheet current through the ionosphere.

Figure 8-31 illustrates a close-in view of the field aligned currents. Region-1 currents are at high latitudes; Region-2 currents are at low latitudes. Figure 8-32 gives a behind-the-Earth perspective.

The ionosphere must have a horizontal current system to close the FAC circuit. The ionospheric closure current, flowing along the electric field, is the *Pederson current* (J_P), which is the horizontal orange sheet in Fig. 8-31 and the current crossing the polar cap in Fig. 8-32. It connects the entering and exiting FACs in the ionosphere. Pederson currents provide an energy sink for the magnetosphere. The closure occurs at a range of heights where the combination of neutral collisions and plasma density supports current flow across ionospheric magnetic field lines. For simplicity, Fig. 8-31 depicts the horizontal closure current as a height-integrated flat sheet.

Chapter 8 Earth's Quiescent Atmosphere and Its Role in the Space Environment and Weather

Fig. 8-31. Current Systems Linking the High-latitude Ionosphere and the Magnetosphere. The ionosphere and magnetosphere are coupled by field-aligned currents. The Sun is off the lower left of the figure. The orange bands indicate Pederson currents aligned with the electric field. The Pederson current system consists of field-aligned currents from the magnetosphere and the closing horizontal currents in the ionosphere. The yellow curves are Hall currents flowing perpendicular to both the electric field and the local magnetic field. The auroral zone, which is not depicted, lies between 65° and 75° latitude. For simplicity the geomagnetic field lines are not shown. *(Courtesy of the COMET Program at the University Corporation for Atmospheric Research)*

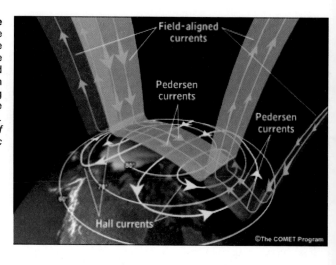

Fig. 8-32. Major Currents Linking the High-latitude Ionosphere to the Magnetosphere. This figure gives the view from above and slightly behind Earth looking down on the North Pole. The white area represents the polar cap. The bright yellow area represents the auroral zone. The pale yellow area represents the neutral sheet. The cusp currents are highly variable, and during active times may have a direction opposite that shown. Region-1 currents flow from the dawn side of the magnetosphere, across the polar cap, as Pederson current, to the duskside of the magnetosphere. Some of the Pederson current diverges into the auroral zone and may close with Region-2 currents. At the magnetopause flanks the Region-1 currents join with and contribute to the magnetopause currents shown in light green (also shown as the dotted green curves in Fig. 8-30). Region-2 currents arise from the divergence of the ring current. The divergence develops from uneven pressure in the ring current caused by enhanced convection. Region-2 currents close in the auroral zone and to a lesser degree across the polar cap. The substorm current wedge is a highly variable current that develops during storm time when the magnetotail does not have sufficient current density to close currents in the neutral sheet. The Southern Hemisphere has a similar current configuration.

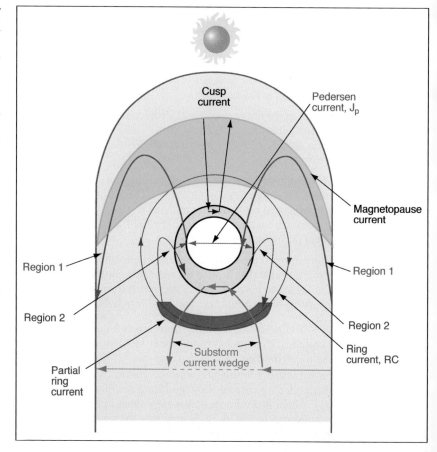

The Lorentz force causes charged particles moving perpendicular to the geomagnetic field lines to deflect. Thus, closure of the ionospheric Pederson current creates an additional current system called the *Hall current (J_H)* system (yellow curves in Fig. 8-31). We determine the direction of the Hall current flow by remembering that the Lorentz force accelerates charged particles in the direction of $J_P \times B$, producing a particle drift consistent with $E \times B$ motion. The resulting plasma flow traces equipotentials of the projected electric field, forming a two-celled pattern with foci in the dawn and dusk high-latitude ionosphere. Because the plasma is collisional, the electrons drift further and faster than the massive ions. A current aligned in the $-E \times B$ direction results. This Hall current creates magnetic signatures at the ground that we use to estimate the degree of magnetosphere-ionosphere coupling.

Pause for Inquiry 8-14

Assume that the electrons are the charge carriers of the FAC. Explain why the Hall currents are Sunward over the central polar cap, given that the electric fields are directed from dawn to dusk (Fig. 7-28b).

Pause for Inquiry 8-15

Assume that the electrons are the charge carriers of the FAC. In Fig. 8-31, in which directions are the electrons moving in the yellow curves on the dawnside and in the yellow curves on the duskside?

The ionosphere is a current generator. In the ionosphere, neutral-gas motions drag charged particles across magnetic field lines. These particles are also affected by a $J \times B$ force and deviate from motion parallel to electric fields. Much of the ionospherically generated current closes within the ionosphere. However, any localized charge buildup resulting from dynamics or conductivity gradients will support field-aligned currents that extend to and close in the magnetosphere. Thus, conditions in the ionosphere may demand that FACs flow into and out of the magnetosphere. These temporal changes are communicated between the regions as Alfvén waves. We recall from Sec. 5.1 that Alfvén waves come from transverse perturbations of the background magnetic field by an induced electric field that causes plasma to oscillate as if tied (frozen in) to the magnetic field. The small oscillations of the local magnetic field lines cause current flow.

A certain element of natural reciprocity exists between the ionosphere and magnetosphere. Closed current loops, whether driven by the magnetosphere, the ionosphere, or both, require a set of consistent conditions at both ends. Both regions must be able to modify electric fields, plasma populations, and conductivity to support the requirements of the other. Polarization electric fields that modify and control current circuits, and magnetic Alfvén waves that transmit the changes along magnetic field lines, are two necessary features of both regions. Electric fields and particle impacts modify the ionospheric conductivity and thus the horizontal current distribution in the ionosphere. This activity assures that the horizontal ionospheric current is consistent with the FAC that is delivering electromagnetic and particle energy from the magnetosphere. If there are insufficient charge carriers at either end, nature acts to build electric field strength until charge flow is balanced and currents can close. Charged particles from the magnetosphere carry FACs that change the ionospheric conductivity that facilitates current closure. The ionosphere, with its large reservoir of charge created by solar photo-ionization, also provides current carriers to the magnetosphere.

Joule Heating Rate (Joule Power). Field-aligned currents (FACs) provide channels for electromagnetic energy exchange between the ionosphere and magnetosphere. The fields and currents produce Joule heating at high latitudes. Electric fields also transfer momentum to the neutral atmosphere. Joule and particle heating change the plasma pressure in the ionosphere and impart upward field-aligned motions to the plasma. The outward ionospheric ion fluxes are further accelerated into the magnetosphere, where they serve as a dominant source of plasma to the inner magnetosphere.

When electric fields force particles through a resistive medium, energy dissipates as heat. This *Joule rate* accounts for the dissipation of roughly 90% of the energy deposited in the upper atmosphere. In its simplest form the heating rate depends on the Pederson conductivity and the square of the electric field.

$$Joule\ power = \sigma_p E^2$$

where

Joule power = energy flux between the ionosphere and magnetosphere [W/m^2]

σ_p = Pederson conductivity [Siemans = 1/Ω]

E = electric field magnitude [N/C]

The hemispherically integrated Joule power provides a measure of the total power dissipation at high latitudes. The Joule dissipation is on the order of 10 GW during quiet times. During extreme geomagnetic disturbances, it exceeds 1000 GW.

Joule heating varies because of changes in the Pederson conductivity, changes in the electric field mapped from the magnetosphere, and ion interactions with the neutral wind. In regions where the ion and neutral velocities differ significantly, the ions are frictionally heated. The heating is proportional to the square of the velocity difference between ions and neutrals. When ions and neutrals travel together, Joule power is minimized. The ionosphere exerts some control over the energy input, reducing the magnitude when the neutral wind conforms to the ion drift. In the auroral zone, where electric fields are stronger and more variable, the ion and neutral velocities are frequently mismatched. Joule power tends to be a maximum in the auroral zone, where it raises the neutral and plasma temperatures, changes the neutral pressure and associated wind field, raises the plasma scale height, and drives outward field-aligned plasma flows. Storm-driven mass enhancements flow out of the auroral zone in wave-like structures (Chap. 12). The disturbances ultimately lead to additional satellite drag.

Under steady-state conditions, ion-neutral collisions impose the convective motion of the ions upon the neutral particles in the ionosphere. Typically about 10% of the energy deposited in the ionosphere is devoted to changes in kinetic energy resulting from *momentum exchange* that largely depends on the same factors as the Joule heating rate. Above the polar cap, the ions drag the neutrals across the polar cap from the dayside to the night side. Dayside radiant heating also pushes the neutrals toward the cooler nightside. Thus the polar cap tends to be a region of lower momentum exchange than the auroral zones.

Particle Energy Flux. Because of their mass, ions dominate the thermal energy in the magnetosphere. The lighter electrons accelerate more easily, so they are the favored particle for magnetosphere-ionosphere coupling during quiet times. Some of the electrons that flow along the geomagnetic field lines collide with neutrals in the ionosphere and release their energy to ionize the neutrals and excite oxygen and nitrogen molecules. Auroral electrons are accelerated out of Earth's plasma sheet by electric fields parallel to Earth's magnetic field lines. These particles are accelerated to energies between 10 eV and 30 keV. Alfvén waves may energize other auroral electrons. When the accelerated electrons strike the atmosphere, they relinquish their energy, resulting in a heat gain for the ionosphere. Particle heating occurs throughout much of the auroral zone. During quiescent times, less than 20% of the magnetospheric energy dissipated in the ionosphere is in the form of particle heating, but this particle heating is important. The ionospheric density is modulated strongly by the local enhancement of ionization, which in turn changes the evolution of the magnetosphere-ionosphere current system. Figure 8-33 identifies the Northern Hemisphere locations that are most susceptible to particle heating during quiet times. The Southern Hemisphere has a similar region.

Just as electrons accelerate downward, ions are heated and accelerate upward. Collisions keep most of the ions locked in the atmosphere. They too are a local heat source. Because of the magnetic mirror force that reflects charged particles away from a stronger magnetic field region, the heated ions that do escape acquire parallel, as well as perpendicular, velocity components as they move upward from the auroral regions.

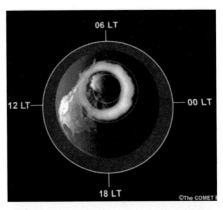

Fig. 8-33. Preferred Locations for Particle Heating during Geomagnetically Quiet Times. In this artist's rendition, the light-green ring corresponds to the heating region. The ring is thinner on the sunlit side and skews toward that side. *(Courtesy of the COMET Program at the University Corporation for Atmospheric Research)*

EXAMPLE 8.6

Auroral Zone Heating

- *Problem Statement:* Determine the amount of energy deposited in the auroral zones during an eight-hour storm in which Joule heating occurs at the rate of 20 mW/m².

- *Given:* Joule heating at the rate of 20 mW/m².

- *Assume:* The inner edge of the auroral zone is at 75° and the outer edge is at 60°. Each degree of latitude is 110 km in length. The Joule heating is uniformly distributed.

- *Solution:* First find the area of the auroral zone using a simple model of the zone being an annulus 15° wide. Subtract the area of a circle with a radius of 15° latitude (inner edge at 75° or 15° from the pole) from that of a circle with a radius 30° latitude.

    Annulus area = $\pi [(30 \times 110 \times 10^3 \text{ m})^2 - (15 \times 110 \times 10^3 \text{ m})^2]$
    = 2.57×10^{13} m²

    Power impinging on one auroral zone =
    $(2.57 \times 10^{13}$ m²$) \times 0.020$ W/m² $= 5.14 \times 10^{11}$ W

    Energy deposited in one auroral zone =
    $(5.14 \times 10^{11}$ W$) \times 2.8 \times 10^4$ s $= 1.48 \times 10^{16}$ J

    Global auroral zone energy deposition =
    1.48×10^{16} J/auroral zone $\times$ 2 auroral zones $\approx 3 \times 10^{16}$ J

- *Interpretation:* This is roughly the amount of energy dissipated by a super cell thunderstorm (Fig. 2-3). However, the auroral storm energy is dissipated in a more tenuous atmosphere. Each particle in the upper atmosphere is ultimately heated and lifted by this energy. Thus, during long-lived storms the upper atmosphere expands. Upper-atmosphere Joule heating tends to be non-uniform, with some regions receiving much more Joule heating and some regions much less. Global geomagnetic storms that last for tens of hours spread energy throughout the magnetosphere and Earth's upper atmosphere. This energy deposition can be as high as 10^{18} J.

8.4.2 Ionospheric Mass Outflows

Objective: After reading this section, you should be able to...

♦ Describe how the ionosphere contributes mass to the magnetosphere

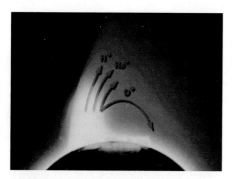

Fig. 8-34. Dayside, High-latitude Ionospheric Mass Contribution to the Geospace. This artist's rendition portrays the flow of H^+, He^+, and O^+ ions into the ionosphere above the low-altitude aurora (light green). *(Courtesy of SCIFER Team and Arnoldy [1996])*

Earth's upper atmosphere leaks neutral atoms as well as oxygen, helium, and hydrogen ions into space from the polar regions. Much of the leakage happens because the wake of the solar wind imposes a near-vacuum boundary condition on the magnetized polar ionospheres of planets, such as Earth, leading to an enhancement of atmospheric outflow at high latitudes (Fig. 8-34). These outflows deposit mass in the plasmasphere, plasma sheet, and tail lobes, contributing a significant and sometimes dominant proportion of ions to the magnetosphere, at least in the near-Earth regions. Solar wind energy that dissipates as heat in the auroral ionosphere adds to the outgoing mass flux. Some plasma and all fast neutral atoms are lost from the ionosphere to the downstream solar wind, but much of the plasma outflow recirculates through the magnetotail neutral sheet and accelerates to form energetic plasmas. Escaping neutral gas produces a geocoronal charge exchange medium that helps image the energetic ion populations. The composition and mass density of the outflow vary widely over the range of solar and magnetospheric activity. During quiet conditions, most of the outflow consists of hydrogen and helium atoms. The total mass of this quiet-time contribution is on the order of one kilogram per second.

The outflow of ions takes a variety of forms: the supersonic polar wind, subsonic polar cap outflows, ion upwelling from the cusp, and upward ion beams from the auroral zone. In addition to these high-latitude sources, strong oxygen atom outflows from the auroral ionosphere develop at times of intense geomagnetic activity.

Pause for Inquiry 8-16

Determine the hydrogen, helium, and oxygen ion flux needed to support a mass flux of 1 kg/s out of the ionosphere. Assume the upper atmosphere mass mixture is 85% hydrogen, 5% helium, and 10% oxygen. Also assume that the ions are atoms, not molecules.

EXAMPLE 8.7

Low-latitude Ionospheric Electric Field

- *Problem Statement:* Determine the electric field strength 200 km above the equator caused by the upward drift of plasma in Earth's magnetic field.
- *Concept:* Electric fields are created by plasma motion in a magnetic field. ($E = -v \times B$)
- *Given:* Plasma drift speed = 20 m/s, surface magnetic field strength = 31000 nT.
- *Assume:* Earth's magnetic field is dipolar
- *Solution:* First, determine the magnetic field strength at 200 km at the equator, using the dipole field equation from Chap. 7.

$$B_{total} = \sqrt{B_r^2 + B_\lambda^2} = B_S \frac{R_E^3}{r^3}\sqrt{1 + 3\sin^2(\lambda)}$$

$$= 3.1 \times 10^{-5} T \left(\frac{6378 \text{ km}}{6578 \text{ km}}\right)^3 \sqrt{1 + 3\sin^2(0)} = 2.8 \times 10^{-5} T$$

directed northward. At the equator $\lambda = 0$, the field has no radial component, and so the magnetic field is horizontal to Earth's surface.

Now calculate E

$$E = -20 \text{ m/s (up)} \times 2.8 \times 10^{-5} \text{ T (north)} = 5.6 \times 10^{-4}$$
V/m east = 0.56 mV/m east

- *Interpretation:* This small electric field is present in the equatorial dayside ionosphere as a result of plasma set into motion by solar heating. But the plasma is not free to accelerate along the electric field. It's constrained to spiral around the magnetic field lines. As it does so, its density increases within 10°–15° of the equator, as shown in Fig. 8-18.

Summary

The thermosphere is the outermost bound portion of Earth's neutral atmosphere. Because collisions are infrequent, thermospheric particles stratify by mass, with heavier particles more likely at the thermosphere base. As a neutral body, Earth's upper atmosphere is energized by waves from the lower atmosphere, absorption of solar photons, and by currents and particles from the magnetosphere. The thermosphere contributes energy to the mesosphere below and mass to the geocorona above. The thermosphere goes through cyclic expansions and contractions. One cycle is linked to solar dayside heating, another to the 11-year solar cycle. Such expansions cause variations in satellite drag and in the behavior of the embedded ionosphere.

Transient luminous events are recently discovered, short-lived phenomena associated with lightning. They appear to be fast sub-visual energy releases in and through the upper atmosphere. Scientists are not yet sure if they are the result of or initiated by lower atmospheric lightning. These events likely create very short-lived ionization channels in the middle and upper atmosphere.

Electron density profiles and total electron content (TEC) are two standard ways to describe the strength of the ionosphere. Electron density profiles can be quantified by Chapman layer functions that vary with the solar zenith angle. They reveal discrete layers of ionization; the D, E, and F layers. The quiescent D layer is a daytime feature created by strong lines in the EUV solar output. The E layer is also a solar-controlled ionization peak, created by X-ray and EUV radiation, ionizing molecular and atomic oxygen. Both layers disappear at night when the ionized constituents recombine. The lower portion of the F layer suffers a similar fate. The upper portion of the F layer (F2 layer and topside) persists through the night due to the relatively low collision rate and low recombination rate for electrons with atomic species. Total electron content is a measure of the ionization in a column of the atmosphere. High values of total electron content correspond to ionization instabilities that perturb radio signals.

The ionosphere exhibits significant variation with latitude. The high latitude regions cycle through extended periods of light and dark. In addition, low-, medium-, and high-energy particles precipitate into the auroral region, exciting auroral light and enhancing ionization and conductivity. Most of the energetic particles are energized in magnetospheric processes. In a few instances, the energetic particles come from the solar wind and solar disturbances. The low-latitude ionosphere is home to ionization crests (TEC enhancements) created by a combination of solar heating and geomagnetic control of ionization. The crests or anomalies develop in the afternoon and evening longitudes. At the crest edges, ionization instabilities develop.

Radio waves are strongly influenced by ionospheric electron density gradients. At low frequencies, the radio waves interact with and heat the D layer. Such waves are likely to be absorbed. Radio waves in specific bands transmitted from ground antennas are refracted by ion density gradients to the point that they bend back toward Earth and so are reflected from the ionosphere. Higher frequency radio waves penetrate the ionosphere and travel into space. However, strong gradients in electron density can bend the path of and retard the propagation of high-frequency radio signals.

The ionosphere is linked to other regions of geospace by currents and waves. The strongest linkage is via field-aligned currents (FACs). Region 1 FACs connect the distant magnetosphere to the high-latitude ionosphere and auroral zone. Region 2 FACs connect the inner magnetosphere to the lower latitude portion of the auroral zone. Cusp currents and substorm currents are the most variable class of FACs. Field-aligned currents and strong particle precipitation events deposit energy in the ionosphere and thermosphere, most of which is dissipated as Joule heat. Ultimately this heating causes atmospheric expansion. The expansion may appear as localized uplift, in traveling waves, in polar ion outflows, and in global lift of the thermosphere.

Key Terms

airglow emission
auroral zones
auroras
blue jets
critical frequency
cusp currents
D layer
diffuse aurora
diffusion
E layer
electron density profiles (EDPs)
elf
equatorial ionosphere
evanescent waves
exospheric temperature
extraordinary wave (x-wave)
F layer
F1 layer
F2 layer
geocorona
group speed
HF radio propagation window
hop transmission
ionograms
ionosphere
ionospheric ducting
ionospheric plasma
Joule rate
lowest usable frequency (LUF)
magnetosphere-ionosphere coupling
maximum usable frequency (MUF)
mid-latitude trough
momentum exchange
noctilucent ("night-shining") clouds
ordinary wave (o-wave)
Pederson current (J_P)
phase speed
polar hole
polar mesospheric clouds (PMCs)
polar wind
red sprites
refractive index n
Region-1 currents
Region-2 currents
skip transmission
sky wave
Snell's Law
substorm current wedges
terrestrial gamma ray flashes (TGFs)
thermosphere
thermospheric winds
topside ionosphere
total electron content (TEC)
transient luminous events (TLEs)
trans-ionospheric wave

Notations

Local rate of change of electron density

$$\frac{\partial n_e}{\partial t} = -\nabla \bullet (n_e \mathbf{v}) + R - L$$

Ionization rate as a function of height

$$R(z) = \sigma \left[n_0 e^{\left(-\frac{\Delta z}{H}\right)} \right] \left[I_0 e^{(-\sigma H \sec(\chi) n_0) e^{-\frac{\Delta z}{H}}} \right]$$

$$R(z) = \sigma I_0 n_0 e^{\left(-\frac{\Delta z}{H} - \tau_z\right)}$$

Maximum ionization rate (1st derivative = 0)

$$\frac{dR(z)}{dz} = \sigma I_0 n_0 \left(-1 + (\sigma H \sec(\chi) n_0) e^{\left(-\frac{\Delta z}{H}\right)} \right)$$

$$e^{\left(-\frac{\Delta z}{H} + (\sigma H \sec(\chi) n_0) e^{\left(-\frac{\Delta z}{H}\right)}\right)} = 0$$

$$L_{recombination} = \alpha n_e n_i \approx \alpha n_e^2$$

$$L_{attach} = \beta n_e n$$

Definition of refractive index $n = c/v_n$

Refractive index for a plasma

$$n = \sqrt{1 - \frac{N_e e^2}{\varepsilon_0 m_e \omega^2}} \approx \sqrt{1 - \frac{80.6 N_e}{f^2}}$$

Plasma frequency $f_p = \sqrt{80.6 N_e}$ or

Plasma angular frequency $\omega_p = \sqrt{\dfrac{N_e e^2}{\varepsilon_0 m_e}}$

Snell's Law $n_i \sin\theta_i = n_r \sin\theta_r$
$n \sin\theta = $ constant

Minimum refractive index for reflection $n_m = \sin\theta_0$

Minimum electron density for reflection of an oblique wave $N_{e\,min} = \dfrac{f^2}{80.6} \cos^2\theta_0$

Special case for vertical incidence and $f = f_p$

$$n = \sqrt{1 - \dfrac{80.6 N_0}{f^2}} = \sqrt{1 - \dfrac{f_p^2}{f^2}} = 0$$

Relation of oblique wave frequency to vertical reflected frequency $f_\theta = \dfrac{f_p}{\cos\theta} = f_p \sec\theta$

Critical frequency for vertical incidence
$f_c = \sqrt{80.6 N_{e\,max}} \approx 9\sqrt{N_{e\,max}}$

Critical frequency for oblique incidence $f_{c\theta} = f_c \sec\theta$

Answers to Pause for Inquiry

PFI 8-1: Atomic oxygen O at solar maximum. Helium He at solar minimum.

PFI 8-2: In the upper atmosphere (500 km), the densities of all the constituent species (O, H, N_2, O_2, He, and Ar) increase during solar maximum, because the temperature at that altitude more than doubles, causing particles from lower altitudes to expand upward.

PFI 8-3: Scale height is $k_B T/mg$. Because scale height depends directly on temperature and exospheric temperature is high during solar maximum, the scale height is largest during solar maximum. Figure 8-2 shows that the density falls off more slowly during solar maximum, consistent with a larger scale height during that period.

PFI 8-4: c) An order of magnitude from 10^{13} to 10^{14}.

PFI 8-5: The exosphere is part of the atmosphere that is only weakly bound, if at all, to Earth by gravity. The material at the base of the geocorona and plasmasphere may be considered part of the exosphere. The geocorona and plasmasphere occupy much of the same region of space. The geocorona consists of neutral atoms. The plasmasphere consists of ionized atoms and is thus subject to electric and magnetic forces.

PFI 8-6: The ionization energy for molecular nitrogen is 15.5 eV. We know that $E = hc/\lambda$. The wavelength λ required for ionization is $\lambda = hc/15.5$ eV $= [(6.62 \times 10^{-34}$ Js$)(3 \times 10^8$ m/s$)]/[(15.5\,eV)(1.6 \times 10^{-19}$ J/eV$)] = 80$ nm.

PFI 8-7: a) At lower density and greater heights.

PFI 8-8: The optical depth = unity surface occurs at greater heights.

PFI 8-9: The F layer.

PFI 8-10: The MUF and LUF both increase. Generally the MUF increases more than the LUF, so the HF propagation window tends to be wider during solar maximum.

PFI 8-11: You should shift the frequency up, looking for a frequency that exceeds the LUF, which has risen because of increased radio absorption in the D layer. The propagation window thus becomes narrower. The LUF may even exceed the MUF and completely close the propagation window.

PFI 8-12: b) Because the F layer height rises.

PFI 8-13: a) Increasing concentration of electrons and decreasing refractive index.

PFI 8-14: In the polar cap, ions and electrons both attempt to drift in the $\mathbf{E} \times \mathbf{B}$ direction, which is anti-Sunward. However, the ions are slowed by collisions, so the electrons drift anti-Sunward faster. This drift produces a current in the opposite (Sunward) direction.

PFI 8-15: Anti-Sunward

PFI 8-16: The outflowing mass flux has 0.85 kg/s of hydrogen, 0.05 kg/s of helium and 0.1 kg/s of oxygen.

The # of hydrogen ions/s needed to provide the mass = $(0.85 \text{ kg/s})/[(1.67 \times 10^{-27} \text{ kg/amu})(1 \text{ amu/hydrogen ion})] = 5.09 \times 10^{26}$ hydrogen ions/s

The # of helium ions/s needed to provide the mass = $(0.05 \text{ kg/s})/[(1.67 \times 10^{-27} \text{ kg/amu})(4 \text{ amu/helium ion}] = 7.49 \times 10^{24}$ helium ions/s

The # of oxygen ions/s needed to provide the mass = $(0.10 \text{ kg/s})/[(1.67 \times 10^{-27} \text{ kg/amu})(16 \text{ amu/oxygen ion}] = 3.74 \times 10^{24}$ oxygen ions/s

Figure References

Arnoldy, Roger L., K. A. Lynch, P. M. Kintner, J. Bonnell, T. E. Moore, and C. J. Pollock. 1996. SCIFER—Structure of the cleft ion fountain at 1400 km altitude. *Geophysical Research Letters*. Vol. 23, No. 14. American Geophysical Union. Washington, DC.

Chamberlain, Joseph W. 1978. *Theory of Planetary Atmospheres: An Introduction to their Physics and Chemistry.* International Geophysical Series, Vol. 22. New York, NY: Academic Press.

Pasko, Victor P. 2003. Atmospheric Physics: Electric Jets. *Nature*. Vol. 423. Nature Publishing Group. New York, NY.

Roble, Ray G. 1977. The Thermosphere, Chapter 3 in the 'Upper Atmosphere and Magnetosphere' monograph for the Geophysical Research Board of the National Academy of Sciences. National Academy of Sciences. Washington, DC.

Further Reading

Cravens, Thomas E. 1997. *Physics of Solar System Plasmas.* Cambridge, UK: Cambridge University Press.

Hargreaves, John K. 1992. *The Solar Terrestrial Environment.* Cambridge, UK: Cambridge University Press.

Hines, Colin O., Irvine Paghis, Theodore R. Hartz, and Jules A. Fejer, eds. 1965. *Physics of the Earth's Upper Atmosphere.* Englewood Cliffs, NJ: Prentice-Hall.

Kelley, Michael, C. 2009. *The Earth's Ionosphere.* Burlington, MA: Academic Press.

Kivelson, Margaret G. and Christopher T. Russell, eds. 1995. *Introduction to Space Physics.* Cambridge University Press. Cambridge, UK.

Prölss, Gerd W. 2004. *Physics of the Earth's Space Environment.* Berlin: Springer-Verlag.

Silverman, Samule M. 1970. Night Airglow Phenomenology. *Space Science Reviews*. Vol. 11. Springer. Dordrecht, Netherlands.

Chapter 8 Earth's Quiescent Atmosphere and Its Role in the Space Environment and Weather

The Active Sun and Other Stars: Sources of Space Weather

9

UNIT 1. SPACE WEATHER AND ITS PHYSICS

Contributions by the AF Weather Agency

You should already know about...

- Cartesian and polar coordinates
- Spectral and thermal emissions (Chap. 2)
- Kinetic energy of relativistic particles (Chap. 2)
- Energy density (Chaps. 2 and 4)
- Solar coordinates (Chap. 3)
- Convective and dynamo processes (Chap. 3)
- The quiescent solar magnetic field (Chap. 3)
- Maxwell's Equations (Chap. 4)
- Co-rotating interaction regions (Chap. 5)
- Plasma characteristics and behavior (Chap. 6)

In this chapter you will learn about...

- Emergent magnetic flux in the forms of sunspots and active regions
- Solar cycles with period of 11 years and longer
- Differential rotations and circulation
- Solar flares and radio bursts and their impacts
- The genesis and impacts of coronal mass ejections
- The genesis and impacts of solar energetic particle events
- The impacts of energetic particles from beyond our solar system

Outline

9.1 The Solar Cycle and Its Roots
    9.1.1 The Solar Dynamo and Its Supporting Motions
    9.1.2 Active Regions and Their Components

9.2 Active Sun Emissions
    9.2.1 Energy Paths of Magnetic Reconfiguration
    9.2.2 Flares and Radio Bursts
    9.2.3 Coronal Mass Ejections (CMEs)
    9.2.4 Solar Energetic Particles

9.3 Space Weather Effects from Other Stars
    9.3.1 Low- and High-energy Particles from the Cosmos
    9.3.2 High-energy Photons from the Cosmos

Chapter 9 The Active Sun and Other Stars: Sources of Space Weather

9.1 The Solar Cycle and Its Roots

The Sun exhibits cyclical activity that relates to a variety of motions in its convective envelope. These motions intensify magnetic fields, causing them to bubble through the convective zone outward to space. The field emerges on a wide range of time scales extending from a few minutes to multiples of the 11-year cycle.

9.1.1 The Solar Dynamo and Its Supporting Motions

Objectives: After reading this section, you should be able to...

- Relate the solar dynamo to vertical and meridional circulations and magnetic flux emergence
- Relate torsional oscillations to the solar cycle

Twisting and Concentrating Magnetic Flux

The Sun is a magnetic star that creates and twists its magnetic field and then expels the twisted field into space. The solar dynamo converts kinetic energy of plasma convection, meridional circulation, and differential rotation into electromagnetic energy. It constantly replenishes and concentrates the magnetism that would otherwise decay via Ohmic dissipation in a matter of decades. The solar dynamo is responsible for the 11-year sunspot cycle with its well-defined characteristics of field polarity and sunspot location (Sec. 3.3.2).

From previous chapters we know that during the solar cycle, differential rotation creates strong toroidal fields from the poloidal field of solar minimum. In the convective zone, velocity shears cause the field to wrap into magnetic flux ropes (Fig. 9-1a). The strong field excludes plasma and becomes buoyant while maintaining pressure balance. The magnetic ropes twist from the Coriolis influences as they rise. A combination of rising and twisting motion puts a component of local pole-to-pole (poloidal) field back into the system, but the field orientation is opposite that of the original dipole field (Fig. 9-1a). This vertical twisting is essential to the 11-year dipolar reversal (field reorientation).

When a strong magnetic flux tube breaks through the photosphere, it creates sunspots, often in pairs, from which the field expands into the less-dense atmosphere as a small dipole. The leading spot of a pair often forms at a slightly lower latitude than the following spot, so the pair has a latitudinal tilt. The westward spots bear the field polarity of the hemisphere in which they originate. The following spots diffuse and drift poleward with other trailing regions to form large unipolar regions near the pole.

We recall from Chap. 4 that magnetic field piercing a surface is magnetic flux. Some of the enhanced magnetic field protrudes through the solar surface in a process called *magnetic flux emergence* (Fig. 9-1b). Convection allows magnetic flux to constantly emerge in quiet and active regions. However, emergence from active regions is more deeply rooted and energetic and therefore more interesting to space weather observers and forecasters. Even as some magnetic flux emerges upward into the solar atmosphere, convective motions below the photosphere further increase the density of the field lines, twisting and hammering strands of field downward toward the tachocline. In the tachocline the field organizes into long, concentrated ropes. As field strength (and hence the magnetic pressure) becomes larger, the ropes (flux tubes) at the base of the convective zone become buoyant. Over the course of three to ten months, the buoyant tubes rise and burst through the surface. Convection aids in the lifting (Example 9.1). Although the rate of magnetic flux emergence from the Sun's quiet regions exceeds that of active regions, emerging active regions contribute most to the large-scale magnetic structure. Figure 9-1c shows an emerged active region in profile on the eastern limb of the Sun.

In the convective zone and photosphere, the thermal energy density is much higher than the magnetic field energy density and the magnetic field is "frozen into" the plasma. But above the photosphere, where density decreases exponentially, magnetic fields more easily affect the plasma. Moderate-strength magnetic fields generally confine plasma in the upper photosphere and chromosphere. In the corona, where plasma is fully ionized, and the interaction between the plasma and coronal magnetic fields is strong, the magnetic field dominates the dynamics.

Active regions extend vertically into the corona, arching over the smaller, dense, random threads of the magnetic carpet. These expansive magnetic loops form canopies that can be imaged because the plasma trapped on the closed-loop field lines radiates in wavelengths characteristic of its temperatures. Cool loops have temperatures less than 10^5 K. Hot loops have temperatures significantly greater than 10^6 K. Substantial energy sources drive the plasma trapped on the closed field lines to coronal heights and maintain the high plasma temperature. Wave energy dissipation and magnetic merging amid the random loops of the low lying magnetic carpet are two likely sources of the heating.

Regions of open field lines also thread the corona. The magnetic flux in these regions is sometimes called unsigned flux. The field lines, with no visible closure on the Sun, allow plasma to exit the Sun in high-speed flows. Such regions are the source of high-speed solar wind that creates co-rotating interaction regions and high-speed flows in the heliosphere during the declining portions of the solar cycle. The tenuous, free-flowing plasma emits virtually no light; hence the regions are seen in X ray as dark coronal holes (top of Fig. 9-1c). Active regions adjoining coronal holes may experience instabilities that result in magnetic reconnection and subsequent coronal mass ejections.

Circulations and Oscillations

As we describe in Chap. 3, data from ground and space instruments reveal *meridional circulations* of solar plasma similar to fluid atmospheric and oceanic circulations on Earth. This circulation along meridians transports magnetic flux in multi-year cycles and supports the 22-year magnetic cycle that reverses and re-establishes the solar main field orientation. Flow in the outer convective zone moves gas and plasma from low latitudes to high latitudes at a rate of 10 m/s–20 m/s. We find hints of upward motion at low latitudes and downward motion at high latitudes. To complete the circulation and maintain mass continuity, a return high-to-low latitude circulation must flow deep in the convective zone. Because of the higher densities at the base of the convective zone, the flow needs a speed of only about 2 m/s–5 m/s to close the circulation.

Helioseismology observations reveal variations of the internal poleward flow during the solar cycle. Accelerations (or decelerations) change the onset of a new solar cycle by several months to a year. Bunching of the flow may also modulate the strength and duration of the activity cycle. Some solar modelers believe the meridional flows act as a system time-keeper to maintain the rhythm of the solar

Chapter 9 The Active Sun and Other Stars: Sources of Space Weather

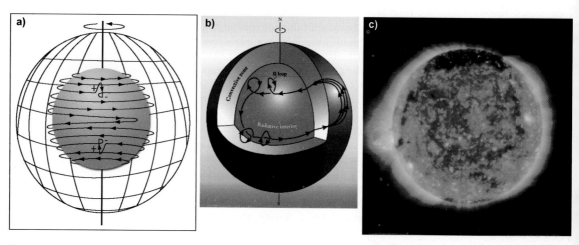

Fig. 9-1.   Dynamo Field Behavior. In panel **a)** the view is near the base of the convective zone and illustrates the motion of a single field line acted upon by differential rotation and convection [Dikpati and Gilman, 2008]. Diagram **b)** shows the emergence of magnetic flux into the convective zone [Parker, 2000]. **c)** This is an extreme ultraviolet image of a limb active region. On the eastern (left) limb is an active region that has ruptured into the corona, forming a coronal loop. The northern pole is covered by a coronal hole. *(Courtesy of ESA/NASA – the Solar and Heliospheric Observatory (SOHO) Program)*

cycle. Their models now include flow-memory terms from previous solar cycles. These memory terms account for the influence of very-long-term flow perturbations near the base of the convective zone that may take two (or more) solar cycles to fully exert their influence.

Scientists have known for decades that some patches of the Sun surge ahead or fall behind surrounding regions while elsewhere the rotation remains substantially steady at the same latitude. *Torsional oscillations* are latitude bands of slightly faster and slower rotation that propagate on a solar cycle time-scale, producing an alternating prograde and retrograde motion at a given latitude (Fig. 9-2). The amplitude of the oscillation is about 5 m/s, which is on the order of 1% of the background solar rotation rate.

The oscillations appear to have two branches, one that propagates poleward and the other equatorward. The stronger equatorward branch originates 2–3 years before the first sunspot eruptions of a new cycle. Sunspots and active regions tend to form where the latitudinal shear is enhanced—poleward of the fast streams of the oscillation. Helioseismology reveals the oscillations are present at significant depths in the convective zone. Torsional oscillations are likely from the effects of the Lorentz force acting on magnetic flux tubes that thread the convective zone and rotate at a different rate than the ambient plasma.

9.1 The Solar Cycle and Its Roots

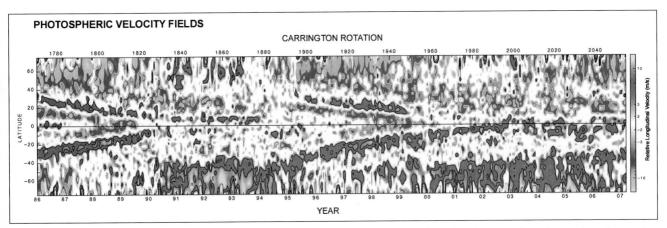

Fig. 9-2. Torsional Oscillations. The vertical axis is solar latitude. The horizontal axis is time in years. Superimposed on the random convective motions are long-lived waves of slightly higher or lower east-west velocity. The waves move from high latitudes to the equator with a period of about 11 years. The retrograde motions are seen most clearly as the red-colored V patterns pointing to the right. Blue indicates prograde motion of ~5 m/s. Red indicates retrograde motion of ~5 m/s. *(Courtesy of Roger Ulrich at UCLA)*

EXAMPLE 9.1

Flux Tube Motion in the Convective Zone

- *Problem Statement:* Determine the average acceleration of a magnetic flux tube that starts at rest at the base of the convective zone and rises to the solar surface in 200 days.

- *Concept:* Kinematics

- *Given:* Time = 200 days, convective zone depth = $0.3 \, R_S$

- *Assume:* Constant acceleration

- *Solution:*

$$y_f = y_0 + v_0 t + \frac{1}{2}at^2 = y_0 + \frac{1}{2}at^2 \quad (\text{since } v_0 = 0)$$

$$a = \frac{2(y_f - y_0)}{t^2} = \frac{2(0.3)(7 \times 10^8 \, \text{m})}{(200 \times 86{,}400 \, \text{s})^2} = \frac{4.2 \times 10^8 \, \text{m}}{2.99 \times 10^{14} \, \text{s}^2}$$

$$= 1.4 \times 10^{-6} \, \text{m/s}^2$$

- *Implication and Interpretation:* The acceleration of a buoyant flux tube through the convective zone seems imperceptibly small. However, a small acceleration over a long interval still produces significant speeds. In fact, the acceleration is probably not constant, but rather grows because of instabilities in magnetic field line configurations. Further, the flux tube moves through an environment with exponentially decreasing mass, which allows increasing acceleration.

Follow-on Exercise: Assuming the constant acceleration found above, what is the speed with which the flux tube bursts through the surface?

EXAMPLE 9.2

Plasma Beta Comparisons

- *Problem Statement:* Use typical temperature, density, and magnetic field strength values to determine the plasma-β in the photosphere and compare that value to the plasma-β in the corona.
- *Concept:* Plasma-β compares thermal energy density to magnetic energy density.
- *Assume:* Photospheric temperature, density, and magnetic field strength are $T = 5800$ K, $n = 1.0 \times 10^{23}$ particles/m^3, $B = 0.003$ T. Coronal temperature, density, and magnetic field strength are $T = 1 \times 10^6$ K, $n = 1.0 \times 10^{14}$ particles/m^3, $B = 0.1$ T.
- *Solution:*

$$\beta = \frac{nk_B T}{B^2/2\mu_o}$$

$$\beta_{ph} = \frac{nk_B T}{B^2/2\mu_o} = \frac{10^{23}\text{particles/m}^3 (1.38 \times 10^{-23}\text{J/K})5800\text{ K}}{(0.003\text{ T})^2/2(1.28 \times 10^{-6}\text{N/A}^2)}$$

$$= 2.27 \times 10^3$$

$$\beta_{co} = \frac{nk_B T}{B^2/2\mu_o} = \frac{10^{14}\text{particles/m}^3 (1.38 \times 10^{-23}\text{J/K})1 \times 10^6\text{K}}{(0.1\text{ T})^2/2(1.28 \times 10^{-6}\text{N/A}^2)}$$

$$= 3.53 \times 10^{-7}$$

- *Implication and Interpretation:* The solar photosphere is a high-beta plasma where thermal particle motion controls the plasma behavior. The solar corona is a low-beta plasma where the magnetic field controls the dynamics of the plasma behavior. The large space weather structures called coronal mass ejections originate in the upper chromosphere and in the corona, where magnetic fields dominate the plasma.

Follow-on Exercise: Compare the plasma-β in a sunspot to that of the corona. You will need to find or determine an appropriate particle number density for the sunspot.

9.1.2 Active Regions and Their Components

> **Objectives:** After reading this section, you should be able to...
> - Explain the origin of active regions
> - Draw elements of active regions in vertical and horizontal diagrams
> - Relate active regions to sunspots, faculae, plage, and magnetic flux emergence

Active Regions in the Solar Atmosphere

Active regions (ARs) are volumes of disturbed solar atmosphere threaded by magnetic fields. Figure 9-3 is a multi-spectral image of AR 9393 that displays an overlay of AR radiative emissions. These disturbed regions exist for one week to several months. In the early 1970s, the Skylab mission produced a series of observations that proved that the common threads for solar active phenomena are active regions. The volumes of emerging or emerged magnetic field extend from the convective zone to the corona as depicted in cross-section in Figs. 3-19 and 9-4. In the photosphere the concentrated fields appear in the form of short-lived pores, magnetic elements, ephemeral regions, and sunspots. Longer-lived sunspots and loops of field arching above them are imaged regularly by various ground-based and satellite-borne sensors.

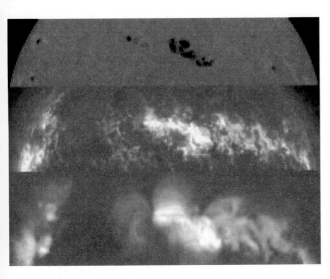

Fig. 9-3. Active Regions Overlie Sunspots. Here we see Active Region 9393 in visible, extreme ultraviolet, and X-ray light. These views correspond (top to bottom) to the photosphere, the chromosphere and transition region, and the corona. *(Courtesy of the Solar and Heliospheric Observatory's MDI/EIT Consortiums and the Yohkoh Space X-ray Telescope Project Astronomical Picture of the Day)*

As sunspots form, the magnetic field erupts horizontally at first but then loops into the upper photosphere and chromosphere, forming an *arch system* (Fig. 9-4). The emerging magnetic fields corresponding to an active region stretch horizontally for great distances because of the differential rotation of the photosphere. Sometimes the span is even cross-equatorial. Most active regions consist of at least two primary sunspots and surrounding faculae and plages. (Section 3.2.1 describes faculae and plages.) More complex regions may be multi-polar (Fig. 9-5). Bipolar regions are initially connected by low-lying magnetic field line arches that may balloon vertically, forming loops (Fig. 3-19). The arch system persists until additional spots develop and complicate the field line structure. Faculae and plages may be present in the vicinity of the arches.

Chapter 9 The Active Sun and Other Stars: Sources of Space Weather

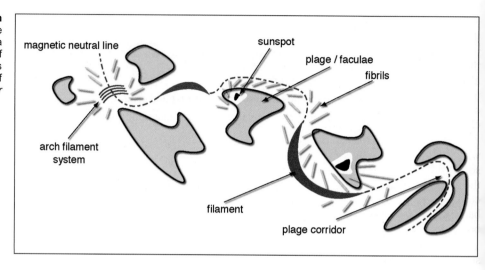

Fig. 9-4. Horizontal Layout of an Active Region. Here we show the elements of Figs. 9-3 and 9-5 in a bird's-eye perspective. The extent of such regions is likely to be thousands to hundreds of thousands of kilometers. *(Courtesy of AF Weather Agency)*

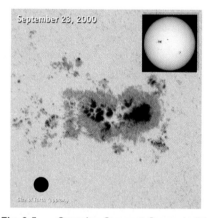

Fig. 9-5. Complex Sunspot Group. In this image, numerous umbral features of different polarities are surrounded by a complex penumbral region. The complexity leads to the likelihood of impulsive eruptions in the form of flares and CMEs. The inset at the top right gives the location of the active region on the Sun. The inset at the lower left is Earth's relative size. Sunspot motion across the solar surface led to the discovery of the Sun's rotation. *(Courtesy of ESA/NASA – the Solar and Heliospheric Observatory (SOHO) Program)*

Some active regions are ephemeral, existing for such short times that they are noticed, but not tagged with a number. In numbered regions the enhanced field persists in time and space long enough to be identified as an active region. Across less dynamic, but mature, active regions, the emergent magnetic field loops may sag in the middle, trapping chromospheric plasma that cools radiatively. These cool, dense plasma clouds are filaments (viewed against the disk of the bright Sun) and prominences (viewed against cool space). Quiescent filaments are associated with large-scale solar sector boundaries, which separate broad regions of opposite and relatively weak magnetic polarity. Dynamic changes related to twisted magnetic field lines juxtaposed with active regions may turn quiescent filaments into erupting filaments and disappearing prominences. We say more about this in the section on coronal mass ejections.

Because sunspots indicate stronger magnetic fields than do plages, they generally form after plages and dissipate before the plages fade. Plages may be present at high solar latitudes (>40° N or S), whereas sunspots are usually seen at latitudes within 40° of the solar equator. This difference indicates that while magnetic fields erupt all over the Sun, causing active regions, the stronger magnetic fields are generally more confined to lower latitudes, where they are strong enough to organize the plasma. The higher radiative EUV and UV output of plages and faculae allows the active Sun to be a hotter body, even when a substantial number of sunspots darken the photosphere.

Individual sunspots last from a few hours to a few weeks. A spot group (active region) may persist for several months. Sunspots appear as irregular blemishes on the Sun's surface (Fig. 9-5). The dark inner spot region, the *umbra*, harbors concentrated field strengths of 0.2 T to 0.4 T. A more luminous fringe, the *penumbra*, has numerous thin, radially extended, bright and dark tubes carrying outward gas flows. The field normally weakens to ~0.1 T at the spot boundary, where the field lines become more horizontal. A typical spot is about 10^4 km across, but spots have been observed in excess of 10^5 km in diameter. They often form as bipolar pairs and these pairs appear in groups. The size of the bipole regions may be uneven, to the point where one pole is so diffuse that it loses its distinction as a spot in the photosphere. Such diffuse regions are tracked more easily by their enhanced chromospheric emissions.

Observations of sunspots near the solar limb reveal that they are depressions in the Sun's surface. Because the magnetic field exerts pressure, and sunspots are in pressure equilibrium with their surroundings, the spot must have a lower thermal pressure (Example 6.11). Thus the density is lower, and less material absorbs the light coming from the spot, so we see further into the spot. At the umbral walls the temperature rises sharply, creating excess emission regions called *faculae*. These bright regions, which extend from the photosphere to the chromosphere, are more easily viewed when sunspots are near the solar limb.

The strong, nearly vertical magnetic field inhibits motions across the field lines, which in turn reduces convection inside the spot. Because convection is the main source of energy transport just below the photospheric surface, less energy reaches the surface through the spot. Thus, spots appear dark because they are relatively cool (~3500 K). If a spot were removed from the Sun and placed against the backdrop of cold space, it would glow with the intensity of a ~3000 K–4000 K body (about as bright as the full moon).

Figure 9-6 illustrates the significant and complicated subsurface flows that scientists now believe challenge the integrity of sunspot structures. If the sunspots are to remain long-lived and coherent, the magnetic flux tubes that create them must withstand the convective motions between spots that continually buffet the tubes. Only the most concentrated fields survive the pummeling. Three-dimensional modeling of the convective zone outside of spots suggests that convection is strong between the opposed-polarity flux tubes and may help to keep them separate, inhibiting turbulent decay.

Active regions may emerge through the convective zone in pulses, often at or near the site of a previous eruption, as shown in Figs. 9-7 and 9-8. Multiple active regions cause complexes (nests) of activity that tend to appear at or near the same magnetic longitude for several months, so they are *active longitudes*. Recent deep dynamo modeling efforts suggest that the convective zone is organized at a very deep level into longitudinal roles that support such structures. This behavior allows a long-lived accumulation of magnetic flux along some longitudes that is ultimately created and destroyed by large-scale dynamo processes.

After a few weeks to months, decaying active regions are subsumed into the background solar field. The poleward portions of the decaying regions drift under the influence of meridional flows, slowly dragging the enhanced field network to high latitudes. These remnants of active region fields are the seed fields leading to the polar reversal of the subsequent solar cycle. The concentrated equatorward portions of decaying active regions drift to lower latitudes, allowing fields of opposite polarities from the two hemispheres to interact in the vicinity of the streamer belt. The interactions probably cause many of the mass ejections accompanying the maximum phase of the solar cycle.

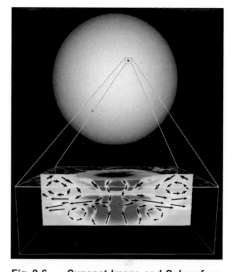

Fig. 9-6. Sunspot Image and Subsurface Temperature and Flow Diagram. Sunspot magnetic fields block the flow of heat to a depth of about 5,000 km below the surface. The hotter zone is in red, while the blue zone is cooler. At the solar surface, plasma flows out of sunspots. Deeper in the Sun, material rushes inward to maintain the pressure in the spot. This inflow is strong enough to pull the magnetic fields together and reduce the amount of heat that normally flows from inside the Sun. Thus sunspots are cooler and appear darker than the surrounding surface. *(Courtesy of NASA/ESA and Philip Scherrer at Stanford University)*

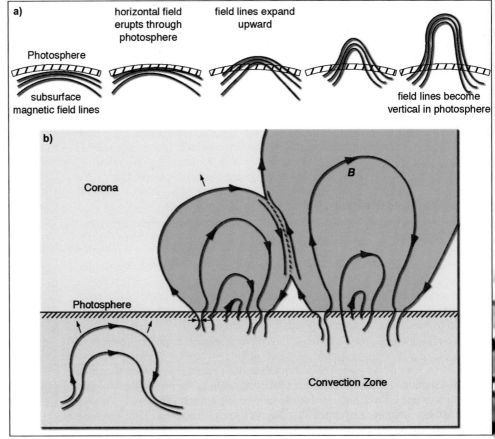

Fig. 9-7.   Schematic Cross Sections of a Developing Active Region. a) This diagram depicts the emergence of magnetic flux from the convective zone to the solar atmosphere. **b)** Here we show continued flux emergence and merging of oppositely directed magnetic fields from two active regions in the corona.

Pause for Inquiry 9-1

Determine the amount of magnetic flux in an average sunspot.

Pause for Inquiry 9-2

Verify that the direction of the magnetic field in the emergent loops of Fig. 9-7 produces a configuration for magnetic merging in locations where the loops juxtapose.

Pause for Inquiry 9-3

Visit the world wide web. Determine the approximate date of the image in Fig. 9-3. Was it made during the rise, maximum, decline, or minimum of the 11-year solar activity cycle?

9.1 The Solar Cycle and Its Roots

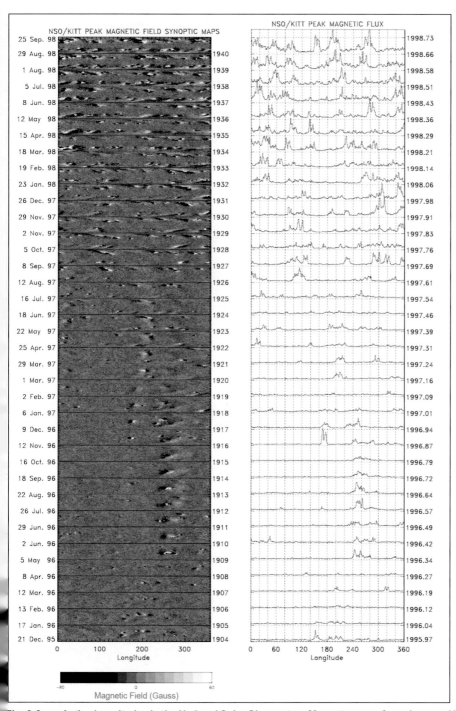

Fig. 9-8. **Active Longitudes in the National Solar Observatory Magnetograms from January 1996 to September 1998.** Time progresses from bottom to top. Each small panel represents a magnetogram image for a single Carrington rotation. The rotation number is shown on the left. The images are cropped at plus and minus 50 deg solar latitude. Black and white regions indicate active regions where the magnetic flux density averaged over a pixel exceeds 60 gauss (6 mT), with black representing negative polarity and white positive. From mid-1996 to late-1996, the Southern Hemisphere had a single active longitude at about 250°. From early-1997 to mid-1997, the Northern Hemisphere had a single active longitude at about 200°. [de Toma et al., 2000]

Chapter 9 The Active Sun and Other Stars: Sources of Space Weather

> **Pause for Inquiry 9-4**
>
> Consider Fig. 9-8. Compare the latitudes of the active regions in mid-1996 and mid-1998. Do the active regions in mid-1996 belong to solar cycle 22 or solar cycle 24?

Focus Box 9.1: Sunspot Counts and Complexity

(Contributed by the National Geophysical Data Center)

The visible indicators of active regions (sunspots) have been counted and sized for centuries. Although many sunspots are so small they are not evident to the unaided human eye, some sunspots are sensational. A telling comparison comes from ancient Korean literature, wherein sunspots were compared to the size of hen's eggs and pears. The literary reference comes from a time when auroras were viewed near the equator. In 1848 Rudolph Wolf devised a daily method of estimating solar activity by counting the number of individual spots and groups of spots on the face of the Sun. He computed his sunspot number by adding 10 times the number of groups to the total count of individual spots, because neither quantity alone completely captured the level of activity. For example, if we see two spots and three groups (with 2, 3, and 4 spots respectively) then the Wolf number is 41 (10 × 3 groups, plus 11 individual sunspots). Today, Wolf sunspot counts continue, because no other index of the Sun's activity reaches continuously so far into the past. Wolf confirmed the cycle in sunspot numbers. He also determined the cycle's length to be 11.1 years by using early historical records. Wolf, who became director of the Zurich Observatory in Switzerland, discovered independently the coincidence of the sunspot cycle with disturbances in the Earth's magnetic field.

Today observers compute a daily sunspot number in the same way. Many refer to the sunspot number as a Wolf number (or as a Zurich Sunspot Number). Results, however, vary greatly, because the measurement depends strongly on observer interpretation and experience and on the stability of Earth's atmosphere above the observing site. The use of Earth as a platform from which to record these numbers contributes to their variability, because the sun rotates and the evolving spot groups are distributed unevenly across solar longitudes. To compensate, each daily international number is computed as a weighted average of measurements made from a network of cooperating observatories. The international community has agreed upon other methods for counting the spots that factor in aerial coverage and groupings of spots.

Observers also categorize sunspots by their magnetic complexity. Table 9-1 describes sunspots in terms of increasing levels of magnetic complexity described by the Mount Wilson Observatory. Other classification schemes exist, some with as many as 60 entries.

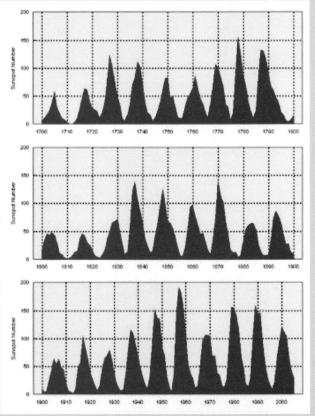

Fig. 9-9. Annual Mean Sunspot Numbers 1700–2005. Sunspot counts rise and fall approximately every 11.1 years. The cycle, though, is not symmetrical. The spot count takes on the average about 4.8 years to rise from a minimum to a maximum and another 6.2 years to fall to a minimum once again. The largest annual mean number (190.2) occurred in 1957. Very often sunspot activity dips slightly during the maximum phase, resulting in a double hump appearance for some solar maxima. *(Courtesy of the National Geophysical Data Center)*

9.1 The Solar Cycle and Its Roots

Table 9-1. Mount Wilson Sunspot Classification Scheme. This table summarizes a commonly used classification for describing sunspots in terms of magnetic complexity and spot orientation. Here "p" means "preceeding" and "f" means "following."

Unipolar α_p	All the magnetic measures in the group are of the same magnetic polarity, which corresponds to the preceding spots in that hemisphere for that cycle
α_f	All the magnetic measures in the group are of the same magnetic polarity, which corresponds to the following spots in that hemisphere for that cycle
Bipolar β	A bipolar group in which the magnetic measures indicate a balance between the preceding and following spots
β_p	A bipolar group in which the magnetic measures indicate that the preceding spots are dominant (factor of 2 or more)
β_f	A bipolar group in which the magnetic measures indicate that the following spots are dominant (factor of 2 or more)
Complex β_γ	A group that has general beta characteristics but in which the polarity of one or more spots is out of place
γ	A group in which the polarities are completely mixed
δ_γ	Will be added to the above classifications if the group is also of the "delta-configuration"; that is, spot umbrae are of opposite polarity within 2 degrees of one another and within the same penumbra
+ reverse polarity group	A plus sign is added after the magnetic classification if the group polarities do not follow the general law of sunspot polarity for the hemisphere.

Pause for Inquiry 9-5

Visit the internet. In about what month and year did solar cycle 24 begin? At what latitude were the first sunspots observed?

Pause for Inquiry 9-6

Assume that the southern hemisphere has a positive magnetic polarity. What is the direction of the leading spots in the southern hemisphere? What is the direction of the leading spots in the northern hemisphere?

Magnetism During the Solar Cycle

Most regions of strong photospheric magnetic fields are grouped in sunspot pairs of opposite polarities. Usually the ordering of positive and negative regions with respect to the Sun's rotational direction is the same in a given hemisphere, but is reversed from northern to southern hemispheres. A few sunspot pairs (5%–10%) have a polarity opposite that of the prevailing polarity; this happens most often in complex sunspot groups.

From one sunspot cycle to the next, the magnetic polarities of pairs undergo a reversal in each hemisphere. Sunspots seldom form poleward of 40° solar latitude and never over the solar equator. Because the latitude of the most intense shearing moves from high to low latitudes over the solar cycle, the formation regions for solar active phenomena also migrate equatorward as the solar cycle progresses from minimum to maximum. The increased local magnetic energy density gradually dissipates via active phenomena such as flares, returning the Sun to a quiescent state by the end of the cycle. Figure 9-10 gives a history of sunspot development and motion constructed by assembling vertical stripes that represent the solar rotation average of the magnetic field at a specific latitude. When assembled as a time series, the spot snapshots look like butterflies flying in formation.

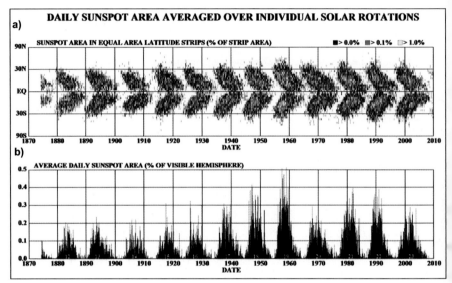

Fig. 9-10. The 11-Year Activity Cycle in Sunspots. a) Here we show the sunspot latitudinal position for 12 solar cycles. With each solar cycle, spot formation begins at high latitudes and proceeds to low latitudes. The cycle length varies, extending anywhere from 8 to 15 years between peak activity. Amplitude variation is often as large as a factor of three. **b)** This figure plots the areal coverage for 12 activity cycles. Each spot cycle exhibits clear temporal asymmetry—the number of sunspots rises more quickly from sunspot minimum to sunspot maximum and decays more slowly. Cycles with more spots tend to have a faster rise time. *(Courtesy of David Hathaway at NASA)*

Coriolis forces on the emerging bipolar magnetic regions of a given 11-year cycle cause the westward or leading spots to be slightly equatorward of their eastward or trailing spots. The north-south dipole moments associated with these tilts are opposite to that of the initial polar fields. The cumulative effect over the cycle is to reverse the Sun's net dipole vector. At the beginning of a cycle, the polar field has the same sign as the leading spots and the opposite sign from trailing spots. Cancellation of the polar fields requires a net trailing-polarity flux be transported from the activity zones to the poles. The lower latitude leader flux preferentially diffuses across the equator and annihilates its opposite-hemisphere counterpart. Surface meridional flow carries the decaying remnants of trailing-polarity flux in each hemisphere as it migrates to the poles. The polar field reversal occurs at solar maximum. The result of spot and diffuse-region migration is a global magnetic field cycle that is ~22 years long.

Increases in radiative emissions accompany active region formation. In the photosphere, faculae are more apparent during solar maximum. And when emerging magnetic fields become stronger and more vertical, plasma migrates upward along magnetic field lines into the upper photosphere and chromosphere. The enhanced plasma radiates more energy and makes its presence known as plages. Plages may appear even if the field is not strong enough to create a sunspot. Figure 9-11 illustrates one of the paradoxes of solar cycle behavior. When more sunspots appear in the photosphere, reducing visible emissions, the total solar irradiance actually increases. The extra, shorter wavelength emissions from faculae and plages exceed the reduction in visible wavelengths.

9.1 The Solar Cycle and Its Roots

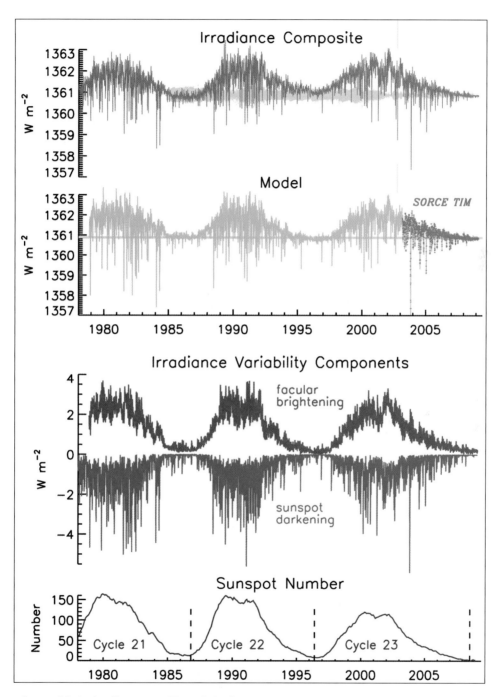

Fig. 9-11. Comparisons of Solar Irradiance over Three Solar Cycles. The upper panel is a record of total solar irradiance (TSI) obtained as an average of three observational composites. The gray curve indicates the standard deviation in the measurements. In the second panel are irradiance variations estimated from an empirical model that combines the influences of facular brightening and sunspot darkening. The blue overlay at the end of the model record shows the comparison with observations from the TIM instrument of the SORCE mission. The third and fourth panels represent the increase in TSI from facular brightening and the decrement due to sunspot darkening. During solar cycle maximum, the facular brightening is about 2.5 W/m². During the same intervals the sunspot darkening, although variable, averages about −1.5 W/m². The lower panel gives the annual mean sunspot numbers for solar cycles 21, 22, and 23, with times of minima indicated by the dashed lines. [Lean, 2010]

9.2 Active Sun Emissions

Magnetic flux rising into the solar atmosphere is how the Sun energizes space weather storms.

9.2.1 Energy Paths of Magnetic Reconfiguration

Objective: After reading this section, you should be able to...

♦ Describe how magnetic merging contributes to eruptive solar emissions

Magnetic reconfiguration is at the root of most space weather events. Figure 9-12 illustrates the paths for energy release from large-scale solar magnetic reconfiguration via flares, energetic particles, and coronal mass ejections. A *flare* is an explosive release of energy stored in the strong chromospheric and coronal magnetic fields in and above active regions. The radiative effects from solar flares arrive at Earth first. Enhanced shortwave photon flux ionizes and heats the dayside atmosphere and often changes the radio propagation characteristics of the ionosphere. Some terrestrial high frequency (HF) radio waves may not propagate at all during such events. Heat and regions of enhanced ionization eventually spread around the globe, affecting satellite orbits and radio communication. Coincidentally, radio bursts from the vicinity of the flare site may jam radio-dependent technologies on Earth's dayside.

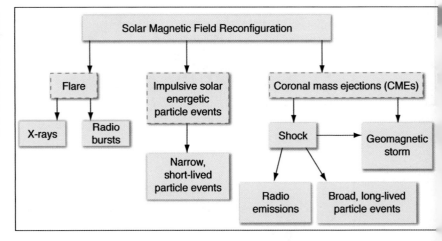

Fig. 9-12. Solar Activity Affecting Earth. The flow goes from the fundamental magnetic flux emergence and reconfiguration at the Sun to the observable processes associated with photons, particles, and fields, shown in the dashed boxes. About half of the reconfiguration energy goes to flares and half to mass ejections. Although the individual particles in energetic particle events tend to have energies in the MeV range, such events are rather rare and put only small amounts of energy into the heliosphere.

Back at the Sun, at the site of magnetic reconfiguration, individual particles accelerate to very high speeds. Some of them accelerate into the lower solar atmosphere, where they create the flare ribbons observed in white light. In the opposite direction, accelerated particles travel along interplanetary magnetic field lines that may connect the flare site to Earth. These particles arrive in narrow impulsive beams. Coronal mass ejections also develop during many solar magnetic merging events (Fig. 9-13). These clouds of plasma and magnetic field have diverse effects on Earth's magnetic field. The effects depend on the CME field orientation with respect to Earth's dayside north-pointing dipole field, a well as the CME speed and the plasma density. Some high-speed coronal mass ejections may drive shocks that energize particles along a wide swath in the interplanetary medium. These particles may arrive at Earth over extended intervals. High-energy protons (1 MeV–1 GeV) often penetrate Earth's magnetic umbrella. They may collide with satellite materials; become resident in the magnetosphere; or penetrate into the polar ionosphere, where they shut down radio communications by changing radio wave propagation characteristics.

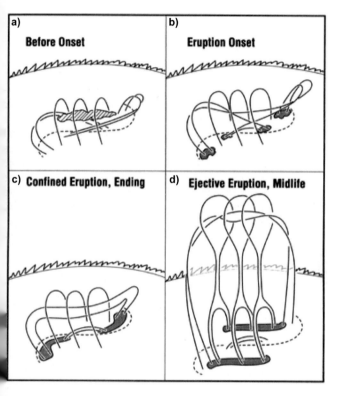

Fig. 9-13. **Development of a Solar Eruption.** Eruptions consist of flares, coronal mass ejections, and energetic particles. This series of drawings shows magnetic field lines that stretch and reconfigure during a solar eruption. **a)** Before eruption onset, a set of arched, quiescent field lines overlie other stretched and perhaps emerging field lines. These arched field lines maintain equilibrium in the system and are sometimes called strapping fields or tethering fields. The strapping fields also overlay a filament (shown in diagonal stripes) and a magnetic neutral line (dashed). **b)** During eruption onset, the stretched field lines beneath the arch begin to merge, thus forming extended ropes of magnetic flux. At the footpoints of the merging field lines, energetic particles, accelerated downward in the merging process, produce chromospheric hot spots (dark patches). The filament has been removed for simplicity. **c)** Some merging events are not energetic enough to break through the over-arching field lines. These eruptions may produce flare signatures (shown as dark patches), but do not produce an eruption. **d)** Energetic eruptions stretch the strapping field lines and allow further merging in the pinched regions beneath. Loops of magnetic field and mass from the pre-existing filament escape to space. At the footpoints of the pinched field lines, energetic particles collide with the chromosphere particles to produce flare ribbon signatures. [Moore et al., 2001]

9.2.2 Flares and Radio Bursts

Objectives: After reading this section, you should be able to...

- Discuss the general characteristics of flares and radio bursts
- Explain how flares develop
- Describe flare categories
- Relate radio bursts to other solar phenomena
- Describe the origins and impacts of radio frequency interference (RFI)

Flare Characteristics. During flares the magnetic field line geometry is rearranged, and a portion of the stored energy is ultimately released as photons. Strong electric fields accelerate charged particles outward to the corona and downward into the denser layers of the Sun (Fig. 9-14). Roughly 10^{36} electrons accelerate to an average energy of 30 keV each second (Example 9.3). These large

Chapter 9   The Active Sun and Other Stars: Sources of Space Weather

fluxes of charged particles, in turn, produce the high levels of electromagnetic radiation emitted by the flare. Most large solar flares occur in mature active regions. Active region size is strongly correlated with flux in the 0.1 nm–0.8 nm soft X-ray band. Although flares may form any time during the development and dissipation of an active region, flare activity peaks when sunspots in active regions attain their maximum coverage. The rate of energy release during a flare is generally on the order of 10^{20} W. Large flares release up to 10^{22} W.

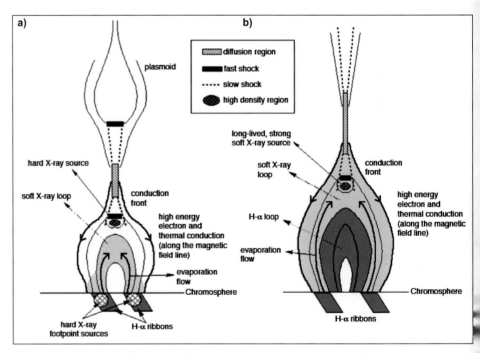

Fig. 9-14. Solar Flare Concepts. Most flares have a sharp rise time (impulsive phase) and a prolonged decay time (gradual phase). **a)** During flares, electrons are heated and accelerated by magnetic merging processes in diffusion regions where oppositely oriented magnetic fields join. Impulsive flare events occur when magnetic energy converts to kinetic and thermal energy. Very fast jets of plasma from the diffusion region create fronts of hot plasma. These fronts, along with individual highly accelerated electrons, interact with nearby matter. Hard X rays develop where the accelerated electrons produce braking radiation, as they slam into regions of denser gas. Hot electrons from the plasma fronts heat the thermal plasma in the lower magnetic loops. This heated gas rises (evaporates) from the chromosphere and also heats the thermal plasma in the overlying loop. Above the diffusion region, magnetic field and matter are released into interplanetary space as a plasmoid (coronal mass ejection). **b)** In less powerful, but prolonged events (long-duration events) or in the subsiding interval of an impulsive flare, radiative cooling of the previously heated gases produces soft X rays in the corona and H-α emission in the chromosphere. These events, called long-duration events (LDEs), span tens of minutes to hours (Figs. 9-16 and 9-17). [Magara et al., 1996]

Flares emit radiation over the entire electromagnetic spectrum. The energy of a typical flare is ~10^{-5} of the Sun's total output. However, its spectrum differs considerably from that of the Sun's background emission. While a flare may produce noticeable brightening in certain visible wavelengths, its greatest output above the background solar radiation is in X rays, extreme ultraviolet (EUV), and radio waves (Fig. 9-15). A solar flare enhances X rays and EUV and radio radiation by a factor of 100 to 10,000. In the radio and X-ray portions of the spectrum, flare emission from relatively small regions often exceeds that of the rest of the solar disk in those wavebands. Large flares heat the local chromosphere and corona to extremely high (10^7 K) temperatures and accelerate particles to relativistic speeds (5%–87% of light speed or kinetic energies of 1 MeV–1 GeV). The most intense flares produce enough enhancements in white light to be visible at Earth's surface. We discuss such an event in Chap. 1—the Carrington white-light flare of 1859.

EXAMPLE 9.3

Flare Power and Energy

- *Problem Statement:* Determine the flare power produced by accelerating 10^{36} electrons to 30 keV each second. This event would be a moderate-to-large flare.
- *Concept:* Power
- *Given:* 10^{36} electrons accelerated to 30 keV each second.
- *Assume:* Energy is distributed evenly to each electron.

- *Solution:* power = energy/time

   total energy = 30,000 eV/electron × 10^{36} electrons

   = 30,000 eV/electron × (1.6×10^{-19} J/eV) × 10^{36} electrons

   total energy = 4.8×10^{21} J

   power = 4.8×10^{21} J/s = 4.8×10^{21} W

- *Implication and Interpretation:* This amount is a typical flare power. The energy is not evenly distributed as assumed; rather some of the particles achieve higher energies (Fig. 9-15) and some achieve lower energies.

Follow-on Exercise 1: If a 30 keV electron suddenly decelerates, what is the wavelength of the photon emitted?

Follow-on Exercise 2: Given the same flare power, if all of the electrons were instead accelerated to the energy of 0.4 nm photons, how many electrons would be accelerated?

Flare Genesis. Large flares usually begin with instability in the coronal magnetic fields overlying complex photospheric sunspots. This instability lets the magnetic structures stretch. Eventually the field lines reconnect (merge) along the neutral line, releasing energy and further amplifying the initial instability (Fig. 9-13). The merging process grows explosively, heating the plasma and initiating the main phase of the flare. Plasma accelerates from the merging line (near the pinch), into the chromosphere below (leading to compressional and shock heating) and into the corona above (leading to short-wave radiation and radio bursts). Electron acceleration provides most of the flare X-ray increases.

Electrons near the merging site heat up and some of them also accelerate away from the flare site. The accelerated electrons attain energies that are much higher than the mean energy of those in the ambient plasma. The heated thermal plasma radiates soft X rays with temperatures on the order of 10 MK–30 MK (~1.5 nm–0.5 nm). Individual spectral lines from the plasma constituents and thermal bremsstrahlung radiation contribute to the soft X-ray enhancement. The hard X-ray to gamma-ray spectrum (wavelength of < 0.1 nm) is dominated by bremsstrahlung from the accelerated electrons and thermal bremsstrahlung from plasma with a temperature above 30 MK. Protons are heated and accelerated also, but because of their relatively large mass, they are much less responsive to energy input. Figure 9-16 shows a three-day time series of soft and hard X-ray emissions from the Sun during the storms of late 2003.

Chapter 9 The Active Sun and Other Stars: Sources of Space Weather

Fig. 9-15. Active Sun Radiative Output. The active Sun is a prodigious emitter of X rays and radio waves. Spikes in gamma rays also occur. The extra X-ray emission is associated with flares, usually from magnetic merging. Radio emissions come from coronal plasma oscillations and from flares and CMEs. Excited atomic nuclei energized by high-energy flare particles emit gamma rays in specific wavelengths based on their atomic number. [White, 1977 and Nicholson, 1982]

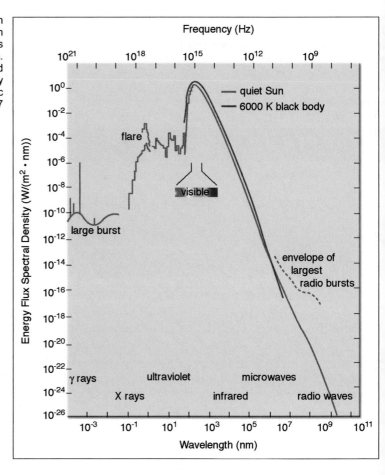

Fig. 9-16. Solar Flare X-ray Flux from the Entire Sun Observed at Geosynchronous Orbit. This diagram shows a multi-day history of emissions in the 0.05 nm to 0.8 nm wavelength bands measured by two GOES satellites. The plots often overlie each other. Flare categories are on the right-hand scale. Soft X rays are in the top curve, hard X rays are in the bottom curve. The sharp rise takes place during the impulsive flare phase. The slower reduction in output coincides with the decay or gradual phase. Flares emit energy in wavebands other than those shown. Constraints on technology limit the wavelengths that can be efficiently monitored from space. *(Courtesy of NOAA's Space Weather Prediction Center)*

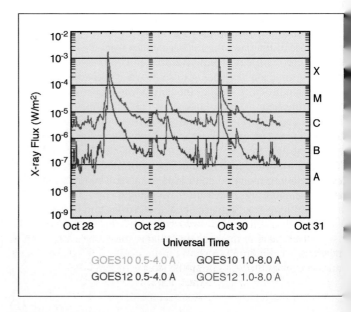

Some flares emit copious amounts of hard X rays and even gamma rays (left side of Fig. 9-13.). Gamma-ray emission is an indicator of high energy nuclear interactions. Ion spectral lines are observed at gamma-ray energies between ~400 keV and 8 MeV. These spectral lines develop when an ion with energy greater than 1 MeV collides with an atomic nucleus in the solar atmosphere. The excited nucleus then radiates a photon at a discrete energy characteristic of the nucleus. The most intense radiation emanates from low in the solar atmosphere, where the plasma density is high. A fraction of the protons accelerate outward into the heliosphere and may intercept Earth.

Many high-energy flares happen in conjunction with a dynamic structure called a *loop prominence system (LPS)*. An LPS seems to be material that's ejected into the corona, then falls back to form a bright structure above the flaring region. From there it flows down to the chromosphere along the magnetic arches on either side of the bright region. An arcade (fan) of such loops forms the bright prominence system and bridges the magnetic neutral line (Fig. 9-17). An LPS generally forms several minutes after the flare peak time, and may persist in soft X-ray and EUV images for one or more hours. The EUV emission often peaks after the X-ray emission, indicating that the emitting material is slowly cooling.

Flares seen in longer wavelengths are the low-altitude signatures of magnetic merging. Traditional H-α flares are chromospheric footpoints of flares, while white-light flares locate the footpoints of extreme X-ray flares that have accelerated particles downward to collide with the photosphere. The Carrington Flare of 1859 (Fig. 1-16) was a white-light flare.

Flare Timing. Flares develop in stages, each lasting from a few seconds to about one hour. Power output usually rises sharply and decays much more slowly. A likely scenario for radiative emissions from large flares is as follows:

- *Precursor* **stage:** A magnetic field reconfiguration releases magnetic energy. Hot chromospheric-coronal plasma radiates thermally at EUV and soft X-ray wavelengths, corresponding to photons emitted when ions capture electrons with kinetic energy <10 keV.

- *Impulsive* **stage:** Gamma rays arise from only the most energetic electrons (kinetic energy >1 MeV) and from fusion reactions, including electron-positron annihilation. Hard X rays are emitted when energetic electrons (kinetic energy >20 keV) accelerate away from the neutral point and collide with denser plasma below. Electron collisions in the hot plasma energize the hydrogen to the n = 3 state (Balmer series). Subsequent n=3→n=2 electron transitions release H-α photons (656.3 nm). The H-α emissions may appear as ribbons of light in the chromosphere, outlining the footpoints of field lines on which plasma is streaming downward. Plasma accelerated into the corona causes radio bursts.

- *Decay* **stage:** Soft X rays gradually cease.

The total energy released during a flare ranges from 10^{21} to 10^{25} J. Although large, even the greater amount is less than one-tenth of the total energy emitted by the Sun every second.

Fig. 9-17. Flare Arcade. This close-up of Active Region 9077 was made shortly after a large flare erupted. The false-color image from a long duration event was made by the Transition Region and Coronal Explorer (TRACE) satellite. TRACE satellite records the million-degree solar plasma cooling while suspended in an arcade of magnetic loops. The image covers an expansive 230,000 km by 170,000 km area on the Sun's surface and was recorded in extreme ultraviolet light. Resembling a "slinky" toy, the enormous loops are actually magnetic field lines that trap the glowing, cooling plasma above the cooler and relatively darker (in X ray) solar surface. *(Courtesy of NASA and the TRACE Satellite System)*

Pause for Inquiry 9-7

Compare flare energy output to that of an extremely large (level 9 on the Richter scale) earthquake.

Chapter 9   The Active Sun and Other Stars: Sources of Space Weather

> **Pause for Inquiry 9-8**
>
> Verify the total energy values stated in Example 9.3 and the flare duration shown in Fig. 9-16. Flare duration is usually reported as the time that power output exceeds 150% of the background value.

Flares occur preferentially near or on magnetic neutral lines (Figs. 9-4 and 9-13), especially if the neutral line contains many bends and kinks, indicating a complex magnetic structure. Although most flares arise in a particular active region, some develop far from active regions, but in association with the sudden disruption or disappearance of a dark filament. The visible components of such flares are called *Hyder flares*; they indicate a change in the local magnetic field structure (Fig. 9-18). These flares correlate strongly with CMEs and solar energetic particle events.

Fig. 9-18. Hyder Flare Mechanism. A quiescent filament may be supported in a magnetic field trough for a long time. If the trough is suddenly disrupted, some of the material may accelerate downward along field lines in a matter of minutes, generating a Hyder flare. The flare name is associated with C. L. Hyder who first proposed the mechanism in 1967. *(Courtesy of Australian Ionospheric Prediction Service)*

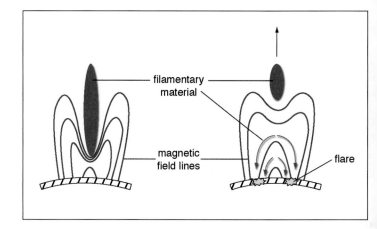

> **Pause for Inquiry 9-9**
>
> Assume that a magnetic field is directed outward from the surface of the Sun on the left side of the elements in Fig. 9-18 and inward on the right side. What is the direction of the current in the left filament? What direction must the disturbance current be to disrupt the filament? Consider Ampere's Law.

High-energy flares may be accompanied by a shock wave that spreads out from the flaring regions in the chromosphere. Viewed in time lapse, these waves appear as an expanding circular ring, sometimes called a *Moreton wave*. Often only part of the circle is apparent. The velocity of expansion is generally around 1000 km/s. Occasionally two or more widely separated flares begin nearly simultaneously. This combination requires a magnetic connection between active regions. The connection may be in the photosphere below, in the corona above, or via Moreton waves that initiate flares in an adjacent region through which they pass. Moreton waves correlate strongly with type II radio bursts, which we discuss in the next section.

A different type of sunquake wave is produced by high-energy particles accelerated from the corona into the underlying chromosphere of the active region. The extreme heating causes a pressure transient that drives a seismic wave into the interior of the Sun. In powerful hard-X-ray flares these particles are almost invariably focused into the penumbra of a sunspot. The sources of seismic emission from flares are now known to be regions of sudden, strong, visible continuum emission white-light flares.

Flare Classification. Early observations of solar flares focused on the white-light classification of flares, because this value could be determined from the ground. We classify flares in the optical wavelengths according to their size and intensity. The size (*importance*) is based on flare area measured in millionths of solar disk area, and in the Doppler velocity shift found in the H-α observations. Optical flare duration also relates directly to its importance, as shown in Table 9-2.

Table 9-2. **Optical Flare Importance and Duration.** Here we list values for optical flare importance, size, duration, and relative frequency. The size of the solar hemisphere is taken to be $2\pi \times R_S^2 \approx 2\pi \times (696{,}000 \text{ km})^2 = 3.04 \times 10^{12} \text{ km}^2$. *(Courtesy of AF Weather Agency)*

Importance	Size in Millionths of the Solar Hemisphere (Disk)	Size ($\times 10^6$ km^2)	Average Duration (min)	Percent of All Flares	Typical Corresponding X-ray Class
0	≥ 10 – <100	30–300	~15	75	C
1	≥ 100 – <250	300–760	~30	19	M-2
2	≥ 250 – <600	760–1800	~70	5	X-1
3	≥ 600 – <1200	1800–3700	>120	<1	X-5
4	≥ 1200	>3700	>120	<1	X-9

With the advent of continuous satellite monitoring, X-ray flux has replaced optical measurements as the determinant of flare strength. The X-ray flux is an integrated measurement, covering the entire solar disk, so visible observations still help us to distinguish among simultaneous flares. Operational measurements are made in X-ray bands between 0.05 nm and 0.8 nm. Flare reports are categorized by flux intensity (W/m^2) of soft X rays (Levels A, B, C, M, and X) in the 0.1 nm to 0.8 nm range as shown in Fig. 9-16 and described in Appendix A. Short-wavelength imagers are being fielded on new satellites and will soon create a new standard for X-ray observations.

Solar Radio Bursts. Several mechanisms produce microwaves and radio waves in the corona, where matter is completely ionized. Coronal plasma produces radio waves at the plasma frequency associated with the local electron density. Additionally, the free electrons emit bremsstrahlung as they accelerate in the vicinity of ions. Blackbody radiation, plasma frequency waves, and bremsstrahlung are thermal radio sources. Because the corona is permeated with magnetic fields, the charged particles that constitute the coronal plasma spiral around these magnetic field lines, continuously emitting cyclotron radiation. Changes in magnetic field strength or plasma density change the nature and intensity of this cyclotron radiation. Relatively fast particles produce other radio emissions. If a charged particle is traveling at an appreciable fraction of light speed, *Cerenkov plasma waves* form. Cerenkov radiation comes from charged particles moving through a medium faster than the phase speed of light in that medium. Charged particles are surrounded by electric fields. When a charged

Chapter 9 The Active Sun and Other Stars: Sources of Space Weather

> **Pause for Inquiry 9-10**
>
> Use the information from Fig. 9-16 and the NOAA Space Weather Scales (Appendix A) to fill in the flare table below.
>
> | X-ray Category | Flux Intensity | NOAA Scale | Frequency |
> |---|---|---|---|
> | X-10 | | | |
> | | | R-3 | |
> | | | | 2000/cycle |
> | | 10^{-6} W/m^2 | Below scale | Thousands/cycle |

particle moves, the electrical field moves along with the particle. However, since the electrical field is carried by photons, it can only travel at light speed. If the particle is travelling faster than light speed in an ambient medium, then in a sense it out-runs its electric field. Cyclotron radiation and Cerenkov plasma waves are non-thermal radio radiation.

The radio output of the Sun yields unique radio signatures measured at Earth. These signatures convey important information about activity in the solar corona.

Noise storms and *gradual rise and fall events* consist of longer duration enhanced continuous emissions above the ambient radio output. Sometimes a large active region produces slightly elevated radio noise levels, primarily on frequencies below 400 MHz. This noise may persist for days. Solar active regions occasionally produce intense flashes of radio waves called *radio bursts*. There are two types: sweep-frequency bursts and microwave bursts.

A *sweep frequency burst* is emitted when plasma shoots upward into the corona during magnetic reconfiguration. This plasma emits radio waves at a frequency characteristic of the electron density at its location, so the frequency emitted depends on height. As the height of the ejected plasma increases, the electron density decreases. Thus the plasma emits at successively lower frequencies—a "sweep" from higher to lower frequencies.

Some *microwave bursts* originate from several processes in the chromosphere and lower corona, and are usually related to downward-moving plasma. They last from a few minutes to hours. Intensities range from a few solar flux units (1 SFU = 10^{-22} W/(m$^2 \cdot$ Hz)) to 10,000 SFU above the ambient emission. Radio bursts greater than 1500 SFU generally accompany large flares, have a more complex profile, last longer, and are less common. *Impulsive microwave bursts* originate with cyclotron radiation from downward-moving electrons in a single burst. In rare cases, protons also contribute a flare radiation component.

Electrostatic bursts occur when electron beams flowing parallel to the magnetic field outrun slower electrons. These beams drive bursty electrostatic Langmuir waves. Table 9-3 relates burst phenomena to the descriptions commonly provided in space weather forecasts.

Table 9-3. Solar Radio Bursts by Type. This table summarizes solar emissions in radio wavelengths from centimeters to dekameters, under both quiet and disturbed conditions. *(Courtesy of Solar Influences Data Analysis Center, Royal Observatory of Belgium, Brussels)*

Type	Description
Type I	A noise storm composed of many short, narrow-band bursts in the meter range (300 MHz–50 MHz). *(Noise storm)*
Type II	Narrow-band emission that begins in the meter range (300 MHz) and sweeps slowly (tens of minutes) toward dekameter wavelengths (10 MHz). Type II emissions occur along with major flares and indicate a shock wave moving through the solar atmosphere. *(Sweep frequency burst)*
Type III	Narrow-band bursts that sweep rapidly (seconds) from decimeter to dekameter wavelengths (500 MHz–0.5 MHz). They often occur in groups and are an occasional feature of complex solar active regions. *(Electrostatic microwave bursts)*
Type IV	A smooth continuum of broad-band bursts primarily in the meter range (300 MHz–30 MHz). They accompany some major flare events beginning 10 to 20 minutes after the flare maximum, and can last for hours. *(Noise storm)*

Focus Box 9.2: Impacts of Active-Sun Radiative Emissions at Earth

Because they propagate at light speed, flare effects are usually underway by the time the flare is observed and categorized. The immediate impacts are mostly limited to Earth's sunlit hemisphere, where they tend to subside shortly after the flare ends. System effects include: absorption of HF (3 MHz–30 MHz) radio communication, scintillation of satellite communications (SATCOM) and radar interference from enhanced background noise.

Radio Communication. Flare X-ray and EUV radiation immediately increases the ionization rate of the ionosphere, especially in the lower layers. As a result, electron density increases in most layers of the ionosphere, producing a *sudden ionospheric disturbance (SID)* or a sudden increase in total electron content (SITEC). Radio signals are absorbed by the enhanced ionization, thereby reducing communication reliability. Low radio frequency disturbance events occur within minutes of an X-ray flare and subside within 30 minutes to several hours. Disturbance strength depends on the solar angle, with an overhead Sun producing a more dramatic disturbance. Table 9-4 provides an empirical correlation between HF blackout durations and the affected radio frequencies with flare magnitudes.

Short wave fades (SWFs) result from abnormally high absorption (signal fade) of an HF radio signal in the D layer. Earth-based radio waves must travel through the D layer to reach the F layer; but each passage through the D layer reduces signal strength. Enhanced absorption may at times be strong enough to completely close the HF (3 MHz–30 MHz) shortwave propagation window, causing a *short wave blackout* (Sec. 8.3). The amount of signal loss depends on a flare's X-ray intensity, location of the HF path

Table 9-4. **X-ray Flare Intensity.** Here we list typical flare intensities with the affected radio frequencies and disruption duration. X-ray intensity is denoted with M and X scales (Fig. 9-15). The optical intensity is rated on the B (brilliance) scale. *(Courtesy of AF Weather Agency)*

X-ray Intensity (Optical)	Lowest Usable Frequency	Approximate Duration
M1 (0B)	12 MHz	20 min
M5 (1B)	15 MHz	30 min
X1 (2B)	20 MHz	40 min
X5 (3B)	28 MHz	90 min
X9	32 MHz	120 min

relative to the Sun, and design characteristics of the system. Some SWFs are accompanied by effects that may cause the altitude of the D-layer base to lower slightly, producing a *sudden phase anomaly (SPA)* that affects very-low-frequency (VLF, 3 kHz–30 kHz) and low-frequency (LF, 30 kHz–300 kHz) transmissions.

Dayside Satellite Communication and Radio Receiver Interference. Flare effects increase solar radio wave energy transmitted to Earth's dayside by a factor of tens of thousands in the 30 MHz to 30 GHz (super-high-frequency) range. The lower frequency portion of the radio burst is absorbed or reflected, but higher frequencies penetrate through the atmosphere to the ground. If the Sun produces a sufficiently intense radio burst at a suitable frequency in the field of view, then radio frequency interference (RFI) will occur on a SATCOM link or in a radar's detection and tracking circuits. Radio interference tends to persist only for a few minutes to tens of minutes. Knowledge of a solar radio burst allows a SATCOM or radar operator to isolate the RFI cause, and avoid time-consuming investigation of possible equipment malfunction or jamming.

GPS receivers are susceptible to particularly strong radio bursts. In December 2006, a number of radio noise bursts occurred with X-class flares. The burst recorded at the Learmonth Solar Observatory on 13 December saturated the receiver at levels exceeding 100,000 SFU. Loss of lock with GPS satellites was experienced for around two hours following the event, and some receivers sustained a complete loss of navigation solution for 6–10 minutes.

Forecasters can not yet forecast a specific disturbance event; rather they predict the likelihood of an event based on the probability of flare occurrence, which in turn is determined by an overall analysis of solar features and past activity. After a flare observation, forecasters quickly issue a warning that predicts the affected frequencies and the duration of signal absorption.

9.2.3 Coronal Mass Ejections (CMEs)

Objectives: After reading this section, you should be able to...

- Describe the origin and structure of coronal mass ejections
- Describe the association between flares and CMEs
- Describe the signature of CMEs

Coronal Mass Ejections (CMEs) Characteristics and Sources

Coronal Mass Ejection Characteristics. *Coronal mass ejections (CMEs)* erupt from the Sun during large-scale reorganization of the magnetic fields in the solar atmosphere (Figs. 9-13 and 9-19). The ejections are huge, transient, magnetized bubbles of plasma from closed field domains near filaments and active regions. They often happen in conjunction with active regions, however about 20% of CMEs erupt from a filament on a spotless Sun. Although initially observed in the corona, some of the erupting material may be dredged up from the chromosphere. A mass release (10^{12} kg to 10^{13} kg) carries with it a substantial magnetic field that is oriented out of the ecliptic (Fig. 9-19b). The ejecta usually have expanding curvilinear shapes that resemble cross sections of loops, shells, or filled bubbles. The magnetic field threads the plasma and confines it as it moves into interplanetary space.

These magnetized bubbles spread through the interplanetary medium and eventually, if moving in the right direction, engulf Earth's magnetic field. Fast ejections create interplanetary shock waves that accelerate ambient protons, causing solar energetic particle events and geomagnetic storms. Well over 10,000 CMEs have been observed from the ground and space since the beginning of the space age (1958). Table 9-5 provides statistics on CME behavior over the course of a solar cycle. Solar maximum ejections are more massive, wider, more energetic, and decelerate faster in the solar wind than do solar minimum ejections.

The structure of CMEs exhibits a high degree of variability, but the simplest have three visible parts: a bright outer rim, a dark central cavity, and a bright inner core. We see these in Fig. 9-19a. The hot outer rim is bright because of the bunched matter that reflects light. The dark cavity may contain a magnetic flux rope with field strength as high as 10^{-4} T to 10^{-3} T, while the inner core may contain the dense prominence or filament material from the chromosphere. The inner core material is often cooler to start (~10^4 K); however, it may be heated by flare processes and then cool as the CME expands. The viewing geometry affects whether we see any or all of the parts. And the brightness of the outer rim and inner core depends on the density of material, which varies. For example, if the CME is moving slowly, it may not cause a large outer rim density increase.

The temperature in CME source regions can exceed 10 MK. The outer rim region usually has a temperature of several MK. Matter exposed to these high temperatures tends to be highly ionized. Chemical elements inside CME structures have higher ionization states than those in the ambient solar wind. *In situ* solar wind observations reveal the presence of ejecta by comparing charge states.

CME Genesis. The origin of CMEs is still a topic of debate and research. Sub-photospheric fluxes store magnetic fields and energy. In most cases the emerging flux reaches a quasi-static equilibrium. Figure 9-20 shows a relatively stable emerging flux rope in the chromosphere and low corona. As long as twisting and

Chapter 9 The Active Sun and Other Stars: Sources of Space Weather

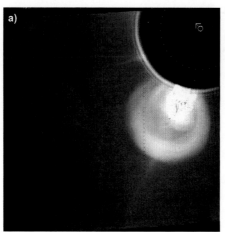
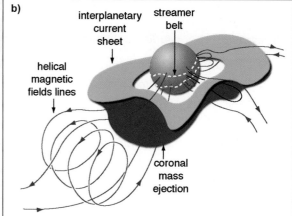

Fig. 9-19. **a) Coronal Mass Ejection (CME) Observed by a Ground-based Coronagraph.** Many CMEs have a bright outer edge encasing a dimmer cavity that surrounds a bright central core. The inner structure may be a rising solar filament. The CME has moved about two solar radii outward, and is headed roughly 90 degrees away from Earth. The uniform disk at upper right in the image is where the occulter has been placed, blocking all direct sunlight. *(Courtesy of the National Center for Atmospheric Research [NCAR] High Altitude Observatory [HAO])* **b) Interplanetary Coronal Mass Ejections.** This artist's rendition of a CME from the streamer belt emphasizes the helical nature of the ejected magnetic field. *(Courtesy of Nancy Crooker at Boston University)*

Table 9-5. **Solar Cycle Coronal Mass Ejection (CME) Statistics.** Here we list several characteristics of solar phenomena during solar minimum, solar maximum, and solar average. The minus sign on the average acceleration means the CME moves away from the Sun. *(Courtesy of Steve Kahler at the AF Research Laboratory)*

	Solar Minimum	Solar Average	Solar Maximum	Over the Solar Cycle
Rate	< 1/day	3/day	6/day	↑ by factor of 6
Mass		1.5×10^{12} kg		↑ by factor of 2
Speed	300 km/s		550 km/s	↑ by factor of 1.8
Width	43°		58°	↑ by factor of 1.4
Kinetic Energy		10^{24} J		↑ by factor of 4
Acceleration		-11 m/s^2		unclear
Latitude of Departure	22°		63°	↑ by factor of 3

Pause for Inquiry 9-11

Based on Table 9-5, are coronal mass ejections occurring at solar minimum more or less likely to be associated with the streamer belt?

---

shearing stay below threshold values, the magnetic structure remains quiescent. Differential motion and convection can destabilize the flux.

Explosive magnetic merging in the solar atmosphere is energetically favored when twisted flux tubes produce complex sunspot groups in the lower portions of active regions, and magnetic flux ropes (active filaments) in the middle and upper layers. During coronal eruptions the general state of force balance (Fig. 9-20) is interrupted by some types of triggering mechanism. Eruptions may be initiated by photospheric motions that shear and twist the coronal magnetic field. We know

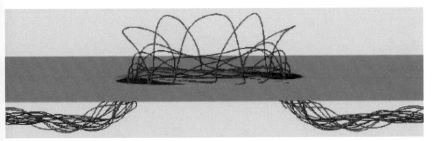

Fig. 9-20. **Model Output of a Quiescent Magnetic Flux Rope.** The subsurface field pierces the photosphere through bipolar sunspots. The expanded loops contain the structure within the atmosphere. In this simulation the overarching strapping fields confine the emerging flux rope. If flux continues to emerge from below, or if the flux rope significantly twists, the strapping field may not be strong enough to contain the emerging flux. [Manchester et al., 2004]

from Chap. 4 that magnetic merging occurs when anti-parallel magnetic field lines created by twisting and shearing reconfigure at an X line and reconnect with new partners (Fig. 9-14). The process rapidly converts magnetic energy into radiation and kinetic energy and results in new topological configurations of the magnetic field lines. In two dimensions an X line is the origin for the energy. In three dimensions it becomes an intersection of surfaces of magnetic field lines. By Ampere's Law, thin current sheets must form in these regions. The magnetic fields move toward the current sheet and merge at the magnetic null. Magnetic merging at a current sheet (pinched regions in Fig. 9-13d) plays an important role in triggering a CME onset and in sustaining the eruption.

Filaments form along magnetic polarity reversal lines and are supported by the magnetic field against gravity and confined by overhead coronal fields. Kinks in the filaments lead to bulging and destabilization that bring oppositely directed fields together. Further, the filament fields may twist so that they contain an orientation opposite that of the confining dipole coronal field.

For some CMEs the weight of the filament mass plays an important role in amplifying the magnetic energy in excess of a threshold value. If the filament mass is disrupted, perhaps by a shock wave or by some kinking instability, and begins to drain toward the surface, the sudden reduction of the prominence weight may trigger the eruption. One model of field reconfiguration suggests that magnetic merging occurs below the filament material. We depict the general structure, described by the streamer model, in Fig. 9-21a. Such eruptions generally correspond to slow-to-moderate speed ejections that the solar wind carries outward without significant acceleration. The degree of field twist and the amount of ejected mass is less than that associated with faster CMEs. In some cases the mass does not successfully eject. Gravitational forces and the confining overhead magnetic fields may bind the mass to the Sun despite magnetic merging.

In a paradigm called "breakout," the overhead, juxtaposed, oppositely directed field weakens the overlying coronal fields that confine the rising filaments (Fig. 9-21b). Sufficient weakening leads to a run-away expansion of the central field. The eruption is triggered by a loss of force balance of the magnetic field and then sustained by subsequent formation of a current sheet and magnetic merging beneath filaments. The process allows twisted magnetic flux ropes to escape or break out into the heliosphere. The release helps the corona shed twisted magnetic field lines and thus locally reduces magnetic energy density.

Chapter 9 The Active Sun and Other Stars: Sources of Space Weather

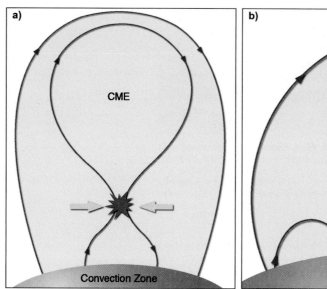 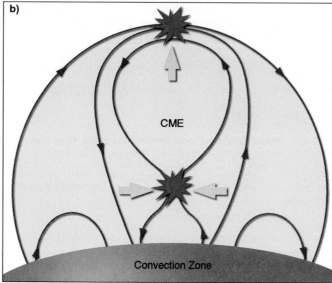

Fig. 9-21. Coronal Mass Ejection (CME) Structure. a) This two-dimensional diagram portrays low-altitude elements of a nascent slow-to-moderate speed CME. We don't show the overlying and confining coronal magnetic field lines. The tops of these structures may glow in soft X-ray wavelengths as magnetic merging commences. The feet of the field lines tie to regions that may produce flare ribbons if particles are accelerated downward. **b)** This diagram depicts additional field merging at a cusp (null) that allows the underlying field to expand outward at high speeds. The erupting field may carry with it filamentary material. *(Courtesy of Spiro Antiochos at the Naval Research Laboratory)*

Numerous other shearing and merging topologies have been proposed. No matter what the source, when complex magnetic field structures develop, nature finds it energetically favorable to simplify the field geometry and dissipate the current sheet by merging the fields at an X line or surface. The process converts magnetic energy to radiation, heat, and mass motion.

Restructuring of the coronal magnetic field produces flares and CMEs. Although the relation between the two phenomena is strong, they are not synonymous. Either may occur without the other. Slightly fewer than half of all flares are associated with CMEs. For flares that lack CMEs, the released energy seems to go mostly into heating, which suggests that small flares contribute proportionally more to coronal heating. In larger eruptive events about half of the energy goes to radiation and half to mass motion. Higher intensity flares correspond to more energetic CMEs. Generally the peak in X-ray output coincides with the peak CME acceleration. Several decades of observations suggest that two types of X-ray flares are associated with CMEs. *Impulsive X-ray flares* of short duration coincide with rapid merging. *Long-duration flares* (sometimes called long-duration events) have a wide range of intensities and last for tens of minutes to hours. They are observed in the vicinity of the arcade loop formation in the solar atmosphere and flux-rope formation in the lifting CME (Fig. 9-17). The long-duration events are the atmospheric embers of a magnetic merging.

Sometimes solar eruptions develop in a series that originates in the same main footpoints defined by visible or EUV images. These rapid-fire events are called homologous events. Eruptions (up to five in a sequence) may occur at the rate of 1 every 10 hours. Each instance of the homologous CME and flares resembles previous ones in the series. The similarity suggests a common root of magnetic flux below the solar surface.

Focus Box 9.3: Soft X-ray and Visible Signatures of CMEs

Some active regions are more eruptive than others. When magnetic fields from adjoining active regions interact, the potential for magnetic reconnection and large-scale flares increases. Hence forecasters want to know how, where, and when they are likely to form, strengthen, and expand. Solar observers watch for regions of growing magnetic complexity, where bipolar regions give way to multi-polar regions in close proximity. Extreme twists in the underlying fields produce a *sigmoid* appearance in soft X-ray images (Fig. 9-22). Such S- or reversed S-shaped signatures are sometimes noted in forecast discussions as connected with high eruption probability.

The northern solar hemisphere preferentially displays reverse-S shapes and the southern hemisphere preferentially shows forward-S shapes. Chromospheric filaments may also have this characteristic. Vector magnetic field observations in the photosphere reveal a small but statistically significant trend for left-handed (sinistral) twist in the northern hemisphere and right-handed (dextral) twist in the southern hemisphere. Scientists believe these sigmoid signatures are CME precursors. The configurations suggest constrained magnetic energy associated with dipped field lines supporting prominence material against gravity (Fig. 9-13). The sigmoid represents current sheet formation along a surface of a twisted flux rope. Hot dense plasma outlines the current sheet and appears as a sigmoid. The signature is not sufficient cause to forecast a CME; rather it is a notice to monitor the area for activity. If the sigmoid activates and is replaced by a cusp or an arcade X-ray signature, then a mass ejection has probably occurred.

In many instances a CME relieves the twist in the magnetic field and the active region becomes less twisted. However, if magnetic flux continues to emerge from below the solar surface, the sigmoid may reform in a few hours or days and again provide the potential for eruption.

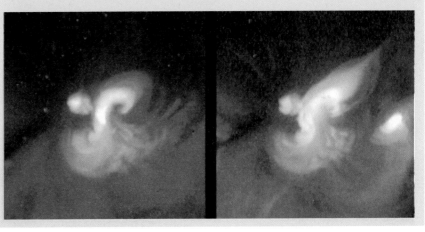

Fig. 9-22. Soft X-ray Signature of a CME. These images made by the Yohkoh spacecraft highlight regions where a twisted magnetic field acted as a root of CME activity. The cusp on the right is consistent with the X-ray loop geometry of Fig. 9-20 and the diagram in Fig. 9-21. *(Courtesy of Yohkoh team: the Lockheed Palo Alto Research Laboratory, the National Astronomical Observatory of Japan, and the University of Tokyo with the support of NASA and the Institute of Space and Astronautical Science)*

CME Origin Characteristics. Scientists determine the speed, angular width, and acceleration from sequences of eruption imagery. The speed measured in the sky plane varies from <20 km/s to >3000 km/s—speeds that range from well below sound speed in the lower corona to highly supersonic. The average speed is ~470 km/s. The highest speed recorded is ~3387 km/s for a CME on November 10, 2004. Average CME speed rises with increasing sunspot activity, consistent with more twisted and energy-laden magnetic fields in active regions. Some active regions become hyperactive. For example, region 10720 in January 2005 produced 11 CMEs as it passed from just east of the central meridian to the west limb. Two of the CMEs intercepted Earth at very high speed, creating major magnetic storms.

Projection effects and viewing angle influence our perception of CMEs. Ejecta width ranges from <5° to 360°. CMEs with apparent widths of 360° are called *halo CMEs*. They are more energetic than average CMEs. When they originate on the frontside of the Sun, they can directly impact Earth. Halo CMEs constitute

<5% of all CMEs, while CMEs with widths ≥120° account for ~10%. The latter category is sometimes called *partial halo CMEs*. Ejecta that are clearly viewed from the side and are moving nearly perpendicular to the Earth-Sun line are called *limb CMEs*.

The speed of CMEs often changes near the Sun because they are subject to propelling and retarding forces. The medium into which the CME is expelled exercises some control on its travel. For example, if a fast CME must push through relatively dense streamer belt material, it will likely slow down, perhaps even to the speed of the ambient solar wind. Slower CMEs accelerate to ambient solar wind speeds. Ejecta ahead of a high-speed stream are likely to accelerate, deflect, and distort. On average, the acceleration is almost zero.

Two general classes of CMEs appear in the data: gradually accelerating CMEs often corresponding to prominence eruptions, and fast CMEs associated with large flares and active regions. The acceleration for most CMEs takes place below $4R_S$. Accelerations and decelerations are typically in the range of 10 m/s²–30 m/s².

Pause for Inquiry 9-12

In Fig. 9-13 identify the following elements:

- Flare ribbons
- CME
- Footpoint motion leading to shearing
- Filament
- Magnetic neutral line
- Arcade
- Merging line or surface

Gradually accelerating CMEs are balloon-like, with central cores that accelerate more slowly than the leading edge. They tend to appear as smooth structures that accelerate to a limiting speed typical of the solar wind. Though their accelerations are less than those of fast CMEs, gradual CMEs may travel fast enough for shock production and particle acceleration.

Fast CMEs often have a ragged appearance in coronagraph imagery (Fig. 9-23). This results from a combination of the explosive magnetic reconnection that creates them and the violent interaction with their surroundings as they accelerate in the low corona. After accelerating, fast CMEs move at constant speed even as far out as $30R_S$, but then decelerate as they plow through the interplanetary medium. Even though decelerated, fast CMEs continue to have interactions with the ambient solar wind that produce shock waves that in turn accelerate individual solar wind particles to high energies (Sec. 10.3).

In the interplanetary medium, most ejections form large loops with ends extending back toward the Sun. Satellite data suggest that the ejected plasma is entrained on expanding helical magnetic field lines. Hot electrons bounce back and forth along the legs of the loops, if the loops remain attached to the Sun. The hot counter-streaming particles help identify such structures when they pass Earth.

Ejecta from near the central meridian usually move radially outward. At Earth the organized wrappings of the magnetic field are seen as *interplanetary flux ropes* or *magnetic clouds*. A magnetic cloud is a transient in the solar wind defined by relatively strong magnetic fields, a large and smooth rotation of the

Fig. 9-23. A Spectacular Coronal Mass Ejection (CME) on January 4, 2002. The event began as a filament eruption. Although the center of the CME was directed almost a full 90 degrees away from the Earth-Sun line, the event appeared as a (weak) full halo event. *(Courtesy of ESA/NASA – the Solar and Heliospheric Observatory (SOHO) Program)*

magnetic field direction over approximately 0.25 AU at 1 AU, and a low proton temperature. They often have extended intervals of southward and northward magnetic fields. There are significant exceptions, but non-cloud CME sources are mostly at larger central meridian distances.

Mass ejections usually arrive at Earth within two to four days of their launch from the Sun. The fastest have arrived within 24 hours. A typical halo CME arrives in 3.5 days. During solar minimum, the Sun shoots out an ejection about once every other day, but during solar maximum, the rate rises to about four per day. A small fraction of these intercept Earth.

EXAMPLE 9.4

CME Kinetic Energy

- *Problem Statement:* How much of the energy in a large solar eruption goes into kinetic energy?
- *Concept:* Kinetic energy = $mv^2/2$
- *Given:* Large CME mass = 10^{13} kg
- *Assume:* Large solar eruptions are associated with fast CMEs. A reasonable speed is 1000 km/s.

- *Solution:*

$$KE = 1/2 \; mv^2$$
$$KE = 1/2 \; (10^{13} \text{ kg})(10^6 \text{ m/s})^2$$
$$= 5 \times 10^{24} \text{ J}$$

- *Implication and Interpretation:* Large, fast CMEs take away an energy similar to that associated with flares. About half of solar eruption energy goes to radiation and half to mass motion.

Follow-on Exercise: What average magnetic field strength is required to provide sufficient energy density in 10^7 km^3 to provide the eruption energy?

Magnetic Topology of Coronal Mass Ejections

Here we provide a conceptual view of the development of a flux rope CME. Figure 9-24 is a replication of some elements of Figs. 9-13 and 9-14 that places more emphasis on the magnetic topology in two and three dimensions. Photospheric motion beneath an active region shears and twists the arcade of field lines overlying a photospheric neutral line. The result is stored magnetic energy. The shear aligns the fields so that oppositely directed fields are in close proximity in a current sheet. Merging rearranges the fields to create 1) low-lying closed field lines and 2) overlying field line loops in the form of a plasmoid or flux rope, whose ejection forms a CME. The recently merged field lines map down to chromospheric ribbons, that appear to separate during the ongoing merging. Accelerated electrons on these field lines collide with the atmosphere producing ribbons of emission in the hydrogen Lyman alpha (L-α) line. The upper portions of the closed field lines may glow with X-ray and or EUV emission (Fig. 9-17) for minutes to hours after the plasmoid or flux rope ejects into the heliosphere. When viewed from the side as a limb CME, the ejected flux rope can take on the appearance of magnetized fireball (Fig. 9-23).

Heated electrons populate the field lines of the ejecta. Electrons on these still-connected-to-the-Sun field lines bounce back and forth, producing a counterstreaming electron flow. Measurements of electron fluxes in the solar wind indicate that CME field lines near and beyond 1 AU are usually connected (~60%) to the Sun at both ends as suggested in Fig 9-19b. On occasion some field lines embedded within CMEs appear to be connected to the Sun at only one

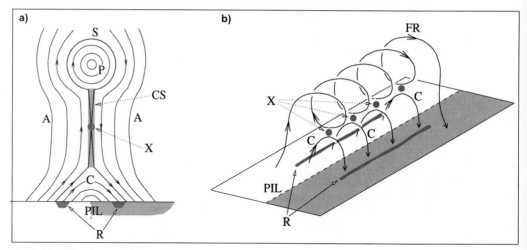

Fig. 9-24. Topology of Magnetic Field Lines Leading to a Flux Rope CME. a) Two-dimensional View. Magnetic merging takes place as oppositely directed arcade field lines flow toward the elongated current sheet and produce closed field lines below and a magnetic flux rope above. If the flux rope completely disconnects from the Sun, it is called a plasmoid. The field line direction changes at and above the neutral line. Flare ribbons show the location were accelerated particles impact the denser chromosphere and photosphere below. **b) Three-dimensional View.** This schematic shows the post-merging eruption of field (and attached) plasma. The merging has taken place along a line of arcade field lines indicated by the dots. The end lines remain attached to the Sun, but under some circumstances can later disconnect. (PIL is polarity inversion line; CS is current sheet; A is arcade field lines; X is X-line reconnection; C is closed field lines; P is plasmoid; R is flare ribbon; S is separatrix; FR is flux rope.) [Longcope and Beveridge, 2007].

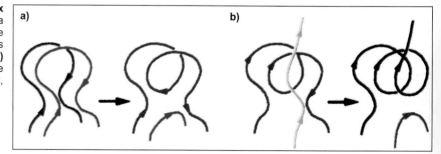

Fig. 9-25. Connected and Disconnected Flux Ropes. a) Merging configuration that produces a flux rope with both legs remaining attached to the Sun. Satellite data show that ~60% of flux ropes that reach Earth have this configuration. **b)** Merging configuration that produces a flux rope with one leg attached to the Sun. [Gosling et al, 1995].

end. Figure 9-25 shows field line topologies that account for various levels of connectedness. The topology in Fig. 9-25a is consistent with that of Fig 9-24b. A partial disconnection occurs at the merging locations, but the field line ends are still attached to the Sun. If a full disconnection occurs, the CME is ejected as a plasmoid (not shown). In Fig. 9-25b an interchange reconnection occurs. Merging between a field line open to the heliosphere and a closed loop in the corona produces and interchange of field topologies. No field lines disconnect from the Sun, yet the process reduces the number of field lines extending into the corona and heliosphere from three to one. Merging between another open field line (not shown) with the partially disconnected field could produce a fully disconnected flux rope (plasmoid) and an additional closed field line.

9.2.4 Solar Energetic Particles

Objectives: After reading this section, you should be able to...

- Explain the origins of solar energetic particle events (SEPs)
- Distinguish between gradual and impulsive particle events
- Describe solar energetic particle effects at Earth

Solar Energetic Particle Characteristics. A less-frequently observed class of solar emissions is the *solar energetic particle (SEP)* event. Energetic solar particles that originate in flares and ahead of CMEs are not part of the solar wind plasma population. Their speed and means of acceleration put them in a different category. Because SEPs are so fast, their arrival time at Earth is measured in minutes to hours. For instance, protons with energies of about 60 MeV take roughly half an hour to reach Earth. Protons with lower energies take several hours (Fig. 9-26). Solar energetic particles have much higher kinetic energies than ambient solar wind particles, but generally lower energies than galactic cosmic rays. On the scale of cosmic radiation, solar-produced ions have relatively low energies, generally below 1 GeV and rarely above 10 GeV.

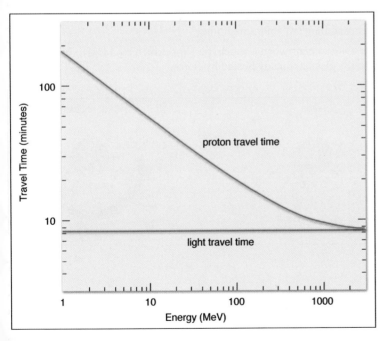

Fig. 9-26. Sun-to-Earth Travel Time for Energetic Protons. Protons with 10 MeV of energy reach Earth in less than an hour.

These fast-arriving particles may harm humans in space, damage space hardware, and hamper communication via the ionosphere (NOAA S-scale in Appendix A). Chapter 13 discusses this topic at length. Figure 2-13 shows the effect of an SEP event on a space-based imaging system. Solar panels suffer serious degradation from protons with energies of a few to 10 MeV. During the largest SEP events, the energetic protons can penetrate to Earth's surface, producing a *ground-level event (GLE)*. On rare occasions, energized solar particles create new radiation belts in Earth's magnetosphere.

Chapter 9 The Active Sun and Other Stars: Sources of Space Weather

Genesis of Solar Energetic Particles. Explosive reconfiguration of magnetic fields in the solar atmosphere accelerates particles to very high speeds. Parker's spiral model of the interplanetary magnetic field (Chap. 5) suggests that particles accelerated in the corona reach Earth, if they are injected into interplanetary magnetic flux tubes near 50°–60° western solar longitude. A fraction of SEP events, called impulsive short-duration events, conform to this model. Other events come from a broader range of longitudes and are longer-lived. In recent years, scientists have sorted the events into two primary forms (Fig. 9-27) and a hybrid form:

- The *impulsive short-duration events* are usually observed from magnetically well-connected locations on the Sun. These particles tend to travel along Parker-spiral field lines that link the Sun's western hemisphere to Earth. The events are electron-rich, and have a strong relationship to impulsive H-α and X-ray flares, and radio bursts. The high ion charge state, with many electrons stripped from heavy ions, indicates their origin in plasma heated by flares. Impulsive events have lifetimes of hours and come from a narrow swath of longitude (< 30°).

- The *gradual events* last for several days. They are proton rich, having about the same element composition and ionization states as those events in the low-density ambient plasma of the high corona or solar wind. Gradual events are associated with gradual X-ray flares, coronal sweep frequency radio emissions, and coronal mass ejections (CMEs). They are observed over a broad range of solar longitudes.

- *Hybrid events* occur when a shock, perhaps driven by a CME, further accelerates plasma already heated by flares.

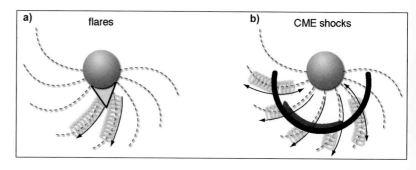

Fig. 9-27. Impulsive and Gradual Energetic Particle Events. a) This diagram depicts particles from an impulsive event. The yellow triangle represents a flare. The narrow beam tends to be rich with electrons accelerated at the flare site. Heavy ions in the mix are mostly highly charged, indicating the high temperatures associated with the flare. Impulsive events are less frequent than gradual ones. The particle beams cover ~30° of longitude and therefore intercept Earth for only a few hours. **b)** This diagram illustrates a gradual SEP event, driven by a CME (red curve). Such events cover broad swaths of longitude. Particles are constantly accelerated. Particle flux decreases as $1/r^2$, starting at the event origin. *(Courtesy of Donald Reames at NASA)*

We discuss impulsive flares in more detail here and treat the gradual events in Chap 10. Most impulsive SEPs come from regions a degree or two in diameter. In the 1990s, the Yohkoh satellite observed compact hard X-ray sources above soft X-ray loops in a region where magnetic fields formed a cusp. The electrons appear to be accelerated in the pointed cusp above the loop. The field lines that continue above the cusp may eventually return to the Sun, or they may open up to the interplanetary medium. If they open to the interplanetary medium, they provide a direct route for energetic flare particles to escape the Sun. Occasionally, impulsive high-energy particles reach Earth even though the parent flaring region is several tens of degrees away from the footpoint of the nominal Parker spiral. Scientists have explored the possibility of diffusion across magnetic field lines and longitudinal transport along coronal magnetic fields above the photosphere. Particles could diffuse across field lines if the field lines are in rapid motion near the source region. Improved solar imaging techniques do reveal instances of such rapid motion. Longitudinal transport is possible, if the field line configuration above an active region is strongly divergent.

Scientists have explored the latter prospect using radio emission from electron beams as a tracer of open coronal flux tubes. Electron beams transiting the solar corona excite Langmuir waves near the local electron plasma frequency. These waves are converted into electromagnetic emissions near the local electron plasma frequency or its harmonic. As the electron beams proceed from the low corona to interplanetary space, they emit at a frequency that decreases with height. Such emission is frequently observed between tens of kHz and hundreds of MHz, and is called a type III burst (Sec. 9.2.3).

The observations suggest that open field lines rooted in active regions, which occupy a narrow range of longitudes in the photosphere and low corona, fan out rapidly with increasing height. The magnetic field extrapolations and radio maps show that extreme excursions (up to 50° from the flare source) can be explained by the outward fanning, where the diverging field lines bend toward a field line connected to Earth in a spiral configuration. Flare particles are introduced into the spiral configuration from a distant active region source. In a few cases the distant source has proven to be several tens of degrees behind the Sun's western limb. The results also suggest that high-latitude flares that should not be well connected to the ecliptic plane can contribute impulsive energetic particles to Earth's geospace.

Recently, scientists have used observations from the Solar and Heliospheric Observatory (SOHO) satellite to verify that many flare-associated particles arise in a thin electrified "sheet" of gas (sometimes called the diffusion region) that stretches from the flare site to the base of the coronal mass ejection. This current sheet acts like an Earth-bound particle accelerator, pushing atomic particles to a large fraction of light speed. Such sheets have been observed by instruments on the SOHO satellite. Fig. 9-28 details the geometry surrounding them.

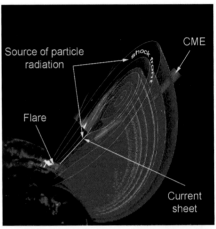

Fig. 9-28. Two Sources of Solar Energetic Particles. The Ultraviolet Coronagraph Spectrometer (UVCS) on board the SOHO spacecraft reveals an extremely narrow region where the gas temperature jumped from <1 MK to >6 MK. This intense heating is one of the hallmarks of the current sheet particle acceleration model. *(Courtesy of Jun Lin at the Harvard-Smithsonian Center for Astrophysics, Terry Forbes at the University of New Hampshire, and ESA/NASA – the Solar and Heliospheric Observatory (SOHO) Program)*

EXAMPLE 9.5

Time of Flight for a Solar Energetic Particle

- *Problem Statement:* Determine how long it takes for a 100-MeV proton to travel from the Sun to Earth.
- *Concept:* Kinetic energy = $(\gamma - 1) m_0 c^2$
- *Given:* 10^7 eV = proton kinetic energy
- *Assume:* Proton rest mass, $m_0 c^2$, is 938.27 MeV and the particle travels along a magnetic field line linking the solar central meridian and Earth.
- *Solution:*

   KE = 100 MeV = $(\gamma - 1)$ 938.27 MeV

   $\gamma - 1 = 0.1066$ (unitless) = $\dfrac{100 \text{ MeV}}{938.27 \text{ MeV}}$

   $\gamma = 1.1066$ (unitless) = $1/\sqrt{(1 - v^2/c^2)}$

Solving for v results in:

   $v = 0.428c = 0.428 \times (3 \times 10^8 \text{ m/s}) = 1.28 \times 10^8$ m/s

Travel time from the Sun to Earth = $(1.5 \times 10^{11}$ m$)/$ $(1.28 \times 10^8$ m/s$) \approx 1168$ s ≈ 19 min

- *Implication and Interpretation:* A 100 MeV particle travels along a magnetic field line directly connected to Earth from the Sun in less than twenty minutes. Particles linked by field lines that connect to more westerly longitudes take slightly longer to reach Earth.

Follow-on Exercise: Determine the speed of a 1 TeV proton.

9.3 Space Weather Effects from Other Stars

Some high-energy particles and photons come from sources other than the Sun. Each of these is characterized by a different energy, flux, and composition of particle species. The particle sources are modulated by variations in the heliospheric magnetic field over the course of the solar cycle.

9.3.1 Low- and High-energy Particles from the Cosmos

Objectives: After reading this section, you should be able to...

- Define the following terms: pick-up ion, anomalous cosmic ray, galactic cosmic ray
- Explain how pick-up ions are created and energized
- Describe the space environment impacts of high-energy charged and neutral particles

Anomalous Cosmic Rays Sources: Charged particles with anomalously high energy compared to normal solar wind ions, but low energy compared to cosmic ray particles, are *anomalous cosmic rays (ACRs)*. They have low charge states and relatively low energies (<100 MeV) compared to cosmic rays. This component of energetic charged particles consists of elements that are difficult to ionize, including helium, nitrogen, oxygen, neon, and argon.

Anomalous cosmic rays have several sources. Interstellar neutral particles pass unimpeded through the heliospheric magnetic field. An interstellar breeze of neutral helium atoms streams through our solar system at ~25 km/s with a density of ~0.01 atoms/cm^3. As neutral particles pass close to our Sun, some are ionized by a solar photon or exchange charge with solar wind ions, becoming singly charged ions. Additionally, the remnant disk from which the planets formed still contains dust and asteroids that contribute neutral particles to the mix. These low-energy *pick-up ions* represent material from inside and outside of the solar system that the solar wind collects and carries outward toward the heliopause. Shock regions associated with especially energetic co-rotating interaction regions (Sec. 5.1) may also produce lower-energy ACRs. During repeated collisions with magnetic irregularities at the flanks of the termination shock that may occur over the course of months or years, the pick-up ions gain energy. Some energized particles (5 MeV–50 MeV) escape from the shock and diffuse toward the inner heliosphere, as depicted in Fig. 9-29.

Galactic Cosmic Radiation Sources and Characteristics. *Galactic cosmic rays (GCRs)* are energetic charged particles (typically GeV) from galactic collisions, and stellar nurseries, death beds, and graveyards. For historical reasons these particles are called cosmic "rays", even though they have mass. In the interstellar medium, stellar winds and shock regions also energize ions to high energies and thus contribute to the cosmic ray population. Most galactic cosmic rays are probably accelerated in supernova explosions within the Milky Way Galaxy. These explosions produce long-lasting plasma clouds threaded by magnetic fields. Charged particles bounce back and forth, gaining energy in the chaotic fields.

9.3 Space Weather Effects from Other Stars

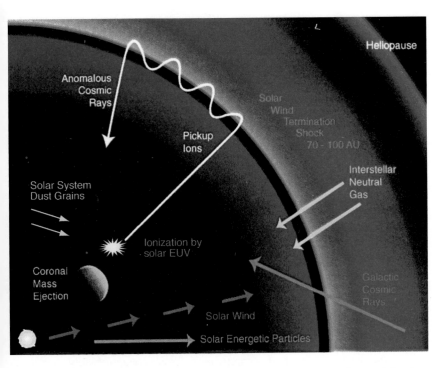

Fig. 9-29. Anomalous Cosmic Rays (ACRs). Anomalous cosmic rays may begin as neutral dust in the remnant materials out of which the planets formed, or as interstellar neutral atoms. The particles convert to pick-up ions and ultimately to ACRs. Scientists now believe that most of the acceleration occurs near the termination shock flanks rather than at the shock nose. *(Courtesy of the Advanced Composition Explorer (ACE) Team and NASA)*

Magnetic field strength, along with the size of the accelerating region and the time held there, determines the maximum energy of the cosmic rays. Particles with sufficient speed escape into the galaxy. But some cosmic rays have been observed at much higher energies than supernova remnants can generate. A few of these have been linked to regions of suspected galactic collisions.

Galactic cosmic rays have extremely high kinetic energies—as high as 10^{21} eV (10^{12} GeV), but more typically in the GeV range. The particle population consists of a uniform flux of charged atomic nuclei composed of hydrogen (87%), helium (12%), and heavier ions (trace). These particles have a relatively low flux (~4 particles/(cm$^2 \cdot$s)), with an energy density of ~0.5 to 1.0 eV/cm^3, similar to that of starlight. When cosmic rays interact with other matter, they usually decelerate and emit gamma radiation or X rays. Cosmic rays are hazardous to humans and radiation-sensitive systems. At Earth's surface, where most biological and technological activities occur, we have minimal interaction with truly energetic cosmic rays.

Once free of their source regions, cosmic particles experience tortuous randomizing trips within the galactic magnetic field. Some of them make their way to the heliopause. Many, if not most, of these are reflected by magnetic structures on the edge of the Sun's domain. Those particles that penetrate must still navigate the magnetic fields distributed by CMEs and solar wind interaction regions in the heliosphere. Only the most energetic particles penetrate into the inner heliosphere.

Cosmic rays have enormous velocities and, as they stream collectively, they excite magnetic irregularities (Alfvén waves). They follow helical orbits along the large-scale heliospheric or galactic magnetic field (the Lorentz force influences their motion). However, self-induced and pre-existing magnetic irregularities cause the helical orbits to change continuously in small increments. Because magnetic irregularities move with the large-scale flow of the solar wind, the cosmic rays, although very energetic and fast, experience constant buffeting.

Chapter 9   The Active Sun and Other Stars: Sources of Space Weather

Fig. 9-30. Cosmic Ray Shower. In this artist's rendering, a cosmic ray (yellow) hits the upper atmosphere and produces a shower of other particles (green). Some of these particles (mostly pions) decay into muons (red). Only a small fraction of the muons reach Earth's surface, because most of them decay in flight. Therefore, we find more muons at higher altitudes, because fewer have decayed. At sea level, each fingernail-size area receives about one muon per minute. About once per week a cosmic particle with energy at or above 10^{20} eV impacts each 3000 km^2 area at Earth's surface. *(Courtesy of Terry Anderson at the Stanford Linear Accelerator Center)*

Low-energy cosmic rays have great difficulty "swimming upstream" into the solar wind, whereas more energetic particles travel much easier. As a result, the spectrum of particles is always less dense at low energies than at high energies.

Earth's atmosphere protects humans and technology systems at Earth's surface from most GCR interactions. Because of their high energy, cosmic rays collide with atoms from the upper atmosphere and create a cascade of secondary cosmic ray particles that reach Earth's surface, as drawn in Fig. 9-30. The process by which a primary cosmic ray particle collides with atmospheric nuclei to create many secondary particles is a *cosmic ray shower* (air shower). Thousands pass through our bodies every minute. These secondary cosmic rays, with an average flux of about 100 particles/(m$^2 \cdot$s) at sea level, represent a few percent of the natural background radiation. As we note in Sec. 7.4, some of the cosmic ray by-products escape outward. If the by-product is a charged particle, it may become part of the inner radiation belt.

The 11-year Variation. One of the most easily monitored by-products of cosmic ray showers is slow neutrons. Figure 9-31 shows the history of slow neutron detections at Moscow, Russia. The neutrons (and hence the cosmic rays that create them) are anti-correlated with the sunspot number. This anti-correlation relates to the magnetic shield provided by the heliospheric magnetic field. When the Sun is active, the enhanced magnetic field scatters more cosmic rays than do the weaker fields of solar minimum. During solar maximum, integrated cosmic ray intensity decreases. During solar minimum, when the field is weaker, Earth is more vulnerable to these cosmic invaders. Lower-energy cosmic flux decreases, but higher energy particles are not appreciably attenuated. The integral cosmic ray dose rate in interplanetary space is approximately 2.5 times higher at solar minimum than at solar maximum. Inner heliospheric disturbances caused by CMEs exert short-term influence on the fluxes of these particles. Hence, galactic cosmic rays are space weather tracers.

The 22-year Variation. In Fig. 9-31, we see that some of the cosmic ray maxima have narrow peaks and some have broad, long-lasting peaks. Recent simulations of the heliospheric magnetic field that account for its orientation show a clear 22-year cycle for cosmic ray penetration into the inner heliosphere. The orientation of the Sun's magnetic field modulates the cosmic ray flux such that every 22 years the maxima in the cosmic ray counts are alternately broad or peaked. These interactions may have important implications. Cosmic rays may influence cloud cover by providing ionization that allows tiny cloud condensation nuclei to form. Scientists are actively looking for the indications of a 22-year climate cycle associated with solar modulation of cosmic rays.

Scientists have long known of the risk to astronauts and space systems from cosmic rays and streams of protons from the Sun. Airline electronic systems, passengers, and crews face a similar, but much reduced, risk. Quantifying this risk is difficult, however, because systematic data are lacking on the number of rays and how many charged particles and neutrons they create in Earth's atmosphere that are encountered during typical flights. Researchers are now collecting that information, thanks to newly developed instrumentation. The need to know the precise level of cosmic and solar radiation along air routes has become more important, as recent generations of commercial aircraft use "fly-by-wire" control systems, managed by on-board computers. And more commercial flights over the poles expose aircrews and frequent flyers to radiation levels that are not well documented. Future aircraft will employ even more sensitive technologies, and will require more redundancy in aviation systems to maintain safety. We need data to make informed decisions about tolerable levels of exposure.

9.3 Space Weather Effects from Other Stars

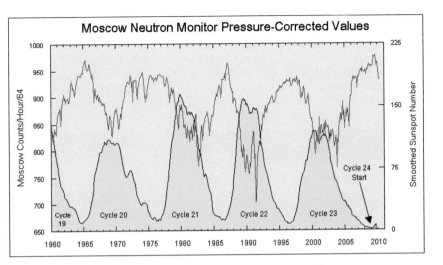

Fig. 9-31. Cosmic Ray Counts Compared to Sunspot Number. This is a plot of data from the Moscow, Russia, neutron monitor. The cosmic ray trace (blue) has an inverse relationship to the sunspot cycle (yellow), because the heliosphere's magnetic field is stronger during sunspot maximum and shields Earth from cosmic rays. *(Courtesy of NOAA's National Geophysical Data Center)*

Focus Box 9.4: Earth and Near-Earth Impact of High-energy Particles

The highly energetic cosmic rays in space pose serious threats to spacecraft and high-altitude airplanes. Computer hardware and sensitive electronic instruments on spacecraft require shielding to prevent cosmic rays from passing through electronic chips and "flipping" the logic states. Astronauts are also at risk of high-level radiation from cosmic rays. The risk is greatest during solar minimum, when the Sun's protective magnetic field is weaker, thus allowing more penetration of energetic particles from the cosmos. Astronauts on long-duration missions are particularly at risk. Figure 9-32 suggests how cosmic rays could produce biological damage. They also create electronic noise in the manufacturing process for computer chips.

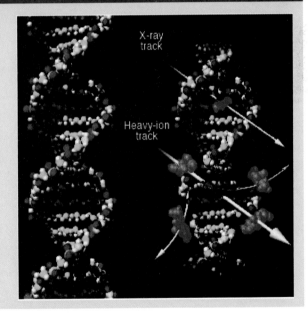

Fig. 9-32. Cosmic Rays Interacting with DNA. This artist's rendering depicts the susceptibility of a strand of DNA to energetic cosmic rays. Particles with energies above 100 MeV (heavy ions) are likely to do the most damage. *(Courtesy of the NASA Office of Biological and Physical Research)*

Pause for Inquiry 9-13

Estimate the cross-sectional area of your body. About how many secondary particles from cosmic ray interactions pass through you each second?

Chapter 9 The Active Sun and Other Stars: Sources of Space Weather

The preliminary tests have confirmed that doses of radiation from cosmic rays and the particles they create are more intense at higher altitudes and at higher latitudes; that is, they are strongest at high altitudes in the Arctic and Antarctic regions. The intensity of lower energy solar- and heliospheric-generated energetic particles is particularly high during solar storms, when large quantities of charged particles reach Earth's atmosphere. New studies are planned to expand this field and facilitate new models of real use for aircraft route planners.

EXAMPLE 9.6

Cosmic Ray Flux

- *Problem Statement:* Determine the power delivered to Earth by cosmic rays with energy of the order of 0.1 TeV
- *Given:* Flux of 0.1 TeV cosmic rays is $1/(m^2 \cdot s)$
- *Assume:* Isotropic flux
- *Solution:* Earth's surface area is $4\pi r^2$

    $= 4\pi (6378 \times 10^3 \text{ m})^2 = 5.1 \times 10^{14} \text{ m}^2$

    Total power delivered to Earth $= [(0.1 \times 10^{12} \text{ eV} \times 5.1 \times 10^{14} \text{ m}^2) \times 1.6 \times 10^{-19} \text{ J/eV}] \times 1/(m^2 \cdot s)$

    $= 8.16 \times 10^6 \text{ W}$

- *Interpretation:* The cosmic ray energy flux at 0.1 TeV is approximately 8 MW. This is a small power compared to that of an average thunderstorm (~100 MW).

Follow-on Exercise: Cosmic rays with energies of $\sim 10^{19}$ eV arrive at a rate of $1/(km^2 \cdot yr)$. How much total energy arrives via these particles each year?

9.3.2 High-energy Photons from the Cosmos

> **Objectives:** After reading this section, you should be able to...
> - Explain how high-energy photons are produced
> - Describe the space environment impacts of high-energy photons

Hot material created under some of the most extreme conditions in the cosmos produces tera-electron volt (TeV) gamma rays. One possible mechanism for this production begins with electrons accelerated to extremely high energy by ultra-strong electromagnetic fields such as those found in supernovae and neutron stars. If an energized electron collides with a low-energy photon, the electron transfers much of its energy to the photon via an inelastic collision. In the process, the low-energy photon transforms into a TeV gamma ray. A variation on the theme involves the interaction of a fast-moving electron with an extremely strong magnetic field. The magnetic field accelerates the electron into cyclotron motion that produces high-energy photons.

Some TeV gamma rays appear to originate in stellar nurseries where groups of young stars produce strong stellar winds (and gusts of stellar winds) that interact to form shocks whose magnetic fields trap and accelerate charged particles to very high speeds. Some of the charged particles are by-products of supernovae explosions: cosmic rays consisting of bare nuclei of iron and other heavy elements. Young stars also expel high fluxes of extreme ultraviolet (EUV) photons that penetrate the bare nuclei and excite nuclear resonances. For a bare (highly charged) nucleus with sufficient speed, an oncoming EUV photon (having energy in its source reference frame of a few eV) has an energy of a few MeV, sufficient to excite the resonance. When the resonance decays, the ejected photon receives a further energy boost, again by the same factor of a million, from MeV to TeV energies. In some cases the excited nuclei will disintegrate creating additional high-energy photons and lighter elements that contribute to the cosmic ray population. [Anchordoqui et al., 2007]

On average, one cosmic TeV gamma ray falls on each square kilometer of Earth per second. Virtually all the rest of them collide with air molecules and produce a cascade of energetic particles in the upper atmosphere. As we discuss in Chap. 8, such cascades have been implicated in the formation of transient luminous events associated with lightning.

On extremely rare occasions a much larger burst of gamma rays floods the near-Earth environment. *Magnetars* (short for magnetic neutron stars) are the magnetic remains of large stars that live fast and die young. Spinning once every few seconds, they emit beams of X rays and soft gamma rays. If their rigid crystal crust buckles, they also emit an occasional powerful burst of gamma rays. In March 1979, August 1998, and December 2004, waves of gamma rays from magnetars saturated space-based detectors. Some instruments switched to a protective safe mode. During such events, the ionized upper atmosphere may experience abrupt changes. The lower edge of the ionosphere may temporarily move to low altitudes, say from 85 km to 60 km or even lower. Radio propagation experiences serious disruption under such circumstances. We discuss magnetar effects in greater detail in Chap. 14.

Chapter 9   The Active Sun and Other Stars: Sources of Space Weather

Summary

Convection, meridional circulation, and differential rotation concentrate magnetic fields into bundles in the solar convective zone. Magnetic field lines, originally in a pole-to-pole formation, twist across longitude and concentrate in buoyant ropes of magnetic flux that rise through the solar surface. This dynamo action occurs on several space and time scales, the most notable of which are the 11-year sunspot cycle and the 22-year magnetic cycle. As the magnetic flux concentrations emerge through the solar surface, they produce active regions that billow into the corona. The photospheric manifestations of active regions are faculae and bipolar sunspots. Higher in the solar atmosphere, active regions produce plages, and may support filaments of chromospheric material. At the highest levels of the solar atmosphere, the active regions manifest large loop structures. Active regions tend to form repeatedly in favored longitudes, suggesting deep-rooted instabilities below the convective zone. Active regions also follow a latitudinal progression that suggests a butterfly pattern over successive solar cycles. The butterfly pattern reveals the equatorward march of oppositely poled leading magnetic sunspots in each hemisphere. A more subtle poleward streaming of oppositely poled following sunspots produces the polar reversal during the maximum of each 11-year solar cycle.

Concentrated and twisted magnetic fields store energy in unstable configurations. Nature releases this energy to other forms during magnetic merging events. The results are solar flares, coronal mass ejections, and solar energetic particle events. The radiative emissions (solar flares) develop when electrons and other charged particles accelerate to high energies at or near the merging site. Collisions and interactions with local magnetic fields produce the radiation sensed at Earth about eight minutes after the explosive merging event. Individual particles also accelerate at the flare site. Some drive deep into the solar atmosphere, where they produce additional radiation; others spew into the heliosphere as solar energetic particles. If magnetically connected to Earth, such particles may arrive within 15 to 20 minutes after flare initiation. In many cases magnetic energy release is also accompanied by the acceleration of a plasma cloud from the corona. This massive, magnetically confined, and relatively cool plasma bubble is a coronal mass ejection (CME). CMEs that intercept Earth at high speed spawn major magnetic storms. High-speed CMEs also create shocks that accelerate solar wind plasma particles to relativistic speeds, thereby creating a secondary but more prolonged source of energetic particles.

Other stars are also sources of energetic particles and photons. Cosmic rays are charged, energetic particles that enter the heliosphere from all directions. To some degree these particles are guided, and even scattered, by the heliospheric magnetic field. When frequent CMEs disturb the heliospheric magnetic field during solar maximum intervals, cosmic rays are less efficient at penetrating to Earth's orbit. Cosmic ray fluxes are in anti-phase with the solar cycle. These energetic particles produce a form of space environment disturbance that peaks during solar minimum.

In addition to cosmic ray particles, Earth's upper atmosphere intercepts energetic photons from regions of stellar death and birth and perhaps from regions of galactic interactions. High fluxes of energetic photons from magnetars appear episodically, perhaps about once per decade. Magnetar events produce fronts of energy that disturb the ionosphere.

Keywords

active longitudes
active regions (ARs)
anomalous cosmic rays (ACRs)
arch system
Cerenkov plasma waves
coronal mass ejections (CMEs)
cosmic ray shower
electrostatic bursts
faculae
flare
galactic cosmic rays (GCRs)
gradual events
gradual rise and fall events
ground-level event (GLE)
halo CMEs
hybrid events
Hyder flares
importance
impulsive microwave bursts
impulsive short-duration events
impulsive X-ray flares
interplanetary flux ropes
limb CMEs
long-duration flares
loop prominence system (LPS)
magnetars
magnetic clouds
magnetic flux emergence
meridional circulations
microwave bursts
Moreton wave
noise storms
partial halo CMEs
penumbra
pick-up ions
radio bursts
short wave blackout
short wave fades (SWFs)
sigmoid
solar energetic particle (SEP)
sudden ionospheric disturbance (SID)
sudden phase anomaly (SPA)
sweep frequency burst
torsional oscillations
umbra

Answers to Pause for Inquiry Questions

PFI 9-1: Since the magnetic field at sunspots is normal to the surface, we can write: magnetic flux = magnetic field × area

Average sunspot magnetic field strength is 0.3 T; typical sunspot diameter is 10^4 km, yielding an area of $\pi r^2 = \pi(5 \times 10^6 \text{ m})^2 = 7.85 \times 10^{13}$ m^2. Magnetic flux = 2.36×10^{13} Wb

PFI 9-2: When a rising loop with downward-directed field juxtaposes with a loop with upward-directed field lines, the field has components in opposite directions, which is the proper configuration for magnetic merging.

PFI 9-3: Fig. 9-3 shows active region number 9393 that was observed in April 2001, near the peak of solar cycle 23.

PFI 9-4: In mid-1996 the active regions were near the equator. In mid-1998 the active regions appeared at higher latitude. The 1996 active regions were the remnants of solar cycle 22. The new cycle 23 sunspots began forming at higher latitudes in 1997.

PFI 9-5: The first reverse-polarity sunspot, marking the beginning of solar cycle 24, appeared on January 4, 2008, at 30° N latitude.

PFI 9-6: The leading spots in the southern hemisphere have a positive polarity; the leading spots in the northern hemisphere have a negative polarity.

PFI 9-7: An earthquake that reaches 9 on the Richter scale releases $\sim 10^{18}$ J. A moderate-to-large solar flare releases 10^{21} to 10^{25} J.

PFI 9-8: The flare power from Ex. 9.3 is $\sim 5 \times 10^{21}$ W. Applying the criterium of flare power in excess of 150% of background to Fig. 9-16 suggests that a typical flare is about three hours in duration. Multiplying power by duration gives total energy values in the 10^{25} J–10^{26} J range.

PFI 9-9: By Ampere's Law, the current in the filament must be directed out of the page. The disturbance current must be directed into the page to weaken or relax the magnetic field.

PFI 9-10: Refer to Fig. 9-16 and Appendix A.

X-ray Category	Flux Intensity	NOAA Scale	Frequency
X-10	10^{-3} W/m^2	R-3	8/cycle
X-1	10^{-4} W/m^2	R-3	175/cycle
M-1	10^{-5} W/m^2	R-1	2000/cycle
C-1	10^{-6} W/m^2	Below scale	Thousands/cycle

PFI 9-11: Coronal mass ejections occurring at solar minimum are more likely to be within the streamer belt. They are confined in latitude close to the equator. During solar maximum CMEs are much more widely distributed.

PFI 9-12:
- Flare ribbons: bottom of lower-right quadrant
- CME: top of the lower-right quadrant
- Footpoint motion leading to shearing: top-right quadrant
- Filaments: all quadrants
- Magnetic neutral line: dashed line in all quadrants
- Arcade: lower half of lower-right quadrant
- Merging line (surface): middle of lower-right quadrant

PFI 9-13: The cross section is about 1 square meter. About 100 secondary particles pass through you each second.

References

Anchordoqui, Luis A., John F. Beacom, Haim Goldberg, Sergio Palomares-Ruiz, and Thomas J. Weiler. 2007. TeV Gamma Rays from Photodisintegration and Daughter Deexcitation of Cosmic Ray Nuclei. *Physical Review Letters.* No. 98. American Physical Society. College Park, MD.

Figure References

de Toma, Giuliani, Oran R. White, and Karen L. Harvey. 2000. A Picture of Solar Minimum and the Onset of Solar Cycle 23. *The Astrophysical Journal.* Vol. 529. American Astronomical Society. Washington, DC.

Dikpati, Mausumi and Peter A. Gilman. 2008. Global Solar Dynamo Models: Simulations and Predictions, *Journal of Astrophysics and Astronomy.* Vol. 29. Indian Academy of Sciences. Bangalore, India.

Gosling, John T., Joachim Birn, and Michael Hesse. 1995. Three-Dimensional Magnetic Reconnection and the Magnetic Topology of Coronal Mass Ejection Events. *Geophysical Research Letters.* Vol. 22. American Geophysical Union, Washington, DC.

Lean, Judith L. 2010. Cycles and trends in solar irradiance and climate. Wiley Interdisciplinary Reviews: *Climate Change.* Vol. 1. Issue 1. John Wiley & Sons, Ltd. Malden, MA.

Longcope, Dana W. and Colin Beveridge. 2007. A Quantitative Topological Model Of Reconnection and Flux Rope Formation in a Two-ribbon Flare. *The Astrophysical Journal.* Vol. 669. American Astronomical Society. Washington, DC.

Magara, Tetsuya, Shin Mineshige, Takaaki Yokoyama, and Kazunari Shibata. 1996. Numerical Simulation of Magnetic Reconnection in Eruptive Flares, *The Astrophysical Journal.* Vol. 466. American Astronomical Society. Washington, DC.

Manchester, Ward IV, Tamas Gombosi, Darren L. DeZeeuw, and Yuhong Fan. 2004. Eruption of a Buoyantly Emerging Magnetic Flux Rope. *The Astrophysical Journal.* Vol. 610. Issue 1. American Astronomical Society. Washington, DC.

Moore, Ronald L., Alphonse C. Sterling, Hugh S. Hudson, and James R. Lemen. 2001. Onset of the Magnetic Explosion in Solar Flares and Coronal Mass Ejections. *The Astrophysical Journal.* Vol. 552. American Astronomical Society. Washington, DC.

Nicholson, Iain. 1982. *The Sun.* Published in association with the Royal Astronomical Society [by] Rand McNally, New York.

Parker, Eugene N. 2000. The physics of the Sun and the gateway to the stars. *Physics Today.* Vol. 53. American Institute of Physics. College Park, MD.

White, Oran R. (ed). 1977. *The Solar Output and Its Variation.* Boulder, CO: Colorado Associated Press.

Further Reading

Aschwanden, Markus J. 2005. *Physics of the Solar Corona: An Introduction with Problems and Solutions.* Berlin: Springer Verlag.

Benz, Arnold O. 2007. Flare Observations. *Living Reviews in Solar Physics.* No. 3. Max-Planck Institute, Katlenburg-Lindau, Germany.

Golub, Leon and Jay M. Pasachoff. 2002. *Nearest Star: The Surprising Science of Our Sun.* Cambridge, MA: Harvard University Press.

The Active Interplanetary Medium: Conduit for Space Weather

10

UNIT 2. ACTIVE SPACE WEATHER AND ITS PHYSICS

You should already know about...

- High-speed streams (Chaps. 1, 5, and 10)
- High-energy particles (Chap. 2)
- Solar filaments and prominences (Chap. 3)
- Coronal holes (Chaps. 3, 5, and 10)
- Magnetic flux emergence from the Sun (Chaps. 3 and 9)
- Heliospheric current sheet (Chaps. 4 and 5)
- How magnetic fields control plasma motion and vice versa (Chaps. 4 and 6)
- Interplanetary magnetic field (IMF) and sector structures (Chap. 5)
- Alfvén waves (Chaps. 5 and 6)
- Heliospheric streamer belt (Chap. 9)
- Coronal mass ejections (CMEs) (Chap. 9)

In this chapter you will learn about...

- Disturbances in the quasi-stationary solar wind
- Coronal hole influence on the interplanetary medium
- Co-rotating and merged interaction regions (CIRs and MIRs)
- Transient disturbances in the solar wind
- Interplanetary signatures of CMEs
- Magnetic clouds and effects of IMF orientation in CMEs
- Effects of fast CMEs
- Shock-energized particles

Outline

10.1 Disturbances in the Quasi-stationary Solar Wind
  - 10.1.1 Quasi-stationary Structures in the Slow Solar Wind
  - 10.1.2 Co-rotating Structures in the Solar Wind

10.2 Transient Disturbances in the Solar Wind
  - 10.2.1 Characteristics of Transients in the Interplanetary Medium
  - 10.2.2 Geoeffectiveness of Transients in the Interplanetary Medium

10.3 Solar Energetic Particles from Interplanetary Shocks
  - 10.3.1 Traveling Shocks and Solar Energetic Particles
  - 10.3.2 Shock Acceleration Mechanisms

Chapter 10 The Active Interplanetary Medium: Conduit for Space Weather

10.1 Disturbances in the Quasi-stationary Solar Wind

10.1.1 Quasi-stationary Structures in the Slow Solar Wind

> **Objectives:** After reading this section, you should be able to...
>
> ♦ Describe sources of minor non-recurrent geomagnetic disturbances in the quasi-stationary solar wind
>
> ♦ Relate the disturbance sources to solar wind speed, pressure, and field variations

Disturbances in Slow Flow. As we describe in Chap. 5, the quiescent outflow of solar wind is bimodal: high speed and low speed. In some literature, the slow flow is called streamer flow and the high-speed flow is called non-streamer flow for reasons we discuss below. In general, the high-speed flow comes from the open field regions (coronal holes) that usually originate at high solar latitudes (Fig. 5-23a and left image of Fig. 5-23). In a simple solar minimum configuration, the open field regions, and thus the high-speed flow zones, expand latitudinally. Most of the expansion occurs between the inner corona and the location where the magnetic field lines become predominantly vertical—the source surface (Fig. 1-10a). The expansion factor f of the magnetic flux tubes often has a value between four and ten. As the high-speed zones expand, they constrict the oppositely directed magnetic fields from each hemisphere to a cusp-like streamer (Fig. 10-1). Streamer plasma forms a relatively dense envelope around the heliospheric current sheet (HCS).

The quiescent slow flow comes from the equatorial streamer regions (Fig. 5-23), but the origin of the flow is not well understood. At the steamer tips (stalks), discrete blobs of material move outward but their masses are too small to fully account for the solar wind flux. SOHO observes three-to-four such emissions each day, with initial speeds of 0 km/s–100 km/s. These structures accelerate to slow solar wind speeds within $3 R_S$–$4 R_S$. Boundary plasma at the edges of some active regions and some coronal holes also contributes to slow flow, especially during solar maximum. SOHO observed these outflows along the outer edges and from the tops of streamers. Observations from multiple spacecraft in 2007 revealed small transients embedded in the slow solar wind. They had smaller scale sizes, smaller speed changes, and lower magnetic field magnitudes than the average solar minimum interplanetary CME at 1 AU. The average size of these structures was less than 0.1 AU, compared to 0.25 AU for large-scale CMEs. Most of the small-scale transients appear to have developed at the edges of coronal holes and flowed into the interplanetary medium at slow solar wind speeds.

With increasing activity, the orderly bimodal character of the corona and the solar wind breaks down. The polar holes shrink and streamers appear at ever higher heliographic latitudes. The right image of Fig. 5-23 shows the multi-streamer Sun during solar maximum. At these times, the slow and fast wind structure gives way to a complex mixture of fast flows from smaller coronal holes and transients, embedded in a slow-to-moderate speed wind from all latitudes.

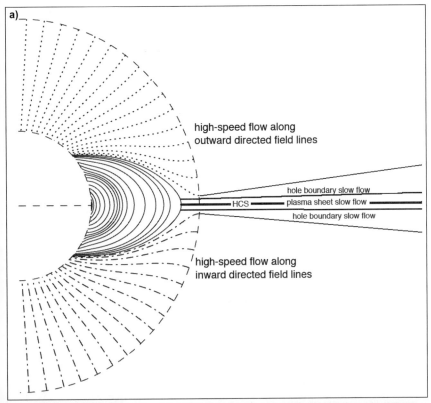

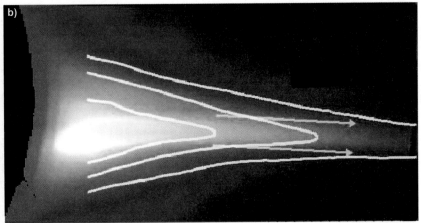

Fig. 10-1. a) Simulation Showing Formation of the Streamer Belt. The model output shows a cross section of the solar corona. The inner dashed curve represents the solar surface. The outer dashed curve is the source surface (2.5 R_S), where the field lines are treated as radial. Field lines beyond the source surface are not shown. Magnetic flux tubes in the open regions expand, forcing the closed flux region into a narrow cone-sheet configuration at low latitudes. The expansion factor is largest at the open/closed boundary. The cusp-like region is thought to be the source of most of the slow solar wind. The observed structure is often broader and more warped than the idealized view presented here. [Wang and Sheeley, 2003] **b) False Color Streamer.** The photosphere and bright inner corona have been masked. Multiply-ionized oxygen appears purple, neutral hydrogen blue, and electron scattering green. Contours of the solar wind outflow speeds are plotted at 0, 50, and 100 km/s, increasing outward. Blobs of outflowing material accelerate along the sides of streamers in the locations indicated by arrows. Material from the streamer tip (stalk) contributes to the high-density regions surrounding the heliospheric current sheet. *(Courtesy of ESA/NASA – the Solar and Heliospheric Observatory (SOHO) Program)*

The solar equator is tipped with respect to Earth's orbital plane, so the HCS crosses geospace regularly. Even at low solar wind speeds, the ~10 R_E-thick current sheet passes Earth in a matter of minutes. At solar minimum, Earth often sits in or at the edge of the current sheet for long times. Sometimes flapping or rippling in the current sheet keep Earth in a broad transition zone with multiple sheet crossings in a single day. These boundaries produce minor variations in solar wind speed and density and small fluctuations in the interplanetary magnetic field (IMF). Small, short-lived southward field excursions add energy to the geospace

Chapter 10 The Active Interplanetary Medium: Conduit for Space Weather

system that dissipates in substorms, as we describe in Chaps. 8 and 11. The increased density from the heliospheric plasmasheet may amplify the effects of the HCS crossings.

Magnetic field fluctuations, with wave-like characteristics, pervade the interplanetary medium. Structures at the largest spatial scale, such as stream structures (Sec. 10.1.2), are sources of fluctuation at frequencies near 10^{-6} Hz. Magnetic flux tubes originating in the chromosphere and extending into the solar wind generate fluctuations with frequencies from 5×10^{-6} Hz to 10^{-5} Hz. Alfvénic fluctuations with frequencies of 1 Hz to 10^{-4} Hz are present in the tubes (Fig. 10-2). Some of these waves are associated with pressure perturbations from solar granulation; other waves are related to MHD turbulence. Fluctuations above 1 Hz are associated with ion motion (cyclotron and acoustic waves).

Data from space-based solar wind monitors suggest that the meso-scale solar wind flow arrives at Earth in bundles of individual flux tubes that probably originate at super granule boundaries. Magnetic field data from the Advanced Composition Explorer (ACE) spacecraft indicate that Earth samples a different flux tube every 15–20 minutes. The flux tubes have an average background field direction consistent with the solar wind (Parker) spiral direction, but each tube exhibits its own motion, plasma properties, and deviation from the background flow (Fig. 10-2). The median diameter is ~5×10^5 km.

Fig. 10-2. Artist's Rendering of Individual Magnetic Flux Tubes in the Solar Wind. Each flux tube has its own local field direction, and within each tube Alfvénic fluctuations make the magnetic field vector wander randomly about this direction. On average, the flux tubes are directed along the Parker spiral. The faster the wind speed, the higher the variability. The field line motion past Earth introduces an element of small-scale randomness in the interplanetary magnetic field record. [Bruno et al., 2001]

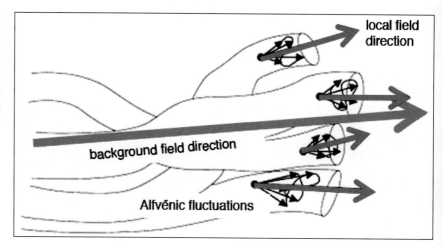

In general, the frozen-in flux condition prevents plasma migration across tube walls. Within the tubes, turbulence produces small-scale random fluctuations of the magnetic field and plasma. Motions of the tube boundaries produce magnetic field discontinuities that may trigger minor disturbances in Earth's geomagnetic field. These disturbances are random and nonrecurrent.

The slow solar wind also exhibits other types of disturbances: pressure pulses, density enhancements, and perhaps magnetic merging events. Periodic density variations may develop in the solar corona where open flux tubes merge with closed flux tubes, leading to an influx of new plasma on the boundary open field lines. Scientists have speculated that compression waves in the convective zone could regulate this reconnection.

Some quiescent solar wind fluctuations may develop after the parcels depart the Sun. Mounting evidence indicates that magnetic merging with scale sizes of millions of kilometers occurs in the slow solar wind. When this happens, a part of the IMF disconnects from its solar roots. Jets created during merging contain accelerated plasma with increased density. At the merging site in the interplanetary medium, the magnetic field shears and Alfvénic fluctuations develop. These discontinuities and fluctuations propagate in the solar wind and sometimes intercept Earth. Scientists have identified several dozen such events in the ACE solar wind record. Events with signatures masked by other processes may be present on a more regular basis. The scale sizes of the merging events are under investigation.

Pause for Inquiry 10-1

a) Compare the diameter of a solar wind flux tube at Earth to the diameter of a supergranule. Explain any differences you find.

b) Compare the diameter of a solar wind flux tube to the diameter of Earth's magnetosphere.

c) Compare the width of the heliospheric plasma sheet to the diameter of Earth's magnetosphere.

EXAMPLE 10.1

Solar Wind Flux Tube Speed

- *Problem Statement:* Use information about the solar wind flux tubes to determine a characteristic flow speed of the solar wind bundles.
- *Concept:* Characteristic size and passage times reflect the speed of the tubes.
- *Assume:* Bundle diameter = 5×10^5 km; Earth passage time = 20 min

- *Solution:*

$$speed = \frac{distance}{time} = \frac{5 \times 10^5 \text{ km}}{(20 \text{ min})(60 \text{ s/min})} \approx 417 \text{ km/s}$$

- *Implication and Interpretation:* The solar wind carries embedded tubes of magnetic field and their attending plasma past Earth at speeds typical of the solar wind flow.

Follow-on Exercise: How long does Earth dwell in the flux tubes of fast solar wind?

Chapter 10 The Active Interplanetary Medium: Conduit for Space Weather

10.1.2 Co-rotating Structures in the Solar Wind

Objectives: After reading this section, you should be able to...

♦ Describe the source of co-rotating interaction regions (CIRs) observed by Skylab instruments in 1973
♦ State which portions of the solar cycle are associated with CIRs
♦ Explain how CIRs become geoeffective
♦ Describe the source of merged interaction regions (MIRs)

Coronal holes grow and often extend to low latitudes (Fig. 10-3), where they form deep corrugations in the heliographic current sheet (Fig. 10-4). Such structures may be present for multiple solar rotations. Recurrent high-speed streams originate in coronal holes. As we note in Chap 5, these streams have low plasma density and vast areas of unipolar radial magnetic field. The flow from each stream is asymmetric with a rapid acceleration, followed by a long interval of decreasing speed. The interaction with the surrounding solar wind produces disruptions in the flow that may lead to moderate geomagnetic storms at Earth.

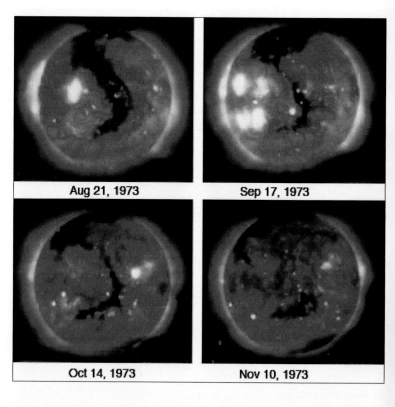

Fig. 10-3. Six-month History of a Coronal Hole Observed by Skylab Instruments in 1973. On August 21, 1973, a trans-equatorial coronal hole extended over a latitudinal distance of a million kilometers. In time, neighboring coronal arches encroached into the hole, slowly closing it and leaving only the persistent hole at the Sun's north pole by mid November 1973. Magnetic field lines diverge at boundaries of the holes; they do not cross the chasm. The magnetically "open" features allow plasma to escape from the Sun at high speeds. Two to three days after departing the Sun, the plasma arrives at Earth as a high-speed stream. The temperature within a coronal hole is about 10^6 K, about a third to a half that in active regions and about 50 percent lower than in surrounding "quiet" areas. Hence in X rays these regions appear dark. Density is a third that of the neighboring corona. The "closed" or arched-over field lines, seen elsewhere on the Sun, give the structured corona its distinctive form. Curiously, coronal holes rotate as though attached to a solid Sun, unlike the slipping layers of the photosphere and chromosphere, which rotate much faster at the equator than at the poles. *(Courtesy of NASA)*

A strong interaction develops when solar wind streams of different speeds collide. *Stream interfaces* form when outward propagating high-speed solar wind streams from coronal holes overtake slow solar wind streams from the streamer belt (Figs. 5-4 and 10-4). The merged structures, called stream interaction regions

10.1 Disturbances in the Quasi-stationary Solar Wind

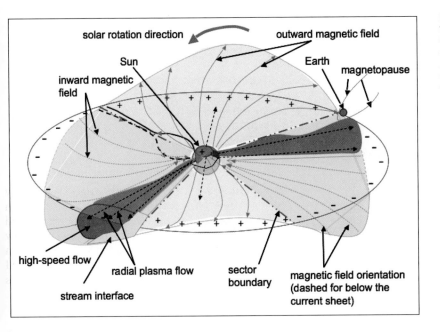

Fig. 10-4. Three-dimensional Concept of the Current Sheet Disturbed by a High-speed Stream. The Sun is at the center, with counterclockwise rotation and positive polarity at the north pole. The current sheet is in a 4-sector structure. Gray curves indicate magnetic field orientation above the current sheet. One set of gray field lines on the left side of the Sun gives a cross sectional view of the current sheet. One set of field lines is slightly stretched from the dipolar configuration. The other set is strongly stretched to form a current sheet. Earth's orbit is the solid red curve. Earth is represented as a blue circle with two curves for the magnetopause. Sector boundaries are red dash-dot lines. Away sectors with outward directed magnetic field (+) are shown in pink. Toward sectors with inward directed magnetic field (−) are shown in blue. Two regions of high-speed flow are in brown. One is approaching Earth, and has a trailing edge shown in a lighter shade of brown. The other is in the foreground with flow directed approximately out of the page. The yellow region in advance of that high speed is the stream interface. [After Alfvén, 1977]

(SIRs), pass Earth at high speed. The interface has several signatures, including flow angle sign changes, compression of the magnetic field, and steep thermal and flow speed increases. A plasma density maximum may occur just ahead of or at the interface. If the solar corona is quasi-stable, the solar rotation produces a series of SIRs that become *co-rotating interaction regions (CIRs)*. Multiple coronal holes may contribute to frequent disturbances during a single 27-day rotation period.

Coronal holes develop and persist at any time, but are most prevalent during the declining and minimum phases of the solar cycle. When persistent coronal holes and their associated high-speed solar outflow are present at low heliographic latitudes, a recurrent disturbance pattern of the HCS and surrounding plasma dominates the solar wind. Although coronal holes normally take up less than 20% of the corona, Earth may be immersed in high-speed flow up to 50% of the time during the declining phase of the solar cycle, sometimes for as long as a week. During solar maximum, up to 30% of the flow is from high-speed regimes that extend to, or form at, low and middle latitudes. The intervals of high-speed flow outside of the declining phase tend to be more sporadic, shorter in duration, and produce speed maxima slightly lower than those in the declining phase.

Pause for Inquiry 10-2

CMEs often accompany stream interface regions. Why do you think this is so?

Near the ecliptic plane, CIRs organize the interplanetary medium by producing pressure enhancements and outward-moving waves. At the leading edge of the stream, we usually find an increase in plasma density, temperature, field strength, and pressure, while the plasma and field on the trailing edge become increasingly rarefied (Fig. 5-5). The high-speed wind transfers momentum and energy to the low-speed wind as the buildup of pressure on the stream's leading edge

EXAMPLE 10.2

Arrival and Duration of a High-speed Stream at Earth

- *Problem Statement:* Determine the arrival time and duration of a high-speed stream at Earth, given a coronal hole map of the Sun, an average expansion factor of the magnetic flux tubes near the Sun, and solar wind speed measurements.

- *Concept:* Arrival time $t_{arrival}$ depends on the movement (rotation) time of the hole's western (leading) edge from the central meridian (t_{rot}), and the travel time of the solar wind to Earth ($t_{Sun\text{-}Earth}$). The solar wind expands from the coronal hole source into the heliosphere.

- *Given:* Coronal hole flow speed is 600 km/s, and solar rotation period τ is 27 days. The solar wind's expansion factor f is 8. The coronal hole's western edge is at 5° E, and the eastern edge is at 12° E.

- *Assumptions:* The hole is static (no expansion or contraction after the flow departs the source surface region at 2.5 R_S.)

- *Solution:*
  1. Arrival time = time for coronal hole rotation to central meridian + travel time to Earth

  $$t_{arrival} = t_{rot} + t_{Sun\text{-}Earth}$$

  $$t_{arrival} = \frac{\Delta\varphi}{\omega} + \frac{d}{v}$$

  where

  $\Delta\varphi$ is the original angle between the leading edge of the hole and the central meridian (deg)
  ω = rotation rate of the Sun (deg/day)
  d = distance from the Sun to Earth (km)
  v = speed of the stream (km/s)

  $$= \frac{5 \text{ deg}}{13.3 \text{ deg/day}} + \frac{1.5 \times 10^{11} \text{ m}}{(6 \times 10^5 \text{ m/s})}$$

  $$= (0.375 \text{ days} \times 86{,}400 \text{ s/day}) + 2.5 \times 10^5 \text{ s}$$

  $$= 2.82 \times 10^5 \text{ s} \approx 3.3 \text{ days}$$

  2. Duration of high-speed flow = expansion factor multiplied by fraction of rotation interval dominated by coronal hole

  $$\Delta\tau = f\tau \, (\Delta\varphi/\varphi)$$

  $$\Delta\tau = 8(27 \text{ days})(7°/360°) = 4.2 \text{ days}$$

  where

  τ = solar rotation period
  $\Delta\varphi$ = coronal hole width

- *Interpretation:* Flow from a coronal hole 7° wide with its western most edge 5° east of the central meridian and a flow speed of 600 km/s will reach Earth in approximately 3 days. The duration of the flow at Earth from this rather narrow hole will be ~4 days.

- **Follow-on Exercise:** The larger coronal holes of the declining phase of the solar cycle have greater widths and higher speed flow. Assume that the coronal hole's western edge remains unchanged, but its width is 15° and the flow speed is 800 km/s. Recalculate the arrival and duration times of such a high-speed flow.

---

accelerates the low-speed wind ahead. The momentum loss slows the high-speed wind. In the fast solar wind behind the interface, the kinetic energy of the plasma converts to thermal energy, resulting in plasma heating and lower density. Figure 10-5 shows a record of compressions and rarefactions, as well as the stream interface, from a high-speed stream passage on 18 May 1999.

As the stream passes a given point, the solar wind speed typically remains high for a few days, but the density and magnetic field strength drop below normal values in the rarefied regions behind the CIR. When the compression passes, it slightly deflects the radial solar plasma flow. We recall from Chap. 5 that the magnetic field lines of the ambient slow solar wind are more curved, whereas those of the high-speed streams are more radial. With increasing distance from the Sun, the mismatch in geometry supports more interaction between streams. Discontinuities steepen into shocks.

10.1 Disturbances in the Quasi-stationary Solar Wind

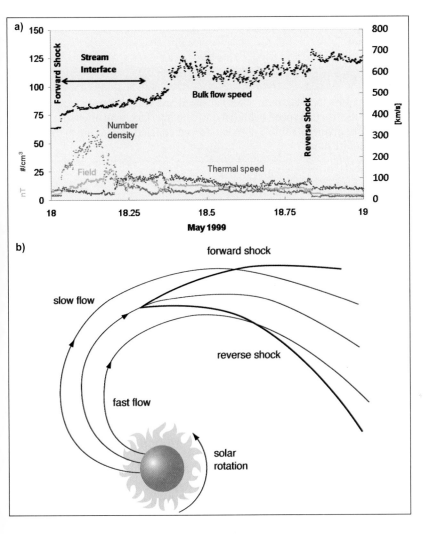

Fig. 10-5. **Shocks Associated with a High-speed Stream. a)** On May 18, 1999, sensors collected these data on the solar wind. The influence of the stream interface extends forward into the slow solar wind and backward into the high-speed stream. In many cases, all three parameters in these plots are needed to identify a shock occurrence. We plot the bulk flow speed in black; the thermal speed in red; the number density in light blue; and the magnitude of the magnetic field strength in green. Solar wind density rises at the leading edge of the stream interface (forward shock). At the trailing edge of the density perturbation, the magnetic field and thermal speed increase. High-speed flow arrives at the trailing edge of the stream interface at 19 UT (reverse shock). *(Courtesy of ESA/NASA – the Solar and Heliospheric Observatory Program and the University of Maryland)* **b)** Black curves show the interplanetary magnetic field direction. Different flow regimes are labeled, as are the forward shock and a reverse shock that expand the flow across the Parker spiral.

If the difference in flow speed between the stream crest and the trough ahead is greater than about twice the speed of small-amplitude pressure signals, then ordinary pressure signals do not propagate sufficiently fast to move the slow wind out of the path of the oncoming high-speed stream. Waves are driven away from the pressure increase in both directions, resulting in a wave in which a leading (forward) wave propagates away from the Sun, while a trailing (reverse) wave propagates toward the Sun but is carried out with the solar wind flow. The pressure eventually increases nonlinearly and a pair of shocks form on either side of the high-pressure region. Both shocks convect away from the Sun by the high-bulk flow of the wind.

Compression-rarefaction couplets often grow into shocks in the flow beyond Earth. Shock formation in the waves occurs via nonlinear amplitude increases (steepening) of waves, thereby requiring several steepening times to elapse before a shock forms. Satellite data reveal occasional CIR-driven shocks at 1 AU. Most shocks happen between 1 AU and 2 AU. Virtually all large-amplitude solar wind streams have formed shock-wave structures at heliocentric distances beyond ~3 AU. Interplanetary shocks are easily identified as simultaneous, abrupt jumps in the speed, density, and temperature of protons, ionized helium, and electrons.

> **Pause for Inquiry 10-3**
>
> Scientists often treat the terms thermal speed and temperature as equivalent. Compare the thermal speed behind the stream interface with high-speed wind temperature given in Table 5-1. What do you conclude about the thermal speed in a fast flow?

> **Pause for Inquiry 10-4**
>
> During a recent 11-year interval, approximately 350 CIRs were recorded in the vicinity of Earth. Verify that 350 CIRs at Earth per solar cycle is a reasonable number.

> **Pause for Inquiry 10-5**
>
> Argue that because most CIRs do not have shocks at 1 AU, but have steepened into shocks by 2 AU, the nonlinear steepening time must be on the order of five days.

At Earth, CIRs are important space weather producers. Strong compressed fields, commonly present within CIRs, create recurrent moderate geomagnetic storms, particularly when the field contains a southward component. When combined with the high-speed flow, the southward field results in a strong interplanetary electric field that transfers energy to the magnetosphere via magnetic merging (Sec. 4.4, and Chaps. 8 and 11). While the typical IMF intensity is about 5 nT–8 nT at Earth, in the ambient solar wind it rises to 10 nT–15 nT within CIRs. For some CIRs, it can reach ~30 nT.

Geomagnetic activity begins with passage of the stream interface and peaks within a matter of hours. In many cases the passage of a CIR is followed by several days of sustained IMF fluctuations. These fluctuations are wave trains (Alfvén waves) excited by the high-speed flow from coronal holes. Over several days they produce mild-to-moderate disturbances in the magnetosphere and deposit 10^{16} J–10^{17} J of energy in the thermosphere and ionosphere. They buffet Earth's magnetosphere and excite waves inside the magnetospheric cavity. Resonances with magnetospheric electrons then energize the low-mass particles to MeV levels. Long-duration fluxes of high-energy electrons damage spacecraft and instruments in the inner magnetosphere.

> **Pause for Inquiry 10-6**
>
> IMF data are not shown in Fig. 10-5. When do you think the peak in magnetic field intensity occurs with respect to the stream interface? Is your answer consistent with Fig. 10-5?

EXAMPLE 10.3

A Faster Solar Wind Stream Overtakes a Slower One

- *Problem Statement:* Determine the meeting point of two solar wind streams traveling at different speeds. The slow flow starts in the streamer region. A second stream originates at 30° east of the first stream, and travels twice as fast.

- *Concept:* The solar magnetic field expands outward following the garden hose or Parker spiral. The streamlines for the Parker spiral are described by the ratio: $v_r/v_\theta = v_{sw}/v_\theta$ (See the Parker spiral equation given by Eq. (5-5)). The faster stream catches up to the slower one.

- *Given:* The rotation rate of the Sun is 2.69×10^{-6} rad/s. The streams are separated by 30° of solar longitude. The speed of the first stream is v_{sw}. The speed of the second stream is $2v_{sw}$.

- *Assume:* $v_{sw} = 450$ km/s

- *Solution:*

$$\frac{dr}{d\theta} = \frac{dr}{dt}\frac{dt}{d\theta} = \frac{v_{sw}}{\omega}$$

Each stream behaves according to

$$dr = \frac{v_{sw}}{\omega}d\theta$$

$r_1 = r_2$ at the meeting point of the two streams.

We first solve for θ:

$$\frac{v_{sw}}{\omega}\theta_1 = \frac{2v_{sw}}{\omega}d\theta_2$$

$$\theta_1 = 2\theta_2$$

From the problem statement,

$$\theta_2 = \theta_1 - 30°$$

Solving for θ_1 gives

$$\theta_1 = 60° = 1.047 \text{ rad}$$

Now we solve for radial distance traveled as the angle advances

$$dr = \frac{v_{sw}}{\omega}d\theta_1$$

$$r - R_S = \frac{v_{sw}}{\omega}\theta_1$$

Substituting

$$r = \frac{4.5 \times 10^5 \text{ m/s}}{2.69 \times 10^{-6} \text{ rad/s}}(1.046 \text{ rad}) + 7 \times 10^8 \text{ m}$$

$$r = 1.76 \times 10^{11} \text{ m} \approx 1.17 \text{ AU}$$

- *Implication and Interpretation:* The meeting point of the two streams with the characteristics described in the problem statement is just beyond Earth. Spacecraft observations reveal that most shock development from stream interactions occurs just beyond Earth.

Follow-on Exercise: Imagine that a spacecraft encounters the meeting point of two streams described in the example at 1.25 AU. Determine the initial separation of the two streams at the Sun.

Merged interaction regions (MIRs) develop as two or more flows interact. At 1 AU, these structures are important, because their large size and strong magnetic fields produce major geomagnetic storms and strongly influence energetic particles and cosmic rays. Colliding CMEs and CIRs can produce MIRs within 1 AU. Interactions between CIRs usually develop beyond 1 AU. Data from the Voyager spacecraft reveal that CIRs continue to develop as they travel past Earth, often forming spiral shapes that may wrap multiple times around the Sun. The CIRs eventually catch one another to form MIRs beyond 10 AU. Beyond about 20 AU, the identity of individual high-speed streams disappears, having been converted to pressure waves and shocks. Consistent with the Second Law of Thermodynamics, such regions convert organized structures from the parent CIRs into heat and turbulence. MIRs are thought to be a partial source of solar wind heating in the distant heliosphere. These complex interactions and structures persist to great distances in the heliosphere. Figure 10-6 shows the extension of Fig 5-18 to large heliospheric distances. The ridges are likely to support MIRs.

Scientists estimate that almost all the equatorial solar wind passes through CIRs by approximately 25 AU. The distant heliosphere is dominated by CIR and MIR structures that often act as particle accelerators and deflectors. Low-energy ions caught at the edges of these structures may be accelerated up to a few tens of MeV. The lumps and bumps created by MIRs in the interplanetary medium may scatter some inbound energetic cosmic particles. Periodic pressure ridges extend as far as 50 AU (Fig. 10-6).

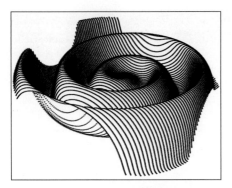

Fig. 10-6. A Computer Simulation of an Ideal Parker Spiral Solar Wind Configuration. When the Sun's magnetic dipole is tilted relative to the rotation axis, the heliospheric current sheet is warped and resembles a ballerina's twirling skirt. In this simple picture, each ridge in the skirt corresponds to a different solar rotation; the ridges are separated radially from one another by about 4.7 AU. Beyond Earth, but within the inner solar system, individual co-rotating interaction regions (CIRs) merge to create merged interaction regions (MIRs). Near the edge of the solar system, the MIRs decay into pressure ridges. [Jokipii and Thomas, 1981]

10.2 Transient Disturbances in the Solar Wind

10.2.1 Characteristics of Transients in the Interplanetary Medium

Objectives: After reading this section, you should be able to...

- Distinguish between large and small interplanetary transients
- Relate coronal mass ejections (CMEs) to the heliospheric streamer belt
- Identify the interplanetary signatures of a CME

Interplanetary Transients

Scale Sizes. The Sun releases stored energy and magnetic helicity in the form of CMEs. Some scientists distinguish between CMEs close to the Sun and those that have escaped into the interplanetary medium. For convenience we use the generic term CME. These large transient structures in the solar wind receive most of the attention from space weather forecasters. However, improved observing capabilities reveal a broad range of transient emissions from the Sun. A campaign involving multiple spacecraft in 2007 showed small transients embedded in the slow solar wind. These have smaller scale sizes, smaller speed changes, and smaller magnetic fields than the average solar minimum interplanetary CME at 1 AU. They also tend to have less longitudinal spread. The average scale size of these structures was less than 0.1 AU, compared to 0.27 AU for large-scale CMEs. Most of the small-scale transients appeared to develop at the edges of coronal holes and flow into the interplanetary medium at slow solar wind speeds. Table 10-1 compares the sizes of small scale transients observed in the streamer belt with larger-scale CMEs.

Table 10-1. **Average Values for Solar Wind Transients.** This table gives values for magnetic fields, speed changes, and diameters of small and large solar wind transients. Data provided by Jian et al. [2006] and Kilpua et al. [2009].

Characteristic	B_{max} (nT)	$|\Delta v|$ (km/s)	Diameter (AU)
Small transients (solar min)	7.3	20	0.072
CMEs (solar min)	13.9	80	0.27
CMEs (solar max)	19.1	154	0.40

Hints of these structures have been noted in the data since the late 1990s. However, their characteristics are close to that of the background solar wind flow; thus they often go undetected in routine solar wind monitoring. If these transients embed in the high-speed flow, they can produce moderate to strong magnetic storms.

Chapter 10 The Active Interplanetary Medium: Conduit for Space Weather

Interplanetary Signatures of CMEs

Most large transient disturbances originate from closed magnetic field regions on the Sun: polar crown filament regions, low-latitude filament regions, and active regions. Within hours of leaving the Sun, a CME blossoms into a structure many times larger than the Sun. By the time it arrives at Earth, it may have expanded to a cross-sectional size of about 1/4 AU. Ejected solar material is marked by some or all of the distinctive signatures we discuss below. Figures 10-7 and 10-10 illustrate several of these characteristics.

Loop-like (Flux Rope) Topologies of Magnetic Fields. Magnetic fields in CMEs are often helical (Fig. 9-19). Ejecta that rise slowly from the Sun may maintain a smooth magnetic field and nearly symmetric configuration during their Earth (or spacecraft) passage. The passage of such a structure often appears as a sinusoidal trace in the interplanetary magnetic field record. These CMEs are *magnetic clouds*. Ejecta that rise rapidly from the Sun tend to have more chaotic fields with much more distorted loops. During solar minimum, the majority of CMEs are magnetic clouds. During solar maximum, they account for only about 20% of the CME population.

Bi-directional Streaming of Particles. Heated (suprathermal) electrons tend to be trapped in narrow cones on the field lines of closed-loop CMEs. When satellite instruments with multiple look angles observe electrons streaming in both directions along the field, we interpret them as signatures of closed field lines and, thus, as a good proxy for CMEs. Figure 10-7 depicts a closed loop on which the electrons stream in either direction. We discuss additional aspects of these particles shortly.

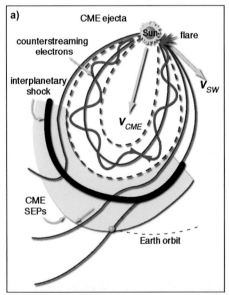
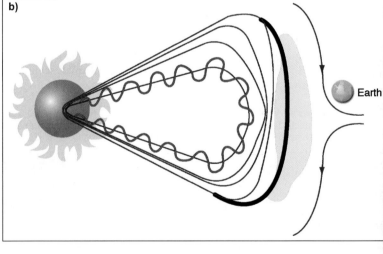

Fig. 10-7.   a) Details of an Interplanetary Coronal Mass Ejection Viewed from above the Ecliptic Plane. This artist's rendition of a CME depicts the plasma and magnetic field frozen together. The ejecta are not constrained to move along the Parker spiral. The ejected flux rope between the dashed curves expands until it reaches pressure equilibrium with the surrounding solar wind. Often both ends of the field lines are rooted in the solar atmosphere, allowing counterstreaming hot electrons to move to and fro on the outward expanding magnetic rope. If the CME is moving fast enough, shock-created energetic particles will precede it (blue region). [Zurbuchen and Richardson, 2006] **b) A Coronal Mass Ejection Interacting with the Plasma and the Interplanetary Magnetic Field (IMF).** This image depicts the plasma and magnetic field in the regions ahead of the CME. The ambient plasma and field are compressed. The CME may distort the ambient IMF. If the ambient magnetic field experiences a southward deflection, magnetic storms may develop well in advance of the CME.

Transient Interplanetary Shocks and Energetic Particle Events. Fast traveling ejecta compress upstream regions of the solar wind, forming a shock that disturbs and concentrates the interplanetary medium ahead of it. The ambient interplanetary magnetic field stretches, distorts, or drapes around the fast-moving ejecta, forming a relatively dense sheath ahead of the CME body (Fig. 10-7b). This dense material arriving at Earth compresses and re-configures Earth's magnetic field. These distorted fields are an additional source of geomagnetic disturbance in advance of the CME at Earth. Fast moving, intense magnetic fields momentarily trap and energize charged particles, producing solar energetic particle (SEP) events. We describe these phenomena more in Sec. 10.3.

Depressed Proton Temperatures and Plasma Beta (β). Expansion of the CME material as it moves through the ambient, quasi-stationary solar wind plasma reduces the CME temperature and plasma β parameter. CMEs that interact strongly with other structures in the solar wind are less likely to expand and thus may not exhibit this characteristic.

Unusual Compositions of Ions, Elements, and Charge States. In the corona, the electron density decreases so rapidly that the plasma becomes collisionless and the relative ionization states for coronal constituents become constant, thus reflecting the conditions in the corona where this change occurs. The charge states of minor ions (those other than hydrogen or helium) in CME flows usually suggest slightly hotter than normal coronal conditions (i.e., $T > 2 \times 10^6$ K) at the "frozen-in" location. In some CMEs, an erupting filament could contribute to a dense plug of material or an extended flow of cool plasma in the trailing edge of the ejecta. This filament material often lags well behind the leading edge of the related CMEs. Enhanced helium flows have also been detected in many large CMEs.

Cosmic Ray Depressions (Forbush Decreases). As CMEs pass Earth, their strong magnetic fields deflect cosmic rays, and so decrease cosmic ray intensity. (Focus Box 10.3)

For a variety of reasons, some CMEs don't exhibit all, or even any, of the signatures in the previous list. Interactions between CMEs and other structures in the solar wind can modify or eliminate the flux rope signature. These same interactions can modify the CME's plasma β. Normally only the fastest CMEs have shocks that drive energetic particles. Noticeable cosmic ray depletions usually accompany the largest CMEs.

It is possible for one CME foot to detach from the Sun via interchange reconnection. This process allows merging between an open field line, perhaps near a coronal hole, and a closed loop (Fig. 10-8). One leg of the CME then detaches from the Sun. This allows the electrons to flow outward, thus eliminating the counterstreaming electron signature. Roughly half of the CMEs arriving at Earth appear to have undergone interchange reconnection.

Suprathermal Electrons as CME Tracers

Measurements of electron energy distributions in the solar wind reveal that a significant fraction of electrons are much hotter than expected for a distribution in thermal equilibrium. These suprathermal electrons are nearly collisionless, and tend to align along the IMF. They thus conduct coronal heat to the heliosphere in beam-like flows with energies in the hundreds of eV. The cooler (thermal) electron population (tens of eV) usually has more energy in gyromotion about the field. The thermal population is sometimes called halo electrons because they move around the field lines (Fig. 10-9). The suprathermal electrons are very helpful in studying the roots and topology of solar wind magnetic field lines. Scientists often use them to trace the roots of coronal mass ejections (Fig. 10-11). When beams of

Fig. 10-8. Interchange Reconnection. Here a closed flux rope with both feet attached to the Sun merges with an open field line. A shorter closed loop reforms close to the Sun. The original loop becomes part of the open heliospheric magnetic flux. [Crooker et al., 2006]

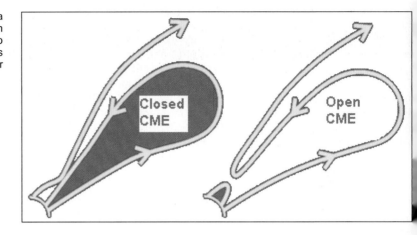

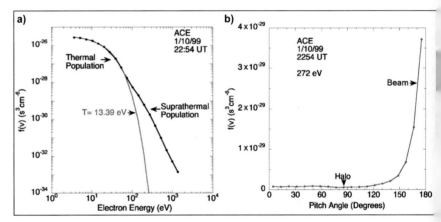

Fig. 10-9. a) Suprathermal vs. Thermal Electrons. Measurements of electron energy distributions in the solar wind reveal the presence of both thermal (halo) and suprathermal (field-aligned beam) populations. The vertical axis is a function of the probability of finding a particle with velocity in the infinitesimal volume element about a specified velocity increment. The red curve represents a thermal Maxwell–Boltzmann velocity distribution with a characteristic energy of 13.39 eV plotted on a log-log scale. Roughly half of the electrons have energies below 13.39 eV. The other half are distributed in a long tail above 13.39 eV. The suprathermal component of the electron distribution is shown by the black curve. The high energy tail of this population is at significantly higher values than those of the thermal population. The characteristic energy of suprathermal population is ~272 eV. **b) Electron Pitch Angle Distribution.** The suprathermal population (272 eV) flows away from the Sun along the field line (pitch angles near 180 deg). Only a tiny fraction gyrate around the field line. (Courtesy of Jack Gosling at the Laboratory for Atmospheric and Space Physics, University of Colorado, Boulder)

suprathermal electrons move in opposite directions on coronal mass ejection field lines that intercept Earth, it indicates that the lines still have both feet connected to the Sun. Beams of particles moving in only one direction mean that the field line has one foot in the Sun and one foot open to the heliosphere (Fig. 10-11). On rare occasions, these open field lines connect directly to Earth's polar cap. When this happens, the suprathermal electrons slam into Earth's upper atmosphere and create an X-ray glow that satellite sensors can detect.

Focus Box 10.1: Using Interplanetary Scintillations to Detect CMEs

Contributed by John Kennewell and Bernard Jackson

Space-based satellite imagery provides significant insight into the origins and dynamics of CMEs. Other ground- and space-based remote sensing techniques are under development. By focusing radio telescopes on powerful sources of natural radio emissions (such as quasars), scientists infer the location of solar ejecta by the intensity fluctuations (scintillation) they produce in these emissions. More material along the line of sight of the radio beam corresponds to more scintillation.

In the absence of any disturbances in the interplanetary medium, the signals received from these point radio sources are constant in amplitude. However, if the radio signal passes through a plasma cloud (CME) on its way to Earth, it is absorbed and refracted by the plasma. The amplitude and phase of the signal as received at Earth vary from one moment to the next—this is scintillation. The signal from the quasar "twinkles", just as the light from visible stars twinkles as it passes through the atmosphere (Fig. 10-10).

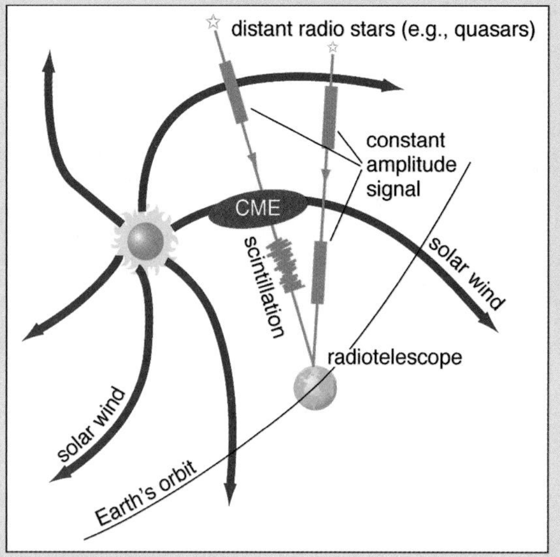

The frequencies used by radio telescopes generally range from about 80 MHz to 300 MHz. Lower frequencies are subject to more scintillation, and are thus better able to detect smaller plasma densities. Higher frequencies see closer to the Sun and so can theoretically provide greater warning time. At these frequencies, the radio telescopes need to be very large to provide a beam narrow enough to resolve closely separated radio stars. For instance, at 100 MHz (a wavelength of 3 m), a linear dimension of 200 m is required to form a beam of 1° in that dimension. The antenna generally employed at these frequencies is an array of many hundreds of dipoles that are electronically combined and steered to scan the sky.

Using a network of four radio telescopes at the Solar-Terrestrial Environment Laboratory, Nagoya University, Japan, researchers have developed a method of detecting the movements of these powerful phenomena in the vast region of space between the Sun and Earth. They can detect the direction and velocity of solar wind disturbances by precisely measuring when a particular fluctuation or twinkle reaches each of the four radio telescopes. The telescopes are separated, one in each of four Japanese radio observatories. The appropriate degree of separation lets scientists correlate the time the scintillation pattern goes from one telescope to the other. Combining all of the information in a computer program produces a three-dimensional picture of the region between the Sun and Earth, sort of a CAT-scan of the solar wind.

An extension of CAT-scan technology is being applied to data from the Solar Mass Ejection Imager (SMEI) satellite. Sky maps from SMEI, when background light is removed, consist of sunlight scattered from heliospheric electrons. Gathering information on the total electron content of the interplanetary medium and its variation allows researchers to locate CIRs and CMEs. Efforts are underway to determine the motions and dynamic development of these structures.

Fig. 10-10. Interplanetary Scintillation Associated with a Coronal Mass Ejection (CME). The extra density within the CME changes the radio signal propagation relative to similar signals received along paths not disturbed by the CME. *(Courtesy of John Kennewell and Andrew McDonald at the Radio and Space Services of the Australian Ionospheric Prediction Service)*

Chapter 10   The Active Interplanetary Medium: Conduit for Space Weather

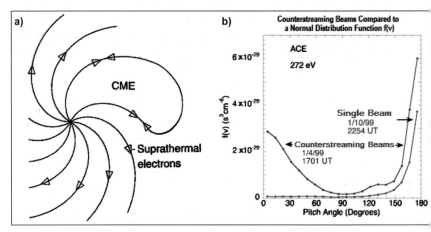

Fig. 10-11.   **Counterstreaming Suprathermal Electrons as Tracers of Closed Magnetic Field Lines in CMEs. a)** In the normal solar wind, field lines are open to the outer boundary of the heliosphere and a single field-aligned, anti-Sunward-directed beam of suprathermal electrons is observed. Coronal mass ejections originate in closed field regions in the corona and field lines within them are at least initially connected to the Sun at both ends. b) Counterstreaming beams (shown in blue) of suprathermal electrons are commonly observed on closed field lines and help identify CMEs in the solar wind. *(Courtesy of Jack Gosling at the Laboratory for Atmospheric and Space Physics, University of Colorado, Boulder)*

10.2.2   Geoeffectiveness of Transients in the Interplanetary Medium

Objectives: After reading this section, you should be able to...

- Describe a geoeffective CME
- Explain why CMEs have preferred solar cycle orientations
- Describe typical propagation times for a CME to 1 AU
- Describe the impact of the interplanetary medium on CME propagation
- Describe the origin of globally merged interaction regions (GMIRs)

Geoeffective Coronal Mass Ejections. To set geomagnetic storms in motion, solar ejecta (or their shocks) need to reach Earth. A tilted streamer belt gives adjacent, but differing, flows access to the ecliptic plane (Fig. 10-4). Ejecta in the slow flow immediately preceding the leading edge of a high-speed stream are channeled or even trapped by the co-rotating interaction region. The probability of encountering a CME in the ecliptic plane is higher near sector boundaries. Average CMEs with widths of 45°–50° usually pass 1 AU within about three days, centered on the sector boundary.

Earth-intercepting CMEs have five elements that produce geospace disturbances: 1) high speed, 2) increased density, 3) high magnetic field strength, 4) magnetic field orientation, and 5) duration of an enhanced field. Launch conditions at the Sun, as well as the nature of the interplanetary medium through which the CME travels, determine which, if any, of these will influence Earth. When space weather forecasters speak of *geoeffective* CMEs, they mean those that produce geomagnetic storms with a disturbance storm time (Dst) index <−50 nT

(Chap. 11). These storms are associated with a large southward component of the interplanetary magnetic field (IMF $-B_z$) and high-speed solar wind flow (v_{sw}). Geoeffectiveness can also refer to the ability of a CME to produce broad and long-lived energetic particle enhancements. Fast CMEs that drive shocks do this. We discuss this latter category in Sec. 10.3.

Figures 10-12a and b illustrate a magnetohydrodynamic simulation of a CME propagating from the corona to near 1 AU. Density is normally the highest at the leading edge. The equatorial portion of the flux rope moves through dense low-speed material, while the high-latitude portions move in a more tenuous, higher speed flow. Thus the shape of the flux rope evolves from circular to elliptical to a "pancake" structure. The pancake becomes more like a boomerang if the equatorial portion of the CME is slowed by the dense material in the heliospheric plasma sheet and the mid-latitude portions interact with high-speed flow. The rapidly moving CME propagates waves along its boundaries, which tend to move poleward. Flux rope CMEs transiting more complicated solar wind structures at solar maximum are likely to become highly distorted.

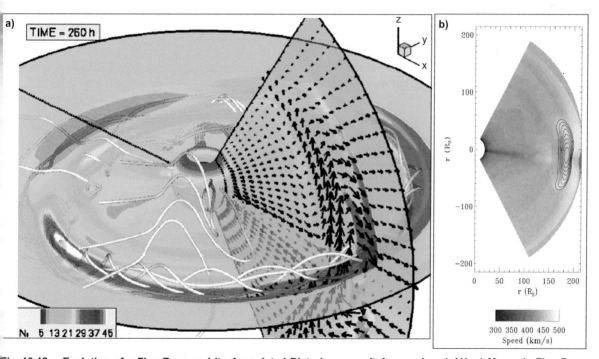

Fig. 10-12. Evolution of a Flux Rope and Its Associated Disturbance as It Approaches 1 AU. a) Magnetic Flux Rope Simulation. Here we see the magnetic flux rope in the ecliptic plane. The cross section shows solar wind velocity vectors. We depict number density n in color. The highest particle number density shown in the ecliptic plane, indicated by white, is in excess of 45/cm^3. *(Courtesy of Dusan Odstrcil at the University of Colorado, Boulder)* **b) Simulated Coronal Mass Ejection Cross Section.** With time, a CME flux rope with a circular cross section becomes a flattened, pancake structure. [Riley et al., 2003]

Nearly 80% of all instances with $B_z < -10$ nT and duration of at least three hours relate to CMEs. Wide, fast CMEs compress and displace the ambient IMF through which they travel. If the ambient field drapes over the disturbance to create an enhanced sheath of southward IMF, the disturbance lasts longer. Strong southward fields often occur either in CMEs or in the sheath regions ahead of them, or both. Fast CMEs also compress the ambient solar wind plasma, sometimes driving a pressure front (shock) in advance of the main disturbance. Dense plasma comes from other sources as well: a CME may carry solar filament material, or it may have interacted with a CIR or another CME, creating a region of higher density. Statistically, those with pressure pulses are more geoeffective. Although the exact influence of enhanced density in geoeffective events is hard to pinpoint, the combination of high speed and density compresses Earth's field, putting it into a higher-than-normal energy state.

To arrive at Earth and create a geomagnetic disturbance, a CME must be launched near enough to the Sun-Earth line that part of the ejecta intercepts Earth. The average CME width is about 45°, so most geoeffective ones originate within a few tens of degrees of the central meridian. Wider ones are more likely to intercept Earth no matter what their source longitude. They also tend to be faster, causing more compression of the ambient interplanetary medium ahead of them. Limb CMEs usually do not intercept Earth directly, but the shocks and disturbed sheath regions generated on the leading edge of high-speed limb CMEs sometimes do hit Earth.

Pause for Inquiry 10-7

Why do CMEs expand into the interplanetary medium?

Hint: Think about magnetic pressure and the proximity of field lines in the corona.

Pause for Inquiry 10-8

Why does the plasma β decrease in a CME?

Two types of CMEs are especially geoeffective: halo CMEs and magnetic clouds. Some ejections produce a halo of light around the Sun (and around the occulting disk used for observation). These are called *halo CMEs*. They tend to be wide (>60°), thus offering Earth-based observers good viewing geometry to see the visible light scattered by electrons in the ejecta (Fig. 1-19b). Although ejections that produce full halos around the Sun constitute less than 5% of the CME population, those that develop on the front side of the Sun usually travel along the Sun-Earth line at higher than average speed. Because of their speeds, halo CMEs are especially effective in producing interplanetary shocks, solar energetic particle events, and geomagnetic storms. At 1 AU, about 70% of them are associated with shocks and counterstreaming electrons or other ejecta signatures.

Front-side full-halo CMEs are the most likely to produce Earth-crossing events. Scientists see the source regions of halo events in greater detail than those observed near the limb. And since halo CMEs tend to travel along the Sun-Earth line, their internal material can be sampled *in situ* by satellites monitoring the solar wind in front of Earth. Partial (elliptical) halo CMEs originate away from the central meridian, but they may be wide enough to create a glancing blow at Earth (Fig. 1-20b). About 70% of the most intense storms of solar cycle 23 involved full or partial halo CMEs with rope-like structures in their magnetic field. In many cases, the feet of the flux rope remain attached to the Sun, so we frequently observed counterstreaming electrons with magnetic clouds.

10.2 Transient Disturbances in the Solar Wind

SOHO's extreme ultraviolet imaging telescope (EIT) and Yohkoh's soft X-ray telescope (SXT) instrument reveal that activity in the low corona, associated with frontside halo CMEs, includes the formation of:

- Dimming regions near the feet of the flux rope (most)
- Filament eruptions (most)
- Long-lived coronal magnetic loop arcades
- Flaring active regions
- Large-scale coronal waves propagating outward from the CME source region
- Decameter radio bursts

Magnetic cloud CMEs have smooth, large rotations of the magnetic field vector and magnetic field strengths greater than the ambient IMF. In a magnetic cloud, the field lines form a helical flux rope (Figs. 10-12a and 10-13). In a flux rope, the magnetic field concentrates in the center (axial field) and weakens with radial distance, producing an outward-directed magnetic pressure force. With increasing distance from the center, the field is more twisted (Fig. 10-13). In such a structure, currents must be present to decrease the field strength and to twist it. Often the field in the outer edge of the flux rope is nearly perpendicular to the axial field, consistent with an inward-directed force. The forces often counterbalance. If plasma pressure in the rope is weak, counterbalancing creates a force-free flux rope. Although force-free, a magnetic flux rope tends to expand in the interplanetary medium as it moves into regions of lower pressure.

Figure 10-13a depicts a magnetic cloud flux rope that brings a southward magnetic field to Earth when it arrives. The field on the back side of the structure is northward. The magnetic signature for Fig. 10-13b is opposite. Either of these will produce geomagnetic storms. High-speed clouds with a south orientation are the most efficient storm producers. Most CMEs are magnetic flux ropes as they depart the Sun. However, due to projection effects and interplanetary interactions, only about one third of them arrive at Earth with a clean flux rope signature like that shown in Fig. 10-13b. Flux ropes may be highly inclined to the ecliptic plane, as in Figs. 10-13c and d. They bring a large IMF B_y field to Earth (east-west field). Sometimes CMEs that develop in or near a highly tilted current sheet have large east- or west-oriented fields. Until recently, these structures had been thought of as non-geoeffective. But mounting evidence suggests that the geospace connections for these out-of-ecliptic structures produce focused-energy deposition at very high latitudes in Earth's magnetosphere and ionosphere.

Solar Cycle Dependence of CMEs. The location from which CMEs originate is linked to the Sun's magnetic field cycle and roughly follows a magnetic butterfly diagram. During solar minimum and the early rising phase of the solar cycle, CMEs tend to originate far from the equator. However, the strong dipole field guides them into or along the current sheet, and a high percentage of these arrive at Earth. As the rising phase yields to solar maximum, numerous CMEs originate at lower latitudes. Even if they are initially Earth-bound, though, many are fully or partially deflected away from Earth by the structured solar wind. During the declining cycle phase, CMEs interact with coronal holes and CIRs. Some are guided to Earth and some are driven away. Success in forecasting CME intercepts requires adequate knowledge of the structure of the interplanetary medium.

The cycling also leads to a preferred orientation of vertical (north–south) field arriving at Earth. The orientation of magnetic clouds arriving at Earth during the transition from even numbered cycles to just after the peak of odd solar cycles

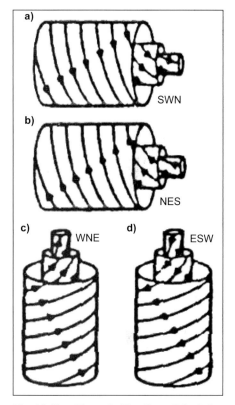

Fig. 10-13. Magnetic Flux Rope Orientations in CMEs. These views are highly idealized versions of a flux rope, like that shown in Fig. 10-12. Magnetic flux ropes are continuously layered structures of magnetic field and current. The outward-directed magnetic pressure gradient force is balanced by the inward directed curvature force of the twisted field around the axis. Currents flow to maintain the twist. The strong, central, axial field is surrounded by sheaths of magnetic field that become nearly perpendicular to the interior field. **a)** The rope travels in or near the ecliptic plane, which would be horizontal in these images. The southward field ($-B_z$) arrives first, followed by westward field and then northward field on the trailing edge. **b)** The northward field arrives first ($+B_z$), followed by the eastward field and then the northward field on the trailing edge. **c)** The rope is inclined 90° to the ecliptic plane. The westward field ($-B_y$) arrives first, followed by the northward field and then the eastward field on the trailing edge. **d)** The eastward field ($+B_y$) arrives first, followed by the southward field and then the westward field on the trailing edge. [After Russell and Elphic, 1979]

Chapter 10 The Active Interplanetary Medium: Conduit for Space Weather

(e.g., 20–21 and 22–23) tends to be as shown in Fig.10-13a. Because the dipole field changes polarity at solar maximum, the direction of the leading field in CMEs and clouds should also change at maximum. Between the peaks of odd-to-even cycle transitions (21–22 and 23–24), the magnetic clouds tend to appear with the orientation of Fig. 10-13b. In most solar cycles a one- to two-year lag in the polarity switch is observed. About 75% of magnetic cloud CMEs are oriented in this manner.

This preferred orientation arises from the strapping or arcade fields shown in Figs. 9-13 and 9-17, respectively. These overarching coronal fields usually align with the dipole. They form the outermost portions of CMEs and are likely to be the leading fields encountered by a spacecraft or a solar system body. This characteristic implies that CMEs carry an imprint of the solar dipole orientation into the heliosphere. Of course, many other factors enter into CME field orientation. The ejecta may come from extremely complex magnetic regions on the Sun and thus not have a typical orientation upon departure from the Sun. Interactions with CIRs and other CMEs cause field deflection and reorientation on the way to Earth.

Some CMEs are generated outside of active regions. Shock waves launched from solar flares may also cause CMEs elsewhere in the solar corona. Such mass ejections do not emanate from structures directly above the eruptive flare, but rather from one side. Interconnecting X-ray loops may disappear suddenly, ejecting huge amounts of material. For these unusual CMEs, a shock wave generated by the flare crosses a large interconnecting loop, causing it to become unstable and erupt. This eruption ejects hot material that represents a significant fraction of the CME mass.

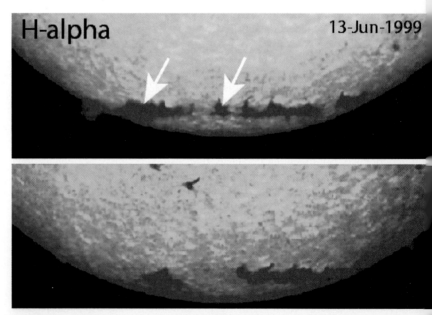

Fig. 10-14. Hydrogen-α Image of the Southern Polar Crown Filament on 13 June 1999. The filament traversed the entire southern polar region. Later that day, the easternmost portion of the filament erupted. *(Courtesy of Big Bear Solar Observatory / New Jersey Institute of Technology)*

10.2 Transient Disturbances in the Solar Wind

A few CMEs originate outside the streamer belt in spectacular eruptions from polar crown filaments (Fig. 10-14). Polar crown filaments are located above the neutral lines that separate the polar magnetic regions from the large-scale active regions in the streamer belt. These extended filaments are occasionally found encircling either polar region of the Sun (latitudes >50°) in a nearly continuous ring. In some events, much of the polar crown filament erupts, producing a wide CME whose field may be significantly distorted on its way to 1 AU. About 20% of polar crown filament eruptions are geoeffective. The low percentage is probably a geometry effect, since these ejecta come from high latitudes and often miss Earth.

Figure 10-15 gives the history of a fast CME passage on 20 November 2003 that produced a super geomagnetic storm. The vertical magnetic field component was southward ($-B_z$) for ~12 hr. The high-speed solar wind combined with the southward field to produce an interplanetary electric field almost an order of magnitude larger than normal. As the CME passed Earth, one of the largest storms of solar cycle 23 developed.

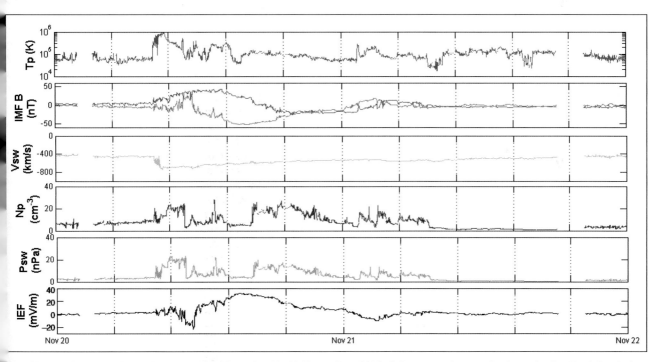

Fig. 10-15. Solar Wind and Magnetic Field Data for 19–21 November, 2003. A fast coronal mass ejection arrived at Earth just before midday on 20 Nov. Top curve: Solar wind proton temperature; second curve: IMF B_z in red and IMF B_y in blue; IMF (B_y and B_z) third curve: solar wind speed; fourth curve: plasma density; fifth curve: solar wind temperature; fifth curve: solar wind dynamic pressure; bottom curve: interplanetary electric field (IEF). The hallmark signature of the magnetic cloud CME was the steady, smooth rotation of the IMF. The combination of high-speed flow and large IMF produced a major geomagnetic disturbance at Earth. *(Data from the ACE Spacecraft and NASA Omniweb)*

Chapter 10 The Active Interplanetary Medium: Conduit for Space Weather

> **Pause for Inquiry 10-9**
>
> Which characteristics in Fig. 10-15 identify the structure as a CME? Which characteristics suggest that it was geoeffective?

> **Pause for Inquiry 10-10**
>
> In Fig. 10-15, verify the value of the peak interplanetary electric field.

> **Pause for Inquiry 10-11**
>
> Which flux rope diagram in Fig. 10-13a or Fig. 10-13b fits the data in Fig. 10-15?

> **Pause for Inquiry 10-12**
>
> Estimate the time it took for the CME indicated in the data of Fig. 10-15 to travel from the Sun to Earth.

Interacting CMEs

Collisions between CMEs may play a key role in determining the interplanetary behavior and distribution of solar ejecta in the heliosphere. Observations from the SOHO spacecraft reveal that fast-moving CMEs overtake slower ones in their paths (Fig. 10-16). The result is a single complex merged interaction region. This violent merging allows high-energy electrons to interact with the local plasma environment. The result is unusual radio bursts in the vicinity of the collision. Less energetic events do not produce a radio outburst detectable with current instrumentation.

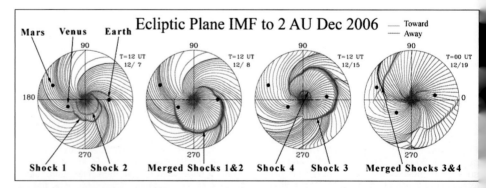

Fig. 10-16. A Merged Interaction Region (MIR) Develops. A single active region ejected several CMEs in sequence. One set emerged in early December 2006 and created a merged interaction region. The same active region ejected two more CMEs in mid-December. A new interaction region formed. By late December, the MIRs were interacting beyond Earth to form a globally merged interaction region (GMIR). We discuss GMIRs further in the section. [Intriligator et al., 2008]

10.2 Transient Disturbances in the Solar Wind

EXAMPLE 10.4

Alfvén Mach Number and Plasma Beta in a CME

- *Problem Statement:* Determine the Alfvén Mach number and plasma beta at the time of peak IMF in the CME on Nov 20 2003. Compare these values to those of the typical solar wind.
- *Concept:* Alfvén Mach number is the ratio of the solar wind speed to the Alfvén speed. The plasma beta is the ratio of thermal pressure to magnetic pressure in the solar wind.
- *Given:* The Advanced Composition Explorer data in Fig. 10-15.
- *Assume:* The IMF B_z value is the largest component of the total magnetic field value. The average number density of the CME interval is 10 particles/cm^3.
- *Solution:*

a) Alfvén speed = $v_A = \sqrt{\dfrac{B^2}{\mu_0 \rho}}$

$$v_A = \sqrt{\dfrac{(5 \times 10^{-8}\,\text{T})^2}{(4\pi \times 10^{-7}\,\text{C}^2/(\text{N} \cdot \text{m}^2))\left(\dfrac{10 \times 10^6\,\text{particles}}{\text{m}^3}\right)\left(\dfrac{1.67 \times 10^{-27}\,\text{kg}}{\text{particle}}\right)}}$$

$= 3.45 \times 10^5$ m/s

Alfvén Mach number
$= v_{SW}/v_A = (6 \times 10^5 \text{ m/s})/(3.45 \times 10^5 \text{ m/s}) \approx 1.7$

b) *Plasma Beta:* $\beta = \dfrac{n k_B T}{(B^2/2\mu_0)}$

$$\beta = \dfrac{(10 \times 10^6 \text{ particles/m}^3)(1.38 \times 10^{-23}\text{ J/K})(1 \times 10^5 \text{ K})}{\left(\dfrac{(5 \times 10^{-8}\,\text{T})^2}{2 \times 4\pi \times 10^{-7}\,\text{C}^2/(\text{N} \cdot \text{m}^2)}\right)}$$

≈ 0.014

- *Implication and Interpretation:* For this strong magnetic field, the Alfvénic Mach number was 1.7. Typical solar wind Alfvénic Mach numbers are close to 10. (Example 5.5).

Further, the strong magnetic field dominated the plasma. Plasma beta values <<1 indicate the magnetic field is in control of the plasma dynamics. A typical solar wind plasma beta value near Earth is ~1.

Follow-on Exercise 1: Verify that the Alfvénic Mach number of the solar wind can be approximated as: $M_A = v_{SW} n^{1/2}/(22B)$, with speed in km/s, density in protons/cm^3, and magnetic field in nT.

Follow-on Exercise 2: Verify that the plasma beta can be approximated as: $\beta \approx 35\, nT/B^2$, with density in protons/cm^3, T in mega-Kelvin, and magnetic field in nT.

Follow-on Thought Exercise 3: What is the solar cycle variation in Alfvénic Mach number and plasma beta?

Chapter 10   The Active Interplanetary Medium: Conduit for Space Weather

Focus Box 10.2: Transient Voids in the Solar Wind

On occasion the solar wind manages to break the "rules of thumb" derived from 40-plus years of observations. It doesn't break any physical laws; rather it reveals a new aspect of behavior that sends scientists on a hunt for better ways of explaining the Sun. The case of May 11–12, 1999 was one that baffled scientists for several years. The solar wind seemed to disappear. It didn't really disappear, but it became so slow and so tenuous that it exerted little pressure on the magnetosphere. The magnetosphere ballooned outward, approaching the moon orbiting on Earth's sunlit side (~60 R_E).

The near void in the solar wind was ultimately tracked to a transient coronal hole (TCH). Transient coronal holes are different from polar coronal holes and come in at least two forms—ones with high-speed flow and ones with very low-speed flow. The ones with high-speed flow have magnetic fields that expand into space over confined regions. The field expands, but not excessively. In low-speed TCHs, the expansion factor is large, perhaps 10 to 100 times that of more normal coronal holes. Sometimes locations close to a large low-latitude active region support open solar magnetic field lines that expand modestly into space, while proximate field lines are tightly bound and closed in the active region. The field lines and plasma have a large spatial void to fill because the nearby active region is holding plasma and field tightly to the Sun. Such a transient has a unipolar field but low speed and very low density. Some of these transients pinch off from the source region. They may head toward Earth, delivering solar wind speed less than 350 km/s and density less than 0.4 ions/cm^3.

These events typically have lifetimes of less than 48 hours. They tend to move slowly, with a four-to-five day travel time to Earth. This makes them rather difficult to trace back to the Sun. It appears that the convoluted current sheet structure of solar maximum supports development of the magnetic field structure needed for such transients, so they are observed more often in solar maximum than in other phases of the solar cycle. The effect at Earth of a low-speed TCH is to greatly reduce geomagnetic activity, especially if the IMF is northward. The magnetosphere expands and becomes quiescent.

Researchers believe merged structures may be the source of CMEs that are larger and more complex in structure than usual eruptions and rather common during solar maximum. These traits cause complex CMEs to trigger protracted magnetic storms when they envelop Earth. Such ejecta often consist of high-speed flows with shocks and other CME signatures, but poorly defined magnetic structures with tangled fields. These collisions change the speed of the eruption, so predicting these events and their impacts at Earth is quite challenging.

Fast-moving CMEs cause space weather events even if they have not merged yet. Energized ions bounce between CME structures and acquire energy with each bounce. If Earth is sandwiched between such structures, it is caught in the middle of a CME-driven energetic particle event. We say more about these unusual storms in Sec. 10.3.

A variant of MIRs created by CIRs develops at the boundaries of transient ejecta and co-rotating flows. Interplanetary signatures of CMEs at Earth tend to cluster around the sector boundary crossings and CIRs during the declining phase of solar cycles. Interaction regions channel ejected material from the Sun. This happens as ejecta from the closed field line regions of the Sun evolve and expand within the slow-flow corridor connected with the streamer belt.

At Earth, MIRs boost magnetic field strength and pressure. The largest recurrent storms linked to high-speed flow are often caused by CMEs at or near the leading edges of high-speed streams. The high-speed flow compresses the rear flank of the CMEs. Conversely, fast CMEs behind a CIR may amplify the field and plasma density in the CIR. Some CIRs and CMEs so disrupt the ambient flow in their path that they create a sheath region that takes on disturbed characteristics of the solar wind. Sheath regions are frequently sandwiched in such a way that they too become part of the MIR. Such MIRs produce major geomagnetic storms and strongly influence energetic particles and cosmic rays, as we describe in Sec. 10.3.

Sometimes during solar maximum, numerous shock waves are driven off all parts of the Sun in a short time. These shocks, propagating in all directions away from the Sun, eventually merge to form a structure of enhanced magnetic field and compressed plasma that almost completely encircles the Sun. These structures are *globally merged interaction regions (GMIRs)*. Such systems, sometimes several AU in width, have been observed in the distant heliosphere by the Voyager 2 spacecraft. Increases in the magnetic field strength, density, temperature, and speed were observed at the leading edge of the system, consistent with the presence of a shock.

The vast extent of GMIRs modifies the heliosphere (Fig. 10-16). In what seems like a paradox, a MIR that accelerates particles to high speeds as it heads toward Earth may transform into a GMIR that blocks the access of energetic cosmic rays to Earth. Further, after forming, the GMIR propagates toward the asymmetric termination shock, colliding first with the nose and later with the tail. The collision with the termination shock sets the entire heliosphere "ringing" as it shakes back and forth, driving pressure waves, and eventually shocks, into the interstellar medium.

Pause for Inquiry 10-13

Suppose an active region produces two CMEs a few hours apart. If both are ejected at about the same speed, the second one may catch the first. Suggest what might happen in the interplanetary medium to allow this to happen.

Pause for Inquiry 10-14

Review Fig. 9-31. In what year did the largest recorded Forbush decrease occur?

Focus Box 10.3: Forbush Decreases

Contributed by Tony Phillips and Frank Cucinotta, NASA

In early September 2005, solar active regions numbered 798 and 808 became extraordinary sources of energetic particles—a local cosmic ray factory. X rays ionized Earth's upper atmosphere. Solar protons peppered the Moon. Amazingly, on board the International Space Station (ISS), radiation levels dropped by 30%.

Scientists have long known about this phenomenon, called a *Forbush decrease* (Fig. 10-17), after American physicist Scott E. Forbush, who studied cosmic rays in the 1930s and 1940s. He found that galactic cosmic ray doses dropped when solar activity was high. Solar ejecta associated with large magnetic reconnection events contain plasma and magnetic fields, knots of magnetism ripped away from the Sun by the emission. These fields deflect charged particles, so when a CME sweeps past Earth, it also sweeps away many of the electrically-charged galactic cosmic rays that would otherwise strike Earth. A single CME suppresses cosmic rays for a few days, while sustained solar activity suppresses them for a much longer time.

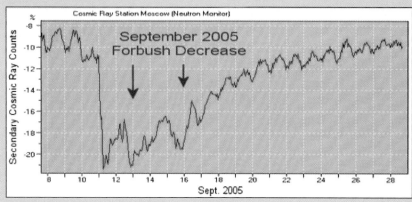

Fig. 10-17. Neutron Counts from a Cosmic Ray Monitoring Station in Moscow, Russia. Radiation levels dropped in early September 2005 during a period of intense solar activity. The entire year was a surprisingly active one for the Sun. Observers counted 14 powerful X-class solar flares and an even greater number of CMEs. As a result, the ISS crew absorbed fewer cosmic rays all year long. *(Courtesy of Science@NASA and the Moscow Neutron Observatory)*

Wherever CMEs go, cosmic rays are deflected. Forbush decreases have been observed on Earth and in Earth orbit on Mir and the ISS. The Pioneer 10 and 11 and Voyager 1 and 2 spacecraft have experienced them, too, beyond the orbit of Neptune. The situation is ironic because flares and CMEs are themselves sources of deadly radiation. CMEs, in particular, cause proton storms or solar energetic particle (SEP) events. En route to Earth, CMEs race through the Sun's outer atmosphere, plowing through the hot gas at shock-generating speeds. Protons caught in the path of one accelerate to dangerous energies. Astronauts, who are particularly at risk from SEP events, must minimize their exposure and stay relatively safe inside their spacecraft.

Cosmic rays are different—and worse. Cosmic rays are supercharged subatomic particles coming mainly from outside our solar system. Unlike solar protons, which are relatively easy to stop with materials such as aluminum or plastic, cosmic rays cannot be stopped completely by any known shielding. Even inside their ships, astronauts are exposed to many dangerous cosmic rays coming through the hull. The particles penetrate flesh, damaging tissue at the microscopic level. One possible side-effect is broken DNA that over the course of time causes cancer, cataracts, and other maladies.

Except during brief trips to the Moon in the 1960s and 1970s, astronauts have never been exposed fully to galactic cosmic rays. Close to Earth, where the ISS orbits, crews are protected by their ship's hull as well as Earth's magnetic field and Earth's gigantic solid body. A six-month trip to Mars, far from these natural shields, would be more hazardous. What are the long-term risks? How much shielding will keep astronauts safe? NASA researchers are grappling with these questions, but one thing is clear—reducing exposure is important.

The Sun also helps. Every 11 years, solar activity reaches its maximum. During solar max, it produces CMEs daily, and the solar wind routinely blows knotty magnetic fields through the inner solar system. These fields provide a measure of extra protection for trips to the Moon and Mars, reducing cosmic ray fluxes in the biologically dangerous energy range of 100 MeV–1000 MeV by 30% or more. Space mission planners of the future will need to determine the costs and benefits of scheduling interplanetary trips to coincide roughly with solar maximum, to exploit the decline in cosmic rays. The trade-off is an interesting one. The CMEs and CIRs that provide the protective covering are themselves copious accelerators of lower-energy particles. We describe these local energetic particles next.

10.3 Solar Energetic Particles from Interplanetary Shocks

10.3.1 Traveling Shocks and Solar Energetic Particles

Objectives: After reading this section, you should be able to...

- Describe the relation between propagating shocks in the solar wind, CMEs, and energized particles
- Distinguish between characteristics of solar energetic particles and energetic storm particles

Solar Energetic Particles, CMEs, and Shocks. Geospace is bathed in a keV-solar wind particle flux and a much lower flux of GeV cosmic rays. Earth is bombarded sporadically by solar particle storms—swarms of electrons, protons, and heavy ions accelerated to high speed by solar flares and CME shocks. These MeV-level *solar energetic particle (SEP)* events produce relatively high fluxes of charged particles, some of which become trapped in the magnetosphere, while others pass through the magnetosphere and penetrate spacecraft, the atmosphere, and aircraft. Solar radiation storm alerts are issued when the flux of 10 MeV particles exceeds 10 particles/($cm^2 \cdot s \cdot sr$) (corresponding to an S1 level solar radiation storm). In this section, we provide an overview of CME-driven traveling shocks in the interplanetary medium that create such geospace disturbances.

We describe in Chap. 9 how SEPs accelerate in solar flare processes low in the solar corona. Here we discuss the long-lived, longitudinally-broad SEP events generated by CME-driven shocks in the upper corona and inner heliosphere. The onset of a CME-driven SEP event usually occurs when the CME is still very close to the Sun (<$4R_S$), where the associated shock has access to the relatively high density of particles in the corona. These events tend to show a fast, intense onset, as well as particle velocity dispersion, with higher-energy particles arriving at the observation point before the low-energy ones. Figure 10-18 shows the flare region related to the large flux of CME-driven SEPs in late January 2005. For that event, high-energy particle fluxes at 1 AU increased by four orders of magnitude in a matter of minutes, as the ejecta ripped through the lower corona.

In a gas, the signal speed (information transfer speed) is the speed of sound. A magnetized plasma has two important signal speeds: the Alfvén speed and the sound speed. A disturbance moving faster than the signal speed in the medium forms a discontinuity separating the slow and fast regimes in an otherwise continuous medium. At the discontinuity, the properties of the medium change abruptly (Chap. 5).

Within 1 AU, most interplanetary shocks are driven by CMEs. Traveling shocks between 0.3 AU–1.0 AU typically have the following properties *(shock parameters)*:

- Shock speed between 300 km/s–700 km/s (occasionally speeds >2000 km/s are observed)
- Shock angular extent between a few tens of degrees to 180°
- Density compression ratio between 1 and 8
- Magnetic compression ratio between 1 and 7

Chapter 10 The Active Interplanetary Medium: Conduit for Space Weather

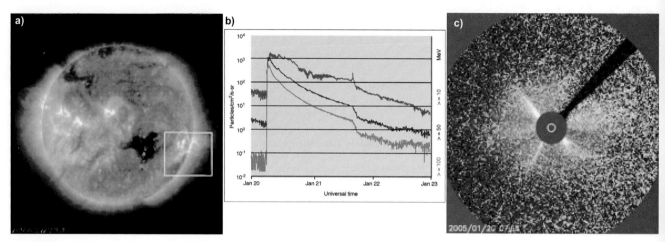

Fig. 10-18. **Solar Energetic Particle (Proton) Event of 20 January 2005. a) The Event Commenced on the Far West Solar Limb Early on 20 January 2005.** A combination of a flare and a fast coronal mass ejection were responsible for the sudden rise in the flux of very energetic protons at Earth. **b) Particle Energy Plots for Three Levels of Particle Flux.** This plot shows the history of energetic particles in three channels (≥ 100 MeV, ≥ 50 MeV, ≥ 10 MeV) observed by the Geostationary Operational Environmental Satellite spacecraft. The sharp peak in the flux of >10 MeV particles late on 21 Jan corresponds to the shock arrival. The sharp rise is called an energetic storm particle event. *(Data courtesy of Space Weather Prediction Center)* **c) Energetic Particles Affect the Solar and Heliospheric Observatory (SOHO) Spacecraft.** The particles penetrated the telescope and their impact was electronically recorded as dots of light or "false stars." *(Images courtesy of SOHO [ESA and NASA])*

> **Pause for Inquiry 10-15**
>
> Compare the duration of the SEP event in Fig. 10-18b to the length of a typical flare. Is it likely that the event could have been exclusively flare driven? Why or why not?

Traveling shocks are of great interest to space weather forecasters because they accelerate individual charged particles to very high energies. We briefly describe the acceleration mechanism in Sec. 10.3.2.

CME-driven shocks expand and propagate through the interplanetary medium (Fig. 10-7). Strong shocks may range across 180° of longitude. If the shock is strong enough, it accelerates particles that propagate along the interplanetary magnetic field lines (Fig. 9-27). SEP events reaching Earth are very likely to arise from regions west of the central meridian. Field lines from the western portion of the Sun often connect to Earth via the Parker spiral, so they provide a natural path for SEPs. Traveling shocks may produce ions with energies up to the GeV level (with tens to hundreds of MeV being typical) and electrons with energies of ~100 MeV. The size and longevity of SEP events are related to the fact that ions accelerated at the expanding CME-driven shock populate magnetic field lines over a broad range of longitudes. Roughly 10% of CMEs produce SEPs; only the fastest 1% of CMEs generate large SEP events. These sudden events pose significant radiation hazards for unprepared humans and technical systems in space (Focus Box 10.4). The vast majority of accelerated particles arriving at Earth are ions. Electrons are also accelerated by solar events, but, being more mobile, tend to scatter easily from shocks. They escape with somewhat lower energies in advance of the ions. Thus, accelerated electrons are less geoeffective than ions. However, they may provide warnings of impending SEP ion arrival (Focus Box 10.6).

CME-related SEP events last several tens of hours to several days and have large fluences. The particles arrive within minutes to hours (depending on particle energy) of the CME lift-off. Advanced warning is very limited.

SEP ionization of the polar atmosphere also produces nitrates that precipitate to become trapped in the polar ice. Observations of nitrate deposits in ice cores reveal individual large SEP events and extend back ~400 years.

Many large SEP events are accompanied by further increases in the intensities of ions in the ~10 MeV–100 MeV energy range (and occasionally higher) that peak when the corresponding CME-driven shock arrives at the observer (e.g., the peak in the 10-MeV particles just after 17 UT on 21 Jan 2005 in Fig. 10-18b). These events are called *energetic storm particle (ESP)* events because of their strong association with geomagnetic storms. Except in advance of the most extreme events, ESP events are characterized by a gradual rise in intensity, often several hours before shock arrival. Space weather forecasters use the signature of rising energetic ion intensities as a signature of an approaching shock with a CME likely to follow. The record in Fig. 10-18b reveals that a CME-driven SEP event developed in the solar atmosphere early in the day of 20 Jan. As the CME traveled in the interplanetary medium, it continued to accelerate particles that sped Earthward. A few minutes before the arrival of the CME, the GOES satellite observed a brief enhancement of 10-MeV particles consistent with an ESP enhancement.

During extreme conditions at 1 AU, ESP events occasionally achieve very high energies, creating a risk of additional radiation dose to astronauts and space hardware. For example, during the large SEP events of October 1989 the (already very intense) flux of >100 MeV protons suddenly went up by a factor of ten for several hours as the shock passed 1 AU. The effect of these shock-associated particle increases on space operations is often magnified because the shock may also cause a geomagnetic storm that suddenly allows the particles access to Space Station latitudes.

Shock Speed. Speed is an important characteristic of geoeffective CMEs; hence space weather observers put significant effort into determining CME and shock speed. When we use coronagraphs to block out the bright solar disk, we see ejecta departing from the solar limbs against the dark surrounding space (plane of the sky). Measurements of height versus time provide a rough estimate of CME speed.

The plane-of-sky technique does not work for CMEs ejected from the central disk (usually as halo CMEs), because Earth-based visual observations are inhibited by the bright Sun. Observers use a dynamic radio measurement across a broad frequency range to estimate the speeds of Earth-directed CMEs. The shock generated by fast-moving CMEs disturbs the coronal plasma, causing a radio emission from electrons accelerated at the leading edge of the disturbance. The electrons generate plasma oscillations that subsequently transform into radio waves that escape the shock. A radio spectrograph displays the variation in radio emissions over time (Fig. 10-19). Example 10.5 shows how information from a radio spectrograph can be used to chart the speed of a disturbance moving through the corona (or the interplanetary medium).

The frequency of the plasma oscillations is given by the electron plasma frequency (Eq. (6-6)) as

$$f \approx 9\sqrt{n_e}$$

where

f = plasma frequency [Hz]

n_e = electron number density [#/m^3]

Chapter 10 The Active Interplanetary Medium: Conduit for Space Weather

Fig. 10-19. Solar Radio Spectrograph. Emissions appear yellow or white against a blue background. The black curve highlights the sweep of frequencies. The continuous interference at several frequencies comes from human-made sources. The transient radio emissions originate from solar activity. A type II burst (sweep) begins at 06:38 (the broad swath slanted down to the right starting at that time). We determine the CME speed from the gradient of the frequency sweep (Example 10.5). The speed then provides an estimated time to arrive at 1 AU. *(Courtesy of the Australian Ionospheric Prediction Service)*

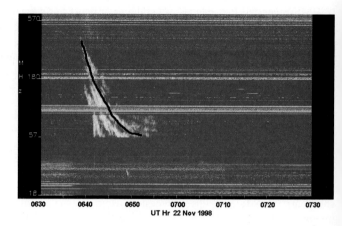

Low in the corona, where electron density is high, the radio frequency is high. As the shock travels through regions of lower density (at greater heights), the radio frequency decreases. The short-duration falling frequency is a *type II radio burst*. Type II radio bursts consist of one, two, or even more bands of enhanced radiation. The emissions are observed at multiple frequencies because the plasma oscillations interact either with low-frequency plasma waves, generating the fundamental emission, or with each other, generating the radio waves of the harmonic emission band.

If the density distribution as a function of radius is known in the corona, i.e., $n_e = n_e(r)$, the observed frequency in the spectrograph corresponds to a height in the solar corona. The frequency drift (slope) corresponds to the speed of the shock in the direction of the density gradient. We combine the time rate of change with models of solar electron density to yield an estimate of shock speed (Example 10.5).

Most SEPs are observed by space instruments and are related to satellite single-event upsets and instrument degradation. However, during the most energetic events, even ground-based instruments observe effects as *ground-level enhancements* (GLEs). A GLE is a sharp, short-duration increase in the counting rate of ground-based cosmic ray detectors. Accelerated charged particles, mostly protons from flares and CMEs, arrive at Earth with energies sufficiently high to penetrate the geomagnetic field into Earth's atmosphere. Significant numbers of particles with energies above ~1 GeV produce a background radiation increase observable on the ground. Solar cycle 23 had 16 such events out of a total of only about 70 since their discovery in 1942 by Scott Forbush.

On 23 February 1956, a GLE was so intense that the ground-level background radiation (normally about 50% of the naturally occurring dose) went up by nearly 50-fold for 15 minutes and remained at levels over ten times the normal rate for more than an hour. The GLE lasted over 24 hours.

On 20 January 2005, a GLE comparable in magnitude (at some energies) to the one in 1956 registered in the neutron monitors of the worldwide network. With the current arsenal of space- and ground-based observatories, scientists were able to deduce how the January 2005 super-GLE developed. An X7.1 solar flare from NOAA Active Region (AR) 10720 near the west limb started at 06:39 UT on 20 January and had a peak emission at 07:01 UT (Fig. 10-18a). Satellite images suggest that before the peak emission, an explosive CME developed with an estimated speed in excess of 3000 km/s. Radio emissions connected with the

10.3 Solar Energetic Particles from Interplanetary Shocks

EXAMPLE 10.5

Calculating Electron Density and Shock Speed

- *Problem Statement:* Determine the electron density between 1.03 and 1.30 R_S using the frequencies from Fig. 10-19 and given in the data table below. Also determine the shock speed over this distance.

Time (s)	Frequency f (MHz)	Electron Density n_e (#/m³)	Height Above Sun (R_S)	Shock Speed (km/s)
42	240.7		1.03	
96	189.3		1.09	
168	79.8		1.15	
254	116.4		1.22	
358	88.4		1.30	

- *Concept:* The emerging material causes the plasma to oscillate at its natural plasma frequency, and so to emit radio waves.
- *Given:* Values from the table and $R_S = 7 \times 10^8$ m
- *Solution:* To solve for electron density, use Eq. (6-6): $f(\text{Hz}) = 9\sqrt{n_e}$

Time (s)	Frequency f (MHz)	Electron Density n_e (#/m³)	Height Above Sun (R_S)	Shock Speed (km/s)
42	240.7	7.15×10^{14}	1.03	--
96	189.3	4.42×10^{14}	1.09	777 km/s
168	79.8	2.66×10^{14}	1.15	583 km/s
254	116.4	5.96×10^{13}	1.22	570 km/s
358	88.4	4.18×10^{13}	1.30	538 km/s

- *Implication and Interpretation:* Spectragraph records in the radio range can be used to track CMEs moving in the solar atmosphere. Such records can warn of high-speed CMEs.

Pause for Inquiry 10-16

Use Fig. 10-19 to determine a typical radio wavelength associated with a type II radio burst.

CME-driven shock commenced at 06:44 UT. The intensity of >100-MeV protons measured by NOAA's GOES-11 spacecraft rose to a maximum within 1/2 hour (Fig. 10-18b). Indeed, observations of the CME were largely lost because of the solar particle background in the SOHO/LASCO charge-coupled device (Fig. 10-18c). The rapid onset was caused in part by the event originating at 67° W, where we expect good magnetic connection with Earth. The ground level background radiation at some observatories exceeded 250% of normal.

Chapter 10   The Active Interplanetary Medium: Conduit for Space Weather

Focus Box 10.4: Energetic Particle Impacts at Earth

High-energy protons represent a direct radiation danger to astronauts and high-altitude aircraft crews. These effects are felt on flights at high altitude and high latitude. International Space Station crew members are advised to take shelter in their shielded sleeping quarters during SEP events. Very-high-energy protons or other heavy ions are capable of penetrating spacecraft skin. As they pass through, they ionize particles deep inside the satellite. A single proton or cosmic ray deposits enough charge to cause an electrical upset (circuit switch, spurious command, or memory change or loss) or serious physical damage to onboard computers or other components. Hence, these occurrences are termed *single-event upsets*. They often drive the design of spacecraft exposed to these particles. Section 13.1 treats of this topic in detail.

Charged particle bombardment during a proton event produces direct collisional damage to a launch vehicle or its payload, or it deposits an electrical charge on or inside the spacecraft. The electrostatic charge deposited may discharge (leading to arcing) during onboard electrical activity, such as during vehicle commands. Satellite solar cells are often particularly hard hit by energetic-particle events. Thus the particles from solar events pose a concern for spacecraft designers.

Many satellites rely on electro-optical sensors to maintain their orientation in space. These sensors lock onto certain patterns in the background stars and use them to achieve precise pointing accuracy. They are vulnerable to cosmic rays and high-energy protons that produce flashes of light as they impact a sensor. The bright spot produced on the sensor may be falsely interpreted as a star (Fig. 10-18c). When computer software fails to find this false star in its star catalogue or incorrectly identifies it, the satellite loses attitude lock with respect to Earth. This loss of lock (*disorientation*) may require human intervention to recover. Directional communication antennas, sensors, and solar-cell panels then drift away from their intended targets. The result may be loss of communication with the satellite; loss of satellite power; and in extreme cases, loss of the satellite because of drained batteries. (Gradual star sensor degradation also occurs under constant radiation exposure.) Disorientation occurs primarily when solar activity is high, and on geosynchronous or polar-orbiting satellites.

At high latitudes the intensified ionization of D-layer atoms and molecules (which produce signal absorption) is caused by particle bombardment from space. SEP-driven events last for hours to several days, and usually occur simultaneously with other radio transmission problems such as non-great circle propagation and multipath fading or distortion. High-frequency polar cap absorption (PCA) events occur when the high-energy protons penetrate into the polar ionosphere's lowest region and collide with atmospheric atoms and molecules. This significantly heightens levels of ionization, resulting in severe absorption of HF radio waves used for communication and some radar systems. This phenomenon, sometimes referred to as a "polar cap blackout", may last for several days and is often accompanied by widespread geomagnetic and ionospheric disturbances as lower-energy solar protons and electrons arrive. Because the ionosphere's base tends to lower during PCA events, concurrent errors on low-frequency navigational systems are also normally observed.

Pause for Inquiry 10-17

Which of the of the images in Fig. 10-20 would be most closely linked to ground-level enhancements?

Why do we record more shock and type II radio CME signatures than other CME associations?

10.3 Solar Energetic Particles from Interplanetary Shocks

Focus Box 10.5: Solar Sources of Disturbances

Figure 10-20 shows the source regions of several types of interplanetary disturbances. The left column compares magnetic cloud CMEs to other types. Clear magnetic cloud signatures tend to originate closer to the central meridian. The middle column shows the origination points of CMEs that generate type II radio bursts (top) and interplanetary shocks (bottom). These signatures often share a common source of fast CMEs. In the right column are the locations of CMEs creating major SEP events and major geomagnetic storms. The major magnetic storm category corresponds well with the magnetic cloud category. The SEP CME distribution is skewed toward the western limb, where field lines from the Parker spiral trace to Earth. Some SEP events even come from behind the west limb.

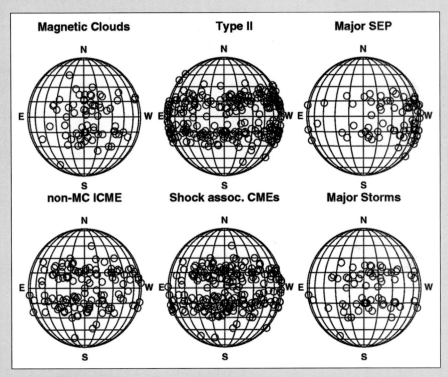

Fig. 10-20. Sources of Solar and Interplanetary Disturbances. Sources of solar disturbances tend to cluster at low latitudes. Sources of major solar energetic particle events at Earth tend to cluster in the western solar hemisphere. [Gopalswamy et al., 2010]

10.3.2 Shock Acceleration Mechanisms

> **Objectives:** After reading this section, you should be able to...
> - Describe how interplanetary shocks energize particles
> - Distinguish between scattering and induction acceleration

Energizing by Turbulence, Resonances, and Reflections

The sharp changes in fluid and magnetic properties at an interplanetary shock boundary tend to concentrate magnetic fields and create small-scale random structures in the field. Charged particles gain energy as they move back and forth across the shock.

Space physicists describe shocks with respect to the orientation between the magnetic field lines in the medium and an imaginary vector that is perpendicular to the shock structure. The perpendicular vector is called the *shock normal*. Generally, the shock normal vector is oriented so that it points into the unshocked medium. In the shock frame of reference, the normal vector points upstream (Fig. 10-21). The physics of collisionless shocks depends strongly on the angle θ_{Bn} between the upstream magnetic field and the shock normal. Shocks are classified in the following way relative to this angle, as shown in Fig. 10-21:

- Perpendicular shocks: $\theta_{Bn} = 90°$
- Parallel shocks: $\theta_{Bn} = 0°$
- Oblique shocks: $0° < \theta_{Bn} < 90°$
- Quasi-parallel shocks: $0° < \theta_{Bn} < 45°$
- Quasi-perpendicular shocks: $45° < \theta_{Bn} < 90°$

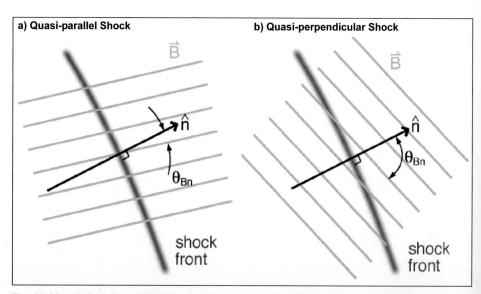

Fig. 10-21. A Quasi-parallel and a Quasi-perpendicular Shock. a) In a quasi-parallel shock, the shock normal vector is approximately aligned with the magnetic field direction. **b)** In a quasi-perpendicular shock, the shock normal is nearly perpendicular to the magnetic field. *(Courtesy of Allan Tylka at the Naval Research Laboratory)*

10.3 Solar Energetic Particles from Interplanetary Shocks

In 1949, Enrico Fermi recognized that charged particles accelerate as they interact with randomly located magnetic scattering centers moving relative to one another. The centers act as magnetic mirrors. His ideas describe two distinct types of acceleration called stochastic (random) and diffusive processes. We discuss these in terms of a parallel shock (Fig. 10-22a). The region disturbed by a shock contains turbulent fields and accelerated particles that generate waves. Both of these produce irregularities that scatter particles. In a *stochastic process*, charged particles encountering these irregularities tend to gain energy. Head-on collisions with a moving magnetic mirror, created by turbulence or waves, increases particle energy upon reflection. The opposite holds if the mirror is receding. Because head-on collisions are more likely than head-to-tail collisions, the net result is an energy increase. This process is rather inefficient, but may act as a pre-accelerator, creating a seed population for other shock processes. Some particles may resonate with waves in the shocked plasma and gain more energy. Suprathermal particles in the solar wind are good candidates for pre-acceleration. Stochastic processes are sometimes called second-order processes, because the energy depends on the square of the mirror velocity.

A more efficient form of particle acceleration comes from repeated reflections in the plasmas converging at the shock front. The turbulent convergence creates random magnetic field concentrations as well as field gradients. A charged particle, traveling upstream-to-downstream through the shock wave, encounters a moving change in the magnetic field, which acts as a local magnetic mirror. A particle reflected back through the shock (downstream to upstream) gains velocity. If a similar process occurs upstream, the particle will gain energy again. Multiple reflections greatly increase its energy. This process is quicker than the purely stochastic process. This is *diffusive shock acceleration*.

The diffusive acceleration time in the interplanetary medium is typically hours to tens of hours. Sometimes acceleration times are reduced by self-excited waves in the magnetic field that create local irregularities from which the particles scatter back to the shock. Diffusive processes are sometimes called first-order processes, because the energy gain per shock crossing is proportional to the velocity of the shock.

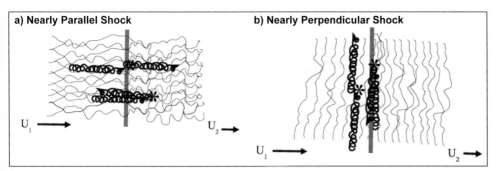

Fig. 10-22. Accelerated Ions at Parallel and Perpendicular Shocks. The speed upstream of the shock, U_1, is higher than the speed downstream of the shock, U_2. The light curves represent the magnetic field; the dark curves are the particle trajectories; the stars indicate a change of trajectory. **a)** At a parallel shock, particles travel across the shock and either gain energy as they scatter from approaching magnetic irregularities in the upstream flow, or lose energy as they scatter from retreating fluctuations in the downstream flow. The particles experience a net energy gain, because the upstream flow speeds are larger than the downstream flow speeds. **b)** At a perpendicular shock, the gradient in the magnetic field causes a gyrating particle to drift along the thin shock front and cross it many times. This drift is in the same direction as the electric field, so the particles gain energy. [Jokipii and Thomas, 1981]

Under some circumstances, the scattering centers are far from the advancing shock. Energized particles escape the shock, move to a distant scattering location, perhaps one arising from a previous CME or a MIR, and then bounce back to the shock. In simulations, protons accelerated by a shock propagating between the Sun and Earth flow outward to an interplanetary disturbance past Earth, and return approximately four times to acquire approximately 1 MeV of energy.

Some recent energetic particle events have been linked to episodic solar eruptions that fill the inner heliosphere with seed particles and shock structures advancing on each other. Earth may be caught in the acceleration zone for hours to days. Large SEP events appear to be preferentially associated with successive rather than isolated CMEs. Also, fast, wide CMEs preceded within 12 hours by slower ones tend to be connected with large SEP events. The propagation of the leading CME may add to the suprathermal seed population, heighten the level of turbulence, increase the density and consequently decrease the Alfvénic speed, and have other effects. These could all help explain the presence of enhanced SEPs.

Energizing by Electric Field Induction

A very rapid acceleration (*shock drift acceleration*) occurs when the shock's normal vector is perpendicular to the magnetic field. In the rest frame of the shock, the upstream solar wind flows into the shock with velocity v_{sw}. The motion produces an induced electric field, $\boldsymbol{E} = -\boldsymbol{v}_{sw} \times \boldsymbol{B}_{sw}$, along the shock front. When a charged particle impinges on the shock, it starts to drift along the induction electric field, gaining energy. Depending on pitch angle (Sec. 7.3) and phase of the gyration, some upstream particles drift along the shock front for long distances before they cross the shock. So, quasi-parallel shocks accelerate predominantly solar wind ions, whereas quasi-perpendicular shocks accelerate predominantly ambient energetic ions, which include material from earlier impulsive events rich in heavy ions. This feature of shock acceleration appears to account for the extreme compositional variations often observed between events at high energies.

After entering the disturbed downstream region, some particles gyrate and scatter in such a way that they turn back to the upstream region and experience further acceleration as they drift, or surf, along the shock again. Figure 10-23 depicts an idealized situation in which a particle gyrates in a large orbit in the upstream region and a smaller orbit in the downstream region, where the magnetic field is stronger. Shock drift acceleration occurs as the induction electric field accelerates the particle during its large gyration in the upstream region and decelerates it during its small gyration in the downstream region. With increasing energy, the ion's velocity increases. Finally, when the velocity component becomes larger than the shock speed, the particle escapes the shock. Upstream irregularities or magnetic mirroring in convergent field regions may scatter some of the energized particles back into the shock, where they undergo further acceleration. In reality, the magnetic field and the shock structure are very turbulent, leading to many small, random-walk segments with bends and turns that create opportunities for particle loss. Thus, only a few particles gain the large energies associated with SEP events.

Figure 10-24 shows the spectra of the proton flux created by a shock on September 29, 1989, which produced a ground-level event. Particle spectra tend to have a large number of low energy particles and a small number of high energy particles. The shape of the curve follows a power spectral form. That is, the flux varies as a function of energy raised to some power. Forecasters are interested in understanding and describing energetic particle power laws in an effort to

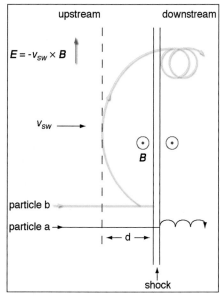

Fig. 10-23. Shock Drift Acceleration. In the shock frame of reference, the upstream solar wind brings ions to the shock. Particle "a" proceeds through the shock without interaction. Particle "b" reflects off the shock, perhaps caused by some irregularity. It accelerates by the solar wind electric field. In the presence of a magnetic field, the motion creates an induced electric field that accelerates particles along the shock. The time during which the particle drifts parallel to the shock depends on the particle's speed perpendicular to the shock. If the perpendicular speed is small, the particle travels along the shock for an extended interval of net acceleration; otherwise it escapes before it gains a large amount of energy. If the perpendicular speed is high, a particle may traverse the shock with little energy change. But if it scatters back to the shock through interaction with magnetic island structures in the turbulent downstream plasma, it may re-approach the shock with a parallel component of motion and be accelerated. [Paschmann et al., 1982]

forecast the flux of particles at the highest energies. The flux of particles above 50 MeV is of concern because these particles penetrate astronaut space suites. Higher energy particles damage spacecraft and penetrate into the atmosphere.

If the particle drifts along the induction field for a longer time, it gains more energy. The drift time depends on the particle's speed perpendicular to the shock. If it is small, the particle sticks to the shock; if it is big, the particle escapes before it can gain a large amount of energy. The particle energy relative to the shock depends on particle speed, shock speed, pitch angle, and θ_{Bn}.

After the acceleration, an initially isotropic particle distribution usually ends up with a very strong magnetic field-aligned beam in the upstream region and a smaller beam perpendicular to the magnetic field in the downstream region. Acceleration at perpendicular shocks is very rapid. Particles with several keV of energy are excited to 1 MeV in a few tens of minutes. Perpendicular shocks are the structures most likely to accelerate particles to the GeV-level associated with extreme SEP events.

The energy for particle acceleration must come from some reservoir. Recent research suggests that CMEs give up 1%–15% of their energy to the particle acceleration process, but usually less than 10%. This energy transfer means that CMEs decelerate, even if only slightly, to accelerate these high-energy particles. In concert with the Second Law of Thermodynamics, large organized motion gives way to random motion.

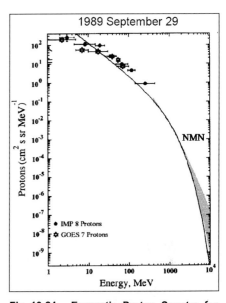

Fig. 10-24. Energetic Proton Spectra for the Ground-level Event of September 1989. Proton flux and energy were measured by the GOES-7 Satellite and the Interplanetary Monitoring Platform-8 Satellite. Inferences about the flux of extremely high-energy protons were made from ground-based neutron monitors. The curve is a model fit to the data. (NMN is neutron monitor network.) [Lovell et al., 1998]

Chapter 10 The Active Interplanetary Medium: Conduit for Space Weather

Focus Box 10.6: Forecasting Solar Energetic Particle Events

Contributed by Arik Posner

Solar radiation storms develop rapidly and are notoriously difficult to predict—often taking forecasters by surprise. Earth and astronauts in low-Earth orbit are protected from these particles by Earth's atmosphere and magnetic field. The danger begins when astronauts travel far from Earth during missions to the Moon or Mars. The type of particle most feared by astronaut safety experts is energetic protons. They damage tissue and break strands of DNA, causing health problems ranging from nausea to cataracts to cancer.

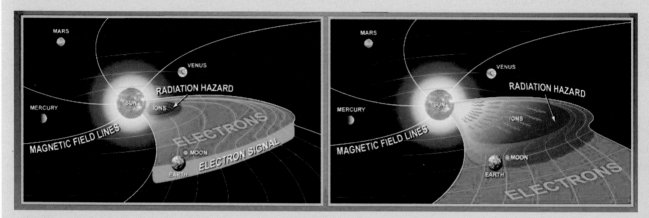

Fig. 10-25. **Radiation Storm Approaching Earth. a)** In this artist's sketch, photons from any flare that precedes particle acceleration have already arrived at Earth. The electron signal is intermediate to the photons and the soon-to-arrive ions. **b)** Within a few tens of minutes the energetic ions arrive at 1 AU. The electron signal continues to lead the ions. *(Courtesy of Arik Posner at NASA)*

Data from the Solar and Heliospheric Observatory (SOHO) offer new opportunities to forecast intense solar radiation storms. Up to one-hour advance warnings are possible, giving astronauts time to seek shelter and ground controllers time to safeguard their satellites. Subatomic particles striking CPUs and other electronics cause onboard computers to suddenly reboot or issue nonsense commands. A satellite operator, aware that a storm is coming, can put the spacecraft in a protective "safe mode" until the storm passes.

The newly recognized key to ion prediction is electron arrival. Every radiation storm is a mix of electrons, protons, and heavier ions. The electrons, being lighter, race ahead, heralding the arrival of energetic ions. One approach developed with data from hundreds of radiation storms recorded by the Comprehensive Suprathermal and Energetic Particle (COSTEP) Analyzer on board SOHO allowed a NASA investigator to construct and test an empirical method for ion forecasts. The rise time and intensity of the initial electron surge suggest how many ions are following and when they should arrive. However, the method is not yet perfect. In some cases the warning time is too short to be useful. The method also generates false alarms that could send astronauts dashing to safety unnecessarily.

EXAMPLE 10.6

CMEs as Energy Sources for Energetic Particles

- *Problem Statement:* Scientists estimate that the solar event on 20 January 2005 accelerated about 2×10^{34} 30-MeV particles into interplanetary space. How much energy was needed to accelerate just these extremely high-energy particles? How does this compare to the kinetic energy of a 3000 km/s CME close to the Sun?

- *Concept:* The particle energy comes from the CME shock process.

- *Assumption:* All the particle energy came from the CME; mass of the CME was 10^{13} kg.

- *Given:* Number of 30-MeV particles and CME speed.

- *Solution:* Total energy of the particles was $(2 \times 10^{34}$ particles) $(3.0 \times 10^7$ eV/particle$)(1.6 \times 10^{-19}$ J/Ev$) \approx 10^{23}$ J

   The kinetic energy of the CME was $(1/2\, mv^2) = 0.5(1 \times 10^{13}$ kg$)(3 \times 10^6$ m/s$)^2 \approx 5 \times 10^{25}$ J. Thus the very-high-energy particles represented ~0.002 of the CME kinetic energy. In fact, the particles probably were not at rest to start, so the energy gain is slightly less than 10^{23} J. On the other hand, particles with lower energies also extracted energy from the CME.

- *Implication and Interpretation:* The deceleration of the CME due to accelerating the very-high-energy particles was negligible. Even if the CME gave up 10% of its energy to particles of all energies, its speed would have decreased by only about 5%.

Follow-on Exercise: How many 1-MeV particles would have been necessary to extract 5% of the CME energy?

EXAMPLE 10.7

Shock Particle Energy and Power

- *Problem Statement:* Compare the relative energy increase per unit mass for a suprathermal solar wind proton that becomes a 10 MeV proton at a CME-driven shock to that experienced by a car in going from 0.5 mph to 60 mph in 4 s. What power is needed in each case to achieve the higher energy?

- *Concept:* Power = energy/time; individual particles experience a large acceleration in the vicinity of a shock

- *Assume:* The solar wind proton is suprathermal with an energy on the order of 100 eV; the proton acceleration occurs over a ten minute interval; vehicle mass = 1500 kg

- *Given:* Proton mass = 1.6×10^{-27} kg, 60 mph = 26.8 m/s; 0.5 mph = 0.224 m/s

- *Solution:*

   Relative change in proton kinetic energy = $(KE_f - KE_i)/KE_i = (10^7$ eV $- 10^2$ eV$)/10^2$ eV $= 10^5$, a 100,000-fold energy increase. The increase per unit mass for the proton is $10^5/(1.67 \times 10^{-27}$ kg$) \approx 6 \times 10^{31}$/kg.

   Relative change in vehicle kinetic energy = $(KE_f - KE_i)/KE_i =$

   $$\left(\tfrac{1}{2}mv_f^2 - \tfrac{1}{2}mv_i^2\right) / \left(\tfrac{1}{2}mv_i^2\right)$$

   $= [(60 \text{ mph} \times 0.447 \text{ m/(s·mph)})^2 - (0.5 \text{ mph} \times 0.447 \text{ m (s·mph)})^2]/$ $(0.5 \text{ mph} \times 0.447 \text{ m (s·mph)})^2 \approx 14,400$ energy increase

   The energy increase per unit mass for the car is 14,400 energy increase /1500 kg, or a factor of ~9.6 /kg

   The relative power per unit mass for the proton is $(6 \times 10^{31}$ /kg$)/600$ s, or a factor of $\sim 10^{29}$.

   The relative power per unit mass for the car is (9.6 /kg) /4 s, or a factor of ~2.4.

- *Implication and Interpretation:* The small fraction of suprathermal solar wind protons that stay in contact with the shock, or repeatedly visit the shock, experience phenomenal gains in individual particle energy. These particles lose minimal energy in their transit to 1 AU, so they are capable of penetrating satellite components and solar panel arrays. As they decelerate, they deliver substantial energy to individual atoms and molecules in the material with which they interact. In many cases they ionize or dissociate atoms and molecules. Thus, the particles are sometimes called ionizing radiation.

Follow-on Exercise: Assume (unrealistically) that the proton is accelerated uniformly between 10 R_S and Earth. Estimate the value of the acceleration and compare it to the value of g on Earth.

Chapter 10 The Active Interplanetary Medium: Conduit for Space Weather

Summary

On average, the quasi-stationary solar wind brings to Earth several small-scale disturbances each day. Most of them are associated with turbulence-driven magnetic flux tube motions in the solar wind. Sector boundary crossings also contribute to solar wind variability. Some sector boundary crossings are followed by high-speed flow from coronal holes. Outflow from these persistent open field regions may repeatedly pass Earth as co-rotating interaction regions (CIRs). CIRs frequently produce repeated moderate storms during the declining and minimum phase of the solar cycle. Beyond 1 AU, these disturbances often create merged interaction regions and pressure ridges.

Transients in the form of CMEs are highly disruptive to the ambient solar wind and to geospace. Two classes of geoeffective CMEs, halo CMEs and magnetic cloud CMEs, have geometries, speeds, and magnetic characteristics that magnify their geomagnetic storm-producing capabilities. Ejecta with large, out-of-ecliptic magnetic fields create the opportunity for dayside merging between the IMF and Earth's magnetic field. Long intervals of such coupling add substantial energy to the geospace and Earth's upper atmosphere. Further, ejecta that emerge rapidly from the Sun often create shocks, accelerate energetic particles in the low solar corona and in the interplanetary medium, and merge with previously ejected materials. Shocks generated on the western limb of the Sun are likely to be magnetically well-connected to Earth via the Parker spiral.

CMEs also create merged interaction regions. Super-Alfvénic CMEs act as long-lasting acceleration mechanisms in the interplanetary medium. In turn, the shock structures behave like imperfect magnetic mirrors moving through the ambient plasma. Two or more successive shocks in the interplanetary medium amplify the mirror effect for ions that become trapped like ping-pong balls between the structures. The shock speed and plasma parameters determine the energy gain for individual particles. A single shock encounter typically boosts a particle's energy by a factor of four. Solar wind keV ions must have several encounters with a shock to accelerate to MeV energies. The seed particles excited by shocks are a mix of ambient solar wind particles and remnants of previous flares and SEP events. They are subject to particle transport and acceleration by the large-scale IMF and by particle plasma wave interactions.

The relative contribution of the shock acceleration mechanisms depends on the properties of the shock. Stochastic acceleration requires a strong increase in downstream turbulence to be effective, while diffusive acceleration requires a sufficient amount of scattering in upstream and downstream media. For perpendicular shocks, the greatest acceleration occurs where the electric induction field is at its maximum value. The shock parameters, such as compression ratio and speed, determine the efficiency of the acceleration mechanism. The energized particles can penetrate spacecraft components, injure human tissue, and create particle showers in Earth's atmosphere. Rarely, some events accelerate particles to such high energies that the resultant interaction particles can be observed at Earth's surface. These are called ground-level events (GLEs).

Nancy Crooker provided useful discussions and insight related to the material in this chapter.

Key Words

co-rotating interaction regions (CIRs)
diffusive shock acceleration
disorientation
energetic storm particle (ESP)
Forbush decrease
geoeffective
globally merged interaction regions (GMIRs)
ground-level enhancements
halo CMEs
magnetic clouds
merged interaction regions (MIRs)
shock drift acceleration
shock normal
shock parameters
single-event upsets
stochastic process
stream interfaces
type II radio burst

Answers to Pause for Inquiry Questions

PFI 10-1: a) A supergranule is approximately 30 Mm in diameter. A solar wind flux tube at Earth is about 10 times larger. We know the solar wind expands as it flows outward, so the scale size should grow. Earth tends to sample the tubes near the heliospheric current sheet, where the tube expansion is inhibited and does not achieve the R^2 growth we would otherwise expect.

b) Earth's magnetosphere has a diameter of about 30 R_E (~1.9×10^8 m). A solar wind flux tube at Earth is about 2.5 times the size of the magnetospheric diameter.

c) The heliospheric plasmasheet has a thickness of about 100 R_E. It is about three times wider than Earth's magnetosphere at its widest point.

PFI 10-2: Stream interface regions bound regions of closed fields. Closed field regions are the source of most CMEs. If a CME has erupted just ahead of a high-speed stream, it is likely to be swept up in the expanding flow from the high-speed region. These complex structures often produce geomagnetic storms of above-average strength.

PFI 10-3: The average post-interface thermal speed for ions is ~70 km/s. Using $(1/2)mv^2 = k_B T$ and solving for temperature, we get: $T \approx 3 \times 10^5$ K. This thermal speed (or equivalently, temperature) much higher than the typical solar wind temperature and even higher than the average high-speed flow thermal temperature.

PFI 10-4: 350 CIRs per solar cycle corresponds to approximately 2.5 CIR passages per month. Because of the solar equatorial tilt, Earth should experience one sector boundary crossing per month during solar minimum. Many CIR passages occur during the declining phase of the solar cycle, so 2.5 per month is reasonable.

PFI 10-5: CIR travel time to 1 AU is roughly 3 days. The shocks must be developing after they arrive at 1 AU but before they arrive at 2 AU. Thus 5 days after departing the Sun is a reasonable time for the pressure waves to steepen into shocks.

PFI 10-6: The maximum magnetic field occurs just ahead of or at the stream interface where the field is being compressed by the interacting flows.

PFI 10-7: CMEs harbor concentrated magnetic fields that exert a pressure. If the magnetic pressure is greater than the thermal pressure in the surrounding solar wind, the CME will expand to achieve pressure balance.

PFI 10-8: The plasma β is the ratio of thermal pressure to magnetic pressure. In an expanding CME the temperature and density are decreasing, so the plasma β decreases.

PFI 10-9: The long-duration sinusoid structure (flux rope) in the IMF identifies the magnetic structure on 20 November 2003 as a CME. The long, smooth rotation of the IMF components identifies it as a magnetic cloud. The long interval of south-directed field suggests it was geoeffective. The associated plasma speed indicates this was a fast CME. Fast CMEs with southward IMF are the most geoeffective structures.

PFI 10-10: The interplanetary electric field (IEF) is $E = -v \times B$. Using $v = -600$ km/s and $B_z = -50$ nT gives a value of +30 mV/m for the IEF. This field points in the positive y direction in magnetic coordinates centered at Earth.

PFI 10-11: The CME arrives with a leading positive z component. The diagram that best fits this scenario is from Fig. 10-13b.

PFI 10-12: The CME arrived with a speed, v_{sw}, of 700 km/s. Dividing this value into 1.5×10^{11} m gives a travel time of 2.14×10^5 s or ~2.5 days.

PFI 10-13: The first CME may sweep away the solar wind material. Consequently the second CME travels through a less dense medium and travels faster and catches the first one.

PFI 10-14: 1991. In that year, the Sun produced a large number of flares and CMEs.

PFI 10-15: The SEP event in Fig. 10-17 lasts for nearly two days. Solar flares have time scales of minutes to hours. Therefore, it is not likely that the event could have been exclusively flare driven. The solar flare would not have sufficient longevity to continue to accelerate particles.

PFI 10-16: The median value of the burst is near 40 MHz. Using $c = \lambda f$ and solving for λ gives $\lambda \sim 7.5$ m.

PFI 10-17: The GLEs would be associated with the major SEP events. However, only a small fraction of major SEP events produce ground-level enhancements.

There are more shocks and type II radio bursts than CMEs because a shock or burst generated by a CME can arrive at Earth even if the CME is deflected or on a trajectory that does not intercept Earth.

References

Jian, Lan, Christopher T. Russell, Janet G. Luhmann, and Ruth M. Skoug. 2006. Properties of interplanetary coronal mass ejections at one AU during 1995–2004. *Solar Physics.* Vol. 239. Springer. Dordrecht, Netherlands.

Kilpua, Emilia K. J., Janet G. Luhmann, Jack Gosling, Yan Li, Heather Elliott, Christopher T. Russell, Lan Jian, Antoinette B. Galvin, Davin Larson, Peter Schroeder, Kristin Simunac, and Gordon Petrie. 2009. Small Solar Wind Transients and Their Connection to the Large-Scale Coronal Structure. *Solar Physics.* Vol. 256. Springer. Dordrecht, Netherlands.

Figure References

Alfvén, Hannes. 1977. Electric Currents in the Cosmic Plasmas. *Reviews of Geophysics and Space Physics.* Vol. 15. American Geophysical Union. Washington, DC.

Bruno, Roberto, Vincenzo Carbone, Pierluigi Veltri, Ermanno Pietropaolo, and Bruno Bavassano. 2001. Identifying intermittency events in the solar wind. *Planetary and Space Science.* Vol. 49. Elsevier. Amsterdam, Netherlands.

Crooker, Nancy U. and Timothy S. Horbury. (Kunow et al., eds.). 2006. Solar imprint on ICMEs, their magnetic connectivity, and heliospheric evolution, in Coronal Mass Ejections. *ISSI Space Science Series.* Vol. 21. Springer. Dordrecht, NL.

Gopalswamy, Natchimuthuk, Sachiko Akiyama, Seiji Yashiro, and Pertti Mäkelä. 2010. Coronal Mass Ejections from Sunspot and non-Sunspot Regions. in "Magnetic Coupling between the Interior and the Atmosphere of the Sun," eds. S. S. Hasan and R. J. Rutten. *Astrophysics and Space Science Proceedings.* Springer-Verlag. Heidelberg, Germany.

Gosling, John T. 2007. "The Solar Wind." The Encyclopedia of the Solar System, 2nd Edition. (Eds. Lucy-Ann McFadden, Paul R. Weissman, and Torrence V. Johnson). Academic Press. New York, NY.

Gosling, John T. 1996. Magnetic topologies of coronal mass ejections: Effects of 3-dimensional reconnection, in Solar Wind Eight. (Eds. Daniel Winterhalter, John T. Gosling, Shadia R. Habbal, William S. Kurth, and Marcia Neugebauer). AIP Conference Proceedings 382. Woodbury, NY.

Intriligator, Devrie S., Wei Sun, Murray Dryer, Craig D. Fry, Charles Deehr, and James Intriligator. 2005. From the Sun to the outer heliosphere: Modeling and analyses of the interplanetary propagation of the October/November (Halloween) 2003 solar events. *Journal of Geophysical Research.* Vol. 110. American Geophysical Union. Washington, DC.

Jokipii, Randy and Barry Thomas. 1981. Effects of Drift on the Transport of Cosmic Rays. IV – Modulation by a Wavy Interplanetary Current Sheet. *Astrophysical Journal.* Vol. 243. The American Astronomical Society. Washington, DC.

Lovell, Jenny, Marc L. Duldig, and John E. Humble. 1998. An Extended Analysis of the September 1989 Cosmic Ray Ground Level Event. *Journal of Geophysical Research.* Vol. 103. American Geophysical Union. Washington, DC.

Paschmann, Göetz, Gerhard Haerendel, Ioannis Papamastorakis, Norbert Sckopke, Samuel J. Bame, John T. Gosling, and Christopher T. Russell. 1982. Plasma and Magnetic Field Characteristics of Magnetic Flux Transfer Events. *Journal of Geophysical Research.* Vol. 87. American Geophysical Union. Washington, DC.

Riley, Pete, Jon A. Linker, Zoran Mikić, Drusan Odstrcil, Thomas H. Zurbuchen, David Lario, and Ronald P. Lepping. 2003. Using an MHD simulation to interpret the global context of a coronal mass ejection observed by two spacecraft. *Journal of Geophysical Research, Space Physics.* Vol. 108. Issue A7. American Geophysical Union. Washington, DC.

Russell, Christopher T. and Richard C. Elphic. 1979. Observations of magnetic flux ropes in the Venus ionosphere. *Nature.* Vol. 279. Nature Publishing Group. New York, NY.

Wang, Yi-Ming and Neil R. Sheeley, Jr. 2003. On the Topological Evolution of the Coronal Magnetic Field During the Solar Cycle. *The Astrophysical Journal.* Vol. 599. The American Astronomical Society. Washington, DC.

Zurbuchen, Thomas H., and Ian G. Richardson. 2006. In-situ solar wind and magnetic field signatures of interplanetary coronal mass ejections. *Space Science Reviews.* Vol. 123. Springer. Dordrecht, Netherlands.

Further Reading

Phillips, John L., John T. Gosling, David J. McComas, Samuel J. Bame, and William C. Feldman. 1992. Magnetic topology of coronal mass ejections based on ISEE-3 observations of bidirectional electron fluxes at 1 AU, in *Proceedings of the First SOLTIP Symposium.* Vol. 2. S. Fischer and M. Vandes, eds. Astronomical Institute of the Czechoslovak Academy of Sciences. Prague, Czech Republic.

Russell, Christopher T. and Richard C. Elphic. 1979. ISEE observations of flux transfer events at the dayside magnetopause. *Geophysical Research Letters.* Vol. 6. American Geophysical Union. Washington, DC.

Chapter 10 The Active Interplanetary Medium: Conduit for Space Weather

The Disturbed Magnetosphere and Linkages Above and Below

11

UNIT 2. ACTIVE SPACE WEATHER AND ITS PHYSICS

With Contributions from Robert McPherron

You should already know about...

- High-energy particles (Chap. 2)
- Magnetic merging (Chaps. 3 and 4)
- The co-rotating interactions regions (Chaps. 5 and 10)
- Particle drifts in various magnetic field configurations, including conservation of the invariants of motion: μ, J, and Φ (Chaps. 6 and 7)
- Magnetospheric configuration (Chap. 7)
- The L-shell coordinate system (Chap. 7)
- Particle populations in the magnetosphere (Chaps. 8 and 10)
- Structures in the solar wind that create a southward interplanetary magnetic field (IMF) (Chap. 10)

In this chapter you will learn about...

- Storms in the magnetosphere, including their phases and intensities
- Solar wind Poynting flux
- Storm indices and parameters
- Stormtime enhancement of the convection electric field, including driven and unloading processes
- The role of energized particles in enhancing the ring current, the radiation belts, and the aurora
- Plasmaspheric response to storm dynamics
- Field-aligned currents during magnetic storms
- Stormtime shielding of the inner magnetosphere

Outline

11.1 Describing Geomagnetic Storms
 11.1.1 The Interplanetary Medium and Magnetic Storms
 11.1.2 The Dst Index and Its Relation to the Ring Current and Storm Phases

11.2 Large-scale Magnetic Disturbances
 11.2.1 Power from the Solar Wind
 11.2.2 Energy Dissipation Processes with Links to the Ionosphere
 11.2.3 Unloading Energy to the Trapping Regions

11.3 Field and Current Coupling during Magnetic Storms
 11.3.1 High-latitude Magnetosphere-Ionosphere Coupling
 11.3.2 Mid- and Low-latitude Magnetosphere Shielding

11.1 Describing Geomagnetic Storms

11.1.1 The Interplanetary Medium and Magnetic Storms

> **Objectives:** After reading this section, you should be able to...
>
> ♦ State the key elements of the solar wind prerequisite to a magnetic storm
> ♦ Explain why some storms are transient and others are recurrent
> ♦ List the general characteristics of a magnetic storm

Geomagnetic storms are large, multi-day disturbances in the near-Earth environment caused by solar wind and interplanetary field structures moving past Earth. We classify them as *recurrent* or *transient*. Recurrent storms arrive roughly every 27 days. These periodic events are most likely produced by flow from coronal holes that form stream interface regions in the solar wind; if the region returns about every 27 days, it is called a co-rotating interaction region (CIR). Geomagnetic storms arise when Earth encounters a southward-oriented magnetic field and high-dynamic pressure regions of CIRs. Recurrent storms occur most frequently in the declining and minimum phases of the solar cycle and are typically moderate in intensity, but rather long-lived. Transient geomagnetic storms, on the other hand, occur most frequently near solar maximum. They are caused by interplanetary disturbances driven by CMEs and usually involve an encounter with the interplanetary shock and the CME that drives it.

Chapter 8 discusses how the dynamic pressure of the solar wind shapes the magnetosphere, but the interplanetary magnetic field (IMF) controls the details of the shape and the level of magnetospheric excitation (storminess). During the strong southward ($-B_z$) IMF and high-speed flow that typifies stormy conditions in the solar wind, the interplanetary medium energizes the magnetosphere. More energy is available and the magnetosphere is more efficient in extracting the energy. Earth's field converts some of the solar wind's kinetic energy to potential energy by stretching from a quasi-dipolar structure to an elongated configuration. The geomagnetic field acts as a reservoir of potential energy awaiting release by internal and external triggering processes. Under routine stressing, the magnetosphere releases energy in substorms and steady magnetospheric convection events. Under rapidly changing and severe stressing, additional modes of energy storage and release appear. In total, the modes create a more global disturbance called a geomagnetic storm.

A key element of the disturbances is a strong southward IMF (10 nT–15 nT) lasting three or more hours. From Chap. 5 we know that the undisturbed IMF tends to align with the Parker (garden hose) spiral with minimal excursions out of the ecliptic plane. Thus the spiral IMF has primarily B_x and B_y components. As Chap. 9 describes, large-scale disturbances in the solar wind, such as CMEs and CIRs, disturb the spiral configuration and introduce significant out-of-the-ecliptic IMF (large B_z values). This appreciable B_z component remains in the solar wind for several hours.

Magnetic storms are multi-faceted. The identifying features depend on the observer's location. Space-based observations reveal that a distinguishing feature of a magnetic storm is an enhancement of the energetic particle environment in the inner magnetosphere that corresponds to energy storage in heated plasma there. Ground-based observations emphasize a different feature: global magnetic perturbations recorded at magnetic observatories over a wide range of longitudes and latitudes. This corresponds to energy storage in magnetic fields and energy dissipation via currents and particle precipitation. Although general agreement obtains on storm drivers, the term "geomagnetic storm" has slightly different meanings in different communities. The differences arise because of various observing techniques, operational impacts, and methods of categorizing storm intensity. In this chapter, we address several of these different meanings and provide a background that promotes an understanding of geomagnetic storms from a physical perspective.

The general physical characteristics of magnetospheric storms include:

- Enhanced coupling of the solar wind-magnetosphere-atmosphere system
- Close-in acceleration and trapping of charged particles (keV–MeV plasma)
- Global magnetic disturbances
- Intensified currents in space and on the ground
- Profound changes in ionospheric dynamics and linkages to the magnetosphere
- A broadening of the polar cap and equatorward extension of auroral displays

A major goal of the space physics community is to predict the occurrence of these magnetic storms based on knowledge of solar activity and how that activity couples to Earth's magnetosphere. We describe the better-known coupling aspects next.

11.1.2 The Dst Index and Its Relation to the Ring Current and Storm Phases

Objectives: After reading this section, you should be able to...

- Explain the association between the Dst index and the ring current
- Distinguish between storm phases using the trend in the Dst index
- Describe the levels of magnetic storm intensity with the Dst index

Geomagnetic Storm Index. In its most quiescent state the magnetosphere is a closed, stretched dipole that allows little transfer of mass, energy, or momentum across its boundary. The solar wind stresses Earth's magnetic field, producing a magnetopause sheet current that confines the geomagnetic field to a certain volume. The volume is compressed on the dayside and stretched on the nightside. When the IMF turns southward, the energy of the situation changes. Field topologies in the IMF that oppose terrestrial field lines "open" a portion of the magnetosphere to mass, energy, and momentum transfer in a magnetic merging process that becomes more efficient during magnetic storms. Inside the magnetosphere, magnetic energy transforms into kinetic energy. Energy deposition heats and accelerates plasma particles and creates changing internal field configurations that drive or enhance current systems.

Within the magnetopause, the plasma motion is dominated by Earth's magnetic field. The magnetosphere has no native plasma population; rather, it accepts a small amount of plasma from the solar wind and, especially during storm-time,

Chapter 11 The Disturbed Magnetosphere and Linkages Above and Below

extracts plasma from Earth's cold ionosphere. During stronger storms, more of each type of plasma is present, but often the percentage of Earth plasma soars.

Energetic particles also exhibit systematic variations during geomagnetic storms. Measurements of particle populations reveal that they vary, but with different cadences. Energetic electrons that form the outer radiation belt are more likely to be enhanced by moderate, long-lasting storms driven by high-speed solar wind streams. Sudden impulsive enhancements of particles that create new radiation belts or fill in the radiation belt slot region are most likely to be driven by high-speed CMEs jolting the magnetosphere. These impulsive events are also more likely to create deep red auroral displays at low latitudes.

Although we categorize magnetic storms through various measurements, for the greater part of the 20th Century and continuing to the present, the hallmark of a geomagnetic storm has been a systematic global decrease and recovery of the surface geomagnetic field strength at low latitudes. This variation is measured with the *disturbance storm time (Dst) index*, expressed in units of nT. The Dst index often serves as a single measure of magnetic storm energy, although it is by no means comprehensive.

From early satellite measurements, investigators learned that the Dst signature was primarily a measure of the energy density of low-to-medium energy ions (tens of keV) that sweep toward Earth from the magnetotail, under the influence of electric fields enhanced by the merging process. As the energized particles approach the inner magnetosphere, the gradient and curvature of Earth's dipole field diverts ions westward and electrons eastward (Sec. 4.4 and Figs. 7-22 and 7-23). The resulting high-altitude ring of current, centered near 4 R_E, causes a depression worldwide in the horizontal (H-component) magnetic field during large magnetic storms (Fig. 11-1).

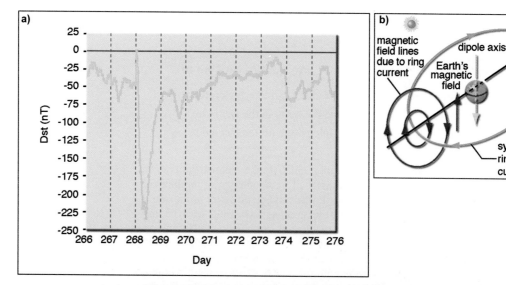

Fig. 11-1. a) The Dst Index for Ten Days in Late 1998. The extremum of the depression in the equatorial surface magnetic field strength was ~230 nT. The ring current was active prior to onset of the storm on day 268 (October 25). A sudden commencement (positive spike) preceded the sharp drop in field strength. *(Courtesy of the World Data Center for Geomagnetism, Kyoto, Japan)* **b) The Current Disturbance Contributing to the Dst Decrease.** The yellow arrow represents Earth's dipole, the green oval represents an idealized ring current at about 4 R_E. The dark gray ovals represent the magnetic perturbation surrounding the ring current at a single longitude. The slanted line represents dawn and dusk. The dark gray arrow represents the depression of the magnetic field inside the ring current. This decrease is manifest at Earth's equatorial surface as a reduction in the horizontal component of the magnetic field.

For strong storms, the horizontal magnetic field reduction is 0.5%–1.0% of the surface field strength (Table 11-1). During a geomagnetic storm's main phase, which typically lasts about a day (but sometimes lasts as long as 2.5 days in the case of a severe storm), charged particles in the near-Earth plasmasheet are energized and injected deeper into the inner magnetosphere. The magnetic perturbation created by this current is shown in Fig. 11-1b. Negative Dst values indicate that work is being done to drive charged particles into Earth's inner magnetosphere. The more negative the Dst is, the more work done and the more intense the magnetic storm. Values below –50 nT indicate a storm-level disturbance.

The Dst index is determined by the World Data Center-C2 for Geomagnetism, Kyoto University, Kyoto, Japan. Traditionally, data from four stations have contributed hourly scalings of the surface horizontal geomagnetic field (Fig. 11-2). Analysts remove quiet-day variations and annual and secular trends from the hourly horizontal-component magnetic variations. A cosine factor of the site latitude transforms residual variations to their equatorial equivalents. The reference level (0 nT) is set so that the index averages to zero on quiet days (designated after the fact by an international team). At any given time, the Dst index is the average of variations over the longitudes covered by the magnetic stations.

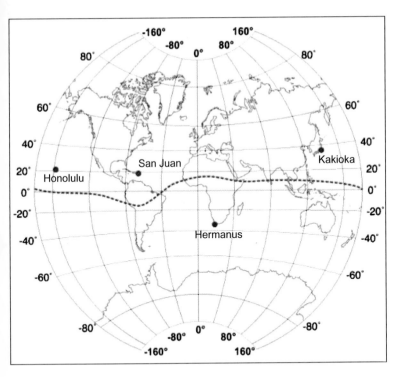

Fig. 11-2. Stations That Contribute to Calculating the Dst Index. The stations are (from left to right) Honolulu, HI, USA; San Juan, Puerto Rico, USA; Hermanus, South Africa; and Kakioka, Japan. These stations are located off the magnetic equator to avoid the effects of equatorial electrojet contribution to the Dst index. *(Courtesy of The World Data Center for Geomagnetism, Kyoto, Japan)*

Electric current measurements from space reveal that near-equatorial geomagnetic variations also receive contributions from other current systems. The Dst index is sometimes contaminated by as much as 30% with magnetic signatures of other current systems. The index has thus come under some criticism as not being a true measure of geomagnetic storm intensity. Nonetheless, the long record of Dst (beginning in 1932) makes it a valuable comparative tool in measuring geomagnetic storms.

Geomagnetic Storm Phases. Magnetic storms have distinct Dst phases associated with the dynamics of the solar wind coupling. In the *initial phase* (onset) an interplanetary plasma structure arrives at Earth. If the arrival is impulsive, the disturbance compresses the magnetosphere, increasing the magnetopause current and the horizontal component of the magnetic field in the inner magnetosphere. This horizontal component increase appears as a spike (*sudden impulse (SI)*) at the ground magnetometers of mid-latitude stations. Increases of 5 nT–30 nT are precursors to many, but not all, magnetic storms. For the strongest events, 50 nT field increases are possible. The compression may last several minutes to an hour. If the IMF is oriented northward during and after the SI, then the ring current is not enhanced. On the other hand, if the IMF is oriented southward, then energy begins to enter the magnetosphere, ions sweep into motion, and a *sudden storm commencement (SSC)* appears. Figure 11-1a shows an SI-SSC at the beginning of day 268 (October 25th) of 1998.

During an SI with accompanying southward IMF, the dayside magnetopause attempts to reach a new equilibrium state, while plasma and fields inside the magnetosphere go into action. The signature of internal magnetospheric response to the new state of solar wind-magnetosphere coupling is the *main phase*. In this several-hour to day-long phase, the Dst rapidly decreases to a minimum. As energy is transferred and stored in the inner magnetosphere, the ring current develops. Stronger storms bring more ring current ions closer to Earth. The larger the westward current flow around Earth, the greater the equatorial surface field reduction. We say more about this in Sec. 11.2.

Nature reduces the energy storage via several energy and ion loss mechanisms. These include ion loss to the atmosphere and magnetopause, charge exchange with hot neutral atoms, and particle wave interactions. When energy release exceeds energy storage in the ring current, the ring current begins to decay (Dst values begin to increase, but remain negative) in the *recovery phase*. This phase is often punctuated by additional small impulses of energy delivered by substorms. We recall that substorms usually deposit most of their energy in the auroral zone, but in stronger substorms some energy apparently leaks to lower latitudes. The recovery phase may be a day to several days long and may be interrupted by another geomagnetic storm onset if conditions are right in the solar wind. In fact, the strongest geomagnetic storms often have double Dst minima, which indicate complex solar wind structures or a special state of the magnetosphere that allows a second ring current enhancement. Table 11-1 shows the Dst values for a typical strong geomagnetic storm.

Table 11-1. **Geomagnetic Storm Phases.** This table lists magnetic signatures of geomagnetic storms near the equator.

	Pre-Storm	Sudden Commencement	Initial Phase	Main Phase	Recovery
$B_{equatorial}$	"Normal" ~30,000 nT	Slight rise 5 nT –30 nT (may not always be present)	Slow ramp up 5 nT–30 nT	Large drop 50 nT–100 nT More during larger storms	Rise back to pre-storm value
Cause	N/A	Arrival of strong solar wind perturbation	Compression of Earth's B field	Large ring current particle "injection"	Slow decay of ring current
Duration	N/A	1 minute–6 minutes	1 hour–10 hours	~1 day	~1 day–5 days

11.1 Describing Geomagnetic Storms

Pause for Inquiry 11-1

Match the storm phases in Table 11-1 to the variations in Dst in Fig. 11-1a.

Pause for Inquiry 11-2

Verify the direction of the magnetic field perturbation inside and outside of the idealized ring current of Fig. 11-1b.

Investigators also use other indices, shown in Table 11-2, to categorize storms, but these tend to lack the temporal resolution and clear link with a storm mechanism that characterizes Dst. The Ap and Kp indices are measures of the more global level of magnetic disturbance and contain information about the state of the auroral currents rather than the ring current. In subsequent sections, we present more quantitative measures of storm intensity.

Table 11-2. Intensity of Geomagnetic Storms Categorized by Various Indices. Here we show the relative values for the disturbance storm time (Dst), the Ap, the NOAA scale, and the geomagnetic level indices.

Dst_{min} hourly (nT)	Storm Category	Ap Daily (Unitless)	Ap Storm Category
		$Ap < 8$	quiet
$Dst_{min} \geq -19$	quiescent	$8 \leq Ap \leq 15$	quiescent
$-20 \geq Dst_{min} \geq -49$	minor storm	$16 \leq Ap \leq 29$	active
$-50 \geq Dst_{min} \geq -99$	moderate storm	$30 \leq Ap \leq 49$	minor storm
$-100 \geq Dst_{min} \geq -249$	intense storm	$50 \leq Ap \leq 99$	major storm
$-250 \geq Dst_{min}$	super storm*	$100 \leq Ap \leq 400$	severe storm
NOAA Scale (unitless) (Appendix A)		**Geomagnetic Level**	
G1	minor storm	$Kp = 5$	minor storm
G2	moderate storm	$Kp = 6$	moderate storm
G3	strong storm	$Kp = 7$	strong storm
G4	severe storm	$Kp = 8$	severe storm
G5	extreme storm	$Kp = 9$	extreme storm

* Sometimes $-300 \geq Dst_{min}$ are referred to as great storms.

EXAMPLE 11.1

- **Problem Statement:** Find the total ring current (I_{ring}) that would cause the main phase decrease in magnetic field shown in Fig. 11-1.
- **Given:** The description of the ring current in this section.
- **Assume:** The ring is concentrated at $L = 4.5$. Ring current magnitude is 230 nT.
- **Relevant Concept:** The Biot-Savart Law describes the influence of a small current segment, $I\,d\mathbf{l}$, on the local magnetic field, $\mathbf{B} = \int d\mathbf{B} = \frac{\mu_0}{4\pi}\frac{I\,d\mathbf{l}\times\hat{\mathbf{r}}}{r^2}$. For a ring current with radius r, the simplification is:

$$B = \frac{\mu_0}{4\pi}I\frac{(2\pi r)}{(r^2)}$$

- **Solution:** The magnitude of the magnetic field at the center of a ring of radius $4.5\,R_E$, carrying a current I, is $B_{loop} = \frac{\mu_0 I}{2(4.5 R_E)}$

Solving for I gives

$$I = \frac{2(4.5\times R_E)B_{loop}}{\mu_0} = \frac{2\times[4.5\times(6378\times10^3\,\text{m})]\times(230\times10^{-9}\,\text{T})}{4\pi\times10^{-7}\,\text{N/A}^2}$$

$$= \frac{9\times(6378\times10^3\,\text{m})\times(230\times10^{-9}\,\text{T})}{4\pi\times10^{-7}\,\text{N/A}^2}$$

$$= 1.1\times10^7\,\text{A}$$

- **Interpretation:** During an intense storm about 11 MA of current flows in the inner magnetosphere as a result of the ring current.

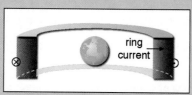

Fig. 11-3. Idealized Schematic of the Ring Current around Earth. In this view from the Sun, the ring current flows into the page on the left side and out of the page on the right side.

Follow-on Exercise: Determine the current density in A/m² in a system modeled as a rectangular ring of height $2\,R_E$ and cross width $1\,R_E$ (Fig. 11-3).

Follow-on Question: If each radiation belt particle carries a single charge, how many particles are in the radiation belts? What thermal pressure do these particles exert?

11.2 Large-scale Magnetic Disturbances

11.2.1 Power from the Solar Wind

Objectives: After reading this section, you should be able to...

- Describe how the solar wind delivers (injects) energy to the magnetosphere
- Distinguish between kinetic energy flux and Poynting flux
- Relate the Akasofu parameter to the energy flux delivered by the solar wind

As we describe in Chaps. 4, 7, and 8, enhanced interaction between the solar wind and the magnetosphere produces strong magnetospheric currents and increases the magnetic energy density in the magnetosphere. The energy produces: 1) reshaping of the magnetosphere, 2) faster magnetospheric convection in the plasmasheet, 3) particle heating and acceleration, and 4) stronger currents. All of these are hallmarks of large-scale magnetic disturbances.

Energy Flux from the Interplanetary Medium. One important question related to large-scale geomagnetic disturbances is, "How much energy is needed to configure and power a disturbed magnetosphere and its storm-time dynamics?" This question is difficult to answer because global sampling of the energy inputs is lacking and because many energy paths exist. However, scientists have settled on two general measures of solar wind power that can be estimated from satellite observations upstream of Earth's bowshock: kinetic energy flux and Poynting flux. The motion of the solar wind particles relative to the Earth produces kinetic energy flux. This plasma energy flux is responsible for shaping the magnetosphere, while the influence of the interplanetary magnetic field on the magnetosphere is characterized by Poynting flux.

Kinetic energy flux, $v(\rho v^2)$, is more closely associated with shaping the magnetosphere than with direct energy injection. This is because to a good approximation the frozen-in magnetic flux limits solar wind-Earth interaction. The fields and plasmas of the two regimes do not mix easily. Many studies reveal that kinetic energy entry to the magnetosphere is very inefficient, ~0.3% during quiet times and perhaps an order of magnitude larger during storms (Example 11.2). Nonetheless, kinetic energy flux plays a significant role in magnetospheric dynamics because the shape of the magnetosphere controls the efficiency of magnetic merging.

Poynting flux is the power delivered by electromagnetic sources. Akasofu [1981] produced an electromagnetic power flux estimate that represents the solar wind power delivered by magnetic merging (Eq. (11-1)). His assumption was that only the tangential component, B_t, of the IMF delivers energy, so the B_x component does not play a role. The elements in the square bracket of Eq. (11-1) constitute the Poynting flux. The factor outside of the brackets accounts for the efficiency of energy entry by an anti-parallel component of the IMF.

$$P = [(1/\mu_0)(E_{SW} \times B_{SW})](\sin(\theta/2))^4$$
$$= (1/\mu_0) v_{SW} |B_t|^2 (\sin(\theta/2))^4 \qquad (11\text{-}1)$$

Chapter 11   The Disturbed Magnetosphere and Linkages Above and Below

where

- P = effective solar wind Poynting flux [W/m^2]
- μ_0 = permeability of free space (1.26×10^{-6} N/A^2)
- v_{SW} = solar wind velocity [m/s]
- $|B_t|$ = magnitude of the tangential component of the IMF = $\sqrt{B_y^2 + B_z^2}$ [T]
- θ = IMF polar angle in the Y-Z GSE plane ($\tan^{-1}(B_y/B_z)$) [deg or rad]

For many storm-time applications, space weather scientists are most interested in the power delivered to the cross section of the dayside magnetosphere. One version of the power estimate, called the *Akasofu epsilon parameter (ε)*, comes from multiplying the effective solar wind Poynting flux by a factor proportional to the cross-sectional area of the magnetopause that intercepts the solar wind:

$$\varepsilon = (1/\mu_0)\, v_{SW}\, |B_t|^2\, (\sin(\theta/2))^4\, l^2 \qquad (11\text{-}2)$$

where

- ε = Akasofu epsilon parameter (W)
- l^2 = effective cross-sectional area of the merging region $\approx (7\, R_E)^2$

Scientists often use ε as a starting point for the energy budget of magnetic storms. However, during severe storms the estimate may go awry because the cross-sectional area of the magnetopause changes drastically, or other poorly understood factors affect the efficiency of the energy transfer. Table 11-2 shows some order-of-magnitude power estimates for quiescent and storm conditions.

Pause for Inquiry 11-3

Consider Eq. (11-1). Justify the substitution made in going from line 1 to line 2 of the equation.

Pause for Inquiry 11-4

Verify that the effective solar wind Poynting flux described in Eq. (11-1) and the kinetic energy flux have units of W/m^2.

Energy Input from Solar Wind Alfvén Waves. In its high-speed mode, the solar wind also contains Alfvénic waves that may buffet and resonate with the magnetosphere. Multi-hour to multi-day passage of high-speed Alfvénic flow causes moderate magnetic storms by virtue of cyclical southward IMF delivered every 20–30 minutes to the magnetopause. The oscillations produce internal ultra-low-frequency (ULF) waves that resonate with magnetospheric electrons in the middle and inner magnetosphere. Electrons with motions synchronized to the ULF oscillations rapidly gain energy. During the recovery phase of moderate storms created by such high-speed stream interactions, ongoing periodic substorms sometimes occur because of north-south oscillations of the IMF. The resulting time-varying fields may be an especially prolific energy source for electrons in the inner magnetosphere.

11.2 Large-scale Magnetic Disturbances

Table 11-3. Power Flux to the Geosphere and Terrestrial Systems. Here we list various energy inputs to the geosphere and terrestrial systems during quiescent and storm times.

Source: Storm or Quiescent	Power Available W/m²	Total Available Power (W)	Location	Result/Comments
Quiet total solar irradiance	1366	~1.7×10^{17}	Top of sunlit atmosphere	Drives the atmosphere
Shortwave enhanced irradiance	0.005	~7×10^{14}		Shortwave enhancement heats the upper atmosphere
Quiet solar wind kinetic energy flux $v(\rho v^2)$	0.0001–0.001	~1×10^{10} (Assume efficiency $\approx 0.3\%$)	Magnetosphere disk $(20 R_E)^2$	Helps to shape magnetosphere
Storm solar wind kinetic energy flux $v(\rho v^2)$	0.001–0.01	$1–5 \times 10^{12}$ (Assume efficiency $\approx 0.1\%$)		Alters magnetosphere's shape
Quiet effective solar wind Poynting flux	0.0–1.0	$1–5 \times 10^{11}$	Magnetopause disk $(7 R_E)^2$	Drives convection and substorms
Storm effective solar wind Poynting flux	1.0–10.0	$5–9 \times 10^{12}$ or more		Drives convection and geomagnetic storms

EXAMPLE 11.2

Kinetic Power From the Solar Wind

- *Problem Statement:* Estimate the actual kinetic power penetrating the magnetosphere during storm time, given a solar wind number density of 10 ions/cm³, a solar wind speed of 800 km/s, and a 1% energy entry efficiency.
- *Relevant Concepts:* Kinetic energy flux = ρv^3
- *Given:* n = 10 ions/cm³, v = 800 km/s; 1% energy efficiency
- *Assume:* The magnetosphere presents an effective area of $(20R_E)^2$ to the solar wind.
- *Solution:* Kinetic energy flux = (10 ions/cm³) (10^6 cm³/m³) (1.67×10^{-27} kg/ion) (8×10^5 m/s)³

  Kinetic energy flux = 8.55×10^{-3} W/m²

  Magnetospheric area = $(20R_E)^2$ = $(20 \times 6378 \times 10^3$ m$)^2 \approx 1.63 \times 10^{16}$ m²

  Kinetic power = (8.55×10^{-3} W/m²) × (1.63×10^{16} m²) = 1.39×10^{14} W

  Using 0.01 efficiency gives: 1.39×10^{14} W × 0.01 = 1.39×10^{12} W

- *Implications and Interpretation:* During storm time the solar wind delivers about a terawatt of power to the magnetosphere. According to Fig. 2-3, a large geomagnetic storm has a power of ten terawatts. Some other energy injection method, such as Poynting flux, is needed for powering geomagnetic storms. This does not mean, however, that kinetic energy flux is unimportant. The shape of the magnetosphere determines many aspects of merging efficiency associated with Poynting flux entry. In this realm, kinetic energy flux is important.

EXAMPLE 11.3

IMF Control of Electromagnetic Power Entry into the Magnetosphere

- **Problem Statement:** Compare the power available to the magnetosphere when the IMF is oriented southward with a value of $B_z = -5$ nT with the power available when its components are $B_z = -4$ nT and $B_y = 3$ nT.
- **Relevant Concepts:** Epsilon parameter; IMF clock angle $= \tan^{-1}(B_y/B_z)$.
- **Given:** Case I: $B_z = -5$ nT; Case II: $B_z = -4$ nT and $B_y = 3$ nT.
- **Assume:** The solar wind speed is the same in both cases; a due north field corresponds to a clock angle of zero.

- **Solution:**

$$\varepsilon = (4\pi/\mu_0)v_{SW}|B_t|^2(\sin\theta/2)^4 l^2$$

In both cases, $4\pi/\mu_0$, v_{SW}, and l^2 are the same. Further, $B_{t1}^2 = (-5\text{ nT})^2 = 2.5 \times 10^{-17}\text{ T}^2$, and $B_{t2}^2 = ((-4\text{ nT})^2 + (3\text{ nT})^2) = 2.5 \times 10^{-17}\text{ T}^2$. We form a ratio of only the values that vary:

$$\frac{[(\sin\theta_1/2)^4]}{[(\sin\theta_2/2)^4]}$$

$\theta_1 = \tan^{-1}(0\text{ nT}/(-5\text{ nT})) = 180°$; $(\sin(180°/2))^4 = 1$

$\theta_2 = \tan^{-1}(3\text{ nT}/(-4\text{ nT})) = 143°$; $(\sin(143°/2))^4 = 0.81$

$(\sin\theta_1/2)^4/(\sin\theta_2/2)^4 = 1.0/0.81 = 1.23$

- **Interpretation:** When the IMF is oriented southward (180°), the energy available to the magnetosphere is 23% greater than when it has the same magnitude, but an orientation of 143°.

11.2.2 Energy Dissipation Processes with Links to the Ionosphere

Objectives: After reading this section, you should be able to...

- Describe the means of energy storage and dissipation in the magnetosphere
- Distinguish between steady magnetospheric convection, substorms, and periodic oscillations
- Describe how substorms contribute to rapid, chaotic energy dissipation
- Describe the differences and similarities between substorms and oscillations

Energy Dissipation. As we mention in Sec. 11.1, geomagnetic storms last several days, during which the perturbed magnetic field that creates the ring current also produces a noticeable long-lived depression in Earth's surface equatorial magnetic field. The main phase of the storm is linked to the dissipation mechanisms we describe in the next topic. The stormtime ring current development occurs during long periods of southward IMF or of fluctuating IMF, accompanied by high solar wind speed. Solar wind-magnetosphere coupling from such an event creates unsteady magnetotail reconnection, global convection enhancements with bursty flow characteristics and deep impulsive penetration of energized particles to geosynchronous orbit, and bright auroras. Ring current particles and radiation belt particles grow in number and energy. Substorms probably play a significant role in energizing these particles.

In the quiet magnetosphere, convection and aperiodic substorms dissipate the energy provided by the solar-wind-imposed electric field. Substorms are a common type of geomagnetic activity, and last two to four hours. They circulate energy through the magnetosphere via a cycle of merging at the dayside magnetopause, storing energy in the magnetotail, and releasing the tail energy during nightside reconnection. Part of the energy release involves reconfiguring the stretched magnetotail into a more dipolar shape (dipolarization). The Earthward convective motion of the plasma is directed from the tail merging (reconnection) line toward and around Earth and then to the dayside magnetopause. The tailward convective motion is associated with plasmoid ejection and post-plasmoid flow. Typical downtail energy loss is on the order of 10^{15} J.

As we describe in Sec. 7.4, when dayside reconnection is continuously enhanced by a southward IMF, the magnetosphere responds in one of several ways: 1) events with steady magnetospheric convection (Fig. 11-4a); 2) intervals of individual aperiodic substorms; 3) strong periodic substorms with growth of the ring current (Fig. 11-4b); and 4) strongly enhanced convection, intermittent large substorms, and a build-up of the ring current (Fig. 11-4c). The last category covers most of the activity and is often referred to as a magnetic storm, although any of the responses may produce a magnetic storm.

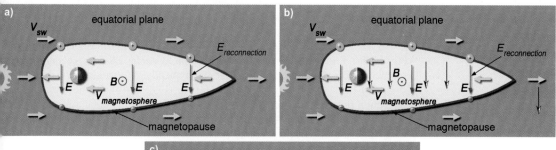

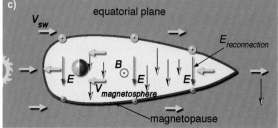

Fig. 11-4. **The Convection Electric Field in the Equatorial Cross Section for Different Modes of Magnetospheric Storm Behavior.** Reconnection generates plasma motion through a cross-magnetotail electric field. High-speed solar wind with southward IMF produces reconnection that varies in strength and location. The result is large-scale structuring of the electric field or a small-scale, and often turbulent, electric field and deposition of energy deep in the inner magnetosphere. The motion ranges from **a)** fairly steady drifts of low-energy plasma that slowly extracts energy from convective motion as it crosses gradients in the magnetic field (steady magnetospheric convection) to **b)** a periodic substorm-driven field that energizes plasma, allowing it to move close to Earth (sawtooth oscillations) to **c)** storm-driven impulses of higher energy plasma that penetrates into the inner magnetosphere.

Convective or Driven Dissipation. A *steady magnetospheric convection (SMC)* event is a global magnetospheric disturbance that develops during stable, continuously southward IMF periods with low, steady, solar wind speeds. The associated interplanetary electric field (IEF) is usually less than 1 mV/m. Such events fall into the category of driven events (Fig. 11-4a). They last for 4–6 hours or more. While they last, the convection system is associated with quasi-steady, balanced reconnection in the mid-tail region (30 R_E–45 R_E beyond Earth), which drives steady Earthward flows, $v_{drift} = (E \times B)/B^2$, suggested in Fig. 11-4a. The flows divert around the inner magnetosphere, leaving it undisturbed. Although plasma drift amplifies convection, the plasma pressure increases only modestly and the cross-tail current remains steady. Particle variations at geosynchronous orbit are small or non-existent. During the SMC, substorms and plasmoids do not develop, although SMC events tend to begin and end with substorms. Steady magnetospheric convective dissipation often leads to minor or moderate magnetic storms. Observations of auroral dynamics suggest that SMCs direct a significant portion of their energy to the ionosphere. They broaden and intensify the auroral oval and current systems that connect to the oval. The ground magnetic disturbance accompanying the auroral current systems normally has a magnitude of ~200 nT. SMCs are rather unusual events; the magnetosphere rarely makes a smooth transition from a state of slow convection and low dissipation to a state of faster convection and higher dissipation. Rather, the transition is most often a series of steps or phases that define a substorm.

Non-steady (Unloading) Dissipation. Substorms are called *isolated non-stormtime substorms* when they occur outside storm periods (Dst > –50 nT) and develop during relatively quiet magnetic conditions. On average the IEF is 1 mV/m. If energy density increases too much or too rapidly, then even the relatively stable storage regions in the magnetotail lobes will destabilize, yielding sporadic reconnection. Stormtime dayside merging allows a very rapid build-up of magnetic flux in the lobes. Lobe magnetic pressure squeezes the plasmasheet until it thins (Fig. 11-4a). The plasmasheet currents become unstable, allowing reconnection to commence. The release regions vary, but are typically in the middle and distant tail, 20 R_E–60 R_E and sometimes beyond 100 R_E (Fig. 11-4b). In these regions, the convection flow is uneven and bursty. Reconnection of the field converts magnetic energy into kinetic energy that accelerates and heats plasma and generally unloads energy into the distant magnetotail (plasmoid), the inner magnetosphere, and the auroral ionosphere. Nightside geosynchronous satellite electron measurements usually display only slight changes in the particle environment during isolated substorms. The aurora expands and brightens noticeably. The magnetic perturbation from the auroral electrojet is usually about 250 nT.

Substorms progress through a series of phases: growth, expansion, and recovery. After dayside merging begins, energy accumulates in the magnetotail during the growth phase (Fig. 11-4a), until it releases explosively in the expansion phase. The expansion phase onset (Fig. 11-5b) signals a switch from relatively slow storage of energy in the magnetotail to a very rapid dissipation and release. During the energy release, stretched magnetic fields dipolarize. During substorms, the inductive fields created by the changing magnetic field add to the quasi-steady convection electric field or even replace it altogether. The associated perturbations and wave-particle interactions transfer energy to drifting plasma during the expansion phase (Fig. 11-5c). During the storm recovery phase, the electric field fluctuations decrease and the magnetosphere returns to a more balanced state (Fig. 11-5d). During the early and middle stages of substorms, fluctuating global-scale electric and magnetic fields energize particles by inward radial motion, which incrementally drives and traps charged particles into regions of stronger magnetic fields (Sec. 11.2.3).

11.2 Large-scale Magnetic Disturbances

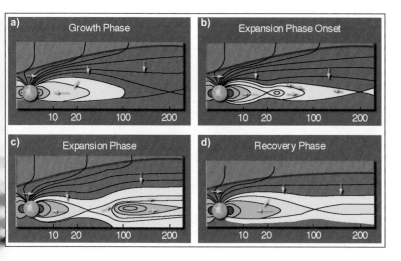

Fig. 11-5. Sequence of Growth, Expansion, and Recovery for a Magnetospheric Substorm. The horizontal axis is in Earth radii. This is the normal sequence for a substorm. During the growth phase, magnetic flux is added to the magnetosphere. Expansion onset is the beginning of the energy release. During the expansion phase, dipolarization occurs, currents flow, particles are energized, and a plasmoid is released. In the recovery phase, the field returns to a more quiet state. [Adapted from Cowley, 1981]

During the expansion phase, dipolarizing magnetic fields produce electric fields that energize particles. These energetic particles drive Earthward into regions of stronger magnetic fields. The dipolarization is also related to a diversion of the cross-tail current into Earth-directed (downward) field-aligned current (FAC) on the dawnside of the magnetosphere, and upward (away-from Earth) FAC on the duskside (Fig. 8-31). These currents close across the auroral ionosphere, creating jets of current (auroral electrojets) and causing significant magnetic disturbances on the ground. At the same time, auroral arcs brighten and expand, creating colorful auroral displays visible from the ground and space. The injection of energetic particles, the Joule heating by the increased ionospheric currents, and the auroral activity account for the dissipation of about 50% of the substorm energy budget ($\sim 10^{15}$ J). The remaining 50% dissipates in plasmoids occurring tailward of the near-Earth neutral line.

Sawtooth events. Sometimes the magnetosphere goes into a quasi-convective state, exhibiting convective and sporadic dissipation characteristics. Strong periodic substorms are related to intermittent and patchy nightside magnetic reconnection and bursts of plasma flow in the mid-tail region. The driving IEF is often near 3 mV/m. Some of the resultant flow-bursts penetrate into the inner magnetosphere, creating plasma injections and field dipolarizations. Plasma observations at geosynchronous altitude often reveal a sudden sharp rise in the number of particles (injections) followed by a slow reduction. This pattern recurs every two to four hours. Several periods of such particle fluctuations look like a series of teeth on a saw and are sometimes called *sawtooth events* (Fig. 11-6). In these global oscillations, the magnetic field in the inner regions stretches strongly and then collapses (crashes) rapidly to a more dipolar state (Fig. 11-7). The field oscillation has a period of two to four hours, with the collapse phase lasting 5–15 minutes. Often the geosynchronous magnetic field is highly stretched at longitudes ranging from dusk to dawn. An enhanced partial ring current is frequently present in the same region.

Chapter 11 The Disturbed Magnetosphere and Linkages Above and Below

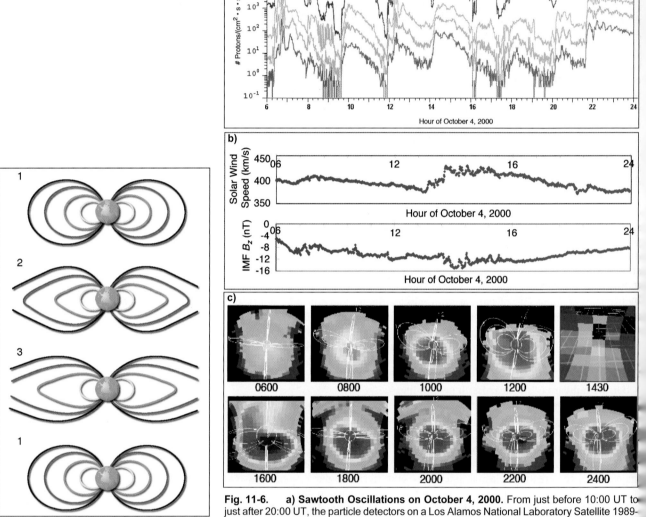

Fig. 11-7. **Global Oscillation of the Geomagnetic Field.** This is a schematic of the dipole-stretch-dipole process during sawtooth oscillations. *(Courtesy of Joe Borovsky at the Los Alamos National Laboratory)*

Fig. 11-6. a) **Sawtooth Oscillations on October 4, 2000.** From just before 10:00 UT to just after 20:00 UT, the particle detectors on a Los Alamos National Laboratory Satellite 1989-046 recorded build-up and decay of various particle populations. The top curve is the flux of particles in the 76 eV–113 eV range in units of protons/(cm$^2 \cdot$s$\cdot$sr). The lower curves correspond to 113 eV–172 eV, 172 eV–260 eV, and 260 eV–500 eV, respectively. **b)** The blue lines plot the solar wind speed and the IMF B_z component measured by the ACE spacecraft. **c)** Ring currents develop during the storm main phase on October 4, 2000. Energetic Neutral Atoms (ENA) images from the IMAGE and POLAR spacecraft from 06 UT to 24 UT on October 4, 2000. Red and brown colors indicate the highest ring current particle flux. The ENA images show fluxes for 16 keV–60 keV particles integrated for 10 minutes and displayed using a logarithmic color scale. One 10-min image is shown each two hours. All images except the one at 1430 are from IMAGE. Between the 12 UT and 16 UT images, IMAGE was in the radiation belts and unable to make ENA observations, but POLAR was well-positioned to make coarser ENA observations (E>37.5 keV) from a similar vantage point. All images are from the northern hemisphere and local noon is located near the top of each image. Pixels with known contamination from scattered sunlight have been blacked out but some photon contamination remains in pixels on the dayside at L>8 and those fluxes should be ignored. [Reeves et al., 2003]

Because of the extremely large amplitudes of the particle and field oscillations, nearly all measurements taken in the magnetosphere show simultaneous signatures of the sawtooth oscillations—that is, the collapses of the field and the particle perturbations are observed at almost all local times. The field-line stretching is sometimes so great that lobe field lines are encountered at the equator at geosynchronous altitude. The sawtooth patterns are normally observed in intense storms (Dst < −100). The oscillations impose periodic impulsive electric fields because of magnetic field polarizations on the background convection electric field. They have a cumulative effect on the ring current development by building high-pressure plasmas in the inner magnetosphere—hence the association with intense storms. Despite having much in common with substorms, these events seem to produce proportionately less energy dissipation in the ionosphere. The ionospheric currents are weaker and the aurora less bright than might be expected for the level of magnetospheric disturbance. Most of the energy appears to be directed into the ring current. Sawtooth events generally occur during geomagnetic storms when the solar wind driving is strong and the IMF is continuously southward for a long time.

A sequence as described above may accompany a more general category of dissipation that includes *aperiodic stormtime substorms*. Fig. 11-8 illustrates a series of substorms within a geomagnetic storm. In the first substorm of the main phase, the aurora expands poleward and equatorward. These events tend to dissipate more energy deeper in the magnetosphere than do isolated non-stormtime substorms. It is not clear that the two categories are really different in terms of their behavior, except that the stormtime substorms are probably driven by a more variable IMF and higher speed solar wind flow. During many geomagnetic storms, the substorms are so rapid that their individual influences cannot easily be determined.

Pause for Inquiry 11-5

Based on the need for stable continuous southward IMF, SMCs are most likely to be associated with

a) High-speed streams

b) Average speed CMEs

c) Slow flow near the heliospheric current sheet

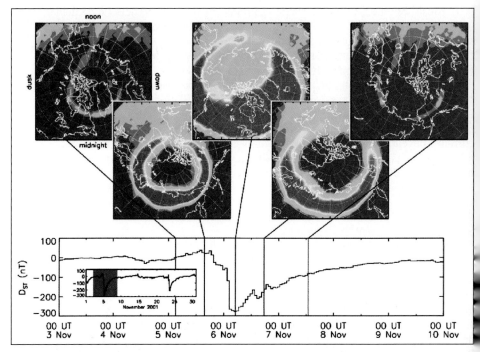

Fig. 11-8. **Aperiodic Substorms within a Geomagnetic Storm.** A sequence of global auroral images from the superstorm of November 2001, captured by NASA's IMAGE spacecraft, traces the development of the auroral oval during a geomagnetic storm and multiple substorms. The progress correlates with the Dst index, a measure of ring current activity. The inset gives the full month of Dst values. [Milan et al., 2008]

11.2.3 Unloading Energy to the Trapping Regions

Objectives: After reading this section, you should be able to...

- Describe how particles become trapped in the inner magnetosphere
- Describe how trapping contributes to the ring current
- Describe how trapping contributes to the outer radiation belt
- Describe how the plasmasphere responds to storm-time convection

Variable Electric and Magnetic Field Contribution to the Trapping Region

The storm-time combination of static and varying magnetic, and convective and inductive electric fields helps trap charged particles in the *trapping region* which includes the radiation belts, ring currents, and plasmasphere (Figs. 7-13 and 7-14). These individual regions also store and dissipate storm energy, as we describe in the next section. Figure 11-9 provides a large-scale view of the trapping region. The solar wind is an important, if not the dominant, source for particles in the trapping regions. However, during intense storms, atomic oxygen from the ionosphere becomes an important constituent.

11.2 Large-scale Magnetic Disturbances

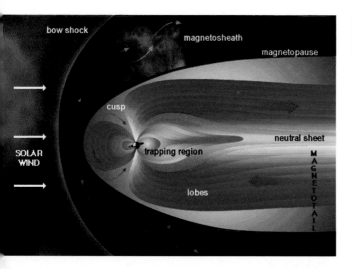

Fig. 11-9. **The Regions Where Particles Are Strongly Influenced or Trapped by Variable Electric and Magnetic Fields.** The orange area represents the region of strong influence. *(Courtesy of the European Space Agency)*

From the previous section, we know that storage and then energy release via bursts of reconnection may heat plasma, accelerate plasma in bulk, and force charged particles to cross gradients of magnetic field lines. Some of the energy from reconnection travels to the innermost portions of the magnetosphere, where the magnetic field is strong and the plasma energy density is not yet saturated. The stronger the storm, the deeper in the magnetosphere this energy storage occurs. Energetic plasma particles may be trapped in Earth's curved field as either ring current particles or radiation belt particles, depending on their energies. Some of these particles are eventually lost to the ionosphere. Not all of the plasma is trapped. However, even the free plasma that drifts toward the dayside and exits the magnetosphere does so at higher speed, thus recycling energy and mass to the boundaries of the magnetosphere.

Particle Drift in Earth's Magnetic Field

Charged particles drifting close to Earth experience gradients in the magnetic field that cause their paths to deviate from straight flow. Where the field stretches, but remains relatively uniform, the drift effect is the same for positive ions and electrons. The more energetic the plasma, the closer the penetration to Earth and the more deviated the paths become. Some particles reach the vicinity of the plasmapause and become trapped in a flow trajectory around Earth. Upon reaching the inner magnetosphere, the charged particles experience drifts from the curvature and gradient in the magnetic field that separate the ions from the electrons. The separation produces the ring current signature associated with short-term particle trapping.

Figures 7-22 and 11-10 provide a perspective on single-particle trapping in Earth's inner magnetic field. Consistent with a quasi-dipole structure, the field strength increases close to Earth. Charged particles driven Earthward by convection are influenced by the field gradient and separated according to sign. These particles, under the right conditions, may completely encircle Earth, becoming trapped for some interval of time that depends on their mass, energy, and momentum and on the variability of the local magnetic field. During storm time, medium-energy particles (10s of keV) are often present in sufficient quantity to produce a net charge flow associated with the ring current. Hotter particles (100 eV–few MeV) also participate in this charge separation, but their low density and more erratic motion do not produce a noticeable current signature.

Chapter 11 The Disturbed Magnetosphere and Linkages Above and Below

Fig. 11-10. A Duskside View of the Gradient Drift Motion in the Magnetosphere in Fig. 7-23. Particles flow toward Earth from the right. Ions follow the green path, and electrons follow the blue path. Some particles completely encircle Earth, while others experience severe path deviation. We don't show the inner magnetospheric field lines and details of other particle motions.

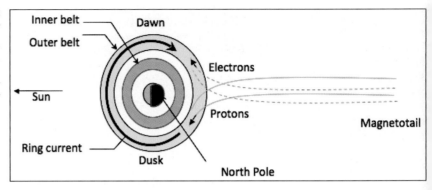

We now quantify the drift description in discussions surrounding Eqs. (6-14b) and (6-15a). Equation (11-3) consolidates these:

$$v_D = \underbrace{\frac{E \times B}{B^2}}_{(1)} + \underbrace{\frac{m(2v_\parallel^2 + v_\perp^2)}{2q} \frac{B \times \nabla B}{B^3}}_{(2)} \quad (11\text{-}3)$$

where

v_D = charged particle drift velocity [m/s]
E = electric field [V/m]
B = magnetic field [T]
m = particle mass [kg]
q = particle charge [C]
$v_\parallel$ = particle velocity component parallel to the magnetic field [m/s]
$v_\perp$ = particle velocity component perpendicular to the magnetic field [m/s]

Drift motion has two contributors: 1) the $E \times B$ convective drift and 2) drift caused by energetic particle motion in magnetic field gradients. If electric fields are absent or if their effect is small compared to the motion of the hot particles, we can drop the first term. If the particles are of low energy (cold), we can drop the second term and the particle drift is dominated by E. Enhanced electric fields and other factors that increase particle energy increase the drift speed of charged particles and allow them to penetrate into regions near Earth that they would otherwise not get to. Co-rotating electric fields trap cold particles from the ionosphere. The bulk of the high energy particles are trapped in the magnetic bottles of the dipole field lines. By energy conservation, these particles must travel on the shells of constant B fields.

Pause for Inquiry 11-6

Which region of Fig. 11-10 is dominated by term 1 of Eq. (11-3)? Which part is dominated by term 2?

We recall from Chap. 7 that an adiabatic invariant is a property of a physical system that stays constant when the system changes slowly. We say the adiabatic invariants are violated during the rapid variations of a magnetic storm. Conservation of the magnetic moment μ, the longitudinal invariant J, and the enclosed magnetic flux invariant Φ, break down whenever the magnetosphere is disturbed on time scales shorter than or comparable to the gyro, bounce, and drift periods respectively.

Geomagnetic disturbances with periods ranging from milliseconds to hours are measured deep in the magnetospheric trapping regions. Violation of the invariants is most likely under resonant conditions; that is, when the period of a particular geomagnetic disturbance coincides with one of the characteristic times associated with the gyromotion, the bounce motion, or drifting in the magnetosphere.

Long-period waves ($\tau \sim$ hours) produce a violation of Φ that allows particles to radially diffuse across field lines. *Radial diffusion* is a result of conserving the ratio of energy to magnetic field strength. As particles move inward to regions of stronger magnetic fields, their energy increases, so they must tap some external energy source. The convection electric field serves as this source. This process is thought to be an important source mechanism for injecting particles from the magnetopause boundary into the region of trapping.

Shorter period waves generally maintain the Φ invariant because their effects average out over a particle drift period. They may however, change μ or J (depending on the wave frequency), which in turn changes the mirror altitude. If the altitude is lowered sufficiently, the trapped particles undergo collisions in the denser regions of the upper atmosphere and leave the radiation belt. This loss process is important for high-energy particles in the radiation zone. Trapped particles continuously precipitate into the atmosphere, causing auroral emissions at high magnetic latitudes. This allows for radiation belt monitoring from LEO.

Local-scale processes also lead to diffusion and loss. These include rapid local acceleration by the large inductive electric fields caused by rapidly changing substorm magnetic fields, and energy exchange between waves generated in the inner magnetosphere and charged particles trapped along the magnetic field lines. Resonant interactions may lead to an energy exchange between waves and particles. Typically, the trapped particles provide the energy source to excite geomagnetic wave phenomena. However, under certain circumstances, the waves pass energy to or accelerate the particles. Table 11-4 describes the variety of resonant processes in the magnetosphere.

Convection and radial diffusion inject the particles into the trapping region. Inward convection is generally restricted to the region outside the plasmapause. Even though the plasmapause moves toward Earth during a geomagnetic storm, the freshly injected energetic radiation belt protons and electrons are effectively excluded from the high-density plasmasphere. Radial diffusion subsequently carries energetic particles across drift orbits and thus acts as an effective source for hot particles in the trapping zone. But the time scale for this transport process is very long, accounting for the long-term stability of the inner-zone particle population.

The Ring Current During Magnetic Storms

The ring current is the transient feature most readily associated with magnetic storms. Scientists still argue over the exact nature and extent of this current system, but the general configuration is understood well enough to explain the penetration of the solar wind energy deep in the magnetosphere and to serve as the physical basis of the Dst index (Sec. 11.1).

Chapter 11 The Disturbed Magnetosphere and Linkages Above and Below

Table 11-4. Acceleration-Diffusion Mechanisms. This table lists various mechanisms that affect the charged particles in Earth's magnetosphere and create changes to the invariants of particle motion. We recall that the pitch angle describes the ratio of motion perpendicular to the magnetic field line to the motion parallel to the magnetic field line ($v_\parallel / v_\perp$). Here μ is the magnetic moment; J is the longitudinal invariant; Φ is the enclosed magnetic flux invariant; v is velocity; VLF is very low frequency; ELF is extremely low frequency. [after Roederer, 1970]

Effect	Violates	Changes	Causes
Magnetospheric compression	Φ	v	Radial diffusion
Drift acceleration by recurrent ionospheric or magnetospheric electric fields	Φ	v	Radial diffusion
Bounce acceleration by micropulsations or electric space charge	J	v	Radial and pitch angle diffusion
Resonant cyclotron acceleration by VLF or ELF waves	μ and J	Pitch angle	Pitch angle diffusion
Random coulomb interactions with the atmosphere	μ and J	Pitch angle	Pitch angle diffusion

EXAMPLE 11.4

Energy Requirements for Crossing L Shells

- *Problem Statement:* Estimate the ratio of energy increase for a charged particle moving across L shells toward Earth in the inner magnetosphere.

- *Relevant Concepts:* L shell, spatial variation of a dipole field, the first adiabatic invariant

$$|\mu| = \frac{\frac{1}{2}mv_\perp^2}{B} = \text{constant}$$

  where $v_\perp$ is the component of velocity perpendicular to the magnetic field.

- *Conceptual Solution:* If B increases, the velocity must increase to maintain μ as a constant.

In the radial direction, B varies as $1/r^3$.

As the particles move to lower L shells, the field magnitude rises by r^3.

To conserve μ, $v_\perp$ must rise by a factor of 3/2 power.

- *Implications:* Particles drifting across dipole field lines must have a significant source of energy. During storm time, this energy comes from the convective electric field that acts as an accelerating force or from inductive electric fields that act as accelerating forces or compressional forces.

In this example, we recognize that charged particle velocity perpendicular to the magnetic field may be different than the charged particle velocity parallel to the magnetic field. This situation develops because force(s) acting on charged particles differ along and perpendicular to the magnetic field.

Follow-on Exercise: Does the parallel velocity component have any role in this process?

Main Phase, Particle Build-up. The source for 10 keV–100 keV storm-time ring current ions is the acceleration of plasmasheet particles by the local-scale inductive and global-scale convective electric fields. During quiet time, the plasma for the ring current is present in the plasmasheet, but at different locations and a lower energies. It tends to drift from the magnetotail past Earth on open field lines Occasional substorms deliver energy that may produce a slight ring current signature. Under southward IMF conditions, moderately energized particles acted on by the imposed electric field are brought into the trapping region. The trapping effect is largely created by the gradient in the magnetic field. Because of fluctuating fields, some particles approach *L*-values of 4 or less (Fig. 11-11). Very strong electric fields move plasma Earthward from the magnetotail and draw heavy ions (primarily oxygen) from the ionosphere to populate the ring current. Thus begins the expansion of the ring current in a storm's main phase.

Pause for Inquiry 11-7

Why is the ring current considered to be a diamagnetic current?

Recovery Phase, Particle Loss. The storm recovery phase begins when the southward component of the IMF weakens, and the ring current decays. The rate of energy increase and inward transport slows, and the various loss processes that remove plasma from the ring current begin to restore it to its pre-storm state. As Fig. 11-1a shows, the storm recovery phase is typically much longer than the main phase. With the build-up of medium-energy particles around Earth, the thermal energy density rises ($P = nk_BT$), and nature tries to overcome the pile-up with a pressure gradient force directed away from the high-density regions. However, the ring current particles are trapped in gyromotion around magnetic field lines and only move out of the region under the influence of a large enough force to break the magnetic field's hold on them. This force may be a pressure gradient force, a collisional force, or a transient force from a wavy field. One other possibility—the charged plasma particle exchanges its charge with a more massive neutral particle. For example, a ring current ion acquires an electron from a cooler neutral hydrogen or oxygen atom, and so becomes an energetic neutral hydrogen atom, while the heavier particle becomes ionized. In doing so, the hotter particle becomes neutral and is no longer bound to the field. The cooler particle is less bound to the magnetic field and also escapes.

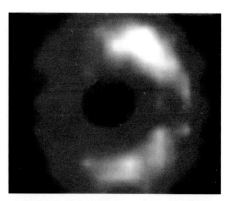

Fig. 11-11. Emissions for Neutral Atoms Escaping Earth's Ring Current. An image of Earth is superimposed. The image was taken by the High Energy Neutral Atom (HENA) imager onboard the NASA IMAGE spacecraft. The current is not smooth, and often does not completely encircle Earth's equatorial zone. It is more prominent on the nightside (right in this figure), and as it moves into the dayside, it breaks up and vanishes, possibly by losing particles into the magnetopause region. *(Courtesy of NASA)*

Pause for Inquiry 11-8

Return to Eq. (11-3) and convince yourself that the second term on the right side becomes less important if particle energy is low.

Asymmetries and Irregularities. Although the ring current is often displayed as circular (Fig. 11-3), the disturbance field is generally not axially symmetric. It is often lumpy and asymmetric as indicated by the bright regions of Fig.11-11. The asymmetry or lumpiness of the ring current creates pressure increases that nature tries to eliminate. Along some longitudes the plasma pressure may be sufficient for some of the lower energy particles to begin flowing along field lines into the low-latitude auroral zones, producing localized currents and heating. The influx of these charged particles changes the conductivity and the response of the ionosphere to storm-time electric fields.

Further, magnetic merging, drift, and particle heating and acceleration are unsteady. Certainly the primary driver of the main storm phase, the interplanetary electric field responsible for the magnetosphere convection electric field, is subject to tremendous fluctuations. Additionally, activity in the magnetotail is punctuated by intense substorms that superpose intense electric fields in time and space onto the storm-time convection electric field. Non-homogeneous plasma, as well as stretching and relaxing of the magnetic field, adds to the level of storm-time variations.

The Plasmasphere during Magnetic Storms

The plasmasphere was directly observed as a separate entity by satellite particle detectors in the early 1960s. In the quiet-time picture arising from many studies, the enclosed volume confines cold (few eV) plasma in a torus. Figure 7-31 shows views of the plasmasphere. The plasma confinement results from a co-rotating electric field, as we describe in Chap. 7. The plasmasphere dominates the mass density of the inner magnetosphere, with densities as high as 1000 particles/cm^3. The mass (~80% ionized hydrogen and 20% singly ionized helium) comes from the outflow of ionospheric plasma along subauroral magnetic field lines at magnetic latitudes of ~60° and less. Most of the time the plasmasphere occupies the same region of the inner magnetosphere as the ring current and radiation belts (between L = 2–7), with the strongest concentration of plasma inside L = 4. During quiet times, the plasmasphere is distinguished from the ring current and radiation by particle energy.

The edge of the co-rotation region is the boundary for the convecting magnetospheric plasma (Fig. 7-31). At this boundary, the interaction between these two flow regimes, one flowing Sunward, the other co-rotating, together with the outflow of plasma from the ionosphere, determines the size, shape, and dynamics of the plasmasphere. This varies strongly according to the level of magnetospheric activity. Recent satellite observations clearly reveal that the spatial extent of the plasmasphere is governed by magnetospheric electric field time history.

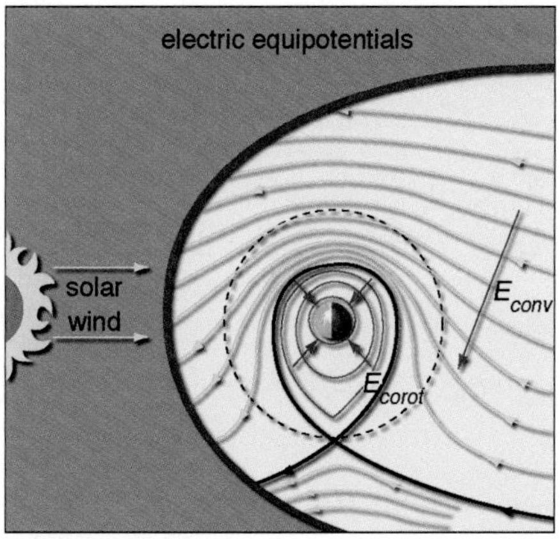

We recall from Chap. 7 that in the inner magnetosphere we algebraically add the electric fields produced by convection and co-rotation to get the total electric field and its equipotentials. The addition reveals two types of equipotential curves, as shown in Fig. 11-12. Some curves are closed loops around Earth, where cold plasma is trapped, and the others are lines beginning in the tail and curving around Earth to the dayside magnetopause, where warmer plasma drifts. The confinement of cold plasma results from trapping on magnetic field lines that rotate with Earth. The singular line (bold black in Fig. 11-12) separating the two types of stream lines is the *electric field separatrix*. This curve has a null on the dusk meridian where the convection and co-rotation electric fields are equal and opposite. At this null, $E = 0$ and plasma doesn't drift, so we call it the *stagnation point*. The duskside stagnation region is sometimes a short-term plasma build-up region. Streamlines toward dusk from the separatrix go around Earth on the duskside, while the opposite is true for all streamlines on the dawnside of the separatrix. Most of the cold plasma in the dayside magnetopause from the tail goes around the dawnside.

Data from multiple spacecraft, especially the IMAGE satellite launched in 2000, along with a ground-based comparison of GPS signals, brought a new perspective on a region that scientists thought was rather inert. The data revealed that the plasmasphere often sheds some of its mass to the magnetosphere and even to the solar wind in long structures called *drainage plumes*. Figure 11-13 shows the development of a storm-compressed magnetosphere with a drainage plume diverting plasma to the dayside magnetopause.

Satellite data in the early 1970s suggested that the plasmasphere deformed depending on the strength of the convection electric field. When the field increases in strength, the flowing plasma invades regions of space formerly dominated by the plasmasphere and squeezes the plasmasphere closer to Earth. The region occupied by the plasmasphere indicated in Fig. 11-13a may be reduced by 50% or more. And some of the plasma at the plasmapause is sheared off into orphaned streams or plumes of cold dense plasma (Figs. 11-13b and c). Immediately adjacent to the plume may be a trough of plasma. These streams are

Fig. 11-12. Equatorial Cross Section of the Global Magnetosphere with Earthward Flowing Magnetoplasma. The view is from above the North Pole. Densities are 100s–1000s particles/cm^3. Earth's magnetic field points out of the page. Curves of electric equipotential are shown in yellow. An electric field, E_{conv}, causes particles to drift Sunward. Inside the separatrix (black line) that loops around Earth is the plasmasphere. This region is dominated by the co-rotation electric field, E_{corot} (red arrows). In this region, the magnetic field and attached plasma co-rotate with Earth. The dashed circle indicates the geostationary orbit. [After Kavanagh, 1968]

plumes may wrap around each other, forming intricate plasma structures that scintillate radio signals, such as those from GPS. Statistical studies indicate that plumes and tails are present more than one third of the time, suggesting that merely enhancing convection is sufficient to redistribute plasmaspheric mass.

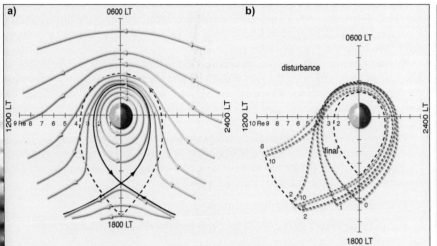
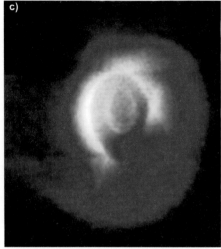

Fig. 11-13. a) and b) Plume Development and Plasma Loss from the Plasmasphere. Here we see the development of a dayside plasma drainage plume. In **a)** the quiescent plasmasphere has plasma trapped in the closed curves and a flow stagnation region near the dusk flank. In **b)** an energized plasma moves toward the dayside under the influence of strong convection. [Grebowsky, 1970] **c) The Plasmasphere during Storm Time.** This ultraviolet image of the plasmasphere viewed from above Earth's North Pole by the IMAGE spacecraft shows the glowing dayside of Earth and an arm of the plasmasphere pointing roughly toward the Sun. The Sun is off the top right of the image. Earth's shadow points away from the Sun. The bright regions in the center are auroral rings over northern Earth, visible as northern lights from the ground. The plasma arm has swung Sunward with convecting plasma from the magnetotail. Inside the arm is a plasma trough. *(Courtesy of NASA)*

Plumes are long-lived (days) with large longitudinal (> 90°) and latitudinal (observed at 50°+) extent near the main plasmasphere. Further from the main structure, plumes exhibit plasmaspheric-like densities and plasmapause-like gradients with large density variability (fine-scale structure) and occupy less than 30° of longitude at geostationary altitudes.

Storm-time behavior of the plasmasphere creates clearly distinguishing characteristics in addition to plumes and tails. Waves also form at the convection co-rotation boundary. In the region of overlap between the ring current and the plasmasphere, two plasma populations interact with each other via multiple processes, such as wave-particle interactions and Coulomb collisions. The ring current ions are a source of free energy to excite plasma waves. This energy, in turn, is redistributed among the thermal and energetic populations as the plasma waves undergo damping. Energy contained in the ring current also transfers to the plasmasphere through Coulomb collisions. Magnetic field lines, threading the disturbed plasma, begin to vibrate in response to plasma turbulence. These vibrations can be sensed at the ground.

The energy addition heats thermal ions in the plasmasphere during a storm. The region of high heating rate corresponds to the location of the peak density of <10-keV ring current ions that provide the main contribution to the plasmaspheric heating. Heating from the ring current increases the ion temperature at altitudes above about 500 km. The ion temperature may continue to rise for more than two days after storm onset. A two-to-five-fold increase in temperature occurs compared with no heating from the ring current. The ring current is also affected by the interaction. As the warm ring current ions drift through the plumes, they may scatter away from their drift paths and precipitate through the plasmasphere and into the atmosphere.

Only recently have researchers discovered the storm-time impacts of this relatively dense region. Extreme structuring of the plasmasphere adds and subtracts total electron content (TEC) between GPS satellites and ground receivers. When the TEC fluctuates, so do the GPS signal strength and propagation properties. Plasmaspheric drainage plumes have been implicated in some severe GPS signal scintillations.

Waves generated at the plasmasphere's boundary influence the behavior of plasma in the ionosphere, the plasmasphere, and the ring current and may cause energized plasma from the ring current to precipitate into and heat the ionosphere. The resulting effects on ionospheric chemistry and dynamics have yet to be fully explored.

Data provided by the IMAGE spacecraft and by GPS signal analysis reveal that sometimes the plasmasphere is smooth, sometimes it is corrugated, and other times it is notched. Whatever its structure, it is certainly a player in storm-time dynamics. The plasmasphere is now known to

- Contribute particles to the plasmasheet
- Modulate fluxes of energetic particles and the intensity of spacecraft charging
- Influence propagation and generation of ultra-low-frequency (ULF) waves
- Influence particle-particle and wave-particle interactions
- Contribute up to 50% of total electron content (TEC) and therefore influence GPS signal paths
- Vary its composition with geomagnetic activity by adding and losing oxygen ions (O^+)

The Radiation Belts During Magnetic Storms

For the most part the radiation belts must be studied from space, so radiation belt science is relatively young (since 1958). Polar LEO satellites cut through the low altitude cusps of the belts (Fig. 11-14). HEO and GEO satellites study the radiation belt particles *in situ*. Data from LEO, HEO, and GEO satellites produce the static view of the belts shown in Fig. 11-14.

Inner Radiation Belt Storm Response. The permanent inner radiation belt is generated by cosmic rays whose journeys begin with the births and deaths of distant stars. These energetic particles collide with atoms in Earth's atmosphere and produce showers of secondary products, some of which are directed back to space as albedo particles. Neutrons created in the collisions subsequently decay into energetic protons that are easily trapped on geomagnetic field lines close to Earth. The inner zone's stability results from a combination of long particle lifetimes in the strong magnetic field, and the slowly varying cosmic ray inputs. On short time scales, the inner belt is relatively insensitive to all but the most extreme of solar wind forcings.

Trapped energetic protons (> 10 MeV and located at < 2.5 R_E) vary on the time scale of the solar cycle. In very rare circumstances, an extreme shock in the solar wind enhances the particle population on the timescale of a minute (Focus Box 11.2). Scientists think the process for such events is prompt acceleration by interplanetary solar wind shocks that hit the magnetosphere and induce intense wave fronts that

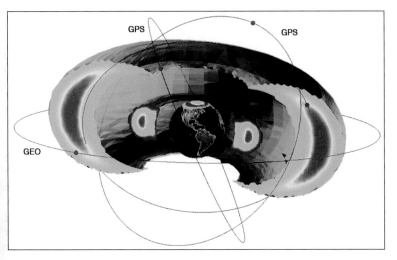

Fig. 11-14. Climatological View of Inner and Outer Radiation Belts. Here we show how the belts encircle Earth. The cross-sectional areas show the relative charged particle density. Higher particle density is indicated in red; lower particle density in green. Between the inner and outer belts is the low-density slot region. During storms, the outer belt expands inward and outward. The outermost region often includes geosynchronous orbit. The constellation of GPS satellites frequently fly through the heart of the outer radiation belt. The bright bands in the northern hemisphere polar region are where particles from the outer belt precipitate into the atmosphere along the loss cone of the geomagnetic field lines. The southern hemisphere has a similar region. *(Courtesy of the AF Research Laboratory)*

surge through the inner magnetosphere (line 1 of Table 7-2). The excited particle populations form new radiation belts whose lifetimes extend from months to years.

Less energetic ions from the Sun or shocked solar wind create a transient population of inner radiation belt ions (energies 5 MeV–10 MeV) that may be trapped for months at a time following energetic eruptions from the Sun. These slightly lower energy particles are more weakly trapped than most of the cosmic ray-induced ions. Weakly trapped energetic protons above an altitude of about 3 R_E respond to storm energy input associated with variable convection electric fields, and may take several weeks to months to return to pre-storm levels (line 2 of Table 7-2).

Particles from the inner belts are so damaging that astronauts and spacecraft avoid, or rapidly transit, the region to the best of their ability. However, some of these particles precipitate into Earth's atmosphere in regions where the field is weak, mostly in the South Atlantic Anomaly (SAA) region. Storm-time enhancements of the inner radiation belts cause excess radiation flux to LEO spacecraft orbiting through the SAA (line 5 of Table 7-2).

Outer Radiation Belt Storm Response. Variations in solar wind-magnetosphere coupling influence the trapped energetic electrons and protons (hundreds of keV to a few MeV) in the outer radiation belt. The outer belt is usually associated with a peak of energetic electron flux at 4 R_E–5 R_E (Fig. 11-14). This belt varies with the solar cycle, season, solar rotation, and most notably with magnetospheric storms. The slot region between the inner belt and outer belt, where fewer particles reside because of constant interactions with waves, may be filled during active times. Because activity is higher during solar maximum, the tenuous particle population in the slot region depends on the solar cycle.

Figure 11-15 shows an 18-month record during the rise of solar cycle 23 of electrons at three energy levels in the outer radiation belt. The early portion of the record displays modest increases in radiation belt fluxes associated with substorm enhancements during the passage of high-speed streams in 1997. Following seven coronal mass ejections and two X-class flares in early May 1998, the population of relativistic electrons in Earth's radiation belts rose dramatically.

Particles trapped in the ring current and the outer radiation belt have a common energizing source in the storm-time convection electric field that drives plasmasheet particles inward toward Earth (line 2 Table 7-2). Data from recent satellite missions indicate that the potential outer radiation belt particles have

Chapter 11   The Disturbed Magnetosphere and Linkages Above and Below

Fig. 11-15. Relative Fluences of Energetic Electrons from January 1977 to June 1998. Before May 1998, the fluences of all energies of relativistic electrons are low, indicative of solar minimum. The lower energy electrons have larger fluences. As a result of the consecutive doses of radiation in the May 1998 event, NASA's Polar spacecraft was upset to the point of being shut down for several hours. [Baker et al., 1998]

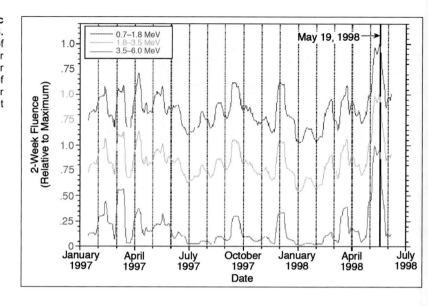

slightly more energy to begin with and are more susceptible to energizing in the middle and far magnetotail. The radiation belt particles in L shells 4–5 are very energetic and less dense than ring current particles. Low density and highly random motions make radiation belt particles incapable of carrying a substantial current. Thus, even though these energetic particles may occupy some of the same region as the ring current particles, they are usually not current carriers.

The intensity and the structure of the relativistic electron belts are controlled by a balance of transport, acceleration, and losses. The Earthward rush of electrons created in the substorm process (substorm injection events) moves particles from open to closed drift orbits. Assuming the magnetic moment μ is conserved, particles moving into regions of stronger magnetic field require an increase in the perpendicular velocity component, which translates to an increase in particle energy (Example 11.4). This comes at the expense of work done by rapidly changing the overall magnetospheric field configuration. Inward radial transport (*radial diffusion*) of electrons increases the flux of 90° pitch angle electrons. Waves created by the motion of the hot electrons produce electron scattering to other pitch angle values. Some are further heated by the waves. Because of their wave-like nature, substorm and sawtooth oscillation behavior is more likely to drive radial diffusion than is steady convection. The ring current may act as an additional source of particles for the outer belt by inflating the magnetic field outside of the ring current. The result is an increase in field value to which the particles must adjust. As the field strengthens, particles drift and shells move outward. Particles originally drifting at 4 R_E may move to 5 R_E as they attempt to remain at a fixed value of ***B***.

Loss mechanisms include precipitation into the atmosphere or escape through the magnetopause (lines 1 and 5 of Table 7-2). The latter situation may develop when the outer radiation belt inflates because of storm processes associated with the passing of a CME sheath region with substantial southward IMF. Arrival of the main body of the CME or an additional shock compresses the magnetopause into the region occupied by some of the radiation belt particles and they escape into the solar wind.

Another loss mechanism involves hot particles interacting with plumes of cool particles in the plasmasphere. The interaction produces scattering waves. Waves inside the plasmasphere scatter electrons into the pitch angle loss cone and thus into Earth's upper atmosphere. This loss controls the inner edge of the outer electron belt. Figure 11-16 shows how a LEO satellite could monitor outer radiation belt particles that scatter into the loss cones that map to polar latitudes.

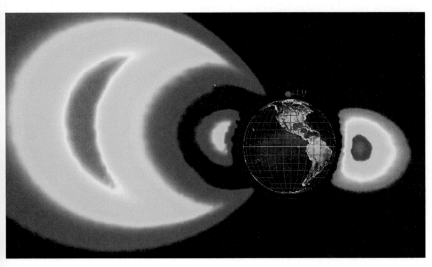

Fig. 11-16. Monitoring Earth's Radiation Belt Particles with Low-Earth Orbiting Satellites. Earth's electron belt (left) and proton belt (right) as simulated by the empirical radiation belt models available from NASA. NASA's SAMPEX satellite provided this type of monitoring for over a decade (Focus Box 11.1). Minor-to-moderate disturbances enhance the drizzle of outer radiation belt particles at high latitudes. Only extreme storms enhance the drizzle of the inner radiation belt particles at low latitudes. *(Generated using the AF-GEOSpace software of the AF Research Laboratory. Courtesy of Munther Hindi at the Lockheed Martin Company.)*

We understand each of these processes in a fairly straightforward manner when we view them as isolated energizing mechanisms. If, however, all of the processes respond to the transient energy input from the solar wind during geomagnetic activity by varying degrees, our understanding of which mechanism dominates is severely challenged.

Asymmetries and Irregularities. The hot particles of the radiation belt tend to have smoother distributions than those in the ring current. However, during some storms, a global dropout of sections of the outer radiation belt develops and lasts 1–2 days. Some electrons previously mirroring in the inner magnetosphere travel on perturbed magnetic field lines that dump the electrons into the upper atmosphere or scatter them back to the magnetosphere. In storm events that begin with a sudden compression of the magnetosphere, the outer belt electrons may be forced into the solar wind as the magnetopause establishes a new equilibrium position closer to Earth. The lost particles are quickly replaced by the processes described above. If the dropout is followed by an event that contains prolonged follow-on, high-speed solar wind flow, then the energetic electrons may reappear with even higher fluences and higher energies. Such events require much vigilance on the part of space weather forecasters and satellite operators whose spacecraft may be subject to severe charging events. Focus Box 11.1 relates an event that was so severe it surprised even veteran space weather observers.

Pause for Inquiry 11-9

Figure 11-17 shows a record of 2-MeV electrons from 1992–2001. Pick the solar minimum in this radiation belt record based on the Dst index shown in black. Estimate when solar cycle 23 began. Find intervals when the slot region was filled.

Chapter 11   The Disturbed Magnetosphere and Linkages Above and Below

Focus Box 11.1:   Reading the Radiation Belt Record

As we discuss in Chap. 7, the L-shell parameter is an important organizing tool for studying the trapping region and also useful for displaying data. Below is a record of the radiation belt history from the SAMPEX spacecraft from 1992 to 2001. We show the approximate orbit geometry for SAMPEX in Fig. 11-16. Each day, SAMPEX made multiple trips through the high-latitude regions where a small fraction of energetic particles from the radiation belts bounce to an altitude at or lower than the spacecraft's altitude and are measured by the particle detectors on board. Most of these detectors measure particles that precipitate into the atmosphere.

In the top panel of Fig. 11-17 is information about the fluence of energetic protons (19 MeV–27.4 MeV) in the inner radiation belt. This display is a simple line plot over time. The proton flux is for L values between $L = 1.33$ and $L = 1.42$. In the same plot, but with a scale on the right, is the sunspot number plotted in black.

In the figure's bottom panel, data from the detector that is sensitive to the 2-MeV electrons are shown. The record begins as the spacecraft transits from the North Pole to the equator. It first encounters 2-MeV particles precipitating from high L values, (distant inner magnetotail), next from lower L values, then from $L \approx 2.5$; it finally moves into L values so low that few energetic electrons are present for measuring. The satellite stores the electron flux record for this partial orbit. Then it proceeds to the southern hemisphere where it again samples through the L shell but in reverse order. The sampling repeats twice more on the return to the North Pole. After all orbits for the day have been completed, the flux values for all L shells are processed and averaged and then mapped in a strip, top to bottom, as in Fig. 11-17. This process proceeds day after day and slowly a picture begins to emerge.

Usually the most intense flux of 2-MeV electrons is between $L = 3.5$ and $L = 5$. Figure 11-17 shows this flux in red. Reduced fluxes are shown in other colors. Occasionally, we see intervals when the high fluxes extend beyond $L = 5$ and as deep as $L = 2$. In late 1996, the 2-MeV particles were absent from the radiation belt, but lower energy particles (not shown) were still there.

The black curve in the lower panel of Fig. 11-17 is the daily Dst index. The closer correspondence between the flux of 2-MeV electrons and the Dst index suggests that these have a common driver in the storm-time magnetosphere.

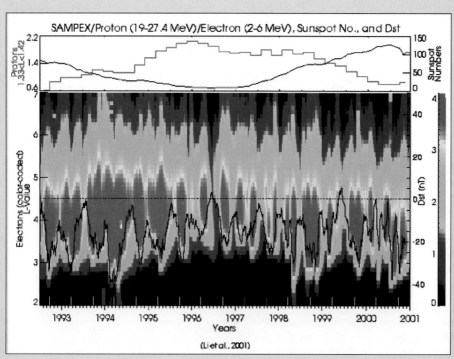

Fig. 11-17.   Ten Years of Radiation Belt Particle Data from SAMPEX. The top panel shows proton flux in the inner radiation belt from 1992–2001 in red and the monthly sunspot number in black. The proton energy range is 19 MeV–27.4 MeV. The bottom panel gives the electron flux that SAMPEX satellite observed in polar orbit during the same years. The electron energy range is 2 MeV–6 MeV. The L shell of the observation is shown on the left vertical axis. The color bar on the right shows the logarithm of electron number flux in units of particles per cubic centimeter per second. The satellite sampled in the latitude regime shown in Fig. 11-16. The Dst index is plotted in black. Its value in nT is shown on the right vertical axis. *(Courtesy of Xinlin Li at the Laboratory for Atmospheric and Space Physics, University of Colorado, Boulder)*

Focus Box 11.2: Shock Radiation Belt Behavior

(Courtesy of Xinlin Li at the University of Colorado and Mary Hudson at Dartmouth University)

One of the most notorious space weather events of the modern era developed from a solar active region at 26° S, 28° E at 2246 UT on March 22, 1991. The bright optical flare (3B) was accompanied by radio, X-ray and gamma-ray emissions, and a large coronal mass ejection. A few hours later at 0245 UT on March 23, a long-duration, soft X-ray event occurred in the same solar active region. Energetic solar particles began to arrive at Earth around 0730 UT on March 23, ahead of a shock moving in excess of 1400 km/s. A strong shock reached Earth just before 0342 UT on March 24, defining an extreme sudden storm commencement (SSC) that drove the magnetopause to deep within the geosynchronous orbit. Within nine hours, the Dst index dropped below −300 nT.

Less than one minute after the SSC, a new radiation belt with electron energies in excess of 13 MeV formed in the slot region (Fig. 11-18b). A 13-MeV electron gyrates at ~99.9% of light speed. Lower energy, 6-MeV electrons filled much of the inner magnetosphere (Fig. 11-18a). Additionally, energetic heavy and light ions of coronal origin were observed in the inner magnetosphere for an extended interval after the SSC.

The interplanetary shock severely compressed the magnetosphere and resulted in a huge inductive electric field that energized pre-existing electrons (1 MeV–2 MeV) at larger L levels and brought them into the lower L (=2.5) region (the slot region) (Fig. 11-18).

The magnetic field measurements from the event show mainly one pulse, corresponding to compression and relaxation. The electric field shows mainly two pulses. The first is associated with compression and the second one with relaxation. As the magnetic field varied, it generated an

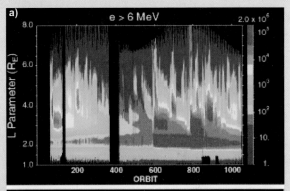

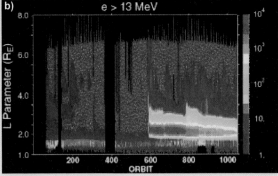

Fig. 11-18. **Radiation Doses from 6-MeV and 13-MeV Electrons.** In each plot the vertical axis is L value and the horizontal axis is days after launch. The highest dose rates are in red. [Blake et al., 1992]

inductive electric field that also propagated through the magnetosphere. The electric field selectively accelerated some pre-existing particles whose drift speed happened to be comparable to the wave propagation speed. The combination of magnetic and electric field variations inside the magnetosphere was probably the energy source for the accelerated electrons. Energetic protons were also injected into the low L region. These protons had a different energy source. They are mostly the solar energetic particles that temporally filled the magnetosphere at larger L shells. Solar energetic particle (SEP) events observed in the magnetosphere are usually caused by transient particles. These energetic ions have large gyro-radii and cannot be trapped by the magnetosphere. However, if a strong interplanetary shock follows them to Earth, some of them will be pushed into much smaller L shells and become trapped.

The whole process of new radiation belt formation took about one minute. Some particles remained trapped for years. Clearly, satellites that operate in the inner magnetosphere, those in HEO, MEO, and GEO orbits, need to be sufficiently radiation-hardened to survive such long-lived storms. Modeling the dynamics of such an extraordinary events remains at the forefront of space weather research.

Chapter 11 The Disturbed Magnetosphere and Linkages Above and Below

11.3 Field and Current Coupling during Magnetic Storms

11.3.1 High-latitude Magnetosphere-Ionosphere Coupling

> **Objectives:** After reading this section, you should be able to...
> - Explain the importance of field-aligned currents during magnetic disturbances
> - Distinguish between Region-1 and Region-2 currents
> - Relate field-aligned currents to horizontal currents in the ionosphere
> - Distinguish between Hall and Pederson currents
> - Explain the origin of the auroral electrojets
> - Explain the origin of the substorm current wedge
> - Distinguish between the substorm current wedge and the partial ring current

The configuration of the geomagnetic field requires three major current systems: the magnetopause current, the ring current, and the tail current. In general, these currents are perpendicular to the local magnetic field. During storm time, all of them connect to the atmosphere via Birkeland (field-aligned) currents (FACs). The storm-time connections are of interest to us here.

We already know that the magnetospheric generator builds excess positive charges on the dawnside of the magnetosphere and negative charges on the duskside. Charges with access to highly conducting magnetic field lines seek to relieve the charge build-up by diverging from the source region. Current continuity requires that a divergence in the horizontal current must be accompanied by a change in the field-aligned currents. Magnetic merging leads immediately to field-aligned currents between the magnetopause and ionosphere. As we show in Figs. 8-30 and 8-31, the FACs originate

- Near the equatorial edge of the magnetopause—the Region-1 (R-1) currents
- At the magnetopause at high latitudes on the dayside—the cusp currents
- In the plasmasheet where the ring current has a divergence—Region 2 (R-2)

These currents close in the ionosphere via several horizontal current systems.

Region-1 and -2 Field-aligned Currents. In Secs. 4.4.1, 7.4.2, and 8.4.1, we describe the configuration of electric fields and field-aligned currents created by charge accumulation and the convection dynamo. Figure 11-19 provides a simplified view of the electrostatic picture. Region-1 (R-1) currents are largely driven by magnetopause charge separation, which we first describe in Fig. 1-25. These currents exist regardless of the direction of the interplanetary magnetic field. When the IMF is southward, the R-1 currents tend to be stronger, consistent with an enhanced merging electric field in the magnetotail.

Another set of field-aligned currents, Region-2 (R-2) currents have their source in the inner magnetosphere. Although their flow pattern is similar to the R-1 currents, their polarity is opposite and they intercept the ionosphere at a lower latitude (Fig. 11-19). The R-2 current system arises from a charge distribution that is

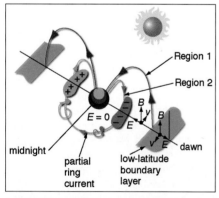

Fig. 11-19. Summary of Region-1 and Region-2 Current Systems. The Region-1 current system is depicted by the red lines. The Region-2 current is shown in green and blue. Current flowing out of the ionosphere at dawn closes through a partial ring current across midnight and then along field lines near dusk. It completes the circuit, continuing as Region-2 current. [McPherron, 1991]

11.3 Field and Current Coupling during Magnetic Storms

a consequence of the drift of charged particles in Earth's dipole field. The magnetotail gradient and curvature drift of charged particles divert electrons toward dawn and ions towards dusk. Charge accumulations produce a dusk-to-dawn electric field that cancels the dawn-to-dusk convection electric field inside the inner edge of the charged regions. Because the electric field is zero inside and nonzero outside, it diverges at this boundary. A field-aligned current must feed or drain the divergence. The particle drifts feeding the R-2 current are much stronger during magnetic storms, so the R-2 current system is often considered a storm-time current.

In the inner magnetotail, the R-2 current is closed by particle drifts across the night side. Effects of the R-2 current and its connection with the R-1 current across the auroral oval tend to cancel each other. Thus the R-2 current is difficult to sense at the ground.

Pause for Inquiry 11-10

In Fig. 11-19, draw the convection electric field vector that is exterior to the R-2 and partial ring currents.

Figure 11-20 is a semi-empirical model of the field-aligned current systems intercepting the ionosphere at the onset of a geomagnetic storm. It shows the projection of the R-1 and R-2 systems onto the northern hemisphere. The R-1 currents tend to project at ~70° magnetic latitude while the R-2 currents are closer to ~60° magnetic latitude. During quiet times, both current systems are present, but at higher latitudes. The R-2 current intensity is noticeably reduced during non-storm time.

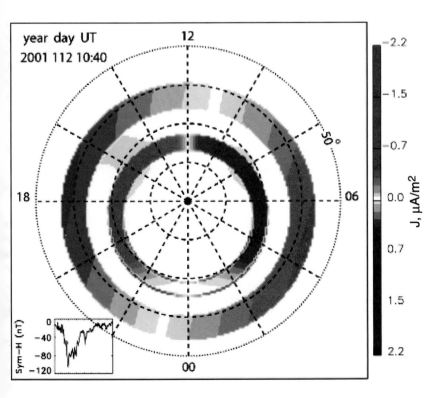

Fig. 11-20. Semi-empirical Mapping of the Region 1 and 2 Currents to the Northern Ionosphere for April 22, 2001. Region-1 currents flow into the ionosphere at dawn and out of the ionosphere at dusk. Region-2 currents have the opposite polarity. Typical current strengths are ~1 µA/m². The inset at the lower left shows the Sym-H index, which is a one-minute version of the Dst index. The current pattern shown is for the early stage of the main phase of a geomagnetic storm, using dynamical data-based modeling of the storm-time geomagnetic field with enhanced spatial resolution. [Sitnov et al., 2008]

Chapter 11 The Disturbed Magnetosphere and Linkages Above and Below

Pause for Inquiry 11-11

Consider a diagram similar to Fig. 11-20 for the southern hemisphere. What is the inward and outward sense of the R-1 and R-2 currents in the southern hemisphere?

Pause for Inquiry 11-12

What are the location and polarity of the cusp currents in Fig. 11-20? Assume a negligible IMF B_y component.

Pause for Inquiry 11-13

On Fig. 11-20, show the direction of the Pederson current flow, the links, and the field-aligned currents.

Additionally, field-aligned *cusp currents* connecting the dayside magnetopause to the ionosphere enter the dayside ionosphere near noon. Often these currents are sensed at latitudes slightly above the R-1 currents. Cusp currents are highly variable and strongly influenced by even small changes in the IMF direction. Figure 11-21 shows how the addition of an IMF B_y component influences the direction of the current flow. When geomagnetic field lines connect to an east- or west-directed IMF, the convection pattern initially twists in the direction of the IMF component but then adjusts to be more aligned with the anti-Sunward flow. The enhanced field-aligned cusp current associated with the IMF B_y influence usually intercepts the ionosphere poleward of the R-1 current. Cusp current variability is an active area of research. Recent studies suggest that some of the most intense field-aligned currents observed at high latitudes flow in very narrow zones when the IMF B_y component is large.

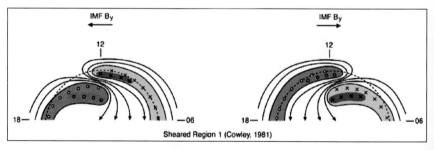

Fig. 11-21. Cusp Field-aligned Current System Response to IMF B_y Variations. Region-1 upward currents are shown in orange; Region-1 downward currents are shown in teal. The B_y component of the interplanetary magnetic field twists the newly opened field lines. The shear requires a supporting field-aligned current. [Strangeway et al., 2000 and Cowley, 1981]

Auroral Electrojets: Ionospheric Effects of Magnetospheric Currents. In the ionosphere, an electric field drives a current in the field direction. This electric-field-aligned current is the Pederson current shown as the horizontal pink arrows in Fig. 8-31. Because the current must be continuous, the only source and closure for the Pederson current is along field lines from the magnetosphere.

The ionospheric electric field drives another horizontal current, the *Hall current*. The magnetospheric origin of this current is the Lorentz force, but it is often referred to as an ionospheric $E \times B$ drift current. In the presence of crossed electric and magnetic fields, electrons and ions drift in the same direction, orthogonal to both E and B. This drift is from noon to midnight across the polar cap and then from midnight to noon along the auroral ovals.

11.3 Field and Current Coupling during Magnetic Storms

EXAMPLE 11.5

Current Carried by Field-aligned Currents

- *Problem Statement:* Estimate the total upward current carried in the R-2 current system shown in Fig. 11-20.
- *Relevant Concepts:* Current density
- *Given:* Data in Fig. 11-20; R-2 current is centered on 60°; Earth radius = 6378 km; 1° latitude spans 110 km, average R-2 current density is 0.5 µA/m².
- *Assume:* The current density is uniform and the R-2 current is 5° in latitudinal width

- *Solution:* The circumference of the R-2 field-aligned current foot print = (2π) 6378 km $(\cos 60°) \approx 2 \times 10^7$ m. The upward current covers one half of this length $\sim 1 \times 10^7$ m

  The area covered by the footprint of the R-2 field-aligned current is width × length
  $= (550 \times 10^3 \text{ m}) (1 \times 10^7 \text{ m}) \approx 0.5 \times 10^{13} \text{ m}^2$

  The total upward current is $(0.5 \text{ µA/m}^2) (0.5 \times 10^{13} \text{ m}^2) = 0.25 \times 10^7 \text{ A} = 2.5$ MA

- *Implications and Interpretation:* The downward current in the R-2 system has the same value. During even more intense portions of some storms the currents are stronger, sometimes exceeding 10 MA.

Follow-on Exercise: Compute the downward current in the R-1 current system.

If the electrons and ions drifted with the same velocity as they do in the magnetosphere, no current would appear. In the ionosphere, however, the ions collide with neutral atoms more frequently than electrons and so they drift more slowly. Thus, a current flows opposite to the drift direction. The idealized Hall current flows from midnight to noon and then back along the auroral ovals (Fig. 11-22). It flows in both polar caps and auroral zones, but tends to concentrate in the auroral zones, where the conductivity is higher.

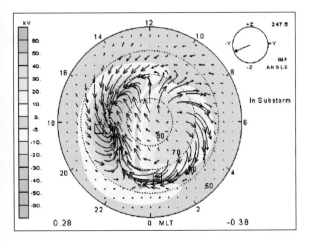

Fig. 11-22. Semi-empirical Model of the Horizontal Current during a Substorm. The horizontal current vectors are superimposed on the electric potential contour map. Peak currents in A/m are indicated below the graph, and the locations of the peaks are marked with the black-line squares. In the westward electrojet near midnight, a peak current of ~0.4 A/m is present. The eastward electrojet has a peak current of ~0.3 A/m near dusk. For this event, the IMF is southward ($-B_z$) and westward ($-B_y$). The pattern is often skewed away from the noon-midnight meridian. Asymmetries in electron precipitation, conductivity, and the IMF B_y component produce the skewing. Magnetic variations on the ground, under the eastward electrojet, are northward (positive). Under the westward electrojet, they are southward (negative). (Courtesy of Dan Weimer at Virginia Tech University)

These current channels are called *auroral electrojets*. Figure 11-23 shows a nightside (toward the Sun view) of the Region-1 and -2 currents and the Hall current auroral electrojets. One electrojet flows westward from roughly dawn to midnight, while the other flows eastward from roughly dusk to midnight. The electrojets appear near and poleward of approximately 67° N and S corrected geomagnetic latitude, during quiet conditions. The strength of the current flow is directly proportional to the product of the conductivity and the horizontal electric field, (σ*E*). Both are at a maximum in the auroral region.

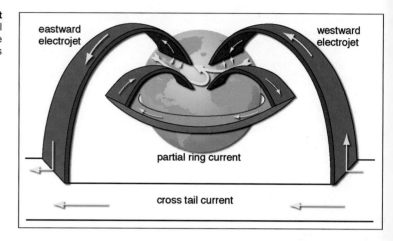

Fig. 11-23. Auroral Electrojets and the Currents that Feed Them. For clarity, this artist's rendering of the auroral electrojets (yellow arrows on blue arcs) shows that they are usually confined to longitudinal channels, with their cores being 5°–8° wide. [After Swift, 1979]

> **Pause for Inquiry 11-14**
>
> Use Fig. 11-22 or 11-23 to verify that magnetic variations on the ground under the eastward electrojet are northward (positive) in the H or X component. Under the westward electrojet, they are southward (negative).

The auroral electrojet is not co-located with the brightest areas of auroral luminosity because such regions correspond to the strongest ionospheric conductivities. Hence, strong electric currents are unable to flow there because the high conductivities short-circuit the electric fields (which produce electric currents). Lower electric fields imply smaller electric currents (even though conductivities are large). Thus, a requirement for strong magnetic perturbations within the auroral electrojet is the presence of strong electric fields, and those fields are only present poleward of the area of strong auroral luminosity.

Substorm Current Wedge. A substorm current wedge is shown schematically in Fig. 11-24. The *current wedge* is a "short-circuit" of the tail current through the midnight ionosphere. The current is downward in the early-morning hours, westward through the ionosphere at midnight, and upward in the late-evening hours. The cause of the short circuit is a subject of current research. Many scientists believe it is a consequence of the pileup of flow from reconnection in the inner tail. In the auroral zone, the effect of the substorm current wedge is a current pattern with a strong enhancement of the westward current across midnight, as illustrated in Fig. 11-23. The current produces a sharp decrease in the horizontal component of the magnetic field.

Partial Ring Current. Another current evident during substorms and the development of the main phase of storms is the partial ring current. In Fig. 11-25, the variable partial ring current is shown in green near midnight, and blue near to dusk. The representation of the partial ring current is similar to the substorm current wedge except that it is reversed polarity. The partial ring current causes a negative perturbation on the ground. When this effect combines with the effect of the ring current, the perturbation pattern is asymmetric, with more negative

11.3 Field and Current Coupling during Magnetic Storms

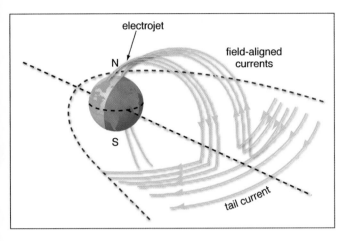

Fig. 11-24. Substorm Current Wedge. The wedge starts in the tail current. Excess current flows out of the cross-tail current, along field lines, and then into and out of the auroral zone. Currents flowing in the resistive ionosphere cause heating of the thermosphere. Some of the current in the auroral zone may flow into the partial ring current (as shown in Fig. 11-25), which is shown near dusk, although it often extends to midnight. [After McPherron et al., 1973]

disturbance at dusk than at dawn. In reality, the ring current and the partial ring current are parts of the same current system. The partial ring current is more likely to be present at storm onset. It is caused by the drift of positive ions around the dusk side. As the ions drift from dawn toward dusk in the dipole field, they gain energy from the cross-tail electric field, produced by convection. This added energy causes the current to be more intense near dusk.

Figure 11-25 summarizes the storm-time magnetosphere current systems. The cusp current circulates from the dayside magnetopause, via field-aligned currents, through the ionosphere, and back to the dayside magnetopause. The inner magnetosphere currents have several elements: 1) on the dawnside, R-1 current flows downward from the inner edge of the low-latitude boundary region. On the duskside, the R-1 current flows upward. This current splits, with some charge flowing over the polar cap and the rest flowing equatorward. 2) Region-2 currents originate in the inner magnetosphere. Their direction is in a sense opposite that of the R-1 currents. The partial ring current also develops in the inner magnetosphere, during the early stages of a geomagnetic storm. The ring currents are the result of particle gradient and curvature drifts. The tail current carries charge from the dawnside to duskside magnetosphere. During rapid storm-time buildup, some of the current may divert to the substorm current wedge in the midnight auroral region. All of these currents contribute to ionospheric currents.

Energy Deposition from Currents. The intensity of the field-aligned currents (FACs) and the closing horizontal currents is sensitive to the magnetic activity level that the solar wind drives directly. The R-1 and R-2 systems expand in space and increase in density during the geomagnetic storm main phase. FACs provide channels into the ionosphere for the magnetosphere's Poynting flux. So an electromagnetic energy flux flows to the ionosphere from the magnetosphere. At high latitudes, the Poynting flux is predominantly downward in the auroral zones and polar cap.

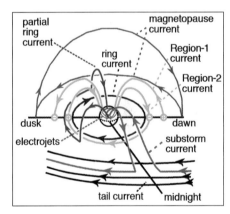

Fig. 11-25. Storm-time Current Systems. These include the magnetopause current at equator and above polar cusp that feed the Region-1 and cusp currents; the Region-1 field-aligned current; the Region-2 field-aligned current. The symmetric ring current and the partial ring current that feed the Region-2 current; and the plasmasheet current that feeds the substorm current wedge. [McPherron, 1991].

Electrojets, driven by field-aligned currents, create substantial resistive heating in the auroral zones. Global Joule heating rates may be in the hundreds of gigawatts during storm time. Thus storm-time currents represent a large energy input to the ionosphere and upper atmosphere. Substorms create temporary current paths to carry extra current generated by reconnection bursts. The paths (substorm current wedges) provide a local diversion of the tail current through the upper atmosphere and are confined to the local time around midnight (Fig. 11-24). During moderate storms, the current wedge carries about 10^6 A. The temporary addition of this new current system adds to the heating in the auroral zone and the expansion of the atmosphere.

As we discuss in Chap. 12, storm-time energy input to the ionosphere has two primary forms: electromagnetic waves and particles. The electromagnetic energy is around four times greater than the particle heating. It is further divided into Joule heating (>90%) and momentum exchange (<10%). Joule heating raises the neutral and plasma temperatures, changes the neutral pressure and associated wind field, raises the plasma scale height, and imparts outward field-aligned plasma flows. Momentum exchange puts the neutral gases into motion.

Focus Box 11.3: The Auroral Electrojet Index

Adapted from World Data Center for Geomagnetism, Kyoto, Japan. Auroral Electrojet (AE) Index Service

The auroral electrojet (AE) index was originally introduced by Davis and Sugiura in 1966, as a measure of global electrojet activity in the auroral zone. The AE index is derived from geomagnetic variations in the horizontal component observed at selected (10–13) observatories along the auroral zone in the northern hemisphere. To normalize the data, a base value for each station is first calculated for each month by averaging all the data from the station on the five international quietest days. This base value is subtracted from each value of one-minute data obtained at the station during that month. Then, among the data from all the stations at each given time (UT), the largest and smallest values are selected. The AU and AL indices are defined respectively by these values. The symbols AU and AL come from the fact that these values form the upper and lower envelopes of the superposed plots of all the data from these stations as functions of UT. The difference (AU − AL) defines the AE index, and the mean value of the AU and AL, i.e., (AU+AL)/2, defines the AO index. The term "AE indices" represents these four indices (AU, AL, AE and AO). The AU and AL indices express the strongest current intensity of the eastward and westward auroral electrojets, respectively. The AE index represents the overall activity of the electrojets, and the AO index provides a measure of the equivalent zonal current. An AE index less than 100 nT represents quiet times. Figure 11-26 shows the AU, AE, and AL indices for 2–8 May 1998.

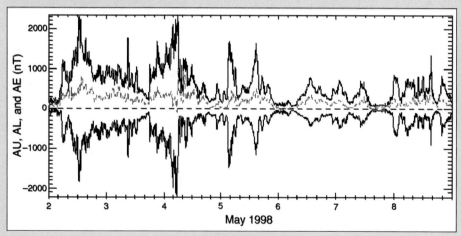

Fig. 11-26. AU, AE, and AL Indices for 2–8 May 1998. The upper black curve is the AU index, the middle red curve is the AE index, and the lower black curve is the AL index. Extreme auroral electrojet activity was recorded on 2–4 May 1998, as several fast coronal mass ejections arrived at Earth. *(Courtesy of Gang Lu at the National Center for Atmospheric Research)*

11.3.2 Mid- and Low-latitude Magnetosphere Shielding

Objectives: After reading this section, you should be able to...

- Describe how undershielding and overshielding develop
- Relate undershielding and overshielding to Region-2 currents

Under equilibrium conditions with little or no convection, magnetospheric particles gradient and curvature drift (Fig. 11-27a). They follow contours of the magnetospheric electric potential Φ. When a cross-tail electric field E from a convection enhancement develops, the plasmasheet surges Sunward and a partial westward ring current develops (Fig. 11-27b). The dusk edge of the plasmasheet is positively charged and the dawn edge is negatively charged. The excess charge creates a dusk-to-dawn polarization electric field (Fig. 11-27b). The superposition of the convection field and the polarization electric field tends to shield the near-Earth region from the enhanced convection dawn-dusk electric field. The shielding configuration usually prevents widespread surges of electric field into the inner magnetosphere and the mid- and low-latitudes ionosphere.

However, the shielding layer and field do not develop instantaneously. Thus, for short intervals, the inner magnetosphere and low- and mid-latitude ionosphere may be exposed to the enhanced convection electric field. This condition is called undershielding. *Undershielding* is the temporary penetration of the dawn-dusk electric field into the inner magnetosphere during times of increasing convection. During undershielding, the dusk-to-dawn polarization field is insufficient to cancel the enhanced convection electric field. The stronger convection field penetrates along field lines into the ionosphere. Additionally, some of the excess charge may flow along the field lines as part of the Region-2 current system. On the other hand, when the driving convection field suddenly decreases, as a result of a northward turning of the IMF, then the dusk-to-dawn polarization electric field dominates across the inner magnetosphere (Fig. 11-27c). *Overshielding* is the temporary dominance of the dusk-to-dawn polarization field over the reduced dawn-to-dusk convection field.

Undershielding affects the pre-midnight equatorward edge of the auroral zone. In this region, electrons usually control the auroral conductance. In the pre-midnight sector, during intervals of heated plasmasheet, ions drift closer to Earth than electrons. The inner edge of the plasmasheet ions forms at a low L-shell. The electrons are confined to enter the auroral zone at higher L-shells and thus do contribute conductivity to the region where the positive ions are attempting to create a current path. Ions that create most of the pressure in the inner magnetosphere also drive most of the R-2 current. Therefore, excess R-2 current flows into the low-conductance, sub-auroral ionospheric region in the pre-midnight sector. As a result, a strong electric field develops at the dusk equatorward edge of the auroral zone.

These fields enhance auroral and sub-auroral plasma flows. Radars that monitor auroral plasma report sub-auroral jets of ionospheric plasma streaming Sunward during the early stages of large geomagnetic storms. These pre-midnight, sub-auroral plasma streams (SAPS) are a hallmark of strong magnetosphere-ionosphere coupling at the onset of large magnetic storms.

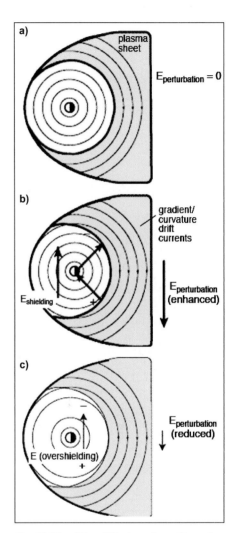

Fig. 11-27. The Effects of a Changing Convection Electric Field on the Inner Magnetosphere. a) Here we show an equilibrium condition with particles drifting (gradient and curvature drift) on curves of electric potential. **b)** This diagram depicts an enhanced convection field that pushes the edge of the plasmasheet inward and sets up an opposing polarization (shielding) field. While the shielding field is setting up, the inner magnetosphere is exposed to penetration electric fields. **c)** The convection electric field weakens, allowing the pre-existing polarization field to dominate for a short time. The inner magnetosphere is temporarily overshielded from the convection electric field. *(Courtesy of Robert Spiro at Rice University)*

Chapter 11 The Disturbed Magnetosphere and Linkages Above and Below

The electric field enhancements that produce shielding and overshielding tend to have an eastward orientation. As we describe in Chap. 12, eastward-directed electric fields that penetrate to the low-latitude ionosphere create plasma uplift. In turn, the uplifted plasma is subject to instabilities that juxtapose high- and low-density plasma plumes. The uneven distribution of plasma ultimately affects radio communication in the low-latitude regions. Observations and predictions of electric field shielding and penetration are an active area of research and have particular application to precision satellite navigation signals that are strongly affected by irregularities in mid- and low-latitude ionospheric plasma.

Summary

When the solar wind carries an interplanetary magnetic field (IMF) that opposes Earth's magnetic field, the dayside terrestrial magnetosphere opens to allow substantial energy transfer. Over several hours, the energy input excites the magnetosphere, creating a geomagnetic storm. Storm intensity is reported with the disturbance storm time (Dst) index—a measure of the depression of the near-equatorial surface field, caused by hot ions drifting around Earth to form a westward ring current. The magnetic field of the ring current decreases the field at Earth's surface, and this depression contributes to the Dst index.

A typical storm consists of three phases: an initial phase that includes a compression of the magnetosphere and often a positive Dst; a main phase when the ring current is growing and causing a rapid depression of Dst; and a recovery phase in which the ring current decays and Dst becomes less negative, tending toward zero. Of primary interest to space forecasters is the storm main phase tendency for close-in acceleration and trapping of charged particles (keV–MeV plasma); intensifying currents in space and on the ground; and enhanced coupling of the magnetosphere-ionosphere system, which usually takes the form of a broadening of the polar cap and equatorward extension of auroral displays.

The interplanetary medium influences and controls the magnetosphere by shaping it and by imposing an electric field. We often use kinetic energy flux to characterize the shaping, and Poynting flux to describe the power delivered by the electric field. The magnetosphere has several ways to dissipate the solar-wind-delivered energy. Under relatively calm conditions, cool particles drift Earthward and many convect to the dayside magnetopause, where they are lost to the magnetosheath. When large amounts of energy are delivered quickly, the most efficient dissipation scheme is a global magnetic storm. Convection dissipates energy during storms, but other modes of dissipation enter the picture. Smaller scale mechanisms, such as strong periodic substorms, steady magnetospheric convection, and isolated substorms, dissipate energy within and outside of storms. Any or all of these mechanisms energize particles. Energized Earthward-moving particles cross magnetic field gradients, and as they do so, they undergo gradient and curvature drift.

Depending on their energy, energized particles may become part of the ring current or the outer radiation belt. Enhancement of ring current and the outer radiation belt is a hallmark of a geomagnetic storm. The ring current develops at the edge of the plasmapause. Storm dynamics that increase the ring current also influence the plasmapause, causing it to contract and contort. The outer radiation belts respond readily to storm dynamics. Their electron density generally increases with increasing storm activity. The belts also broaden across L shells. The inner radiation belt is usually perturbed only by very intense activity. Extreme shocks in the solar wind add new particles to the inner belt or create a new radiation belt in the slot region.

Another feature of geomagnetic storms is increased coupling between the ionosphere and magnetosphere. Much of this coupling occurs via field-aligned currents. Region-1 currents, which connect the flank regions of the magnetopause to the ionosphere, are a direct response to solar wind forcing. Region-2 currents, which connect the inner magnetosphere to the ionosphere, are driven by magnetospheric dynamics, responding to solar wind energy input. Region-1 and -2 currents contribute to and are fed by the auroral electrojets. These horizontal current jets flow in the highly conducting auroral zone. Particles flowing along the electric field (Pederson current) and perpendicular to the field (Hall current) contribute to the electrojets. Most of the ground signature of the jets is from the Hall current. Surges in the electrojets may arise from impulsive contributions of current flowing out of the mid-magnetotail region in the form of a substorm current wedge. The partial ring current also contributes to asymmetries in the electrojets. They flow through the auroral zone, heating the surrounding neutrals, and dissipating some of the storm energy provided by the field-aligned currents from the magnetosphere.

Impulsive changes in magnetospheric convection produce an electric field effect at Earth as well as current effects. Undershielding is the temporary penetration of the enhanced dawn-dusk electric field into the inner magnetosphere, during times of increasing convection. Prevalent at storm onset, undershielding allows for significant magnetosphere-ionosphere coupling at the onset of major geomagnetic storms. Most of the signatures of this coupling appear in the pre-midnight regions at sub-auroral latitudes. Overshielding is the temporary dominance of the dusk-to-dawn polarization field over the reduced dawn-to-dusk convection field. Undershielding and overshielding electric fields contribute to low-latitude ionospheric instabilities.

Chapter 11 The Disturbed Magnetosphere and Linkages Above and Below

Key Words

Akasofu epsilon parameter (ε)
aperiodic stormtime substorms
auroral electrojets
current wedge
cusp currents
disturbance storm time (Dst) index
drainage plumes
electric field separatrix
geomagnetic storms
Hall current
initial phase
isolated non-stormtime substorms
kinetic energy flux $v(\rho v^2)$
main phase
overshielding
Poynting flux
radial diffusion
recovery phase
recurrent
sawtooth events
stagnation point
steady magnetospheric convection (SMC)
sudden impulse (SI)
sudden storm commencement (SSC)
transient
trapping region
undershielding

Notations

Effective Solar Wind Poynting Flux to the Magnetopause

$$P = [(1/\mu_0)(E_{SW} \times B_{SW})](\sin(\theta/2))^4$$
$$= (1/\mu_0) v_{SW} |B_t|^2 (\sin(\theta/2))^4$$

Akasofu epsilon parameter

$$\varepsilon = (1/\mu_0) v_{SW} |B_t|^2 (\sin(\theta/2))^4 l^2$$

Charged particle drift velocity

$$v_D = \frac{E \times B}{B^2} + \frac{m(2v_\parallel^2 + v_\perp^2)}{2q} \frac{B \times \nabla B}{B^3}$$

(1)   (2)

Answers for Pause for Inquiries

PFI 11-1: The sudden impulse (SI) is indicated by the sharp increase in the Dst value (in this case the value rises to ~0 nT). The main phase is associated with a sharp decrease in Dst values (~0 nT to~ –235 nT). The recovery phase is indicated by the slow rise in Dst values from the most negative value (~ –235 nT) back to pre-storm values (~ –30 nT).

PFI 11-2: The ring current flows around Earth in a clockwise direction (looking down on the North Pole (from East to West)). The magnetic perturbation for such a ring of current is downward inside of the ring of current and upward on the outside of the ring of current.

PFI 11-3: We recall that the electric field imposed by the solar wind is given by $E_{SW} = -(v_{SW} \times B_{SW})$. Assuming that the solar wind has only a radial (x) component and the x component of the electric field does not contribute to the energy deposition, we obtain the electric field by evaluating the determinant:

$$E_{SW} = -(v_{SW} \times B_{SW}) = \begin{vmatrix} x & y & z \\ v_x & 0 & 0 \\ 0 & B_y & B_z \end{vmatrix} = E_y = (v_x B_z)$$

Now put the result from the electric field into the determinant for $E_{SW} \times B_{SW}$

$$(E_{SW} \times B_{SW}) = \begin{vmatrix} x & y & z \\ 0 & v_x B_z & 0 \\ 0 & B_y & B_z \end{vmatrix} = (v_x B_z) B_z = v_{SW} |B_t|^2$$

PFI 11-4: The dimensioned quantities of Eq. (11-1) are vB^2/μ_0 with units of $((m/s)(N/A \cdot m)^2)/(N/A^2)$

Manipulating these units produces $((m/s)/(N/m^2)) = (J)/(s \cdot m^2) = W/m^2$

The dimensioned quantities for kinetic power are ρv^3 with units of $((kg/m^3)(m/s)^3)$

Manipulating these units produces $((kg \cdot m^2/s^2)(1/s)(m/m^3)) = (J)/(s \cdot m^2) = W/m^2$

PFI 11-5: b) Average speed CMEs

PFI 11-6: The right-most region, where the particle motion is Earthward, is dominated by term 1, the $E \times B$ term. The region in which the particles begin to circle around Earth is dominated by term 2, the gradient and curvature drift term.

PFI 11-7: The ring current is considered diamagnetic because it produces a magnetic field that opposes the source field that created the current.

PFI 11-8: When the particle energy is low, the parallel and perpendicular speeds, which are elements of the second term, becomes smaller. Therefore, the second term on the right side becomes less important.

PFI 11-9: In Fig. 11-17, the Dst values are shown in the vertical scale on the right. During solar minimum, the ring current is less active so the Dst index tends to be less negative. This is also the interval when the outer radiation belt has reduced energetic particle flux and these particles are observed at large L-values. These characteristics are evident in 1996–1997, the timeframe that coincides with the minimum number of sunspots as indicated by the black trace at the top of Fig. 11-17. Solar cycle 23 began in 1998, according to the Dst index record and the more active outer radiation belt. The slot region filled once in 1994 and several times in 1998.

PFI 11-10: The convection electric field vector points from dawn to dusk.

PFI 11-11: The Region-1 current sense in the southern hemisphere is inward (toward Earth) on the dawn side and outward (away from Earth) and the dusk side. The Region-2 current sense in the southern hemisphere is inward (toward Earth) on the dusk side and outward (away from Earth) and the dawn side. This is the same as in the northern hemisphere.

PFI 11-12: Under conditions of little or no IMF B_y influence, the dayside cusp currents form part of the Region-1 dayside current system. The current flow is toward Earth in the pre-noon sector and away from Earth in the post-noon sector.

PFI 11-13: Pederson currents flow across the polar cap from dawn to dusk. It also flows from high latitude to low latitude between the Region 1 and Region 2 currents in the post midnight and early morning sectors. In the evening and pre-midnight sector of the auroral zone, the Pederson current flows from low latitude to high latitude.

PFI 11-14: Using the right hand rule with the thumb along the direction of the current and the fingers following the direction of the magnetic field, the magnetic field associated with eastward electrojet points northward. Similar reasoning produces a southward magnetic field associated with the westward electrojet.

References

Akasofu, Syun-Ichi. 1981. Energy coupling between the solar wind and the magnetosphere. *Space Science Review.* No. 28. Springer. Dordrecht, Netherlands.

Cowley, Stanley W. H. 1995. The Earth's Magnetosphere: A brief beginners guide. *Eos. Transactions - American Geophysical Union.* Vol. 76. American Geophysical Union. Washington, DC.

Davis, T. Neil and Masahisa Sugiura. 1966. Auroral electrojet activity index AE and its universal time variations. *Journal of Geophysical Research.* Vol. 71. American Geophysical Union. Washington, DC.

Figure References

Baker, Daniel, Joseph H. Allen, Shri G. Kanekal, and Geoff D. Reeves. 1998. Space environmental conditions during April and May 1998: An indicator for the upcoming solar maximum. *EOS.* Vol. 79. American Geophysical Union. Washington, DC.

Blake J. Bernard, Wojciech A. Kolasinski, R. Walker Fillius, and E. Gary Mullen. 1992. Injection of electrons and protons with energies of tens of MeV into L<3 on March 24, 1991. *Geophysical Research Letters.* Vol. 19, No. 821. American Geophysical Union. Washington, DC.

Grebowsky, Joseph M. 1970. Model study of plasmapause motion. *Journal of Geophysical Research.* Vol. 75. American Geophysical Union. Washington, DC.

Kavanagh, L. D., Jr., J. W. Freeman Jr., and A. J. Chen. 1968. Plasma Flow in the Magnetosphere. *Journal of Geophysical Research*. Vol. 73. American Geophysical Union. Washington, DC.

McPherron, Robert L., Christopher T. Russell, and Michael P. Aubry. 1973. Satellite studies of magnetospheric substorms on August 15, 1978, 9, Phenomenological model for substorms. *Journal of Geophysical Research*. Vol. 78. American Geophysical Union. Washington, DC.

McPherron, Robert L. 1991. Physical processes producing magnetospheric substorms and magnetic storms. *Geomagnetism*. Vol. 4. Edited by J. Jacobs. Academic Press, Ltd. London, England.

Milan, Steve, Adrian F. Grocott, Colin Forsyth, Suzanne M. Imber, Peter D. Boakes, and Benoit Hubert. 2008. Looking through the oval window. *Astronomy and Geophysics*. Vol. 48. Royal Astronomical Society. West Sussex, UK.

Reeves, Geoff D., Michael G. Henderson, Ruth M. Skoug, Michelle F. Thomsen, Joseph E. Borovsky, H. O. Funsten, Pontius C. Brandt, Donald J. Mitchell, Joerg-Micha Jahn, C. J. Pollock, David. J. McComas, and Steven B. Mende. 2003. IMAGE, POLAR, and Geosynchronous Observations of Substorm and Ring Current Ion Injection. *Geophysical Monograph*. Vol. 142. American Geophysical Union. Washington, DC.

Roederer, Juan G. 1970. Dynamics of Geomagnetically Trapped Radiation. Vol. 2. *Physics and Chemistry in Space*. Springer-Verlag. New York, NY.

Sitnov, Mikhail I., Nikolai A. Tsyganenko, Aleksandr Y. Ukhorskiy, and Pontus C. Brandt. 2008. Dynamical data-based modeling of the storm-time geomagnetic field with enhanced spatial resolution. *Journal of Geophysical Research*. Vol. 113. American Geophysical Union. Washington, DC.

Strangeway, Robert J., Christopher T. Russell, Charles W. Carlson, James P. McFadden, Robert E. Ergun, Michael A. Temerin, David M. Klumpar, William K. Peterson, and Thomas E. Moore. 2000. Cusp field-aligned currents and ion outflows. *Journal of Geophysical Research*. Vol. 105df. American Geophysical Union. Washington, DC.

Swift, Daniel W. 1979. Auroral Mechanism and Morphology. *Reviews of Geophysics and Space Physics*. Vol. 17. No. 4. American Geophysical Union. Washington, DC.

Further Reading

Cowley, Stanley W. H. 1981. Magnetospheric asymmetries associated with the Y component of the IMF. *Planetary Space Science*. Vol. 29. Elsevier Science, Ltd. Amsterdam, Netherlands.

Hargreaves, John K. 1992. *The Solar Terrestrial Environment*. Cambridge Atmospheric and Space Science Series. Cambridge University Press.

Kamide, Yohsuke and Abraham Chian (Editors). 2007. Handbook of Solar-Terrestrial Environment. Springer-Verlag. New York.

Kivelson, Margaret G. and Christopher T. Russell. 1995. *Introduction to Space Physics*. Cambridge University Press. Cambridge, UK.

Prölss, Gerd W. 2004. *Physics of the Earth's Space Environment*. Springer Verlag. Dordrecht, Netherlands.

Space Weather Disturbances in Earth's Atmosphere

12

UNIT 2. ACTIVE SPACE WEATHER AND ITS PHYSICS

Contributions by Tim Fuller Rowell and Devin Della-Rose

You should already know about...
- Field and current concepts (Chap. 4)
- Plasma concepts (Chap. 6)
- The quiescent upper atmosphere (Chaps. 7 and 8)
- Flares and solar energetic particles (Chaps. 9 and 10)
- The storm-time magnetosphere (Chap. 11)

In this chapter you will learn about...
- Energy flow into the upper atmosphere from the magnetosphere
- Solar cycle variations in the upper atmosphere
- Effects of flares and solar energetic particles on the upper atmosphere
- Energy deposition by particles and Poynting flux into the polar atmosphere
- Contributions of diffuse, discrete, and broadband aurora
- Influences of traveling atmospheric and ionospheric disturbances
- Auroral zone physics and structure
- The definition of an ionospheric storm, including positive and negative phases
- Ionosphere-thermosphere interactions during ionospheric storms
- Plasma chemistry and dynamics during ionospheric storms
- Ionospheric disturbance phenomena at high, middle, and low latitudes

Outline

12.1 **Upper-atmosphere Disturbance Drivers**
- 12.1.1 Characteristic Solar Cycle Behavior
- 12.1.2 Flare and Fast CME Energy Response
- 12.1.3 Magnetospheric Energy Sources
- 12.1.4 Poynting's Theorem, Poynting Flux, and Joule Heat

12.2 **Thermospheric Disturbance Effects**
- 12.2.1 Internal Disturbances: Thermospheric Tides and Winds
- 12.2.2 External Disturbances From High Latitudes
- 12.2.3 Outflowing, Upwelling, and Composition Changes in the Thermosphere

12.3 **Ionospheric Storm Effects**
- 12.3.1 Ionospheric Polar Cap Absorption Events
- 12.3.2 Ionospheric Storms and Disturbance Features

Chapter 12 Space Weather Disturbances in Earth's Atmosphere

12.1 Upper-atmosphere Disturbance Drivers

Upper-atmospheric storms develop in the thermosphere and ionosphere as a result of solar activity and geomagnetic storms. Response times range from minutes for individual solar flares to years or more for solar cycle forcing. The cumulative energy deposition during solar and geomagnetic storms changes the thermospheric temperature, density, global circulation, and neutral composition. Storm processes involve currents; aurora; plasma transport; drag forces; heating; dynamo effects; and perturbations in neutral winds, waves, and composition. Disturbances in the neutral atmosphere are shared with the embedded ionized atmosphere. Conversely, strong interaction between the ionosphere and magnetosphere drives the thermosphere into a disturbed state. Storm effects also include variations in satellite drag and disturbances in radio propagation.

12.1.1 Characteristic Solar Cycle Behavior

> **Objectives:** After reading this section, you should be able to...
>
> ♦ Describe the changes in upper atmosphere density and temperature over the solar cycle
> ♦ Describe and differentiate between sources of thermospheric energy flux

Varying solar and magnetospheric energy inputs produce vigorous response in Earth's upper atmosphere. Figure 12-1 shows the variations in neutral temperature and density and electron density over the course of the solar cycle. Regions above 150 km are most affected. Much of the variation in Fig. 12-1 comes from the solar cycle changes of solar X-ray and extreme ultraviolet radiation that photoionizes N_2 ($\lambda < 79.6$ nm), O ($\lambda < 91.1$ nm), and O_2 ($\lambda < 102.5$ nm). Photo-dissociation occurs at $\lambda > 102.5$ nm. The absorbed energy is redistributed, with about half going to break electron-ion bonds and half to the energy of the ejecta, including the photoelectron. Roughly 60% of the absorbed energy ultimately heats neutral particles. About half of the remainder dissociates molecular oxygen into atomic oxygen, which is the chief component of the upper atmosphere. The other half appears as ultraviolet airglow. Extra energy associated with stormtime dynamics can excite nitric oxide (NO), which in turn radiates energy in infrared wavelengths. This IR radiation may be present for 1–3 days after storm onset.

> **Pause for Inquiry 12-1**
>
> By how much do the quantities in Fig. 12-1 change over a solar cycle? Use the values at 300 km to state your answers in terms of (max value–min value)/average value.

The Sun's deposition of energy to Earth's dayside contributes on average approximately 80% of the upper atmosphere's total energy. Table 12-1 shows energy flux contributions from various sources in the heliosphere and geospace. Energy flux from particles and electromagnetic sources primarily occurs in the auroral zones. When integrated over time and space, the electromagnetic energy deposited as Poynting flux accounts for ~15% of thermospheric energy, and particles account for ~5%. During severe geomagnetic storms the non-solar energy input may surpass the Sun's input.

12.1 Upper-atmosphere Disturbance Drivers

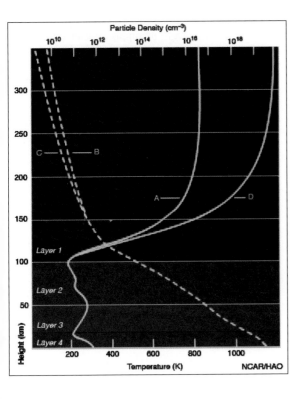

Fig. 12-1. **Fundamental Effects of Storming in the Upper Atmosphere.** Energy input from solar and magnetospheric sources ionizes and heats the constituents of the upper atmosphere. The heated atmosphere expands upward, increasing the density in regions where low-Earth orbiting satellites operate. Layer 1 represents the thermosphere. Curve D is a solar maximum temperature profile, while curve A is a solar minimum profile. Curve B is the solar maximum density profile. Curve C is the solar minimum density profile. The values at the top of the image show the scale for thermospheric density. *(Courtesy of the COMET Program at the University Corporation for Atmospheric Research)*

Table 12-1. **Energy Sources for Geospace.** The sources in the left column contribute energy to the locations shown in the two right columns. Highlighted sources produce significant ionization (Fig. 12-2). The second column gives energy flux. Values preceded by ± are the solar cycle amplitude of the flux. In the X-ray and EUV ranges, higher amplitude variations occur during flares. Magnetospheric quantities are shown as a range of quiet-to-storm values. *(Courtesy of Judith Lean at the Naval Research Laboratory)*

Source	Energy Flux (W/m^2)	Deposition Altitude (km)	Latitude/Longitude
Solar Radiation			Dayside
X ray–EUV 5 nm–120 nm	0.0032 ± 0.0009	100–500	
EUV-UV 121 nm–300 nm	14.9 ± 0.1	30–120	
UV-RF 301 nm–10000 nm	1350.5 ± 0.5	Surface	
Stellar Radiation			Global
Total starlight	0.0000018	Surface–120	
UV 123 nm–135 nm	0.000001	90–120	
Particles			
Solar energetic protons	0.002 ± 0.002	10–90	Polar cap
Magnetospheric protons	0.001–0.006	100–130	High latitude
Magnetospheric electrons	0.003–0.03	30–130	High latitude
Galactic cosmic rays	0.000007	0–90	
Magnetospheric Poynting Flux	0.000015–0.15	100–500	High latitude
Tidal Forcing from the Lower Atmosphere	0.00005	100–125	Dayside

Chapter 12 Space Weather Disturbances in Earth's Atmosphere

Figure 12-2 depicts the ranges of ionization rates from solar photons and energetic particles. Solar photons and auroral electrons supply significant amounts of energy to the tenuous upper atmosphere. Energetic electrons, accelerated in the middle and inner magnetosphere, and solar protons, accelerated in the corona and or interplanetary medium, reach the mesosphere and stratosphere. Chemical changes are the atmospheric hallmark of these deeply penetrating particles.

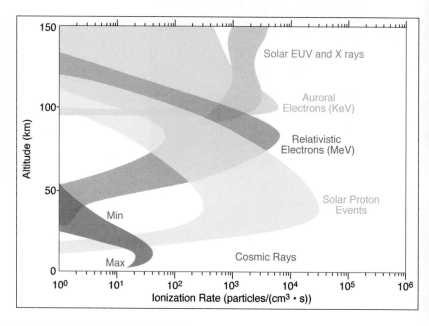

Fig. 12-2. Solar Cycle Ionization Rates. The ranges shown are typical solar cycle ranges. The ionization rate from auroral protons is not shown. Extreme quiet or extreme storms produce values outside of the ranges shown. On very rare occasions, extreme solar proton events create ionization all the way to the ground. For most curves the solar maximum value is on the right and the solar minimum values are on the left. However, cosmic rays have better access to Earth during solar minimum, thus for cosmic rays the solar minimum value is on the right and the solar maximum values are on the left. [Based on data from Richmond, 1987 and after Baker, 2001]

Pause for Inquiry 12-2

Of the categories shown in Fig. 12-2, which are present only as the result of storms?

12.1.2 Flare and Fast CME Energy Response

Objectives: After reading this section, you should be able to...

♦ Describe ionospheric and thermospheric effects of impulsive solar events
♦ Discuss characteristic response times for impulsive events

Flares and fast CMEs produce responses in some aspects of the coupled ionosphere-thermosphere that are not well captured in Fig. 12-1. During extreme flares, soft (2 nm < λ < 5 nm) and hard (0.1 nm < λ < 1 nm) X rays and photons from the hydrogen Lyman-α transition (121 nm) penetrate the lower dayside thermosphere and mesosphere and produce E-layer and D-layer enhancements (Fig. 12-3). These ionization enhancements result in sudden ionospheric disturbances (SIDs) and sudden frequency deviations (SFDs) during radio transmissions, and other radio frequency propagation disruptions related to the E layer and F layer (Chap. 14). Figure 12-4 displays the predicted regions of radio propagation disruption for an X-3 level event on July 14, 2000.

12.1 Upper-atmosphere Disturbance Drivers

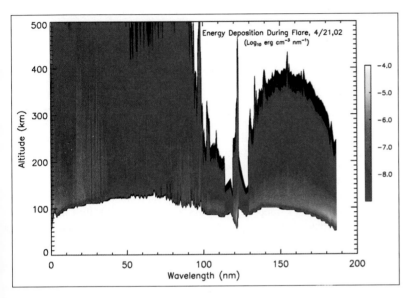

Fig. 12-3. Simulated Ionization Rates for a Solar Flare on 21 April 2002. The simulation covers the wavelength range from less than 1 nm to 185 nm. The color bar shows the logarithm of energy deposition in units of erg/(cm³·nm). Energy deposition spans more than 5 orders of magnitude. Regions shown in white experience negligible energy deposition. Very-short-wavelength X rays ($\lambda < 4$ nm) and EUV radiation with wavelengths near 121 nm penetrate deep into the lower ionosphere and upper mesosphere. *(Courtesy of Stan Solomon at the National Center for Atmospheric Research)*

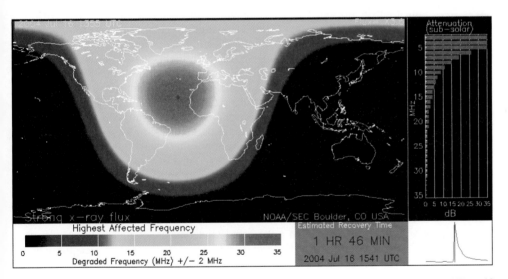

Fig. 12-4. Prediction of Ionospheric D-Region Absorption. The X-3 flare peaked at 1355 UT on 16 July 2004. In the D layer, free electrons absorb low-frequency radio waves and collide with neutrals before they reradiate the radio energy. A large region of the middle Atlantic experienced radio wave absorption in frequency bands below 10 MHz. Some effects on radio signals were predicted at frequencies as high as 32 MHz. The predicted recovery time for all frequencies was 1 hr 46 min after the flare peak. *(Courtesy of NOAA's Space Weather Prediction Center)*

The excess ionization also allows enhanced current flow in the subsolar region. During the most extreme of these events (impulsive hard X-ray flares), dayside ionospheric current perturbations can be detected in magnetic signals at the ground. The signals indicate sudden, large, overhead currents that develop as streams of ionized flow in the equatorial electrojets. In magnetometer records, we see a sharp rise and a long decay—thus plots look like crochet hooks (Fig. 1-17). Such events are called magnetic or flare crochets.

> **Pause for Inquiry 12-3**
>
> Consider Fig. 12-4. Why was the North Pole region predicted to have more degradation than the South Pole region?

Additionally, in the ionosphere's E and F layers, flares produce episodic sudden increases in total electron content (SITECs) and rapid heating of the upper thermosphere. Figures 12-5a-c show model output of the electron density variations at 110 km altitude, produced by a large solar flare. The subsequent thermal expansion (Figs. 12-5d-i) changes the mix of constituents at fixed heights and creates pressure gradients that drive neutral winds.

Solar Energetic Particle (SEP) Interactions. As we describe in Chaps. 9 and 10, solar flares and CMEs energize particles in the ionosphere. A small fraction of these penetrate into Earth's upper atmosphere, along open field lines in the polar caps. These hot particles have enough energy to create multiple ionizations along their path. Most penetrating SEPs are ions with energies between 1 MeV and 100 MeV. Their penetration depths range from 90 km to 30 km—stretching from the ionosphere to the stratosphere (Fig. 12-5). The most energetic ions (>1000 MeV) breach the troposphere and create ground-level events.

More important than the heating effects of this flux are the immediate excess ionization in the polar cap and longer term chemical changes that affect the mesosphere and stratosphere. Sudden changes in polar cap total electron content can inhibit polar cap radio communications for hours. As a result, radio communication blackouts cause over-the-pole flights to be re-routed.

In the hours and days after an SEP event, some of the ions produced by SEP impact create ozone-destroying radicals, such as NO_x and HO_x. These catalytic substances convert ozone to molecular oxygen. Recent simulations of SEP events suggest that their effects extend further into the atmosphere than previously thought, because secondary products, in particular X rays produced by electron bremsstrahlung, also ionize deep in the atmosphere.

One of the great ironies of large space weather events is the interplay between Forbush cosmic ray decreases and SEP enhancements. Large CMEs effectively limit the Earth access of ionizing cosmic rays. However a large fast-moving CME can create high fluxes of SEPs. Although these locally accelerated particles tend to have lower energy than cosmic rays, the SEP flux is higher and often directed into the magnetically open polar caps. Modeling the complex energetic particle flux variation during flare-CME events is a great challenge for space physicists.

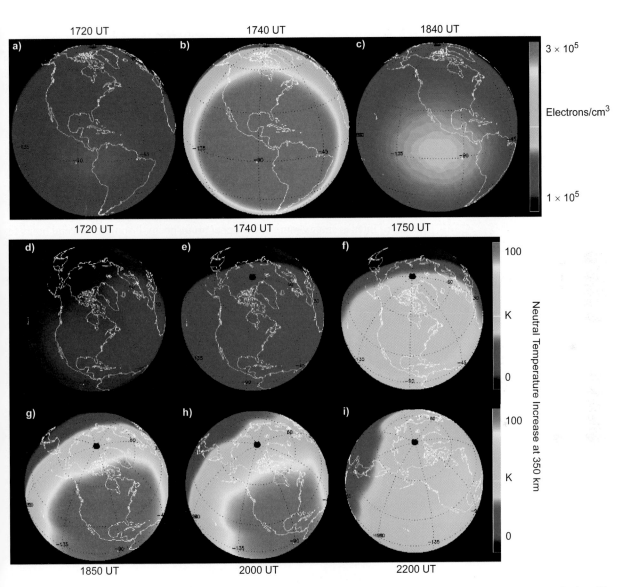

Fig. 12-5. Simulations of a Solar Flare on Sept 7, 2005. a–c) Electron Density at 110 km. The electron density is displayed in the color bar on the right. For a short time, the X-17 flare approximately tripled the dayside electron density in the E layer. **d–i) Neutral Temperature Change at 350 km.** The neutral temperature change (~100 K increase) produced by the flare is slower to evolve. Effects of the temperature change are conveyed to the nightside via pressure gradient forces that drive neutral wind surges. *(Courtesy of Gang Lu at the National Center for Atmospheric Research)*

12.1.3 Magnetospheric Energy Sources

> **Objectives:** After reading this section, you should be able to...
> ♦ Recognize space weather phenomena in the auroral zone
> ♦ Describe the sources of auroral precipitation and their variation with geomagnetic activity
> ♦ Explain how auroral particles energize the ionosphere and thermosphere

In addition to flare effects, thermospheric disturbances are often created by magnetospheric consequences of coronal mass ejections or high-speed solar wind streams impinging on Earth's geomagnetic field. Geomagnetic storms produce a period of intense energy input to the upper atmosphere, lasting for several hours to more than a day. Thermospheric and ionospheric storms develop within an hour or two of a storm commencement in the magnetosphere. Global upper atmosphere response lags local response by two to eight hours.

During magnetic storms, energized particles precipitate to the lower thermosphere and below, expanding the auroral zone, and increasing ionospheric conductivities. Electrons (and sometimes protons), energized by magnetotail processes, spiral down along magnetic field lines, bombard Earth's upper atmosphere, and then interact with the air molecules, producing auroral emissions. Satellite data reveal several types of auroral energy deposition. Unaccelerated ions and electrons create *diffuse auroras*. In *discrete auroras*, most of the energy flux is from the electrons, because their light mass, and thus high speed, delivers more energy than the sluggish, more massive ions. Accelerated, discrete electrons arrive in two forms: 1) mono-energetic beams, accelerated by quasi-static electric fields, and 2) deposition of broadband electron accelerated by Alfvén waves. An image from a low-Earth orbiting spacecraft (Fig. 12-6) reveals diffuse and discrete auroral forms.

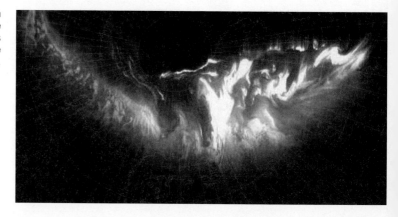

Fig. 12-6. High-latitude Processes Associated with Upper-atmospheric Storms. This image from a Defense Meteorological Satellite Program (DMSP) satellite reveals two auroral forms. The featureless glow relates to diffuse auroras. The bright features are part of the discrete aurora. *(Courtesy of the US Air Force)*

Diffuse Auroras. During quiet times, most of the particles arriving at high latitudes are mirrored by Earth's dipole field and return to the magnetosphere. Only those with small pitch angles penetrate the atmosphere and produce the diffuse aurora and quiet-time particle energy deposits. The diffuse aurora typically forms toward the equatorward-boundary of the oval, where field lines connect to the plasmasheet in Earth's magnetotail. Mirroring electrons (and some protons) with energies of a few eV–1 keV may be scattered into the loss cone via interactions with magnetospheric plasma waves. These particles then funnel into the high-latitude atmosphere, where they collide with upper-atmospheric gases to produce auroral light. Because this process spreads over a relatively wide latitudinal area, diffuse auroral intensities are normally weak. As a result, a quiet-time, diffuse aurora is not usually visible from the ground, even at night. Figure 12-7 shows the spatial distribution of energy flux from diffuse electron precipitation for low- and high-geomagnetic activity. When integrated over the auroral oval, the quiet diffuse energy deposition rate is ~7 GW, roughly a third of the rate for active times.

Some protons take a different path to influencing the aurora. Precipitating protons capture free electrons in the upper atmosphere and become neutral hydrogen. The hydrogen atoms then drift in any direction, no longer bound to follow magnetic field lines. They can re-ionize and re-neutralize many times before their original energy is spent. The proton auroral light is thus spread over a large area, and is usually very faint.

Discrete Auroras. We identify the discrete auroral regions by their emissions of light visible to the human eye, though auroral photons range across the EM spectrum from radio waves to ultraviolet light to X rays. Visible discrete auroras range across the spectrum from red to green to blue-violet. Auroral brightness is measured in *rayleighs*

$$1 \text{ rayleigh} = 10^6 \text{ photons}/(\text{cm}^2 \cdot \text{s})$$

One kilorayleigh (kR) is about the same brightness as the Milky Way on a moonless night. By comparison, during a space weather storm, the auroral brilliance reaches several hundred kilorayleighs.

Pause for Inquiry 12-4

Convert the maximum energy deposition shown in Fig. 12-7 to W/m^2.

In mono-energetic discrete auroras, the energy dispersion of the precipitating electrons is small, meaning most electrons in the beam have nearly the same energy. The quiet and active coverage of the mono-energetic flux is shown in Fig. 12-8a. During active times, discrete structures dominate most of the dusk (evening) region. Auroras in which the precipitating electrons cover broader ranges of energy (broadband) tend to be associated with time-varying fields. They are found generally near midnight (Fig. 12-8b).

The structures connected with discrete auroras are brighter than diffuse auroras, and usually occur at higher latitudes within the oval. Upper atmospheric magnetic field lines in this region map to the outer (boundary layer) plasmasheet, or perhaps to magnetotail reconnection sites during storm times. The patches and curtains of light are between 10 km and 100 km in vertical extent, usually with a base near 100-km altitude. Satellite measurements indicate that precipitating electrons with energies >1 keV cause discrete auroras, requiring an energizing process to raise their energies above the thermal background. Many scientists believe that discrete auroral shapes develop at the base of a "localized" current circuit (i.e., not connected to the Region-1 and Region-2 current system) connecting the magnetosphere and upper atmosphere. The circuit may be electrostatic (direct current) or time varying (alternating current).

Fig. 12-7. Diffuse Auroral Electron Energy Deposits for Quiet and Active Conditions in the Winter Hemisphere. These data come from observations made by the Defense Meteorological Satellite Program (DMSP) satellites over 11 years. Local noon is at the top of both images. The data have been binned and fitted to show the variations in energy deposition for quiet and active geomagnetic conditions. As geomagnetic activity increases, the polar cap becomes broader and the auroral oval becomes wider. The energy of the precipitation electrons also rises. As a result, energy deposition increases. [Newell et al., 2009]

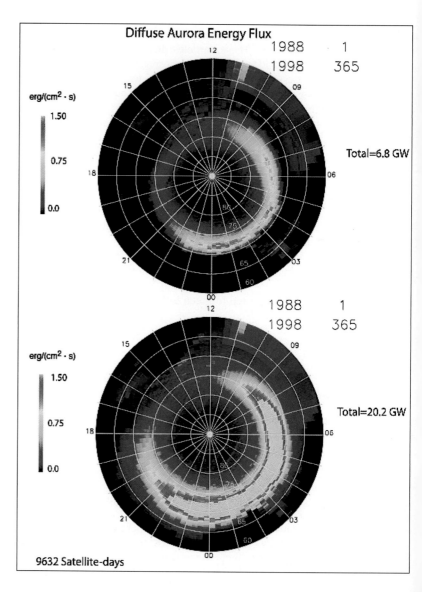

The physics of the local current circuit that creates discrete auroras and its association with the distant magnetotail is still the subject of research. However, theory predicts intense radio wave emissions from the vicinity of the arcs. Such emissions, termed *auroral kilometric radiation (AKR)*, have been observed and are one of the brightest radio sources in our solar system. Spacecraft in the Cluster constellation, launched in 2000 by the European Space Agency, have pinpointed the sources of the 540 kHz–550 kHz radio noise in bright spots in the auroral oval. Downward accelerating electrons emit sporadic bursts of radio frequency whistles and squawks. Emission strengths may exceed 10^9 W, which is about ten thousand times stronger than Earth-based commercial radio signals. Fortunately, this noise is reflected by the topside ionosphere and does not interfere with Earth-based broadcasts.

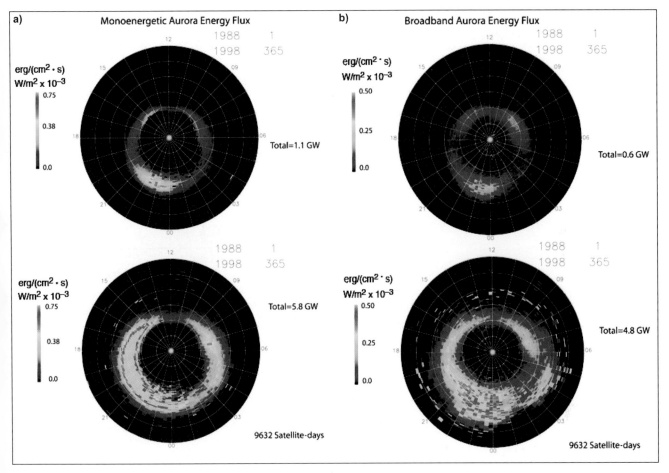

Fig. 12-8. **Discrete Auroral Electron Energy Deposition for Quiet and Active Conditions.** Local noon is at the top in all four images. **a) The Distribution of Monoenergetic Auroras during Quiet and Active Times. b) The Distribution of Broad-band Auroras during Quiet and Active Times.** These data are derived from observations made by the Defense Meteorological Satellite Program (DMSP) satellites over 11 years. The data have been binned and fitted to show the variations in energy deposition for quiet (top) and active (bottom) geomagnetic conditions. As geomagnetic activity increases, mono-energetic auroral beams cover much of the dusk region. Broadband auroras intensify in the pre-midnight region. The energy of the precipitating electrons also rises, increasing energy deposition. [Newell et al., 2009]

The accelerating mechanism probably develops at altitudes between 0.2 R_E–2 R_E as potential differences along the magnetic field line. Figure 12-9 shows upward (downward) directed electric fields with downward (upward) accelerating electrons. Downward accelerated electrons strike the neutral atmosphere, creating the aurora. The narrow, parallel electron beams, energized by a parallel potential drop of ~1 V/m, are typically 3 km–10 km wide. Some satellite observations suggest that the parallel electric fields are a series of small discontinuous potential steps created by local plasma instabilities.

In structures with the opposite electric field direction (middle segment of Fig. 12-9), the electrons move out of the ionosphere. This behavior creates gaps between auroral curtains, sometimes called *black auroras*. Data from the Cluster spacecraft show that the potential structures that create the black auroras extend to altitudes greater than 20,000 km and that they grow and intensify, then

Fig. 12-9. Electrostatic Structure of a Set of Auroral Arcs. In the outermost structures, the potential is lower closer to the central field line. The electric field and current are upward-directed and the electron motion is downward. The electrons are accelerated to energies of a few keV. When they strike the neutral atmosphere, they ionize and excite neutral particles. In the middle structure, the potential is highest near the central field line. The electric field and current are directed downward and the electrons flow up. The more massive ions are less accelerated and do not penetrate the ionosphere. With fewer particle impacts, a void in the auroral emissions develops. This structuring explains the alternating patterns of curtains separated by non-emitting regions. Other structures related to time-varying fields are often observed. [After Marklund et al., 2001]

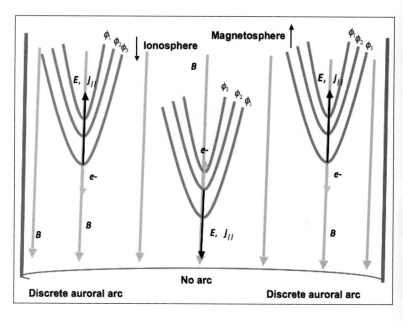

collapse over timescales of a few minutes. These anti-auroras have electric fields as strong as or stronger than the adjacent auroral structures. Some evidence suggests that the potential structures extend deeper into the atmosphere.

Substorm Auroral Evolution. During the International Geophysical Year (1957–1958), space scientists around the world coordinated their efforts to record the aurora from many places at the same time. Shun-ichi Akasofu studied the dynamics of auroral substorms from images made by more than 100 all-sky cameras. He determined that substorms develop in a series of phases (Fig. 12-10). Each substorm is different, but most have many common features we describe below.

During the quiet phase, auroral structures (*arches* and *arcs*) are long, unstructured, motionless ribbons of light extending across the sky. The *growth phase* of a substorm begins with enhanced dayside merging that transports magnetic flux into the tail (Fig. 11-5a). During the growth phase of an auroral substorm, a previously quiet arc slowly brightens and drifts equatorward.

The transition from a slowly moving arc in the growth phase to rapidly moving auroras with more structured curtains and rays is known as *substorm onset* (Fig. 11-5b). Onset occurs in a matter of seconds after a 60–90 minute interval during which magnetic flux, and thus magnetic energy, accumulates in the tail. With its surplus energy, the magnetotail stretches, increases the cross-tail current, and thins the plasmasheet. During this reconfiguration, the most equatorward auroral arc often brightens and moves poleward.

In the *substorm expansion phase* (also known as the *break-up phase*) a bulge of auroral emissions expands poleward and along the oval. From space, the auroral oval thickens along its north-south extent. A surge of auroral emissions propagates westward. From the ground, shooting rays appear to flow down the curtain structures to lower altitudes, and groupings of these rays may flow westward or eastward along the curtain. Other times, *bands* of aurora with red at high altitudes, green at middle altitudes, and purple at low altitudes move rapidly across the sky (Fig. 12-11). As a curtain passes directly overhead, light seems to rain from the sky.

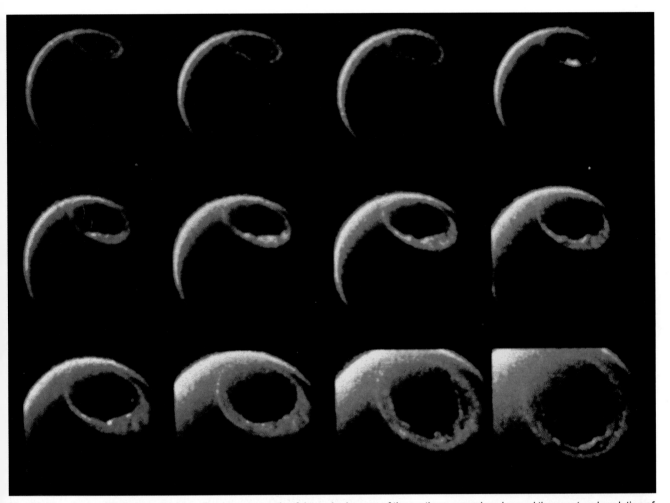

Fig. 12-10. **Auroral Substorm Phases.** Twelve consecutive false-color images of the northern auroral oval record the onset and evolution of a substorm as the spacecraft Dynamics Explorer 1 descends from apogee position during the period 0529 through 0755 UT on 2 April 1982. The image cadence is twelve minutes. The poleward bulge of the surge is first seen in the fourth image, which begins at 0605 UT. A westward traveling surge then propagates along the poleward edge of the oval, as highly structured, eastward-moving auroral forms develop in the post-midnight sector. The sunlit atmosphere is visible at the upper left in each image. The passband of the optical filter is 123 nm–155 nm, for which the dominant responses are from emissions of atomic oxygen at about 130.4 and 135.6 nm and bands of molecular nitrogen. The spacecraft location for the first image at the upper left corner of the image sequence is 23° N geographic latitude and 22 hours local time, and the altitude is 3.67 R_E. The spacecraft is directly over the auroral oval as the last image frame is telemetered, at an altitude of 2.17 R_E. *(Courtesy of the University of Iowa and NASA)*

After about thirty minutes, the activity fades and retreats poleward. Diffuse glows are the most common at the end of an auroral display, during the substorm recovery phase. At high latitudes, the displays are usually greenish. At lower latitudes during intense storms, red diffuse auroras accompanying proton precipitation are common. Diffuse auroras are often very faint to the human eye. From the ground, large patches of diffuse aurora that brighten and fade within seconds are visible, called *pulsating auroras*. From space, the recovery phase is a thick oval, often with two bright regions to the north and south. An uninterrupted recovery phase lasts several hours. At any time during this sequence, the aurora may erupt into another auroral display, if Earth's magnetotail region disturbances trigger a new magnetic substorm event.

Chapter 12 Space Weather Disturbances in Earth's Atmosphere

Fig. 12-11. Substorm Expansion of the Aurora Australis (Southern Lights). In this image made by astronauts aboard the International Space Station in May 2010, auroral curtains and rays are evident. Excited nitrogen molecules interact with the atomic oxygen, causing a green emission at a wavelength of 558 nm. This auroral green-line emission is seen throughout the aurora, as low as about 100 kilometers, giving the aurora its dominant green appearance. Oxygen atoms emit 630 nm auroral red-line light only at very high altitudes. *(Courtesy of NASA)*

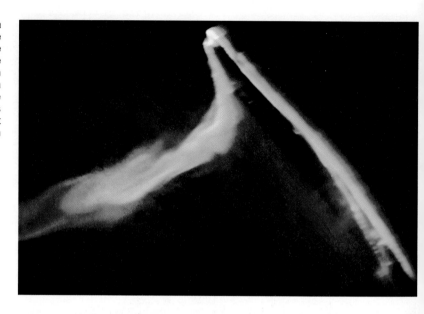

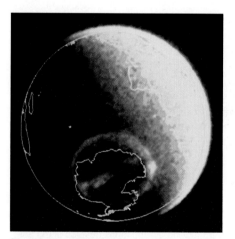

Fig. 12-12. Transpolar Arc Observed by the Dynamics Explorer 1 Spacecraft in the Southern Hemisphere. This image, obtained at 0022 UT on 11 May 1983, shows spatial distribution comprising the auroral oval and a transpolar arc. The arc extends into the polar cap from local midnight to the auroral oval at local noon. Motion of the transpolar arc in the dawn-dusk direction appears to be controlled by the interplanetary magnetic field direction along the dawn-dusk direction. In the Northern Hemisphere, the arc moves in the direction of the IMF B_y component. *(Courtesy of the University of Iowa and NASA)*

Transpolar Arcs. Transpolar arcs (TPAs) are found in the polar cap during intervals of northward interplanetary magnetic field (IMF) (Fig. 12-12). When the IMF B_y component is large, these discrete arcs develop and migrate across the polar cap. The arcs are typically 100 kilometers or more wide and can span from midnight to noon. When present in the central polar cap, TPAs look like the Greek letter theta (θ). Luminosity along the arc is usually less intense than the average luminosity elsewhere along the auroral oval. Particle characteristics (spectra and ion composition) observed in the arcs indicate that theta auroras are created on closed field lines. Arcs in the summer hemisphere are more pronounced. Because ionospheric conductivity depends on solar illumination, the TPA luminosity is also strongly influenced by Earth's dipole tilt angle. TPAs have been associated with auroral kilometric radiation.

Auroral Storm Power. During disturbances, precipitating auroral electrons with an average energy of about 10 keV deposit energy at lower auroral zone altitudes. This energy enables three main processes: *ionization* of the upper atmosphere, heating of the upper atmosphere, and production of auroral light. Simulations suggest that about 35 eV of energy is consumed every time an ion-electron pair forms. Ionization of the most common thermospheric constituents requires ~15 eV. Thus the actual ionization consumes slightly less than half of the energy. Most of the remainder heats the upper atmosphere. Only a few percent of the energy goes to auroral emissions.

This deposited auroral electron energy, also called *hemispheric power*, along with the EM energy transfer (next section), represents a significant energy reservoir to fuel upper atmospheric storms. Table 12-2 summarizes the power provided by auroral particles under quiet and active conditions. The data represent average conditions within each category. During extreme forcing, the hemispheric power exceeds 100 GW. Auroral and EM heating of the polar upper atmosphere establishes strong temperature gradients that mobilize substantial neutral winds. These winds, along with the storm-time increase in auroral ionization, alter electric current flowing in the dynamo region. In turn, the new electric current pattern feeds back on, and changes, the original EM energy transfer rate.

Table 12-2. **Power Delivered to Each Hemisphere by Auroral Particles for Quiet and Active Geomagnetic Conditions.** The data were derived from measurements made by the Defense Meteorological Satellite Program (DMSP) low-Earth orbiting spacecraft. Percents are percentages of hemispheric power. [Newell et al., 2009]

Aurora Type	Hemispheric Power: Quiet (Gigawatts)	Hemispheric Power: Active (Gigawatts)	Hemispheric Power: All Conditions (Gigawatts)
Diffuse (e–)	6.8 (63%)	20.2 (57.5%)	12.6 (61.5%)
Diffuse (ion)	2.3 (21%)	4.9 (14%)	3.4 (16%)
Monoenergetic	1.1 (10%)	5.8 (15.5%)	3.3 (16.5%)
Broadband	0.6 (6%)	4.8 (13%)	1.5 (6%)

EXAMPLE 12.1

Heating Rate from Particle Energy Deposition

- *Problem Statement:* Estimate the volumetric heating rate of the ionosphere near 150 km from particle ionization, using the facts that about 35 eV of energy is expended for each ionization and 50% of the energy ultimately goes to heating.

- *Given:* Heating efficiency = 0.5; energy needed for ion-electron pair production = 35 eV; electron density profile in Fig. 8-11; recombination coefficient = 9×10^{-14} m^3/s.

- *Assume:* The location of interest at 150 km is in photochemical equilibrium.

- *Concepts:* The energy deposition rate is proportional to the ionization rate ($W_{ionization} \propto P$). The ionization rate is proportional to the square of the ionization density $P \propto n^2$ (Eq. 8-5a).

- *Solution:* From Fig. 8-11, the electron density at 150 km is $\sim 2 \times 10^{11}$ electrons/m^3

$$W_{ionization} = 0.5\,(35\text{ eV})\,\alpha n^2$$
$$= 0.5\,(35\text{ eV})(1.6 \times 10^{-19}\text{ J/eV})\,(9 \times 10^{-14}\text{ m}^3/\text{s})$$
$$(2 \times 10^{11}\text{ electrons/m}^3)^2$$
$$= 1.008 \times 10^{-8}\text{ W/m}^3 \approx 10^{-8}\text{ W/m}^3$$

- *Interpretation:* The particle heating rate at 150 km is $\sim 1 \times 10^{-8}$ W/m^3. The photon heating rate at 150 km is $\sim 2 \times 10^{-8}$ W/m^3. Thus the volumetric heating rate from ionization is half that from solar photons at the same level. Further, solar photons are incident on the entire dayside, while particle ionization is limited to the auroral zone, and primarily the nightside auroral zone. Therefore, photons contribute more total heating. Nonetheless, auroral heating can be an intense heating source for the nightside thermosphere.

Follow-on Exercise: Determine the heating rate per neutral particle at 150 km.

Focus Box 12.1: Auroral Zone Collisions and Light

Auroral light spans much of the EM spectrum. The complete auroral emission spectrum is very complex, and quantum mechanics helps us achieve a detailed understanding of all the processes involved. We highlight some of the main results here.

As energetic (up to about 10 keV) electrons descend into the auroral zone, they lose energy in a way similar to electrons striking the inside of a TV tube—in both cases, they emit light. However, the upper atmosphere is much less dense than the material in a TV tube, and so the electrons lose their energy in a series of upper-atmospheric collisions. These interactions excite, dissociate, and ionize atoms and molecules in the path of the "primary" auroral electrons (e_p). Ionizing reactions lead to additional energetic secondary electrons, (e_s) that lead to a cascade of upper-atmospheric collisions. Eventually, the original electron energy spreads among many upper-atmospheric particles. Surprisingly, only a small fraction of this energy converts to auroral light, even during a strong magnetic storm (more on this later). For the emitted light, the main emission process should be de-excitation of excited states (formed directly from an auroral electron collision). However, numerical models predict that chemical reactions occurring after the initial electron impact create most of the excited particles that subsequently emit auroral light as they de-excite. For instance, the yellow-green photons (λ = 557.7 nm), emitted by atomic oxygen (Fig. 12-13), form the emission line via the reaction chain

$$N^+ + O_2 \Rightarrow NO^+ + O^*$$

followed by

$$O^* \Rightarrow O + \text{photon (557.7 nm)}$$

Here, O^* denotes an excited state of oxygen. We make two interesting points about this chemical reaction. First, atomic nitrogen (N) is a trace gas in Earth's atmosphere, so how could enough nitrogen ions (N^+) exist to result in bright green auroral light? The answer is that N^+ forms when an energetic electron (either primary or secondary; we'll assume it's a primary electron for clarity) collides with N_2, and a dissociative ionization occurs

$$N_2 + e_p \Rightarrow N + N^+ + e_p + e_s$$

where the secondary electron e_s has been stripped away from its parent nitrogen atom. The second point is that the quantum probability of emitting the 557.7-nm photon is relatively small, and initially, scientists were stumped over the identity of the emitting atom. As late as the 1920s, a new element, "geocoronium," was proposed to explain the dilemma. Figure 12-13 lists the other major auroral emissions in the visible spectrum. Of note are the oxygen red lines at 630 and 630.4 nm. Spontaneous photon emissions from those excited states of oxygen are so unlikely that the emissions must come mainly from altitudes above 200 kilometers. At lower altitudes, the higher density means that collisions between excited oxygen atoms and other particles de-excite the oxygen before it has time to spontaneously emit a photon (in such cases, the excess energy appears as particle kinetic energy).

As a final example, we recall that auroral light even includes X rays. Neither collisional excitations nor chemical reactions are responsible for this light. Rather, the primary electrons are sometimes decelerated rapidly in the collision process, and *bremsstrahlung* (German for "braking") radiation results, producing a high-energy X-ray photon.

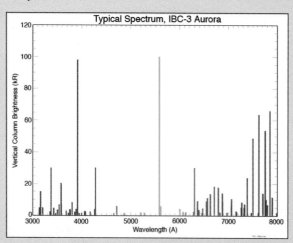

Fig. 12-13. Main Auroral Emission Lines. This diagram shows the wavelengths of emissions from the aurora in the range from 300 nm to 800 nm. The dominant visible auroral emissions are at 557.7 nm, 630 nm, and 630.4 nm. *(Courtesy of Stan Solomon at the National Center for Atmospheric Research)*

12.1.4 Poynting's Theorem, Poynting Flux, and Joule Heat

Objectives: After reading this section, you should be able to...

- Relate electromagnetic wave energy deposition to Joule heating
- Calculate Poynting flux given an electric field and perturbation magnetic field

In addition to auroral input, energy is transferred from the magnetosphere to the upper atmosphere via enhanced Poynting flux from field-aligned currents and Alfvén waves and via charged-particle deposition (previous section). All sources increase in magnitude and expand in latitude (usually equatorward) during intervals of southward IMF. Poynting flux delivers energy to large portions of the auroral zones, where it dissipates predominantly as Joule heat.

In Chap. 4, we show that the *dynamo process*, the conversion of mechanical (solar wind) energy into electrical energy, strengthens electric fields during space weather disturbances. Highly conducting space plasmas carry currents as a result, and these currents generate their own disturbance magnetic fields (which superpose on Earth's dipole magnetic field). Table 4-3 gives the approximate strengths of these fields. The dynamo electric fields produced by the solar wind generator connect into the high-latitude ionosphere, and together with Earth's dipole magnetic field, drive $\boldsymbol{E} \times \boldsymbol{B}$ plasma drifts (*convection*), as we explain in Sec. 4.4.2. We describe electromagnetic wave energy in terms of the *Poynting vector S* (Eq. (4-6)) and the *electromagnetic energy density* u_{EB} (Eq. (4-9)). These relations, rewritten below for electromagnetic waves, are valid in any region of space where electric and magnetic fields co-exist, even if waves aren't the sole source of the fields.

$$\boldsymbol{S} = \frac{\boldsymbol{E} \times \boldsymbol{B}}{\mu_0}$$

$$u_{EB} = \frac{\varepsilon_0 E^2}{2} + \frac{B^2}{2\mu_0}$$

where

- $\boldsymbol{S}$ = Poynting vector energy flux, carried by an electromagnetic wave [W/m^2]
- $\boldsymbol{E}$ = electric field strength vector [N/C]
- $\boldsymbol{B}$ = magnetic field strength vector [T]
- μ_0 = permeability of free space, ($4\pi \times 10^{-7}$ N/A^2 = 1.26×10^{-6} N/A^2)
- u_{EB} = electromagnetic energy density [J/m^3]
- ε_0 = permittivity of free space (8.85×10^{-12} C^2/(N·m^2))

In this section, we modify Eq. (4-6) to account for Poynting flux delivered by interactions in the form of convection fields $\boldsymbol{E}_{conv}$, and perturbation magnetic fields $\Delta \boldsymbol{B}$, from field-aligned currents. The magnitudes of $\boldsymbol{E}_{conv}$ and $\Delta \boldsymbol{B}$ relate directly to the strength of the space weather event that produces them. This form of the

Poynting flux is sometimes called the *perturbation Poynting flux* (S_p). Figure 12-14 shows a cross section of the system that deposits energy at high latitudes.

$$S_p = \frac{E_{conv} \times \Delta B}{\mu_0} \qquad (12\text{-}1)$$

$$u_{EB} = \frac{\varepsilon_0 E_{conv}^2}{2} + \frac{(\Delta B)^2}{2\mu_0} \qquad (12\text{-}2)$$

where

E_{conv} = convection electric field vector [N/C or V/m]

ΔB = magnetic field perturbation caused by field-aligned currents [T]

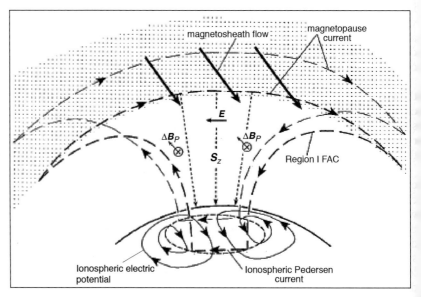

Fig. 12-14. Energy Deposition to the High-latitude Regions. The solar wind flowing in the magnetosheath is shown with heavy arrows. Currents are depicted with dashed curves. The ionospheric electric potential, which traces out the convection pattern, is shown in light curves. The dotted vertical arrows represent the perturbation Poynting flux. In this idealized view, the polar cap magnetic perturbations ΔB, created by field-aligned currents (Region-1 currents in this case), interact with the convection electric field E to produce Poynting flux directed into the ionosphere S_z (S_p in Eq. (12-1)), where the subscript z indicates vertical perturbation Poynting flux. Although the Region-1 current is emphasized here, Poynting flux is also generated by other field-aligned currents. The magnetic perturbations between the Region-1 and Region-2 currents (not shown) interact with the convection electric field in the auroral zone to produce significant Poynting flux deposition in the auroral zones. [Cowley, 2000]

As suggested by Fig. 12-14, electromagnetic energy from the magnetosphere is deposited (converges) into the upper atmosphere, where it heats the upper atmospheric gases (neutral and charged particles). We quantify these effects with the conservation-of-energy relation for electromagnetic fields, called *Poynting's Theorem*, Eq. (12-3). Poynting's Theorem provides a relationship between energy density and the flux of an electromagnetic (EM) field.

$$\frac{\partial u_{EB}}{\partial t} + \nabla \bullet S = -J \bullet E \qquad (12\text{-}3)$$

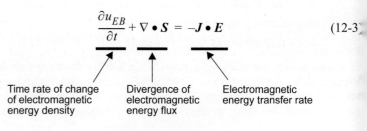

where

J = volume current density vector [A/m^2] (Eq. (4-12a))

The first term on the left side represents the time rate of change of EM energy density, and the second term is the divergence of EM energy flux. The right side is the *electromagnetic (EM) energy transfer rate*, which quantifies the sum of the heat energy change and the particle accelerations produced by changes in EM energy.

Using a vector calculus identity, we expand the electromagnetic energy transfer rate on the right side to account for Joule heating and particle acceleration.

$$\boldsymbol{J} \bullet \boldsymbol{E} = \boldsymbol{J} \bullet (\boldsymbol{E} + \boldsymbol{v} \times \boldsymbol{B}) + (\boldsymbol{J} \times \boldsymbol{B}) \bullet \boldsymbol{v} \qquad (12\text{-}4)$$

where

v = plasma velocity [km/s]

The first term on the right side of Eq. (12-4) is *Joule heating* and the second term is the scalar product of the force per unit volume on the medium, $\boldsymbol{J} \times \boldsymbol{B}$, with the velocity, and represents the kinetic energy generation. Usually Joule heating accounts for more than 90% of the electromagnetic energy transfer rate. This form of energy deposition (and dissipation) greatly influences winds and atmospheric composition changes during space weather storms.

When $\boldsymbol{J} \bullet \boldsymbol{E}$ is positive (current and electric field in the same direction), then the electric field performs positive work on the charged particles that carry the current, and so EM energy is lost from the driving (usually the magnetosphere) system. The energy appears in the ionosphere-thermosphere as heat (temperature rises) and mechanical energy (kinetic energy increases). When $\boldsymbol{J} \bullet \boldsymbol{E}$ is negative, the opposite occurs, and system EM energy increases with time.

Using a circuit analogy, we model space weather storms as part of an electric circuit, in which Earth's upper atmosphere functions as a huge resistor. A resistor converts electrical energy to heat. Electromagnetic forces also perform mechanical work to accelerate charged particles (which, in turn, accelerate the neutral gas via drag forces). Both of these processes drive a multitude of upper atmospheric disturbances that span all latitudes (Secs. 12.2 and 12.3). Figure 12-15 provides a statistical view of the magnitude and location of Poynting flux deposition mapped to 110 km. The plot is organized by direction of the IMF. Most of the energy deposition is in the auroral zones between the Region-1 and Region-2 currents. Intervals of southward IMF have the strongest deposition.

Pause for Inquiry 12-5

Show that the dimensions of each term in Poynting's Theorem are energy per volume per time.

Poynting's Theorem helps us understand the thermodynamics in a volume of space containing EM energy and matter. First, we consider a situation where on the whole, EM energy flow neither enters nor leaves the volume, which means that $\nabla \bullet \boldsymbol{S}$ is zero (very quiet conditions). If $\boldsymbol{J} \bullet \boldsymbol{E}$ is positive, then the right side of Eq. (12-3) is negative. Therefore, the electromagnetic energy density (u_{EB}) in the driving system decreases in time, consistent with our comments in the previous paragraph. We apply similar reasoning for the case when $\boldsymbol{J} \bullet \boldsymbol{E}$ is negative.

Chapter 12 Space Weather Disturbances in Earth's Atmosphere

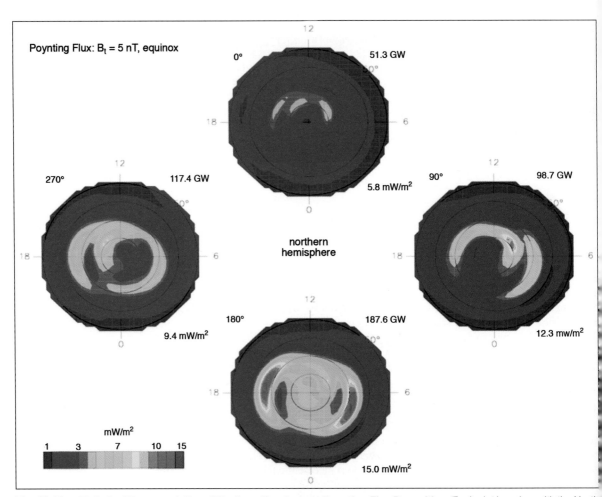

Fig. 12-15. Statistical Representation of Northern Hemisphere Poynting Flux Deposition. Each plot is a view with the North Pole at the center and the Sun at the top. The data are extracted from measurements made by the Dynamics Explorer 2 Satellite. The plots are organized by IMF orientation. The top plot is for northward IMF (0°) and the right plot is for duskward IMF (90°). Most of the Poynting flux is deposited in the auroral zone. The value at the top right of each plot is the hemispherically integrated Poynting flux deposition. *(Courtesy of Astrid Maute at the National Center for Atmospheric Research)*

For a second scenario, we assume the volume of matter and electromagnetic energy is in steady state, so the time derivative of u_{EB} is zero. However, we now permit $\nabla \cdot S$ to be non-zero (active conditions). Because u_{EB} is constant in time, any convergence or divergence of S must be counter-balanced by Joule heating. If, for instance, electromagnetic energy flux converges into our volume, the magnitude of $\nabla \cdot S$ must exactly equal the Joule heating plus particle acceleration. Put another way, if $\nabla \cdot S < 0$, Eq. (12-1) indicates that $J \cdot E > 0$, meaning that electromagnetic energy converts to thermal and kinetic energy in our volume. We reason in a comparable fashion for the situation where Poynting flux diverges from the volume. To use Poynting's Theorem for space weather computations, we need to specify further the high-latitude electric and magnetic field structure during a storm; that is, we need values to go with the illustration in Fig. 12-15.

Example 12.2 gives some representative values for energy deposition in the ionosphere-thermosphere system. The magnetic fields associated with space weather disturbances are from the field-aligned current (FAC) system (Figs. 4-26

12.1 Upper-atmosphere Disturbance Drivers

and 4-29). We recall from Sec. 4.4 that the FAC system connects the upper atmosphere to the dynamo regions of the magnetosphere. We can show that Earth's dipole magnetic field does not contribute to the divergence of the Poynting flux in this case—all we need to know is the storm current magnetic field.

EXAMPLE 12.2

Poynting Flux at High Latitudes

- *Problem Statement:* Research indicates that moderate space weather storms deposit $\sim 10^{11}$ W (100 gigawatts) of power in the high latitudes of the upper atmosphere. Use Poynting's Theorem to assess the accuracy of this power estimate.

- *Relevant Concept or Equations:* Perturbation Poynting flux (Eq. (12-1)), convection electric field E_{conv} and magnetic perturbations ΔB from field-aligned currents, and scale analysis from Chap. 4.

- *Given:* Approximate numerical values for E and ΔB during a space weather storm: $|E_{conv}| \approx 50$ mV/m and $|\Delta B| \approx 500$ nT

  The auroral zone is 50 km thick and covers a horizontal area of about 5×10^6 km^2

- *Assume:* All the deposited EM energy in the space weather storm transforms into upper-atmospheric heat energy. The ionospheric electromagnetic fields vary slowly enough to neglect the time derivative of electromagnetic energy density (u_{EB}). Thus $\nabla \cdot S_p = -J \cdot E$

- *Solution:* To compute the perturbation Poynting flux S_p we return to Eq. (12-1)

$$S_p = \frac{E_{conv} \times \Delta B}{\mu_0}$$

To find the cross product magnitude, we need to understand the directions of E and ΔB, not just their magnitudes. The electric field direction is shown in Fig. 12-16. The magnetic field is created by a set of field-aligned currents like those shown in Fig. 12-16. Between the Region-1 and Region-2 current sheets, the FAC magnetic field is approximately parallel to Earth's surface, and points anti-Sunward. This fact means that E and ΔB are about perpendicular to each other, so the magnitude of the cross product reduces to $|E||\Delta B|$. The Poynting vector points downward, toward Earth's center (Fig. 12-14), consistent with our notion that electromagnetic energy flux converges into the upper atmosphere.

$|S_p| = (|E||B|)/\mu_0 \approx (50 \times 10^{-3}$ V/m$)(500 \times 10^{-9}$ T$) /$
$(4\pi \times 10^{-7}$ N/A$^2) \approx 2 \times 10^{-2}$ W/m^2

Now, using scale analysis, we replace the divergence of the Poynting flux $\nabla \cdot S_p$ with the Poynting flux divided by a characteristic scale length L.

$$\frac{S_p}{L} = -J \cdot E$$

For L, we use the auroral zone thickness, 50 km. Finally, the electromagnetic energy transferred to Joule heat is

$|-J \cdot E| \approx S_p/L = (2 \times 10^{-2}$ W/m$^2) / (5 \times 10^4$ m$)$
$= 4 \times 10^{-7}$ W/m$^3 = 0.4$ µW/m^3

Auroral zone volume = (area) (thickness) = $(5 \times 10^{12}$ m$^2)$
$(5 \times 10^4$ m$) \approx 3 \times 10^{17}$ m^3

The storm-time power deposition is $(0.4 \times 10^{-6}$ W/m$^3)$
$(3 \times 10^{17}$ m$^3) = 1.2 \times 10^{11}$ W

- *Interpretation:* A value of ~100 GW of energy transfer between an electromagnetic source and heat deposition is consistent with the auroral storm-time values of $|E_{conv}| \approx 50$ mV/m and $|\Delta B| \approx 500$ nT.

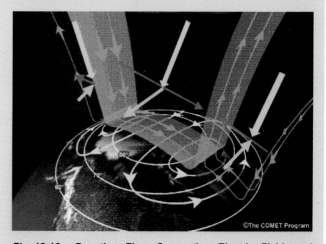

Fig. 12-16. **Poynting Flux, Convection Electric Field, and Field-aligned Current Magnetic Perturbation.** The electric field (red arrows) points from dawn to dusk in the polar cap and from dusk to dawn in the auroral regions. The perturbation magnetic field, shown with green arrows, is Sunward in the polar cap and anti-Sunward in the auroral zones. Ribbons of aligned current are shown in orange. The height-integrated Hall current circuit is displayed in yellow. The Poynting flux (blue arrows) in all regions is directed toward Earth. In other high-latitude regions, the vectors have slightly different directions, but the Poynting flux is usually directed Earthward. *(Background image courtesy of the COMET Program at the University Corporation for Atmospheric Research)*

Chapter 12 Space Weather Disturbances in Earth's Atmosphere

Satellites with particle detectors and magnetometers have been used for many years to measure field properties in space. Figure 12-16 (Focus Box above) depicts the physical situation. A polar orbiting satellite with electrometers and magnetometers, crossing the dawn-dusk meridian, observes the magnetic field of the FAC currents ΔB and the dynamo electric field E. A dawn-to-dusk electric field is usually observed in the polar cap field, while dusk-to-dawn fields are observed in the auroral zone. The FAC magnetic field has two anti-Sunward regions (near the vertical lines in the figure) where $|\Delta B|$ is largest. These maxima occur when the satellite passes between the *Region-1* and *Region-2* FAC current sheets in the dawn or dusk sectors.

Observations show that convection electric fields and large-scale, direct electric currents flowing along magnetic field lines into Earth's high-latitude regions are normally the main power sources for the auroral zones. However, evidence from recent spacecraft missions suggests an additional contribution from an electromagnetic wave disturbance propagating along magnetic field lines (Alfvén waves). At auroral altitudes, an Alfvén wave normally has frequencies of several Hertz, corresponding to wavelengths a few times larger than an Earth radius. These waves propagate on closed field lines that link the hemispheres. The form of this energy is similar to that of the perturbation Poynting flux, but it arises from alternating current. The difference is attributable to the wave nature of the energy. Instead of a relatively static convection electric field, we are dealing with a wave perturbation electric field δE and a wave perturbation magnetic field δB (Fig. 12-17). The Alfvén Poynting flux takes the form

$$S_A = \frac{\delta E \times \delta B}{\mu_0} \tag{12-5}$$

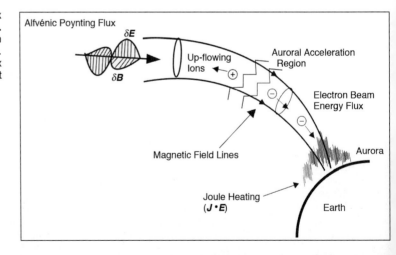

Fig. 12-17. Schematic of Alfvén Poynting Flux Flowing from the Magnetosphere to the Ionosphere. Alfvén waves originating in the magnetosphere, with periods of 6 s to 180 s, are the source of the perturbations. Some of the energy from the Alfvén Poynting flux probably goes into auroral particle acceleration. [Wygant et al., 2000]

Data from the POLAR spacecraft (altitude 25,000 km to 38,000 km) reveal that the flow of this wave electromagnetic energy focuses into the auroral oval. Typical fluxes mapped to ionospheric heights range from 1 mW/m²–5 mW/m² and tend to be the strongest in the post-noon and pre-midnight regions. Thus the alternating current (Alfvén wave) Poynting flux is about 10% of the direct current Poynting flux delivered by quasi-static field-aligned currents. Researchers think that some of the wave energy goes into accelerating particles in the auroral acceleration region (previous section) as well as into Joule heat in the ionosphere.

EXAMPLE 12.3

Thermospheric Heating by Poynting Flux

- *Problem Statement:* Verify that 100 gigawatts of power deposition can raise the local temperature of the thermosphere by several hundred degrees during the early stages of a geomagnetic storm.

- *Relevant Concept or Equations:* Energy density, thermospheric neutral particle density; $1 \text{ eV} \approx 11{,}600 \text{ K}$

- *Given:* From Example 12.2, we know that 100 GW of power deposition is about $0.4 \times 10^{-6} \text{ W/m}^3$; initial storm duration of 3 hours; density in the lower thermosphere is about 10^{18} particles/m^3

- *Assume:* Energy is distributed equally among all the neutral particles

- *Solution:* Energy density deposited over the three hour interval $= (0.4 \times 10^{-6} \text{ W/m}^3)(10{,}800 \text{ s}) = 0.00432 \text{ J/m}^3$

Individual particle energy addition is $(0.00432 \text{ J/m}^3) / (10^{18}$ particles/m$^3) = 4.32 \times 10^{-21}$ J/particle

Recalling that $1 \text{ eV} = 1.6 \times 10^{-19}$ J, this is roughly equivalent to 0.03 eV/particle.

Therefore, if we could insulate the auroral zone particles from their surroundings, the added Joule heating from the storm would be sufficient to raise the auroral zone gas temperature by $(0.03)(11{,}600 \text{ K}) \approx 350 \text{ K}$

- *Interpretation:* About 100 GW of energy deposition is roughly equivalent to a 350 K temperature increase, if all of the energy goes to Joule heating. In fact, some of the energy goes to acceleration, increasing the speed of the particles, which translates to increased temperature. However, particle energy deposition should also be included in the energy budget and many storms last longer than 3 hours. Temperature increases of more than 350 K are often observed during geomagnetic storms.

Follow-on Exercise: Use the data in Table 12-2 to estimate the temperature increase associated with particle deposition.

12.2 Thermospheric Disturbance Effects

12.2.1 Internal Disturbances: Thermospheric Tides and Winds

Objectives: After reading this section, you should be able to...

- Distinguish between migrating and non-migrating perturbations
- Describe sources and effects of Sun-synchronous (migrating) tides in the upper atmosphere
- Describe sources and effects of non-migrating tides in the upper atmosphere

Absorption of solar radiation excites thermal tides in Earth's atmosphere, creating upheavals and bulges. Strong tidal signatures in the troposphere may be further modulated by land-air and land-sea interactions. Latent heat release in thunderstorm complexes and seasonal convergence zones contributes to a time-lagged tidal forcing. In the stratosphere (~50 km), ozone is a strong absorber of UV solar radiation, while molecular nitrogen and oxygen at ~150 km are good EUV absorbers in the thermosphere. The immediate tidal effect is a Sun-synchronous atmospheric bulge in post-noon longitudes. Figure 8-4 shows the thermal high pressure centered near the post-noon equator, caused by solar heating.

Space-based observers see the bulge fixed with respect to the Sun as the planet rotates beneath it. To a ground-based observer, the bulge moves westward with the apparent motion of the Sun once a day. This pressure bulge drives thermospheric winds toward Earth's night side; that is, air moves poleward in the mid-morning to mid-afternoon local-time sectors in both hemispheres. Additionally, westward winds blow near dawn, and eastward winds blow near dusk. On the night side, these winds continue to flow toward the thermal low centered near the equator at midnight. Typical wind speeds are 50 m/s–100 m/s. Other atmospheric processes produce tides that do not migrate with the Sun. Figure 12-18 shows the influence of upward, non-migrating waves on the crests of F-layer ionization on either side of the magnetic equator.

Upward-propagating tides contribute to the momentum budget of the lower thermosphere. In general, tides are capable of propagating vertically to higher, less dense, regions of the atmosphere. Because of the exponential decrease in density, wave amplitudes grow with height. The waves become large and unstable, generate turbulence, and deposit heat and momentum into the atmosphere. Momentum deposition produces a net meridional circulation, and associated rising motions (cooling) at high latitudes during summer, and sinking motions (heating) during winter. Low-Earth orbiting satellites experience a small drag effect from upward-propagating disturbances with long wavelengths.

Recent research has focused on regional effects of gravity waves excited by deep convection in tropical regions. These primary up-and-down motions excite large-scale secondary gravity waves in the thermosphere at altitudes near the bottom of the F layer (~250 km). The secondary waves have long wavelengths (200 km–4000 km) and produce wind perturbations of 70 m/s–150 m/s and density perturbations of 10%–15%. Researchers are investigating the role that such disturbances may have on plasma instabilities that cause GPS signal scintillation.

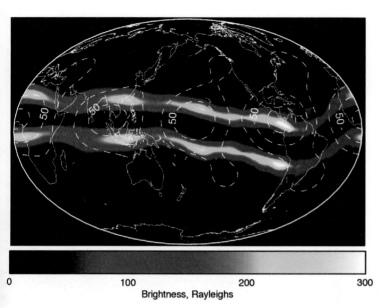

Fig. 12-18. **Equatorial Airglow Amplified by Lower Atmospheric Tides.** This image is a reconstruction of nighttime ionospheric equatorial anomaly emissions over 30 days (March 20–April 20, 2002). The airglow enhancements have a 1000-km scale longitudinal variation that matches a pattern of strengthened upward propagating tides, associated with tropical convection. The observations were made with the Far Ultraviolet Imager on the Imager for Magnetosphere-to-Aurora Global Exploration (IMAGE) spacecraft and the Thermosphere Ionosphere Mesosphere Energetics and Dynamics (TIMED) satellite. This image is representative of the local ionospheric properties at 20:00 LT. Overlaid on this figure in white dashed contours is the amplitude of the diurnal temperature variation at 115 km from upward-propagating lower-atmospheric tides, as reported by the Global Scale Wave Model. The temperature contours are in 25 K increments. [Immel et al., 2006]

12.2.2 External Disturbances From High Latitudes

Objectives: After reading this section, you should be able to...

- State the time scales from ion acceleration of neutral winds
- Explain the source, directions, and magnitudes of storm-driven neutral circulations
- Determine the source and speed of traveling atmospheric disturbances

Winds from External Forcing. During upper-atmospheric storms, mechanical forcing and strong heating from electrodynamic interactions and auroral sources accelerate thermospheric winds (Figs. 12-20a and b). Strong electric fields enhance the magnetic field-aligned current connection with the ionosphere. These currents convey Poynting flux that dissipates as heat, accelerates ions, and transfers momentum to the neutrals in the thermosphere. In concert with the strengthened fields is an expansion and intensification of auroral particle energy deposition.

At high latitudes and high altitudes, ions embedded in the thermosphere drift in response to the electric field ($E \times B$) and, by colliding with neutrals, drive horizontal winds. The wind speeds are usually a fraction of a kilometer per second, but during storm time, narrow channels with speeds in excess of a kilometer per second may develop. The higher ion velocities mean more momentum is transferred from the ions to the neutrals. The ions stir the neutrals into motion via collision, and thus the electric field becomes a momentum source to the neutral particles. However, pressure gradients and the Coriolis force prevent the neutral atmosphere from completely mimicking the motion of the ions. The resultant wind field is the superposition of the quiet- and storm-time patterns (Fig. 12-20). Wind flow is generally anti-Sunward in the polar cap and Sunward at lower latitudes. Neutral particle acceleration usually dissipates only 1/10th of the energy dissipated by Joule heating.

Chapter 12 Space Weather Disturbances in Earth's Atmosphere

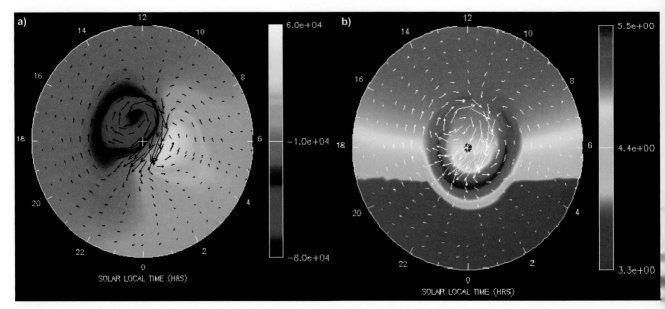

Fig. 12-19. Simulated Motion of Lower Thermosphere Neutral Particles Driven by $E \times B$ Drifting Ions for a Storm in April 2002. a) The view is from above the northern polar cap and extends to a latitude of 27 deg. Noon is at the top. The dawn-to-dusk electric field crossed with Earth's downward magnetic field produces anti-Sunward ion drift in the polar cap. We note the similarity between the circulation in this image and that in Figs. 7-28b and 12-14. In this image, background colors depict the electric potential pattern, which is slightly skewed away from noon by a large IMF B_y component in the solar wind. The highest potential is ~60 kV, and the lowest potential is ~–80 kV. During this storm, ion-neutral collisions created winds in excess of 100 m/s (the longest arrows). **b)** Here we show the same motion superimposed on an electron density map. The colors represent logarithms of ion density (#/cm^3). In this image, the highest electron density is in the auroral zones. The entire dayside is also strongly ionized. The wind pattern pulls dayside ionization into the polar cap. *(Courtesy of the High Altitude Observatory of the National Center for Atmospheric Research)*

The ion flow becomes stronger as the ionization rate increases. Storm-driven plasma convection, given enough time, creates thermospheric winds comparable to those created by thermal heating. The effect is time-lagged, because the plasma density is at least 1000 times less than the neutral density. As a result, the time required to generate storm neutral winds is from 30 minutes up to several hours. Ions are the main plasma participants in this interaction, because electrons are much too light to effectively transfer momentum to the much heavier thermospheric particles.

At low altitudes, frequent collisions allow the ions to heat the neutral particles. Auroral heating creates high-pressure bulges in the polar regions of both hemispheres, and this pressure drives equatorward blowing winds at all longitudes, at speeds of 50 m/s–200 m/s (Fig. 12-20). Some storm energy is dissipated in regions outside of the thermosphere. Molecular and turbulent thermal conduction moves heat from the thermosphere to the mesosphere, where collision frequencies are high enough that polyatomic molecules such as CO_2, O_3, and H_2O radiate infrared energy to help balance storm energy input.

In the daytime, storm winds oppose the circulation created by solar heating (in Fig. 8-5a). Additionally, if the magnetic pole is tilted toward the Sun, the neutral winds are slowed as neutral particles collide with excess ions created in the sunlit sector. The equatorward winds are usually stronger at night, because they add to the background day-to-night circulation and because they are reinforced by anti-Sunward ion drag caused by magnetospheric convection $E \times B$ drifts. The longitude sector with the strongest response tends to be on the nightside. In the

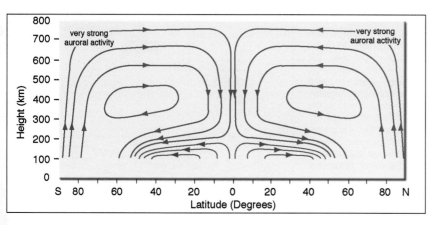

Fig. 12-20. Dayside Thermospheric Circulation Forced by Strong Storm Heating in the Auroral Zone. Joule heating in the auroral zone raises the neutral air temperature, making the air more buoyant and allowing it to rise. Air above 125 km displaces horizontally as well as vertically. Return circulation at high altitudes enhances the mid-latitude poleward circulation. [Roble, 1977]

midnight sector, the winds blow material far away from the auroral oval. The "winner" of this opposing-wind struggle depends on the severity of the storm, and perhaps other factors such as season and timing in the solar cycle.

Some of the auroral energy dissipates in impulsive waves of *traveling atmospheric disturbances (TADs)* that launch from the auroral zones and propagate equatorward (Fig. 12-21). Gravity is the restoring force for the vertically displaced parcels, so these disturbances are also called thermospheric *gravity waves*. In some instances, the density enhancements exceed an order of magnitude at satellite altitudes. Figure 12-21 is a schematic of propagating density enhancements from both auroral zones toward the equator, after the sudden injection of auroral energy during a geomagnetic storm. During such events, dayside TADs with typical density amplitudes ~20%–30% propagate toward the equator from the Northern Hemisphere and Southern Hemisphere auroral regions with phase speeds on the order of 500 m/s to 700 m/s.

EXAMPLE 12.4

Traveling Atmospheric Disturbances

- *Problem Statement:* How long does a TAD wave, generated at 70°, take to reach the equator?

- *Given:* The wave must travel a distance equivalent to 70° of latitude. The wave speed is 700 m/s.

- *Assume:* The wave travels at an altitude of 300 km. At that altitude, each degree of latitude equals ~115 km.

- *Solution:*

$$\frac{70° \times 115 \text{ km}/°}{(0.70 \text{ km/s})} = 1.15 \times 10^4 \text{ s} \approx 3.2 \text{ hr}$$

- *Interpretation:* About three hours after initial deposition at storm onset, neutral atmospheric energy redistributes globally. Waves, originating in both hemispheres, are likely to launch. At locations near the equator, a superposition of these waves may create a significant density upheaval three to four hours after storm onset.

**Fig. 12-21. Schematic for a Traveling Atmospheric Disturbance (TAD).
a)** Density residuals (enhancements and depletions) associated with a TAD appear as lighter and darker patches. **b)** This schematic illustrates two hypothetical, elongated density enhancements propagating away from the auroral regions. The first event propagates from both auroral zones, the second event from the southern auroral zone only. The density disturbance, caused by a simultaneous, impulsive Joule heating event in both hemispheres, is assumed to be aligned along lines of constant magnetic latitude. The almost-vertical dotted lines approximate the orbital path of the CHAMP satellite. At points of intersection between CHAMP and a density enhancement, positive density residuals are expected, flanked by negative residuals on either side. Each black arrow traces the path of a disturbance minimum or maximum. The heavy black curves represent disturbance maxima at subsequent times (t_1, t_2, t_3) from the start of the event. [Bruinsma and Forbes, 2007]

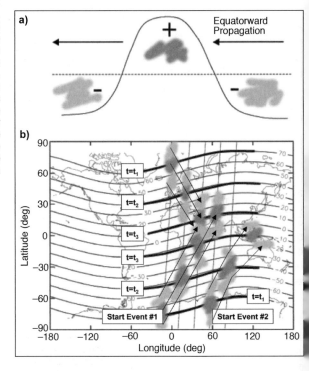

12.2.3   Outflowing, Upwelling, and Composition Changes in the Thermosphere

Objectives: After reading this section, you should be able to...

- State the effects of upwelling
- Describe the influence of production, loss, and transport on thermospheric composition
- Describe changes to Earth's atmospheric composition during space weather events

Outflow. The neutral atmosphere gains heat and momentum from storm energy deposition. In the high-temperature regime, thermospheric hydrogen and helium are bound weakly by gravity. Increasing temperature allows these atoms to escape (outflow) into space, enhancing the hydrogen and helium density in the geocorona but decreasing it at lower altitudes of the thermosphere. The flow to the geocorona is a tenuous polar breeze. Escaping neutral hydrogen and helium reduce the amount of hydrogen ions in the ionosphere, when thermospheric temperatures rise. A more energetic neutral outflow develops out of the storm-driven auroral zone. Neutral outflow of heavier matter (mostly atomic oxygen) develops when an energized outflowing of ions undergoes charge-exchange with the background neutral atoms. Outflow rates, primarily from the energized auroral zone, are on the order of 10^9 particles/(cm$^2 \cdot$ s). Over the extended periods of geomagnetic storms these particles change the mass and momentum balance in the inner and mid magnetosphere. Mass exchange between the atmosphere and magnetosphere is an active area of storm research.

12.2 Thermospheric Disturbance Effects

Upwelling and Composition Changes. Most storm energy is deposited deep in the thermosphere (100 km–150 km), causing the thermosphere to expand upward (but not escape), in an attempt to restore hydrostatic balance. This *upwelling* (air moving through constant pressure surfaces) results in molecular-rich air rising at high latitudes. To compensate, sinking occurs at lower latitudes. At higher altitudes, the upflowing mass from the large mass reservoir in the 100 km–120 km range provides a source of thermospheric density increase in the middle thermosphere.

Upwelling also produces departures from diffusive equilibrium and an increase in the mean molecular mass. Thermospheric heating produces an increase in oxygen atom density in the middle and upper thermosphere. Although oxygen atoms are held by gravity, rising temperature increases the scale height of atomic oxygen. Figure 12-22 shows simplified height profiles of the primary neutral molecular nitrogen and atomic oxygen in Earth's thermosphere. During quiet times, atomic oxygen accounts for 4% of the neutral density at the base of the thermosphere and 96% at 500 km. In strong storms, upwelling enhances oxygen density at higher altitudes. The ratio of atomic oxygen density to heavier background gases, such as molecular nitrogen, rises. One of the most common measures of thermospheric disturbance is the ratio $[O]/[N_2]$.

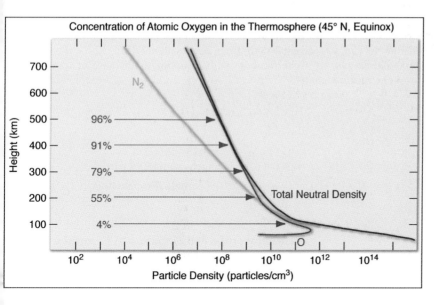

Fig. 12-22. Undisturbed Total and Relative Neutral Density Contributions from Atomic Oxygen and Molecular Nitrogen. Number densities are shown in units of particles per cubic centimeter. The results are from the Mass Spectrometer and Incoherent Scatter Model. Atomic oxygen is in red; molecular nitrogen is in green; and the total is in blue. The percentages represent the fraction of total neutral density composed of atomic oxygen. (Courtesy of the COMET Program at the University Corporation for Atmospheric Research)

We consider the dynamics of upwelling by assuming each gas species is in hydrostatic equilibrium, with the density of each species "s" decreasing exponentially with height according to the Law of Atmospheres, Eq. (3-6b). We rewrite the law here, allowing for multiple species in the thermosphere:

$$n_s(z) = n_{so} \exp\left(-\frac{z-z_o}{H_s}\right) \quad (12\text{-}6)$$

where

$H_s = k_B T/m_s g$ = scale height of species "s" (we assume that a single constant temperature T is valid for all species)

n_{so} = number density of species "s" at the height z_o (base of thermosphere)

Because species mass is in the denominator of the scale height, more massive molecules have smaller scale heights. In accordance with the negative sign in Eq. (12-6), the heavier thermospheric constituents have the larger exponential density decrease with height. As suggested by Fig. 8-2a, they are present only in tiny amounts at high altitudes, while the lighter elements are more abundant. Atomic oxygen dominates during quiet times above 200 km. Things change during storm time.

A mass continuity equation describes how the amount of a species of gas s changes in a volume. Gas may be added by chemical reactions (production P_s) and consumed by other chemical reactions (loss L_s). Further, the gas may move into or out of the volume (transport). The transport term is the product of the number density of gas species n_s and the drift velocity $\boldsymbol{u}_s$. (Here we use the vector $\boldsymbol{u}$ for neutral motion, to distinguish it from the velocity $\boldsymbol{v}$ of plasma particles). The product $n_s\boldsymbol{u}_s$ defines the particle flow rate (number flux) of the gas, and has dimensions of # particles per area per time. If we multiply $n_s\boldsymbol{u}_s$ by the mass per particle, we get mass per area per time (mass flux).

In general, all three processes—chemical production, chemical loss, and transport (number or mass flux)—act at the same time. The net effect of these is a change in the gas density with time in a volume (Eq. 12-7). For a given atmospheric gas, we write the continuity equation as the sum of transport and local chemical reactions:

$$\frac{\partial n_s}{\partial t} = -\nabla \bullet (n_s \boldsymbol{u}_s) + P_s - L_s \qquad (12\text{-}7)$$

where

$\partial n_s/\partial t$ = gas density change with time [# particles/(m³·s)]

$-\nabla \bullet (n_s \boldsymbol{u}_s)$ = divergence of the particle flow rate (transport) [#particles /(m³·s)]

u_s = particle drift speed [m/s]

P_s = chemical production of species "s" [# particles /(m³·s)]

L_s = chemical loss of species "s" [# particles /(m³·s)]

The first term on the equation's right side is the negative of the number flux divergence and represents the number density change of the gas of species "s" via transport. Transport changes the gas number density in a volume by importing more gas than exits or vice versa.

Pause for Inquiry 12-6

Verify that the first term on the right side of Eq. (12-7) has units of [# particles/(m³·s)]

Pause for Inquiry 12-7

Compare Eqs. (8-1) and (12-7). How are they the same? How are they different?

Focus Box 12.2: Storm-time Thermospheric Density Changes

Very often, we assume a balance of production and loss and focus only on gas motion. To illustrate this aspect of Eq. (12-7), we consider polar-orbiting satellite data from an altitude range of 231 km to 340 km over many orbits. Figure 12-23 shows relative gas density variations for argon, molecular nitrogen, atomic oxygen, and helium along the satellite's track as a function of magnetic latitude. Before the storm, all values are near unity. With storm onset, the atomic oxygen density remains steady and then decreases slightly. However, the heavier minor gases show a steep rise in relative density at high latitudes (factors of 80 for argon and 10 for molecular nitrogen), whereas the relative density of the lighter gas, helium, decreases at high latitudes by a factor of 10. The result is a composition bulge.

Assuming that $P_s - L_s \approx 0$, and that spatial changes in number flux occur primarily in the vertical direction, we rewrite Eq. (12-7) as

$$\frac{\partial n_s}{\partial t} = -\frac{\partial}{\partial z}(n_s u_s)$$

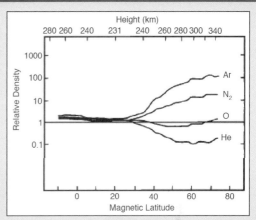

Fig. 12-23. **Ratio of Neutral Gas as a Function of Latitude before and during a Magnetic Storm.** The gas ratio relative to quiet time is shown on the vertical axis. The top horizontal axis shows the spacecraft perigee where the observation was made. The bottom horizontal axis shows the magnetic latitude of the observation. Although the observations are discrete, they have been connected by lines to guide the eye. The local time of the observations is 0900. Prior to storm onset, all species were at their expected density for the altitudes shown. After storm onset, atomic oxygen maintained its expected density, while the N_2 density increased. The factor of ~5 increase in N_2 density brought N_2 into competition with atomic oxygen as the dominant species at ~250 km. At higher altitudes (>300 km), N_2 had a larger relative increase, but actual N_2 density was still only a fraction of that of atomic oxygen (see Fig. 12-22). [Prölss, 1980]

From Fig. 12-23, we reasonably assume, to within a factor of two or so, that the density of atomic oxygen (the major neutral gas) is unchanged by the storm. Thus the number flux of oxygen $n_O u_O$ is approximately constant. But the thermospheric storm heating certainly causes atomic oxygen to flow upward ($u_O > 0$). Therefore, for the oxygen number flux to remain constant, the exponential decrease in number density n must be counterbalanced by an increase with height in speed u_O, as follows:

$$u_O(z) \propto exp\left(+\frac{z-z_o}{H_O}\right)$$

Strong vertical motions develop in the rarified regions of the thermosphere. Further, atomic oxygen drags the minor neutral particles upward at roughly the same speed. However, the minor species number flux is not constant with height, and thus for the minor species:

$$n_M u_M = n_M u_O \propto exp\left(-\frac{z-z_o}{H_M}\right) \cdot exp\left(+\frac{z-z_o}{H_O}\right)$$

where the subscript M indicates the minor gas number density and scale height.

To discover how the minor gas number density changes with time, we apply the vertical continuity equation, Eq. (12-7), to the minor species number flux expression:

$$\frac{\partial n_s}{\partial t} = -\frac{\partial}{\partial z} n_M u_O \left(exp\left[\frac{H_M(z-z_0) - H_O(z-z_0)}{H_M H_O}\right]\right) =$$

$$-\frac{\partial}{\partial z} n_M u_O \left(exp\left[-\frac{(z-z_0)}{(H_M H_O)/(H_O - H_M)}\right]\right)$$

After differentiating, the result is

$$\frac{\partial n_s}{\partial t} = n_M u_O \frac{1}{H'} \left(exp\left[\frac{-(z-z_0)}{H'}\right]\right)$$

where $H' = \frac{H_O H_M}{H_O - H_M}$ is a modified scale height involving the mass of atomic oxygen and the minor species gas being considered.

H' is positive if the minor gas is heavier than atomic oxygen (more massive gases have smaller scale heights). In this case, the minor gas density increases with time, which is just what happens to argon and molecular nitrogen in Fig. 12-23. Conversely, H' is negative for lighter gases such as helium, and so the helium density decreases with time. In the height range 231 km–280 km, heating and expansion processes bring molecular nitrogen and oxygen into competition for the dominant thermospheric species, and raise the overall atmospheric density by a factor of three or more.

Meridional Changes. Storm-time composition changes extend beyond the high-latitude thermosphere. Because of composition modifications, total thermospheric storm length may be several days. The region of altered composition is sometimes called a *composition bulge*; it represents a volume of increased average mass (Fig. 12-23). The excess mass results in higher than average pressure, which then causes pressure gradients that modify the global thermospheric circulation. Changes in temperature communicate forces to the air, which produce regional circulation cells and waves that locally displace parcels, but have a global long-range effect. Sudden upwelling generates a large-scale wave, a traveling atmospheric disturbance (TAD), early in the storm, which can propagate globally (Fig. 12-20). Later, large-scale changes in Joule heating generate further large-scale waves, and more local changes generate smaller-scale waves. As the storm continues, the horizontal, polar-cap anti-Sunward winds accelerate, and upwelling provides sufficient molecular-rich material for horizontal advection to cause large increases in the density of the molecular species over a wide area. Enhanced equatorward winds transport the composition changes to lower latitudes. The storm-time neutral winds transport thermospheric composition changes toward mid-latitudes. This effect is evident at night, where strong equatorward winds blow. Sometimes composition changes are also observed on the dayside.

During a geomagnetic storm, the twin-celled neutral convection pattern produces convergence near midnight, as shown in Figs. 12-19 and 12-24. Gas, rich in molecular species, may form a plume that co-rotation moves towards dawn. This leads to downward winds and heating. Where the air moves directly from the auroral oval, these downward winds do not compensate for the composition changes caused by upwelling in the auroral regions. The heating at midnight sets up a pressure gradient force that weakens the diurnal tide. This implies increased convergence, and heating, at other local times as well. The entire thermosphere heats up. Storm heating also creates winds in the thermosphere that transport the composition changes to other latitude regions.

Simulations suggest that after the composition bulge is created, it can be transported to middle latitudes by the nightside equatorward winds and brought onto the dayside as Earth rotates. Seasonal effects also arise in the simulations. The prevailing summer-to-winter circulation at solstice transports the molecular-rich gas to mid- and low-latitudes in the summer hemisphere over the day or two following the storm. In the winter hemisphere, poleward winds restrict the equatorward movement of the composition. The seasonal migration of the bulge is superimposed on the diurnal oscillation.

Our simple reasoning in Example 12.4 is consistent with model calculations, indicating that TAD impulses propagate with sufficient speed to reach from 70° latitude to the equator in less than four hours. The first signs of storm-related composition changes usually appear at low latitudes, less than four hours after a storm begins. The often uneven expansion of the thermosphere also produces pressure gradients, which drive strong neutral winds. Wind surges propagate from high latitudes around the globe and transport plasma along magnetic field lines to regions of altered chemical composition, changing ion recombination rates and total electron content (TEC). The disturbed thermospheric circulation alters the neutral composition, and moves the plasma up and down magnetic field lines, changing rates of production and recombination of the ionized species. At the same time, the disturbed neutral winds produce polarization electric fields, by a dynamo effect, as they collide with the plasma in the presence of Earth's magnetic field. These electric fields, in turn, affect the plasma and the neutrals.

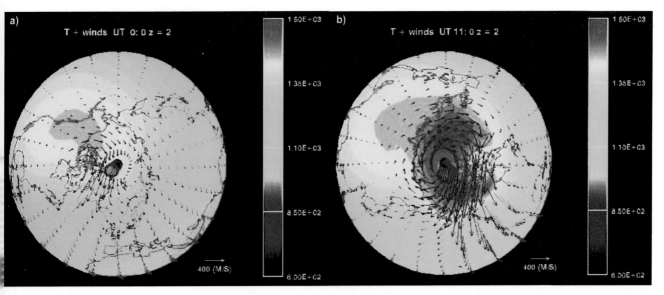

Fig. 12-24. Simulated Thermospheric Neutral Temperature and Winds at ~250 km. This simulation output is from the Thermosphere Ionosphere Electrodynamics General Circulation Model. The view is from above the geographic northern pole. Noon is at the top. **a) Quiet Conditions.** Neutral winds flow from the dayside to the nightside across the polar cap at low speed. **b) Active Conditions.** Higher temperatures (~1600 K) are shown in red, and cooler temperatures (~850 K) are in blue. Neutral winds flow predominantly from the dayside to the nightside across the polar cap. Storm-time winds exceed 700 m/s. Convergence of winds in the midnight region leads to vertical motion and neutral composition changes. *(Courtesy of Alan Burns at the National Center for Atmospheric Research)*

Thermospheric Cooling. We've shown many paths for energy input, transport, and dissipation in the thermosphere. Ultimately, much of the energy goes to heat. During solar minimum, molecular conduction to low altitudes is the main thermospheric cooling agent. Additionally, some of the heat goes into mechanical expansion of the atmosphere. Radiative cooling to space by nitric oxide (NO_x) at 5.3 μm and CO_2 at 15 μm is also important. In terms of radiative loss, NO_x is the larger cooling agent in the global average. Nitric oxide (NO) is extremely important in storm events.

The maximum NO density, which occurs near 110 km, is highly variable. Some of the variability is driven by solar X-ray flux in the 2 nm–10 nm band. Solar soft X rays ionize the major constituents (molecular nitrogen and atomic oxygen) of the lower thermosphere, producing energetic photoelectrons. In a cascade, these photoelectrons produce additional ionization and more photoelectrons. These in turn dissociate molecular nitrogen and excite the product nitrogen atoms. These then react with molecular oxygen to produce nitric oxide. Satellite observations over the tropics, where solar irradiance is most intense, reveal order of magnitude increases in nitric oxide density, after an intense solar flare.

Solar X-ray flux is only one source of NO variation. Numerous observational and modeling studies have linked auroral energy input with thermospheric NO concentration. The high-latitude abundance of NO is directly affected by precipitating electron flux that ionizes and dissociates molecular nitrogen. These processes enhance concentrations of excited-state atomic nitrogen, which in turn leads to greater NO production in the lower thermosphere. Joule heating also contributes to increased NO production, via a temperature-sensitive reaction between ground-state atomic nitrogen and molecular oxygen. Enhancements in temperature and molecular oxygen density lead to excess NO at altitudes above 140 km.

Further, enhanced auroral energy deposition influences the collisional excitation of NO, which occurs primarily via inelastic collisions with atomic oxygen. A change in the concentration of atomic oxygen is thus accompanied by a change in the number of excited NO molecules. Rising background temperatures, associated with Joule heating, increase the rate of collisional excitation. Recent observations of the infrared radiance instrument onboard the Thermosphere Ionosphere Mesosphere Energetics and Dynamics (TIMED) satellite show dramatic increases in 5.3-µm emissions during periods of enhanced solar and geomagnetic activity. Increases in 5.3-µm cooling dampen thermospheric temperature increases arising from increased geomagnetic activity, and reduce timescales of temperature relaxation in the recovery phase of a storm. Recent modeling and observational studies suggest that meridional winds transport excess NO, produced at auroral latitudes, to middle and low latitudes. Enhanced levels of NO, created by solar and high auroral activity, therefore result in an increase in 5.3-µm radiative cooling on a global basis. Figure 12-25 shows before- and after-storm images of nitric oxide radiance. The enhanced NO radiance is the result of strong solar activity on 12–14 April 2002, followed by an intense geomagnetic storm on 17–18 April 2002.

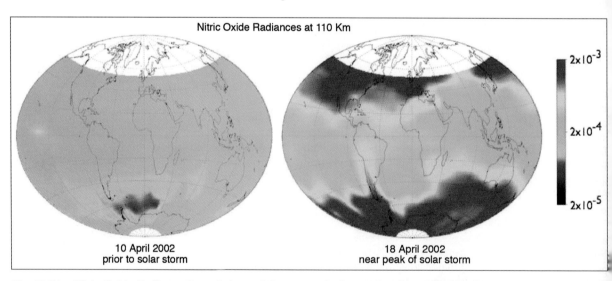

Fig. 12-25. Nitric Oxide Radiance for a Solar and Geomagnetic Storm Sequence in April 2002. Nitric oxide emissions, measured by the Sounding of the Atmosphere using Broadband Emission Radiometry (SABER) instrument aboard the TIMED polar-orbiting spacecraft, increased by an order of magnitude in auroral and sub-auroral regions, during a strong solar-geospace disturbance. Most of the radiance increase on 18 April 2002 was the result of recent auroral activity associated with a geomagnetic storm in which Dst dropped below 100 nT. *(Courtesy of the Space Dynamics Laboratory at Utah State University and the NASA TIMED Team)*

12.3 Ionospheric Storm Effects

We know many causes of variability in the ionospheric plasma. As we describe in Chap. 8, the ionosphere responds to solar EUV radiation, which causes the ionospheric electron concentration to vary diurnally, latitudinally, seasonally, and over the 11-year cycle of solar activity. The polar ionosphere can be deeply disturbed by very energetic particles accelerated in advance of fast-moving CMEs. Internally, the composition of the thermosphere controls electron density by affecting the rate at which ions and electrons re-combine. Thermospheric winds, through collisions, transport the ionospheric plasma along the magnetic field lines, thus raising or lowering the ionosphere. Further, the geometry of the magnetic field and the presence of electric fields cause plasma drift across magnetic field lines that leads to, for example, the equatorial anomaly that consists of an ionization trough at the geomagnetic equator and ionization crests on either side.

12.3.1 Ionospheric Polar Cap Absorption Events

Objectives: After reading this section, you should be able to...

- Describe the sources of polar cap absorption events
- State locations affected by polar cap absorption events

Solar proton events (SPEs) are major (though infrequent) space weather phenomena that inhibit radio communication in the polar regions of the ionosphere. Protons in the MeV energy range are the primary species in SPEs, although lower fluxes of heavier ions (e.g., He, Fe, and O in various charge states) and electrons are also present. The ionospheric effects of SPEs produce large signal losses on high-latitude VHF communication circuits. Increased ionization at low altitudes also creates sudden disturbances of phase and amplitude on LF and VLF radio signals. The signal loss, or absorption, creates blackouts on ionospheric sounding equipment. These events are called *polar cap absorption (PCA)* events.

PCA events originate with fast coronal mass ejections (CMEs) that accelerate a small part of the ambient coronal or solar wind particles to a sizable fraction of light speed (~0.4c for a 100 MeV proton). Some solar flares also accelerate particles that reach Earth. The particles become trapped in Earth's nearly vertical polar field lines and penetrate into the D layer in the polar regions. The protons gain access to the ionosphere along the open magnetic field lines in the polar caps and, because of the high energies involved, cross closed field lines into the auroral zone. The disturbance may extend to 65 degrees latitude. In such events, the lowest usable frequency (LUF) available for radio transmission exceeds the frequency for trans-ionospheric transmission. No radio communication using the E and F layers is possible.

Figure 12-26 shows two time profiles of PCA effects. One is associated with particles accelerated on field lines connected to the solar central meridian, and the other shows CME particles accelerated along field lines from a more westerly solar location. Energetic particles originating near the western solar limb are guided by the magnetic Parker spiral to Earth. Consistent with the top right image in Fig. 10-20, these are the particles with the best connection to the vicinity of Earth. In rare instances, a behind-the-limb CME may create a PCA event.

Fig. 12-26. Relative Intensity and Timing of Polar Cap Absorption (PCA) Events. On the left, the CME-generated energetic proton flux is created near the solar central meridian; much of the flux bypasses Earth. On the right, the particles accelerated by the CME are well connected to Earth. The PCA event onset is sharp, strong, and long-lasting. *(Courtesy of Margaret Shea at the AF Research Laboratory)*

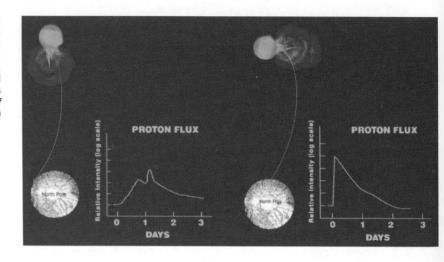

The principal Earth-based method of observing PCAs is with a relative ionospheric opacity meter (RIOMETER), which measures the absorption of the cosmic radio noise at a given high frequency, usually between 20 MHz and 60 MHz. During a PCA, the majority of radio absorption happens at low altitudes because the precipitation of high-energy solar protons (>10 MeV) creates enhanced ionization at low altitudes. The atmospheric penetration of these particles is much greater than that of typical auroral electrons. Figure 12-2 shows the range of heights associated with solar energetic proton penetration. Protons with energies of 100 MeV deposit energy at altitudes as low as 30 km.

Figure 12-27 shows the locations of energetic proton flux observed by a low-Earth orbiting spacecraft, during an SEP storm on July 15, 2000. An X-ray flare and Earth-directed CME were the sources of these particles. All polar transits on that date revealed significant particle penetration into the atmosphere. The white curves show contours of three values of the L coordinate. The outermost contour is $L = 3$, the center is $L = 4$, and the innermost is $L = 6$. The plots provide a picture of the extent of the northern and southern polar areas being affected during a solar energetic proton event. This also provides an indication of those HF radio propagation paths that will be badly degraded because of the signal absorption.

12.3 Ionospheric Storm Effects

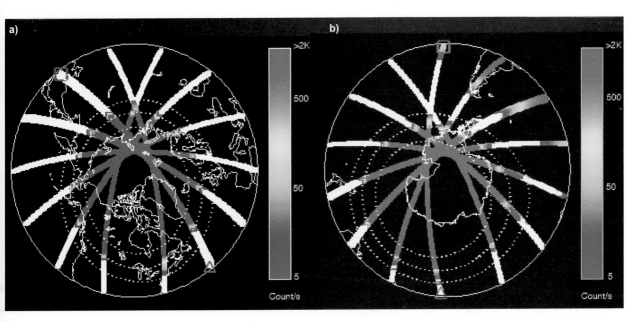

Fig. 12-27. The Polar Extent of Energetic Particle Penetration Associated with a Polar Cap Absorption (PCA) Event. a) Energetic proton counts during 14 northern polar passes of the POES NOAA-15 satellite. **b)** Energetic proton counts during 13 northern polar passes of the POES NOAA-14 satellite. The 16-second averaged count rates from the >15 MeV (nominally 15 MeV–70 MeV) proton detector on each satellite are color-coded and plotted along that satellite's transit. The geographic location of each data point is not the location of the sub-satellite point but rather the location mapped along the geomagnetic field from the satellite to 120-km altitude, where the solar protons enter the atmosphere and produce the excess ionization. The red box shows the satellite location at the beginning of the day. The red triangle indicated its location at the time of the last data download. *(Courtesy of NOAA's Space Weather Prediction Center)*

12.3.2 Ionospheric Storms and Disturbance Features

> **Objectives: After reading this section, you should be able to...**
> - Describe the E-region dynamo
> - Compare and contrast ionospheric disturbances at different latitudes
> - Explain why eastward electric fields are important for low-latitude ionospheric disturbances
> - Describe equatorial spread F and equatorial plasma bubbles
> - Explain the sources of polar cap scintillations
> - Explain the chain of chemistry that leads to ionization loss in a steady-state ionosphere
> - Compare and contrast ionospheric disturbances at different latitudes
> - Compare and contrast negative and positive ionospheric storm phases
> - Describe common properties of ionospheric storms
> - Describe the sources of mid-latitude storm-enhanced density (SED) plumes

Predicting the outcome of a specific geospace storm, as a function of latitude and time, is still on the cutting edge of research, and we don't pursue such advanced topics here. However, we do outline the physics and dynamics that cause the plasma density to change. In preparation for this discussion, Focus Box 12.3 extends the Chap. 8 discussion of ionospheric formation, to explain how ionization is destroyed in the upper atmosphere. The balance between production and destruction then dictates the plasma density in Earth's ionosphere.

Disturbances at high latitudes are ubiquitous. Only weak magnetospheric forcing is needed to disrupt the layering of high-latitude regions. These regions often experience radio signal scintillation because blobs and streaks of plasma can be torn from the dayside and moved across the polar cap by the field lines caught up in magnetospheric convection. At mid-latitudes, the summer and equinox seasons experience decreases in ionization density, especially during strong storms. Density reductions are more common in winter months during weak magnetic activity. The mid-latitude nightside ionosphere tends to be more structured than the dayside ionosphere.

The term *ionospheric storm* refers to changes in plasma densities and dynamics in response to upper atmospheric storm energy. Geomagnetic storms that deposit energy at high latitudes produce atmospheric waves and changes in thermospheric winds and composition. Both increases and decreases in the electron concentration occur. These changes hinder high-frequency (HF) and satellite communications and degrade satellite precision navigation accuracy. Traveling ionospheric disturbances (TIDs) are waves or ripples in the electron density that propagate horizontally. They affect the refraction of radio waves and so degrade directional finders.

Additional (and sometimes separate) storm features of low-density plasma plumes or bubbles develop at low latitudes. These voids or depletions can span the topside ionosphere to heights of 1500 km. Locations on the nightside of the dusk terminator and 10°–20° either side of the equator are particularly productive

Focus Box 12.3: Ionospheric Chemical Processes

Here we explore some of the chemical factors that determine the upper-atmospheric plasma density. In Chaps. 3 and 8, we describe how ionospheric plasma is produced. Although plasma production may be intense and rapid, it does not necessarily follow that plasma density is high. Recombination and chemical destruction may be equally robust. We determine the plasma density by balancing production and chemical destruction of plasma. This kind of equilibrium remains throughout much of the sunlit ionosphere during quiet periods. Storm periods upset this. Before covering this topic, however, we need to say more about the equilibrium plasma density that exists prior to a storm.

Plasma is created in Earth's upper atmosphere when neutral particles absorb solar ultraviolet radiation (or are impacted by energetic electrons), resulting in a positive ion and a free electron from each neutral particle. For example, some neutral species "A" could be ionized as follows:

$$A + hf \Rightarrow A^+ + e^- \qquad \textit{photoionization}$$

Above about 150 km, the species "A" is mainly atomic oxygen. Below this altitude, molecular oxygen and nitrogen (O_2 and N_2) also take part. For simplicity, we don't address this more complicated case. As in Chap. 8, we compute the O^+ (and thus the free-electron) production rate by knowing the density of atomic oxygen and the intensity of the ionizing radiation. To do this computation, we also need to know how efficient this reaction is. In other words, when an oxygen atom and an extreme ultraviolet (EUV) photon come together, how likely is this reaction? This probability is determined by the laws of quantum physics, and is beyond the scope of this text. However, we know the answer is not 100 percent. Although we might think that plasma is destroyed by reversing the direction of the reaction (radiative recombination), the laws of quantum physics predict that it's rather improbable, compared to other types of chemical reactions. In reality, the most likely way to destroy plasma is through a chain of chemical reactions. The first reaction changes the neutral oxygen atom into an ion

$$O + hf \Rightarrow O^+ + e^- \qquad \textit{photoionization}$$

followed by

$$O^+ + N_2 \Rightarrow NO^+ + N \qquad \textit{ion-atom interchange}$$

In this case, the molecular ion is nitric oxide. Finally, the molecular ion and a free electron recombine to destroy the plasma pair

$$NO^+ + e^- \Rightarrow N + O \qquad \textit{dissociative recombination}$$

This process is *dissociative recombination*, because the molecule dissociates into two atoms. This two-step process can also involve molecular oxygen

$$O^+ + O_2 \Rightarrow O_2^+ + O \qquad \textit{ion-atom interchange}$$

followed by a recombination that dissociates the molecular oxygen into atomic oxygen. The requirement for conservation of energy and momentum make the dissociation a far more likely event than the production of neutral oxygen molecules.

$$O_2^+ + e^- \Rightarrow O + O \qquad \textit{dissociative recombination}$$

breeding grounds for low-latitude plasma bubbles. Strong magnetic storm forcing may spread the less-dense plumes to longitudes spanning much of the night side. Radio signals used in the GPS system can experience phase shift and scintillations when they traverse these highly structured plumes.

Predicting ionospheric response to geomagnetic storms and local disturbances is of great practical importance. Originally the greatest need was to forecast storm disturbances for ground-to-ground communication via the ionosphere using HF radio propagation. With the advent of satellites and associated communication paths, storm effects on higher frequencies became a concern. Regional and global navigation systems present new forecasting challenges. Irregularly structured ionospheric regions can cause diffraction and scattering of trans-ionospheric radio signals (ionospheric scintillation). These compromised signals present random temporal fluctuations in both amplitude and phase to a receiver. The disturbances also challenge accurate assessment of the vertical delay calibrations used in aircraft navigation systems. The parameters of most practical consequence are the peak F-region electron density (NmF2), or critical frequency (foF2), which is related to the maximum usable frequency (MUF) for oblique propagation of radio waves, and the total electron content (TEC), which is significant for the phase delay of high-frequency ground-to-satellite navigation signals.

Next we organize the effects of disturbed ionospheric behaviors into three latitude bands: low-, mid- and high-latitudes, and then further discuss the multi-latitude disturbances called ionospheric storms and traveling ionospheric disturbances.

Low Latitudes. Figure 4-26 shows the quiet dayside ionospheric current system. Solar-driven atmospheric heating on the dayside and radiative cooling on the nightside generate tidal winds that move ionospheric plasma vertically through the geomagnetic field. The plasma motion induces an eastward electric field and current in the dayside dynamo region between 100 km and 140 km. This zonal electric field is called a dynamo electric field because it results in conversion of mechanical energy of the winds to electrical energy of the plasma.

Near-equatorial Internal Field Sources. To be consistent with upward plasma drift across a horizontal magnetic field, the electric field must be eastward. (We recall that plasma drifting across magnetic field lines does so because of $E \times B$ interactions, as in Example 12.5). At night, when the atmosphere cools, the neutrals and plasma sink to lower altitudes (called downwelling). Downwelling motion creates a westward electric field. The induced field drives currents (Fig. 4-26 and Example 12.5). Because the conductivity changes near the terminator, these currents diverge on the evening side and converge on the morning side, producing positive and negative charge accumulation, respectively. The dayside electric field results from a ~5 kV electric potential difference.

Figure 12-28 illustrates that the eastward electric field in the E region at off-equatorial latitudes maps along Earth's magnetic field to F-region heights above the magnetic equator. The resulting $E \times B$ drift transports F-region plasma upward at the magnetic equator. The uplifted plasma then moves north to south along the nearly horizontal magnetic field in response to gravity and pressure-gradient forces. As a result, an *equatorial ionization anomaly* forms, with minimum F-region ionization density at the magnetic equator and maxima at the two crests approximately 15° to 20° in magnetic latitude to the north and south. The crests of plasma density contain atomic oxygen that emits ultraviolet radiation (Fig. 12-18). The ionization crests are subject to destabilization from local and distant perturbations.

12.3 Ionospheric Storm Effects

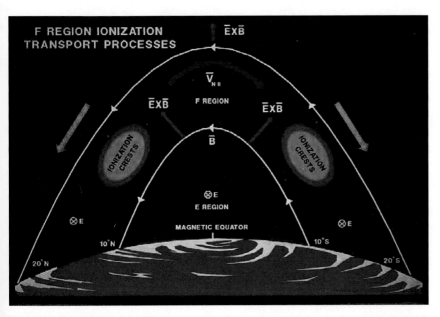

Fig. 12-28. **Formation of the Latitude Variation of Ionization Density in the Equatorial F Region (Equatorial Ionization Anomaly).** The diagram illustrates that, during daytime, the eastward dynamo electric field from the E-region maps along the magnetic field to F-region heights above the magnetic equator. The plasma moves upward because of $E \times B$ drift and then diffuses along the magnetic field to form two crests with maximum ionization density near ±15° magnetic latitude and minimum ionization at the magnetic equator. Gravity and the pressure gradient force move the plasma downward along field lines and out of the region where it lifted. *(Courtesy of the AF Research Laboratory)*

Sudden changes in the electric field in the inner magnetosphere can penetrate into the low-latitude F region and either strongly enhance or completely disrupt the direction of the dynamo electric field. When this happens, the low-latitude plasma undergoes sudden vertical upwelling or downwelling that translates into motion along the bending magnetic field lines and to outright plasma instabilities. This phenomenon is called *spread F*, indicating that the F layer is not found at a single height but spreads vertically. Equatorial spread F (ESF) is one of the most troublesome ionospheric space weather issues. The instabilities instigate the formation of field-aligned plasma depletions, sometimes called *equatorial plasma bubbles (EPBs)*, which rapidly rise to the topside F layer. The sharp gradients in plasma density disrupt the smooth propagation of radio waves. Under the most severe circumstances, UHF signals to and from satellites may go completely astray.

Near sunset, plasma densities and dynamo electric fields in the E region decrease, and the ionization anomaly begins to fade. However, another dynamo develops in the F region. The result is that polarization charges within conductivity gradients at the terminator enhance the eastward electric field for about an hour after sunset. This is also where the dayside eastward and nightside westward currents meet. The sources of the conductivity gradients are the subject of intense research, because this *pre-reversal enhancement (PRE)* can produce excess plasma uplift, which can be further acted upon by other instabilities. The post-sunset (pre-reversal) electric field moves the ionospheric plasma upward, allowing the ionization anomaly crests to intensify.

Often, after sunset, vertical plasma density gradients form in the bottom side of the F layer. This happens as plasma recombines to create neutral gas at low altitudes, but remains ionized in high-altitude regions where the setting Sun is still shining. The upward density gradient is opposite in direction to gravity. This configuration, in which a heavy fluid (high ionization) rests on top of a light fluid (low ionization), is unstable and allows plasma density irregularities to form. In fluid descriptions, this instability is called a *Rayleigh-Taylor instability*. Bubbles of low-density plasma shoot upward, sometimes to heights of more than 1000 km

EXAMPLE 12.5

Eastward Electric Fields and Electrojets

- *Problem Statement:* Verify that the dynamics described in the previous discussion produce eastward equatorial electric fields and an equatorial electrojet in the ionospheric E region.
- *Given:* North-directed, horizontal magnetic field at 100 km above the magnetic equator
- *Assume:* Daytime heating causes uplift in the neutral atmosphere. The uplifting neutral gas drags the plasma along. The electron mobility is larger than the ion mobility.
- *Solution:* Part I
  a) The drift electric field direction is described by $E = -v \times B$

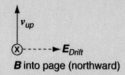

  b) In response to the drift electric field, both species accelerate. However, the electrons are more mobile, so they accelerate westward, which creates an eastward current $j_{primary}$. The mobility of the electrons also produces an upward polarized electric field E_{pole}.

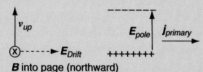

- *Interpretation:* Consistent with the problem statement, the drift electric field and associated current are directed eastward in the E layer. However, the geometry shown in the second sketch suggests that there is also an upward directed polarization electric field, created by the differences in charge mobility. We tackle this next.

- *Solution:* Part II
  c) The westward motion of the electrons across magnetic field lines is consistent with a secondary Hall electric field E_{Hall}, directed vertically downward.

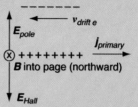

  d) The mobile electrons respond to E_{pole} by accelerating downward, but this motion is across the magnetic field and results in an $E \times B$ drift of the electrons to the west, producing an east-directed secondary current that adds to the primary current.

- *Interpretation:* The mobile electrons are constantly adjusting to the induced and polarization electric fields. The result is an east-directed, primary electric field and the sum of an eastward primary and secondary current that together form the equatorial electrojet, which is a narrow, intense eastward current across the dayside.

Follow-on Thought Exercise: Redo the geometry and thought process for nightside behavior.

(Fig. 12-29). Strong ESF and plasma depletions (plasma bubbles) arise during the PRE of the zonal electric field, when upward $E \times B$ plasma drifts elevate the F layer sufficiently for instability initiation (Focus Box 12.3).

External Field Sources. In addition to internal ionospheric sources of eastward electric field in the post-sunset realm, the atmosphere produces electric fields in other ways. Tropospheric gravity waves, tropical storms, and distant lightning are considered contributors. The upward push by waves from these sources creates local zonal currents and concomitant east and west fields. After an instability develops, the presence of a westward electric field, if it exists, may not be sufficient to damp the growth.

Additionally, external forcing comes from the magnetospheric penetration of electric fields during strong magnetic storms. Currents flowing in the equatorial

12.3 Ionospheric Storm Effects

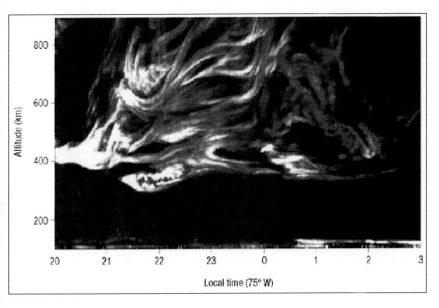

Fig. 12-29. **Plumes of Plasma Turbulence Observed by the Jicamarca Radio Observatory in Peru.** The image is a 7-hour history at 75 W longitude, of radio signal returns from plasma turbulence associated with electron plasma depletions. The turbulence and depletions develop in regions of eastward electric fields and the Rayleigh-Taylor instability. The convective ionospheric storm occurred in October 2006. Eastward (westward) post-sunset electric fields enhance (lessen) the instability. GPS signals that encounter the sharp gradients in electron density scintillate and may undergo loss of lock between receivers and satellites. On some days, E is too small to cause equatorial anomaly crests or ionospheric irregularities before its polarity reverses. [Kelley et al., 2006]

ionosphere are affected by intensifications of the southward interplanetary magnetic field. In turn, the dawn-to-dusk electric field E_y increases throughout the magnetosphere. This electric field causes energetic ions in the inner magnetosphere to gain kinetic energy and flow toward Earth. As a result, the gradient-curvature drift paths of ions and electrons begin to separate and the ring current and partial ring current grow. If strong electric fields continue for more than an hour, space charge builds up to shield the innermost parts of the magnetosphere from the strong E_y.

Magnetometers at low latitudes on Earth detect decreases in the north-south component of the local magnetic field caused by ring current growth. Negative slopes in Dst (dDst/dt < 0) identify times when the interplanetary magnetic field is imposing a dawn-to-dusk electric field in the inner magnetosphere, Earthward of the existing ring current.

Region-1 and -2 field-aligned currents also globally affect the electric field. Depending on the orientation of the zonal field at the equator, these currents either suppress or trigger ESF and equatorial plasma bubbles. Simulations show that Region-1 currents produce eastward electric fields of ~1 mV/m just east of the dusk terminator. The Region-2 currents shield the equatorial region from the Region-1 currents and tend to suppress, if not reverse, the electric field. Therefore, the efficiency of high-latitude currents in creating and suppressing scintillations depends on the phase of the magnetic perturbation.

Recent studies demonstrate that plasma bubbles form frequently in the initial and main phases of geomagnetic storms. Penetration electric fields are the likely triggers for the creation of these structures. Bubbles are seldom detected during local evening in storm recovery phases. They are, however, evident in the post-midnight-to-dawn local time sector. Bubble suppression in the evening sector and growth after midnight is attributed to the action of a counter dynamo driven by penetration E fields or by energy deposition at nighttime auroral latitudes.

High Latitudes. Major effects of geomagnetic storms occur in the high-latitude E and F regions, where particle precipitation in the cusp and aurora produce ionization. Strong magnetospheric convection electric fields map along magnetic

Focus Box 12.4: Physics of Equatorial Plasma Bubbles: Rayleigh-Taylor Instability

Here, we briefly describe the elements of a Rayleigh-Taylor instability as it applies to plasma bubble formation. Unlike most space weather conditions, in this situation, gravity plays a role in the plasma behavior. The geometry is shown in Fig. 12-30. In the diagram, g points down (toward Earth), and B points into the page (northward).

Initially, dense plasma resides on top of significantly less-dense plasma. This situation may develop just after sunset, when the lower ionosphere moves into darkness and plasma begins to recombine. Meanwhile, the F region is still sunlit, and the low densities in the F region ensure that the plasma is long-lived.

The $g \times B$ drift causes ions to flow toward the right (eastward). In the presence of a seed perturbation (e.g., a sinusoidal wave caused by some local or distant disturbance), ions pile up on one side, and electrons on the other. These charge displacements generate a perturbation electric field δB. The local $E \times \delta B$ drift causes ridges to rise and troughs to deepen. The developing ridges carry depleted plasma upward.

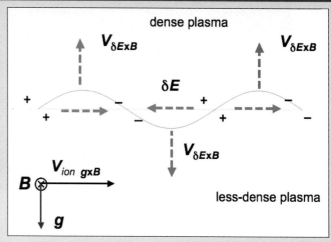

Fig. 12-30. **Fields Associated with a Rayleigh-Taylor Instability.** The view is from the south looking north. Small perturbations of the plasma by local wave sources seed the instabilities.

The growth rate γ of the instability depends on several factors. Among these are the ion-neutral collision frequency ν and the vertical density gradient. The stronger the density gradient, the larger the growth rate. The smaller the ion-neutral collision frequency, the larger the growth rate. The base of the F layer provides ideal conditions for instability growth.

$$\gamma \propto -\frac{g n_0}{\nu}\left(\frac{\partial n_0}{\partial z}\right)$$

Small scale instabilities grow into structures that traverse several hundred vertical kilometers in a matter of 1–2 hours. In turn, these structures cascade to smaller scales (200 km–30 cm). Non-linear growth results in plasma depletions that are a different shape from the initial seed perturbation. The amplitude growth rate may depend on the size of the seed perturbation (Fig. 12-30).

---

field lines to the polar ionosphere, where they channel dayside gas with higher plasma density across the polar cap and into the nightside. Sometimes these plumes appear as tongues of ionization drawn from the dayside into the polar cap (Fig. 12-31). The flows often decay into blobs and streaks in their transit of the polar cap. If the convection speed is moderate to high, the enhanced ionization flows into the nightside, where additional ionization may develop as the result of auroral particle impacts. Some of the flow attempts to return to the dayside via the dusk channel, and some returns via the dawn channel. The co-rotation electric field tends to slow the flow in the dusk return channel and accelerate the flow in the dawn channel. The combination of motions leads to patches of ionization with sharp ionization gradients. These cause large variations in the radio propagation characteristics of the polar and auroral ionosphere. The resultant variation in the number density and height of the F2 layer leads to scintillation in GPS signals and unusual returns from over-the-horizon scanning radars. Although strong convection is a hallmark of the southward IMF, signal scintillation may also be present during northward IMF, if Sun-aligned arcs develop in the polar cap.

12.3 Ionospheric Storm Effects

We recall that during intense magnetic storms, energy exchange between the plasma and the neutral particles maximizes in the E region, producing substantial auroral Joule heating. The effects of this energy deposition are observed well into the low latitudes. Large-scale, equatorward flow (~150 m/s) and traveling atmospheric disturbances (TADs) surge from the auroral regions, producing meridional winds that veer westward under the influence of the Coriolis force. The dynamo action of these winds drives an equatorward Pederson current and a resulting charge buildup at the equator, which in turn creates a poleward electric field. This field, when crossed with the inclined magnetic field shown in Fig. 12-28, produces a current consisting of mobile westward-moving electrons. The current produces negative charge buildup at dawn and a positive charge buildup at dusk. This charge configuration is opposite that of the quiet-day configuration. The resulting zonal electric field from the disturbance dynamo opposes the quiet-time dynamo electric field. In particular, the field disruption on the nightside may instigate upward plasma drifts in the post midnight regions that are normally down-drifting. Simulations show that post-sunset spread-F disturbances tend to dampen and the post-midnight ones grow during strong geomagnetic activity.

Middle Latitudes. The mid-latitude region is generally far from the spread-F dynamics we describe in Sec. 12.3.2. However, the effects of plasma transport, linked to a strong electric field in the overlying plasmasphere, during extreme storms, often overwhelm ionospheric characteristics at auroral and sub-auroral latitudes. With increasing magnetic activity during geomagnetic storms, the equatorward extent of such effects expands to lower latitudes. Large-scale enhancement of the ionospheric convection electric field allows solar-produced F-region ionospheric plasma to be transported Sunward and poleward from a source region at middle and low latitudes in the afternoon sector.

Additionally, storm-induced electric fields penetrate the inner magnetosphere, where they uplift and redistribute the plasma of the low-latitude ionosphere. Eastward electric fields near dusk produce a poleward displacement of the equatorial ionization anomalies and enhancements of TEC in the post-noon plasmasphere and mid-latitude ionosphere. These strong magnetospheric electric fields are generated as storm-injected energetic particles fill the enhanced ring current. In the meantime, sub-auroral polarization electric fields erode the plasmasphere boundary layer, producing plasmaspheric drainage plumes, which carry the high-altitude material toward the dayside magnetopause. The near-Earth footprint of the plasmaspheric erosion events is seen as the mid-latitude streams of *storm-enhanced density (SED)*, which sweep poleward (Fig. 12-32). With time, the ionospheric convection transports the enhanced F-region plasma into the polar cap (Fig. 12-31). These processes produce storm fronts and narrow plumes of dense thermal plasma, which extend continuously from low latitudes into and across the polar regions.

A storm-enhanced density plume is a continuous, large-scale feature spanning the afternoon sector at latitudes between the polar cap and the mid- to low-latitude source region. These streaks account for the pronounced enhancement of ionospheric density near dusk at middle latitudes. SEDs are observed during the early stages of some magnetic storms. They are detected in total electron content (TEC) maps. We recall that TEC is a measure of the total number of electrons in a vertical column that extends from the ground through the ionosphere and into the plasmasphere. Within the SED plume, TEC is greatly enhanced and steep latitude gradients of TEC appear where the SED extends into the dusk sub-auroral regions. Recent investigations have found that SEDs and plumes of greatly elevated TEC are linked to the erosion of the outer plasmasphere by strong sub-auroral electric

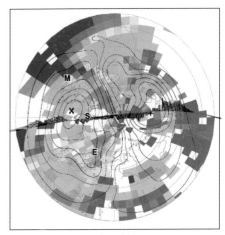

Fig. 12-31. **Multi-instrument Map of a Tongue of Ionization, Overplotted on the Ionospheric Convection Pattern at 1730 UT, 20 Nov 1993.** Noon is at the top, and the north geomagnetic pole is in the middle. Latitude circles are drawn in 10° increments. The convection pattern is provided by the Super Dual Auroral Radar Network and the Defense Meteorological Satellite Program's F-13 spacecraft. Ionization information comes from GPS total electron content observations. The positions of the three Incoherent Scatter radars are indicated (M, S, E). At 1730 UT, a plasma plume extended through the cusp convection convergence zone near noon. The tongue of ionization was drawn from dayside and duskside mid-latitudes across the polar cap. Subsequent measurements suggest the ionization became turbulent and patchy as it moved to the nightside and then cycled back into the auroral and sub-auroral regions. [Foster et al., 2005]

Chapter 12 Space Weather Disturbances in Earth's Atmosphere

Fig. 12-32. Development of a Storm-enhanced Density Plume on April 11, 2001. Maps of total electron content (TEC) before a storm (a) and during a storm (b). Blue colors represent low values of TEC. Red and brown colors represent high values. The interval between the two images is about 4 hours. The maps show dramatic density features developing over the continental US during a large geomagnetic storm. TEC is measured by ground station receivers, distributed across the globe, which continually monitor signals from GPS satellites. The image on the right shows how the plume of high TEC moved northwestward over the south, central, and eastern United States and eastern Canada. *(Courtesy of John Foster and Anthea Costner at the Massachusetts Institute of Technology's Millstone Hill Radar, the National Science Foundation, and NASA)*

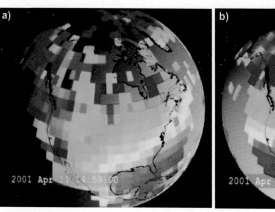
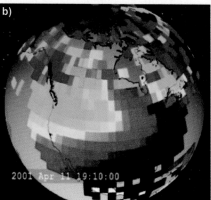

fields. The SED-TEC plumes identified at low altitudes map closely onto the boundaries of the dayside plasmaspheric drainage plume determined by EUV imaging from the IMAGE spacecraft (Fig. 11-13).

Some investigations suggest that the North American continent is a favored region for intense SED activity, because the weak magnetic fields of the South Atlantic Anomaly allow easier transport of plasma out of the equatorial ionization crests.

Multi-Latitude Storm Features. Although complicated, ionospheric storms have several common properties: enhancements (positive phase) and depletions (negative phase) of electron density occur in different regions. The location and timing of these perturbations depend on the strength of magnetic activity and the local time of storm onset. Some storm effects seem to propagate from high to low latitudes. Ionospheric storms bring large changes in ionospheric density distribution, total electron content, and the ionospheric current systems. Ionospheric electron densities in a given latitude band can either increase or decrease over the evolution of a geomagnetic storm. The former condition is called a *positive ionospheric storm*, the latter a *negative ionospheric storm*. These ionospheric storm effects develop in concert with geomagnetic storms. They often depend on local time and the time since storm onset. Figure 12-33 shows the temporal development of the ionospheric storm phases at a mid-latitude station in Rome, Italy.

The positive phase usually occurs in the early stages of a storm in or near the polar region. Joule heating from dissipation of electric currents, and energy deposited by particles flowing from the magnetosphere, cause an increase in the atmospheric pressure at high latitudes. A meridional wind from the polar region, carried by fast-moving traveling atmospheric disturbances (TADs), propagates energy to lower latitudes. The enhanced equatorward winds lift the plasma to higher altitudes in the F-region ionosphere, where fewer molecular species are present and recombination is slow; hence the increase in electron density. This upward drift may also be caused by stormtime magnetospheric convection electric fields that penetrate to Earth's dayside.

Negative storm effects develop when the ratios $[O]/[N_2]$ and $[O]/[O_2]$ decrease. The regions of upwelling (increases of N_2 and O_2) hasten recombination and thus cause the ionosphere to decay faster, creating negative phases. These composition disturbances often result from thermospheric wind

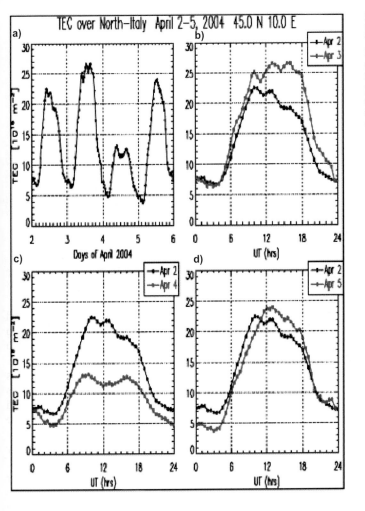

Fig. 12-33. **Total Electron Content (TEC) Positive and Negative Ionospheric Storm Phases on 2–5 April 2004. a)** This plot shows a four-day history of total electron content over Rome, Italy in terms of TEC units. One TEC Unit = 10^{16} electrons/m^2. April 2nd is the quiet reference day; its trace is repeated as the black line in the other three plots. **b)** A positive phase storm develops on April 3rd. **c)** The negative phase storm appears on April 4th. **d)** This plot shows the recovery to normal conditions on April 5th. *(Courtesy of the Istitituto Nazionale di Geofisica e Vulcanologia in Rome, Italy)*

circulation, driven by geomagnetic storms. Wind surges in the midnight sector expand the disturbed composition zone toward middle latitudes. The disturbance zone reaches the lowest latitudes at night, but then rotates with Earth into the morning sector. During the recovery of the geomagnetic activity, the composition disturbance region continues travelling to the forenoon sector. Equatorward of this zone and during the afternoon hours, poleward neutral winds may form, producing convergence and downwelling.

The geographic progression of ionospheric storm phases is shown in Fig. 12-34. Although the [O]/[N_2] ratio recovers during the transit from night to day, the perturbations are still large enough to produce daytime negative ionospheric storm effects. Negative phases normally begin many hours after geomagnetic storm commencement. Sometimes the delay is more than 24 hours. Negative storm phases in the dayside ionosphere are usually delayed until several hours after magnetic storm onset. The ion density perturbations are typically much larger in the F2 layer than the F1 layer.

Fig. 12-34. Spatial Distribution of Ionospheric Storm Features on 19–20 December 1980. a) This diagram shows pre-storm conditions at four mid-latitude stations: Wakkanai (WAK) Japan; Boulder (BOU), USA; Poitier (POI), France; and Rome (ROM), Italy. Shading indicates regions of reduced ionospheric electron density. **b)** Storm onset with weak positive storm effects appear at POI and ROM. The storm surges spill out of the midnight sector. The scalloped curve indicates a traveling atmospheric disturbance (TAD). **c)** Additional geomagnetic activity enhances the TADs. An intense negative phase occurs at BOU as the station rotates into the morning disturbance zone previously generated in the post-midnight sector. Later during this phase, positive storm effects appear at POI and ROM locations around noon. These may have been associated with penetration electric fields. **d)** In the storm recovery phase, the weakened composition disturbance region rotates into the forenoon sector, including POI and ROM. [Tsagouri et al., 2000]

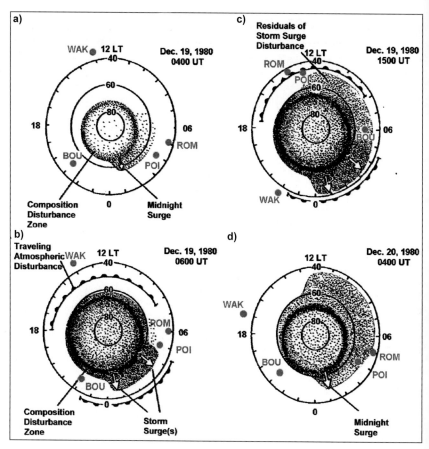

Recent statistical studies suggest that positive phase storms tend to be the largest and last the longest (14 hr–15 hr) in the equatorial anomaly crests and in the vicinity of the auroral oval. Negative phase storms are observed at many latitudes. They are of longest duration at high to mid-latitudes. Seasonal and local time dependencies also exist. Observations suggest that positive phases of storms are more likely in winter at mid-latitudes, and negative phases are more likely in summer. Statistically, the composition disturbance zone reaches lower latitudes in summer than in winter. A prevailing transequatorial summer-to-winter flow restricts the equatorward motion of the composition changes in winter, while allowing them to reach lower latitudes in summer. Further, in winter mid-latitudes, a decrease in molecular species, associated with downwelling, persists and produces a positive phase.

At low latitudes, penetration electric fields also have important effects on the global ionosphere, during magnetic storms. An eastward penetration electric field in the dayside ionosphere lifts plasma to high altitudes, resulting in high electron density and positive phases of ionospheric storms at middle and low latitudes. The field also enhances the equatorial fountain effect and causes electron density decreases over the equator and density increases in the equatorial ionization anomaly region.

Penetration and neutral disturbance dynamo electric fields occur at low latitudes during magnetic storms. Timing is a critical factor in separating the different sources. For the first few (2–3) hours, penetration electric fields can cause ionospheric disturbances simultaneously at low latitudes and dominate the dayside ionospheric evolution. In contrast, large-scale atmospheric gravity waves take 2–3 hours to travel from the auroral zone to the equatorial ionosphere, and a significant propagation delay can be identified at different latitudes. Two to three hours after storm commencement, neutral disturbance dynamo electric fields become important at low latitudes.

Traveling Ionospheric Disturbances (TIDs). Another, more transient, form of cross-latitude ionospheric disturbance is the traveling ionospheric disturbance. Many of the increases in electron density at the peak of the F2 layer and TEC are caused by TADs. At ionospheric heights, the motion of the neutral gas in the TAD sets the ionospheric plasma into motion. The waves displace the plasma, resulting in a *traveling ionospheric disturbance (TID)*. These disturbances appear as traveling corrugations in the ionosphere, and they can seriously affect HF radio communications and surveillance.

Large-scale TIDs, generated in the aurora, with 1000 kilometers or more between wave crests, usually move from a pole to the equator at 300 m/s–600 m/s. Large-scale TIDs lead to total electron content variations of ~20%. Medium and small-scale TIDs with wavelengths of several hundred kilometers are generated by thunderstorms, cold fronts, hurricanes, changes in ionospheric conductivity at the solar terminator, earthquakes, and tsunamis. The TEC perturbation for these smaller scale structures is usually 10% or less. Sometimes TIDs are observed in airflow images and by HF radars. Figure 12-35 depicts a TID over Japan observed by a dense network of GPS receivers. The speed of this small-scale TID was about 0.1 m/s.

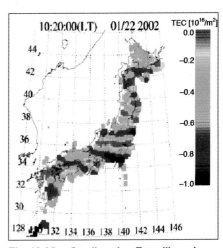

Fig. 12-35. Small-scale Travelling Ionospheric Disturbance (TID) over Japan. The chain of crests and troughs in the TID covered the entire nation. This disturbance had a wavelength of ~100 km and amplitude of 1 TECU. [Sekido et al., 2004 and after Saito et al., 1998]

Summary

Earth's upper atmosphere responds markedly to varying solar and magnetospheric energy inputs. Solar activity affects Earth's neutral atmosphere at virtually all altitudes. The dominant aspect of solar variability that modulates upper atmospheric parameters is emissions in the EUV band. All solar EUV flux incident on Earth's atmosphere is absorbed within the thermosphere. Over an 11-year solar activity cycle, the solar EUV emission varies by a factor of about two in integrated intensity. As a result, the neutral temperature varies by a factor of two and neutral density varies by at least an order of magnitude. Electron density varies even more. The sunlight provides roughly 80% of the heating in the upper atmosphere. Most of the remainder comes from geomagnetic activity, which focuses energy in the auroral zones. At times, the heating in the auroral regions exceeds that from dayside solar input. Solar energetic particles, while not a large energy source, have profound effects on radio communication and ionospheric chemistry in the polar regions, when fluxes are elevated as a result of flares and CMEs.

Geomagnetic effects in Earth's upper atmosphere are most obvious in the high-altitude regions where auroras are visible. Diffuse auroras are present nearly all of the time. Discrete auroras are associated more with active conditions and tend to deliver more energy to the nightside ionosphere. The principal effect of geomagnetic activity on the neutral atmosphere is intense localized warming of the upper atmosphere in the polar and auroral regions. This warming is caused by kinetic heating from the precipitation of energetic charged particles and by Joule heating from enhanced ionospheric currents in the auroral zone.

Chapter 12 Space Weather Disturbances in Earth's Atmosphere

The resulting uneven expansion of the thermosphere produces pressure gradients, which drive strong neutral winds. The disturbed thermospheric circulation alters the neutral composition, and moves the plasma up and down magnetic field lines, changing rates of production and recombination of the ionized species. At the same time, the disturbed neutral winds produce polarization electric fields by a dynamo effect, as they collide with the plasma in the presence of Earth's magnetic field. These electric fields in turn affect the neutrals and the plasma alike, demonstrating that the ionized and neutral species in the upper atmosphere are closely coupled.

Earth's ionosphere is a relatively thin layer of partially ionized magnetized plasma extending from about 90 km to 500 km altitude. Typical plasma densities in the ionosphere range from 10^3 to 10^6 particles per cubic centimeter, but often increase by at least a factor of two during storm time. Because of strong coupling between the ionosphere and the Sun, Earth's atmosphere, and the magnetosphere, the phenomenology of the ionosphere is complex. To categorize the various storm phenomena, the ionosphere is subdivided into two or three zones in latitude, and separated by day and night. The low-latitude ionosphere has received more attention since the advent of global satellite navigation systems that rely on a signal transit in a smoothly varying ionosphere. The equatorial ionosphere, and especially the equatorial ionization anomaly, are prone to plasma instabilities that mix low-density plasma plumes into highly ionized regions overhead. These plumes generally appear near sunset and often last several hours into the night. Strong magnetic storms that produce changes in electric fields that penetrate from the inner magnetosphere into the ionosphere generate plasma bubbles even in the typically stable post-midnight region.

Plasma irregularities also appear in the middle latitudes and polar regions after the onset of strong magnetic storms. During storms, streams of excess plasma are distributed by plasmaspheric and magnetospheric convection into regions that are normally smoothly layered.

Multi-latitudinal disturbances are also created by magnetosphere-ionosphere-thermosphere interactions. Thermospheric composition change and winds driven by storm energy feed back to the ionosphere. Many of the increases in F-layer density and TEC are thought to result from TADs. The divergent nature of the wind causes upwelling of molecular-rich thermospheric gas from lower altitudes. These regions of enhanced molecular species or composition bulges at F-region altitudes are transported by the pre-existing background wind fields and by the storm winds. The regions of upwelling (increases of N_2 and O_2) cause the ionosphere to decay faster and create the negative phases (abnormally low electron density) of ionospheric storms. During storms, ionospheric electron densities can decrease or increase by a factor of two or more.

Traveling atmospheric disturbances may produce undulations in the ionosphere called traveling ionospheric disturbances. The wavelike features cause uneven reflection of radio waves. Smaller scale TIDs are generated by disturbances at the ground and in the lower atmosphere.

Key Words

arches
arcs
auroral kilometric radiation (AKR)
bands
black auroras
break-up phase
bremsstrahlung
composition bulge
convection
diffuse auroras
discrete auroras
dissociative recombination
dynamo process
electromagnetic (EM) energy transfer rate
electromagnetic energy density u_{EB}
equatorial ionization anomaly
equatorial plasma bubbles (EPBs)
gravity waves
growth phase
hemispheric power
ionization
ionospheric storm
Joule heating
negative ionospheric storm
polar cap absorption (PCA)
positive ionospheric storm
Poynting vector S
Poynting's Theorem
pre-reversal enhancement (PRE)
pulsating auroras
rayleighs
Rayleigh-Taylor instability
Region 1
Region 2
spread F
storm-enhanced density (SED)
substorm expansion phase
substorm onset
traveling atmospheric disturbances (TADs)
traveling ionospheric disturbance (TID)
upwelling

Notations

Perturbation Poynting flux $S_p = \dfrac{E_{conv} \times \Delta B}{\mu_0}$

Electromagnetic energy density

$$u_{EB} = \dfrac{\varepsilon_0 E_{conv}^2}{2} + \dfrac{(\Delta B)^2}{2\mu_0}$$

Poynting's Theorem $\dfrac{\partial u_{EB}}{\partial t} + \nabla \bullet S = -J \bullet E$

Electromagnetic energy transfer rate
$J \bullet E = J \bullet (E + v \times B) + (J \times B) \bullet v$

Alfvén Poynting flux $S_A = \dfrac{\delta E \times \delta B}{\mu_0}$

Law of Atmospheres $n_s(z) = n_{so} \exp\left(-\dfrac{z - z_o}{H_s}\right)$

Mass continuity equation $\dfrac{\partial n_s}{\partial t} = -\nabla \bullet (n_s u_s) P_s - L_s$

Answers for Pause for Inquiries

PFI 12-1: Total neutral number density variation at 300 km:
$[(9.5 \times 10^{-9} - 1 \times 10^{-10})/cm^3]/[(9.75 \times 10^{-9})/mm^3] = 0.96 \approx 100\%$

Neutral temperature variation:
$[(1500 - 800) K]/[(1150) K] = 0.61 \approx 60\%$

PFI 12-2: Relativistic electrons and solar protons.

PFI 12-3: According to the figure caption and information contained in the figure, the event occurred in July, when the northern hemisphere is in summertime and thus absorbing significant solar EUV energy.

PFI 12-4: 1.5 erg/(cm$^2 \cdot$s) = $1.5 [((10^{-7}$ J/erg) erg)]/[((1 m^2/10^4 cm^2) cm^2)]s = 1.5×10^{-3} W/m^2 = 1.5 mW/m^2

PFI 12-5: First term on the left side has units of time rate of change of energy density: J/(m$^3 \cdot$s) = W/m^3

Second term on the left side has units of energy flux/length = (W/m^2)/m = W/m^3

The term on the right side has units of current density × electric field = $(A/m^2)(N/C) = (C/m^2 \cdot s)(N/C) = N/(m^2 \cdot s) = J/(m^3 \cdot s) = W/m^3$

PFI 12-6: The grad term has units of $1/m$; therefore the entire term has units of $((\# \text{ particles}/m^3)(m/s))/m = (\# \text{ particles}/(m^3 \cdot s))$.

PFI 12-7: The two continuity equations represent the same physical processes. Equation (8-1) applies to ionized matter; Eq. (12-7) applies to neutral matter.

Figure References

Baker, Daniel N. 2001. Coupling between the solar wind, magnetosphere, ionosphere, and neutral atmosphere. *Encyclopedia of Astronomy and Astrophysics.* Ed. P. Murdin. Institute of Physics Publishing. Bristol, UK.

Bruinsma, Sean L. and Jeffrey M. Forbes. 2007. Global observation of traveling atmospheric disturbances (TADs) in the thermosphere. *Geophysical Research Letters.* Vol. 34. American Geophysical Union. Washington, DC.

Cowley, Stanley W. H. 2000. Magnetosphere-Ionosphere Interactions: A Tutorial, in Magnetosphere Current Systems, Geophysical Monograph 118. American Geophysical Union. Washington, DC.

Foster, John C., Anthea J. Coster, Philip J. Erickson, John M. Holt, Frank D. Lind, William Rideout, Mary McCready, Anthony P. van Eyken, Robin J. Barnes, Raymond A. Greenwald, and Frederick J. Rich. 2005. Multiradar observations of the polar tongue of ionization. *Journal of Geophysical Research.* Vol. 110. American Geophysical Union. Washington, DC.

Immel, Thomas J., Eiichi Sagawa, Scott L. England, Sidney B. Henderson, Maura E. Hagan, Stephen B. Mende, Harold U. Frey, Charles M. Swenson, and Larry J. Paxton. 2006. Control of equatorial ionospheric morphology by atmospheric tides. *Geophysical Research Letters.* Vol. 3308. American Geophysical Union. Washington, DC.

Kelley, Michael C., Jonathan J. Makela, and Odile de la Beaujardiére. 2006. Convective Ionospheric Storms: A Major Space Weather Problem. *Space Weather Quarterly Digest.* Vol. 3. American Geophysical Union. Washington, DC.

Marklund, Göran T., Nickolay Ivchenko, Tomas Karlsson, Andrew Fazakerley, Malcolm Dunlop, P. A. Lindqvist, S. Buchert, C. Owen, Matthew Taylor, A. Vaivalds, P. Carter, M. André and André Balogh. 2001. Temporal evolution of the electric field accelerating electrons away from the auroral ionosphere. *Nature.* Vol. 414. Nature Publishing Group. New York, NY.

Newell, Patrick T., Tom Sotirelis, and Simon Wing. 2009. Diffuse, monoenergetic, and broadband aurora: The global precipitation budget. *Journal of Geophysical Research.* Vol. 114. American Geophysical Union. Washington, DC.

Prölss, Gerd W. 1980. Magnetic Storm Associated Perturbations of the Upper Atmosphere: Recent Results Obtained by Satellite-Borne Gas Analyzers. *Reviews of Geophysics and Space Physics.* Vol. 18. No. 1. American Geophysical Union. Washington, DC.

Richmond, Arthur. 1987. The Ionosphere. *The Solar Wind and the Earth.* Syun-Ichi Akasofu and Yohsuke Kamide, (eds). Dordrecht, Netherlands: Reidel Publishing Co.

Roble, Ray G. 1977. The Thermosphere, Chapter 3 in the 'Upper Atmosphere and Magnetosphere' monograph for the Geophysical Research Board of the National Academy of Sciences. National Academy of Sciences. Washington, DC.

Saito, Akinori, Shoichiro Fukao, and Shin'ichi Miyazaki. 1998. High resolution mapping of TEC perturbations with the GSI GPS network over Japan. *Geophysical Research Letters.* Vol. 25. American Geophysical Union. Washington, DC.

Sekido, Mamoru, Tetsuro Kondo, Eiji Kawai, and Michito Imae. 2004. Evaluation of Global Ionosphere TEC by Comparison with VLBI Data, International. Very Long Base Line Interferometer Services, (IVS). 2004 General Meeting Proceedings, Ottawa, Canada.

Tsagouri, Ioanna, Anna Belehaki, George Moraitis, and Helen Mavromichalaki. 2000. Positive and negative ionospheric disturbances at middle latitudes during geomagnetic storms. *Geophysical Research Letters.* Vol. 27. American Geophysical Union. Washington, DC.

Wygant, John R., Andreas Keiling, Cynthia A. Cattell, Michael T. Johnson, Robert L. Lysak, Michael A. Temerin, Forrest S. Mozer, Craig A. Kletzing, Jack D. Scudder, William K. Peterson, Christopher T. Russell,

George K. Parks, Mitchell J. Brittnacher, Glynn A. Germany, and James Spann. 2000. Polar spacecraft based comparisons of intense electric fields and Poynting flux near and within the plasmasheet-tail lobe boundary to UVI images: An energy source for the aurora. *Journal of Geophysical Research*. Vol. 105. No. A8. American Geophysical Union. Washington, DC.

Further Reading

Cravens, Thomas E. 1997. *Physics of Solar System Plasmas*. Cambridge, UK: Cambridge University Press.

Hargreaves, John K. 1992. *The Solar Terrestrial Environment*. Cambridge, UK: Cambridge University Press.

Hines, Colin O., Irvine Paghis, Theodore R. Hartz, and Jules A. Fejer, eds. 1965. *Physics of the Earth's Upper Atmosphere*. Englewood Cliffs, NJ: Prentice-Hall.

Kamide, Yohsuke and Abraham Chian, eds. 2007. Handbook of Solar-Terrestrial Environment. New York: Springer-Verlag.

Kelley, Michael C. 2009. *The Earth's Ionosphere*. Burlington, MA: Academic Press.

Kivelson, Margaret G. and Christopher T. Russell, eds. 1995. *Introduction to Space Physics*. Cambridge, UK: Cambridge University Press.

Magnetosphere-Ionosphere Interactions: A Tutorial, in Magnetosphere Current Systems. 2000. Geophysical Monograph 118. American Geophysical Union. Washington, DC.

Midlatitude Ionospheric Dynamics and Disturbances. 2008. Geophysical Monograph 181. American Geophysical Union. Washington, DC.

Prölss, Gerd W. 2004. *Physics of the Earth's Space Environment*. Berlin, Germany: Springer-Verlag.

Schunk, Robert and Andrew Nagy. 2000. *Ionospheres*. Cambridge, UK: Cambridge University Press.

Chapter 12 Space Weather Disturbances in Earth's Atmosphere

Near-Earth Is a Place...with Susceptible Hardware and Humans

13

UNIT 3. IMPACTS AND EFFECTS OF SPACE WEATHER AND SPACE ENVIRONMENT

With Contributions by George Davenport, R. Chris Olsen, W. Kent Tobiska, Steve Johnson, Eugene Normand, AF Research Laboratory, and NOAA Space Weather Prediction Center

You should already know about...

- Energy delivered by fields, particles, and photons (Chap. 2)
- Maxwell's Equations (Chap. 4)
- Plasma motions (Chap. 6)
- Magnetospheric current systems (Chaps. 7 and 11)
- Plasma populations of the magnetosphere (Chaps. 7 and 11)
- Geomagnetic field variations (Chaps. 7 and 11)
- Thermospheric variations (Chaps. 8 and 12)
- Drivers of space weather disturbances (Chaps. 9–12)

In this chapter you will learn about...

- How geomagnetic field variations affect space- and ground-based technologies
- Ground-induced currents and their effects
- Radiation terminology and units
- Single-event effects
- How space weather affects orbiting astronauts
- Effects of energetic plasmas
- Satellite drag
- The meteor and space debris environment
- Space weather economic and societal effects

Outline

13.1 **Damage and Impacts from Particles and Photons**
    13.1.1 Particle Radiation Environment
    13.1.2 Energetic Particle Radiation Environment for Humans and Hardware
    13.1.3 Energetic Plasma, Photon, and Neutral Atmosphere Effects on Hardware
    13.1.4 Satellite Drag

13.2 **Damage and Impacts Associated with the Meteor and Artificial Debris Environment**
    13.2.1 The Natural Meteor Environment
    13.2.2 Artificial Space Debris Environment

13.3 **Hardware Damage and Impacts Associated with Field Variations**
    13.3.1 High-energy Electrons
    13.3.2 Geomagnetic Field Interactions with Satellites
    13.3.3 Geomagnetic Field Interactions at the Ground

13.4 **Surveying the Impact of Space Weather on Systems—I**

Chapter 13 Near-Earth Is a Place...with Susceptible Hardware and Humans

13.1 Damage and Impacts from Particles and Photons

Space hazards associated with particles and photons cover a wide range of deleterious effects. Figure 13-1 depicts an organization scheme for discussing this damage. It is organized by energy. Damage at the highest energies is associated with individual particles that cause radiation damage. At slightly lower energies, medium-energy plasma particles create charging effects. At even lower energies is surface degradation, caused by the low-energy plasma environment or individual photons. Chemical interactions in Earth's upper atmosphere produce erosion and loss of surface integrity. Finally, at the end of the spectrum is damage caused by the impact of debris and meteors. The individual particles have low energy, but the motion of a vast number of such particles still delivers undesired energy and momentum to orbiting spacecraft.

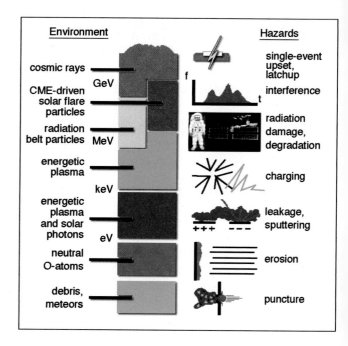

Fig. 13-1. Elements of Space Environment Hazards Sorted by Energy. Example environments are in the left column, and hazards are on the right. *(After the European Space Agency, Space Environment and Effects Analysis Section)*

13.1.1 Particle Radiation Environment

Objectives: After reading this section, you should be able to...

- Understand radiation terminology that describes the space environment
- Describe how radiation affects astronauts and high-altitude flyers
- Describe radiation effects associated with single-event upsets, sensor degradations, and launch delays

The Nature of Space Radiation

Definitions. This section describes space radiation, a known hazard in the space environment. In its most generic meaning, *radiation* is the transfer of energy from one entity to another without a local medium. Very energetic particles and photons travel through the local medium but do not use it to transfer energy. Energetic particles of cosmic, solar, or radiation belt origin are capable of penetrating materials, causing biological effects on human space operations, solar cell degradation, detector malfunction or degradation, optical system degradation, memory system alteration, and control system malfunction or failure. Stellar collapse, shock interactions, magnetic merging, cyclotrons, and local accelerations in Earth's magnetotail are sources of space radiation.

Ionizing radiation removes electrons from parent atoms or molecules. The threshold ionization energy for most materials depends on the target atom, but typically ranges from ~5 to 25 eV. Usually the quoted ionization energy is for the most loosely bound electron. Removing the inner electrons requires up to several thousand electron volts. Very thin layers of matter absorb radiation of only a few tens of eV (a piece of paper is sufficient to seriously attenuate UV energy). Except for causing superficial skin reactions, most radiation must be at least a few tens of keV to be biologically significant. Neutrons are an exception to this rule, because they are uncharged. These neutral particles "sneak" past the electromagnetic potential barrier of the atomic electron cloud and induce radioactivity in the nuclei of the materials they penetrate.

Highly energetic particles penetrate deep into materials, ionizing the constituent material. Under some circumstances, ions and neutrons arrive with sufficient energy to displace atoms from crystal lattice structures, or even create nuclear reactions and energetic by-products of nuclear reactions. The by-products are *nuclear radiation*.

Table 13-1 shows radiation types relevant to the space environment. Each has a characteristic way of interacting with matter. One engineering approach to preventing radiation damage uses shielding provided by dense materials. However, for the highest energy particle radiation, even massive shielding fails. The highly structured heliospheric magnetic field and Earth's magnetic field effectively shield against many types of particle radiation.

Orbital inclination and geomagnetic cutoff determine the amount of radiation a spacecraft or astronaut receives. For Earth-orbiting spacecraft, orbital inclination is the angle between the orbital and equatorial planes as shown in Fig. 13-2. Low-Earth, low-inclination orbits experience high geomagnetic shielding except when the orbit passes through the low-altitude cusp of the inner radiation belt in the South Atlantic Anomaly. High-inclination orbits traverse the polar region, encountering field lines open to the heliosphere. Hence, the natural shielding is lower for these orbits. High-energy solar particles and cosmic rays spiral along the open field lines and deposit energy in the materials that stop them.

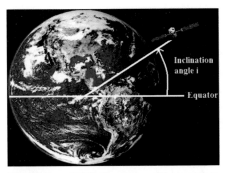

Fig. 13-2. Orbit Geometry. We measure an orbit's inclination angle from Earth's geographic equator.

Chapter 13 Near-Earth Is a Place...with Susceptible Hardware and Humans

Table 13-1. **Radiation Terminology.** Here we list radiation types associated with the space environment and space weather effects.

Energy Carrier	Radiation Terminology
High-energy photons – (Energy > 4 eV) UV-EUV	Ionizing radiation
Very high-energy photons, X rays, and γ rays	Ionizing X-ray and gamma radiation
Neutral particles—mostly neutrons	High-energy neutrons
Electrons and positrons	Beta (β) radiation
Energetic protons	Z = 1
Energetic helium atoms, stripped of electrons	Z = 2, alpha (α) radiation
Energetic heavier atoms, many electrons missing*	High charge and energy (HZE)
By-products from nuclear reactions	Nuclear radiation or radioactivity

* The most biologically damaging component of space radiation.

Geomagnetic cutoff rigidities are a quantitative measure of the shielding provided by Earth's magnetic field. Rigidity is momentum per unit charge. The rigidity value (in volts) predicts the energetic charged particle transmission through the magnetosphere to a specific location as a function of direction. We recall that moving charged particles experience a Lorentz force in the presence of a magnetic field and orbit about the field lines. Charged particles with high momentum require a stronger field to be deflected from their original trajectory. We display the link between such points and the atmosphere in Fig. 13-3. If the particle lacks the requisite momentum, its access to a location in geospace is "cut off." The closed field lines at low latitudes are effective in limiting energetic charged particle access. Because the magnetosphere is asymmetric, the cutoff for a given particle is a function of latitude, longitude, magnetic local time, geomagnetic activity, and particle arrival direction.

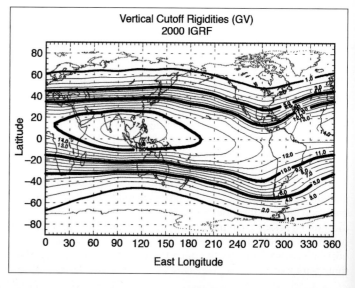

Fig. 13-3. **Global Geomagnetic Cutoff Rigidity.** This plot shows values for geomagnetic cutoff. Longitude is on the horizontal axis and latitude is on the vertical axis. The calculation assumes a dipole geomagnetic field. The unit of rigidity is volts (V) or as in the plot, gigavolts (GV). The higher the rigidity, the lower the probability that primary particles will penetrate the magnetic field. *(Courtesy of Margaret Ann Shea and Don Smart at the AF Research Laboratory)*

Focus Box 13.1: Units for High Energy Particles

The units used to describe high-energy particles may seem strange compared to SI units. The choice is a matter of convenience. In particle physics, the standard unit of energy is the electron volt (eV). It is the amount of energy that an electron gains when it moves through a potential difference of 1 volt (in a vacuum). A GeV is 10^9 eV. The rest-energy of a proton or neutron is approximately 1 GeV.

Instead of using kilograms to measure mass, high-energy particle physicists use the mass-energy relationship, $E = mc^2$, to relate mass to units of energy in eV. One eV is 1.6×10^{-19} J. The proton rest mass is 1.67×10^{-27} kg. Using $E = mc^2$ we get:

$$E = 1.67 \times 10^{-27} \text{ kg} \times (3 \times 10^8 \text{ m/s})^2 = 1.503 \times 10^{-10} \text{ J} = 9.39 \times 10^8 \text{ eV}$$

Similarly the rest mass energy of an electron is 511 keV.

If we solve $E = mc^2$ for m, we get $m = E/c^2$, one might think the unit of mass should be eV/(m/s)^2. What happened to c^2? For convenience, particle physicists set light speed = 1. This system of units is sometimes called the *natural system* of units. As long as everyone understands the system, we only need a conversion factor to get back the "real" (i.e., SI) units, if we want them.

We apply a similar line of reasoning to momentum. Thus $p = E/c$ and momentum per unit charge is $p/q = E/qc$. High-energy-particle momentum (per unit charge) is often stated in units of volts. The energy value in eV is divided by the charge (in units of e) of the particle, leaving V/c. Using the unit of light speed = 1, the natural system unit for momentum becomes volt (with the necessary prefix, usually giga). Most cosmic rays have energies in the GeV range.

Charged Particle Penetration and Displacement

Ordinary matter (atoms and molecules) is very open when viewed on an atomic scale. If we represent a nucleus by a 0.1 mm dot (just visible), the next nucleus in a typical solid will be about 10 meters away. So atoms have plenty of room for particles with sufficient kinetic energy to penetrate into their volume. The space between the nuclei is occupied by electron clouds that have little mass but generate strong electromagnetic fields.

Inelastic collisions of electrons and nuclei produce X-ray bremsstrahlung over a continuous range as the incident charged particles interact with the Coulomb fields of nuclei (Chap. 2). The electrons lose energy at a high rate in a single encounter, with much of their energy going to photons. The nucleus energy state can rise too. High-energy electrons impinging on the upper atmosphere produce a measurable X-ray glow. A more likely interaction is elastic scattering, where an electron interacts with a nucleus. The electron trajectory changes, but its kinetic energy and speed remain constant. This process is known as electron backscattering.

These backscattered electrons can also interact with the electron clouds of the target material. Because they carry an electric charge, they have an electric field and interact with other electrons without touching them. As they pass through the material, the incident electrons, in effect, push the other electrons away from their paths. If the forces on the target electrons are sufficient to remove them from their atoms, the result is ionization. In other cases, the atom rises to a higher energy level (excited state). Regardless of the interaction type, the moving electrons lose some of their energy. The electrons are thermalized (reduced to the ambient temperature) as they continue in a sequence of energy-loss interactions.

The rate at which an electron transfers energy to a material is the linear energy transfer (LET), and is expressed in terms of the energy transferred per unit distance traveled. Typical units are kiloelectron volts per micrometer (keV/μm). In some material, such as human tissue, the LET value depends on the kinetic energy of the electrons. The total distance a particle travels in a material before losing all its energy is its *range*. The two factors that determine the range are (1) the particle's initial energy and (2) the material's density. One important characteristic of electron interactions is that all electrons of the same energy have the same range in a specific material. A 250-keV electron travels slightly more than 0.5 mm in water or soft tissue before stopping.

When a heavy charged particle such as a proton or an alpha particle (He^{++}) enters a solid, it interacts almost entirely with the distributed electrons. We know from mechanics that in collisions of heavy projectiles with light targets (e.g., a bowling ball hitting a ping pong ball) only a small fraction of the projectile energy transfers to the target. Thus, as a relatively massive particle travels through the electron cloud, it undergoes many collisions, losing a little of its kinetic energy each time, until it finally stops. The "lost" (transferred) energy is often enough to free one or more of the electrons. Massive charged particles have well-defined ranges.

Penetrating energetic particles also cause *displacement* of constituent atoms from their proper crystal lattice locations, creating defects in the crystal structure that appear as low points (wells) in the electric potential. These wells trap conduction electrons, increasing the material's resistance. The minimum kinetic energy needed to displace an atom from its lattice site is called the *threshold displacement energy*. Displacement is especially harmful to solar cells, where the accumulated displacement damage and increased resistance gradually reduce power output.

Neutrons carry no charge, so their primary interaction is with the compact, massive nuclei. The target nucleus sometimes captures the neutron and becomes unstable, leading to a nuclear reaction. The secondary radiation particles from these individual nuclear reactions often have sufficient energy to start a cascade (shower) of further reactions. As an example, calculations suggest that a single extremely energetic neutron striking the top of Earth's atmosphere creates 10^{11} radiation particles at sea level. When very-high-energy particles strike shielding material in a spacecraft, the shielding becomes a source of more radiation.

Radiation Units

Energy, Rates, and Effects on Matter. In 1978, metric (SI) units were introduced to describe various types of radiation and their effects. However, the old set of units still regularly appears in publications, alongside the SI units. In particular, energy is still described in multiples of eV (keV, MeV, GeV, TeV) rather than in fractional joules. An approved SI prefix is available (attojoule = 10^{-18} J), but so far it is not used widely.

For radioactive materials, a unit of activity describes the number of disintegrations per second. Radioactivity was formerly measured by the number of disintegrations per second in one gram of radium—one curie (Ci), which is 3.7 × 10^{10}. The SI unit is the becquerel (Bq) (one disintegration per second).

Table 13-2 shows some of the conversions from one radiation measurement system to the other. The remaining units we need deal with the effects of radiation on matter. The *dose* refers to the amount of energy deposited by radiation in specific materials through ionization. An older unit refers to *radiation absorbed dose (rad)*, which is the amount of any radiation that deposits 10 mJ of energy in a kilogram of material. The SI unit of absorbed dose is the gray (Gy) and equals one J/kg (100 rad).

Different types of radiation have differing biological effects. A dose of alpha radiation is much more damaging to living tissue than the same dose from *gamma radiation* because the alpha particle has many more interactions with its target material. To estimate the potential harm to humans from ionizing radiation, scientists use *dose equivalent* (biological dose) for radiation types. A given amount of dose equivalent produced by two different radiation types produces the same risk of cancer mortality. The old unit for dose equivalent was the *rem (rad equivalent mammal/man)*, which had to be related to rads by a relative biological effectiveness (RBE) factor for the specific radiation. The factor compares the effectiveness of different radiations to a common endpoint (cell death, cancer inductions, etc.). The SI unit for biological dose is the *sievert (Sv)*. The *quality factor (QF)* considers the type of radiation and its energy that together alter the biological effect in terms of the cancer risk. The relationship between the old and SI biological units is

$$1 \text{ sievert} \approx 100 \text{ rem}$$

The use of the approximate equality sign indicates that QF and RBE are not exactly the same. A chest X ray is about 20 µSv. Now and in the future, the measure of human interaction with radiation will be quantified in sieverts.

Table 13-2. Old and SI Units of Radiation. This comparison is useful, because many older (but still valuable) references use the old units.

Old unit	SI unit
Activity rate 1 curie (Ci) = 3.7 × 10^{10} disintegrations/s	Activity rate 1 becquerel (Bq) = 1 disintegration/s = 27.03 pCi
Deposition 1 rad = 10 mJ/kg	Deposition 1 gray (Gy) = 100 rad = 1 J/kg
Equivalent dose 1 rem = relative biological effectiveness (RBE) × rad	Equivalent dose 1 sievert = quality factor (QF) × gray

Radiation Sources

High-energy particles come from several sources, each characterized by a different composition of particle species and different energy and flux. Many of these sources vary during the course of the solar cycle. We illustrate the terminology and the relationships in Fig. 13-4 and Table 13-3.

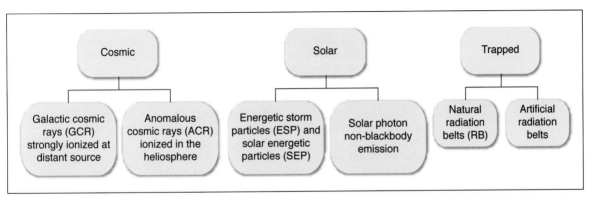

Fig. 13-4. Radiation Sources for the Space Environment. These are the three main sources of radiation that affect our lives on and near Earth.

Table 13-3. **Properties of Energetic Particle and Photon Populations.** This table lists the key properties of energetic particles and photons. (GCR is galactic cosmic ray; ACR is anomalous cosmic ray; SEP is solar energetic particle; ESP is energetic storm particle; CIR is co-rotating interaction region.) The spatial scales are referenced to the size of the heliosphere. Global scales involve the entire heliosphere, local scale is on the order of interplanetary distances. *(Courtesy of Margaret Ann Shea at the AF Research Laboratory)*

Population	Temporal Scales	Spatial Scales	Energy Range*	Acceleration Mechanism
GCR	Continuous	Global	GeV–TeV	Supernova shock
ACR	Continuous	Global	10 MeV–100 MeV	Diffusive shock—heliosheath
SEP	Seconds to hours	Local to large	keV–100 MeV	Reconnection, shock stochastic heating
ESP	Days	Large	keV–10 MeV	Diffusive shock and shock drift
CIR	27 days	Large	keV–10 MeV	Diffusive shock
Bow Shock	Continuous	Local	keV–MeV	Shock drift
Artificial	Rare	Local	keV–MeV	Nuclear detonation
X ray, γ ray	Minutes	Local (solar)	keV–MeV	Reconnection, nuclear reactions

* Room temperature particles have energies of about 1/40 eV.

Galactic Cosmic Radiation. This radiation originates from outside the solar system. The *galactic cosmic ray (GCR)* particle population consists of ionized, charged atomic nuclei from hydrogen (87%) and helium (12%), to uranium (trace) that are characterized by extremely large kinetic energies for each nucleon—as high as 10^{21} eV, but more usually in the GeV range. These particles are distributed isotropically and found in relatively low fluxes. The GCR energy density is ~0.5 eV/cm^3 to 1.0 eV/cm^3, similar to the energy density of starlight. The integral flux of four isotopes of GCRs at solar minimum and maximum conditions is depicted in Fig. 13-5. The particle flux drops sharply beyond one GeV per nucleon. Although not shown in Fig. 13-5, electrons are also part of the GCR mix. They play a larger role in cosmic radio waves than in matter interaction.

During solar maximum, the interplanetary magnetic field generated by the Sun provides some protection to the inner solar system, decreasing the integral cosmic ray intensity (Fig. 9-31). The lower energy GCR particle fluxes are in antiphase with respect to the solar-generated energetic particles. Lower energy cosmic ray flux decreases, but higher energy particles are not appreciably attenuated. Fortunately, the flux of GCRs decreases exponentially as energy increases (Fig. 13-5).

In 2009, cosmic ray intensities rose 19% beyond previous space age measurements (Fig. 13-6). The long solar minimum between solar cycles 23 and 24 appears to be the cause. The Sun's magnetic field decreased to 4 nT from normal values of 6 nT to 8 nT. Further, the solar wind pressure was at a 50-year low and the heliospheric current sheet was unusually flat. The current sheet is important because cosmic rays tend to be guided by its folds. A flat current sheet allows cosmic rays more direct access to the inner solar system.

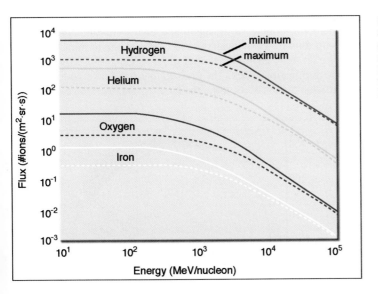

Fig. 13-5. Integral Energy Spectra for Major Ions at Solar Maximum and Solar Minimum. The curves show the galactic cosmic ray (GCR) fluxes of hydrogen, helium, oxygen, and iron ions, for solar cycle maximum and minimum. For all isotopes the flux is higher at solar minimum, when the Sun's field provides a less effective shield. [NASA, 1993 and Badhwar and O'Neill, 1994]

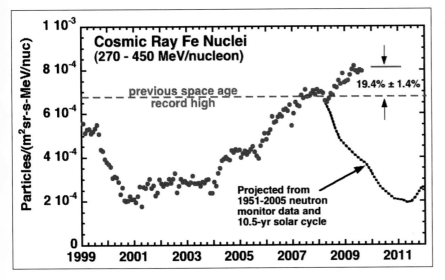

Fig. 13-6. Galactic Cosmic Ray Variation from 1999 to 2009. The number of energetic iron nuclei, counted by the Cosmic Ray Isotope Spectrometer on NASA's ACE spacecraft, reveals that cosmic ray levels jumped 19% above the previous space age high. The dotted curve on the lower right shows the predicted flux for iron nuclei based on climatology. *(Courtesy of Richard Mewaldt at CalTech and NASA)*

Anomalous Cosmic Rays. *Anomalous cosmic rays (ACRs)* are charged particles with low charge states and energies that are typically no higher than 10^8 eV. This component of the cosmic ray population consists only of those elements that are difficult to ionize, including helium, nitrogen, oxygen, neon, and argon—hence the term "anomalous." One theory suggests that high-energy GCRs that interact with the heliopause shock front might be decelerated, resulting in their transformation into lower energy anomalous cosmic rays. Another theory suggests that ACRs are created from electrically neutral particles of galactic origin. Because they are unaffected by magnetic fields, galactic neutral atoms are able to enter the heliosphere, where solar EUV photons ionize them. Subsequently, the Sun's outflowing magnetic field could channel these ionized particles (pick-up ions) and the solar wind termination shock that marks the inner edge of the heliosheath. The production of anomalous cosmic rays requires a magnetic connection to the

Chapter 13 Near-Earth Is a Place...with Susceptible Hardware and Humans

termination shock. The blunt nose of the shock allows the pick-up ions to remain in contact with the shock for months, slowly gaining energy. With time, the magnetic field and the pick-up ions flow backward to the flanks of the termination shock, where the particles gain more energy in shorter times. Figure 13-7 provides a not-to-scale view of the termination shock. The pick-up ions are delivered to the shock nose, but largely accelerated to ACRs at the flanks.

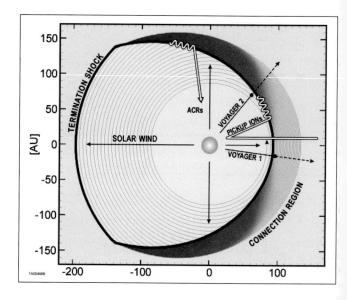

Fig. 13-7. The Heliosphere's Termination Shock. This artist's view cuts through the termination shock at the equator. The Sun is near the center of the image. Also shown are the approximate 2004 positions of Voyager 1 at the nose of the termination shock and Voyager 2 still inside the shock. Voyager 1 has exited to the connection region in the heliosheath. The blunt shape is created by the motion of the heliosphere through the interstellar medium. The Sun's outflowing magnetic field connects to the termination shock at the nose, where it delivers pick-up ions for acceleration. Most of the acceleration occurs as the field and ions flow backward and reach the flanks. With sufficient energy, a fraction of these particles diffuse inward and return to the vicinity of Earth where they are sensed as anomalous cosmic rays. [McComas and Schwadron, 2006]

Energetic Particles from the Sun and Its Atmosphere. Periodically, the Sun emits significant numbers of high-energy particles during severe solar disturbances. These events inject energetic electrons, protons, alpha particles, and heavier particles into interplanetary space. The frequency of such events varies with the solar activity cycle, as shown in Fig. 13-8. The energy of the protons and neutrons reaches 10^{11} eV but is mostly in the range of 10^7 eV–10^9 eV. They come from solar flares as *solar energetic particles (SEPs)* or from interplanetary shocks associated with CMEs as *energetic storm particles (ESPs)*. During solar maximum, the frequency and intensity of solar eruptions increase. Most eruptions don't present a significant hazard, because they are either too small to inject significant numbers of energetic particles, or they occur at solar longitudinal positions that are unfavorable for directly transferring particles to Earth along interplanetary magnetic field lines. However, eruptions of great intensity or rapid sequences of large eruptive events that drive interacting CMEs are particularly likely to generate very large particle events. Very large solar proton events, those having an omnidirectional solar proton fluence $>1 \times 10^9$ protons/cm^2 at one AU with energies >30 MeV, impose operational constraints on human space missions and equipment. These may include storm shelters with sufficient shielding to reduce the radiation dose to tolerable levels or the need to protect or actually turn off sensitive equipment that may be susceptible to soft errors or radiation damage. These events generally take place only once or twice in a solar cycle.

The multi-source and directional nature of solar emissions complicates prediction. Particle events from a flare site are directionally confined by the magnetic field into rather narrow cones. If the particle path is well-connected to Earth, they will arrive promptly and impulsively. A flare erupting 50°–60° west

13.1 Damage and Impacts from Particles and Photons

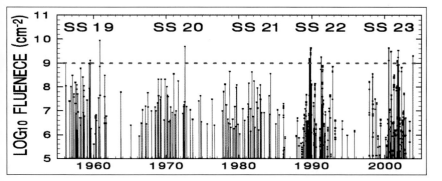

Fig. 13-8. Solar Proton Events with Energies Greater than 30 MeV since Solar Cycle 19. The dashed line indicates events with extreme fluence (integration over the interval of the event) that occur about once per solar cycle at one AU. *(Courtesy of Don Smart at the AF Research Laboratory)*

of the central meridian is optimum for high-energy solar flare particles to intercept Earth most quickly. If the reconnection event that creates the culpable solar flare also drives a fast CME, then shock acceleration of particles along an expanding front is possible. Thus, energetic charged particles from shock interactions, even though they tend to follow the interplanetary magnetic field (IMF) lines, may approach Earth from broad longitude bands in advance of CMEs. The most energetic ones arrive within less than an hour, but the majority take several hours to days. We observe more of the shock-drive event particles simply because they have broader and longer-lasting sources. The worst-case scenario is a combination of the two sources in a well-connected geometry. On occasion, a large reconnection event occurs behind the west limb of the Sun and creates a shock acceleration region that connects to Earth, but has no signs of an associated solar flare. Such events are virtually unpredictable.

Trapped Radiation. Earth's magnetic field forms the trapped radiation belts that surround it. As we describe in Chaps. 6, 7, and 11, electrons and protons are naturally trapped in torus-shaped regions at distances of 1.5 R_E–10 R_E. Particle energies are as high as 10^9 eV for protons and 10^7 eV for electrons under extreme conditions. Charged particles travel through these zones, spiraling around the magnetic field lines, and oscillating between "mirror points" located in opposite hemispheres. Figure 13-12 illustrates the distribution of trapped protons and electrons around Earth. As a result of Earth's tilted geomagnetic field and the anomalous reduction of the field strength in the South Atlantic, the trapped proton belt extends down to the atmosphere in the South Atlantic Anomaly (SAA) region (Sec. 11.2).

The flux of high-energy protons is high in the region close to Earth's atmosphere because the particles are created (and trapped) there from cosmic-ray-neutral-atmosphere interactions. This is the inner Van Allen radiation belt. The flux of energetic electrons is farther from Earth (outer radiation belt) because their primary source is in the magnetospheric plasmasheet that is subject to considerable external forcing from the solar wind. Energetic protons are largely absent from the outer belt because they penetrate deeper into the atmosphere, where they're mostly removed by atmospheric collisions.

New Radiation Belts. In rare circumstances, new radiation belts form. On March 24, 1991, a new electron belt of >13 MeV electrons formed in the slot region in less than one minute (Focus Box 11.2). An enhancement of the inner proton belt was also observed. Activity on the Sun about a day before resulted in a fast interplanetary shock traveling >1400 km/s. The shock compressed the magnetosphere and produced an inductive electric field, which energized electrons by bringing pre-existing 1 Mev–2 MeV electrons from $L = 8$ to $L = 2.5$.

Chapter 13 Near-Earth Is a Place...with Susceptible Hardware and Humans

On June 19, 1962, the US detonated its first nuclear explosion in space, code-named STARFISH. The explosion, at an altitude of 400 km, injected energetic electrons into Earth's magnetic field, where they formed an artificial radiation belt. This artificial nuclear-effects belt lasted until the early 1970s. The radiation produced by STARFISH destroyed seven satellites within seven months, primarily from solar-cell damage. The losses motivated spacecraft manufacturers to harden many of their systems against radiation effects.

The interaction of nuclear detonations with the space environment is of significant concern for all space-faring nations. Some research is underway to determine how space assets might be protected in the event of a high altitude nuclear detonation. The objective is to remove trapped particles—a process called radiation belt remediation. In 1967, an international treaty banned nuclear explosions in space. If such treaties are honored, only natural processes will henceforth make new radiation belts.

13.1.2 Energetic Particle Radiation Environment for Humans and Hardware

Objectives: After reading this section, you should be able to...

- Describe radiation effects on living tissue
- Explain why more mass does not equate to more shielding on orbit
- Compare the relative advantages of low-Earth orbit (LEO) and high- and low-inclination orbits for radiation shielding
- Describe radiation hazards to humans in the polar aviation environment

Particle Radiation Environment—Physical Damage and Impacts to Humans

Spaceflight Environment for Humans. Because some high-energy particles inflict greater biological damage than that resulting from typical terrestrial radiation hazards, exposures to spacecraft crews sometimes exceed those received by terrestrial radiation workers (Table 13-4). An energetic charged particle passing through a cell produces a region of dense ionization along its track. The ionization of water and other cell components damages DNA molecules near the particle path and otherwise compromises cell chemistry, thus inhibiting cell function. Direct hits to DNA molecules do even more damage. Problems with cell functions are particularly noticeable for blood cells and organs that use or produce blood cells.

Radiation exposure sometimes results in acute, delayed, or chronic illnesses. Radiation sickness symptoms are mild, transitory, or severe. As a result of damage to blood-forming and -processing organs, a person may suffer loss of appetite, digestive failure, brain damage, or death. Even small doses of radiation are a concern, because of the possibility of long-term genetic effects and increased cancer rates. However, the risk of acute effects during a Space Shuttle flight or during a stay on the International Space Station is very small, because Earth's atmosphere and magnetic field provide adequate shielding.

NASA tracks the amount of radiation each astronaut accumulates during orbital missions on the Space Shuttle and the International Space Station, and an astronaut is not permitted to fly in space when a specific limit is reached. Thus

Table 13-4. Recommended Limits to Radiation Exposure in Equivalent Dose. Here we list the maximum allowed dose of radiation for humans in various types of exposure. [US Nuclear Regulatory Commission, Part 20]

Exposure	Maximum Dose (Sv)	Equivalent Dose (rem)
Occupational exposure	50 mSv in one year	~5 rem in one year
Astronaut exposure	50 mSv in one year	~5 rem in one year
Public exposure in the US	5 mSv in one year	~0.5 rem in one year
Early-fetus exposure	0.5 mSv/month	~0.05 rem in one month
Average chest X ray	0.01 mSv–0.05 mSv	~0.001 rem–0.005 rem
Natural background	~3 mSv in one year	~0.3 rem in one year

far the longest stay in orbit is a few months. The possibility of long-term spaceflight makes this cumulative dose issue much more important than it has been in the past. Many former astronauts have suffered some form of cataracts after flying in space, according to a NASA 2001 study. Of these, most had flown on high-radiation missions such as the Apollo Moon landings. Radiation damage to fiber cells at the center of the lens causes the usually transparent cells to cloud, forming the cataracts.

In 1989, the US National Council of Radiation Protection and Measurements issued a report with a recommendation for limits to radiation exposure for humans. Table 13-4 summarizes these limits. The basis for the occupational dose limits is that a risk to an individual of a fatal cancer from exposure to radiation should be no greater than that of fatal accidents in safe industries. NASA uses a concept for dose limits known as "As Low As Reasonably Achievable" (ALARA). For ALARA, the radiation protection must keep doses low in relationship to the obtainable benefits.

Ionizing radiation sometimes breaks DNA strands. Single strand breaks are common, while double strand breaks are less so, but they are an important component of long-term risk. Strand breaks are often repaired by built-in cell mechanisms, but clustered DNA damage leads to cell death. For most cell types, the death of a single cell is not catastrophic—cells continually die and are replaced by normal processes. A more dangerous event is the non-lethal change of DNA molecules that may lead to cell proliferation, a form of cancer. Also, evidence exists of advanced aging and increased risk of coronary disease and pulmonary problems in the retiring astronaut corps. Studies are underway to track these health changes.

Though the total dose from GCRs is small, their high energies allow them to penetrate most surfaces easily. During solar minimum, the unshielded interplanetary dose to the blood-forming organs (BFO) in astronauts is approximately 0.6 Sv/year, significantly exceeding the acceptable values shown in Table 13-4. Some astronauts reported light flashes inside their eyes, even when their eyes were closed. These flashes are evidence of high-energy particles striking the retina and triggering a false signal that the brain interprets as a light flash. Health specialists are trying to determine if long-term vision is compromised by such events. Some form of shielding is necessary for human interplanetary travel.

Thin-to-moderate shielding is effective in reducing the projected equivalent dose rate, but increasing shield thickness beyond this does little to improve its effectiveness. This is because of the large number of secondary particles, including neutrons, produced from nuclear interactions between the GCRs and shield nuclei. Figure 13-9 illustrates the annual BFO dose equivalent as a function of different shield materials. The total dose from these particles is quite small (a few rads/yr = a few tens of mJ/(kg·yr)).

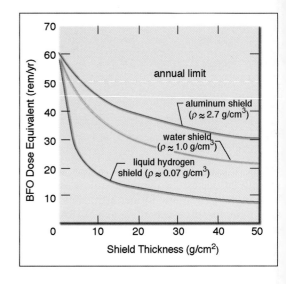

Fig. 13-9. Material Shielding Effectiveness against Galactic Cosmic Rays (GCRs) at Solar Minimum. This plot shows how well aluminum, water, and liquid hydrogen shield blood-forming organs (BFO) against GCRs. [NASA, 1993]

In the International Space Station (ISS) orbit, the geomagnetic field deflects many of the lower energy GCR components, providing a factor of ten reduction in total GCR exposure relative to the free space environment. This protection is primarily a function of latitude (not altitude), as we show in the data from the ISS Tissue Equivalent Proportional Counter (TEPC) dose monitor in Fig. 13-10. Spacecraft in higher inclination orbits are exposed to increased GCR levels as the spacecraft transits the higher latitudes. During geomagnetic storms, GCR exposure often expands at lower latitudes because storm dynamics compress the field and reduce the shielding effectiveness.

Proton spectra and flux are strong functions of altitude and latitude for space operations in Earth's magnetic field—low inclination LEO usually minimizes exposure. At the higher altitudes, most crew exposure comes during transits through the South Atlantic Anomaly (SAA), as a result of greater trapped proton flux levels. At lower altitudes, the protons in the SAA interact with the residual atmosphere and some of the protons fly away and contribute to an anisotropic proton distribution. The proton flux from the west is more than twice as great as that from the east. In addition to altitude, the integrated dose is a function of solar cycle phase. Increases in solar activity expand the atmosphere, enhancing the losses of protons in LEO. Therefore, the trapped radiation dose in LEO decreases during solar maximum and increases during solar minimum.

Although trajectories of very-low-inclination flights do not pass through the regions of maximum intensity within the SAA, the orbits have longer exposure at the edges of the SAA. Low-inclination space flights normally transit a portion of the SAA during six or seven consecutive orbits each day. Moderate-inclination flights transit between north and south 58° and pass through the SAA maximum intensity regions, but spend less time in the SAA than low-inclination flights. Thus, crews in moderate-inclination flights receive less net exposure to trapped radiation than those in low-inclination flights for a given altitude.

13.1 Damage and Impacts from Particles and Photons

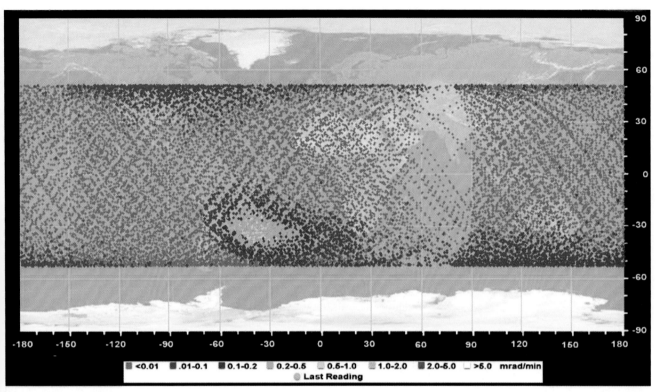

Fig. 13-10. **International Space Station (ISS) Tissue Equivalent Proportional Counter (TEPC) Geospatial Dose Rate Map.** Dose rate is shown in millirads/minute. The map is centered at the equator over the Greenwich meridian (0 longitude, 0 latitude). The dose rate history is from May 1, 2002 to May 27, 2002 and follows the ground track of the ISS. The ISS altitude varied between 385 km and 395 km. Slightly elevated dose rates occur at the northern and southern orbit excursions into geomagnetic high-latitudes. Enhanced dose rates are evident in the vicinity of the South Atlantic Anomaly. *(Courtesy of NASA)*

The intensity and spectral distribution of solar energetic particles affect shielding significantly. Figure 13-11 shows the modeled shielding effectiveness for three large particle events. Fortunately, most solar events are relatively short-lived (1–2 days), which makes small-volume, on-orbit "storm shelters" feasible. To minimize exposure, the crew enters the storm shelter during the most intense portion of the solar proton event, which may last for several hours. Storm-shelter shielding of ~20 g/cm^2 (200 kg/m^2) or more of water equivalent material should protect the crew adequately.

Solar energetic particles pose the greatest short-term threat to unprotected crews in polar or interplanetary orbits. To date, the greatest risk for significant exposures to astronauts existed during the Apollo Program (late 1960s–early 1970s). The Apollo astronauts were fortunate that no significant SEP events occurred during the lunar missions. The extreme solar proton event of August 1972 occurred between missions. Solar monitoring since about 2000 has significantly improved knowledge of the precursors to large particle events, but solar eruptions still occur with little warning. The magnitude and intensity of such events are difficult to determine until they are in progress.

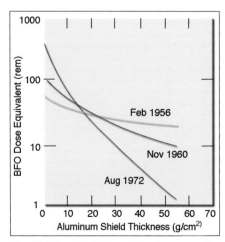

Fig. 13-11. **Shielding Effectiveness for Three Large Solar Proton Events.** This plot shows how shielding reduces the effect of radiation on blood-forming organs, as measured during three solar particle events on the dates shown. [NASA, 1993]

Chapter 13 Near-Earth Is a Place...with Susceptible Hardware and Humans

Aviation Radiation Environment for Humans

Space weather is a concern for airlines that fly commercial flights over the polar cap. Solar radiation storms increase radiation exposure to passengers and crews in jet aircraft in the polar region. Some officials in the airline industry are concerned about radiation risks, and therefore the Federal Aviation Administration (FAA) suggests that airline leaders educate crews about radiation hazards. The European Union (EU) requires airlines to track crewmembers' exposure levels and educate them about the risks. The FAA recommends occupational radiation limits for commercial aircraft crews. These include a five-year average dose of 20 mSv/yr, with no more than 50 mSv in a single year. For a pregnant crew member, the recommended limit for the fetus is an equivalent dose of 1 mSv, with no more than 0.5 mSv in any month.

The FAA's Civil Aeromedical Institute (CAMI) offers information on how much galactic radiation dose is received on a flight between any two airports in the world. For example, for a flight between New York and Tokyo, the effective dose to an individual is approximately 67 µSv. Although doses from SPEs are not yet calculated by the CAMI model, this area is a subject of active research. The Centers for Disease Control (CDC), the National Institute for Occupational Safety and Health (NIOSH), and the FAA/CAMI continue to research the radiation hazards to the aviation industry.

Particle Radiation Environment—Physical Damage and Impacts to Hardware

Particle Penetration into Materials. Energetic particles from three different sources (cosmic, solar, and radiation belts) in space are capable of penetrating most spacecraft shielding and depositing their energy within the man-made components, potentially causing problems called device single events. In general, microelectronics (integrated circuits (ICs)) are the human-made components most susceptible to performance degradation by these highly ionizing particles, but optics and polymeric materials (teflon and others) are also affected. In this section, we deal with damage from a single charged particle. In the next section, we consider problems arising from groupings or collections of energetic particles.

The highest energy cosmic rays are subatomic particles carrying the energy of macroscopic objects. Protons with energy greater than 10 MeV penetrate typical spacecraft shielding and pose energy deposition risks. Ions with energies > 30 MeV are capable of breaching ICs and thereby inducing faults. The primary risk to the ICs is called *single-event effects (SEEs)*. These result from the deposition of charge in an analog, digital, or power circuit caused by the interaction of a single particle (Fig. 13-12).

In the low-Earth orbit (LEO, <500 km) environment, the South Atlantic Anomaly (SAA, over Brazil and the Atlantic Ocean) is the location of the largest number of SEEs in spacecraft because the trapped proton belt extends to its lowest altitudes at this location. At higher altitudes, SEEs occur at all latitudes and longitudes, and they even happen at Earth's surface, although at a much lower rate. Near the polar regions, where Earth's magnetic field is open to the solar wind, GCR particles cause SEEs. Long-term physical damage takes place when a system drops out of operation because of an SEE and remains out of operation until repaired. A service mission in space is expensive and dangerous, so it is hardly ever done; therefore spacecraft systems are designed to account for the possibility of SEEs in the microelectronics. This is done by selecting ICs that are tested on the ground and

13.1 Damage and Impacts from Particles and Photons

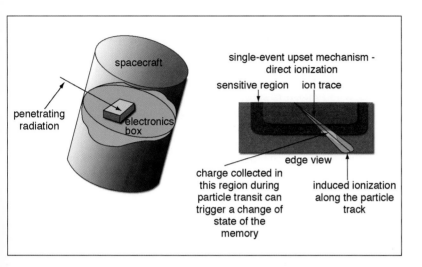

Fig. 13-12. The Role and Path of High-energy Particles in Producing a Single-event Upset (SEU). This diagram shows how solar radiation penetrates a spacecraft and its electronic equipment to cause an SEU, such as a change in the memory state. [After Baker, 2002]

by incorporating fault mitigation features such as redundancy and error detection and correction (EDAC, the correcting of single bit errors).

A *single-event upset (SEU)* in an IC device within a spacecraft or aircraft comes about when a high-energy charged particle penetrates it, causing a burst of free electrons and ions. The proton, alpha particle, or heavy ion causes a short electrical pulse and an unplanned change in the circuit state that sends a false command and flips the logic state of a single bit (in a random access memory). This soft error is recoverable by rewriting to the memory or rebooting the system. Sometimes SEUs are referred to as *bitflips*.

A *single-event latchup (SEL)* is a different condition, caused by energy deposition, which leads to a high-current state in which the device no longer operates properly. Complementary metal-oxide semiconductor (CMOS) devices are the type of ICs most sensitive to latchup. An SEL often results in the destruction of the IC through overheating. Newer CMOS devices are more prone to a non-destructive "mini-latch," although the IC is still not operating correctly. The SEL state in an IC is removed only by recycling power and is thus considered a permanent effect of the penetrating radiation. Power transistors or other high-voltage devices are sometimes damaged by a different mechanism called *single-event burnout (SEB)*, which depends on another parameter, the drain-source voltage. Another SEE effect, known as *single-event functional interrupt (SEFI)*, happens to complex parts in which the IC malfunctions because of errors in control or other specialized registers. The SEFI reveals itself as a lockup or other erroneous operation.

Shielding reduces the probability of SEEs. Satellites in LEO with inclinations less than ~60° benefit from the natural shielding of Earth's magnetic field against solar energetic particles.

Single-event effects are not limited to orbital altitudes. Effects at aircraft altitude and at the ground come primarily from galactic cosmic ray interaction. Neutrons created by cosmic ray interactions with the atmosphere reach a maximum at ~18 km and produce a peak flux of about four neutrons/(cm$^2 \cdot$s). At nine kilometers altitude, the neutron flux reduces to ~1/3 the peak value, and on the ground, the value is 0.0025 times the peak flux. Other particles, such as secondary protons and ions are also created, but for SEUs the neutrons are the most important. As with cosmic rays on orbit, neutron flux is greater at higher latitudes.

Chapter 13 Near-Earth Is a Place...with Susceptible Hardware and Humans

Focus Box 13.2: Energetic Particles on Humans and the Atmosphere

August 1972. During 2–12 August, 1972, between the launches of Apollo 16 and 17, a large solar active region appeared, producing numerous sequential eruptions and a severe proton storm. At 0620 UTC on one day, a large optical flare was observed. Within hours, energetic protons invaded the geospace. By 1400 UTC, an astronaut outside of the command module would have received the 30-day maximum radiation exposure for blood-forming organs and the year dose for eyes. By 1700 UTC, that same astronaut would have exceeded the career exposure limit for skin.

Events such as this have driven research into the shielding capacity of various materials. Shielding capacity is measured in units of mass/area.

- A typical space suit: 0.25 g/cm^2
- Apollo command module: 7 g/cm^2 to 8 g/cm^2
- Modern space shuttle: 10 g/cm^2 to 11 g/cm^2
- ISS (most heavily shielded areas): 15 g/cm^2
- Future moon bases will have storm shelters possibly exceeding 20 g/cm^2

NASA estimates that a hypothetical moon walker caught in the storm might have absorbed 400 rem (4000 mSv) of radiation dose. The Apollo command module would have attenuated the dose from 400 rem to less than 35 rem at the astronaut's blood-forming organs. That's the difference between needing a bone marrow transplant and oral medication.

September 1989. On September 29, 1989, a flare from an active region that had rotated behind the Sun's west limb produced a large *ground-level event (GLE)*. A GLE is a very high-energy solar proton event detectable at Earth's surface. The potential full-body dose of radiation from this event was computed for astronauts in different circumstances. The radiation exposure concerns for an unshielded astronaut exposed during an extra-vehicular activity on the lunar surface or on a deep-space mission, such as to Mars, would have been significant. Some models have suggested lethality would have been about 10% for astronauts inside a lightly shielded deep-space vessel. Other models using different assumptions projected no lethality, but significant radiation sickness.

According to a NOAA study, data from radiation sensors on Concorde supersonic jets cruising just above 18 km show that passengers and crew received a radiation dose equivalent of a chest X ray during the September 1989 event. The US Air Force used data from these events to revise standards of radiation exposure for pilots of high-flying aircraft.

October 1989. NASA launched the Space Shuttle Atlantis on October 18, 1989. Late on October 19, one of the largest energetic particle events of solar cycle 22 commenced (Fig. 13-13). Astronauts experienced irritating "flashes" in their eyes during this mission. These flashes did not subside until the proton event ended.

October 2003. Solar flare activity caused NASA flight controllers to issue contingency directives for the ISS Expedition-8 crew to briefly relocate to the aft portion of the station's Zvezda Service Module and the Temporary Sleep Station (TeSS) in the US Lab. The crew spent brief periods of time in the aft end of the Service Module, which is the most heavily shielded location on the ISS. During Tuesday Oct. 28, the crew spent five 20-minute periods in the aft end of Zvezda. These actions cut the potential radiation exposure of the crew by approximately 50%. Only once before, in November 2000, have NASA officials asked the ISS crew to seek shelter during solar radiation storms.

The Federal Aviation Administration issued the following commentary regarding aviation radiation dose on October 28, 2003:

> "Satellite measurements indicate unusually high levels of ionizing radiation, coming from the sun. This may lead to excessive radiation doses to air travelers at Corrected Geomagnetic (CGM) Latitudes above 35° north or south. Avoiding excessive radiation exposure during pregnancy is particularly important. Reducing flight altitude may significantly reduce flight doses. Available data indicates that lowering flight altitude from 40,000 feet to 36,000 feet should result in about a 30 percent reduction in dose rate. A lowering of latitude may also reduce flight doses but the degree is uncertain..."

Figure 13-14 illustrates the high latitude regions of concern.

January 2005. Event effects extended into Earth's atmosphere. Solar eruptions during 16–21 January 2005 led to a substantial flux of charged particles in the Earth's atmosphere. Enhanced solar proton flux increased production of OH in the mesosphere. The events also led to the production of NO$_x$ (NO, NO$_2$), as the protons and associated secondary electrons dissociated molecular nitrogen (N$_2$). Atmospheric chemists reported long-lived stratospheric ozone changes, causing an ozone decrease that lasted for several weeks.

Focus Box 13.2: Energetic Particles on Humans and the Atmosphere (Continued)

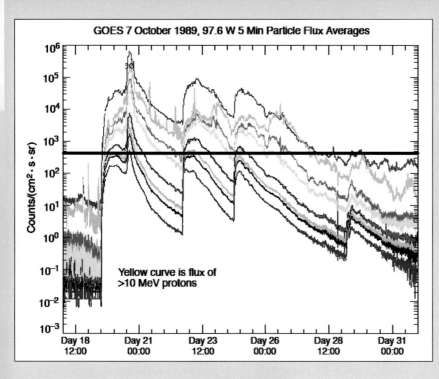

Fig. 13-13. Time Profile of the Flux of Energetic Electrons and Protons during the Latter Half of October 1989. This plot shows how the flux of energetic electrons and protons (in units of counts/(cm$^2 \cdot$ s $\cdot$ sr)) during a series of eruptions on the Sun produced solar energetic particle events. Proton levels remained above alert levels for 10 days. The top curve is the integrated electron flux for >2 MeV electrons. Below that are curves for >1, 5, 10, 30, 50, 60, and 100 MeV protons. Particle flux decreases as energy increases. The yellow curve represents the 10-MeV protons used in NOAA solar radiation alerts. The black line shows the maximum particle flux for a solar proton event in November 1997. *(Courtesy of NOAA)*

Fig. 13-14. FAA Warning Regions for Possible Elevated Radiation Dose during the Solar Radiation Storm of October 28, 2003. The orange shaded regions were most susceptible to increased energetic particle penetration and radiation dose. We note the correspondence between the susceptible regions and the regions of low rigidity values in Fig. 13-3. The FAA advised flight crews to fly at lower altitudes and latitudes to reduce radiation dose-risk in these regions while the storm was in progress. [NOAA, 2004]

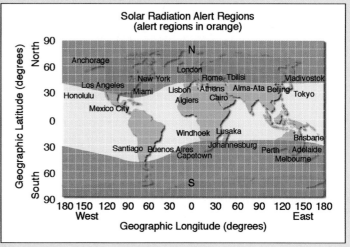

Chapter 13 Near-Earth Is a Place...with Susceptible Hardware and Humans

In recent studies, scientists discovered that the variations in aviation in-flight SEU rates with altitude and latitude correlate with the variation in atmospheric neutron flux. They also found that ground-based computing systems with large memory stores have field upset rates correlated with upset rates based on measurements in a laboratory neutron beam combined with cosmic ray neutron flux at ground level. So the same kind of radiation field is largely responsible for effects on the ground and at aircraft altitude, and the same types of devices and circuit boards are involved. The upset rates for ground and aircraft altitudes, based on ICs in use during the 1990s, are shown in Table 13-5. Because of the rapid changes in IC design and manufacture, the per bit SEU sensitivity of more recent RAMs is lower by several orders of magnitude, even though they contain many more bits. Scientists are trying to determine if the solar cycle modulates these effects and if extreme solar events may contribute to episodic events.

Table 13-5. Electronic Upset Rates for Ground and Airborne Equipment. Here we list how often computers and avionics equipment are upset by high-energy charged particles. *(Courtesy of Eugene Normand at the Boeing Radiation Effects Laboratory)*

Location	Computer Type	Upset Rate
Ground level SEU rate	Large computer memory banks	~2×10^{-12} upset/bit-hr
Aircraft SEU rate (neutron flux ~300 times higher)	Avionics	~6×10^{-10} upset/bit-hr

Avionics engineers mitigate SEU problems in flight controls by combining error detection and correction (EDAC) and system redundancy. Computer engineers also use EDAC and are seeking other mitigation methods.

Another form of penetrating particle effect in spacecraft is internal (*deep dielectric charging*), caused when multiple energetic particles lodge in material, building charge as they accumulate. Charge-deposition into dielectric (insulating, non-conducting) materials occurs when electrons with energies of 2 MeV–10 MeV penetrate deep into spacecraft structures over periods of days. The internal electric field builds up if the charge leak rate is less than the charge collection rate. Although excess charge spreads evenly on conducting surfaces, it produces an uneven potential distribution on dielectrics. Potential differences sometimes become large enough to produce an electrostatic discharge (spark), as depicted in Fig. 13-15. A discharge produces substantial current flow at a low pressure, and in a few microseconds or less. Discharges occur among components inside the spacecraft circuits. Proper grounding and shielding reduces the possibility of internal charging.

Internal discharge is more damaging because it occurs within dielectric materials and well-insulated conductors, close to sensitive electronic circuitry. The probability of a discharge increases dramatically with increasing electron fluence. Normally, a flux of 10^{10} electrons/cm^2–10^{11} electrons/cm^2 (over a period relative to the dielectric leak rate) builds up a sufficient charge to arc. Satellites in geosynchronous orbit (GEO) are particularly susceptible to deep dielectric charging and discharging, because they spend much of their orbits within the electron radiation belts. An Air Force Research Laboratory study based on Combined Release and Radiation Effects Satellite (CRRES) data obtained at geosynchronous orbit indicated that most environmentally induced spacecraft anomalies result from deep dielectric charging and the resulting discharge pulses

13.1 Damage and Impacts from Particles and Photons

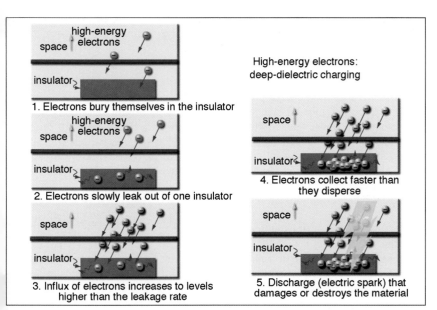

Fig. 13-15. **Deep Dielectric Charging Process.** These diagrams show how electrons cause deep dielectric charging when they penetrate spacecraft electronic components. The problem is most acute during intervals of large high-energy electron flux, when the charges have insufficient time to diffuse away from the location of deposition. *(Courtesy of Geoff Reeves at the Los Alamos National Laboratory)*

and not from surface insulator charging or single-event upsets. The 674 anomalies recorded during the 18-month CRRES mission correlated well with high levels of high-energy electron flux and poorly with every other environmental parameter. Figure 13-16 shows the close correspondence between fluxes of high-energy electrons (> 2 MeV) at geosynchronous orbit and the occurrence of anomalies on geosynchronous spacecraft used for European communications.

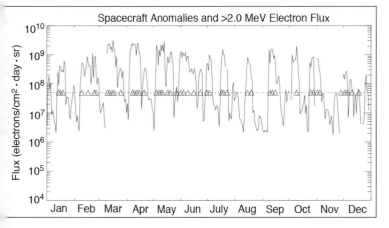

Fig. 13-16. **Daily Averaged GOES Electron Flux (>2 Mev) Compared to Reported Anomalies on DRA-delta Geostationary Spacecraft.** The horizontal axis is month of 1994. The small triangles represent each reported anomaly associated with electrostatic discharge. [Wrenn and Smith, 1996]

Direct Thermal Heating. Particle penetration causes another subtle difficulty—instrument and sensor heating on spacecraft. Energy conservation requires that the kinetic energy lost from the penetrating particles appear in other forms such as *direct heating*. Some spacecraft use the ambient outside temperature to maintain the very cold temperatures required for the proper operation of their onboard infrared (IR) sensors. This technique is inexpensive because it requires no cryogenics, but it does not work well when numerous

Chapter 13 Near-Earth Is a Place...with Susceptible Hardware and Humans

Focus Box 13.3:   Device Single Events

Based on discussions with the NOAA Space Weather Prediction Center staff

November 1982. The visible and infrared spin-scan radiometer on the Geostationary Operational Environmental Satellite (GOES)-4, which maps cloud cover, shut down minutes after high-energy protons arrived from a major solar flare.

March 1989. Sensors on board the GOES-7 spacecraft indicated that the flux of energetic protons was above event levels intermittently for several days. The geostationary Japanese telecommunications satellite CS-3b, operated by Nippon Telephone and Telegraph Corporation, failed during this time. Prior to 2000 UT on March 17, 1989, the particle flux again increased sharply. MARECS-1, a geostationary maritime communications satellite operated by the European Space Agency (ESA) for the International Maritime Satellite Organization (Inmarsat), experienced operational problems just before GOES-7 detected the sharp increase in energetic protons. The International Telecommunications Satellite Organization (Intelsat) reported "hits" on its geostationary communication satellites on March 18, 1989, when energetic proton fluxes were well above event levels. MARECS-1 failed in 1991 after numerous environment-related problems.

September 1989. On September 29, 1989, relativistic solar protons with energies greater than 450 MeV manifested the largest flux increase measured since the start of the satellite age.

Photo-sensitive components on the Magellan spacecraft en route to Venus were damaged. GOES power panel outputs were permanently reduced. GOES-5 and GOES-6 recorded single-event upsets. A series of 13 unidentified geostationary satellites recorded 46 "hits." A NASA Tracking and Data Relay Satellite (TDRS-1) recorded 53 random access memory (RAM) hits. NOAA-10 experienced rare phantom commands. The intensity of the event was verified by its detection by underground cosmic ray detectors at Embudo, New Mexico, which has a threshold rigidity of 19 GV. This was the first detection of a ground-level solar cosmic ray event by any underground cosmic ray muon telescope, where the event was clearly distinguishable above the background cosmic ray intensity.

July 1991. ESA's Earth Resources Satellite, ERS-1, was launched with a number of different experiments. Among them was an active microwave instrument. After about five days of operation, the experiment was shut down following a transient over-current condition and could not be restarted. This occurred during a pass over the SAA. Subsequent proton testing on the ground of the microelectronics board verified that energetic protons induce single-event latchup in the 64 kbit SRAM. [Adams et al., 1992].

September 2001. For two weeks, a variety of problems delayed the first orbital launch from the Kodiak, Alaska site. Operators solved these problems, so the rocket was ready for launch on September 24, 2001. However, a major solar flare (X2.6/2B) and a high-energy solar proton event further delayed the launch. The guidance system in the Athena 1 rocket is particularly sensitive to high fluxes of high-energy protons, so the launch was delayed until the event subsided. On September 29, 2001, the Athena 1 rocket finally launched a quartet of research satellites into space for NASA and the Air Force.

October 2003. The Mars Odyssey Spacecraft entered its safe mode during a severe radiation storm. During downloading on October 29, the spacecraft had a memory error that was corrected with a cold reboot on October 31. Its MARIE instrument had a temperature red alarm leading operators to power off the instrument on October 28. It did not recover. Ironically, MARIE's mission was to assess the radiation environment at Mars to determine the radiation risk that astronauts on a Mars mission may encounter.

energetic particles from a solar flare or coronal mass ejection collide with and heat the IR sensor to unacceptably high temperatures.

Loss of Attitude Control. Many satellites rely on electro-optical sensors to maintain their orientation in space. These sensors lock onto certain patterns in the background stars and use them to achieve precise pointing accuracy. These star sensors are vulnerable to cosmic rays and high-energy protons, which can produce flashes of light as they impact a sensor (Fig. 13-17). The bright spot produced on the sensor may be falsely interpreted as a star. When computer software fails to find this false star in its star catalogue or incorrectly identifies it, the satellite can lose attitude lock with respect to Earth. Directional communications antennas, sensors, and solar cell panels would then fail to see their intended targets. The result may be loss of communication with the satellite loss of satellite power, and, in extreme cases, loss of the satellite because of drained batteries (gradual star sensor degradation can also occur under constant radiation exposure). Disorientation occurs primarily when solar activity is high and on geosynchronous or polar-orbiting satellites.

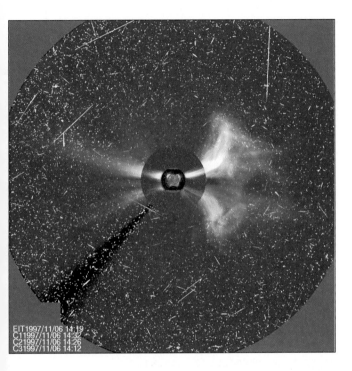

Fig. 13-17. Solar and Heliospheric Observatory (SOHO) Image during the November 6, 1997 Proton Storm. The many speckles are solar protons striking the spacecraft's digital camera. The white shadow near the Sun is the coronal mass ejection that drove the particle acceleration. *(Courtesy of ESA/NASA – the SOHO Program)*

Focus Box 13.4:   Satellite Disorientation

November 2000. The interplanetary spacecraft Stardust was launched in 1999 on a trajectory to rendezvous with Comet P/Wild 2 in 2004. Its mission was to collect samples of dust flying off the comet's nucleus and to collect interstellar particles flowing through our solar system. The spacecraft was only 1.4 AU from the Sun on November 8, 2000, when a powerful solar flare erupted. A cloud of high-energy particles headed for Earth and for Stardust. A stream of high-energy protons hit the spacecraft. The two star cameras that it used to control its orientation were peppered with radiation. Protons from the solar flare electrified pixels in the star cameras, producing dots that they interpreted as stars. The twelve brightest dots, the ones the spacecraft relied on to point its way, had electrified pixels that appeared as false stars. Hundreds of these star-like specks inundated the star camera's field of view, preventing it from computing its attitude.

The spacecraft went into its standby mode, turned its solar panels toward the Sun, and waited for instructions from Earth-bound operators. While it was waiting, the spacecraft tried again to determine its attitude using two different sets of cameras. But the particles repeatedly produced hundreds of bogus stars.

On November 13, 2000, Stardust was commanded to leave its safe mode. The star cameras started working again, controlling the orientation of the spacecraft perfectly. Stardust returned to Earth in 2006 to drop off the samples in a parachute-equipped return capsule.

Spacecraft Power. Solar arrays suffer from a variety of problems associated with the space environment. Solar cells are the most common power source for operational satellites, many of which operate in regions of appreciable ambient radiation and must face the Sun to produce power. The power output of a solar cell declines by 30%–40% or more over the lifetime of the satellite because of radiation-induced damage, as we describe in Focus Box 13.5.

In general, solar proton events are more intense and more frequent at solar maximum. They result in step-like losses in satellite power. Figure 13-18 shows the effect of prolonged radiation exposure and episodic solar proton events on the power output of the SOHO spacecraft solar cells between 1997 and 2002. These multi-hour to day-long events age a satellite's power array by a year or more during a single storm. However, damage is not limited to solar particles. During solar minimum, the solar arrays receive ongoing doses of radiation from galactic cosmic rays, which also reduce the lifetime of solar panels.

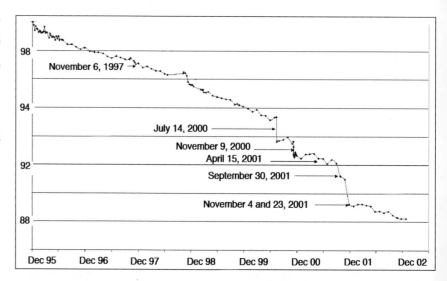

Fig. 13-18. Power Degradation in Solar and Heliospheric Observatory (SOHO) Solar Cells. The vertical axis is percent of power production. The horizontal axis is time. Exposure to cosmic rays and a low level of energetic solar protons slowly degrade the power output from spacecraft solar cells. The November 2001 solar proton events caused nearly a year's worth of degradation in a few hours of energetic proton dose. The data come from two sections of the solar array. *(Courtesy of Paul Brekke, European Space Agency)*

Focus Box 13.5: Solar Panel Degradation

September and October 1989. In September 1989, the Geostationary Operational Environmental Satellite (GOES-7) was the only National Oceanic and Atmospheric Administration (NOAA) fully functional spacecraft for observing terrestrial weather and for monitoring the Sun and the space environment from geostationary altitude. It detected a major X9.8 solar flare with maximum intensity at 1133 UT on September 29, 1989. During the ensuing particle event, the output of its solar panels degraded more in one day than their expected loss over a full year of normal radiation aging.

A few weeks later, during several October 1989 proton events (Fig. 13-13), a succession of large-amplitude energetic particle fluxes bombarded the satellite's power panels. The output degradation produced in a few days was equivalent to half of a satellite's normal lifetime loss.

Total Dose Effects. Total radiation dose is what usually limits the operational lifetime of spacecraft electronics and instruments. The electrical properties of solid-state components change upon exposure to radiation. As the dose accumulates, these changes drive the component parameters outside of their design range for their circuits. Ultimately, these changes cause the circuit to cease functioning properly. For example, several of the solid state detectors in the

Space Environment Monitor aboard the NOAA POES spacecraft experience normal radiation damage over time that reduces their sensitivity.

Focus Box 13.6: Spacecraft Solar Cell Operations

Background information from the National Renewable Energy Laboratory

A typical solar cell consists of special materials joined to produce electron flow. These materials are usually arranged in a layered crystalline structure that transfers the energy of an absorbed photon to an electron in the crystal lattice (Fig. 13-19a). The structure is a p-n junction and is a few tenths of a micrometer below the surface, with the n-type material occupying the region between the surface and the junction. The bulk of the material at the base of the cell is p-type material, where the "p" designates positive charge. The "n" designation indicates that the material donates electrons (negative charge) that flow freely when interacting with the blue light of the visible spectrum. When a photon of appropriate energy state releases its energy to a lattice electron in the valence band, the electron is excited into the conduction band and becomes mobile. These electrons migrate to the top electrode, leaving holes in the depletion zone that attract electrons from the p-type material. The p-type material, on the bottom, intercepts the longer wavelength red light that excites electrons from the valence band to the conduction band. These electrons diffuse upward to the depletion zone, where they are accelerated by the positive charge centers remaining in the n-type material. The net result is negative charges migrating to the top electrode and positive charges migrating toward the bottom electrode, creating a potential difference of about 1 volt. Connecting these electrodes to an external circuit produces a current (Fig. 13-19b). The product of voltage and current is power ($P = IV$). Usual solar cell efficiencies are 15%–20%, meaning that roughly 80% of the incident solar radiation is lost in the form of heat. In fact, a significant design problem is keeping the solar cell running at an optimum temperature.

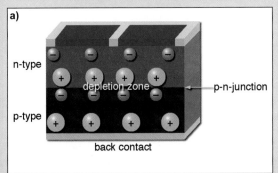
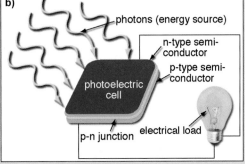

Fig. 13-19. a) Close Up View of p-n Junction. Blue light is 99% absorbed within 0.2 micrometers of the surface in the n-type material, whereas red light penetrates about 200 micrometers before being 99% absorbed in the p-type material. **b) Solar Cell Circuit.** The diode created by the electric field allows current to flow in only one direction across the junction. Connecting the sides of the cell externally causes electrons to flow to their original p side to meet with holes.

For space operations, the solar cells generally have a transparent (to visible light) cover that rejects higher-energy photons to minimize damage to the p-n material. Radiation affects solar cells in two major ways: darkening their coverslides and decreasing charge carrier lifetimes. If a free electron is created in the coverglass, it may be trapped to form a charged defect called a color center. A color center inappropriately absorbs photons that should pass through to the p-n material. Color centers decrease the incident solar flux on the cells underneath and reduce the power output from the cells.

Radiation also displaces atoms in a cell's crystal lattice, thus forming positively charged interstitials. An electron diffusing through the crystal on its way to the n-p junction gets trapped into an orbit around an interstitial atom and then no longer contributes to the current flow. This process decreases the average lifetime of the free charge carriers, meaning that fewer electrons reach the junction and the output current drops. Energetic protons from solar flares and the shock regions of CMEs are the culprits in this type of solar cell degradation.

13.1.3 Energetic Plasma, Photon, and Neutral Atmosphere Effects on Hardware

> **Objectives: After reading this section, you should be able to...**
> - Distinguish between deep dielectric charging and surface charging
> - Distinguish between the physical origins of surface charging and sputtering
> - Explain how ultraviolet (UV), extreme ultraviolet (EUV), and X-ray photons contribute to spacecraft charging and surface deterioration
> - Explain why spacecraft surfaces deteriorate after long exposure in low-Earth orbit (LEO)
> - Describe the likely source of spacecraft glow

In this section we move down the energy ladder (of Fig. 13-1) to levels that are more typically associated with plasma (as opposed to individual penetrating particles). The delineation is not sharp. Energetic particles that are part of a background plasma population still separate themselves from the background to cause damage and impacts. Conducting materials (thermal paints and coatings) as well as electrical resistive coatings often degrade from impacts. Results cascade to a premature end of life for spacecraft or their components.

Damage to Spacecraft Components from Surface Charging, Discharging, and Sputtering

One of the most common anomalies associated with the moderate-energy particles in the space environment is spacecraft electrical charging. Spacecraft surface charging occurs when the electrostatic potential between the spacecraft surface and the surrounding plasma increases. A large potential difference provokes a discharge, which often causes spurious electronic switching, breakdown of thermal coatings, amplifier and solar cell degradation, and optical sensor degradation. Most reported problems are at high altitudes (>5 R_E), where magnetotail particle fluxes are greater.

Surface charging is produced by 1) an object's motion through a medium containing charged particles (called "wake charging"), which is a significant problem for large objects such as the Space Shuttle or the International Space Station; 2) directed particle bombardment that occurs during geomagnetic storms and proton events; or 3) the 'photoelectric effect' of solar illumination that causes electrons to escape from an object's surface. The impact of each phenomenon is strongly influenced by an object's shape and the materials used in its construction. If sufficient differential charge collects, a spontaneous discharge, similar to a small lightning stroke, will occur, damaging or disrupting the operation of components on or near the spacecraft's surface. This leads to a variety of operational anomalies, including degradation of surfaces and sensors and interference in electrical circuits due to arcing.

Under certain conditions, a reservoir of charge called *Debye sheath* surrounds the spacecraft (we recall the discussion of Debye spheres in Chap. 6). This reservoir and the spacecraft's surface support three mechanisms for current flow to and from the reservoir. The charge mobility creates the opportunity for charging, and sometimes a related phenomenon called sputtering. The charge mobility, and thus charging, has three sources:

- Charge flow from the ambient plasma—*plasma induced charging* associated with differences in ion and electron mobility
- Charging by *photoelectron emission* from the spacecraft (*Note:* We treat a photon as a particle here)
- Charging from *secondary electron emission* due to plasma bombardment

We present the general scenario in Fig. 13-20, where we identify the different current sources. These effects generally result in a non-zero charge on the satellite's body (and surfaces), and hence non-zero potentials. The potentials are often quite large—into the kilovolt range.

Plasma-induced Charging. We assume the satellite is initially uncharged. If the two gases, one composed of electrons and one composed of ions, have approximately the same temperature and density, then the electron current will exceed the ion current by at least an order of magnitude. We determine this by considering the temperature-thermal velocity relationship

$$\langle E \rangle = \frac{3}{2}k_B T = \frac{1}{2}mv_i^2 = \frac{1}{2}mv_e^2$$

where

$\langle E \rangle$ = particle energy (assuming thermal and kinetic energies are equal) [kg·m²/s²]

k_B = Boltzmann constant (1.38 × 10⁻²³ J/K)

T = particle temperature [K]

m = particle mass [kg]

v_i = ion speed [m/s]

v_e = electron speed [m/s]

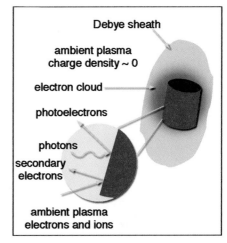

Fig. 13-20. Surface Charging. In a plasma, charge neutrality is maintained by strong, long-range electric forces. This requires that on average, the electron and positive ion densities be equal (quasi-neutrality). The average kinetic energies (temperatures) of ions and electrons are often similar, but in general, their speeds are quite different, because of the significant mass difference. *(Courtesy of R. Chris Olsen at the Naval Post Graduate School)*

Even assuming that most ions are protons, the ratio of thermal velocities is fairly high

$$\frac{v_e}{v_i} = \frac{\sqrt{m_i}}{\sqrt{m_e}} \approx 43$$

Hence the electron speed is much higher than the ion speed for similar temperatures. This identifies the electrons as being highly mobile. Thus, negative charge will accumulate on the satellite first, which will inhibit the incoming electron flow, as well as increase the ion inflow. Equilibrium will be established when the negative satellite potential is the same order of magnitude as the average electron energy in electron volts (eV). This is the situation for a high-altitude satellite that goes into eclipse, where photoelectrons do not exist. Thus, if the electrons are very energetic (keV range), then large negative charging occurs (into the kV range).

In the simplest case (no effects from satellite motion or plasma flow), the current density (current/area) J, incident on a satellite's surface is given by

$$J = qnv$$

where

J = local current density vector [A/m^2]
q = particle charge (1.6 × 10^{-19} C)
n = number of particles [unitless]
v = particle velocity [m/s]

Because the charge magnitude and the density are the same for both species, the current density is much higher for the electrons than the ions (43 times higher for a proton plasma).

Charging from Photoelectric Emissions. When photons of sufficient energy strike a material's surface, electrons are emitted from the surface, leaving the target positively charged. We represent this in Fig. 13-21.

Fig. 13-21. Surface Charging from Photons. A photon impact frees an electron, leaving the surface positively charged. Nature keeps this process partially in check by electrically attracting the freed electrons to the positively charged surface.

The energy balance is

$$hf = \phi + KE_e$$

where

h = Planck's constant (4.136 × 10^{-15} eV-s)
f = photon frequency [1/s]
hf = energy of incident photon [eV]
ϕ = material work function (amount of energy required to dislodge an electron) [eV]
KE_e = ejected electron kinetic energy [eV]

The work functions of most spacecraft materials are about 4 eV–5 eV, which means that only photons in the UV and X-ray region ($\lambda \leq 300$ nm) generate photoelectrons. Nevertheless, because of the intense fluxes of UV and soft X rays (particularly the Lyman-alpha emission) this mechanism is important in positive spacecraft charging. This mechanism operates only when the spacecraft is sunlit, so spacecraft charging goes through cycles for orbiting satellites passing in and out of eclipse.

In general, the flow of photoelectrons away from a satellite is much greater than the ambient electron flux toward the satellite. As the spacecraft charges positively with respect to the ambient plasma, the outgoing photoelectron flux is inhibited (photoelectron energies are generally a few eV at most), and the incoming ambient electron flow increases. Equilibrium occurs when these competing flows balance, and the potential is a few volts (positive) at most. For positive potentials, the incoming ambient ion flux plays no role in establishing equilibrium, because it's small to begin with, and decreases even further because of the positive potential. Even if a large influx of high-energy ambient electrons exists—which could cause a large negative charging event in the absence of photoelectrons (during eclipse)—the large flux of outgoing photoelectrons in sunlight mitigates these large negative voltages.

13.1 Damage and Impacts from Particles and Photons

Charging from Secondary Emissions. Substantial currents are generated by incident electrons with energies of a few hundred electron volts (100 eV–500 eV). These low-to-medium-energy electrons ionize some of the surface material, producing secondary electrons. The latter escape the surface with energies of a few eV. For some materials, more than one secondary electron escapes. This electron flow leads to the incongruous result of having a net positive current produced by the incident electron flux. Nature tends to rein in this process as the sheath of secondary electrons begins to repel incoming electrons.

As the plasma temperature increases, the net secondary yield decreases, normally at temperatures exceeding 10 keV. The energetic primary electrons penetrate sufficiently far into the material that the secondary electrons, though freed from their parent atoms, don't have enough energy to reach the spacecraft's surface. The sheath of secondary electrons does not develop to repel the incoming electrons. As a result, the surface becomes negatively charged with respect to the ambient plasma. In such a case, a substantial number of electrons with energies over 10 keV must be present for a net negative charge to flow to the satellite, which is possible during geomagnetic storms, when electrons energized in the tail drift Earthward and toward dawn. Anomalous charging of this type is often reported by satellites in the magnetic local times between midnight and dawn. Figure 13-22 illustrates the situation for geosynchronous satellites. Satellites in HEO and MEO orbits show similar charging probabilities in the midnight to dawn sector.

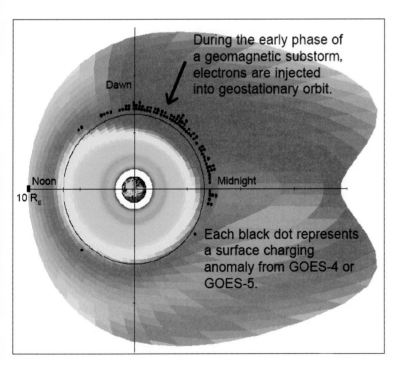

Fig. 13-22. A Model View of the Inner and Middle Magnetosphere from Above Earth's North Pole. Orange and red colors indicate a high population of energetic electrons. Yellow, green, and blue represent low fluxes of energetic electrons. The black circle represents the geosynchronous orbit. Black dots indicate the location of surface charging anomalies from the GOES-4 and GOES-5 spacecraft in geosynchronous orbit. *(Courtesy of National Geophysical Data Center and John Freeman of Rice University)*

Sputtering. When atoms or ions of sufficient energy strike a solid surface, they eject atoms in a process called *sputtering*. Common sputtering thresholds are in the range of a few tens of eV. The energies available in collisions between atmospheric atoms and moving satellite surfaces are generally lower than the sputtering thresholds. Thus, collisions caused by the orbital motion of the satellite do not cause a significant amount of sputtering. However, we know that satellite surfaces become negatively charged to several kilovolts, and these negatively charged surfaces accelerate any positive ions in the atmosphere toward the surface, yielding impact energies of hundreds or even thousands of eV, more than enough to cause sputtering. On extended missions, the sputtering of materials, in particular metals, causes erosion as well as changes in the surface properties. Spacecraft materials erode, exposing underlying regions to atomic oxygen attack. Optical coatings are particularly sensitive to sputtering. They improve the efficiency of solar arrays, but their performance depends in part on their transmittance, which usually degrades after long exposure to the space plasma environment.

Photon Interactions with Spacecraft Surfaces

The process by which photons interact with matter is fundamentally different from that of charged particles. Rather than multiple collisions, photons transfer all of their energy in a single interaction and vanish in the process. This leads to an exponential decrease of intensity I with penetration depth into the absorber, as we describe in Chap. 3

Virtually all solar radiation with a wavelength less than 0.3 μm (~4 eV) is absorbed by the atmosphere before it reaches Earth's surface. A satellite, on the other hand, is exposed to a broad spectrum of solar radiation, from radio waves through X rays. Photons with wavelengths between 0.13 μm and 0.82 μm can break many carbon-, oxygen-, and nitrogen-based chemical bonds and thereby alter the physical properties of spacecraft surface materials. Exposure to EUV photons causes materials, especially paints and thermal coverings, to develop micro-cracks, which in turn leads to brittle structures that can spread into the spacecraft environment. Over time, materials deeper in the spacecraft skin become susceptible to atomic oxygen attack and thermal cycling.

One important quantity for spacecraft thermal control is the solar absorbance α_s of spacecraft materials, which varies from 0 for a perfect reflector to 1 for a perfect absorber. Large changes in the absorbance (on the order of 50%) have been observed because of UV irradiation for certain materials and exposure times on the order of 1000 hours.

Deterioration of Surface Materials, Sensors, and Solar Panels by Neutral Atoms

Atomic oxygen and some other heavier elements react chemically with spacecraft surface materials, sensors, and solar panels. The deterioration effect is cumulative. Damage is most severe on spacecraft exposed to atomic oxygen for months or years in low-Earth orbit.

At 300 km altitude, the ambient atmospheric density is about ten orders of magnitude below that encountered at sea level (~10^{15} oxygen atoms/m³). A one square meter spacecraft surface orbiting at 8 km/s collides with about 10^{19} atoms per orbit, producing an effective collision energy of about 5 eV. Because atomic oxygen is highly reactive, these collisions oxidize and erode surface materials.

Spacecraft Glow Caused by Neutral Atoms

Observations reveal that ram surfaces of LEO spacecraft exhibit a visible luminescence or glow. Such glows have been detected during flights of the Atmospheric Explorer (AE) satellite and the Space Shuttle. Figure 13-23 shows

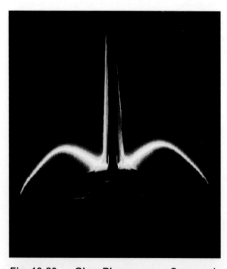

Fig. 13-23. Glow Phenomenon Surrounding the Vertical Stabilizer and Orbital Maneuvering System Pods. This photograph was taken as the Space Shuttle Columbia was orbiting Earth during a "night" pass. This glow is in the red portion of the visible band near 680 nm. *(Courtesy of NASA)*

13.1 Damage and Impacts from Particles and Photons

glow on the Space Shuttle. This emission extends about 0.1 m from orbiting surfaces and peaks at a wavelength of 680 nm. More glow is observed at low altitudes. The most likely explanation is the recombination of fast oxygen atoms in the upper atmosphere with nitric oxide absorbed on the Shuttle's surface or exhaust gases from the Shuttle's thrusters. The recombination creates excited NO_2, which radiates as it desorbs. Spacecraft glow may be a source of interference in space-based spectroscopy; anomalous airglow observations made by the AE spacecraft have been attributed to this contamination.

Focus Box 13.7: Spacecraft Contamination

Aerospace engineers' concerns go beyond those effects created solely by the ambient space environment that we describe in the previous sections. The spacecraft contaminates the environment through which it travels. Therefore, engineers are concerned also with the *spacecraft environment*, which is the sum of the natural space environment and the contamination from the spacecraft.

Contamination comes in two broad categories labeled particulate and molecular. Particulate contamination is visible-sized (micrometer or larger) pieces of matter—dust. Spacecraft are great sources of dust contamination that is nearly impossible to eliminate (even though incredible efforts are often taken to do so). Dust comes from internal components and from the local, transport, and launch environment. During launch, the spacecraft may shake violently, freeing more particulates that ride along to orbit where they are free to float about and interact with the space environment. Table 13-6 lists particulate cleanliness levels. Level 1 is the most stringent.

Molecular contamination is the accumulation of individual molecules on a surface. It comes from a spacecraft's materials. These molecules form films hundreds of micrometers thick on payload sensors, solar arrays, thermal control surfaces, and other spacecraft surfaces and often cause unwanted effects. We can theoretically eliminate particulate contamination, but molecular contamination cannot be eradicated, because it is not foreign matter. The materials of the spacecraft outgas molecules from their surface. *Outgassing* means releasing molecules from the surface by one of three processes: desorption, diffusion, or decomposition. *Desorption* is the release of surface molecules held by chemical forces. *Diffusion* is the motion of molecules to the surface from random thermal motions and the physical process of smoothing gradients of molecular concentrations in the material. Finally, *decomposition* is the chemical reaction of a compound splitting into two (or more) simpler substances that are then free to outgas through desorption and diffusion. Some materials outgas more than others, and aerospace engineers carefully consider material selection when constructing a spacecraft.

Thus we must understand the natural space environment, space weather, and contamination effects. For example, solar power production for early GPS satellites degraded at about twice the rate predicted by radiation damage models. Engineers believe this rate came from photo-enhanced molecular deposits on the solar arrays.

Table 13-6. Particle Cleanliness Levels. Here we list the requirements for a given level of particulate contamination control for spacecraft in Publication IEST-STD-CC1246D from The Institute of Environmental Sciences and Technology.

Cleanliness Level	Particle Size μm	# of Particles of Sizes ≥ Stated Size Per 0.1 m^3
1	1	1
10	1	8
	2	7
	5	3
	10	1
100	15	265
	25	78
	50	11
	100	1
1000	100	42,658
	250	1022
	500	39
	750	4
	1000	1

13.1.4 Satellite Drag

Objectives: After reading this section, you should be able to...
- State the factors that most affect satellite drag
- State the general level of uncertainty in predicting satellite drag
- Explain why satellites accelerate as they experience drag

Variations in solar activity influence thermospheric density levels and the thermal environment. When solar activity is high, solar extreme ultraviolet (EUV) radiation heats and expands Earth's upper atmosphere. Additionally, geomagnetic storms add energy to the auroral zones via particle and Poynting flux deposition. Impulsive energy increases propagate out of the auroral zones as traveling atmospheric disturbances, scale height changes, and composition changes. The associated enhancement of the upper atmosphere's neutral density increases satellite drag and associated orbital decay rate. The underlying cause of atmospheric drag is momentum transfer from atoms or molecules in the atmosphere. The more atoms or molecules the spacecraft encounters, the greater the momentum transfer and thus, the greater the drag.

Forecasting satellite lifetimes depends upon a knowledge of the initial satellite orbit, the satellite drag coefficient, the satellite mass to cross-sectional area (in the direction of travel), and a knowledge of the upper atmospheric density and how the density responds to energy inputs, which must also be predicted. Even when most of the quantities are known, a ~10% uncertainty remains in the satellite lifetime estimate. In other words, the error in predicting the duration of a satellite expected to remain aloft for about 10 years is one year, whereas the demise of a satellite expected to re-enter in 24 hours is accurate to about 2 hours. Improving observations and models decreases the short-term forecasting uncertainty.

Long-term variations in atmospheric density, such as those driven by solar cycle variations in EUV, have order-of-magnitude effects on the lifetime of satellites in low-Earth orbit (LEO). Figure 13-24 provides an example of satellite lifetime as a function of F10.7 solar flux (solar emissions at 10.7 cm wavelength) for circular orbits at various initial altitudes and assuming constant solar emissions over the satellite lifetime. Long-term forecasts of solar EUV variability are used to determine on-board fuel requirements, estimate satellite lifetimes, and plan re-boost maneuvers. Short-term variations in density, which occur during geomagnetic events, perturb the orbital motions of satellites and lead to difficulties in controlling attitude dynamics, precision tracking, and cataloging objects at low-orbit altitudes. These short-term perturbations lead also to uncertainties in position for re-entry of orbiting vehicles and difficulty in providing collision avoidance warnings.

Figure 13-25 illustrates the very large effect a severe geomagnetic storm can have on satellite drag. The black dots show the satellite acceleration during quiet geomagnetic conditions, but variable solar emissions. The change in acceleration is approximately linear with the F10.7 index. However, impulsive energy deposition during magnetic storms creates a radical increase in acceleration, as shown by the open circles.

The dynamics of a satellite experiencing drag can be rather counter-intuitive. As the satellite encounters higher density and hence more drag, its mechanical energy decreases, its altitude decreases, but its speed increases. The speed

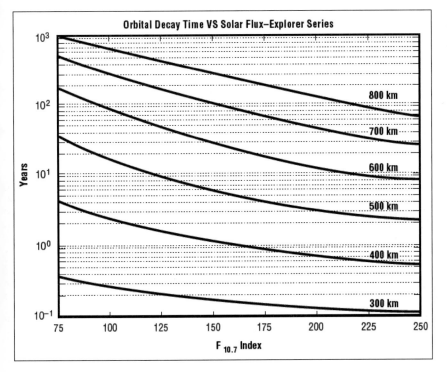

Fig. 13-24. Satellite Orbital Lifetime versus Solar Emissions Described by the F10.7 Index. The horizontal axis is the F10.7 cm radio index in solar flux units. (1 SFU = 10^{-22} W/(m$^2 \cdot$Hz)), which is a proxy for solar extreme ultraviolet radiation heating of the upper atmosphere. The vertical axis is satellite lifetime in years. A satellite in circular orbit is assumed (unrealistically) to experience the same conditions over its entire lifetime. A satellite in a circular orbit at 400 km at the lowest F10.7 value remains in orbit for about 4 years. The same satellite placed in the same orbit, but experiencing the highest F10.7 value for its entire lifetime would stay on orbit for less than a year. [Gorney, 1990]

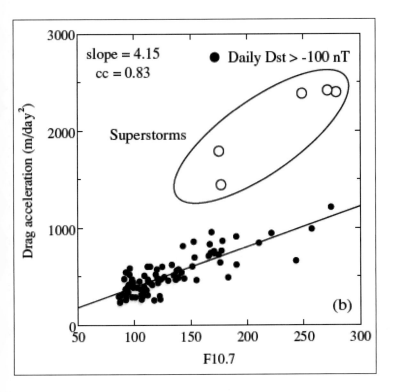

Fig. 13-25. Scatter Diagrams of the KOMPSAT-1 Drag Acceleration versus the F10.7 Index. The filled dots represent measurements made when the Dst index indicated the upper atmosphere was quiet to moderately disturbed (The magnitude of the Dst index was less than 100 nT). The open circles are when the upper atmosphere was greatly disturbed during superstorms (the magnitude of Dst was greater than 250 nT). The numbers in the top left corner refer to the slope of the fit line for the quiet-to-moderately disturbed conditions and to the correlation coefficient (cc) for the fit. KOMPSAT-1 was in a Sun-synchronous orbit at ~685-km altitude starting in 1999. By late 2003, when the superstorms occurred, the satellite orbit altitude was ~660 km. [Kim et al., 2006]

increase comes from the conversion of some of the mechanical energy to kinetic energy, even as the drag force continues to act.

The drag force on a satellite is given by

$$D = (1/2)\rho v^2 A C_d$$

where

- D = drag force [N]
- ρ = atmospheric density [kg/m^3]
- v = satellite speed [m/s]
- A = satellite cross-sectional area perpendicular to the direction of motion [m^2]
- C_d = drag coefficient, usually assumed to be about 2 [unitless]

As a satellite drops in altitude, it encounters exponentially increasing density and its speed increases, which further increases the drag (by the v^2 term). But most satellites are not in circular orbits. Satellite drag makes a non-circular orbit more circular as the satellite passes through the atmosphere. Figure 13-26 shows the height and speed profile for the STARSHINE-1 satellite, placed in a circular orbit near 400 km in 1999. As the satellite altitude decreases, its speed rises.

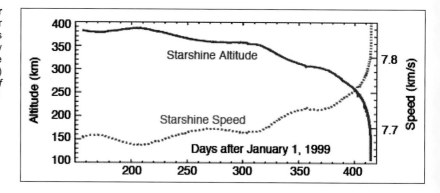

Fig. 13-26. Altitude and Speed Profile for STARSHINE 1. The satellite orbited Earth for eight months before vaporizing in Earth's atmosphere in February 2000. These data show how the satellite fell slowly at first, then more rapidly as it descended into lower (and denser) layers of Earth's atmosphere. *(Courtesy of Judith Lean at the Naval Research Laboratory)*

An important result of the speed increase is that the orbital period shortens. This makes monitoring the locations of satellites difficult because the satellites are moving faster than forecast and thus, are ahead of their expected positions. The errant satellites fly through the ground-based radar beam ahead of schedule. During strong geomagnetic storms, many satellites are behaving in this manner, making satellite tracking and identification difficult, if not impossible. Agencies charged with satellite tracking are increasing their efforts in forecasting atmospheric density variations so that computer models can better predict the likely future locations of satellites. Figure 13-27 illustrates the situation.

13.1 Damage and Impacts from Particles and Photons

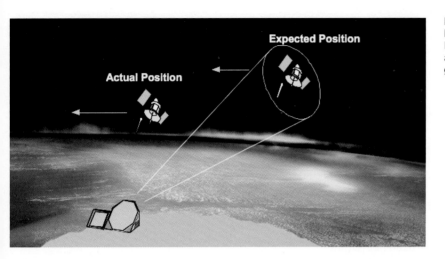

Fig. 13-27. A Satellite that Is Ahead of Its Expected Position because of Satellite Drag. Increased drag causes the satellite to be below and ahead of the position expected by the ground radar.

Chapter 13   Near-Earth Is a Place...with Susceptible Hardware and Humans

Focus Box 13.8:   Satellite Drag

The Skylab Space Station. Skylab was launched on May 14, 1973, from the Kennedy Space Center on a huge Saturn V launch vehicle, the moon rocket of the Apollo Space Program (Fig. 13-28). Upon completion of its mission in 1974, Skylab was positioned into a stable attitude and systems were shut down. Operators expected that it would remain in orbit eight to ten years. However, in the fall of 1977, they determined that it was not in a stable attitude because of greater-than-predicted solar activity that expanded the atmosphere, causing excess air drag. On July 11, 1979, Skylab re-entered the atmosphere and impacted Earth's surface. The debris dispersion area stretched from the southeastern Indian Ocean across a sparsely populated section of western Australia.

April 7, 1984. NASA launched the Long Duration Exposure Facility (LDEF) into low-Earth orbit, at an altitude of about 509 km, aboard Space Shuttle Challenger. The mission of the unmanned laboratory was to study the effects of the natural space environment and orbital debris on spacecraft materials. According to the original plan, LDEF was supposed to ride a Shuttle back to Earth after about one year. Then scientists would

Fig. 13-28. Skylab Space Station. This laboratory orbited for over six years, but reentered prematurely because of increased solar activity that expanded the atmosphere, causing higher atmospheric drag than predicted. *(Courtesy of NASA)*

study the payload. But Shuttle schedules slipped, then Challenger exploded shortly after launch, grounding the remaining fleet of Shuttles, and LDEF was stranded in orbit. LDEF nearly re-entered Earth's atmosphere because of air drag. In January 1990, it was losing about a kilometer of altitude per day. This was close to solar maximum, near the peak of solar cycle 22. Within a few weeks, LDEF would have broken apart in the atmosphere. Instead, the Space Shuttle Columbia captured it on January 12, 1990, after 32,422 orbits, and it provided a wealth of information about the effects of the space environment on spacecraft materials.

March 1989. During this storm, US Space Command had to re-track over 1,000 objects whose orbits had been affected by the increased air resistance. During the great geomagnetic storm of March 13 and 14, 1989, tracking of thousands of space objects was lost and the North American Defense Command (NORAD) took many days to reacquire them in their new, lower, faster orbits. One LEO satellite lost over 30 kilometers of altitude, and hence significant lifetime, during this storm.

July 2000. Increased air drag caused failure of the Advanced Satellite for Cosmology and Astrophysics (ASCA), originally named ASTRO-D, launched by the Japanese Institute of Space and Astronautical Science (ISAS) in February 1993 (Fig. 13-29). The 417-kg spacecraft featured an X-ray telescope and a pair of cameras to record images from the telescope. It was designed to perform studies ranging from searches for black holes to studies of dark matter and the chemical evolution of these. Problems for ASCA started on July 15, 2000, when it encountered strong atmospheric drag. A severe geomagnetic storm (Kp = 9) expanded the tenuous upper atmosphere. This additional drag created a torque on the satellite. Its attitude control system could not compensate for the torque and it started spinning with a period of about three minutes. Spacecraft controllers put it into a protective safe mode, but the torque was strong enough to move the spacecraft's solar panels out of proper alignment with the Sun. This misalignment reduced ASCA's ability to generate power and its batteries discharged. Efforts to restore power and charge the batteries were unsuccessful. The battery cells suffered serious unrecoverable damage, and the satellite mission was lost. ASCA reentered the atmosphere on March 2, 2001

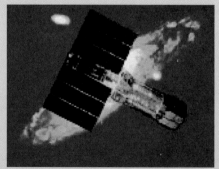

Fig. 13-29. Japanese Advanced Satellite for Cosmology and Astrophysics (ASCA) Spacecraft. Atmospheric drag increased the torque on this spacecraft during a solar storm. The attitude control system couldn't compensate for the torque, so the spacecraft was lost. *(Courtesy of NASA and Institute of Space and Astronautical Science)*

13.2 Damage and Impacts Associated with the Meteor and Artificial Debris Environment

Hardware in space may suffer hypervelocity encounters with natural meteoric and human-made debris. These encounters cause penetration, pitting, perforations, cracks, and spalling, all of which lead to component or system failure. Collisional objects with radii less than 0.5 mm are predominantly meteors, and those with radii greater than 1 mm are mostly debris.

13.2.1 The Natural Meteor Environment

Objectives: After reading this section, you should be able to...

- Distinguish between various forms of meteoric material
- Describe meteoric effects on the atmosphere and spacecraft

Characteristics

Ablation of comets and asteroids produces a natural solar system particle and dust population called *meteoroids*. Table 13-7 lists the sizes for pre-atmosphere entry.

Table 13-7. Size Comparison of Micrometeoroids and Meteoroids Entering Earth's Atmosphere. The larger the meteoroid, the less common it is.

Meteoroids > 10 m	0.01 m < Meteoroids < 10m	Micrometeoroids < 1 cm
Very rare	Rare	Largest flux at < 0.5 mm

These objects enter Earth's atmosphere with speeds ranging from 2 km/s–25 km/s, with a mean value of ~18 km/s. Meteoroids, upon entering the atmosphere, become *meteors* (shooting stars). Large, fast meteors with speeds in excess of 30 km/s sometimes explode as they transit the atmosphere, creating blasts heard and reported by the public. These exploding meteors are *bolides*. Meteors that survive the atmospheric transit and reach the ground are meteorites. Spacecraft collisions with meteors and meteoroids are often catastrophic. Even micrometeoroids are dangerous to spacecraft occupants and components.

Meteor fluxes are often categorized as cosmic dust, sporadic meteors, and showers. Cosmic dust consists of high-flux-rate micrometeoroids with relatively low speeds (~11 km/s). These tiny particles constantly bombard Earth's atmosphere. Sporadic meteors do not belong to a recognized shower or stream. They are by far the dominant component of the meteoroid flux at Earth, and become more frequent at smaller sizes. Sporadic events probably come from the asteroid belts or long-ago perturbed comet debris. Meteor showers originate in Sun-orbiting comets, though a few are linked with asteroids. The meteoroids generally follow the parent-body orbit, and are thought to be relatively recent (<10,000 years) ejecta from the parent. The larger (r > 0.5 m) meteors that result in the intense bolides most likely come from asteroids. Large meteors and their trails are sometimes misidentified as spacecraft, missiles, and unidentified flying objects. Radar operators and individuals involved in tracking satellites and identifying space objects must know about the anticipated meteor showers.

Earth's gravitational field has a focusing effect that enhances the meteor flux near Earth. At any time approximately 200 kg of meteoric mass is within 2000 km of Earth's surface. Most of the particles have low density (0.1 g/cm^3–0.5 g/cm^3) and mass. The mass flux is quasi-isotropic—that is, sensors and spacecraft surfaces that face Earth have a reduced chance of meteor impact.

Atmospheric Effects

Meteoric and dust constituents are responsible for thin sodium and lithium layers in the upper atmosphere. They may also contain and shed to the atmosphere iron, magnesium, aluminum, and calcium ions. These substances are observed with resonance dayglow and lidar techniques. Their high kinetic energies allow them to create ionization trails in the atmosphere with both beneficial and detrimental consequences. Long-lived trails create sporadic E layers (E_s), and they enable meteor burst communications used by some operators for low-volume data transfer. The trails are sensed by radars and may occasionally interfere with radar functions. Dust and micrometeoroid particles provide nucleation centers for forming noctilucent (illuminated at night) clouds. The smallest micrometeoroids (r < 100 μm) do not fully ablate but remain intact and settle to Earth's surface. Section 14.2 describes how meteors affect communications.

Effects on Spacecraft

The high relative speeds between meteor material and spacecraft are the basis for serious impact damage, even from tiny specks of material. NASA has reported meteor damage (cells and blanket punctures) to the Hubble Space Telescope solar array. During meteor showers, spacecraft operators usually close sensitive apertures or (rarely) re-orient spacecraft. Small particles also cause pitting of optical surfaces and mirrors, degrading the performance of critical sensors. Coatings on unprotected surfaces deteriorate or even delaminate.

Further, some of the impacts may create an ionization channel near the spacecraft, permitting a surface electrostatic discharge that causes further damage. Short circuits are one of the greatest dangers to satellites. Meteoroids that hit spacecraft disintegrate, creating a cloud of electrically charged plasma. Under the right conditions, this plasma cloud sets off a chain reaction, causing a massive short circuit. The loss of the European Space Agency's Olympus communication satellite in 1993 was attributed to a strike from a Perseid meteor and the resulting plasma discharge that destroyed the spacecraft's electronics. Table 13-8 compares natural debris (meteoric) with artificial debris, which we describe in the next section.

13.2 Damage and Impacts Associated with the Meteor and Artificial Debris Environment

Table 13-8. Debris Comparisons. Here we list some of the natural and artificial debris located in near-Earth space.

Natural debris—Meteoroids and Dust	Artificial debris—Human made
• Originate from comets, asteroids • 200 kg of mass within 2000 km • Largest flux below size of 0.5 mm • Low densities and mass (0.1 g/cm^3–0.5 g/cm^3) • High velocity: average speed ~18 km/s • Flux steady over long intervals • Affected slightly by solar cycle • Quasi-isotropic flux (some Earth shielding factor)	• >9000 large enough to be tracked • 1.5×10^6 kg–3.0×10^6 kg within 2000 km • Largest flux bigger than 1 mm • Higher densities and mass (2 g/cm^3–9 g/cm^3) • Lower velocity: average speed ~10 km/s • Flux increasing with time • Affected by launch rate, launch operations, and solar cycle • Majority in high-use orbits

Focus Box 13.9: The Leonid Meteor Shower—A Regular November Event

Contributed by Aerospace Corporation and NASA

The Leonid shower results as Earth passes through the path of the comet Tempel-Tuttle's debris cloud. Gravity and other forces nudge the particles within each stream until, over centuries, the trails blur together into a broad swath of dust in the comet's orbit. When Earth passes through the ancient cometary debris each year around November 18, we see a smattering of 10–15 meteors an hour.

The meteor activity associated with the comet Tempel-Tuttle is a "Leonid" event, because the meteors appear to be coming from the direction of the constellation Leo. The Leonids produced extraordinary shows between 1998 and 2002, when Earth encountered material freshly ejected by the comet. The comet swings around the Sun every 33 years, shedding dust at every pass. Outbursts of hundreds of meteors an hour—and storms of thousands an hour—occurred when Earth swept through the dense trails created in the past 200 years (Fig. 13-30).

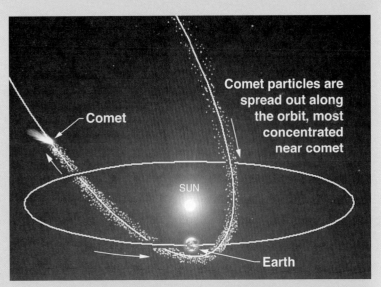

Fig. 13-30. The Comet Tempel-Tuttle's Orbit Crosses Earth's Orbit. Debris from the comet spreads along its orbit and causes the annual Leonid meteor showers when Earth passes through the region of space where the comet has been. *(Courtesy of William H. Ailor at the Aerospace Corporation)*

Chapter 13 Near-Earth Is a Place...with Susceptible Hardware and Humans

Focus Box 13.9: The Leonid Meteor Shower—A Regular November Event *(Continued)*

Contributed by Aerospace Corporation and NASA

The comet Tempel-Tuttle, named after William Tempel and Horace Tuttle, who first discovered it in 1865 and 1866, is about 4 km in diameter. As it makes its closest approach (perihelion) to the Sun, it passes close to Earth's orbit. Perihelion occurred most recently on February 28, 1998. Earth passed through that same region in space on November 17, 1998, and because of the comet's passage (and new cometary debris), we expected an increase in the amount of debris over what Earth would normally encounter. Hence, rather than the meteor shower on or about November 17, the Leonids put on a spectacular meteor storm (Fig. 13-31).

Historically, Earth experiences the most intense meteor activity during November of the year following the comet's perihelion, rather than in the same year (Fig. 13-32). This increased activity occurs when fresh material lies along Earth's orbital path. For example, following perihelion of Tempel-Tuttle in 1965, Earth experienced a major Leonid meteor storm in November of 1966. What is a major storm? In a normal year, we expect to see 15 or so meteors per hour on November 17. On the night of November 17, 1966, the meteor flux was approximately 150,000 meteors per hour—the most intense storm on record. Normally, the encounter with the debris cloud from Tempel-Tuttle lasts for several days, but the most intense part of the encounter, the storm, typically lasts only two to three hours.

Fig. 13-31. The November 1998 "Attack of the Fireballs." This image is of the meteor shower from an all-sky camera at the Modra Astronomical Observatory in the Slavak Republic. [Toth et al., 2000]

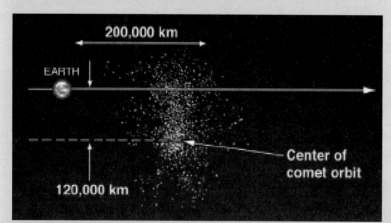

Fig. 13-32. Earth's Encounter with Comet Tempel-Tuttle's Debris Cloud. Earth passes very close to the comet's orbit every year. The center of the comet's orbit passes closer than the Moon, which is about 384,000 km from Earth. The cometary dust and debris cloud is naturally patchy or clumpy, which makes precise predictions of the storm's intensity impossible. *(Courtesy of William H. Ailor at the Aerospace Corporation)*

Although the particles of rock and dust are very small, the speed is enormous—about 71 km/s—over 200 times the speed of sound. Even a grain of sand the size of the head of a pin has the same energy as a .22 caliber bullet. At these speeds, the impact of a particle smaller than the diameter of a human hair creates an electrically charged cloud—a plasma. This plasma, in turn, causes a sudden electrical pulse that upsets sensitive electronics.

13.2 Damage and Impacts Associated with the Meteor and Artificial Debris Environment

13.2.2 Artificial Space Debris Environment

Objectives: After reading this section, you should be able to...

- Distinguish between natural and artificial space debris
- Explain why artificial debris poses a long-term hazard to space operations
- Describe sources of artificial debris
- Explain why artificial debris is more of a hazard in LEO than GEO

Characteristics

Functional spacecraft account for less than 20% of Earth's orbital population. Objects in Earth orbit that are not functional spacecraft are considered debris. The approximately 4000 space missions since 1957 have left thousands of large objects and perhaps tens of millions of medium-sized objects in near-Earth space. Unlike meteoroids and meteors that transit near-Earth space, artificial space debris orbits Earth and often remains in orbit for long periods. More than 19,000 objects with diameters greater than 10 cm are currently in Earth orbit. Most of them are expected to stay in orbit for tens or hundreds of years.

Sources of space debris include

- Non-functional spacecraft
- Rocket bodies
- Intentional release of explosive bolt parts, spring release mechanisms, spin-up devices, camera covers, and waste products from MIR, Shuttle, and ISS
- Mission operations and maneuvers, solid rocket motor effluent, etc.
- Fragments from explosions, deteriorations, and collisions

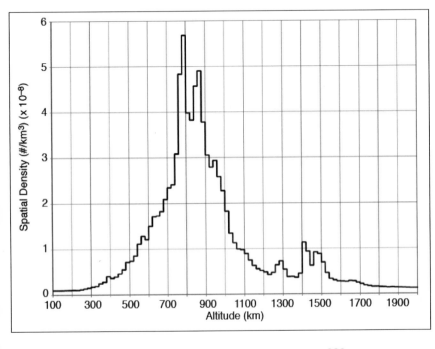

Fig. 13-33. The Spatial Density of Large Orbital Debris. The vertical axis is in units of number/km^3 × 10^{-8}. The horizontal axis is altitude. The debris above 800 km are long-lived. [NASA, 2009]

Chapter 13 Near-Earth Is a Place...with Susceptible Hardware and Humans

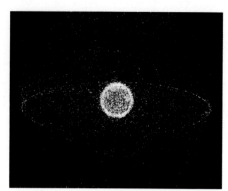

Fig. 13-34. **Objects in Earth Orbit.** In this computer-generated image, the orbital debris dots are scaled according to the image size of the graphic to optimize their visibility and are not scaled to Earth. These images provide a good visualization of where the greatest orbital debris populations exist at low-Earth and geosynchronous altitudes. *(Courtesy of NASA)*

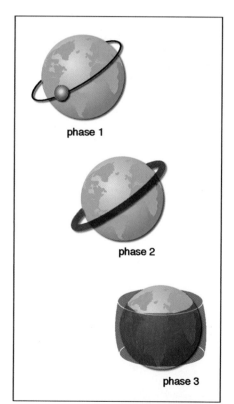

Fig. 13-35. **Orbital Evolution of Fragmentation Debris.** This sequence of diagrams shows that a single satellite breakup progresses into a ring of debris in orbit and eventually into an annulus of debris.

Figure 13-33 shows the spatial density of orbital debris below 2000 km in mid-2009. The peak near 800 km is associated with the collision of the Iridium 33 satellite and Cosmos 2251 in 2009. The peak at 870 km is the result of the intentional destruction of China's Fengyun-1C satellite in 2007. Figure 13-34 shows a computer-generated image of objects in Earth orbit that are currently being tracked. Approximately 95% of the objects in this illustration are orbital debris, i.e., not functional satellites. The dots represent the location of each item and are scaled according to the image size of the graphic to optimize their visibility and are not scaled to Earth.

Fragmentation creates the bulk of space debris that constitutes more than 40% of the US space object catalog (and undoubtedly represents an even larger fraction of non-cataloged objects). The fragmentation debris released from a breakup is ejected at a variety of initial velocities. As a result of their varying velocities, the fragments spread into a toroidal cloud that eventually expands to the limits of the maximum inclinations and altitudes of the debris, as shown in Fig. 13-35. The rate at which the toroidal cloud evolves depends on the original spacecraft's orbital characteristics and the velocity imparted to the fragments. In general, the greater the fragments' initial velocity spread, the faster the cloud evolution.

After entering orbit, debris follows perturbing forces that alter its trajectory and sometimes remove it from orbit. For space objects in middle-altitude orbits (above 800 km), solar and lunar gravity are important factors. Small debris is affected also by solar radiation pressure, plasma drag, and electrodynamic forces, although the effects of the latter two forces are minor.

Other than Earth's gravitational attraction, the primary forces on a space object in lower orbits (below about 800 km) are gravitational perturbations from Earth's oblate shape and atmospheric drag. Generally, gravitational perturbations do not strongly affect orbital lifetimes. Drag forces, however, have significant effects, and depend strongly on solar cycle influences. The rate at which a space object loses altitude is a function of its mass, its average cross-sectional area normal to its direction of motion, and the atmospheric density. As we describe in Chap. 7, although Earth's atmosphere technically extends to great heights, its retarding effect on space objects falls off rapidly with increasing altitude above 800 km. Atmospheric density at a given altitude is not constant and varies considerably (particularly at less than 1,000 km) because of atmospheric heating associated with the day-night cycle, the 11-year solar cycle, and geomagnetic storming. These natural phenomena accelerate the orbital decay of debris during periods of solar maximum. During the peaks in solar cycles 21 and 22, the total cataloged space object population declined, because the rate of orbital decay exceeded the rate of space object generation from new launches and fragmentations. During the extended solar minimum between solar cycles 23 and 24, the rate of orbital decay slowed.

Three important factors affect orbital decay: 1) objects with low ratios of cross-sectional area-to-mass decay much more slowly than objects with high area-to-mass ratios; 2) objects at low altitude experience more rapid orbital decay than objects at high altitude (higher air density at lower altitudes); 3) objects decay much more rapidly during periods of solar maximum than during the solar minimum. The combination of these factors has caused approximately 16,000 cataloged objects to re-enter the atmosphere since the beginning of the space era. In recent years, an average of 2–3 large space objects (cataloged), as well as numerous smaller debris particles, re-enter Earth's atmosphere each day. Over the course of a year, this rate amounts to hundreds of tons of material, composed primarily of large objects that were launched into low orbits (multi-ton rocket

13.2 Damage and Impacts Associated with the Meteor and Artificial Debris Environment

bodies) and small objects with high cross-sectional area-to-mass ratios. Seldom do any larger objects initially placed into orbits higher than 600 km re-enter the atmosphere because of drag.

Debris Effects, Collision Avoidance, and Mitigation

Hypervelocity impacts are the primary effect of space debris. The probability for a collision depends upon the spacecraft size, the debris flux in an orbital environment, and the amount of time spent exposed to the environment. Damage in a collision is a function of kinetic energy, which depends on the debris velocity relative to the spacecraft. Figure 13-36 shows the laboratory results of a high-speed impact between a small sphere of aluminum, travelling at approximately 6.8 km/s, and a block of aluminum.

A collision between a LEO satellite and artificial debris is much more likely to produce major damage than a collision between a GEO satellite and artificial debris, because the relative velocity between artificial debris and a LEO satellite is often very high. Active GEO spacecraft and debris from retired GEO spacecraft are all moving at about the same speed and in roughly the same direction. Their relative velocity is small, making damaging collisions unlikely.

As with meteors, one of the hazards of debris encounters is *spalling*, the ejection of flakes or chunks of material from the parent body. The impact creates a back-spray of material that contaminates sensors or an ionization channel that conducts stray currents.

Particles less than 1 cm pose less of a catastrophic threat, but they do cause surface abrasions and small holes in spacecraft. The greatest challenge is medium-sized particles (objects with a diameter between 1 cm and 10 cm), because they are not easily tracked and are large enough to cause catastrophic damage to spacecraft and satellites. Larger particles (objects greater than 10 cm across) are being tracked and catalogued by USSPACECOM radar. Spacecraft and satellites can avoid collisions by maneuvering around the larger debris. For example, analysts regularly examine the trajectories of orbital debris to identify possible close encounters with the International Space Station (ISS) and Space Shuttle. If a catalogued object is projected to come within a few kilometers of either spacecraft, the latter will normally maneuver away. Other large spacecraft are moved occasionally. NASA's Earth Observing spacecraft has maneuvered three times to avoid collisions.

NASA's main source of data for debris 1 cm to 30 cm across is the Haystack radar, operated by MIT Lincoln Laboratory in Massachusetts. It has been collecting orbital debris data for NASA for two decades under an agreement with the US Department of Defense. Haystack statistically samples the debris population by "staring" at selected pointing angles and detecting debris that fly through its field-of-view. The data are used to characterize the debris population by size, altitude, and inclination. From these measurements, scientists have concluded that over 500,000 debris fragments as small as 1 cm across orbit Earth.

The Department of Defense, other US government agencies, and many international agencies are funding research efforts to better characterize atmospheric neutral density. Some satellites now carry accelerometers to measure forces associated with satellite drag. Better specification of the neutral density improves computerized orbit propagation algorithms, thus allowing spacecraft operators to plan maneuvers for collision avoidance. Spacecraft are also being designed to withstand hypervelocity impacts by untrackable particles. Conducting hypervelocity impact tests on spacecraft and satellite components assesses the effect of debris hitting orbiting spacecraft. New materials and

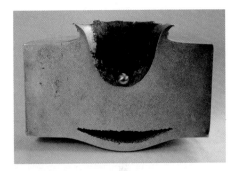

Fig. 13-36. Results of a Hypervelocity Impact. A 1.7-g aluminum sphere, 1.2 cm in diameter, hit a block of aluminum 18 cm thick at approximately 6.8 km/s. This test simulated what happens when a small space debris object hits a spacecraft. The impact diameter is 9 cm and the crater depth is 5.3 cm. Such a collision can produce pressures and temperatures that exceed those found at Earth's center, i.e., greater than 365 GPa and more than 6,000 K. *(Courtesy of ESA)*

designs from hypervelocity impact data help protect spacecraft from the space debris. One type of spacecraft shielding, termed multishock, uses several layers of lightweight ceramic fabric to act as "bumpers," which shock a projectile to such high energy levels that it melts or vaporizes and absorbs debris before it can penetrate a spacecraft's walls. Lightweight shields based on this concept are used on the ISS.

Because of the increasing number of objects in space, NASA and the international aerospace community have adopted guidelines and assessment procedures to reduce the number of non-operational spacecraft and spent rocket upper stages orbiting the Earth. One method of active debris removal is post-mission disposal that allows the reentry of these spacecraft, either from natural orbital decay (uncontrolled) or controlled entry. One way to accelerate orbital decay is to lower the perigee altitude so that atmospheric drag causes the spacecraft to enter Earth's atmosphere more rapidly. However, in such cases the surviving debris impact footprint cannot be guaranteed to avoid inhabited landmasses. Controlled entry normally is achieved by using more propellant with a larger propulsion system to cause the spacecraft to enter the atmosphere at a steeper flight path angle. The vehicle then enters the atmosphere at a more precise latitude and longitude, and the debris footprint can be positioned over an uninhabited region, generally in the ocean.

Focus Box 13.10:   Space Debris Encounters

Tiny pieces of space debris smash into the Space Shuttle during every flight. The debris includes very small natural meteoroids and miniscule remnants from abandoned spacecraft. For example, during Shuttle flights since 1983, at least eighteen windows have been damaged so severely that NASA replaced them. On one occasion, a fleck of paint, the size of a grain of salt, dug a pit into a Shuttle windshield. Each window costs $50,000.

In July 1996, artificial debris damaged the 50-kilogram French Ministry of Defense military research microsatellite named Cerise. A suitcase-sized piece of an old Ariane rocket body struck Cerise. The rocket body had been left in orbit after launching France's Spot 1 remote-sensing satellite in November 1985. The debris sliced the 6-meter long boom that had stabilized Cerise, sending the satellite tumbling end over end. The boom became another piece of space debris.

The Hubble Space Telescope has endured bombardment by tiny bits of debris since its launch in 1990, leaving the spacecraft covered with hundreds of little divots and one gaping hole that illustrate how hostile space is. On average, every square meter of Hubble gets hit by about 5 sand-grain-sized bits each year. Such debris is too small to see and avoid. So far, the impacts have not threatened the telescope's functions.

13.3 Hardware Damage and Impacts Associated with Field Variations

13.3.1 High-energy Electrons

Objectives: After reading this section, you should be able to...

- Describe the likely energy sources for relativistic electrons at geosynchronous altitude
- Relate relativistic electron increases to their drivers in the solar wind

Outer Magnetosphere Ultra-low-frequency Field Fluctuations and "Killer Electrons"

On occasion, the relativistic (energy >0.5 MeV) electron fluxes at and inside geosynchronous orbit rise dramatically. These events are sometimes called *killer electron events* because some of the electrons penetrate deep into satellite bodies and then produce an electrostatic discharge that may disable the satellite. Such events commonly take place one to two days after a prolonged high-speed increase in the solar wind arriving at Earth. A similar, but more rapid, flux increase often occurs with a moderate-speed solar wind, if it's accompanied by a prolonged period of southward interplanetary magnetic field (IMF). A third opportunity for accelerating develops with solar wind shocks. Scientists have proposed two methods by which electrons can be accelerated to killer status. One relies on very-low-frequency (VLF) waves of 3 kHz–30 kHz and the other on ultra-low-frequency (ULF) waves of 0.001 Hz–1 Hz. Both types of waves accelerate electrons in Earth's radiation belts but with different time scales. The ULF waves are more efficient during short intervals than the VLF waves, because their amplitudes tend to be larger. The VLF waves are likely to be more important in the inner magnetosphere near the plasmapause boundary.

From Faraday's Law, we know that a changing magnetic field creates a changing electric field. Simulations suggest that a slowly oscillating magnetic field induces an electric field that changes in step with the motion of some electrons forced into circular motion about Earth by storming processes in the magnetotail (a combination of $E \times B$, gradient, and curvature drifts). When the electric field oscillations align their phase with electron drift motion, the electrons are sped up. Other electrons, out of phase with the oscillation, are slowed. A plot of their flux is in Fig. 13-37 (Focus Box 13.10). Here we describe some root causes of the acceleration.

Killer electrons are trapped in Earth's outer radiation belt (4 R_E to 8 R_E above the planet's surface). Their fluxes may increase by three orders of magnitude during persistent high-speed solar wind flow and after the arrival of some coronal mass ejections. During high-speed flow in the solar wind, the magnetopause and regions inside it are buffeted by fluctuations in the solar wind speed, density, and magnetic field. The result is ULF waves at the magnetopause nose and flanks that penetrate the magnetosphere. Over an interval of one to two days, ULF waves can drive a portion of the moderate-energy electrons, already resident at distances (L values) of 7 R_E–10 R_E inward to lower L values. This radial diffusion requires the particles to extract energy from the waves. As moderately accelerated electrons move to lower L-shells and into a stronger magnetic field,

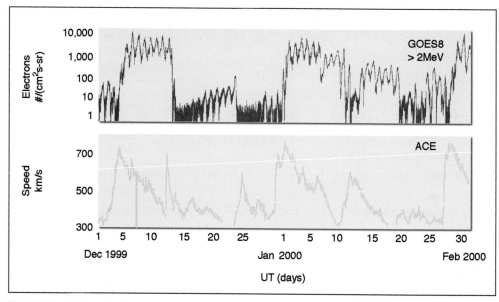

Fig. 13-37. **Energetic Electron Enhancement and Its Relation to Solar Wind Speed Increases.** The upper curve displays the flux of electrons with energies greater than 2 MeV. The lower curve is the solar wind speed recorded by the ACE satellite. *(Courtesy of Terry Onsager at NOAA's Space Weather Prediction Center)*

their motion changes to accommodate curvature and gradients in the field. Thus, they become trapped in approximately circular orbits around Earth in the middle magnetosphere (L-values between 3 and 7). Magnetic field waves, generated inside the magnetosphere with a period of approximately 10 minutes, may lock into phase (resonate) with electrons drifting around Earth at the same frequency. It's similar to a surfer catching a wave. Electrons with such a resonance quickly gain energy (up to several MeV) as they orbit Earth; electrons out of phase with the waves lose energy.

During impulsive storm events, a two-step ULF wave acceleration process may happen. The initial acceleration is caused by the strong shock-related magnetic field compression. Immediately after the impact of the interplanetary shock, Earth's magnetic field lines begin to shudder at ultra-low frequencies. The time-varying magnetic field produces an electric field that accelerates the seed electrons provided by the first step, to become killer electrons.

An additional (and possibly alternative) acceleration mechanism associated with VLF waves has been suggested. The VLF acceleration probably develops as a resonance of a source population of ~100-keV electrons with strong waves created by plasma instabilities near the dawnside plasmapause. This process may become efficient when hot electrons from one of the ULF-driven processes come into the vicinity of the dawnside plasmapause. Further complicating the understanding of VLF acceleration is the fact that some VLF waves scatter electrons into the atmosphere. Thus, VLF waves may energize electrons, but also cause them to disappear from the location where they are energized.

We find a well-established link between high-speed solar wind enhancements and killer electrons. When the solar wind speed exceeds 500 km/s, the energetic electron population grows, with higher speeds creating stronger enhancements. The acceleration mechanism associated with high-speed streams takes about two days to build a significant flux of relativistic electrons in the inner magnetosphere. The flux remains enhanced for days, or even weeks, as in the case of early January 1994 (Focus Box 13.11). During periods of unusually steady and prolonged southward IMF, but moderate-speed solar wind, a similar phenomenon may develop. The flux build-up often happens much more rapidly. However, climatological studies show these cases to be rather unusual, because the IMF is rarely steady at Earth. In extremely rare events that accompany coronal mass ejections (CMEs) traveling at very high speeds, the impulse delivered to the magnetosphere is sufficient to drive a magnetosonic wave through the entire magnetospheric cavity. Such an impulse was observed with an event in March of 1991. The disturbance was so immediate and so pervasive that the accelerated electrons formed a new radiation belt.

No matter what drives the electron acceleration, satellites in GEO orbit or semi-synchronous orbits (GPS constellation) ultimately, and sometimes suddenly, find themselves operating in an environment filled with electrons with enough energy to penetrate deeply into spacecraft electronics. The result is deep dielectric charging and discharging—a very dangerous condition for satellite components that we describe in Sec. 13.1.

Focus Box 13.11: Relativistic (Killer) Electrons

August 1993. During a high-speed solar flow in August 1993, temporary pointing errors in five geosynchronous International Telecommunications Satellite (Intelsat) satellites were reported. These were associated with electrostatic discharges that affected the attitude control system and produced various uncommanded status changes.

January 20–21, 1994. Two Canadian communication satellites (ANIK E-1 and E-2) experienced loss of momentum wheel control during a prolonged period of enhanced energetic electron flux. Charge build-up in the satellite created damaging discharges. Engineers regained control of ANIK E-1 in a few hours. Weeks passed before they created an alternative control scheme for ANIK E-2. The Intelsat-K satellite began to wobble a few hours before the January 20, 1994 event. Intelsat-K also experienced a short outage of service during this time.

November 2004. On 7 November 2004, an interplanetary shock wave followed by a large magnetic cloud swept over the Solar and Heliospheric Observatory (SOHO) and Advanced Composition Explorer (ACE) satellites. The speed of the solar wind suddenly increased from 500 km/s to 700 km/s. Shortly thereafter, the shock wave arrived at Earth's magnetosphere. The impact induced a wave front propagating inside the magnetosphere at more than 1200 km/s at geosynchronous orbit. ESA's satellite constellation Cluster observed a rapid increase in energetic electrons in the outer radiation belt.

Chapter 13 Near-Earth Is a Place...with Susceptible Hardware and Humans

13.3.2 Geomagnetic Field Interactions with Satellites

Objectives: After reading this section, you should be able to...

- Explain what happens to a satellite during a magnetopause crossing
- Describe how geomagnetic storms affect satellites that reference the geomagnetic field for navigation
- Explain the meaning of unloading momentum

Attitude Control Failures and Difficulty Unloading Momentum

Some satellites use Earth's magnetic field as a reference to maintain their attitude. This reference is lost when a satellite suddenly enters the magnetosheath (or even the solar wind during severe compressions). Two mechanisms may cause such an event: magnetopause compression and erosion.

Although the Sunward magnetopause is usually about 8 R_E–10 R_E from Earth's center, variations in solar wind pressure and other factors can change it. During intervals of high velocity and density, this boundary is pushed inside the geosynchronous equatorial orbit (GEO) belt at 6.6 R_E (Fig. 13-38). Similarly, during times of unusually strong southward IMF, the magnetopause boundary may erode by magnetic merging to locations inside of GEO. Magnetosheath visits for GEO spacecraft last for minutes to hours.

In the magnetosheath, sensors sample the compressed and turbulent IMF. Magnetic sensors that measure the usually stable and northward pointing magnetic field become confused as the detected magnetic field drops from ~200 nT to near zero and changes erratically. This is called a *geostationary magnetopause crossing (GMC)*. The GOES spacecraft carry magnetometers that unambiguously identify magnetopause crossings at their positions. However, because magnetopause compression is time-varying, and different spacecraft are at different longitudes, GOES satellites don't all experience the same crossings.

The strength and direction of the geomagnetic field at orbital altitudes often fluctuate during a geomagnetic storm. This poses a problem for spacecraft that normally depend on Earth's magnetic field for orienting their axes. Human controller intervention is often required to properly control spacecraft attitude.

High plasma pressure from a coronal mass ejection (CME) or interplanetary shock is not the only culprit in magnetopause crossings. If the IMF is also coincidentally southward, then reconnection may strip away closed field lines from the dayside more rapidly than they are replenished by field-line convection from the magnetotail. In this case, the magnetopause erodes inward toward the geosynchronous orbits. Such erosion events don't normally last very long. Instead, the magnetosphere speeds its convection process to rebalance the flux distribution and restore the magnetopause to a more Sunward location.

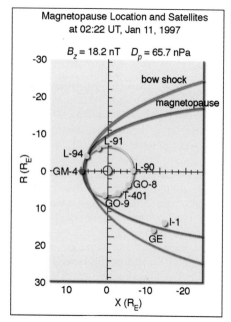

Fig. 13-38. Modeled Magnetopause Location for the January 11, 1997, Space Weather Event. The dynamic pressure in the solar wind was nearly 66 nPa and the IMF had a strong southward component of –18 nT. Satellites on the yellow circle are at geosynchronous orbit. The magnetopause compression was sufficient to cause one dayside geosynchronous satellite and one late evening mid-tail satellite to exit the magnetosphere. *(Courtesy of Jih-Hong Shue at NASA)*

Focus Box 13.12: Geostationary Magnetopause Crossing (GMC)

March 1989. The GOES-7 spacecraft (Fig. 13-39) crossed through the inward moving magnetopause several times on March 13 and 14, 1989. These crossings and the associated magnetic storm caused considerable difficulty for the satellite, because it oriented its antenna by reference to the ambient magnetic field.

October 2003. TV satellite controllers reported problems with maintaining routine operations during the large geomagnetic storms of late 2003. They resorted to "manual attitude control" for 18- to 24-hour periods because of magnetopause crossing events that affected the attitude controllers of two or more of their fleet. A component burned out in one ungrounded circuit box in a newer satellite and required a workaround.

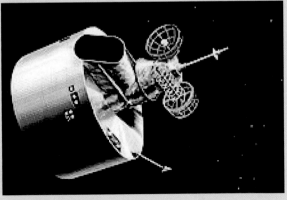

Fig. 13-39. **Geostationary Operational Environmental Satellite (GOES-7).** This weather-forecasting spacecraft in GEO crossed into the magnetopause when a particularly strong geomagnetic storm compressed the bow shock to 4.7 R_E. (Courtesy of NOAA)

Difficulty Unloading Momentum

Some spacecraft use the geomagnetic field for stabilizing and maneuvering procedures. These spacecraft have three perpendicular reaction wheels whose momenta remain zero unless the craft performs an attitude change maneuver. To begin a maneuver, the vehicle control system spins one or more wheels, and by conservation of momentum moves the vehicle in the opposite direction. At the end of the maneuver, the control system reduces the wheel speed to zero and the vehicle stops at the commanded attitude. This zero momentum state is maintained only in the absence of net external torques. Small external torques from gravity-gradients, solar pressure gradients, and atmospheric drag, among other effects, are almost always present and create a net increase in the system's momentum.

Typical spacecraft require momentum unloading every few hours or days and to maintain zero momentum. They do this by supplying a controlled external torque to the vehicle using reaction jets or by pushing against Earth's magnetic field using large electric coils (internal $\boldsymbol{E} \times \boldsymbol{B}$ force). Applying magnetic torque to slow a spinning satellite is difficult or impossible when the geomagnetic field fluctuates wildly during a geomagnetic disturbance.

13.3.3 Geomagnetic Field Interactions at the Ground

Objectives: After reading this section, you should be able to...

- Describe the role of geomagnetic variation in magnetic surveys, drilling operations, and navigation
- Describe the effect of geomagnetically induced currents in power and telecom systems
- Describe the effect of geomagnetically induced currents on pipe and rail lines

Consequences of Storm-driven Geomagnetic Field Fluctuations at Earth's Surface

Although most direction-finding is now done by Global Positioning System (GPS) receivers, some activities still rely on a steady magnetic field.

Surveys. Exploration geophysicists investigate Earth's subsurface features using magnetic surveys. In some cases, these studies are large aerial surveys over ocean and land. Time-varying magnetic fields often hide important details, so surveyors try to avoid and remove time variations from the data. In the case of land surveys, they use nearby magnetometer readings as a baseline for variation removal. Over the oceans, the best approach is to avoid surveying during geomagnetic activity.

Directional Drilling. In oil exploration and extraction, multiple wells are sometimes drilled from a single platform. Geomagnetic measurements guide the direction of the drill. Deep drilling projects that rely on magnetic data for accuracies of a tenth of a degree cease when the geomagnetic field is active and varying by a few degrees.

Ground-induced Currents

As we describe in Chaps. 4, 7, 8, 11, and 12, some of the currents flowing in and between the magnetosphere and ionosphere are sensed at the ground. Rapidly changing current structures are particularly evident at Earth's surface at auroral and sub-auroral latitudes. During severe space weather disturbances, unsteady currents, exceeding megaampere levels, flow through the ionospheric electrojets. At ground level, ionospheric current effects are manifest as a disturbance in the geomagnetic field, which in turn induces a disturbance geoelectric field. The electric field drives electric currents, called *geomagnetically induced currents (GIC)*, in ground-based technological systems. Currents flow through artificial conductors present on the surface (pipelines or cables). In effect, GICs constitute the ground and below-ground end of the space-weather chain in electric power grids, telecommunication cables, extended pipelines, and railway lines. Most developed and developing countries have what we can best describe as rivers of GIC conductors.

The chain of events for GICs begins with a time-varying magnetic field external to Earth that induces electric currents in the conducting material on and under the surface. These currents are usually strongest in the auroral electrojets, but tropical regions could feel a similar short-lived effect caused by an anomalous electrojet created by severe compressions of Earth's dayside magnetic field. These time-varying currents create a secondary (internal) magnetic field. As a consequence of Faraday's Law of Induction, an electric field at Earth's surface is induced by the time variations of the magnetic field. The surface electric field causes the GICs to flow into any conducting material. This GIC-induced electric field (typically measured in V/km) acts as a voltage source in electrical power transmission grids, oil and gas pipelines, undersea communication cables, telephone and telegraph networks, and railways.

Power Systems. Figure 13-40 indicates that these currents enter the grid of transmission lines through their grounding points. The effects on long conducting lines or cables increase where Earth's surface is a poor electrical conductor, such as igneous rock. Regions of North America have significant amounts of igneous rock and thus are particularly susceptible to the effects of GICs. Electric currents flowing in the ocean contribute to GICs by entering along coastlines and are amplified by local changes in conductivity. The problem, however, is not generally in the soil, or the high voltage lines, but rather in the transformers connecting them. Transformers sometimes trip or overheat as a result of GICs.

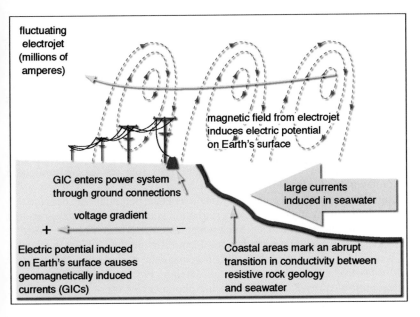

Fig. 13-40. An Ionospheric Time-varying Electric Current. Storm-time dynamics drive this current (large yellow arrow). The current generates a magnetic field of the same frequency and phase. The magnetic field lines penetrate Earth's surface and other conducting bodies. A voltage (an electromotive force shown in blue) grows within the conductor, according to Faraday's Law. Currents flow in the conducting material on and under ground. These currents usually flow through the conductor in planes perpendicular to the lines of the magnetic field from the ionosphere but may be restricted by the conductor's geometry and by irregularities in the conductor. Current flow within Earth's surface generates secondary magnetic field lines at the conductor that oppose those of the primary magnetic field. For simplicity we don't show the secondary field. *(Courtesy of John Kappenman at Storm Analysis Consultant)*

Characteristic times associated with fluctuating geomagnetic variations are on the order of seconds and minutes. Thus one hertz is an important characteristic frequency for GICs. Consequently, GICs flowing in electric power transmission systems are quasi–direct currents (DC), in comparison with the 50/60 Hz alternating current (AC) frequencies generated for the power in many countries. Those GICs flowing through the transformer winding produce extra magnetization, which, during the half-cycles when the AC magnetization is in the same direction, saturates the core of the transformer (Fig. 13-41). A saturated transformer converts energy to heat, thus reducing the energy available for transmission and in turn, reducing the voltage. Excess heat causes internal damage to the transformers. A large GIC event sometimes disrupts the operation of step-up and step-down transformers, relays, and other equipment throughout the network, ultimately leading to problems ranging from trip-outs of individual lines to the collapse of the whole system. Large GIC events are relatively rare; however, they have nearly instantaneous effects over a large region (Focus Box 13.13).

Chapter 13   Near-Earth Is a Place...with Susceptible Hardware and Humans

Fig. 13-41.   A Transformer Core. When a transformer core saturates, extra eddy currents flow in the core and structural supports. These currents heat the transformer. Usually, the large thermal mass of a high-voltage power transformer means that this heating produces a negligible change in the overall transformer temperature. However, localized hot spots sometimes occur and cause damage to the transformer windings.

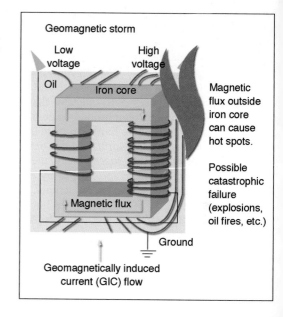

Modern electric power transmission systems consist of generating plants interconnected by electrical circuits that operate at fixed transmission voltages, controlled at substations. The grid voltages largely depend on the path length between these substations. In most systems, voltages of 200 kV are common. We see a trend toward higher voltages and lower line resistances to reduce transmission losses over longer path lengths. Low line resistances produce a situation favorable for the flow of GICs.

Of significant concern to emergency planners is the potential GIC damage that a 4800 nT/min (or greater) magnetic perturbation could have on power grids that extend over continental distances. Such a magnetic perturbation was observed in May of 1921. The magnetic perturbations accompanying the Carrington storm of 1859 were likely even larger than that. The magnetic perturbation that brought down the Hydro Québec power system in March of 1989 was ~500 nT/min.

Telecommunication Cables. Submarine phone cables cross the ocean floor. Their trans-oceanic lengths mean that the potentials induced in them during geomagnetic storms are easily hundreds—or even thousands—of volts, leading to possible problems. Time variations in the magnetic field induce potential differences in cable communication systems. Telecommunication cable systems use the Earth as a ground return for their circuits, and these cables thus provide highly conducting paths for concentrating the electrical currents. The voltages, which are also affected by the flow of seawater, are monitored on trans-Atlantic and trans-Pacific cables and on some shorter ocean cables, as well.

Although modern optical-fiber cables do not carry GICs, they aren't completely immune from GIC problems, because power to repeaters is fed by metallic cables, in which GICs flow. Furthermore, electronic systems, used in telecommunications, are more complicated and miniaturized than previous versions, so even small GICs could have harmful influences not previously recognized.

On August 4, 1972, an outage of a coaxial cable system in the mid-western US occurred during a major geomagnetic disturbance. At the time of the outage, Earth's magnetic field had been severely compressed by a shock in the solar wind. The resulting magnetic disturbance had a peak rate of change of 2200 nT/min,

Focus Box 13.13: Geomagnetically Induced Currents (GICs)

Although transformer failure is a rare event, statistics reveal that the transformers in auroral and sub-auroral locations have significantly shorter lifetimes than those in middle and low latitudes. Researchers have proposed that the shortened lifetime is caused by undocumented effects of repeated exposure to low-level GICs. Statistically, the failure rate follows the solar cycle, but with a three year lag. This failure rate is a major concern to insurance companies that insure transformers, generators, and other electrical equipment for utility companies.

Usually only a few amperes of exciting alternating current (AC) is needed to provide the magnetic flux for the voltage transformation in a power transformer. The transformer operates within the range where the voltage depends linearly on the exciting current. However, a DC-like GIC moves the operating curve to a nonlinear regime, leading to a very large exciting current (hundreds of amperes). Transformer DC current causes half-cycle saturation, harmonics, over-heating, increased audible noise, and mechanical stress. These produce huge reactive power losses in the power system, contributing to a voltage drop. A cascading sequence of harmful processes in the system may quickly result in a complete collapse and blackout.

The harmonics and the reactive power demands sometimes severely stress transformers. They draw excess power from the system and turn it to heat, attaining temperatures high enough to blister paint and decompose internal insulation. The audible noise level of a transformer increases from a hum to a roar because of the vibration of a saturated steel core. Furthermore, the saturation changes the path of the magnetic flux in the transformer, so that flux densities may become very large, causing overheating. The resulting hot spots permanently damage the insulators, and cause out-gassing of transformer oil, thus leading to serious internal failures of the transformer (Fig. 13-42).

On March 13, 1989, one of the very few "great" storms occurred. It caused auroras, normally located within a band or oval between 65° and 75° geomagnetic latitude, to be visible as far south as Cancun, Mexico.

As a result of magnetic variations on the order of 400 nT/min, GIC-driven disruptions caused a nine-hour blackout of the 21 GW Hydro Québec power system, leaving six million customers without power. The "domino effect" in the grid developed in roughly two minutes.

Fig. 13-42. Geomagnetic Storm-induced Transformer Damage. During the storm of March 13–14 1989, overheating caused a catastrophic failure of this generation step-up transformer at a Salem New Jersey facility. *(Courtesy of New Jersey Public Service Electric and Gas)*

At the same time, in Denmark, a ground magnetic observatory recorded a 2000 nT/min disturbance in the geomagnetic field. Simultaneously, in Sweden, seven high-voltage lines tripped, and the temperature on a rotor in a nuclear power plant increased 5° C. In the US, several safety systems tripped, and a large step-up transformer at a nuclear power plant in New Jersey was damaged beyond repair. The transformer cost several million dollars to replace. Such expensive units are usually built to order and the delivery time is normally about one year. Fortunately, in this case, a replacement transformer was available and delivery and installation took only six weeks. Even so, having the transformer out of service restricted the deliverable power from the Salem generating station, and the purchase of replacement power from neighboring utilities cost about $17 million, far more than the cost of the transformer.

In April 2002, active region 9393 produced the largest solar flare in 12 years. The ensuing geomagnetic storm caused transformer safety switches to flip repeatedly in Wisconsin, New York, and other northeastern regions. Improved warning capability allowed power grid operators to take mitigating actions and avert serious damage to regional power grids.

Focus Box 13.13: Geomagnetically Induced Currents (GICs) (Continued)

Another great storm took place in late October 2003. The Scottish power grid authorities reported a peak GIC of 42 A, which they also managed, but which significantly exceeded the 25 A "concern threshold" for grid operations. The peak GIC observed on 30 October 2003 exceeded previously reported values in the United Kingdom in modern times. The geomagnetic superstorm brought down a part of the high-voltage power transmission system in southern Sweden. The blackout lasted for an hour and left about 50,000 customers without electricity.

In South Africa, the discovery of transformer damage in the ESKOM system, Africa's largest electricity producer, was delayed by several days until weekly inspections revealed that transformer cooling oil and insulating paper had heated and transformer windings were failing because of overheating. In all, 15 transformers were damaged during the superstorm. Each transformer was valued at $2–$3 million. Total losses amounted to $73M per day for an extended time period, after officials accounted for the lost revenue. These transformer failures were particularly surprising because they happened in a mid-latitude power grid.

observed at the Geological Survey of Canada's Meanook Magnetic Observatory, near Edmonton, and a rate of change of the magnetic field at the cable location estimated to be 700 nT/min. The induced electric field at the cable was calculated to have been 7.4 V/km, exceeding the 6.5 V/km threshold at which the line experiences a high current shutdown.

Pipelines. In pipelines, GICs exacerbate problems with corrosion at points where the current flows from the pipe into the surrounding soil (or electrons flow to the pipe from the soil). Through tiny holes in the pipeline coating, pipeline steel may come into contact with the soil, water, or moist air and corrode. Most steel pipes are insulated with a protective, high-resistance coating to reduce corrosion. However, at defects in the coating, large pipe-to-soil voltage differentials develop. To minimize the risks, engineers use cathodic protection systems that keep the pipeline at a lower voltage (~1 V) with respect to the soil. The voltage depends on the soil properties. The GIC-related problems arise where currents enter and leave the pipeline. There the pipe-to-soil voltage differentials easily exceed the protection voltage, thus interfering with the protection system. Further, control measurements of pipe-to-soil voltages during magnetic disturbances are often unreliable, resulting in an inappropriate voltage adjustment.

The largest pipeline-to-soil GIC flow is in the vicinity of discontinuities, such as changes in the pipeline material or size, inhomogeneities in Earth's conductivity, bends and branches of the pipeline, etc. The size of an area around a discontinuity that experiences significant pipe-to-soil voltage differentials is characterized by an "adjustment distance," typically tens of kilometers. To mitigate corrosion, engineers try to maintain good electrical continuity throughout the system, sometimes by creating long, uninterrupted pipelines. Figure 13-43 shows the onset of a GIC event measured on a Canadian pipeline during one of the large storms of solar cycle 23.

Railways. Railways also constitute a large network of conductors experiencing GICs. Physically, a railway system is usually somewhere between a discretely grounded power system and a continuously grounded buried pipeline, because the electric wires are isolated from the ground, but the rails have continuous ground contact. On Swedish railways, GICs caused traffic signals to turn unexpectedly red during a magnetic storm in July, 1982. A GIC-associated voltage change caused the automatic safety equipment to falsely sense that a train was short-circuiting the rails.

13.3 Hardware Damage and Impacts Associated with Field Variations

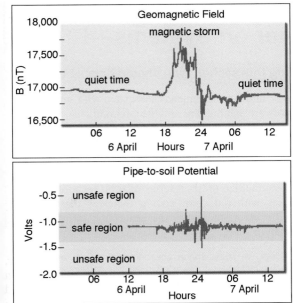

Fig. 13-43. Geomagnetically Induced Currents (GICs) Measured in a Canadian Pipeline during April 2000. The upper plot shows the variation in one component of Earth's magnetic field at Ottawa, Ontario; the lower plot gives the accompanying pipe-to-soil voltage. *(Courtesy of Natural Resources Canada)*

A recent statistical analysis of automatic signaling and train control equipment for 2004–2005 on the East-Siberian Railway, located at mid-latitudes (51°–56° N), revealed an increase of the daily duration of anomalies by 3–4 times during big geomagnetic storms (local geomagnetic index A > 30). The anomalies consist mainly of unstable functioning and false operations in automatic traffic control systems (rail chain, switches, locomotive control devices, etc.), often resulting in false engagement of railway tracks (red signals instead of green). On one train, a seasonal effect was found in the relative numbers of anomalies similar to that observed in geomagnetic activity. Analysis of linkages between these anomalies and geomagnetic conditions in 2004 suggest that the anomalies occur during the main phase of geomagnetic storms.

Chapter 13 Near-Earth Is a Place...with Susceptible Hardware and Humans

13.4 Surveying the Impact of Space Weather on Systems—I

Objective: After reading this section, you should be able to...

♦ Appreciate the economic and societal impacts associated with space weather damage to hardware and humans and the environment

From the previous sections, we know space weather significantly impacts space and terrestrial activities. Here we tabulate those effects. Table 13-9 lists hazards and effects on humans, hardware, and systems. Because particle effects are so widespread, we also summarize these in terms of orbit location in Table 13-10. The number of boxes with red color coding indicates that energetic particles and photons play an important role in the health and status of space vehicles and humans operating in space.

Table 13-9. Space Weather Hazards and Their Effects. Here we show space environmental effects on humans, hardware, and systems. Relevant categories of each effect or event are listed in the subheadings of column 1. (After the appendix to NASA Reference Publication 1396)

Environment Variation	Human, Hardware, and System Effects
Solar Photons • Solar dynamics, flares (X ray and UV) • Solar cycle variation See radio bursts in Chap. 14	**Atmosphere** • Satellite drag affects tracking accuracy, orbit prediction, pointing and torques, mission planning, guidance, navigation and control, and longevity of space debris **Humans** • Causes sunburn and skin cancer **Hardware** • Influences solar cell design • Influences spacecraft design for thermal properties • Degrades thermal, electrical, optical properties • Degrades structural integrity
Ionizing Radiation • Trapped proton/electron radiation • Galactic cosmic rays • Solar energetic particle events • High altitude nuclear effects See influence on signals in Chap. 14	**Atmosphere** • Changes atmospheric chemistry and conductivity **Humans** • Radiation dose limits EVAs, crew time on-orbit • Affects long-duration mission plans • May limit over-the-pole and very-high-altitude flights **Hardware** • Radiation dose degrades solar cells, spacecraft materials, and structures • Deep charging and single-event effects range from temporary to mission-ending • Corrupts star tracker • Causes image noise and data corruption • Damages spacecraft circuits and instruments • Damages aircraft avionics
Plasma Environment • Ionospheric plasma • Auroral plasma • Plasmaspheric plasma • Magnetospheric plasma See influence on signals in Chap. 14	**Hardware** • Arching causes electromagnetic interference (EMI) leading to upsets • Shifts floating potential and pulsing • Causes current losses and power drains • Biases instrument readings • Re-attracts contaminants • Changes absorptance/emittance properties • Causes sputtering contamination

Table 13-9. Space Weather Hazards and Their Effects. (Continued) Here we show space environmental effects on humans, hardware, and systems. Relevant categories of each effect or event are listed in the subheadings of column 1. *(After the appendix to NASA Reference Publication 1396)*

Environment Variation	Human, Hardware, and System Effects
Neutral Environment • Atmospheric density • Density variations • Atmospheric composition (atomic oxygen), winds	**Atmosphere** • Satellite drag affects tracking accuracy, orbit prediction, pointing and torques, mission planning, guidance, navigation and control, and longevity of space debris **Hardware** • Atomic oxygen degrades material • Produces spacecraft glow and interferes with sensors
Micrometeoroids and Meteor Storms • Size distribution • Mass distribution • Velocity distribution • Directionally	**Hardware** • Impacts cause electromagnetic interference • Damages solar cells • Damages surfaces • Adds shielding and propulsion requirements • Affects crew survivability • Possibly ruptures pressurized tanks
Magnetic Field Variation • Time rate of change • Resonance with particles • Drives currents in atmosphere	**Geospace/Magnetosphere** • Produces highly energetic "killer" electrons **Atmosphere** • Heats auroral zone and upper atmosphere **Hardware** • Affects spacecraft attitude
Ground Events	**Hardware or System Effects**
Rapid Magnetic Field Variations • Time rate of change and location	**Ground-induced Currents** • Damages or destroys transformers • Damages or destroys power grids • Causes voltage swings on long communication cables • Corrodes pipelines • Causes spacecraft charging, which degrades surface and interior materials • Increases magnetic survey and navigation uncertainty
Ground-level Energetic Particles • Particle spectrum	**Atmosphere** • Causes chemical changes **Hardware** • Causes manufacturing defects in electronics

The influence of space weather is now discussed in scholarly science journals and in industrial and popular literature. The following are examples of space weather impacts during solar cycles 22 and 23. Much of the following information comes from the NOAA Space Weather Prediction Center and reports issued by the National Research Council.

- The effects of a March 1989 space weather storm cost two large utilities (Hydro Québec in Canada and PSE&G in New Jersey) an estimated $30 million in direct costs. Hydro-Québec's solution to the blackout was to install devices that block storm-induced currents from traveling through its transmission lines. Unfortunately, this solution is extremely complex and expensive ($1.2 billion). Comprehensive real-time protective space weather prediction services could have significantly reduced damages and costs.

Chapter 13 Near-Earth Is a Place...with Susceptible Hardware and Humans

Table 13-10. Space Environmental Hazards. Here we show many hazards in space from energetic particles. (LEO < 60° means low-Earth orbit, less than 60° inclination; LEO > 60° means low-Earth orbit, more than 60° inclination; MEO means medium-Earth orbit; GPS is Global Positioning System; GTO is geosynchronous transfer orbit; GEO is geosynchronous orbit; HEO is highly elliptical orbit; O is atomic oxygen.) *(Courtesy of the Aerospace Corporation)*

Space Hazard	Spacecraft Charging		Single-event Effects			Total Radiation Dose		Surface Degradation	
Specific cause	Surface	Internal	Cosmic rays	Trapped radiation	Solar particle	Trapped radiation	Solar particle	Ion sputtering	O erosion
LEO <60°									
LEO >60°									
MEO									
GPS									
GTO									
GEO									
HEO									
Interplanetary									

Legend: ■ Important   ▨ Relevant   □ Not applicable

- In research supported by the National Science Foundation, Forbes and St. Cyr (2004) used a multivariate economic analysis of one cross-state transmission system to conclude that space weather produces congestion in that system, which transmits power from the generating site to the distribution site. For the interval studied—June 1, 2000, through December 31, 2001—they concluded that solar-initiated geomagnetic activity increased the wholesale price of electricity by approximately 3.7 percent, or approximately $500 million. The economic conclusions of this analysis have occasioned some controversy, and further discussions and analyses are in progress.

- An aviation insurance underwriter has estimated that $500 million in satellite insurance claims from 1994 to 1999 were the direct or indirect result of space weather

- The DOD has estimated that disruptions to government satellites from space weather cost about $100 million a year

- Solar-initiated geomagnetic storms appear to have caused the demise of several communication satellites in solar cycles 22 and 23. Development and launch costs for large telecommunication satellites range from $200 million to $250 million.

- Space weather information is employed commonly as important operational input for determining launch and on-orbit operations (including space walks) of Space Shuttle and International Space Station (ISS) activities.

- In late October and early November 2003, a series of exceptional solar events caused an unprecedented number of physical events and system effects. Numerous deep-space missions in progress during that time were affected by the severe solar activity. Scientists stressed the importance of situational awareness during solar storms. The Stardust team said, "If we had not known about the flare we could have floundered for days and perhaps even sent commands that would have been detrimental."

- **Mars Explorer Rover** – The spacecraft entered its "Sun Idle" mode because of excessive star tracker events. It waited for the event to end and recovered.
- **Microwave Anisotropy Probe** – The spacecraft star tracker reset, and the backup tracker autonomously turned on. The primary tracker recovered.
- **Mars Express** – The spacecraft had to use gyroscopes for stabilization, because a flare interfered with navigation using stars as reference points. The radiation storm blinded the orbiter's star trackers for 15 hours. The flares also delayed a scheduled Beagle 2 checkout procedure.
- **Advanced Composition Explorer (ACE)** – The EPAM Low-energy Magnetic Spectrometer (LEMS 30) was damaged. Noise levels increased in several ion channels and remained abnormally elevated. The instrument has not recovered.
- **Kodama, Data Relay Test Satellite** – This geostationary communication satellite went into its safe mode on the morning of October 29 during the severe (S4) solar radiation storm. It normally relays data among low-Earth orbiting (300 km–1000 km altitude) spacecraft (including the International Space Station) and ground stations. Excessive signal noise coming from the Earth sensor assembly suggested that the satellite was affected by a proton barrage. Kodama was recovered on November 07, 2003.
- **CHIPS** – The satellite computer went offline on October 29, and contact was lost with the spacecraft for 18 hours. It lost its three-axis control because its single on-board computer stopped executing. When contacted, the spacecraft was tumbling, but was recovered successfully. It was offline for a total of 27 hours.
- **GOES-9, 10, and 12** – NOAA saw high bit error rates on 9 and 10 and magnetic torquers disabled 12 because of solar activity.
- **Inmarsat** (a fleet of nine geosynchronous satellites) – Controllers at the satellite control center had to quickly react to the solar activity to control Inmarsat's fleet of geosynchronous satellites. Two of the satellites experienced speed increases in momentum wheels, requiring them to fire thrusters, and one had an outage when its CPU tripped out.
- Other satellites with known anomalies during this time include: **FedSat, POLAR, GALEX, Cluster, RXTE, RHESSI, Integral, Genesis,** and more. On October 28, 2003, the **NOAA-17** spacecraft experienced a significant problem with the scan motors of the AMSU-A1. The instrument was powered down and operators plan no recovery efforts.
- **Pay Radio Satellite Services** – A pay-radio satellite had several short periods where it lost satellite lock.
- **Department of Defense** – Satellite operations over high-interest regions were lost for 29 hours when three spacecraft experienced anomalies or were shut down to avoid damage from space weather.
- **Satellites or instruments "safed" on Aqua, LandSat, Terra, TOMS, TRMM, Upper Atmosphere Research Satellite/HALOE Instrument, SIRTF, and CHANDRA.** The CDS instrument on the SOHO spacecraft at the L1 point went into its safe mode for three days (October 28–30).

- **Geomagnetically Induced Currents (GICs)** in excess of 100 A were observed in the US, and larger GICs were observed globally. Power companies in North America did experience some problems. Effects and actions reported by grid operators included less use and switching between systems; high levels of neutral current observed at stations throughout the country; a capacitor tripped in the northwest (known to be GIC susceptible); transformer heating in the east (precautions were implemented); and a 'growling' transformer that was backed down to help cool it. GIC impacts were more significant in northern Europe, where heating in a nuclear plant transformer was reported and a power system failure occurred on October 30 in Malmo, Sweden, resulting in blackout conditions.
- Some actions taken by the US Nuclear Regulatory Commission were included in their Power Reactor Status Report for October 30, 2003. (Table 13-11)

(From Severe Space Weather Events, 2008 and NOAA, 2004)

Table 13-11. Power Reactor Status Report, October 30, 2003. This table lists the status of six power reactors in the US during the solar storm of October 30–November 7, 2003. *(From the NOAA Service Assessment Report on the October–November 2003 Storms.)*

Unit	Power	Comment
Hope Creek 1	80 MW	Reduced power because of solar magnetic disturbances
Salem 1	80 MW	Reduced power because of solar magnetic disturbances
Braidwood 2	90 MW	Coastdown to refueling outage, reviewing system planning operating guide for solar flare response
Arkansas Nuclear 1	100 MW	Holding off on switchyard maintenance for solar flare
Palo Verde 1	98 MW	T-Hot limited, taking extra readings on plant computer because of solar flare
Point Beach 1	83 MW	Increasing power following a decrease in power caused by grid geomagnetic disturbances

Summary

Hardware Impacts: Time-varying magnetic fields in the space environment affect systems in space and on the ground. Moderate to rapid variations energize particles that interact with spacecraft materials and even create new radiation belts. Field variation in space and time affects spacecraft navigation and the ability of some spacecraft to properly maintain their attitude.

Energetic particles come from stellar and solar emissions and secondarily from geospace. The most energetic particles are responsible for single-event upset (SEU) phenomena, total dose effects, and deep dielectric charging. Single-event effects are sometimes observed in aircraft avionics. Medium-to-low energy particles contribute to surface charging and materials degradation.

Solar photons have influences as individual particles in photonics noise that compromises spacecraft orientation and spacecraft surface charging. Changes in satellite drag are related to solar cycle variations in X-ray and EUV photons, as well as to geomagnetic storms that produce episodic auroral zone heating.

Debris and meteorite impacts are mostly associated with kinetic effects on hardware. High-speed impacts often damage or disable a satellite.

Consequences to Humans: Most spaceflight operations have taken place within the protection of Earth's magnetic field. This shields the space crews from large solar particle-induced radiation events and a significant portion of the galactic cosmic radiation. For the low-Earth orbiting (LEO) missions that have typified the US space program, most radiation exposure is in the South Atlantic Anomaly (SAA), where the intensity of Earth's geomagnetic field is reduced. The remainder of the exposure is from high-energy galactic cosmic radiation. During lunar missions, astronauts rapidly traverse the trapped radiation belts into the unprotected realm of free space outside the magnetosphere. These excursions place astronauts at risk of an episodic energetic particle event emitted by the Sun or accelerated by CMEs. These events are difficult to predict and potentially life-threatening to an inadequately protected crew. Without proper shielding, crewmembers exposed to large solar particle events will likely develop acute health problems and reduced daily performance.

Long-duration, exploratory-class missions are not protected by Earth's geomagnetic field from cosmic radiation and energetic solar protons. Astronaut exposure to galactic cosmic radiation within a thinly shielded spacecraft during solar minimum exceeds current astronaut exposure guidelines.

Key Words

anomalous cosmic rays (ACRs)
bitflips
bolides
contamination
Debye sheath
decomposition
deep dielectric charging
desorption
diffusion
direct heating
displacement
dose
dose equivalent
energetic storm particles (ESPs)
galactic cosmic ray (GCR)
gamma radiation
geomagnetic cutoff rigidities
geomagnetically induced current (GIC)
geostationary magnetopause crossing (GMC)
ground-level event (GLE)
ionizing radiation
killer electron events
meteoroids
meteors
molecular contamination
nuclear radiation
outgassing
photoelectron emission
plasma induced charging
quality factor (QF)
rad equivalent mammal/man
radiation
radiation absorbed dose
secondary electron emission
sievert (Sv)
single-event burnout (SEB)
single-event effects (SEEs)
single-event functional interrupt (SEFI)
single-event latchup (SEL)
single-event upset (SEU)
solar energetic particles (SEPs)
spacecraft environment
spalling
sputtering
threshold displacement energy

References

Adams, L., Eamonn Daly, Reno Harboe-Sorensen, Robert Nickson, James Haines, W. Schafer, M. Conrad, H. Griech, J. Merkel, T. Schwall. 1992. "A Verified Proton Induced Latchup in Space." *IEEE Transactions on Nuclear Science*. Vol. 39, No. 6. Nuclear and Plasma Sciences Society. Los Alamitos, CA.

NOAA Service Assessment, Intense Space Weather Storms October 19–November 7, 2003. 2004. U.S. Department of Commerce. National Oceanic and Atmospheric Administration.

NOAA. 2004. Solar Storms Cause Significant Economic and Other Impacts on Earth. *NOAA Magazine*. April 5, 2004. Archived online at http://www.magazine.noaa.gov/stories/mag131.htm.

Nuclear Regulatory Commission. Title 10, Code of Federal Regulations, Part 20—Standards for Protection Against Radiation. Annual update.

Severe Space Weather Events—Understanding Societal and Economic Impacts: A Workshop Report. Committee on the Societal and Economic Impacts of Severe Space Weather Events. The National Academies Press, 2008.

Vaughan, William W., Keith O. Niehuss, and Margaret B. Alexander. 1996. Spacecraft Environments Interactions: Solar Activity and Effects on Spacecraft. NASA Reference Publication 1396. Marshall Space Flight Center, Alabama.

Figure References

Badhwar, Gautam D. and Patrick M. O'Neill. 1994. Long term modulation of galactic cosmic radiation and its model for space exploration. *Advances in Space Research*. Vol. 14. Elsevier. Amsterdam, NL.

Baker, Daniel N. 2002. How to Cope with Space Weather. *Science*. Vol. 297. Vol. 5586. American Association for the Advancement of Science. Washington, DC.

Gorney, David J. 1990. Solar cycle effects on the near-Earth space environment. *Reviews of Geophysics*. Vol. 28. American Geophysical Union. Washington, DC.

Kim, Kyoung Ho, Yong-Jae Moon, Kyung Suk Cho, H. D. Kim, and Jong- Y. Park. 2006. Atmospheric drag effects on the KOMPSAT-1 satellite during geomagnetic Superstorms. *Earth Planets and Space.* Vol. 58. Terra Scientific Publishing Company. Tokyo, Japan.

McComas, David J. and Nathan A. Schwadron. 2006. An explanation of the Voyager paradox: Particle acceleration at a blunt termination shock. *Geophysical Research Letters.* Vol. 33. American Geophysical Union. Washington, DC.

NASA. 1993. NASA Technical Memorandum 104782, Spaceflight Radiation Health Program at the Lyndon B. Johnson Space Center.

NASA. 2009. *Orbital Debris Quarterly News.* Vol. 13, Issue 3. July 2009. A publication of The NASA Orbital Debris Program Office.

NOAA. 2004. Service Assessment, Intense Space Weather Storms, October 19–November 07, 2003. US Department of Commerce. National Oceanic and Atmospheric Administration. Silver Spring, MD.

Toth, Juraj, Leonard Kornos, and Vladimir Porubcan. 2000. Photographic Leonids 1998 Observed at Modra Observatory. *Earth, Moon and Planets.* Vol. 82-83.

Wrenn, Gordon L. and Rob J. K. Smith. 1996. Probability factors governing ESD effects in geosynchronous orbit. *IEEE Transactions on Nuclear Science.* NS-43. Vol. 6. Nuclear and Plasma Sciences Society. Los Alamitos, CA.

Further Reading

Allen, Joe H. and Daniel C. Wilkinson. 1993. Solar-Terrestrial Affecting Systems in Space and on Earth. *Solar-Terrestrial Predictions – IV,* Volume 1, Proceedings of a Workshop at Ottawa, Canada, May 18–22, 1992, Hruska. J., M.A. Shea, D.F. Smart and G. Heckman (eds.), NOAA, September 1993.

Anderson, C. W., Louis J. Lanzerotti, and Carol G. Maclennan. 1974. Outage of the L-4 system and the geomagnetic disturbances of August 4, 1972, *Bell System Technical Journal*, 53, 1817–1837.

Baker, Daniel N., J. Bernhard Blake, Shrikanth Kanekal, B. Klecker, and Gordon Rostoker. 1994. *Satellite anomalies linked to electron increase in the magnetosphere*, EOS, 75, 401.

Baker, Daniel N., Joe H. Allen, Shrikanth Kanekal, and Geoffrey D. Reeves. 1998. *Disturbed space environment may have been related to pager satellite failure*, EOS, 79, 477.

Blais, Georges and Paul Metsa. 1993. Operating the Hydro-Québec Grid Under Magnetic Storm Conditions Since the Storm of 13 March 1989, *Solar-Terrestrial Predictions – IV,* Volume 1, Proceedings of a Workshop at Ottawa, Canada, May 18-22, 1992, Hruska. J., M.A. Shea, D.F. Smart and G. Heckman (eds.), NOAA, September 1993.

Dorman, L. I., N. G. Ptitsyna, G. Villoresi, V. V. Kasinsky, N. N. Lyakhov, and M. I. Tyasto. 2008. Space storms as natural hazards. *Advances in Geoscience.* Vol. 14.

Jansen, Frank, Risto Pirjola, and René Favre. *Space Weather, Hazard to the Earth?* 2000. Swiss Reinsurance Company, Zurich.

Kappenman, J. G. 1996. Geomagnetic Storms and Their Impact on Power Systems. *IEEE Power Engineering Review.* Vol. 58.

Koskinen, H., E. Tanskanen, R. Pirjola, A. Pulkkinen, C. Dyer, D. Rodgers, P. Cannon J.-C. Mandeville, and D. Boscher. 2001. *Space Weather Effects Catalogue.* European Space Weather Study. FMI-RP-0001. Issue 2.2.

Normand, Eugene. 2004. Single Event Effects in Avionics and on the Ground in Radiation Effects and Soft Errors in Integrated Circuits and Electronic Devices, Selected Topics in Electronics and Systems. Vol. 34, *International Journal of High Speed Electronics and Systems*, Vol. 14, No. 2, Ron Schrimpf and Dan Fleetwood, Editors. World Scientific.

Pirjola, R., A. Viljanen, A. Pulkkinen, and O. Amm. 2000. Space weather risk in power systems and pipelines, Physics and Chemistry of the Earth, Part C. *Solar, Terrestrial & Planetary Science.* Vol. 25, Issue 4.

Rodgers, David J., Lesley M. Murphy, and Clive S. Dyer. 2000. *Benefits of a European Space Weather Programme.* DERA report no. DERA/KIS/SPACE/TR000349. ESWPS-DER-TN-0001. Issue 2.1.

Rodger, Craig J., Mark A. Cliverd, Thomas Ulich, Pekka T. Veronnen, Esa Turunen, and Neil R. Thomson. 2006. The atmospheric implications of radiation belt remediation. *Annuals of Geophysics*, Vol. 24.

Scherer, Klaus, Horst Fichtner, Bernd Herber, and Urs Mall (Eds.). 2005. Space Weather, The Physics Behind A Slogan. Lecture Notes for Physics 656. Springer. Berlin.

Smart, Don F. and Margaret A. Shea. 2005. A review of geomagnetic cutoff rigidities for earth-orbiting spacecraft. *Advances in Space Research.* Vol. 36, Issue 10. Solar Wind-Magnetosphere-Ionosphere Dynamics and Radiation Models.

Space Storms and Space Weather Hazards (NATO SCIENCE SERIES: II: Mathematics, Physics and (NATO Science Series II: Mathematics, Physics and Chemistry) [Hardcover] Ioannis Daglis, Editor, 2000.

Spacecraft System Failures and Anomalies Attributed to the Natural Space Environment. K. L. Bedingfield, R. D. Leach, and M. B. Alexander Editors. NASA Reference Publication 1390, August 1996.

Tribble, Alan C. and James W. Haffner. 1991. Estimates of Photochemically Deposited Contamination on the GPS Satellites. *Journal of Spacecraft and Rockets.* AIAA. Vol. 28, No. 2. March–April 1991.

Tribble, Alan. 2003. *The Space Environment, Implications for Spacecraft Design.* Princeton University Press.

Walker, M., T. Dennis, and J. Kirschvink. *The Magnetic Sense and its Use in Long-distance Navigation by Animals.* Current Opinion in Neurobiology 2002, 12:735–744, Elsevier Science Ltd. Published online 11 November 2002.

Effects of Space Weather and Space Environment on Signals and Systems

14

UNIT 3. IMPACTS AND EFFECTS OF SPACE WEATHER AND SPACE ENVIRONMENT

*With Contributions by George Davenport and Richard C. Olsen,
Air Force Weather Agency and Air Force Research Laboratory*

You should already know about...

- The ionospheric layers and ionizing radiation (Chaps. 3 and 8)
- Radio wave propagation and refractive index (Chap. 8)
- Maximum and minimum usable frequency (Chap. 8)
- Solar radio noise (Chap. 9)
- Drivers of space weather (Chaps. 9–12)
- Spread-F, scintillation, and traveling atmospheric disturbances (Chap. 12)
- Polar cap absorption events (Chap. 12)

In this chapter you will learn about...

- Space environmental effects on wave propagation
- Radio frequency interference
- Scintillation effects
- Radio signal propagation effects from regional anomalies
- Meteor effects on signals
- Magnetar effects on signals

Outline

14.1 **Background and Solar-driven Effects**
    14.1.2 Prompt Signal Effects
    14.1.3 Impulsive Solar Events Generating Polar Cap Absorption (PCA)
    14.1.4 Geomagnetic Storms and Seasonal Effects on Signals

14.2 **Non-solar Effects on Signal Propagation**
    14.2.1 Meteor Effects
    14.2.2 Magnetar Effects
    14.2.3 High-altitude Nuclear Signal Effects

14.3 **Surveying the Impact of Space Weather on Systems—II**

Chapter 14 Effects of Space Weather and Space Environment on Signals and Systems

Navigation, communication, and many forms of surveillance rely on radio wave propagation, which in turn depends on the refractive (bending) index of the ionosphere. Space weather effects contribute to sharp spatial and temporal changes in the refractive index and to changes in the layers of Earth's ionosphere. Indeed, much of what we know about space weather effects on the ionosphere comes from radio probes of plasma behavior.

14.1 Background and Solar-driven Effects

14.1.1 Background Ionosphere Effects

> **Objectives: After reading this section, you should be able to...**
> ♦ Explain why some radio signals do not follow great circle propagation
> ♦ Describe ionospheric terminator effects on radio waves

Before we delve into the harsh effects of solar driven events on communication signals, we provide some additional information on the background ionosphere in which radio waves propagate.

Non-great-circle Propagation

An undisturbed radio signal propagating in the ionosphere follows an undeviated path over Earth's curved surface. Such a signal follows a *great-circle route*. On this type of path, the signal circles the planet and eventually passes its point of origin. The center of this imaginary circle coincides with Earth's center. The distance the wave travels is Earth's full circumference—hence the term "great circle."

Longitude lines are great-circle routes, but latitude lines are not (except for the equator). This is easy to see when considering the latitude lines close to the North or South Pole—the lines are circles whose radii become smaller and smaller as they approach the pole. In general, east and west signal propagations do not remain eastward or westward, unless the path is equatorial. Figure 14-1 illustrates these features.

Considering that an undisturbed radio signal follows a great-circle route, we know that an unperturbed signal directed north or south continues along that course. But if a signal starts east or west, what happens? From Fig. 14-2 (or using a globe), we see that the signal must gradually deviate toward the equator, in order to follow a great-circle route. The radio station A could not transmit a directed signal to B by pointing it directly east even though B is directly east of A. Instead, A must direct the signal a little north of east so that the equatorward deviation along the great-circle route brings it to B.

Non-great-circle propagation is radio wave transmission that deviates from the normal great-circle route, meaning that the signal deviates left or right of its intended path. The ionosphere refracts HF radio waves away from regions of greatest electron density. Downward or upward refractive bending results from

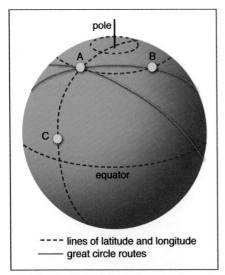

Fig. 14-1. Great Circle Routes Versus Latitude and Longitude Lines. Curve A-C is a great-circle route for communication, as is the equator. A great circle communication route may connect points A and B, but it would not follow the latitude curve that connects the two points. *(Adapted from AF Weather Agency)*

654

vertical gradients in the electron density. Non-great-circle propagation is refractive bending left or right due to horizontal electron density gradients.

When radio operators communicate with each other, they direct the signal along the great-circle route, which may necessitate transmitting the signal through regions with variable electron density. Changes in electron density cause anomalous propagation. Figure 14-2 illustrates the difference between expected and deviated propagation.

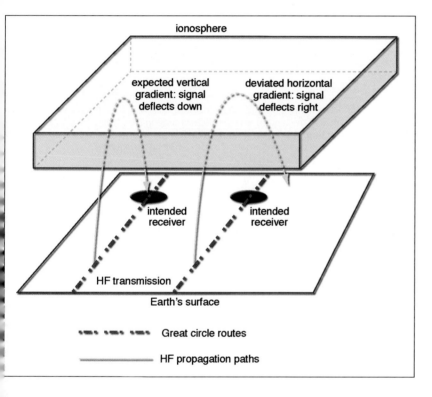

Fig. 14-2. Standard Versus Non-great-circle HF Propagation. An abnormal electron gradient structure bends a radio signal from its intended path. The amount of deviation depends on its frequency. *(Adapted from AF Weather Agency)*

Non-great-circle propagation results in signal fade or loss because the main energy lobe from the transmitting antenna does not reach the intended target. Instead, the weaker edge of the main energy lobe is received. The degree of loss depends on the narrowness of the directed radio beam.

Terminator Effects. Horizontal electron density gradients in the ionosphere at the terminator result in non-great-circle propagation of HF and VHF signals. Figure 14-3a illustrates how such a radio signal nearly parallel to the terminator deflects toward the night side, away from the higher electron densities associated with the dayside ionosphere. Horizontal electron density gradients in the ionosphere at the terminator also cause ionospheric ducting. A signal crossing the terminator into the nightside sometimes becomes trapped between the E and F2 layers, as Fig. 14-3b illustrates.

Chapter 14 Effects of Space Weather and Space Environment on Signals and Systems

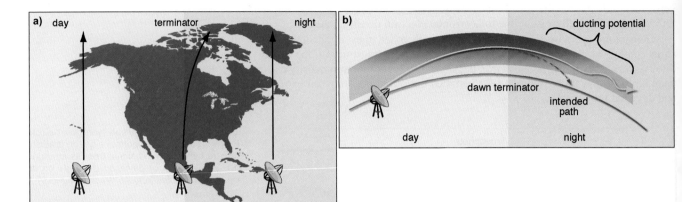

Fig. 14-3.   a) Non-great-circle Propagation at the Terminator. HF antennas (and to a lesser extent, VHF antennas) transmitting roughly northward or southward are vulnerable for a short time every evening and morning to non-great-circle propagation, while horizontal electron density gradients associated with the terminator pass overhead. **b) Ducting at the Terminator.** This diagram shows how signals passing into the nightside sometimes get trapped between the E and F2 layers because of ducting. *(Adapted from AF Weather Agency)*

14.1.2 Prompt Signal Effects

> **Objectives: After reading this section, you should be able to...**
>
> ♦ Describe event-based emissions and their impacts on signals
> ♦ Describe sudden ionospheric disturbances (SIDs)
> ♦ Describe the effects that SIDs have on radio frequency systems
> ♦ Describe how a solar X-ray event affects the total electron content

Space weather impacts on signals usually originate at the Sun. We illustrate the timing and duration of these in Fig. 14-4. Strong magnetic reconfiguration events on the Earth-facing side of the Sun produce the most intense flares and the fastest CMEs. In turn, fast CMEs produce shocks that accelerate energetic particles along magnetic field lines connected to Earth. In general, the stronger and better connected the solar activity, the greater the impact on Earth-based signals and systems.

Some space environmental impacts on signals or signal transmitting-receiving systems occur coincident with the flare produced by solar magnetic reconfiguration. Other impacts are delayed for several minutes, hours, or even days.

Traveling at light speed, the X-ray, visible, and solar radio frequency emissions from flares arrive at Earth in about 8-1/3 minutes. Flare effects include the general category of sudden ionospheric disturbances, sudden signal enhancements, and sudden phase anomalies.

Sudden Ionospheric Disturbances (SIDs)

In the lowest layer of the ionosphere (D layer, normally 80 km–95 km altitude) a large number of neutral air atoms and molecules coexist with the ionized particles. As a passing radio wave causes the ions and free electrons to oscillate, the particles collide with the neutral air particles, and the oscillatory motion dampens and converts to heat. Thus the D layer often absorbs passing radio wave signals. The

14.1 Background and Solar-driven Effects

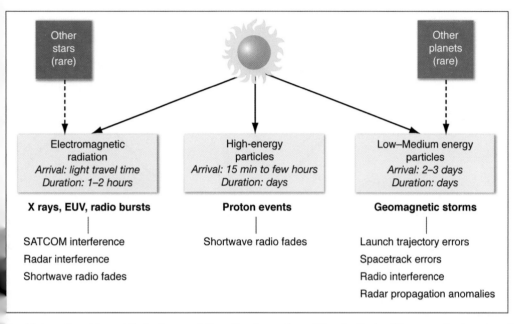

Fig. 14-4. **Event-based Emissions and Resulting Impacts on Signals.** Each of the general categories of emissions has its own characteristics and types of immediate or delayed system influence. Some effects are generated by more than one category of events. Natural sources other than the Sun generate some impacts on signals. We discuss these later in the chapter. (EUV is extreme ultraviolet.) *(Courtesy of the AF Weather Agency)*

lower the frequency, the greater the degree of signal absorption. The *lowest usable frequency (LUF)* is that frequency below which radio signals encounter too much ionospheric absorption to permit them to pass through the D layer. Section 8.3 discusses this topic in detail.

Normally the LUF lies in the lower portion of the HF band. Any event that produces an abnormal increase of ionization in the D layer affects the HF propagation and usually involves a LUF change. Absorption affects skywaves and Earth-to-space-to-Earth waves, but has the greatest effect on the former.

Figure 14-5 illustrates several phenomena that constitute *sudden ionospheric disturbances (SIDs)*—a disturbance to the ionosphere's total electron content and its electron density profile, either locally or regionally. During a solar flare or series of flares like those illustrated in Fig. 9-16, the flux of solar X-ray and extreme ultraviolet (EUV) electromagnetic radiation often increases significantly, sometimes by two, three, or more orders of magnitude. The effects on the ionosphere are dramatic. The lower ionosphere on Earth's sunlit side, especially the D region, is likely to intensify and thicken as very short wavelength energy penetrates and produces enhanced ionization. Often the altitude of the D region decreases slightly. These changes affect propagation conditions throughout the radio frequency spectrum. An X-class flare almost certainly produces a SID.

Chapter 14 Effects of Space Weather and Space Environment on Signals and Systems

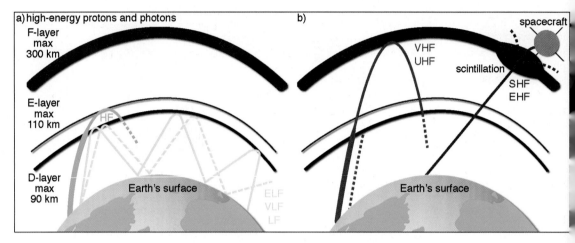

Fig. 14-5. **Sudden Ionospheric Disturbance (SID) Effects on Radio Communications. a)** During SIDs, extremely low- to low-frequency waves (yellow) may travel an altered path as the reflecting layer height varies. High-frequency waves (green) may be absorbed by the enhanced D layer. **b)** High- to ultra-high-frequency waves (red) may also be absorbed by the enhanced D layer or anomalously refracted by variations in the F layer. If radio operators increase the radio frequency to overcome the blockage, they may inadvertently create trans-ionospheric waves that do not return to the surface. Super-high and extremely-high-frequency waves (blue) may suffer unusual refraction or amplitude and phase shifts, rendering them ineffective. Altered blue curves represent transmissions that do not reach their intended destination. Scintillation effects are usually considered as a separate category and are not shown here. *(Adapted from George Davenport)*

Fluctuating Total Electron Content (TEC) from a Solar X-ray Event

A significant solar flare event and a resulting SID cause the total electron content (TEC) in the ionosphere to change sharply over a short time period, on the order of a few minutes, during the solar flare. As the TEC fluctuates, detection and tracking radar systems often experience errors in range, range rate, and direction as the radar signal passes through the ionosphere.

Sudden Ionospheric Disturbance Effects on Low-frequency Systems

During a sudden ionospheric disturbance (SID), some radio waves that are normally trans-ionospheric are absorbed by the enhanced D-region ionization. The radio signal energy is converted to random motion of the electrons—heat.

Thunderstorms produce thousands of lightning strokes on Earth every second, most of which generate LF radio frequency noise. The worldwide amplitude of this radio frequency noise is fairly constant. It sounds like a hum on the loudspeaker of a radio tuned to the incorrect frequency. However, during a solar flare, the reflectivity of the D region increases on Earth's sunlit side, and the waveguide in which the lightning pulse is propagating becomes more efficient, producing *sudden enhancements of atmospherics (SEAs)*. The amplitude of the hum heard in the radio loudspeaker increases, which is usually a mere annoyance.

A *sudden enhancement of signal (SES)* is similar to the SEA. The difference is that the signal source is another radio transmitter. The strength of the signal received from a distant very low-frequency (VLF) transmitter increases during a solar flare.

Observers use SEAs to monitor solar flare activity. They analyze the intensity of a solar flare event simply by monitoring the signal strength of the noise received by an LF radio receiver.

A *sudden phase anomaly (sudden phase advance) (SPA)* is an abrupt shift in the LF or VLF signal phase from a distant transmitter. When a solar flare increases the D region ionization, the effect is to lower the height of the reflecting layer. The path length within the Earth-ionosphere waveguide decreases, and the phase of the signal received from a distant transmitter changes, producing an SPA. The yellow curves in Fig. 14-5 illustrate this.

Ground-based navigation systems such as Loran-C depend upon a stable D region to provide reliable location information. Loran-C transmits radio signals with a frequency of about 100 kHz, and the receiving equipment uses the phase of this LF signal to calculate an accurate position. A phase shift introduces significant errors in the position calculation.

Sudden Ionospheric Disturbance Effects on High-frequency Systems

Short-wave fade and the sudden frequency deviation are SIDs that affect the high-frequency (HF) portion of the radio spectrum. Long-distance HF communication uses radio signal refraction within the ionosphere's F region. The HF radio signal reflects off the F region, returning toward Earth's surface, where it is received or reflected back to the ionosphere for another bounce. A *short-wave fade (SWF)* occurs when the D region blocks the HF radio signal from reaching the F region on its way up or returning to Earth's surface on its way back down.

The dashed and dotted red curves in Fig. 14-5 illustrates an SWF. If a large solar flare intensifies the D region ionization, an HF radio signal may be absorbed during its transit. As a result, the strength of the HF radio signal is reduced on sunlit paths where the LUF has increased.

Figure 14-6 is the propagation window for a hypothetical high-frequency radio path. The frequencies between the LUF and the MUF are the frequencies that propagate between the transmitter and the receiver at that time. In the pre-noon sector, operators normally maintain effective communication with frequencies between the LUF (about 4 MHz) and the MUF (about 8 MHz).

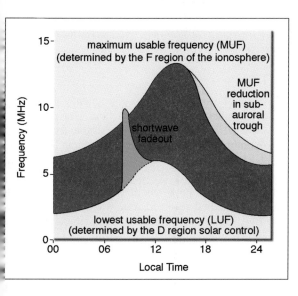

Fig. 14-6. High-frequency (HF) Propagation Window. Over the course of a day, the maximum and lowest usable frequencies vary with Sun angle, as depicted by this wavy band of transmission frequencies. The MUF has a stronger dependence on ionospheric dynamics and tends to peak after mid-day. Solar flares change the strength of the D region, resulting in a higher LUF. For large flares, the LUF may exceed the MUF, effectively closing the HF propagation window. Geomagnetic storms can cause a reduction in ionization in regions just equatorward of the auroral zone, particularly in the evening hours. The loss of ionization reduces the MUF and narrows the HF propagation window. *(Courtesy of AF Weather Agency)*

Chapter 14   Effects of Space Weather and Space Environment on Signals and Systems

Occasionally a SWF causes the LUF to increase so much that it becomes greater than the MUF. The HF radio propagation window closes completely, as is the case for the morning hours in Fig. 14-6. The result is a total SWF (blackout). Such blackouts last for the duration of the solar flare, affecting HF radio paths on Earth's sunlit side. Figure 14-7 shows the HF blackout region for the November 4, 2003, X-17 flare.

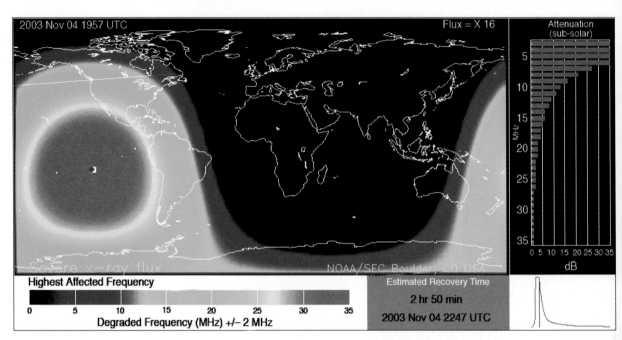

Fig. 14-7.   HF Blackout caused by the November 4, 2003 Extreme Flare Event. The space-based X-ray detector saturated due to extreme radiation between 1943–1956 UTC on 4 November 2003. Best estimates have the maximum flux as X-28, occurring at 1950 UTC on November 4. The image is rendered for an X-16 flare. The bright colors on the map indicate the regions affected by the radio blackout. The Pacific sector was severely affected (more than 10 db signal fade) at frequencies below 10 MHz. The graph at the right quantifies the HF attenuation in decibels. The blue box on the bottom shows the estimated recovery time. The inset on the bottom right gives the time profile of the event intensity in relative values. Time to recovery after the very sharp rise in flare photons was nearly three hours. *(Courtesy of NOAA's Space Weather Prediction Center)*

The *frequency of optimum transmission (FOT)* is that frequency capable of providing the most reliable communications for a given hour and path. This frequency is usually just below the MUF. Because the lowest usable frequency is subject to more spread, radio operators tend to avoid the lower boundary of the radio propagation window (LUF). Transmissions near the LUF are subject to absorption in a dynamic radio environment.

A *sudden frequency deviation (SFD)* is a small, abrupt change in the frequency of a high-frequency radio signal received from a distant transmitter. The increased solar X-ray and EUV flux from a flare often intensifies the F1 and E regions of the ionosphere. The HF radio signal suddenly reflects from one of these regions, instead of the higher portion of the F region, known as the F2 region. Signal reflection from the lower altitude introduces a Doppler shift in the frequency of the received signal. This frequency shift is the SFD.

Focus Box 14.1: Short-wave Fades (SWFs) of High-frequency (HF) Radio Propagation

March 1989. The Space Environment Center received reports of unusual VHF radio propagation during large solar flares on March 9 and 10, 1989. A dispatcher in Colorado reported mobile units sounding distorted and breaking up badly. Radio pagers and cellular telephones had similar problems. An amateur radio operator in Minnesota reported receiving strong transmissions from the California Highway Patrol.

July 2000. Earth experienced severe space weather storm conditions from 1500 UT on July 15, 2000 until 0300 UT on July 16, 2000. The ARINC HF radio ground-to-aircraft voice communications station at Bohemia, NY, reported fadeout at 2200 UT on July 15, 2000 and again from 2235 UT on July 15 until 0120 UT on July 16.

April, 2001. The biggest solar flare in 12 years was observed late on April 2, 2001. X-ray radiation from the flare overpowered satellite sensors, forcing solar scientists to guess its power at about X-22. As the X rays encountered Earth's ionosphere, they generated a blast of radio waves that caused an R-5, or extreme radio blackout, resulting in loud hissing and popping on two-way radios at higher latitudes.

October–November 2003. Airlines and ground controllers experienced communication problems almost daily during the October–November 2003 solar activity outbreak. Initially (October 19–23), elevated X-ray solar emissions caused the degraded HF communications. On October 19, Air Traffic Centers reported moderate-to-severe impacts on all HF groups, and HF service was degraded for over two hours, following an X-1 flare. These periods were the first of several with severely degraded communications. With each major flare, HF communications at low and middle latitudes encountered a range of problems from minor signal degradation to complete HF blackout.

Sudden Ionospheric Disturbance Effects on Very High-, Ultra High-, Super High-, and Extremely High-frequency Systems

Sudden ionospheric disturbances (SIDs) also affect the portion of the radio spectrum with the highest frequencies and shortest wavelengths. These radio signals should pass through the ionosphere and into space. The thicker and more intense D region impedes the transmission of commands from a ground station to a spacecraft.

The D region absorbs downlink telemetry signals from spacecraft to the ground, and it also absorbs uplink commands from the ground to the spacecraft. The dashed blue curve in Fig. 14-5 illustrates degraded telemetry from a spacecraft because of signal absorption within the intensified D region. This phenomenon has not been assigned a name, but the result is degraded *spacecraft commanding and telemetry*. Although these phenomena are rare, spacecraft operators have observed and documented anecdotal occurrences of the problems, and the events have correlated with significant solar flares, usually X-class.

The stars and other galaxies produce measurable radio noise. Astronomers have measured galactic radio noise for many years using relative ionospheric opacity meters (RIOMETERS) that operate at or near 30 MHz. They know the intensity of the radio noise associated with each location in space. As Earth rotates, the strength of the cosmic radio noise is measured and compared with its anticipated intensity. When a solar flare causes the D region to be become thicker and more intense than normal, the strength of the observed cosmic radio noise is less than expected. The D region has absorbed some of it, which is *sudden cosmic noise absorption (SCNA)*.

Chapter 14 Effects of Space Weather and Space Environment on Signals and Systems

Focus Box 14.2: Changes in the Total Electron Content (TEC) Caused by a Solar X-ray Event

July 2000. An X-5.7/3B solar flare commenced at 10:03 UT on 14 July 2000. GPS observations revealed a temporal evolution of the ionospheric TEC values within the latitude range of 30° ~ 45° N and the longitude range of 15° ~ 45° E. The dayside TEC values increased during the flare event by as much as 5 TEC Units (1 TEC Unit = 10^{16} electrons/m^2) in regions with small solar zenith angles. The increases tended to be largest at the sub-ionospheric point but were somewhat asymmetric, being smaller in the local morning than in the afternoon and larger in the Southern Hemisphere than in the Northern Hemisphere for the same solar zenith angle. The asymmetry suggests that the background levels of the ionosphere and thermosphere influenced the TEC increase. The temporal variation of TEC displayed minor correlated disturbances from 10:15 to 10:27 UT, when the solar flare was in the maximum phase. The minor disturbances likely resulted from the evolution of flare emissions in the EUV domain.

October–November 2003. Some of the most intense solar flares measured in 0.1 nm to 0.8 nm X rays in recent history occurred near the end of 2003. High-time resolution, ~30 s ground-based GPS data and satellite far ultraviolet (FUV) dayglow data were used to examine the flare-ionosphere relationship. The data indicate strong spectral EUV and FUV variability from flare to flare. The October 28 TEC ionospheric peak enhancement at the sub-solar point was ~25 TECU (30% above background). In comparison, the November 4, October 29, and Bastille Day events had ~5–7 TECU peak increases above background. The October 28 TEC increase lasted about three hours, far longer than the flare itself. This latter ionospheric feature is consistent with increased electron production by EUV photons in the middle altitude ionosphere, where recombination rates are low. Scientists believe that the location where the flare originates may account for some of the variability. For example, the flare on November 4, 2003 originated at the solar west limb. The solar atmosphere may have absorbed some of the flare-emitted EUV and FUV photons, thus preventing their transmission to Earth.

Radio Frequency Interference (RFI) from the Sun

The atmosphere is awash in radio signals. *Radio frequency interference (RFI)* is the unintended reception of a radio signal. This signal may be natural or artificial. Natural sources include lightning, the aurora, the Sun, some of the planets (Jupiter and Saturn), some stars, and the core of the Milky Way galaxy. Artificial sources include radio and television broadcasts, private radio operators, aircraft, ships, computer devices, phones, satellites, power lines, and radars. RFI obscures the intended signal. During World War II, radar signals were a primary element of British defense against the Luftwaffe bombers. Solar radio noise at times filled the airways, prompting investigations of German jamming capabilities. In fact, particularly severe radar jamming during a large solar event in 1942 led to the discovery of solar radio bursts (reported following the end of the war).

When the Sun produces a radio burst, some of the energy strikes Earth's sunlit hemisphere, causing RFI for antennas aligned too close to the Sun, or whose side lobes are pointing toward the Sun. Radars scanning broad areas in search of targets are the most susceptible to RFI caused by a solar radio burst.

The flux of solar radio frequency emissions increases sharply during a significant solar flare. The Sun emits radio noise over the entire radio spectrum, so a solar radio burst degrades any system that operates at frequencies high enough for its signal to pass through the ionosphere. Sometimes a significant solar radio burst occurs without an accompanying X-ray solar flare event.

Focus Box 3.4 discusses unintended radio reception from the Sun. The Sun interferes with the receipt of radio signals from a spacecraft when the two are in nearly the same line of sight from a ground station. The Sun also introduces interference when it is located in one of the lobes of a ground station antenna pattern during a solar radio burst. Radar operators usually avoid having the Sun in the primary radar beam, so an HF communication system is more likely to experience unexpected RFI when a secondary lobe of the receiving antenna is pointed toward the Sun. Sometimes operators can't avoid having a radar beam aligned with the Sun. A spacecraft command or tracking ground station is also more likely to experience RFI when a lobe of the receiving antenna is pointed toward the Sun. The Sun passes through the main lobe of a ground station antenna pointed at geostationary satellites twice a year near the spring and autumn equinoxes.

Space weather RFI is primarily a space-to-Earth sensor problem. Military radio and radar operators must distinguish between natural and hostile RFI, because hostile RFI is an act of aggression. Mistaking natural RFI for an attack could lead to embarrassing misunderstandings.

Focus Box 14.3: Radio Frequency Interference (RFI)

November 2000. The ARINC HF radio ground-to-aircraft voice communications station at Bohemia, NY, informed the Federal Communications Commission of RFI sufficient to degrade operations on two frequencies at various times throughout the day on November 25, 2000.

On the same day, the solar radio observatory at Sagamore Hill, Massachusetts, reported solar radio bursts of 17,000 solar flux units (SFUs) above background at 245 MHz and 15,000 SFUs above background at 410 MHz (1 SFU = 10^{-22} W/(m$^2 \cdot$Hz)). The solar radio observatory at Palehua, Hawaii, also reported a solar radio burst of 10,000 SFUs above background at 606 MHz, and the solar radio observatory at Learmonth, Australia, reported a radio burst of 11,000 SFUs above background at 2695 MHz.

November 2004. The GOES-9, GOES-10, and GOES-12 satellites experienced radio noise during November 3 and 4, 2004. These satellites have several systems that receive 401 MHz to 406 MHz, including the Satellite Aided Search and Rescue repeater. Of these events, the GOES-12 event was the most severe in magnitude and duration (about 100 minutes). The problem was attributed to solar radio burst emissions, particularly in the low 400 MHz range. This occurred during a 7,200 SFU burst on 410 MHz from a solar flare in Active Region 696. Focus Box 9.2 describes additional effects.

Multi-year. A research team from New Jersey Institute of Technology and Lucent Technologies' Bell Laboratories has correlated four decades of solar data with cellular telephone dropped calls. The problem develops when radio waves associated with solar radio bursts hit cell phone towers, creating static that overwhelms the signal at the tower, where calls are relayed. The researchers found that, on average, solar radio bursts powerful enough to disrupt wireless communications occur 10 to 20 times per year.

14.1.3 Impulsive Solar Events Generating Polar Cap Absorption (PCA)

Objective: After reading this section, you should be able to...

♦ Describe polar cap absorption and polar cap blackout effects on radio frequency systems

Polar Cap Absorption Events. During some events, the Sun ejects energetic charged particles that travel at large fractions of light speed. Shocks driven by high-speed coronal mass ejections also produce populations of energetic protons that reach Earth 20 minutes to several hours after the ejection enters interplanetary space. Earth's magnetic field allows the charged particles to enter near the cusp and over the polar regions, where they penetrate and bombard the atmosphere and sharply increase the ionization of the D region and lower E region (50 km–100 km), creating a *polar cap absorption (PCA) event*, in which radio waves are absorbed by the increased ionization.

The effects of a PCA event on the D region are similar to those of a SID, except that they persist long after events at or near the Sun have ended. Figure 12-27 illustrates the effects of a PCA event on the lower ionosphere. Because the energetic particles are constrained to enter the polar cap region, the effects of a PCA are limited to high latitudes. Often a PCA event occurs simultaneously with other radio transmission problems. Long-lasting PCAs are *polar cap blackouts* and are often followed or accompanied by widespread geomagnetic and ionospheric disturbances as lower-energy plasma and field disturbances associated with a CME arrive at Earth. Because the ionosphere's base tends to be lower during PCA events, concurrent errors on LF navigational systems are also observed.

The D region serves as the upper surface of the waveguide through which ELF, VLF, and LF radio signals propagate. The effects on radio propagation at these frequencies are similar to the effects of a SID. In particular, a PCA event introduces significant errors in the position data indicated by the Loran-C navigation system.

Long-distance HF radio propagation requires the signals to "bounce" off the ionospheric F region and return to Earth's surface well beyond the line of sight. The D region absorbs some of the signal before it ever reaches the F region, or it prevents the signal from returning to the surface. A PCA event eliminates any HF radio propagation across the polar region for days, perhaps up to a week, following a major solar flare. In some circumstances, cross-polar flights are rerouted to lower latitudes, where HF communication can be maintained. The rerouting may require additional stops for refueling.

Focus Box 14.4: Polar Cap Absorption (PCA) Effects

March 1989. High-energy charged particles from a major solar event and probable CME on March 9, 1989 intercepted Earth's orbit and produced a solar proton event. The charged particles caused an unusually long-lasting and intense PCA event. This PCA lasted until March 14, 1989. The Loran system experienced large position errors. A change of 7 km–10 km in the height of the D region produced position errors of 1 km–12 km. Further, an HF monitoring station at Thule, Greenland reported that most frequencies were unusable because of severe absorption. HF propagation was almost nonexistent from 75° north latitude to the North Pole.

Solar Cycle 23. Radio operators in Thule, Greenland had strange multi-day radio outages that were attributed to equipment maintenance issues. Maintainers spent days pulling radios apart, never finding the cause. They then began receiving space weather support and discovered that PCA events were the cause of the long-term radio outages and had nothing to do with the performance of the radios.

October–November 2003. A polar cap absorption event developed during the first geomagnetic storm of the interval. A major carrier rerouted three polar flights to lower latitudes to obtain better data-link and SATCOM reception. This action required an additional 26,600 pounds of fuel and resulted in over 16,500 pounds of cargo being denied. Additional impacts were reported on October 24 following the onset of a second strong geomagnetic storm. Solar radiation remained at background levels, and high-latitude communications were severely degraded. Higher latitudes experienced even more difficulty following the onset of the radiation storms on October 26. Air traffic operators reported minor-to-severe impacts on HF communications every day between October 26 and November 5. Communication was so poor on October 30 that additional staff was necessary to handle air traffic.

14.1.4 Geomagnetic Storms and Seasonal Effects on Signals

Objectives: After reading this section, you should be able to...

- Describe auroral clutter, noise, and interference
- Describe geomagnetic storm effects on radio frequency systems
- Distinguish between sporadic E events and spread F events
- Relate the effects of scintillation to signal and navigation degradation

Most geomagnetic storm effects on signals occur in the HF to EHF domain. The plasmasphere-ionosphere is the affected medium. Discontinuities in the solar wind energize the magnetosphere. During geomagnetic substorms, particles from the magnetotail are accelerated Sunward, and many of them follow the magnetic field into the auroral zones. Those that don't reach their magnetic mirror points slam into the upper atmosphere and produce discrete aurora. The lower energy particles are stopped in the F layers of the ionosphere, while the higher energy particles penetrate to the E layer. Auroral phenomena hamper radio communications and radar operations, causing a number of ionospheric disturbances and direct radio interference. Many of the effects tie to storm-time magnetic and electric field variations that create changes in total electron content (TEC) or significant variability in local electron density profiles (EDPs). Other effects are related to radio frequency interference generated in the auroral zone.

Radar Clutter, Noise, and Interference

Shortly after World War II and during the beginning of the Cold War, radars became powerful enough to send their pulses far over the horizon. Poleward pointing radar beams with frequencies of 200 MHz–600 MHz occasionally received backscatter from unknown targets. Radar operators learned that the strange returns often appeared about 48 hours after a solar flare. Today we know these returns are radar auroral clutter (RAC), produced when pulsed radar echoes backscatter from the irregular plasma structures of the aurora. The clutter return reaches the radar at the same time as echoes from later pulses returned by much closer targets.

Radar clutter can produce false targets and tracks. The effect is most noticeable when the radar beam is perpendicular to geomagnetic field lines around which substantial numbers of auroral particles are gyrating. Detecting and tracking radar systems experience auroral clutter, noise, and interference in radar returns when the signal is incident upon or passes through active auroras. Eliminating the backscattered signals is not feasible. Instead, radar operators identify and map the aurora as part of the space environment situational awareness.

Space-based radars, operating in the VHF frequency range, experience a related problem. They observe backscatter from electron density irregularities that develop in response to unstable current systems in the auroral and equatorial electrojets. Models of ionospheric densities, temperatures, and electric fields are being developed to predict radar electrojet clutter.

Chapter 14 Effects of Space Weather and Space Environment on Signals and Systems

Geomagnetic Storm Effects on High-frequency (HF) Systems

High-frequency radio signal propagation is often degraded on long-distance paths during a geomagnetic storm. The effect is most pronounced at high geomagnetic latitudes, where irregularities in the ionospheric layering make the ionosphere unsuitable for steady HF wave refraction. Interestingly, ham radio operators often report enhanced propagation across the equator during moderate geomagnetic activity. The turbulent ionosphere opens some communication paths not normally present. These paths are not usually well-suited for robust critical communication channels.

Geomagnetic storms lead to large variations and irregularities in electron density that produce changes to signal coverage and strength and in some cases reduce the number of frequencies available for use. Further, geomagnetic storms are one source of traveling ionospheric disturbances that lead to anomalous propagation effects. As we describe in Chap. 12, storms may increase or decrease the local electron density, or even leave it roughly unchanged, but the layer structure becomes deeply disturbed.

Auroral activity excited by geomagnetic storms disturbs the E region of the high-latitude ionosphere and causes radio signals propagating to and from the F region to deviate from normal great-circle paths. HF radio signals, bouncing off or passing through the E region, bend and change direction similar to visible light rays passing through a lens or reflecting off an irregular mirror.

The sources of these path deviations are *sporadic E* conditions associated with strong, irregular enhancement of the E layer. As the name implies, sporadic E (E_s) describes the E layer of the ionosphere as being irregular. This phenomenon is characterized by patches of enhanced E layer electron densities that are transient and irregular. The E_s patches range from tens to hundreds of kilometers wide and are about one kilometer thick. Sporadic E persists from minutes to several hours and strongly refracts HF and occasionally the lower VHF radio band, causing some signals to fade and others to be enhanced. Intense patches of the E region ionosphere develop in the auroral zone or near the geomagnetic equator. Clouds of intense E_s also form at middle latitudes, particularly during the summer months. The clouds usually cover a relatively small geographic area, with a diameter of less than 100 km. They form randomly, and usually dissipate within a few hours. Sporadic E occurs day or night, and varies markedly with latitude. It is associated with

- Solar flares bringing about extreme ionization events and their associated electron density gradients
- Geomagnetic activity causing particle precipitation and electrojet activity in the auroral zone
- Plasma bubble activity—either seasonal or storm driven
- Thunderstorms, hurricanes, and other natural phenomena that generate vertically propagating waves from the ground and neutral atmosphere
- Meteor trails

While the precipitation of particles in the aurora enhances the auroral ionosphere, subauroral regions may lose ionization. Escape of charged particles just equatorward of the aurora is associated with field-aligned currents that connect the ring current with the auroral electrojets. A sudden surge in these currents can initiate a *traveling ionospheric disturbance (TID)*, which can propagate equatorward for thousands of kilometers. Loss of F region electron density depresses the maximum usable frequency (MUF) available for HF propagation and so narrows the window of available frequencies for long-distance communication. The effect is most pronounced during the nighttime hours, as illustrated by the narrowed HF radio propagation window between 1800 and 2400 hours local time in Fig. 14-6. Electron density is unusually low within a structure called the main electron density trough or the *sub-auroral trough*. The trough is located just equatorward of the auroral oval, especially on Earth's nighttime side.

The effect worsens during storm time, when chemistry and dynamics cause ion-electron recombination, thus a decrease in electron density. The auroral oval expands and its equatorward-edge shifts towards the equator. The sub-auroral trough also deepens (the electron density becomes even less than normal) when Earth's magnetic field is disturbed. These effects combine to degrade HF propagation at high latitudes during a geomagnetic storm.

In the region 20°–30° north and south of the geomagnetic equator, the ionosphere responds more to features in Earth's atmosphere and magnetic field structure than to solar illumination. High-frequency propagation is less dependable in the region of the equatorial ionization anomaly than elsewhere. As we describe in Chaps. 8 and 12, the *equatorial ionization anomaly* relates to plasma uplift in and below the F2 region. The equatorial anomaly contains irregularities and structures that are not typical of the ionosphere elsewhere. They may or may not refract HF radio signals back to Earth's surface. Figure 14-8 is a simulation of a vertical total electron content (TEC) contour plot for middle and low latitudes. The equatorial ionization crests appear as regions of enhanced TEC. As we describe in Chap. 12, the plasma in these regions is subject to instability, especially in the dusk sector.

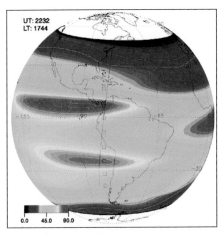

Fig. 14-8. Simulation of Plasma Enhancements in the Equatorial Ionization Anomaly. This image gives color contours of vertical total electron content (TEC). The TEC is the line-integrated electron density through the ionosphere; one TEC unit is 10^{16} electrons/m². The simulation covers ±60° magnetic latitude and clearly shows the ionization crests approximately ±15° from the magnetic equator. The crest regions are subject to plasma instabilities that cause radio signals to go astray. The Sun is to the left. The enhancements develop on the dayside and extend into the evening sector. *(Courtesy of Joseph Huba at the Naval Research Laboratory)*

Geomagnetic Storm Effects on Very High- to Extremely High-frequency (EHF) Systems

Spacecraft operators use 30 MHz–100 GHz (very high to extremely high frequency) radio bands to communicate with satellites and deep space systems. Unlike the HF and lower bands, these signals pass through the ionosphere to reach spacecraft in orbit or deep space. Ionization, created by photons and particles in the upper atmosphere, produces concentrations of free electrons. The total electron content (TEC) is the key parameter for characterizing the ionospheric influence on Earth-space radio links. Radio signals transiting the ionosphere are bent and retarded. The delay results from the integration of the refractive index and varies as a function of the TEC and the signal frequency. The ionospheric path delay is about 0.162 m/TECU at the L1 frequency, 1575.42 MHz.

Perturbations in TEC create changes in the refractive index of the medium and cause signal refraction and delay. In the L-band, which corresponds to the GPS frequencies, the delay may reach several tens of meters. Further, as these signals pass through the disturbed ionosphere, they sometimes change direction or experience a phase shift, change in polarization, or signal fade. Both organized traveling disturbances (Chap. 12) and localized random irregularities reduce signal quality.

The E and F regions have the highest electron density and are responsible for this effect, called scintillation. Scintillation is most common near the geomagnetic equator, where it's associated with plasma depletions, and at high latitudes (65°–75°), where it's related to auroras. On rare occasions, scintillation is observed at middle latitudes during severe geomagnetic storms (Sec. 12.3). Figure 14-9 illustrates the likely locations for the scintillation of a spacecraft's radio signal as it passes through the ionosphere toward a ground station.

Chapter 14 Effects of Space Weather and Space Environment on Signals and Systems

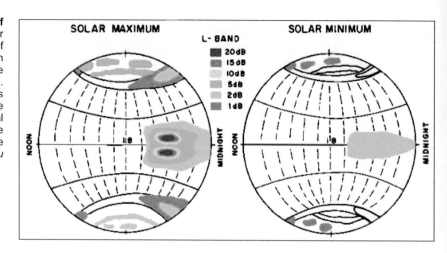

Fig. 14-9. Global Climatology of Ionospheric Scintillation. The polar regions are at the top and bottom of each figure. The view is centered on dusk. The color coding refers to the reduction of the GPS L-band signal. During solar maximum, scintillation is pronounced at high latitudes and in the pre-midnight region. The equatorial ionization crests on either side of the magnetic equator are the largest source of scintillation. *(Courtesy of Santi Basu at the AF Research Laboratory)*

Radio waves propagating through an irregular, ionized medium exhibit turbulent changes in the refraction index and may experience signal scintillation, or a randomization in the signal phase and amplitude, as illustrated in Fig. 14-9. Because the plasma density determines the refraction index for radio waves, the radio waves are particularly sensitive to irregularities in the plasma density within the ionosphere. The electron density change during scintillation events may be as high as 10% in the worst case. Furthermore, the effect of density irregularities as scattering centers is influenced strongly by the relative size of the irregularities compared to the signal's wavelength.

Strong geomagnetic storms stir the magnetic field lines in the magnetosphere and the plasma to which these field lines are linked in the plasmasphere-ionosphere. Sudden storm-driven motions in the plasmasphere are the subject of much recent scientific investigation. The plasmasphere's size and shape is strongly controlled at its outer boundary by the magnetospheric electric field (Fig. 11-13). Sudden field changes expand and contract the plasmasphere, producing a corrugated structure with large spatial and temporal variations in electron density. Just inside the boundary layer (inside the closed black curve of Fig 11-13a) electron content becomes highly variable. This structuring provides high-altitude plasma irregularities from which radio waves may scatter and scintillate (Focus Boxes 14.5 and 14.6).

Shortly after storm onset, cold plasma total electron content (TEC) increases at middle latitudes in the post-noon sector, inside the boundary layer. Near the poleward edges of this region, plumes of enhanced TEC move to higher altitudes and latitudes (Fig. 12-32). In the dusk sector, disturbed electric fields, transmitted from the magnetosphere, cause storm-enhanced density plumes to strip away the outer layers of the plasmasphere-ionosphere. This process produces narrow drainage plumes of cold plasma that extend along magnetic field lines between the plasmasphere and the ionosphere. These streaks of cold plasma flow Sunward ($v \approx E \times B$) under the influences of disturbance electric fields.

Closer to Earth, magnetic fields in the ionosphere draw attached plasma into the circulations associated with magnetic storms and substorms. Excess plasma originally located in the F region's equatorial anomaly is drawn into middle and even high latitudes by convection. Extreme TEC gradients at the edges of these storm fronts produce severe scintillation of GPS signals.

14.1 Background and Solar-driven Effects

Focus Box 14.5: Quantifying Scintillation

Scintillation effects are characterized by changes in amplitude and phase. Here we discuss the amplitude measure of scintillation, the S_4 *index*. The S_4 index indicates the amount of variation in the amplitude of received signal power over an interval of time, typically one minute. Usually the signal reception is measured at the ground, although scintillations do occur on purely space transmission segments. The dimensionless index is given by the formula: (Fig. 14-10):

$$S_4 = \sqrt{\frac{\langle I^2 \rangle - \langle I \rangle^2}{\langle I \rangle^2}}$$

where I is the signal intensity and $\langle I \rangle$ is the mean intensity.

Figure 14-10 shows a time profile of the S_4 index at a single location on a day of low geomagnetic activity, with strong scintillation. The index value ranges from zero to one. Strong scintillation and ionospheric signal scatter is inferred to occur when S_4 is greater than ~0.6. Potential loss of signal lock, as well as degradation of precision

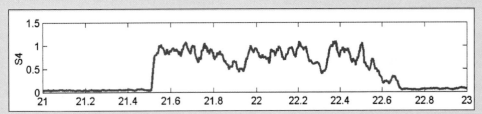

Fig. 14-10. The S_4 Index for a Two Hour Interval at Ascension Island, British Overseas Territories in the South Atlantic Ocean. The highly scintillated signal was recorded from a single satellite on 27 March 2000. Geomagnetic activity was low on this day, however the low latitude plasma irregularities were present in and near the equatorial anomaly. *(Courtesy of the AF Research Laboratory)*

navigation, occurs when S_4 values exceed 0.8. Weak scintillation is associated with S_4 values between 0.3 and 0.6. An S_4 value below 0.3 is insignificant to signal reception and processing. Figure 14-11 shows signal scintillation on three radio frequencies. Higher frequencies are subject to more and stronger scintillation.

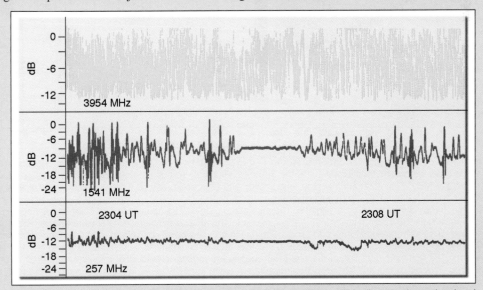

Fig. 14-11. The Effects of a Turbulent Ionosphere on Radio Waves. This diagram shows the signal variability on three frequencies in the UHF and SHF bands related to scintillation. The upper curve is scintillation on UHF SATCOM frequencies. The middle curve is scintillation on GPS-type frequencies. The bottom curve is scintillation on surveillance satellite frequencies. *(Courtesy of the AF Research Laboratory)*

Chapter 14 Effects of Space Weather and Space Environment on Signals and Systems

Focus Box 14.5: Quantifying Scintillation (Continued)

Natural and local reflectors increase the S_4 index by increasing the path length between signal source and reception. In addition to scattering off of ionospheric irregularities, *multipath signals* are produced when a GPS signal bounces off local reflectors such as buildings or uneven canyon walls. Shipyards and regions with wind farm wind mills have a notoriously high number of multipath signals. Signals from low-elevation GPS satellites tend to have higher S4 index values because the signal traverses a longer path in the ionosphere before arriving at the receiver.

For a locally static environment, multipath variations repeat with a period of ~24 hours. Operators identify multipath contamination from local sources and try to eliminate it from the S_4 index by comparing signal behavior on consecutive days. This processing allows the S_4 index to be a better measure of ionospheric scintillation.

Figure 14-12 presents a global model of enhanced S_4 values. Regionally, we see modest increases in S_4 values in the auroral zone and polar cap during geomagnetic activity. The more pronounced and persistent increases in S_4 are in low latitudes of the summer hemisphere, and in the low-to-mid-latitude in the post-sunset to pre-midnight sectors. Except during extreme geomagnetic storming, the S_4 values tend to diminish in the pre-dawn sector.

On a seasonal basis, more geomagnetic disturbances and ionospheric structuring occur in the equinoctial months of September–October and March–April than in other months. During these seasons, periods of degraded GPS performance occur in the late afternoon and especially at dusk and after sunset. These disturbances typically occur at the same time each day and last for several days or even a few weeks.

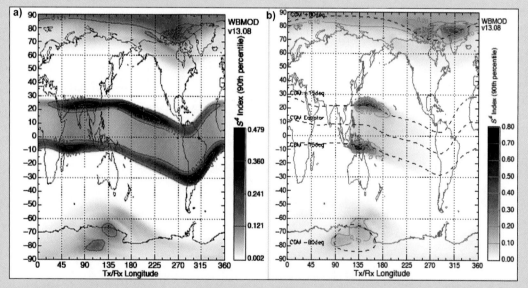

Fig. 14-12. Climatological Location of Radio Signal Scintillation. a) This map depicts the model predictions of a scintillation index at 2300 local time at the Southern Hemisphere autumnal equinox (DOY 091) for the GPS L1 (1575.42 MHz) signal, during low magnetic activity and high solar activity. Each division on the X axis (15°) corresponds to 1 hour. Two strong scintillation bands follow the ~15° geomagnetic latitude contours. Enhanced scintillation also occurs in the polar regions. The mid-latitude regions are usually free of scintillation, especially at GHz frequencies. However at lower frequencies, closer to 100 MHz, significant scintillation activity is occasionally recorded. Severe magnetic storms can produce mid-latitude scintillation. **b)** This map shows model predictions under the same conditions as a) but for 1200 UT, rather than at constant local time (the right and left borders of the plot). The vertical center line of the plot (180° longitude) corresponds to local midnight, and dusk is at 90° longitude. The equatorial scintillation is present at decreasing intensity levels throughout most of the nightside. The scintillation peak in the magnetic equatorial regions occurs between 2100 and 2200 local time (135°–150° longitude). *(Courtesy of the Australian Ionospheric Prediction Service)*

The TEC along the receiver-satellite line of sight influences how radio signals in general, and GPS signals in particular, propagate through the ionosphere. TEC variations change the GPS signal's propagation velocity, so it differs from the model value used to determine range. This velocity change, in turn, affects range measurements. When a GPS signal encounters large TEC gradients, the range measurement error grows.

Spatial and temporal gradients in TEC lead to errors in differentially augmented GPS systems such as the FAA's Wide Area Augmentation System (WAAS). In extreme cases, they cause tracking loss (Focus Box 14.7). At mid-latitudes these gradients reach 10 TECU/min during geomagnetic storms, which WAAS does not accommodate either in accurately measuring the gradients (the spatial density of reference receivers is too small), or in updating the ionospheric correction model (the data rate is inadequate). These gradients cause WAAS and GPS navigation to be unavailable for commercial air travel for up to tens of hours per event. At tropical latitudes, similar TEC gradients produced by different physical processes complicate the delivery of WAAS technology and services. At polar latitudes, the TEC density gradients associated with auroral arcs change so rapidly that the GPS signal phase is not trackable by dual-frequency GPS receivers, leading to cycle slips.

Chapter 14 Effects of Space Weather and Space Environment on Signals and Systems

Focus Box 14.6: Signal Fade

TEC spatial gradients and their associated irregularities are detected in fluctuations in the signal carrier-to-noise ratio. The loss of GPS tracking depends on the amplitude of the carrier-to-noise ratio decrease (fading), the temporal length of the fade- and the receiver-tracking algorithm parameters. The fade amplitude depends on the size of ionospheric irregularities and the TEC gradients, but the fading time scale depends on the reference frame. To first order, fading is modeled as a spatial pattern moving ionospheric irregularities with typical horizontal speeds of 50 m/s–1000 m/s. Hence, if a moving receiver, such as an aircraft, matches the translating pattern velocity, the fade lengths may be several seconds or longer. If the carrier-to-noise ratio is small within the fade, loss of tracking often occurs. The likelihood of loss increases during the period of intense signal or phase dynamics.

In an ionosphere that is permeated with irregularities of scale-size d_o, and a distance D from the transmission source, signals with wavelength λ are most susceptible to scintillation according to the following relationship:

$$\lambda \geq 2\frac{\pi^2 d_o^2}{D}$$

Using this relationship, we determine that a communication signal at 1541 MHz ($\lambda \approx 0.7$ m), originating from a GPS transmitter at ~20,000 km and propagating through highly structured ionosphere, would be most affected by irregularities that have scale sizes on the order of 500 meters.

In creating a forecast scheme for these scintillations, system designers must determine the

- Presence of these irregularities or their statistical likelihood
- Amplitude of these irregularities (are the irregular regions deeply depleted?)
- Distribution in scale size

Studies that provide probability distribution functions in scale size and depletion depth allow forecasters to determine a first-order reduction of signal-to-noise strength of a wave that passes through the disturbed medium.

Scintillation is such a widespread problem that it has prompted the US Air Force Research Laboratory and Aerospace Corporation to develop the Communications/ Navigation Outage Forecasting System (C/NOFS) to monitor and forecast global ionospheric scintillation in real time. The C/NOFS, which consists of a low-inclination LEO satellite and ground-based support, alerts users to impending satellite communication outages, GPS navigation degradations, and space-based radar tracking errors caused by equatorial ionospheric scintillation. Figure 14-13 is an artist's version of the operation of the ground-space C/NOFS that depicts a single satellite moving through a radar field of view.

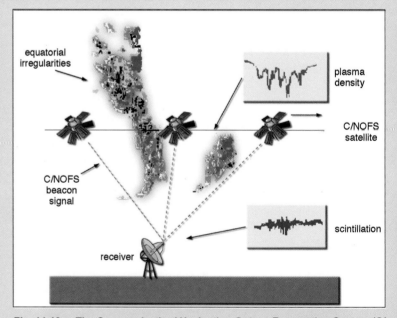

Fig. 14-13. The Communication/ Navigation Outage Forecasting System (C/NOFS) Space and Ground Segments. C/NOFS is a prototype operational system, designed to monitor and forecast ionospheric scintillation in real-time and on a global scale. It includes three critical elements: 1) a LEO space-borne sensor system with seven sensors to provide data for real-time specification, and 4-hour forecast capability; 2) a series of regional ground networks that augment the space based sensors for high-resolution coverage; and 3) a forecasting and decision aid software package that produces space environmental forecasts and outage maps. *(Adapted from AF Weather Agency)*

Focus Box 14.7: Wide Area Augmentation System for the Global Positioning System

Contributed by Patricia Doherty at Boston College

The Wide Area Augmentation System (WAAS) was designed by the Federal Aviation Administration and the Department of Transportation to become the primary means of air navigation. WAAS was commissioned for vertical guidance approach (APV) services in July 2003. APV is a service level that guides an aircraft to 250 feet above a runway in poor visibility conditions. The coverage area for the WAAS APV service is currently limited to the CONtiguous United States (CONUS). Analysis of WAAS performance, since commissioning, has shown limitations to the availability of the APV service during extreme geomagnetic storm events. (Fig. 14-14)

In the WAAS system, the standard GPS service is augmented with corrections for time, GPS satellite orbital errors, and ionospheric delays. These augmentations enable the WAAS system to meet the very stringent aviation requirements for accuracy, availability, and integrity. Incremental improvements for WAAS are planned, with modernization efforts to extend the coverage region and improve availability. These improvements will ultimately lead to a greater level of precision approach services.

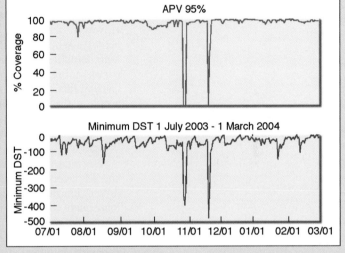

Fig. 14-14. Wide-area Augmentation System (WAAS) Vertical Guidance Approach (APV) Response to Geomagnetic Activity. These plots show how geomagnetic activity interferes with the WAAS APV availability. When a solar storm causes the Dst value to drop to –400 or below, the APV value drops to zero, so airplane pilots flying precision approaches lose that critical information for a short time. *(Courtesy of Patricia Doherty at Boston College)*

Performance reports reveal that one of the greatest challenges for WAAS is maintaining continuous APV availability during extreme geomagnetic storm events. Figure 14-14 illustrates this effect by plotting WAAS availability statistics and magnetic activity for an eight-month period surrounding the extreme storm events of October and November 2003. The top figure displays the percent of CONUS that had APV availability 95% of the time. The bottom figure shows the daily minimum Dst as a proxy for geomagnetic activity. The largest drops in Dst indicate periods of extreme storm activity. Both plots cover the period of July 1, 2003, to March 1, 2004. The figures illustrate that during non-storm days, WAAS generally maintained 95% availability over 95% of the CONUS. During the extremely disturbed days of October 29–30 and November 20, 2003, however, the APV service was unavailable over all of CONUS for approximately 15 and 10 hours respectively.

Seasonal Effects on High-frequency (HF) Systems

F Region Anomalies. When causes other than solar events change the electron density in the F2 region, we refer to the changes as "anomalies." These anomalies result from

- Seasonal and solar-cycle variation of the temperature and consequently the scale height H
- Seasonal variation of the O/O_2 and O/N_2 ratios
- Transport of electrons by diffusion, global circulation, electromagnetic forces ("fountain effect"), etc.
- Transport of electrons between geomagnetic conjugate points

Chapter 14 Effects of Space Weather and Space Environment on Signals and Systems

Seasonal Patterns. We know from Chaps. 3 and 8 that solar heating causes rising motions in the sunlit region of the atmosphere. In turn, these motions cause diurnal, seasonal, and solar cycle variations in the ionosphere. Further, Earth's orientation with respect to the Sun influences some aspects of radio signal propagation.

Equinoctial Patterns. Near the equinoxes (September and March), solar heating occurs most notably along the equator. The rising motion in the equatorial regions results in two circulation cells—one in each hemisphere. However, these two circulation patterns change according to geomagnetic activity. When geomagnetic activity is low, very little thermal input occurs at high latitudes—only a small circulation pattern, as indicated by the dashed curves in Fig. 8-5. As geomagnetic activity increases, so does the thermal input, causing this circulation pattern in the polar regions to dominate a much broader swath at the mid-latitudes as in Fig. 12-20. Figure 14-16 shows a developing storm-time pattern in which circulation from the polar regions is transporting neutrals and plasmas into the mid- and low-altitude regions. The interface region is highly turbulent and often creates regions of plasma instabilities.

When the Sun is directly overhead in the equinoctial equatorial regions, the E and F1 layers reach a maximum electron density because of the intense solar-driven ionization. Large-scale motion of the neutral atmosphere causes the ionosphere to lift. The lifted plasma in this layer follows magnetic field lines north and south to higher geomagnetic latitudes. This motion is the *fountain effect*, as we describe in Chap. 8. The circulation diagram in Fig. 14-15 shows how this fountain effect draws F2-layer plasma away from the equator and deposits it ±20° from the equator during times of average geomagnetic activity (bottom panel of equinox diagrams). If the activity is low or high, the circulation pattern disrupts this effect. For HF and VHF signals, non-great-circle propagation and ducting are possible wherever plasma gradients develop as a result of the circulation.

F2-layer Winter Anomaly. Observations reveal that the F2 layer is slightly stronger during the winter months between the 45° and 55° geomagnetic latitudes, especially during high geomagnetic activity. This anomaly is caused by the transport of E and F layer plasma from the summer hemisphere, where the Sun's emissions strengthen the ionosphere in these layers. The bottom panel of the solstice diagrams in Fig. 14-16 shows thermal circulation in the summer hemisphere from solar heating and thermal circulation in the winter hemisphere from auroral heating (auroral heating also occurs in the summer hemisphere, but it merely adds to the solar heating effect). These two circulation patterns converge over the winter hemisphere and deposit plasma in the mid-latitude F2 layer. For HF and VHF signals, non-great-circle propagation and ducting are possible during equatorial anomaly events.

14.1 Background and Solar-driven Effects

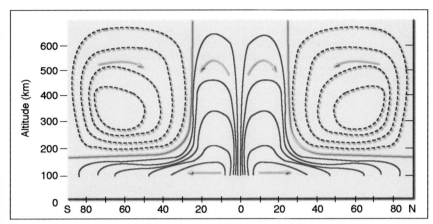

Fig. 14-15. Developing Storm-time Meridional Circulation during the Equinox: Equatorial Anomaly. At the equinoxes, heated plasma expands upward at the equator. As it rises, it also flows poleward. We don't show the return circulation at lower altitudes. The auroral zones, indicated by the dashed curves, are heated by geomagnetic effects and maintain their own circulation that expands equatorward during more active times. The equatorial uplift region is confined by the expanding polar circulation. At the interface of the two circulation systems, plasmas with different characteristics are juxtaposed. Plasma irregularities form that may scatter, scintillate, or duct radio waves. *(Adapted from Ray Roble at National Center for Atmospheric Research)*

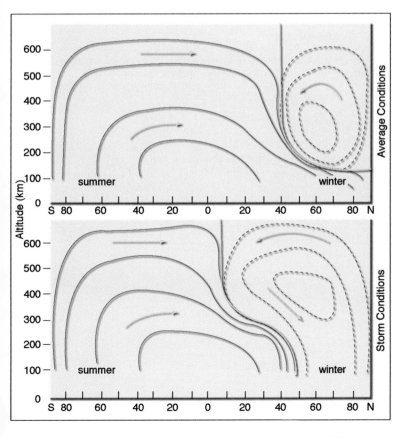

Fig. 14-16. Meridional Circulation during the Solstice: F2-layer Winter Anomaly. At the solstices, heated plasma expands upward from the sunlit hemisphere. As it rises, it flows toward the other hemisphere. We don't show the return circulation at lower altitudes. The auroral zones, indicated by the dashed curves, are heated by geomagnetic effects and maintain their own circulation, which expands equatorward during more active times. At the interface of the two circulation systems, plasmas with different characteristics are juxtaposed. Plasma irregularities form that may scatter, scintillate, or duct radio waves. *(Adapted from Ray Roble at National Center for Atmospheric Research)*

Chapter 14 Effects of Space Weather and Space Environment on Signals and Systems

Focus Box 14.8: Using Radio Scatter and Noise to Probe the Environment

In the late 19th century, scientists learned that free electrons immersed in an electromagnetic field are accelerated by the electric component of the field and reradiate a part of its energy. This process accounts for incoherent scattering of radio waves from plasma. After they knew the mass and charge of the electrons, the scientists computed the electron scattering cross section for electromagnetic waves. This value was so small (on the order of 10^{-24} m^2) that they initially thought the process was of no practical significance.

However, developing high-power VHF and UHF radars during WW II pushed technology to a point where physicists and radar engineers realized that, given a large enough radar, they could detect incoherently backscattered radar returns from the plasma in the ionosphere. From that signal they determined the electron density at altitudes above the F layer peak that normal ionosondes could not reach. Tests performed at different laboratories during the early 1960s showed that backscattered signals could be detected

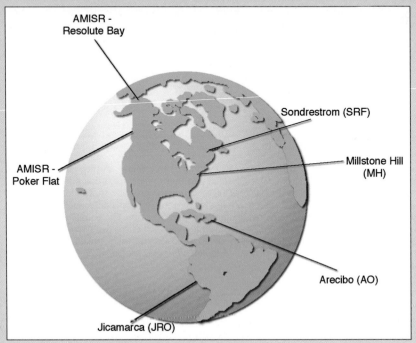

Fig. 14-17. **The North-South Chain of Incoherent Scatter Radars Operated by the National Science Foundation.** These radars operate independently to study local plasma processes or in coordination to study space weather events. *(Adapted from National Science Foundation)*

at the ground. The researchers were surprised that the initial predictions of signal strength and information content were overly restrictive, and in fact the radar echoes contained a wealth of information about the scattering plasma. This discovery led to the establishment of a number of dedicated incoherent scatter radar (ISR) facilities in many areas of the world during the 1960s and 1970s. Figure 14-17 shows the chain of ISR radars operated by the National Science Foundation to support the study of the coupled ionosphere-magnetosphere system. Much of what we describe in this section on space weather effects on communication has been determined or verified by ISR facilities.

14.2 Non-solar Effects on Signal Propagation

14.2.1 Meteor Effects

Objective: After reading this section, you should be able to...

♦ Describe how meteors affect the near-Earth space environment

As a meteor enters Earth's atmosphere, it transfers its energy and particles to the atmosphere via collisions. Part of the ablated material (usually metals) ionizes or exchanges charge with atmospheric molecules, so that much of the meteor's kinetic energy goes into a pronounced trail of dense plasma, which disperses with time. The trail may be tens of kilometers long and is usually between 80 and 120 km above Earth's surface. Depleting or enhancing ionization reflects and scatters radio energy, making meteor plasmas visible to radars. Meteors also create heavy metal layers, sporadic E layers, and global increases of TEC.

Two possibilities exist for trail formation: depleted and enhanced trails. Sub-micrometer meteor dust is distributed along the meteor path, resulting in ion-electron recombination that leaves an ionization depletion in the ionosphere. These trails have lifetimes in the 100 ms–300 ms range. The depletion produces irregular and diffuse reflections from this ionized trail, which change the paths and disrupt the intensity of radio waves. Trails with enhanced ionization form when frictional ablation creates extra ionization (Fig. 14-18). At an altitude of roughly 100 km, frictional heating is so great that electrons are stripped off atmospheric atoms, forming a temporary ion trail. The trails persist for several seconds to several minutes. Polarized electric fields in the ionosphere may extend the lifetime. This brief trail can be used for burst-mode data communications between one point and another within a radius of 2000 km, without the need for repeaters.

Meteor burst communication is one benefit of meteor interaction with the ionosphere. Other effects in the ionosphere, including heavy metal layers and sporadic E layers, are as much a nuisance as a benefit. Some meteor effects are simply detrimental. As an example, model estimates of the ionospheric behavior during an intense meteor shower similar to the Leonid event of 1966, indicate that electron density could rise by a factor of 100 for as long as a day. In such an event, radio wave propagation would be severely disrupted.

Radar reflections from meteor plasma are known as head echoes and specular (mirror-like) and non-specular trail echoes. A transmitted radio signal may be reflected from the cylindrical meteor path or from the meteor head. The hot meteor plasma rapidly thermalizes, producing a column of increased ionization in the ionospheric E region. These meteor columns are unstable to the growth of a special type of wave, a Farley-Buneman gradient-drift wave that becomes turbulent and generates large magnetic-field-aligned irregularities. These irregularities arise when a perturbed electric field interacts with the density perturbations caused by the meteors. They result in radar reflections called non-specular meteor trails. The information gained from radar interrogation of these trails provides data on speed, mass, and direction of the meteors entering Earth's atmosphere.

Chapter 14 Effects of Space Weather and Space Environment on Signals and Systems

Fig. 14-18. Head and Non-specular Echoes from a Meteor. The narrow straight line on the left of the trail is the head echo. The longer-lived broad echo to the right is the non-specular echo resulting from long-lived plasma created by the meteor. *(Courtesy of Meers Oppenheim at Boston University)*

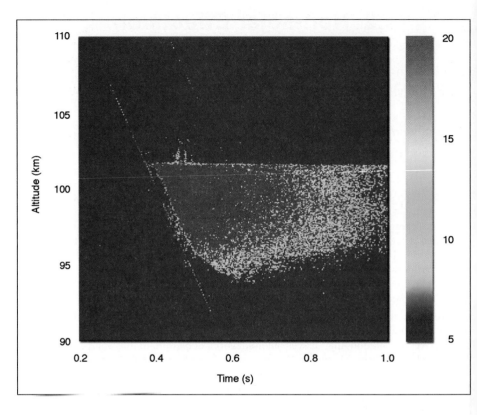

The locations and times of the trails are random, but observations reveal that meteors strike the atmosphere more during the dawn hour, and their incidence is lower during sunset. Meteor incidence also varies seasonally, because the intersections of meteor orbits with Earth's orbit are not uniformly distributed, but are concentrated so as to produce a maximum of intersections in August and a minimum in February.

14.2.2 Magnetar Effects

Objective: After reading this section, you should be able to...

♦ Describe how magnetars affect the near-Earth space environment

On March 5, 1979, gamma-ray detectors on nine spacecraft in the solar system recorded an intense radiation spike lasting 0.2 seconds followed by a 200-second emission that exhibited a clear eight-second pulsation period. Most of the pulse energy was concentrated in gamma rays. Astronomers tied the burst to a supernova remnant in the Large Magellanic Cloud, a small sister galaxy to the Milky Way. They estimated that the emission exceeded the energy emitted by the Sun in over 1000 years.

On August 27, 1998, an even more intense high-energy radiation front arrived at Earth from a star 20,000 light years away. The energy saturated detectors on seven scientific spacecraft. One interplanetary probe, NASA's Comet Rendezvous

Asteroid Flyby, was forced into a protective shutdown mode. The gamma rays arrived at Earth at nighttime in the mid-Pacific region. Fortuitously, at the time, observers from Stanford University were gathering data on the propagation of very low-frequency radio waves around Earth. They noticed an abrupt change in the ionized upper atmosphere. The inner edge of the ionosphere plunged from 85 to 60 kilometers for five minutes and pulsated with a period of 5.16 seconds (Fig. 14-19). The source of the emission was a catastrophic magnetic flare in a rapidly spinning magnetic neutron star (magnetar).

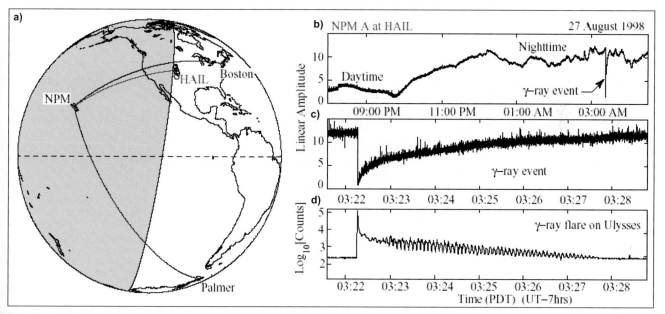

Fig. 14-19.   The Very-low Frequency (VLF) Great-circle Paths from a 21.4 kHz Transmitter in Hawaii (NPM), USA to Boston, MA, USA, Palmer, Antarctica and the HAIL VLF Observing Network in Colorado, USA. a) The part of the globe illuminated by the gamma-ray flare is indicated by shading. b) In the top curve, the amplitude of the 21.4 kHz NPM signal is observed in Trinidad, Colorado. c) In the middle curve, we show an expanded record of the gamma-ray flare event, which occurred at 3:22 a.m. PDT. d) In the lower curve, the intensity of the giant flare is observed by the Ulysses spacecraft. The gamma rays produced enhanced ionization at altitudes of 30 km to 90 km, which was observed as unusually large amplitude and phase changes of very-low frequency (VLF) signals propagating in the Earth-ionosphere waveguide. [Inan et al., 1999 and the Stanford University HAIL team]

A magnetar forms from the supernova explosion of a very large, but otherwise ordinary star. The star collapses under its own gravity into a dense ball of neutrons. The remnant outer layers solidify into a rigid, but highly unstable, crust with a surface of iron and some of the most intense magnetic fields in the universe. When the crust buckles, energy is released in explosive starquakes, then streams into space as intense flashes of gamma rays and X rays. Figure 14-20 is an artist's concept of the magnetar eruption. More than a dozen magnetar candidates have been identified. Most are within the plane of the Milky Way.

On December 27, 2004, more than a dozen spacecraft picked up another powerful burst of gamma rays from the other side of our galaxy (50,000 light years). For two-tenths of a second, it doused Earth with a higher rate of energy than any previously observed object outside the solar system. Almost all of its power was emitted in the form of gamma rays. The event completely saturated the gamma ray telescope on NASA's Swift satellite, which had been launched into orbit just five weeks earlier.

Chapter 14 Effects of Space Weather and Space Environment on Signals and Systems

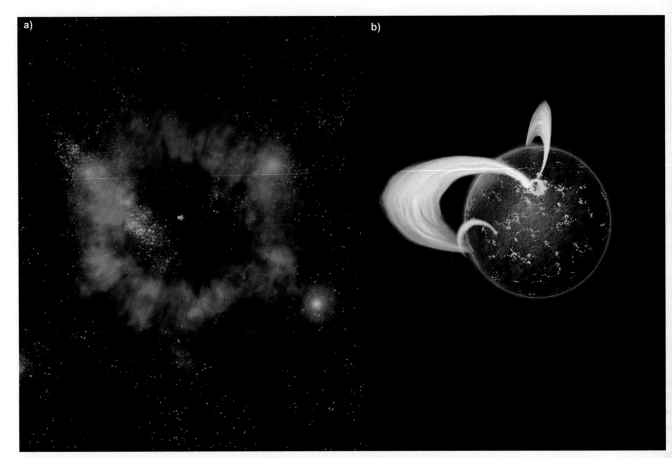

Fig. 14-20. A Magnetar Outburst. a) This artist's rendition shows the shock wave of gamma rays and X rays emanating from a magnetar's crustal explosion. **b)** This artist's drawing shows how loops of dense charged particles trace the star's intense magnetic field. The magnetic field may buckle, producing extremely large currents. Instabilities develop in the strongly magnetized plasma of a magnetar, helping to produce magnetic merging. Merging events on the Sun emit as much as 10^{25} J of energy. Flares from magnetars are about 10^{12} times stronger (~10^{37} J), befitting their more intense magnetic fields. *(Courtesy of NASA)*

The Stanford University radio instruments were again observing, but this time on the dayside, when the gamma rays arrived. Above the US, the effect of the flare was huge—more intense than the Sun in terms of producing ionization. More powerful and brighter than the nighttime flare of 1998, the daytime flare pumped 1000 times as much energy into the atmosphere. For a brief period, it ionized the atmosphere down to an altitude of 20 kilometers, just above where airplanes fly. Its most intense effects in ionizing the atmosphere, called the peak, lasted a few seconds. The second-most-intense effects, the oscillating tail, lasted five minutes. And the least intense effects, the afterglow, lasted an hour. The flare increased the electron density at an altitude of 60 kilometers by six orders of magnitude. Even in full sunlight, the ionosphere took an hour to return to normal from the disturbance. Radio signals across a wide spectrum were disrupted.

14.2.3 High-altitude Nuclear Signal Effects

Objective: After reading this section, you should be able to...

♦ Describe nuclear blast effects on the near-Earth space environment

Although effects associated with high-altitude (exo-atmospheric) nuclear blasts are not space weather events, they are impulsive events that have many of the detrimental effects of extreme space weather storms. Space professionals whose jobs involve situational awareness need to distinguish between space weather effects and high-altitude blast effects.

A nuclear detonation releases large amounts of energy and radiation in a short time and a relatively small space. The altitude of the blast largely determines how this energy and radiation couple into the surrounding environment. *High-altitude nuclear effects (HANE)* seriously degrade communication systems, damage satellites, blanket large geographical areas with damaging radiation and electromagnetic energy, and expose humans to life-threatening doses of radiation. In this section, we focus on the first of these.

High-altitude nuclear bursts occur at altitudes above 30 km. Most of the energy is in the form of the initial burst of radiation (X rays, gamma rays, and neutrons, 75%–80%), the kinetic energy of the debris (15%–20%), and the residual or delayed radiation (gamma rays ~5%). In addition, atmospheric heating creates infrared radiation that persist for minutes to hours.

The electromagnetic radiation associated with high-altitude detonations differs from space weather phenomena in three important ways: 1) the X-ray and gamma-ray signatures of space weather phenomena persist for seconds to minutes compared to nanoseconds for HANEs; 2) radiation from HANEs decreases as the square of the distance from the burst point, while radiation from flares is practically constant throughout the sunlit hemisphere of geospace; 3) radiation from flares affects the entire sunlit hemisphere, while the radiation from HANEs is limited to the line-of-site or magnetic-field-contained footprint, and is much smaller. Table 14-1 provides a comparison of strong space weather events and HANEs.

Ionization Effects in the Atmosphere. Radiation propagates outward from the HANE detonation point. The altitude to which the radiation penetrates is defined as the altitude at which the arriving ionizing radiation deposits most of its energy (usually 90%) by absorption in the atmosphere. The deposition region represents a large volume of the upper atmosphere that is ionized by the prompt X rays and gamma rays emitted by the detonation.

The ionization of the upper regions of the atmosphere seriously affects radar and communication systems. Electromagnetic radio and radar waves that pass through the deposition region experience attenuation, signal distortions, and in some cases complete absorption. Satellite communication systems that rely on phase shifting or frequency shifting techniques experience unwanted shifts in phase and frequency because of ionization-induced changes in the propagation velocities of electromagnetic waves that pass through the deposition region.

Other System Effects. Other systems that may be affected by HANE effects on electromagnetic wave propagation include sensors in the IR, visible, and UV regions, and laser communication, which may be affected by the background IR. A very hot (but transparent) region of the atmosphere can act as a lens to refract a laser communications beam off of its intended receiver. At frequencies above about 300 MHz (UHF, SHF, and EHF), signals may be disrupted by scintillation,

Chapter 14 Effects of Space Weather and Space Environment on Signals and Systems

Table 14-1. Effects of Natural and Human-made Events. Here we compare the probability/location, duration, and energy of natural and human-made events. (HANE is high-altitude nuclear effects; UV is ultraviolet; IR is infrared; EMP is electromagnetic pulse; SGEMP is system-generated EMP.) *(Adapted from National Security Space Agency)*

Photon Type	Solar Flare			Magnetic Storm			HANE		
	Probability	Duration	Energy	Location	Duration	Energy	Probability	Duration	Energy
Gamma	rare	s	0.5 MeV–100 MeV	no			always	10 ns–20 ns	0.5 MeV–10 MeV
X-ray	always	min	~MeV	intense aurora	min	several keV	always	10 μs–20 μs	<15 keV
UV	always	hr	~eV	aurora	days	eV	always	min	~eV
Visible	slight	min		aurora	days		always	min	
IR	slight	min		aurora	days		always	hr	
Radio	always	s–hr	MHz–GHz	aurora noise	days		always	mon*	MHz
EMP	sometimes		SGEMP	power grid	min		always	μs	100 V/m

* Disturbance duration in months

primarily characterized by intermittent fading and multipath transmission. These effects may persist for long periods and can degrade and distort a signal almost beyond recognition (for example, the plasma clouds are dispersive so that the speeds of all frequencies of electromagnetic radiation are not equal in the cloud). Temporal and frequency coherence can be destroyed.

Electromagnetic Pulse (EMP). Nuclear explosions are accompanied by an *electromagnetic pulse (EMP)*, a sharp pulse of electromagnetic radiation characterized by intense electric and magnetic fields with rise times on the order of a few nanoseconds, and decay times of a few tens of nanoseconds. The energy associated with EMP is contained in an electromagnetic wave with a sharp leading edge. This energy impacts and transfers to satellite components and equipment via intense currents that severely damage equipment.

Compton scattering of prompt gamma rays creates the EMP. Compton scattering refers to a high-energy photon interacting with matter, usually an electron. The electron receives part of the energy (making it recoil), and a photon with the remaining energy is emitted in a different direction from the original, so that the overall momentum of the system is conserved. As the gamma rays interact with matter, the emitted electrons gain energies on the order of 0.5 MeV. These electrons propagate away from the blast at near light speed but tend to travel along geomagnetic field lines in a spiral path until they collide with air molecules. The electron motion about the geomagnetic field lines produces a strong electric current transverse to the propagation direction of the gamma rays. This transverse current radiates an electromagnetic wave that is the EMP. The emitted radiation at various altitudes is approximately in phase, thus greatly increasing the intensity of the pulse.

Unlike the localized electromagnetic fields of lightning, the EMP is orders of magnitude higher and widely distributed. The EMP caused by a single high-altitude nuclear burst could cripple electronic systems and power grids for thousands of kilometers from the detonation point. The energetic debris entering the ionosphere produces ionization and heating of the upper atmosphere. In turn, these effects cause the geomagnetic field to heave, producing a delayed EMP, called a *magnetohydrodynamic (MHD) heave signal*.

Initially, because the plasma from a high-altitude burst is slightly conducting, the geomagnetic field cannot penetrate this volume and is displaced instead. The field displacement generates Alfvénic fluctuations in the field. Although the signal variations from this process are not large, systems connected to long lines could suffer effects similar to the geomagnetically induced current effect we describe in Chap. 13. Power lines, telephone wires, and tracking wire antennas are at risk because of the large induced current. The combined effects of the MHD-EMP would damage unprotected civilian and military systems that depend on or use long-line cables. Small, isolated systems tend to be unaffected.

Another significant effect associated with HANE is the electromagnetic pulse that develops within hardware, called the *system-generated EMP/internal EMP (SGEMP/IEMP)*. This type of pulse is generated when the ionizing radiation from a high-altitude nuclear burst, particularly gamma rays and X rays, interacts with a satellite body. The irradiated material releases electrons because of Compton scattering and the photoelectric effect, creating large electron emission currents and intense electromagnetic fields. This electromagnetic environment couples into the interior of the structure and is defined as SGEMP. In addition, very strong electric fields are induced across the spacecraft's interior. These in turn induce currents in the electrical and electronic components onboard the satellite.

Chapter 14 Effects of Space Weather and Space Environment on Signals and Systems

14.3 Surveying the Impact of Space Weather on Systems—II

Objective: After reading this section, you should be able to...

♦ Appreciate the economic and societal impacts associated with space weather effects on signals

In the previous sections, we focus primarily on how space weather influences signal propagation. But it also affects bigger systems in terms of hardware and in terms of command and control. In the description and diagrams below we address the broader issues of systems. Table 14-2 lists some communication systems affected by space weather events.

Table 14-2. Communication Systems Affected by Space Weather Events. This table lists the types of adverse effects that space weather events have on communication systems. (ULF is ultra-low frequency; LF is low frequency; HF is high frequency; VHF is very high frequency; SHF is super high frequency.) *(After Louis Lanzerotti)*

Communication Systems Effects	
ULF-LF Communications • Anomalous propagation • Increased absorption	**Satellite Systems** • Scintillation • Loss of phase lock • Radio frequency interferences (RFI)
HF Communications • Increased absorption • Depressed maximum usable frequencies (MUF) • Increased lowest usable frequencies (LUF) • Increased fading and flutter (scatter)	**Navigation Systems** • Position errors **Radar Surveillance Systems** • Radar energy scatter (auroral interference) • Location error (range, elevation, azimuth angle) • Tracking errors
VHF-SHF Communications • Scintillation	

Whether wired or wireless, beneath the ionosphere or through the ionosphere, signals react to changes in the upper atmosphere and space environments. The impacts of space weather on signal propagation have been a favorite topic of communication specialists for over a century. Below are examples of space weather impacts on communication systems during solar cycle 23.

- Antarctic science groups and staff rely on MacRelay radio operations to provide essential communications between McMurdo Station and remote sites on Antarctica. MacRelay is also responsible for communication links with aircraft and ships supporting the US Antarctic Program. The primary means of communication is HF radio. MacRelay experienced over 130 hours of HF communication blackout during the October–November 2003 solar activity.

- Following an extended solar flare-induced HF outage earlier in solar cycle 23, the McMurdo staff developed a contingency plan to use Iridium satellite phones as backup during HF outages. During the October–November 2003 activity, the LC-130 aircraft that service the remote sites used Iridium phones to communicate with McMurdo and the remote locations. And to ensure safety, take-off and landing restrictions were changed during the HF blackout periods. The 150-meter cloud ceiling with 3.2 km of visibility was increased to 900 meters with 4.8 km of visibility.

- During that same interval, a DoD maritime interdiction mission, requiring 100% effective communications, was cancelled based on scintillation forecasts issued by the USAF Space Weather Operations Center.

- For a 15-hour period on October 29 and an 11-hour interval on October 30, the ionosphere was so disturbed that the GPS WAAS vertical error limit, as defined by the FAA's Lateral Navigation Vertical Navigation (LNAV/VNAV) specification (no more that 50 m), was exceeded. That problem translated into commercial aircraft being unable to use the WAAS for precision approaches.

- During the October–November 2003 activity, companies delayed high-resolution land surveying; postponed airborne and marine survey operations; cancelled drilling operations; and some, as was the case with the C.R. Luigs deep water drill ship, resorted to backup systems to ensure continuity of operations. An international oilfield services company issued an internal "technical alert" via their worldwide network, to alert their surveying and drilling staff to potential impacts from solar storms. They reported six cases of survey instrument interference in late October from sites around the world.

- During the extreme radio flare events in December 2006, solar radio bursts overwhelmed ground-based GPS receivers, causing some systems to lose signal lock. Table 14-3 lists the general categories of communication disturbance associated with space weather and the space environment.

Chapter 14 Effects of Space Weather and Space Environment on Signals and Systems

Table 14-3. Summary of Space Weather and Environment Effects on Communications. Degradation from these events often leads to delays, cancellations, and costly work-arounds to operations that depend on electronic communications. [Lanzerotti et al., 2001]

Event	Direct Signal Effects
Ionospheric Variations • Affects civilian signals, precision navigation, geolocation signals, and command and control signals for national defense, homeland security, and local safety • Wireless signal reflection, propagation, attenuation • Communication satellite signal interference and scintillation	**Equatorial Anomaly** • Scintillation at <20° geomagnetic equator on trans-ionospheric paths • Non-great-circle propagation • Degraded HF communications on trans-equatorial paths **Geomagnetic Disturbance-High Latitude** • Auroral clutter and noise • Non-great-circle propagation through auroral oval • Satellite tracking difficulty • Ionospheric disruption leading to radar errors • Depressed MUF • Degraded communications at high latitudes • Scintillation through auroral zone and polar cap **Geomagnetic Disturbance at Middle Latitude** • Traveling ionospheric disturbances • Scintillation with plumes of storm-enhanced density
X-ray and EUV Flare • Affects civilian and defense signals and aircraft communications	**Dayside** • Ionospheric disruption leading to radar errors • Elevated LUF • Short-wave fade on sunlit paths
Micrometeoroids and Meteor Storms • Affects data transmission and communication	**Worldwide** • Enhances or inhibits reflection from meteor trails
Solar Radio Bursts • Affects cell phone communication, precision navigation and geolocation • Interferes with radar and radio receivers	**Dayside** • Excess noise in wireless communication systems • Radio frequency interference: white-out and target masking • Compromise of GPS receivers in extreme events
Energetic Particle Event • Affects civilian and defense signals and aircraft communications	**High Latitudes** • Polar cap absorption: degraded HF communications at polar latitudes
Event	Effects on Communication Hardware or Systems
Charged Particle Radiation	**Geospace /Magnetosphere** • Spacecraft solar cell damage • Semiconductor device damage and failure • Spacecraft charging of surface and interior materials • Aircraft communications and avionics
Magnetic Field Variations	**Geospace /Magnetosphere** • Attitude control of communications spacecraft
Micrometeoroids and Meteor Storms	**Geospace/Magnetosphere** • Spacecraft solar cell damage • Damage to surfaces, materials, and complete vehicles • Attitude control of communications spacecraft

Summary

The ionosphere enables and hampers radio frequency communications. Even in the relatively undisturbed ionosphere, daily and seasonal variations in ionospheric layering cause radio signals to deviate from anticipated great circle routes. These effects are aggravated during more active geomagnetic conditions. Perturbations to radio signal propagation are produced by many mechanisms, including direct solar photon emissions (solar EUV and X-ray emissions), energetic solar and CME-driven particles (directly impacting the polar region ionosphere), and magnetospheric particles precipitating into the auroral zone during geomagnetic storms. Sudden enhanced shortwave emissions from the Sun are responsible for increases in total electron content, which lead to dayside shortwave fades (SWF) on high-frequency paths. Often accompanying solar flares are intense long-wave radio bursts that produce radio frequency interference (RFI). Such bursts overwhelm the signals from Earth-based systems and saturate radio receivers. Cell tower receivers and transmitters and Global Positioning System (GPS) receivers have experienced such saturation.

Energetic particles from solar flares and, more often, from disturbances in advance of high-speed coronal mass ejections, have easy access to Earth's polar regions. These particles drive deep into the polar ionosphere, creating excess ionization that also inhibits normal radio wave refraction. Often during such events, the lowest usable frequency exceeds the maximum usable frequency, causing any radio transmission that reaches the F-region reflection level to exit the ionosphere into space.

During geomagnetic storms, lower energy particles from the magnetosphere excite auroral emissions and produce irregular excess ionization in the auroral regions. The associated plasma turbulence is responsible for radar auroral clutter (RAC). Disturbed ionosphere storm conditions also cause extra problems at all geomagnetic latitudes in navigation signals from GPS. Electron density gradients scintillate radio signals in the equatorial ionization anomaly, the auroral zones, and, on rare occasions, in storm-enhanced density plumes at middle latitudes. Signal scintillation in the near-equatorial regions is a significant problem in the dusk-to-midnight sector. These ionosphere perturbations degrade the accuracy of positional determinations, thus limiting some uses of space-based navigation techniques for applications ranging from air traffic control to naval navigation to many national security systems.

Some communication disturbances are driven by other than space weather events. Micrometeors produce ionization trails that enable short bursts of enhanced communications, and on the other hand, produce radar clutter. Rare and extraordinary gamma-ray and X-ray bursts from distant stars may cause extraordinary ionization deep into Earth's neutral atmosphere. Extreme events come from high-altitude nuclear detonations that produce regional ionizations and intense heating. Such events also produce electromagnetic pulses that have long-term effects on many forms of space- and ground-based communication systems.

Increasing reliance on and use of wireless signals create vulnerabilities to natural solar and stellar emissions and to emissions from impulsive energy releases from artificial electromagnetic sources. Some of the vulnerability relates directly to variations in the ionized medium that propagates the signal. Other vulnerabilities come from platform and communication hardware sensitivity to particles and field variations that arise in the course of space weather events.

Chapter 14   Effects of Space Weather and Space Environment on Signals and Systems

Key Words

amplitude scintillation index S_4
electromagnetic pulse (EMP)
equatorial ionization anomaly
fountain effect
frequency of optimum transmission (FOT)
great-circle route
high-altitude nuclear effects (HANE)
lowest usable frequency (LUF)
magnetohydrodynamic (MHD) heave signal
non-great-circle propagation
phase scintillation index P_{rms}
polar cap absorption (PCA) event
polar cap blackouts
radio frequency interference (RFI)
short-wave fade (SWF)
spacecraft commanding and telemetry
sporadic E
sub-auroral trough
sudden cosmic noise absorption (SCNA)
sudden enhancement of signal (SES)
sudden enhancements of atmospherics (SEAs)
sudden frequency deviation (SFD)
sudden ionospheric disturbance (SID)
sudden phase anomaly (sudden phase advance) (SPA)
system-generated EMP/internal EMP (SGEMP/IEMP)
traveling ionospheric disturbance (TID)

Reference

Lanzerotti, Louis J. 2001. Space weather effects on communications. In: Daglis, Ioannis A. (ed.) *Space Storms and Space Weather Hazards.* Dordrecht, Netherlands: Kluwer Publishing.

Figure Reference

Inan, Umran S. et al. 1999. Ionization of the Lower Ionosphere by γ-rays from a Magnetar: Detection of a Low Energy (3-10 kev) Component. *Geophysical Research Letters.* Vol. 26, No. 22. American Geophysical Union. Washington, DC.

Further Reading

Daglis, Ioannis A. (ed.) 2000. Space Storms and Space Weather Hazards (NATO Science Series II: Mathematics, Physics, and Chemistry).

Kamide, Y. and A. Chian (eds.). 2007. Handbook of the Solar-Terrestrial Environment. Springer-Verlag. Berlin Heidelberg.

Koskinen, H., E. Tanskanen, R. Pirjola, A. Pulkkinen, C. Dyer, D. Rodgers, P. Cannon J.-C. Mandeville, and D. Boscher. 2001. *Space Weather Effects Catalogue.* European Space Weather Study. FMI-RP-0001. Issue 2.2.

NOAA. 2004. Service Assessment, Intense Space Weather Storms, October 19–November 07, 2003. US Department of Commerce. National Oceanic and Atmospheric Administration. Silver Spring, MD.

Severe Space Weather Events—Understanding Societal and Economic Impacts: A Workshop Report, Committee on the Societal and Economic Impacts of Severe Space Weather Events. 2008. The National Academies Press.

NOAA Space Weather Scales

The NOAA Space Weather Scales were introduced as a way to communicate to the general public the current and future space weather conditions and their possible effects on people and systems. Many of the Space Weather Prediction Center products describe the space environment, but few have described the effects that can be experienced as the result of environmental disturbances. The scales describe the environmental disturbances for three event types: geomagnetic storms (**G** Scale), solar radiation storms (**S** Scale), and radio blackouts (**R** Scale). The scales have numbered levels, analogous to hurricanes, tornadoes, and earthquakes that convey severity. They list possible effects at each level. They also show how often such events happen, and give a measure of the intensity of the associated index or physical causes.

Appendix A   NOAA Space Weather Scales

NOAA Space Weather Scales

Category		Effect	Physical measure	Average Frequency (1 cycle = 11 years)
Scale	Descriptor	Duration of event will influence severity of effects	Kp values* determined every 3 hours	Number of storm events when Kp level was met; (number of storm days)
Geomagnetic Storms				
G 5	Extreme	**Power systems:** widespread voltage control problems and protective system problems can occur, some grid systems may experience complete collapse or blackouts. Transformers may experience damage. **Spacecraft operations:** may experience extensive surface charging, problems with orientation, uplink/downlink and tracking satellites. **Other systems:** pipeline currents can reach hundreds of amps, HF (high frequency) radio propagation may be impossible in many areas for one to two days, satellite navigation may be degraded for days, low-frequency radio navigation can be out for hours, and aurora has been seen as low as Florida and southern Texas (typically 40° geomagnetic lat.)**.	Kp=9	4 per cycle (4 days per cycle)
G 4	Severe	**Power systems:** possible widespread voltage control problems and some protective systems will mistakenly trip out key assets from the grid. **Spacecraft operations:** may experience surface charging and tracking problems, corrections may be needed for orientation problems. **Other systems:** induced pipeline currents affect preventive measures, HF radio propagation sporadic, satellite navigation degraded for hours, low-frequency radio navigation disrupted, and aurora has been seen as low as Alabama and northern California (typically 45° geomagnetic lat.)**.	Kp=8, including a 9-	100 per cycle (60 days per cycle)
G 3	Strong	**Power systems:** voltage corrections may be required, false alarms triggered on some protection devices. **Spacecraft operations:** surface charging may occur on satellite components, drag may increase on low-Earth-orbit satellites, and corrections may be needed for orientation problems. **Other systems:** intermittent satellite navigation and low-frequency radio navigation problems may occur, HF radio may be intermittent, and aurora has been seen as low as Illinois and Oregon (typically 50° geomagnetic lat.)**.	Kp=7	200 per cycle (130 days per cycle)
G 2	Moderate	**Power systems:** high-latitude power systems may experience voltage alarms, long-duration storms may cause transformer damage. **Spacecraft operations:** corrective actions to orientation may be required by ground control; possible changes in drag affect orbit predictions. **Other systems:** HF radio propagation can fade at higher latitudes, and aurora has been seen as low as New York and Idaho (typically 55° geomagnetic lat.)**.	Kp=6	600 per cycle (360 days per cycle)
G 1	Minor	**Power systems:** weak power grid fluctuations can occur. **Spacecraft operations:** minor impact on satellite operations possible. **Other systems:** migratory animals are affected at this and higher levels; aurora is commonly visible at high latitudes (northern Michigan and Maine)**.	Kp=5	1700 per cycle (900 days per cycle)

* Based on this measure, but other physical measures are also considered.
** For specific locations around the globe, use geomagnetic latitude to determine likely sightings (see www.sec.noaa.gov/Aurora)

Appendix A

Solar Radiation Storms

			Flux level of $\geq$ 10 MeV particles (ions)*	Number of events when flux level was met**
S 5	Extreme	**Biological:** unavoidable high radiation hazard to astronauts on EVA (extra-vehicular activity); passengers and crew in high-flying aircraft at high latitudes may be exposed to radiation risk.*** **Satellite operations:** satellites may be rendered useless, memory impacts can cause loss of control, may cause serious noise in image data, star-trackers may be unable to locate sources; permanent damage to solar panels possible. **Other systems:** complete blackout of HF (high frequency) communications possible through the polar regions, and position errors make navigation operations extremely difficult.	10^5	Fewer than 1 per cycle
S 4	Severe	**Biological:** unavoidable radiation hazard to astronauts on EVA; passengers and crew in high-flying aircraft at high latitudes may be exposed to radiation risk.*** **Satellite operations:** may experience memory device problems and noise on imaging systems; star-tracker problems may cause orientation problems, and solar panel efficiency can be degraded. **Other systems:** blackout of HF radio communications through the polar regions and increased navigation errors over several days are likely.	10^4	3 per cycle
S 3	Strong	**Biological:** radiation hazard avoidance recommended for astronauts on EVA; passengers and crew in high-flying aircraft at high latitudes may be exposed to radiation risk.*** **Satellite operations:** single-event upsets, noise in imaging systems, and slight reduction of efficiency in solar panel are likely. **Other systems:** degraded HF radio propagation through the polar regions and navigation position errors likely.	10^3	10 per cycle
S 2	Moderate	**Biological:** passengers and crew in high-flying aircraft at high latitudes may be exposed to elevated radiation risk.*** **Satellite operations:** infrequent single-event upsets possible. **Other systems:** effects on HF propagation through the polar regions, and navigation at polar cap locations possibly affected.	10^2	25 per cycle
S 1	Minor	**Biological:** none. **Satellite operations:** none. **Other systems:** minor impacts on HF radio in the polar regions.	10	50 per cycle

* Flux levels are 5 minute averages. Flux in particles·s⁻¹·ster⁻²·cm⁻². Based on this measure, but other physical measures are also considered.
** These events can last more than one day.
*** High energy particle measurements (>100 MeV) are a better indicator of radiation risk to passenger and crews. Pregnant women are particularly susceptible.

Appendix A NOAA Space Weather Scales

Radio Blackouts

		Description	GOES X-ray peak brightness by class and by flux*	Number of events when flux level was met; (number of storm days)
R 5	Extreme	HF Radio: Complete HF (high frequency**) radio blackout on the entire sunlit side of the Earth lasting for a number of hours. This results in no HF radio contact with mariners and en route aviators in this sector. Navigation: Low-frequency navigation signals used by maritime and general aviation systems experience outages on the sunlit side of the Earth for many hours, causing loss in positioning. Increased satellite navigation errors in positioning for several hours on the sunlit side of Earth, which may spread into the night side.	X20 (2×10^{-3})	Fewer than 1 per cycle
R 4	Severe	HF Radio: HF radio communication blackout on most of the sunlit side of Earth for one to two hours. HF radio contact lost during this time. Navigation: Outages of low-frequency navigation signals cause increased error in positioning for one to two hours. Minor disruptions of satellite navigation possible on the sunlit side of Earth.	X10 (10^{-3})	8 per cycle (8 days per cycle)
R 3	Strong	HF Radio: Wide area blackout of HF radio communication, loss of radio contact for about an hour on sunlit side of Earth. Navigation: Low-frequency navigation signals degraded for about an hour.	X1 (10^{-4})	175 per cycle (140 days per cycle)
R 2	Moderate	HF Radio: Limited blackout of HF radio communication on sunlit side, loss of radio contact for tens of minutes. Navigation: Degradation of low-frequency navigation signals for tens of minutes.	M5 (5×10^{-5})	350 per cycle (300 days per cycle)
R 1	Minor	HF Radio: Weak or minor degradation of HF radio communication on sunlit side, occasional loss of radio contact. Navigation: Low-frequency navigation signals degraded for brief intervals.	M1 (10^{-5})	2000 per cycle (950 days per cycle)

* Flux, measured in the 0.1–0.8 nm range, in $W \cdot m^{-2}$. Based on this measure, but other physical measures are also considered.
** Other frequencies may also be affected by these conditions.

URL: *www.sec.noaa.gov/NOAAScales*

March 1, 2005

Electromagnetic Waves and the Speed of Light

B

Here we show how to manipulate Maxwell's Equations to determine light speed. In a vacuum, Maxwell's equations become

$$\nabla \bullet \boldsymbol{E} = 0 \tag{B-1}$$

$$\nabla \times \boldsymbol{E} = -\partial \boldsymbol{B}/\partial t \tag{B-2}$$

$$\nabla \bullet \boldsymbol{B} = 0 \tag{B-3}$$

$$\nabla \times \boldsymbol{B} = \mu_0 \varepsilon_0 \, \partial \boldsymbol{E}/\partial t \tag{B-4}$$

The right sides simplify because we assume there is no charge density or current density in a vacuum (no particles to carry the charges or currents). To derive the electromagnetic wave equation, we take the curl of Faraday's Law, and then substitute Ampere's Law and Gauss' Law. The result is an equation for $\boldsymbol{E}$ only. The first line of Eq. (B-5) uses vector identity 14 from Appendix C. We note that the gradient of the divergence of the electric field equals zero. The gradient of the divergence produces the last term of the first line. The second line of Eq. (B-5) relates the results of the first line to the negative time rate of change of the curl of the magnetic field. The third line of Eq. (B-5) relates the result of the first two lines to the negative second time derivative of the electric field.

$$\begin{aligned}
\nabla \times (\nabla \times \mathbf{E}) &= \nabla(\nabla \bullet \mathbf{E}) - \nabla^2 \mathbf{E} \\
&= -\nabla \times \frac{\partial \mathbf{B}}{\partial t} = -\frac{\partial}{\partial t}(\nabla \times \mathbf{B}) \\
&= -\frac{\partial}{\partial t}\left(\mu_0 \varepsilon_0 \frac{\partial \mathbf{E}}{\partial t}\right)
\end{aligned} \tag{B-5}$$

Thus:

$$\nabla^2 \mathbf{E} = \mu_0 \varepsilon_0 \frac{\partial^2 \mathbf{E}}{\partial t^2} \tag{B-6}$$

Equation (B-6) has the general form of a 4-dimensional wave equation

$$\nabla^2 \mathbf{Y} = \frac{1}{v_p^2} \frac{\partial^2 \mathbf{Y}}{\partial t^2} \tag{B-7}$$

Appendix B Electromagnetic Waves and the Speed of Light

where the vector quantity that oscillates, producing a disturbance that propagates at the phase speed of the wave, v_p. The phase speed of an electromagnetic wave in a vacuum is

$$c = v_p = \frac{1}{\sqrt{\mu_0 \varepsilon_0}} \quad \text{(B-8)}$$

where

μ_0 = permeability of free space, $4\pi \times 10^{-7}$ [N/A^2] = 1.26×10^{-6} [N/A^2]
ε_0 = permittivity of free space, 8.85×10^{-12} [C^2/N m^2]

The phase speed is light speed in a vacuum expressed in terms of the proportionality constants in Maxwell's equations.

The ratio of light speed in a vacuum to light speed in a medium is the index of refraction

$$n = \frac{c}{v} \quad \text{dimensionless} \quad \text{(B-9)}$$

where

c = light speed in a vacuum [m/s]
v = light speed in a medium [m/s]

Variations in the index of refraction lead to important space environment and space weather effects in the ionosphere.

Vector Identities

The following vector identities are from the 2009 Naval Research Plasma Formulary.(See the reference at the end of Chap. 6.) Notation: f, g, are scalars; $\mathbf{A}$, $\mathbf{B}$, etc. are vectors.

(1) $\mathbf{A} \cdot \mathbf{B} \times \mathbf{C} = \mathbf{A} \times \mathbf{B} \cdot \mathbf{C} = \mathbf{B} \cdot \mathbf{C} \times \mathbf{A} = \mathbf{B} \times \mathbf{C} \cdot \mathbf{A} = \mathbf{C} \cdot \mathbf{A} \times \mathbf{B} = \mathbf{C} \times \mathbf{A} \cdot \mathbf{B}$

(2) $\mathbf{A} \times (\mathbf{B} \times \mathbf{C}) = (\mathbf{C} \times \mathbf{B}) \times \mathbf{A} = (\mathbf{A} \cdot \mathbf{C})\mathbf{B} - (\mathbf{A} \cdot \mathbf{B})\mathbf{C}$

(3) $\mathbf{A} \times (\mathbf{B} \times \mathbf{C}) + \mathbf{B} \times (\mathbf{C} \times \mathbf{A}) + \mathbf{C} \times (\mathbf{A} \times \mathbf{B}) = 0$

(4) $(\mathbf{A} \times \mathbf{B}) \cdot (\mathbf{C} \times \mathbf{D}) = (\mathbf{A} \cdot \mathbf{C})(\mathbf{B} \cdot \mathbf{D}) - (\mathbf{A} \cdot \mathbf{D})(\mathbf{B} \cdot \mathbf{C})$

(5) $(\mathbf{A} \times \mathbf{B}) \cdot (\mathbf{C} \times \mathbf{D}) = (\mathbf{A} \times \mathbf{B} \cdot \mathbf{D})\mathbf{C} - (\mathbf{A} \times \mathbf{B} \cdot \mathbf{C})\mathbf{D}$

(6) $\nabla(fg) = \nabla(gf) = f\nabla g + g\nabla f$

(7) $\nabla \cdot (f\mathbf{A}) = f\nabla \cdot \mathbf{A} + \mathbf{A} \cdot \nabla f$

(8) $\nabla \times (f\mathbf{A}) = f\nabla \times \mathbf{A} + \nabla f \times \mathbf{A}$

(9) $\nabla \cdot (\mathbf{A} \times \mathbf{B}) = \mathbf{B} \cdot \nabla \times \mathbf{A} - \mathbf{A} \cdot \nabla \times \mathbf{B}$

(10) $\nabla \times (\mathbf{A} \times \mathbf{B}) = \mathbf{A}(\nabla \cdot \mathbf{B}) - \mathbf{B}(\nabla \cdot \mathbf{A}) + (\mathbf{B} \cdot \nabla)\mathbf{A} - (\mathbf{A} \cdot \nabla)\mathbf{B}$

(11) $\mathbf{A} \times (\nabla \times \mathbf{B}) = (\nabla \mathbf{B}) \cdot \mathbf{A} - (\mathbf{A} \cdot \nabla)\mathbf{B}$

(12) $\nabla(\mathbf{A} \cdot \mathbf{B}) = \mathbf{A} \times (\nabla \times \mathbf{B}) + \mathbf{B} \times (\nabla \times \mathbf{A}) + (\mathbf{A} \cdot \nabla)\mathbf{B} + (\mathbf{B} \cdot \nabla)\mathbf{A}$

(13) $\nabla^2 f = \nabla \cdot \nabla f$

(14) $\nabla^2 \mathbf{A} = \nabla(\nabla \cdot \mathbf{A}) - \nabla \times \nabla \times \mathbf{A}$

(15) $\nabla \times \nabla f = 0$

(16) $\nabla \cdot \nabla \times \mathbf{A} = 0$

Appendix C Vector Identities

Physical Constants and Heliophysical and Geophysical Values

Constant	Symbol	Value
Speed of light	c	2.998×10^8 m/s
Elementary charge	e	1.6×10^{-19} C
Electron mass	m_e	9.11×10^{-31} kg
Proton mass	m_p	1.67×10^{-27} kg
Gravitational constant	G	6.67×10^{-11} N·m²/kg²
Permittivity constant	ε_0	8.85×10^{-12} C²/(N·m²)
Permeability constant	μ_0	$4\pi \times 10^{-7}$ N/A²
Boltzmann's constant	k	1.38×10^{-23} J/K
Stefan-Boltzmann constant	σ	5.67×10^{-8} W/(m²·K⁴)
Planck's constant	h	6.63×10^{-34} J·s
Universal gas constant	R	8.31 J/(K·mole)
Avogadro's number	N_A	6.02×10^{23} mole⁻¹

Appendix D Physical Constants and Heliophysical and Geophysical Values

Heliophysical and Geophysical Values	Symbol	Value
Solar radius	R_S	6.96×10^8 m
Solar mass	M_S	1.99×10^{30} kg
Solar luminosity	L_S	3.83×10^{26} W
Sun-Earth distance	AU	1.5×10^{11} m
Mean solar synodic period	Ω_S	27.3 days
Earth radius	R_E	6.378×10^6 m
Earth mass	M_E	5.97×10^{24} kg
Earth's magnetic moment	M	8.0×10^{22} A·m^2
Earth's orbital period	Yr	3.16×10^7 s
Earth's equatorial magnetic field	B_0	3.1×10^{-5} T

Index

Numerics

22-year magnetic cycle 438, 442

A

ablation 625
    asteroids 625
    comets 625
absorption cross section, σ 132
absorption, energy 131–135, 137
ACE *See* Advanced Composition Explorer (ACE)
active debris removal 632
    post-mission disposal 632
        atmospheric drag 632
        controlled entry 632
        flight path angle 632
        natural orbital decay 632
active longitudes 407
active regions (ARs) 10, 29, 97, 105, 109, 121, 123, 157, 172, 201, 400–402, 405, 406–410, 412, 414, 416, 420–422, 425, 426, 429, 430, 431, 435, 442, 476, 521, 606, 641
adiabatic expansion 201
adiabatic invariant 511
    longitudinal invariant J 511
    magnetic flux invariant Φ 511
    magnetic moment μ 511
Advanced Composition Explorer (ACE) 200, 215, 238, 448
Advanced Satellite for Cosmology and Astrophysics (ASCA)
    atmospheric drag 624
    attitude control system 624
    safe mode 624
    solar panels 624
    torque 624
Aeronomy of Ice in the Mesosphere (AIM) 349
Air Force Research Laboratory 608
airglow 559
airglow emission 352, 367
Akasofu epsilon parameter ε 500
Alfvén Mach number 469
Alfvén Poynting flux 556

Alfvén speed 213–214
Alfvén waves 213, 214, 284–286, 389, 500, 556
    super-Alfvénic 213
Alfvén, Hannes 213
Alpha Centauri 15
alternating current (AC) 639
Ampere's Law 222, 276, 282, 314, 427
Ampere's Law for Magnetic Fields with Maxwell's addition 162
amplitude (A) 70
ANIK E-1 635
anomalous cosmic rays (ACRs) 436, 597
    argon 597
    neon 597
    nitrogen 597
anti-dynamo 297
Ap index 497
aperiodic stormtime substorms 507
Apollo 200, 606
Aqua 647
arcade loop 428
arcades 122, 419, 431
arch system 405, 546
arcing 614
arcs 546
artificial space debris 626, 629
    fragments 629
    non-functional spacecraft 629
    rocket bodies 629
    spatial density 629
astronauts 601, 606
    Apollo Program 603
    solar proton event 603
astronomical unit (AU) 11
Athena 1 610
atmosphere *See also* Earth, atmosphere 3, 6, 17, 19, 211
    drag 21, 32, 624
    isothermal 211
    mixing 19
    temperature 21
    terrestrial 2
    upper 12, 17–21, 23, 27, 33, 38
atmosphere, layers
    exosphere 7, 21
    ionosphere 7, 21, 33, 39

mesosphere	6
stratosphere	6, 23
thermosphere	6, 21
troposphere	6, 23
atmospheric constituent densities	350
atmospheric density	622
solar flux F10.7	620
atmospheric effects	626
ionization trails	626
lithium layers	626
noctilucent clouds	626
sodium layer	626
sporadic E layers (E_s)	626
atmospheric neutral density	631
atomic oxygen	352, 357, 372, 374, 618
erode surface materials	618
oxidize surface materials	618
sensors	618
solar panels	618
surface materials	618
atoms, ionized	11
carbon	11
helium	11
hydrogen	11
iron	11
oxygen	11, 20, 38
attachment coefficient β	366
attachment rate (L_{attach})	366
attitude change maneuver	637
conservation of momentum	637
aurora australis	548
aurora borealis	200
auroral	335
arcs	546
conductance	529
electrojet (AE) index	528
electrojets	188, 504, 505, 525, 528, 531, 638
emissions	511
kilometric radiation (AKR)	544
lights	335
ovals, auroral zones	189, 335, 372, 504, 508, 523–525, 544–548, 550
substorms	335
zones	372, 391, 527, 528, 531, 620
heating	648
auroras	3, 26, 28, 32, 33, 39, 176, 247, 342, 350, 357, 372, 502, 641
broadband	545, 549
colors	374, 546, 547, 550
diffuse	542–544, 547, 549
discrete	542–545
energy deposition	549, 552–554
monoenergetic	545, 549
pulsating	547
Ayer's Rock	13

B

ballerina skirt	456
Balmer (H-α) radiation	82, 84
band spectra	81
barometric equation	130, 208
Bartels, Julius	35
Beer's Law	132
beta (β) radiation	592
Big Bear Solar Observatory (BBSO)	29, 117, 466
biological dose	
cancer mortality	595
chest X ray	595
damage	439
quality factor (QF)	595
relative biological effectiveness (RBE) factor	595
sievert (Sv)	595
Biot-Savart Law	498
bipolar	
group	411
magnetic regions	123, 405, 412, 429
pairs	406
sunspots	442
Birkeland (field-aligned) currents (FACs)	522
Birkeland currents *See also* field-aligned currents	187
Birkeland, Kristian	200
bitflips	605
black auroras	545
blackbody radiation	75–78, 82, 84
blackbody radiators	75
blackout	641
blood-forming organs (BFO)	601, 606
blue jets	358
bolides	625
Boltzmann's constant	129
bounce motion	316, 319–323, 511
bounce period	319, 511
bow shock	4, 16, 228, 231, 234, 238
collisionless environment	238
foreshock	229

laminar zone	229
magnetic barrier	229
turbulent zone	229
bowshock	16
braking radiation	416
break-up phase	546
bremsstrahlung	87, 550

C

Carrington	
longitude (L)	103
rotation	223, 409
rotation number	103
Julian date	103
storm	640
white-light flare	417
Carrington, Richard	25–28, 103
Center for Integrated Space Weather Modeling	279
Centers for Disease Control (CDC)	604
central meridian (CM)	103
Cerenkov plasma waves	421, 443
Cerise	
Ariane rocket body	632
Spot 1	632
CHAMP satellite	562
CHANDRA	647
Chapman mechanism	138
Chapman, Sydney	138, 208
barometric equation	208
charge conservation	173
charge density	616
charge magnitude	616
charge sheet	153
charged particle density	517
charged particles	4, 22, 36–39, 104, 115, 200, 205, 209, 230, 310, 311, 316, 317, 321, 325, 327, 335, 342, 416, 421, 436, 438, 440, 493, 495, 504, 508, 511, 512, 523, 530, 593, 618
acceleration	252, 255–259
alpha particle	594
gyro-motions	205
proton	594
charged plasma particles	513
chemical energy	296
chromosphere	110, 206, 401, 404, 405, 407, 412, 416, 417, 419–421, 425, 432
chromospheric ribbons	431
circulations	108
closed field lines	308
closed magnetic flux	122
closed system	45
closed-field convection	334
CLUSTER spacecraft	232, 544, 635, 647
collisionless shock	230
comet	200
Comet Hale-Bopp	200
debris cloud	628
perihelion	628
tails	238
Tempel-Tuttle	628
COMET Program	388, 391, 563
Comet Rendezvous Asteroid Flyby	678
communication	
high-frequency	30, 33
long-range	22
loss	34
short-wave	22
communication systems effects	684–686
Communications/Navigation Outage Forecasting System (C/NOFS)	672
composition bulge, composition disturbance	565, 566, 582
Comprehensive Suprathermal and Energetic Particle (COSTEP) Analyzer	484
compression regions	204, 238
compression-rarefaction zone	204
Compton Gamma Ray Observatory	359
conductance Σ	177, 179
conduction	65
conduction currents	156
conductivity σ	177–179, 189, 216, 305
conservation laws	231
conservation of energy	208
conservation of magnetic flux	231
conservation of mass	208
conservation principle	99
equation	99
external forces	100
force density	100
gravity	100
pressure gradient force	100
sinks	100
sources	100

convection   65, 104, 105, 313, 333, 335, 341, 402, 407, 442, 503, 509, 511, 514, 518, 527, 529, 530, 551
    cells   107
    zone
        turbulence   112
convection dynamo   522
convection electric field E_{conv} 339, 342, 503, 504, 513, 514, 517, 523, 552, 555, 556
convection scale   105
convective zone   104–107, 112, 121, 400–408
convective dissipation   504
convective drift   510
convective instability   104
    buoyancy   104
coordinate systems   299
    spherical coordinates   299
Coordinated Universal Time (UTC)   26
core turbulence   298
Coriolis forces   298, 412
corona   12, 114, 204–206, 208–210, 214, 216, 224, 235, 401, 404, 405, 408, 415–422, 425, 427, 430, 432, 434, 442
    plasma beta   181
    radio emissions and   476
    temperature   209
coronagraph   426
coronal flux tubes   435
coronal heating   428
coronal holes   123, 211, 214, 225, 401, 446, 450–452, 492
    polar   225
coronal loop   402
coronal magnetic field   428, 435
coronal mass   528
coronal mass ejections (CMEs)   12, 13, 30, 33, 114, 127, 201, 202, 207, 212, 215, 221, 279, 313, 317, 330, 401, 406, 414–416, 425, 426, 428, 431, 434, 435, 442, 457–459, 461–470, 479, 517, 521, 569, 610, 633, 635, 636
    Forbush decrease   472
    solar cycle dependence   465
    solar energetic particles   473–475, 479, 482, 485
coronal plasma oscillations   418
coronal streamer   211
coronal sweep frequency radio emissions   434
co-rotating electric fields   514
co-rotating interaction flows   330
co-rotating interaction regions (CIRs)   204, 401, 436, 451, 453–456, 470, 492, 596
co-rotation electric field   340, 578
cosmic ray counts   438
cosmic ray energy flux   440
cosmic ray flux   438, 608
cosmic ray muon telescope   610
cosmic ray shower   438
cosmic ray trace   439
cosmic rays   235, 305, 310, 340–341, 438–442, 459, 472, 516, 591, 593, 604
    detectors   610
        Embudo, New Mexico   610
    rigidity   610
Coulomb collisions   206, 207, 515
counterstreaming electrons   458, 462, 464
critical distance   209
critical frequency   285, 383–385
critical pitch angle α 320
cross-polar-cap field   331
cross-tail current   505, 527
cross-tail electric field E 527, 529
cryogenics   609
curl ($\nabla \times$)   159
current continuity   173–175
current density J 174–176, 179, 186, 190, 328, 525, 615
current loops   316
current sheet *See also* heliospheric current sheet (HCS)   122, 152–154, 220, 223, 325, 329, 333, 427, 429, 431, 435, 451, 507
current systems   305, 314
current wedge   526
currents   114, 307, 493
curvature B drift   265, 266, 523, 527, 530
cusp currents   387, 522, 524, 527
cusps   306, 312, 429, 434, 516
cycloidal drift   264
cyclotron frequency   258, 285, 316
cyclotron radiation   88, 421
cyclotron radius   316
cyclotrons   591

D

D layer *See* ionosphere, D layer
dayside magnetopause   337
dayside regions   314

debris	649
human-made	625
meteoric	625
Debye length (λ_D)	248–250, 253
Debye sheath	614
Debye shielding	249–251, 253
Debye sphere	251–253
declination D	301
deep dielectric charging	608
Air Force Research Laboratory	608
anomalies	609
charge	608
CRRES mission	609
Geostationary Operational Environmental Satellite (GOES)	609
geosynchronous orbit (GEO)	608
high-energy electron flux	609
potential distribution	608
current flow	608
electrostatic discharge	608
grounding	608
shielding	608
radiation belts *See* Van Allen radiation belts	
Defense Meteorological Satellite Program (DMSP)	542, 544, 545, 549
F-13 spacecraft	579
desorption	619
diamagnetic current	513
diamagnetism	257
dielectric charging	635
differential rotation	10, 95, 96, 103, 400, 402, 442
diffuse auroras	372, 542
diffusion	349, 511, 619
diffusion region	435
diffusive magnetic flux	278
diffusive shock acceleration	481
dipolar field	318, 341
dipolar shape	503
dipolarization	333, 505
dipole	493, 510, 512, 527, 592
dipole configurations	152
dipole coronal field	427
dipole field	294, 318, 333, 400, 414, 523
dipole tilt angle	304
direct currents (DC)	639
direct heating	609
discrete auroras	542
disorientation	478
dispersion	80
displacement	594
displacement current	159, 161
dissociation	366
dissociative recombination	573
distances in space	4
disturbance field	513
disturbance storm time (Dst) index	494, 497, 511, 520, 521, 530, 621
divergence ($\nabla \cdot$)	159, 173
DNA	
non-lethal change	601
Doppler	108
radar	120
shift	108
Doppler effect	72
dose	594, 602
Tissue Equivalent Proportional Counter (TEPC)	602
dose equivalent	595
downwelling	574
drainage plumes	514, 519
drift	513
collisions with	263
curvature B	265
cycloidal	264, 267
$E \times B$	261–263
gradient	264
gravity	260, 262
guiding center	260–262, 267
other	261
drift motion	322, 511
Dst index *See* disturbance storm time (Dst) index	
ducting of radio signals	656
Dungey, James	328
Dunn Solar Observatory	117
dynamic pressure	237, 324
balance	237
Dynamics Explorer 1	547, 548
Dynamics Explorer 2	554
Dynamics Explorer satellite	357
dynamo electric field	331
dynamo process	551
dynamos	37, 182, 192, 194, 281, 407, 442, 551
ionospheric	178, 179

E

$E \times B$ drift	261, 262

$\boldsymbol{E} \times \boldsymbol{B}$ drift current	524
E layer *See* ionosphere, E layer	
Earth	133–136
atmosphere	348
absorption	137
atomic oxygen	137, 138
emission	137
ionized	358–374
mesosphere	136
neutral	348–357
opaque	133
ozone	135–138
photodissociation	133, 135
photoionization	136
production rate	137
recombination rate	138
temperature	351
thermosphere	136
tides	352–354
transparent	133
troposphere	136
wind flow pattern	353
thermal envelope	21
Earth's Moon	202
Earth-Sun line	304
ECHO 7	259
ecliptic plane	200, 204, 219, 304, 435, 492
elastic scattering	593
electric (Coulomb) force	36–38, 155
electric currents	154, 173, 294, 296, 305, 526, 639
polarization	189
electric dipole	153
electric field separatrix	514
electric fields	36–38, 66, 185–189, 217, 327–330, 332–334, 336–342, 422, 503, 505, 507, 508, 510, 514, 521, 523, 524, 526, 529, 530, 633, 639, 642
compression	521
convection	507
convection dawn-dusk	529
interplanetary	193
ionospheric	393
polarization	389
reference frames, dependence on	191
relaxation	521
upper atmosphere	192
electric merging field	331
electric potential difference ΔV	332
electric potential energy	206, 207
electric potential Φ	529
electric power transmission systems	
geomagnetically induced current (GIC) damage	640
grid voltages	640
low line resistances	640
magnetic perturbation	640
electrical conductance (Σ)	177, 179
electrical conductivity (σ)	177–179, 189, 216, 305
electrojets	372, 527, 638
equatorial	576
electromagnetic (EM) energy	3, 400
electromagnetic (EM) energy density u_{EB}	551
electromagnetic (EM) energy flux	527
electromagnetic (EM) energy transfer rate	553
electromagnetic (EM) fields	441, 593
electromagnetic (EM) power	211
electromagnetic (EM) radiation	69–71, 75–80, 86, 135
amplitude	70
Doppler effect	72
frequency	70
wavelength	70
electromagnetic (EM) spectrum	12, 30, 70, 71, 416
extreme ultraviolet (EUV)	416
radio waves	416
visible wavelengths	416
X rays	416
electromagnetic (EM) waves	22, 164, 169, 284, 311, 342, 528
electromagnetic pulse (EMP)	682
electron backscattering	593
electron belts	518
electron clouds	593
electron density	22, 369, 373–385, 424, 477, 530, 541, 560, 654
gradients	655
electron density profiles (EDPs)	360, 362, 385
electron energy	311
electron mobility	615
electron temperature	310
electron volts (eV)	4
definition	56
electrons	37, 38, 339, 341, 606, 615, 617, 634
electro-optical sensors	
cosmic rays	610
high-energy protons	610
lose attitude lock	610
communications antennas	610

solar cell panels	610
electrostatic bursts	422
electrostatic discharges	
attitude control system	635
uncommanded status changes	635
electrostatic microwave bursts	423
elf	358
emission lines	81
energetic charged particles (ECPs)	363, 592, 599
energetic electron enhancement	634
energetic electrons	340, 494, 517, 519, 520, 607, 617, 635
flux	517
energetic ions	521
energetic neutral hydrogen atom	513
energetic particle events	442, 459, 607, 649
energetic particle motion	510
energetic particles	247, 414, 433, 435, 439, 494, 505, 511, 512, 516, 520, 530, 594, 596, 603, 604, 614, 644, 648
anomalous cosmic ray (ACR)	596
deep dielectric charging	648
discharge	614
electrical charging	614
energetic storm particle (ESP)	596
galactic cosmic rays (GCRs)	596
materials degradation	648
single-event upset (SEU)	648
solar energetic particle (SEP)	596
surface charging	614, 648
total dose effects	648
energetic protons	516, 520, 607, 613
energetic solar particles	521
energetic storm particles (ESPs)	598
events	475
energy	44
categories	51
conservation	51, 53, 61
conversions	57
ionization	58
kinetic	50, 62
mechanical	50
nuclear	57–59
physical concept	44–49
potential	50
solar flare	48
system, in a	44–49
values	47
versus power	49
energy density	67–69, 169–172, 271–273, 341, 437, 494, 504, 558
magnetic	170, 172, 181, 184
energy flux	203, 308, 537
geospace	536
intensity	73–77, 79
non-solar	536
energy release	496
energy state	81
energy storage	496
Energy Storage and Thermal Science Program	172
energy transfer	65–69
conduction	65
convection	65
radiation	65
energy, thermonuclear	4
enhanced networks	111
entropy S	61, 230
ephemeral regions	122, 405
epsilon parameter	502
equatorial electrojet	188, 495
equatorial ionization anomaly	373, 574, 667, 675, 686
equatorial ionization crests	668
equatorial ionosphere	373
equatorial pitch angle	320, 322
equatorial plasma bubbles (EPBs)	
See also plasma bubbles	575, 577, 578
equatorial spread F (ESF)	575–577
error detection and correction (EDAC)	608
escape velocity	21, 53
ESKOM system	
transformer failures	642
European Space Agency (ESA)	126, 610
European Union (EU)	604
evanescent waves	381
exobase	21
exosphere	21, 357
exospheric temperature	350
Explorer 1	309
explosive energy release	333
extra vehicular activity (EVA)	32
extraordinary wave (x-wave)	384
extreme ultraviolet (EUV)	337, 416, 441, 620
extreme ultraviolet imaging telescope (EIT)	465

F

F layer *See* ionosphere, F layer	
F region anomalies	673
equinoctial patterns	674
seasonal patterns	674
F1 layer	370
F2 layer	371
F2-layer winter anomaly	674, 675
Fabricius, Johannes	25
faculae	109, 405, 407, 412, 442
Faraday's Law	162, 233, 277, 633, 639
Farley-Buneman gradient-drift wave	677
fault mitigation	605
Federal Aviation Administration (FAA)	
Civil Aeromedical Institute (CAMI)	604
flight doses	606
ionizing radiation	606
occupational radiation limits	604
FedSat	647
field intensity maps	294
field polarity	400
field-aligned beam	460, 483
field-aligned currents (FACs)	185–189, 330, 386–390, 505, 522–525, 527, 531, 554–556
field-line convection	636
fields	150
electric	115, 150, 152, 155, 164, 166, 190, 194
gravitational	150–152
magnetic	95, 97, 101, 104–115, 150, 152, 155–157, 164, 190
static	151
superposition	151
filaments	52, 111, 406, 420, 425, 426, 427, 430, 442
disintegrating	112
erupting	406
prominences	111
quiescent	406, 420
first adiabatic invariant	512
First Law of Thermodynamics	47, 61
flare crochets	27, 539
flare ribbons	414, 432
flares	412–422, 424, 428–430, 434, 435, 521, 598, 606, 612, 613, 641, 647, 657, 661, 662, 682, 685
bremsstrahlung	417
thermal	417
Carrington	419
compressional heating	417
decay stage	419
soft X rays	419
electron collisions	419
extreme X ray	419
gamma ray	417
H-α	419
hard X rays	417, 419, 421
high energy	419
impulsive stage	419
large-scale	429
optical	421
precursor stage	419
radar interference	424
radio bursts	417
scintillation	424
shock heating	417
shock wave	420
short-wave radiation	417
soft X rays	417
thermal plasma	417
white-light	419, 421
flow continuity	202
fluid dynamics	268, 274–283
flux Φ	157
flux rope	425, 429, 431, 458, 460, 463–465
flux transfer events	329
flux tubes	401, 434
following spot	400
Forbush decrease	459, 471, 472
forced convection	65
fountain effect	674
Fraunhofer lines	82
free convection	65
F-region critical frequency (foF2)	574
F-region electron density (NmF2)	574
frequency (f)	70
frequency of optimum transmission (FOT)	660
frozen-in magnetic field	216, 338
frozen-in magnetic flux	180, 216, 218, 231, 278, 448, 499
frozen-in plasma	329

G

galactic cosmic rays (GCRs)	9, 23, 433, 436, 596, 612, 649
hydrogen	597
iron	597

oxygen	597
GALEX	647
Galilei, Galileo	25
gamma rays	98, 417–419, 437, 441, 521, 594
garden hose angle ψ	218–220
Parker angle	220
garden hose spiral	218
Gauss's Law for Electric Fields	162, 276
Gauss's Law for Magnetic Fields	162, 233
generalized Lorentz force	36, 40
Geocentric Solar Ecliptic Coordinates (GSE)	304
Geocentric Solar Magnetospheric Coordinates (GSM)	304
geocorona	308, 356
geodynamo	296
chemical energy release	296
convection kinetic energy	296
electromagnetic energy	296
internal radioactivity	296
iron-filled liquid outer core	296
magnetic induction	296
natural electric generator	296
rocky mantle	296
solid iron inner core	296
geoeffectiveness	462–464, 467
geoelectric field	638
Geographic Coordinate System (GCS)	301
geomagnetic activity	503
geomagnetic crochet	27
geomagnetic cutoff	591
geomagnetic cutoff rigidities	592
function of	592
geomagnetic activity	592
latitude	592
longitude	592
magnetic local time	592
particle arrival direction	592
geomagnetic disturbances	511
geomagnetic field	22, 33, 35, 294, 306, 336, 341, 492–495, 522, 592, 602, 636, 637, 641
geomagnetic field lines	517, 524
geomagnetic latitude angle Λ	303
geomagnetic poles	297
geomagnetic shielding	591
geomagnetic storms	238, 294, 323, 425, 492–495, 497, 502, 507, 511, 523, 527, 557, 558, 568, 602, 614, 617, 620, 622, 624, 636, 641, 646, 648
automatic traffic control systems	643
impulsive energy deposition	620
geomagnetic superstorm	642
geomagnetically induced currents (GICs)	638, 643, 648
East-Siberian Railway	643
electric power grids	638
extended pipelines	638
pipe-to-soil voltage	643
power system failure	648
railway lines	638
telecommunication cables	638
transformer heating	648
geospace	219, 235, 606
geosphere	501
geostationary magnetopause crossing (GMC)	636
Geostationary Operational Environmental Satellite (GOES)	475, 610, 617, 647, 663
energetic protons	610
flare	610
GOES-11 satellite	477
GOES-7 satellite	483, 612, 637
high-energy protons	610
single-event upsets	610
geostationary orbit	320
geosynchronous altitude	341
geosynchronous orbit (GEO)	303, 308, 502, 504, 517, 521, 636, 646
geosynchronous satellites	17, 120, 312, 315, 617
Hale's Law	123
geosynchronous transfer orbit (GTO)	646
giant convection cells	107–108
Glatzmaier-Roberts Geodynamo Model	298
global magnetic disturbances	493
global magnetic perturbations	493
Global Oscillation Network Group (GONG)	95, 126
Global Positioning System (GPS)	310, 368, 638, 646, 668–672
Global Scale Wave Model	559
globally merged interaction regions (GMIRs)	471
Goddard Space Flight Center (GSFC)	
Space Science Mission Operations Team	30
GPS signal scintillations	516
gradient drift	264, 321, 510, 523, 527
gradient-curvature drift	265
gradual events	422, 434
granules, granulation	105, 121
acoustic (sound) waves	96, 107, 114
low-frequency	112
shock	112

cells	112
magnetic field	107
gravitational drift	262
gravitational field	66, 210, 626
gravitational force	155
gravitational potential energy	209
gravity waves	558, 561
great storms	641
great-circle route	654
ground level enhancements (GLEs)	476, 478
ground magnetic disturbance	504
ground magnetic observatory	641
ground-level event	433, 482, 606
deep-space mission	606
extra-vehicular activity	606
proton event	606
radiation sickness	606
ground-level solar cosmic ray event	610
group speed v_{gr}	284, 378
growth phase	546
guiding center	261
guiding-center drift	267
gyrational kinetic energy	212
gyrofrequency ω	39, 231, 257, 316
gyromotion	316–318, 323, 511
gyroradius r	39, 231, 257, 261, 316, 318, 326
gyroperiod	318, 511

H

H-α emission	416
H-α flare	434
Hale, George	121, 123
Hale's Law	123
Hall current J_H	388, 389, 524, 525, 531, 555
Hall electric field E_{Hall}	576
Halloween storm	661, 662, 664, 684, 685
halo CMEs	429, 431, 464
halo electrons	460
Haystack radar	
Lincoln Laboratory	631
orbital debris data	631
heat	45–47, 428
heavy ions	201
helical trajectories	298
heliographic coordinate system	103
heliopause	14, 15, 17, 200, 234–237, 436
shock front	597

heliopause stagnation point	237
Helios	200
helioseismology	401
acoustic waves	96, 97
granular motion	107
reflection	96
refraction	96
resonance	97
HINODE	126
heliosheath	15, 40, 235, 597
heliosphere	6–8, 15, 19, 200, 209, 216, 220, 227, 234, 238, 334, 401, 414, 419, 427, 431, 436–438, 442, 471, 591, 596, 597
inner	16
heliospheric current sheet (HCS) See also	
current sheet	220, 222, 238, 447, 450, 451, 596
heliotail	236
helium	201, 339, 341, 436, 437
helmet streamers	122
hemispheric power	548, 549
heterosphere	20
HF blackout	660
HF radio propagation window	376
high-altitude nuclear effects (HANE)	681–683
high-energy electrons	310, 314, 593
high-energy neutrons	592
high-energy particles	434, 436, 438, 510, 593, 595, 598, 600–601, 605, 608
biological damage	600
blood cells	600
cell chemistry	600
DNA molecules	600
ionization	600
high-energy protons	415
high-frequency (HF) systems	666, 673
high-inclination orbits	591
highly elliptical orbit (HEO)	315, 646
high-speed CMEs	442
high-speed flow	223, 225, 401, 492, 635
high-speed streams (HSS)	12, 15, 204, 238, 430, 451–453, 507, 517
Hinode mission	126, 201
Hodgson, Richard	26
homosphere	20
hop transmission	376
horizontal magnetic intensity vector H	301
Hubble Telescope	15
hybrid events	434

Hyder flares	420
Hyder, C. L.	420
Hydro Québec power system	24, 640, 641, 645
hydrogen	204, 231, 437, 513
hydrogen Lyman alpha (L-α) line	431
hydrogen-Balmer-α (H-α) line	111
hydrostatic atmosphere (Earth)	210
hydrostatic equilibrium	99, 100, 102, 128, 208
hypervelocity encounters	625
hypervelocity impacts	631

I

Ideal Gas Law	68, 106, 128, 130
idealized dipole field	300
idealized magnetic field	301
IMAGE satellite	337, 513, 515, 516
Imager for Magnetosphere-to-Aurora Global Exploration (IMAGE)	559, 580
impact ionization	246
impulsive microwave bursts	422
impulsive short-duration events	434
impulsive stellar burst magnetar	9
impulsive X-ray flares	428
inclination I	301
incoherent scatter radar (ISR)	676
induction	163
induction electric fields	168, 482
inductive electric fields	504, 511, 512, 521, 599
inelastic collisions	593
infrared (IR) sensors	609
initial phase	496
Inmarsat	647
inner corona	216
inner radiation belt	311, 591
instabilities	22
Institute of Space and Astronautical Science Japan Aerospace Exploration Agency	126 126
intensity (energy flux), I	73, 618
interaction regions	235, 437
interface regions	492
interference	424
dayside satellite communication	424
radio receiver	424
internal EMP (IEMP)	683
internal energy	53
International Geophysical Year	546

International Maritime Satellite Organization (Inmarsat)	610
International Space Station (ISS)	356, 548
Tissue Equivalent Proportional Counter (TEPC)	603
International Telecommunications Satellite Organization (Intelsat)	610
interplanetary current sheet	221
heliospheric current sheet	221
interplanetary disturbances	237, 479
interplanetary electric field (IEF)	193, 503, 513
interplanetary flux ropes	430
interplanetary magnetic field (IMF)	11, 165, 183, 194, 200, 215, 218, 219, 226, 236, 304, 309, 330–331, 341, 414, 447, 454, 458, 492, 499, 522, 530, 596, 598
dipole	218
ecliptic	218
spiral	218
interplanetary magnetic flux tubes	434
interplanetary medium	200, 234, 238, 430, 434
Interplanetary Monitoring Platform 1 (IMP-1)	223
Interplanetary Monitoring Platform-8 satellite	483
interplanetary scintillations	461
interplanetary shocks	480–485, 492, 598, 634–636
classifications	480
shock normal vector	480
interstellar medium	234, 237, 238, 436
ions	235
pressure	237
interstellar neutral particles	236, 436, 437
interstellar space	9, 15
intranetwork field	121
ion acoustic waves	286
ion mobility	615
ion-atom interchange	573
ionization	21, 27, 109, 113, 360–363, 366, 371, 424, 438, 548, 593
radiation	33
trails	22
ionization crests	373, 574
ionization energy	56, 58, 362, 363
ionization loss	365
ionization potential	270, 271
ionization rates $R(z)$	363, 538
off-zenith	364
ionizing radiation	591–592, 594, 601
ionograms	384, 385
ionosonde	385

ionosphere	21, 294, 305, 308, 311, 327, 329, 330, 332, 334, 335, 336, 340, 342, 360–362, 367, 414, 424, 433, 441, 442, 494, 504–508, 513, 516, 522, 523, 525–527, 529, 531, 569
chemical processes	573
D layer	361, 370, 375, 478, 538, 656–658, 661, 664
E layer	361, 370, 385, 538, 576
F layer	361, 370, 376, 574, 578, 579
F1 layer	370, 385
F2 layer	369, 371, 373, 383–385
ionospheric current effects	638
plasma	254, 360
polar	371
radio wave propagation and	375–385
topside	371
ionosphere-magnetosphere coupling	305
ionosphere-plasmasphere boundary	308
ionospheric convection	332
ionospheric currents	187–189, 527
ionospheric disturbances	
high latitude	577–579
low latitude	574–577
middle latitude	579
ionospheric ducting	376
ionospheric dynamo	178, 179
ionospheric layers	361–362, 370
ionospheric mass outflows	392
ionospheric storms	572
negative	580–582
positive	580–582
ions	4, 604
gyration	38, 39
Iridium	630
isolated non-stormtime substorms	504, 507
isolated system	45
isothermal, unaccelerated motion	208

J

Japanese Institute of Space and Astronautical Science (ISAS)	624
Joule heating	391, 505, 527, 551, 553, 557, 567
Joule power	390
Joy, Alfred H.	124
Joy's Sunspot Tilt Law	124
jump conditions	231
Jupiter	220, 223, 225

K

Kelvin scale	4
Kepler, Johannes	25
Kew Observatory	26–28
killer electrons	633–634
high-speed solar wind enhancements	634
high-speed streams	634
plasma instabilities	634
plasmapause	634
kinetic energy (KE)	50, 62, 206, 207, 212, 230, 318, 329, 333, 336, 400, 419, 427, 431, 435, 437, 492, 493, 504, 530, 593, 609, 622, 631
plasma	329
thermal energy	206, 207
kinetic energy flux $v(\rho v^2)$	499–501
kinetic power	211
kinetic speed	212
kinking instability	427
Knudsen Number K_n	279, 281
Kodiak, Alaska	610
Kp index	497

L

L1 frequency	368
L2 frequency	368
Laboratory for Atmospheric and Space Physics, University of Colorado	349
Lagrange points	215
Lagrange, Joseph	215
Landau length (L)	207
Langmuir waves	285–286, 435
Large Magellanic Cloud	9
Larmor radius	257
Lateral Navigation Vertical Navigation (LNAV/VNAV)	685
latitude θ	103
latitudinal shear	402
Law of Atmospheres	129, 131, 138, 269, 563
law of mass action	138
L-B coordinate system	302, 303
leader flux	412
leading spot	400
Learmonth Solar Observatory	424
Lenz's Law	167

Leonid meteor shower	
cometary debris	627
Tempel-Tuttle	627
lifetime	81
light speed c	164, 424
limb CMEs	430
limb darkening	109
line spectra	81, 83
absorption lines	82, 84, 85
emission lines	85
linear energy transfer (LET)	593
lines	314
long-duration events	416
long-duration flares	428
longitudinal drift	316
longitudinal drift period	322
longitudinal invariant	512
long-period waves	511
loop prominence system (LPS)	419
Loran-C	659
Lorentz force	38, 151, 182, 191, 216, 231, 260, 262, 327, 389, 402, 437, 524, 592
loss cone	320, 519
loss mechanisms	518
low radio frequency disturbance	424
low-density slot region	517
low-Earth orbit (LEO)	315, 646
lower energy particles	513, 517, 520
lowest usable frequency (LUF)	376, 377, 569, 657, 659
low-frequency electromagnetic fields	305
luminescence	618
LUNIK	200
Lyman-α radiation	81–82, 357, 370, 538

M

mach number, Alfvén	201
mach number, sound	201
MacRelay radio operations	684
Magellan spacecraft	610
magnetars	9, 441, 442, 678–680
magnetic arches	419
magnetic bottles	320, 510
magnetic butterfly diagram	124
magnetic carpet	122, 401
magnetic clouds	430, 458, 464–466, 479
magnetic collisions	212
magnetic connection	597
magnetic crochet	30
magnetic cusp	330
magnetic cycle	401
magnetic dipole	154, 158, 265, 304
magnetic disturbances	497, 499, 505, 640
magnetic elements	405
magnetic energy	333, 427, 428, 429, 431, 442, 493, 499, 504
magnetic energy density	230, 401, 404, 412, 427
magnetic equatorial plane	303
magnetic field	36–39, 66, 101, 104, 107, 109, 113, 121, 123, 151, 201, 204, 209, 212–214, 216–218, 220, 222, 223, 226, 228–231, 235, 238, 294, 301, 302, 305, 309–312, 314, 315, 317–321, 323, 325, 326, 328–331, 333–335, 337–340, 342, 400, 404, 405, 407, 408, 411–412, 414, 417, 419, 425, 426, 429–431, 434–439, 441, 442, 493, 495–498, 502, 504, 505, 509, 511, 512, 519, 521, 524, 526, 530, 591, 592, 598–600, 602, 604, 633, 636, 639, 642, 648
antipodal asymmetry	295
coronal	235, 417
cycle	412
depression	323
dipole	16, 38, 121, 123, 216, 302
Earth	3, 37
energy density	216
flux tubes	113, 298
frozen field lines	104
horizontal field component	302
inward-directed	226
lines	217, 218, 220, 300, 303, 311, 316, 427, 430, 432, 514, 519
breakout	427
dipole	303
non-uniform	316
loops	406
magnetic dipole	299
magnetic surveys	638
magnetized fluid waves	114
magnetometer readings	638
north geomagnetic pole	295
orientation	223
outward-directed	226
paleomagnetic	296
plages	111
polarizations	507
poloidal	101

reconfiguration	419
shells	302
south geomagnetic pole	295
toroidal	101
toroidal shell	302
trough	420
magnetic flux	157, 158, 168, 336, 342, 401, 408, 415, 428, 442, 504
frozen-in	180
magnetic flux emergence	400
magnetic flux invariant	511
magnetic flux rope See also flux rope	425, 465
magnetic flux tubes	400–402, 403, 407, 447–449
magnetic force	36, 39, 155
magnetic force vectors	299
magnetic induction equation	277
magnetic irregularities	437
Alfvén waves	437
magnetic loops	202, 401, 416
magnetic merging	37, 52, 114, 182, 182–184, 194, 235, 311, 327–329, 331, 333, 336, 340–342, 401, 408, 414, 416, 418, 419, 426–428, 432, 442, 493, 499, 513, 522, 636
magnetic mirroring	318
magnetic moment μ	317, 318, 511, 518
conservation	318
magnetic network	201
magnetic neutral line	419
magnetic neutron stars	441
magnetic observatories	35
magnetic perturbations	219, 495
magnetic polarity	411
magnetic pressure	69, 104, 237, 324, 333, 401
magnetic reconfiguration	414
magnetic reconnection	401, 429, 505
magnetic Reynolds Number R_M	279–281
magnetic ropes	400
magnetic sectors	224
magnetic sensors	636
magnetic star	400
magnetic storms See also upper-atmospheric storms, geomagnetic storms	311, 336, 429, 442, 496, 500, 503, 504, 511, 523, 529, 542, 565, 637, 642
magnetic tension	333
magnetic turbulence	238
magnetic wave speed	212
magnetized planets	227
magneto-convection	104, 105
magneto-coupling	314
magnetogram	409
magnetohydrodynamic (MHD)	114, 274–283, 463
equations	275, 278
waves	114
magnetohydrodynamic (MHD) heave signal	683
magnetometer stations	323
magnetometers	35, 496
magnetonose	305
magnetopause	17, 37, 39, 190, 229, 305, 311, 313, 324–332, 341, 493, 496, 500, 503, 511, 513, 514, 519, 521, 522, 524, 527, 530, 633, 637
boundary	636
compression	636
current	185, 522
current system	305, 306
magnetoplasma	331
magnetosheath	16, 229, 231, 233, 306, 312, 325, 530, 636
magnetosonic waves	287
magnetosphere	8, 16–19, 23, 37, 39, 185, 203, 211, 212, 221, 223, 227–229, 235, 238, 294, 304, 305, 307, 308, 309–313, 316, 317, 324, 327–336, 339, 341, 415, 470, 492–496, 498–504, 505, 507, 509, 511–514, 517, 519–522, 525, 529, 530, 592, 617, 633, 638, 649
bullet shape	18
current systems	527
curvature-gradient drift and	266
energy sources	542
inner	17, 306, 308
ionosphere and	386–390
magnetosonic wave	635
nightside	21
outer	16
quasi-convective state	505
tear-drop shape	17
magnetosphere convection field	331
magnetosphere-ionosphere coupling	386–391, 529, 531
magnetosphere-ionosphere system	332
magnetospheric acceleration	310
magnetospheric convection	342
magnetospheric convection events	492
magnetospheric current systems	294
magnetospheric substorms	333

magnetotail	16, 17, 37, 185, 186, 305–307, 311–314, 329, 330, 332–335, 342, 522, 633, 636
magnetotail gradient	523
magnetotail lobes	17
main phase	496
mantle	298
MARECS-1 satellite	610
Mariner II	200
Mars Odyssey	
safe mode	610
mass conservation	202
mass continuity equation	564
mass ejections, magnetized	12
mass flux	626
mass motion	428, 431
mass spectrometer	258
Mass Spectrometer and Incoherent Scatter Model	563
mass-energy relationship	
proton rest mass	593
matter	4, 36
phases	4
states	5
Mauna Loa	226
K-coronameter	226
Maunder, Edward	25
maximum usable frequency (MUF)	376, 384, 574, 659, 666
Maxwell's Equations	34, 159–164
Ampere's Law	160–165, 167, 174
differential form	160–162
Faraday's Law	162–168
Gauss's Law for electric charge	160–163, 166, 174, 192
Gauss's Law for magnetic fields	162
integral form	160, 162
Maxwell-Boltzmann distribution	268
Maxwell-Boltzmann function	53
McIlwain, Carl	302
McMurdo Station	684
mean free path	205, 207, 230
Meanook Magnetic Observatory	642
mechanical energy	50–53, 622
medium-Earth orbits (MEOs)	315, 646
medium-energy particles	509, 513
merged interaction regions (MIRs)	235, 333, 455, 468, 470
merging	335
merging "X-line"	183
merging region	333
meridional	108
meridional circulations	108, 400, 401, 442, 675
meridional flows	401
mesosphere	136, 145, 348, 606
meteor fluxes	
cosmic dust	625
showers	625
sporadic meteors	625
meteor trail	22
meteorite impacts	649
meteorites	625
meteoroids	22, 625
plasma	626
meteorologists	3
meteors	22, 625, 677, 686
communications effects on	677
head echo	678
impacts	626
electrostatic discharge	626
ionization channel	626
short circuits	626
non-specular echo	678
showers	626
micrometeoroids	626
micropores	106
microwave bursts	422
microwaves	421
middle magnetosphere	305
mid-latitude trough	373
mid-magnetotail current system	294
Milky Way Galaxy	436
mirror points	319, 599
mirroring	543
molecular contamination	619
payload sensors	619
solar arrays	619
thermal control surfaces	619
momentum balance	324
momentum conservation	208, 333
momentum exchange	390, 528
Moon	19, 229
Moreton wave	420
Mount Wilson Observatory	410, 411
multipath signals	670
multi-polar regions	429

N

Nançay Radio Observatory	117
National Aeronautics and Space Administration (NASA)	126
National Council of Radiation Protection and Measurements	
As Low As Reasonably Achievable (ALARA)	601
limits	601
occupational dose	601
National Geophysical Data Center	11
National Institute for Occupational Safety and Health (NIOSH)	604
National Oceanic and Atmospheric Administration (NOAA)	30, 32, 612
National Optical Astronomy Observatory	15
National Solar Observatory	214, 409
natural space debris	626
asteroids	627
comets	627
dust	627
meteoroids	627
solar cycle	627
navigation	32, 33
near-space environment	6–9, 17, 19, 22, 38, 201
negative ionospheric storm	580
neon	436
ionized	202
network	121
active	121
quiet	121
neutral current sheet	314
neutral density ratios	563–565, 580
neutral outflow	562
neutral winds	548, 559, 567
neutrinos	60
neutron flux	608
neutron stars	9, 441
neutrons	438, 594
induce radioactivity	591
Newton, Sir Isaac	200
Newton's Second Law	36
nightside region	314
nitric oxide (NO_x)	567, 619
nitrogen	308, 357, 372, 374, 436, 606
NOAA scale	497
NOAA solar radiation alerts	607
NOAA Space Weather Scales	422
Nobeyama Radioheliograph (NoRH)	118
noctilucent ("night-shining") clouds	349
noise storms	422, 423
non-conservative forces	61
non-great-circle propagation	654–656, 686
non-homogeneous plasma	513
non-hydrostatic atmosphere	210
non-ionized gases	206
non-thermal radio radiation	422
Cerenkov plasma waves	422
cyclotron radiation	422
north geographic pole	294
northern lights (Aurora Borealis)	336–337, 515
northward interplanetary magnetic field (IMF)	330
northward terrestrial field line	334
nuclear explosions	600
artificial radiation belt	600
energetic electrons	600
radiation belt remediation	600
radiation effects	600
solar-cell damage	600
STARFISH	600
nuclear power plant	641
nuclear radiation	591
nuclear reaction	
shielding	591
nucleus energy state	593

O

oblique waves	284
occulting disks	
coronagraph	114
Ohm's Law	177–179
Ohmic dissipation	400
oil exploration	
geomagnetic measurements	638
Olympus	626
open magnetic flux	122
open system	45
open-ended field lines	305
open-field convection	334
optical depth τ	131–133, 137, 364–366
optical-fiber cables	640
orbit propagation algorithms	631
orbital debris	624, 630
annulus of debris	630

atmospheric density	630
11-year solar cycle	630
atmospheric heating	630
geomagnetic storming	630
atmospheric drag	630
fragmentation	630
geosynchronous altitudes	630
low-Earth altitudes	630
lunar gravity	630
oblate shape	630
orbital decay	630
solar cycle	630
solar gravity	630
toroidal cloud	630
orbital inclination	591
orbital lifetime	621
orbital period	622
ordinary wave (o-wave)	384, 385
outer belt	311
outer corona	216
outgassing	619
overshielding	529, 531
oxygen	308, 339, 341, 436, 508, 513, 516
oxygen ions (O+)	516
ozone	23, 606

P

paleomagnetic records	297
parallel waves	284
parcel	210, 227
Parker (garden hose) spiral	218, 448, 453, 455, 456, 474, 479, 492
Parker spiral angle	221, 229
Parker, Eugene	208, 218
modern solar wind theory	208
Parker's spiral model	434
Parker-spiral field	434
partial halo CMEs	430
partial ring currents	505, 523, 526, 527, 531
particle acceleration processes	
diffusive	481
electric field induction	482
stochastic	481
particle deposition	620
particle drifts	523
particle energy flux	390
particle gradient	527
particle heating	513
particle precipitation	493
particle trapping	509
particle-induced radiation events	649
Pederson current (J_P)	387, 390, 524, 531, 579
penetration electric fields	577, 582
Penticton Observatory	116
penumbra	406
perpendicular waves	284
perturbation Poynting flux S_p	552, 555, 556
phase speed	378
phase velocity v_{ph}	70, 284
Phobos	202
phonons	66
photochemical processes	66
photoelectric effect	614–616
photoelectric transfer	66
photoionization	66, 246, 363, 573
photons	4, 70, 436, 441, 442, 596, 615, 616, 618, 644, 648
photosphere	109, 201, 219, 400, 404, 405–407, 412, 419, 427, 429, 432, 435
photothermal energy	66
pick-up ions	436, 598
pipelines	
adjustment distance	642
cathodic protection systems	642
corrosion	642
current	642
discontinuities	642
electrons	642
geomagnetically induced currents (GICs)	642
inhomogeneities	642
insulated	642
magnetic disturbances	642
pipe-to-soil voltage differentials	642
pitch angle α	317, 322, 335, 518
equatorial	319, 322
plages	111, 122, 405, 412, 442
Planck's Radiation Law	75, 76, 78
planetary index, Kp	32, 35
plasma	4, 5, 21, 36, 114–116, 123, 200, 203–206, 209, 212, 214, 216, 228–233, 235, 238, 244–288, 308, 311–314, 324, 330, 332, 333, 334, 335, 336, 337, 339–342, 401, 402, 412, 414, 416, 425, 430, 433–435, 493, 496, 503, 505, 509, 512, 514, 529, 614, 617
characteristics	244, 248

compressed equilibrium state	232
conducting magnetized	238
corona	238
definition	244, 253
downstream state	233
energy density	216, 271–273, 509
frozen-in magnetic flux	180
high-pressure	507
instabilities	205
magnetic energy	216
magnetized	94, 182, 232, 238
natural and artificial	245
nightside oval	336
parcels	104, 105, 217
plumes	21, 530, 572, 574
pressure	504
quasi-ideal magnetized	232
sources	246
state	247
subsonic	232
supersonic	217, 231, 238
temperature	235
temperature and density ranges	245
terrestrial	334
thermal energy	216
two-cell convection pattern	336
uncompressed equilibrium state	231
upstream state	233
wave dissipation	230
waves	86
plasma beta parameter (β)	181, 279–281, 459, 464, 469
plasma bombardment	615
plasma bubbles See also equatorial	
plasma bubbles	572, 574–577
plasma clouds	436, 442
plasma convection	308, 332, 400
plasma density	337, 415, 419
plasma drift	331
plasma energy flux	499
plasma fountain	373
plasma frequency (ω_p)	248, 252, 253, 375, 379, 383, 475–477
plasma induced charging	615
plasma mantle	18, 312, 313
plasma models	
kinetic	248
single-particle	246, 248, 255
thermal (fluid)	246, 248, 268–271
plasma plumes	572, 574, 577, 579
plasma trough	515
plasma turbulence	515
plasma waves	284–288
plasma-β	404
plasmapause	18, 308, 314, 337, 339, 509, 511, 514, 530
plasmasheet	17, 306, 308, 311–314, 335, 495, 499, 504, 512, 516, 517, 522, 529, 599
boundary layer	312
plasmasphere	8, 18, 306–311, 314, 336–340, 357, 508, 511, 514–516, 519
plasmaspheric erosion	337
plasmoid ejection	503
plasmoids	416, 431, 504
POES NOAA-14 satellite	571
POES NOAA-15 satellite	571
Poker Flats Research Range	259
polar cap	168, 188, 390
polar cap absorption (PCA)	
events	33, 478, 569–571, 663, 686
polar cap blackouts	478, 664
polar crown filament	466, 467
polar cusps	311, 335
polar field reversal	412
polar hole	371
polar mesospheric clouds (PMCs)	349
Polar Orbiting Environmental Satellite (POES)	613
polar reversals	297
POLAR spacecraft	518, 556
polar wind	371
polarity reversal	297
polarization electric fields	189, 529
pole-to-pole (poloidal) field	400
polodial magnetic fields	101
pores	107, 405
positive ionospheric storm	580
post-plasmoid flow	503
potential energy (PE)	50, 333, 492
power transmission systems	640
Poynting flux deposition	620
Poynting flux vector S	169–171, 386, 499–501, 527, 530, 551–556
Poynting's Theorem	552–555
pre-reversal enhancement (PRE)	575
pressure	68
pressure gradient	210
pressure perturbations	227
production rate R	137

prominences	111, 406, 429, 430
propagation window	659
proton (ion) energy	311
proton belt	519
proton events	606, 612, 614
proton storms	606
proton temperature	310
proton-proton (PP) fusion	58, 59
protons	37, 39, 339, 341, 604, 606, 615
pulsating auroras	547

Q

quality factor (QF)	595
quasi-dipolar structure	492, 509
quasi-parallel shock	229
quasi-perpendicular or tangential	229
quasi-static magnetic fields	305
quasi-stationary solar wind	446
quiet regions	401

R

R-1 current	527
R-2 current	523, 525, 529
rad equivalent mammal/man	595
radar	22, 26
radar auroral clutter (RAC)	665
radial diffusion	511, 518
radiance	73, 132
radiation	65, 111, 428, 431, 591, 601, 604, 605, 612, 613, 618
absorbance	618
UV irradiation	618
aircraft	604
astronauts	601
chemical bonds	618
damage	601, 619
dose	601, 607, 612
electromagnetic *See* electromagnetic radiation	
extreme ultraviolet	94, 99, 113, 115, 126, 416
gamma ray	98
hazards	604
polar region	604
radio waves	618
types	
by-products from nuclear reactions	592
electrons	592
energetic heavier atoms	592
energetic helium atoms	592
energetic protons	592
γ-ray photons	592
high-energy photons	592
neutral particles	592
positrons	592
very high-energy photons	592
X-ray photons	592
ultraviolet	111
X rays	94, 98, 113, 115, 122, 618
radiation absorbed dose	594
radiation belts	17, 306, 309, 314, 315, 317, 339, 341, 433, 438, 494, 498, 502, 508, 511, 516, 519–521, 530, 599, 633, 635, 648
electrons	599
inner	599
mirror points	599
outer	599
protons	599
slot region	599
radiation damage	619
radiation dose	607, 612
radiation energy	427
radiation sickness	
blood-forming organs	600
cancer	600
genetic effects	600
radiation storms	484, 606, 647
radiation-hardened	521
radiative diffusion	99
radiative power	211
radii of curvature	321
radio	94, 115
blackout	13, 22, 33
burst	33, 35
communication	3
F10.7	116
propagation	33
waves	21, 22, 33
radio bursts	414, 422, 423, 424, 434
electrostatic bursts	422
electrostatic Langmuir waves	422
microwave bursts	422
chromosphere	422

cyclotron radiation	422
sweep frequency bursts	422
magnetic reconfiguration	422
radio waves	422
radio emissions	521
radio frequency interference (RFI)	424, 662, 663, 686
radio maps	435
radio telescopes	461
radio waves	21, 22, 33, 285, 375–378, 382, 418, 421, 618
blackbody radiation	421
bremsstrahlung	421
coronal mass ejections	475–477
electron density	421
high-frequency (HF)	375, 377
plasma frequency waves	421
propagation window	376–382
type II burst	476, 478, 479
radioactive materials	594
radioactivity	
alpha radiation	594
becquerel (Bq)	594
biological effects	594
curie (Ci)	594
gray (Gy)	594
ionizing radiation	594
railways	
network of conductors	642
range	593
Rankine-Hugoniot Equations	231, 233
rayleighs	543
Rayleigh-Taylor instability	575, 577, 578
reaction wheels	637
recombination	619
recombination coefficient a	365
recombination rate ($L_{recombination}$)	365, 371
reconfiguration	427, 434
reconnection	311, 333–336, 340, 502–504, 509, 527
reconnection event	599
recovery phase	496
recurrent storms	492
red sprites	358
refractive index (n)	378–384
Region-1 currents	187, 387, 522, 523, 531, 552, 556, 577
Region-2 currents	187, 387, 522, 527, 531, 556, 577
regional power grids	641
relative ionospheric opacity meters (RIOMETERS)	570, 661
relativistic electron belts	518
relativistic electron fluxes	
electrons	633
electrostatic discharge	633
geosynchronous orbit	633
interplanetary magnetic field (IMF)	633
magnetosphere	633
radiation belts	633
solar wind	633
solar wind shocks	633
relativistic electrons	517, 635
resistive heating	527
rest mass energy	57, 593
reversed-polarity patches	298
rigidity, geomagnetic	592, 607
ring current ions	311
ring currents	187, 314, 321, 323, 341, 342, 494, 496–498, 502, 505, 507–509, 511–515, 518, 522, 526, 527, 530
root-mean-square speed (v_{rms})	54
Royal Astronomical Society	26
Royce, Frederick	28
RXTE satellite	647

S

S_4 index	669
Sabine, Edward	25
Saha equation	268–270
SAMPEX	520
satellite communication (SATCOM)	376
satellite drag	390, 622, 623, 648
atmospheric density	620
drag coefficient	620
low-Earth orbit (LEO)	620
mechanical energy	620
momentum transfer	620
orbital decay	620
satellite lifetimes	620
satellite tracking	622
satellites	600
Advanced Composition Explorer (ACE)	7, 12, 16, 448, 634, 635, 647
Advanced Satellite for Cosmology and Astrophysics (ASCA)	624
Aeronomy of Ice in the Mesosphere (AIM)	349
ANIK E-1	635
Apollo 16	606

Apollo 17	606	MARECS-1	610
Aqua	647	Mars Explorer Rover	647
Atmospheric Explorer (AE)	618	Mars Express	647
Cassini	30	Mars Odyssey	610
Cerise	632	Microwave Anisotropy Probe	647
CHAMP	562	NOAA-10	610
CHIPS	647	Olympus	626
Cluster	544, 635, 647	POES NOAA-14	571

Apollo 17 606
Aqua 647
Atmospheric Explorer (AE) 618
Cassini 30
Cerise 632
CHAMP 562
CHIPS 647
Cluster 544, 635, 647
Combined Release and Radiation Effects
    Satellite (CRRES) 608
Cosmos 2251 630
Defense Meteorological Satellite
    Program (DMSP) 542, 544, 545, 549, 579
DRA-delta 609
Dynamics Explorer 357
Dynamics Explorer 1 547, 548
Dynamics Explorer 2 554
Earth Observing 631
Earth Resources Satellite (ERS-1) 610
Explorer 1 17, 309
FedSat 647
Fengyun-1C 630
GALEX 647
Genesis 647
Geostationary Operational Environmental
    Satellite (GOES) 7, 31, 33, 35, 126, 475,
    609, 610, 612, 617, 637, 647
    GOES-11 477
    GOES-7 483
    Solar X-Ray Imager (SXI) 126
Global Positioning System (GPS) 7, 22, 310, 368, 638
Hubble Space Telescope 626, 632
IMAGE 337, 514, 514–516
Imager for Magnetosphere-to-Aurora
    Global Exploration (IMAGE) 559, 580
Integral 647
International Space Station (ISS) 2, 30, 548, 600,
    602, 603, 614, 631, 646
International Telecommunications Satellite
    (Intelsat) 635
Interplanetary Monitoring Platform-8 483
Iridium 630
Japanese telecommunications satellite CS-3b 610
Kodama 647
KOMPSAT-1 621
LandSat 647
Long Duration Exposure Facility (LDEF) 624
Magellan 610

MARECS-1 610
Mars Explorer Rover 647
Mars Express 647
Mars Odyssey 610
Microwave Anisotropy Probe 647
NOAA-10 610
Olympus 626
POES NOAA-14 571
POES NOAA-15 571
POLAR 647
Polar 518
Polar Orbiting Environmental Satellite (POES) 613
RHESSI 647
RXTE 647
SAMPEX 519, 520
Skylab 450, 624
Solar and Heliospheric Observatory
    (SOHO) 2, 7, 31, 95, 101, 107, 113,
    115, 122, 126, 435, 446, 465,
    468, 474, 484, 611, 612, 635
    Extreme Ultraviolet Imaging Telescope (EIT) 122
    Michelson Doppler Imager (MDI) 95
Solar Mass Ejection Imager (SMEI) 461
Space Shuttle 600, 614, 618, 624, 631, 632, 646
Stardust 611, 647
Starshine 356
STARSHINE-1 622
Television Infrared Observation Satellite 7
Terra 296, 647
Thermosphere Ionosphere Mesosphere
    Energetics and Dynamics (TIMED) 559, 568
Tracking and Data Relay Satellite (TDRS-1) 610
Transition Region and Coronal Explorer
    (TRACE) 113, 126
Upper Atmosphere Research Satellite 647
Voyager 455
Voyager 1 598
Voyager 2 471, 598
Yohkoh 113, 429, 434, 465
Saturn 15, 220, 236
sawtooth events 505
sawtooth oscillations 503, 506, 507
scale analysis of differential equations 192
scale height 130, 369, 563–565, 620
Schwabe, S. Heinrich 25
scintillation 22, 572, 574, 578, 667–672, 686
Scottish power grid authorities 642
second invariant motion 319

Second Law of Thermodynamics	61, 114
secondary electron emission	615
sector boundaries	223
sector, heliospheric current sheet	222–223
seismic wave	421
semi-synchronous orbits	635
sensor heating	609
shielding	591, 600–605, 606, 632, 649
aluminum	602
Apollo command module	606
blood-forming organs (BFO)	602
International Space Station (ISS)	606
liquid hydrogen	602
moon bases	606
multishock	632
Space Shuttle	606
space suit	606
water	602
shock	227, 230, 231, 233, 415, 434, 442, 516, 518, 521, 640
acceleration	599
electric field	233
energy flow	233
interactions	599
magnetic flux	233
mass flow	233
momentum flow	233
quasi-ideal magnetized	233
quasi-parallel	230
tangential	232
vector normal	232
shock acceleration mechanisms	480–485
shock drift acceleration	482
shock fronts	207
shock normal	480
shock parameters	473
shock regions	436, 613
shock waves	425, 427, 430
short wave blackout	424
short-term plasma build-up region	514
short-wave fade (SWF)	424, 659, 661, 686
shortwave photon flux	414
sievert (Sv)	595
sigmoid	429
signal fade	672
single-event burnout (SEB)	605
drain-source voltage	605
single-event effects (SEEs)	604
neutrons	605
pions	605
secondary protons	605
single-event functional interrupt (SEFI)	605
single-event latchup (SEL)	605, 610
complementary metal-oxide semiconductor (CMOS)	605
high-current state	605
mini-latch	605
single-event upset (SEU)	32, 478, 605
alpha particle	605
electrons	605
heavy ion	605
ions	605
protons	605
short electrical pulse	605
single-particle dynamics	255
skip transmission	376
sky wave	376
sky wave radio propagation	376
Skylab	405, 450, 624
slot region	310, 311, 521, 530, 599
Snell's Law	379–381
ionosphere	380
sodium airglow	348
soft X ray	429, 434
Lyman-alpha emission	616
soft X-ray event	521
soft X-ray telescope (SXT)	465
solar	
active region	109
ephemeral	122
atmosphere	94, 113, 114, 131, 208, 238
lower	109–113, 114, 130
physics	128, 131
upper	108, 109, 111, 113, 114–118
atomic composition	94
bipolar regions	101, 123
Carrington rotation number	103
cells	109
characteristics	94
constituents	
calcium	102, 111
carbon	95, 102
free electrons	102
helium	95, 102
hydrogen	95, 111, 112

iron	95, 102
lithium	95
oxygen	95, 102, 111
convective zone	281
coordinate system	103
Carrington longitude	103
central meridian	103
latitude	103
corona	114–115, 117
temperature	95
coronal holes	123, 125
coronal mass ejections (CMEs)	
See also as a separate entry	114, 127, 415
cycle	107, 110, 123–125
density	95, 112
core	94
electron	112, 116
mass	101, 112
mean	94
differential rotation	96, 103
dynamo	101, 104, 107
Earth L1 libration point	126
emissions	94, 113, 115, 119
background component	116
shortwave	117
slowly varying component	116
equator	103
faculae	109
filaments	111
prominences	111
quiescent prominences	112
flux	116, 121
giant cells	107
granulation	105–107, 121
layers	106
chromosphere	107, 116, 121, 134
corona	107, 114–115, 117
inner corona	114
photosphere	107, 109, 114, 115, 121, 131, 134
limb	103, 109
luminosity	94
magnetic carpet	122
magnetic field	95, 107–109, 111, 113, 114, 123
magnetic map	97, 113
maximum See also solar maximum	115, 123, 125
meridional circulation	108
minimum See also solar minimum	116, 123, 125
nuclear reactions	95
optical	117, 120
parcels	101, 104, 105
plages	111
pressure	94, 98–100, 102, 104, 105, 112
balance	105
core	94, 98
gradient	99, 102
thermal	105
quiescent	94, 109, 120
radius	94
rotation	12, 94, 95, 108
shearing	95, 101
sound waves	95, 107, 114
spectrum	94, 109, 111, 115, 117, 121
blackbody	109, 119
stratified	98
supergranulation	105–107, 121
surface gravity	94
tachocline	95, 101, 102, 106
temperature	94, 98, 111–114, 116
blackbody	94
core	94
profile	111
transition region	110–113
wind	94, 115, 123, 125
zonal circulation	108
zones	
convection	107
convective	98, 112
radiative	99, 101
Solar and Heliospheric Observatory (SOHO)	201, 226, 435, 446, 465, 468, 474, 484, 612, 635, 647
C2 Large Angle Spectrometric Coronagraph	226
coronal mass ejection	611
Extreme Ultraviolet Imaging Telescope	226
protons	611
solar arrays	612, 618
radiation-induced damage	612
solar cells	612
transmittance	618
solar atmosphere	208, 238
upper	208
solar cells	594, 612
color center	613
crystal lattice	613
efficiencies	613
electron	613

photon	613
solar cycle	2, 6, 11, 15, 21, 23, 26, 29, 39, 238, 315, 401, 407, 410, 412, 436, 492, 516, 537, 598, 606, 642
effects on the upper atmosphere	536, 538
solar cycle 23	664, 684
solar dipole axis	220
solar disturbances	
energetic alpha particles	598
energetic electrons	598
energetic protons	598
solar dynamo	104, 107, 400
solar emissions *See also* Sun, emissions	
extreme ultraviolet radiation heating	621
solar energetic particle (SEP) events	28, 420, 425, 472–475, 478, 482, 484, 485, 521
forecasting	484
solar energetic particles (SEPs)	12, 23, 64, 415, 433, 442, 540, 598
solar eruption	12, 415
coronal mass ejections *See* coronal mass ejections	
emerging field lines	415
filament	415
magnetic neutral line	415
strapping fields	415
tethering fields	415
energetic eruptions	415
energetic particles	415
flare ribbon signatures	415
energetic particles	415
eruption onset	415
extended ropes	415
flare signatures	415
flares	415
electric fields	415
quiescent field lines	415
solar flares	12, 22, 23, 30, 33, 39, 52, 212, 416, 442
effects on the upper atmosphere	539–541
energy	48
protons	12
solar flux	613
solar flux units (SFUs)	115, 422, 663
solar magnetic orientation	224
solar magnetospheric coordinates (SM)	304
Solar Mass Ejection Imager (SMEI)	461
solar maximum	225, 350–352, 361, 377, 412, 425, 431, 438, 442, 446, 451, 458, 466, 471, 472, 517, 596–598, 602, 612, 624, 630
solar minimum	225, 235, 335, 350–352, 361, 377, 425, 431, 438, 439, 442, 446, 447, 458, 518, 596, 601, 612, 630
solar panels	612
solar particles	310
solar proton events (SPEs)	569, 570, 599
operational constraints	598
radiation damage	598
radiation dose	598
soft errors	598
storm shelters	598
solar quiet (Sq) current system	187
solar radiation	6, 438
extreme ultraviolet	6, 12, 34
infrared	6
ionizing	33
radio	12
scale	33
ultraviolet	6, 23
visible	6
X-ray	6, 12, 22, 30, 33–35
solar rotation	218, 220, 238, 517
solar spectrum	10, 78, 85
solar storms	648
anomalies	647
bit error rates	647
CHANDRA	647
safe mode	647
situational awareness	647
Solar and Heliospheric Observatory (SOHO)	647
solar system	436
solar system, scaling	14
Solar TErrestrial RElations Observatory (STEREO)	127
solar wind	11–18, 22, 23, 37, 39, 69, 94, 115, 123, 125, 193, 194, 200–238, 281, 294, 304–306, 309, 311–315, 324–335, 340–342, 392, 427, 430, 433, 436–438, 442, 446–449, 452, 455, 492–494, 499, 502, 507–508, 511, 518–519, 530, 604, 633, 635, 636, 640
Alfvén speed	214
bimodal	225
bulk flow speed	213
climatology	225
collisionless	200, 230
critical radius	209

density	213, 232, 237
disturbances	17, 457–472
fast	203
flow continuity	230
free-flowing	229
frozen-in magnetic flux	218
high-speed	206, 214, 451, 454
interaction	37
Mach number	214
magnetic field	215, 233
non-transient	201
number density	235
parcels	230
plasma	229
plasma density	236
plasma motion	218
pressure	237, 636
quasi-stationary	201
shocks associated with	453, 459
slow	203, 446–449, 453–455
sound speed	212, 214
sound waves	227
speed	211, 218, 220–221, 225, 231, 233
steady state	202
structure	229
sub-Alfvénic	214
subsonic	209, 229, 235, 238
super-Alfvénic	214
supersonic	16, 201, 209, 213, 214, 215, 227, 229, 234
temperature	213, 214, 233, 236
termination shock	597
thermal speed	213
transients	203, 457–459, 470
solar wind coupling	496
Solar Wind Observations Over the Poles of the Sun (SWOOPS)	226
solar wind particles	311
solar wind pressure	596, 636
solar wind speed	220
solar wind spiral	219
solar wind-magnetosphere-ionosphere-thermosphere connection	336
Solar-Terrestrial Environment Laboratory	461
SORCE mission	413
sound speed c_s	209, 211, 230
Mach number	211
supersonic	211
sound waves	228
high-density turbulent flow	228
subsonic	228
supersonic	228
wave amplitude	228
wave modes	228
source surface	216, 217, 223
solar	223
South Africa ESKOM grid	30
South Atlantic Anomaly (SAA)	294, 296, 310, 315, 517, 580, 591, 599, 602, 603, 604, 649
south geographic pole	294
southern lights (Aurora Australis)	336
southward interplanetary magnetic field (IMF)	330, 332, 334, 336, 635, 636
space climate	3, 40
space debris	632
collision	631
collision avoidance	631
fragments	631
hazards	631
ionization channel	631
sensors	631
space environment	2, 3, 624
disturbances	33
Kp index	35
model	34
space environment interactions	3
space environmental effects	
hazards	644
space hazards	
chemical interactions	590
debris	590
erosion	590
loss of surface integrity	590
meteors	590
particles	590
photons	590
plasma particles	590
radiation damage	590
space object catalog	630
space radiation	
ionizing radiation	591
biologically significant	591
threshold ionization energy	591
sources	
cyclotrons	591
local accelerations	591

magnetic merging	591
shock interactions	591
stellar collapse	591
Space Shuttle	600, 614, 618, 624, 631, 632
windshield	632
space weather	3, 39, 238, 305, 400, 414, 604
disruptions	646
disturbance	123, 638
economic effects	646
environment	39
events	39, 238, 646
forecasters	238, 422
insurance claims	646
model	34
operational input	646
International Space Station (ISS)	646
Space Shuttle	646
storm	13, 21, 26, 38, 39, 95, 414
alerts, watches, warnings	34
geomagnetic	12, 17, 21, 27, 30, 33
magnetic	12, 17, 22, 27, 30
radiation	13, 30, 33
scales	32
system effects	646
space weather activity	337
space weather disturbances	638
space weather event	636
space weather forecasts	422
Space Weather Prediction Center (SWPC)	29, 35, 645
space weather storm 2003	2, 29, 35
space weather tracers	438
space-based spectroscopy	619
spacecraft charging	516
spacecraft commanding and telemetry	661
spacecraft contamination	
molecular	619
particulate	619
spalling	631
spectral emissivity (ε_λ)	78
spectral lines	419
spectrograph	80
spectroscope	80
spherical magnetic coordinates	337
spicules	107
Spirit Mars Rover	172
sporadic E (E_s)	666
spot group	406
spread F	575, 579
sputtering	615, 617
erosion	618
optical coatings	618
Sq current system	187
stagnation point	514
Stardust	611, 647
flare	611
high-energy particles	611
high-energy protons	611
solar panels	611
standby mode	611
STARSHINE-1	356, 622
steady magnetospheric convection (SMC)	336, 503
Stefan-Boltzmann Law	76
stellar collapse	591
stellar nurseries	441
stellar winds	436
stochastic process	481
storm	515, 519, 521, 530
initial phase	530
magnetic	530
main phase	530
recovery phase	530
storm conditions	308
storm energy	508
storm events	634
storm phases	497, 504, 513, 530
storm shelters	603
storm-enhanced density (SED)	579
storm-induced currents	3, 645
storm-level disturbance	495
storm-time currents	527
storm-time disturbance	322, 342
storm-time dynamics	639
storm-time enhancements	517
stratosphere	135
stream interaction regions (SIRs)	450
stream interfaces	450, 453, 454
streamer belt	201, 204, 407, 430, 447
sub-auroral plasma streams (SAPS)	529
sub-auroral trough	667
submarine phone cables	640
sub-photospheric fluxes	425
sub-solar magnetopause	324
substorm current	526
substorm current wedges	387, 526
substorm enhancements	517
substorm expansion phase	546

substorm injection events	518
substorm onset	546
substorm oscillation	518
substorms	333, 336, 492, 496, 502–504, 507, 512, 525, 526, 530
auroral	546
break-up phase	546
expansion phase	504, 546
growth phase	504, 546
onset	546
recovery phase	504, 547
sudden cosmic noise absorption (SCNA)	661
sudden enhancement of signal (SES)	658
sudden enhancements of atmospherics (SEAs)	658
sudden frequency deviation (SFD)	660
sudden impulse (SI)	496
sudden impulsive enhancements	494
sudden increase in total electron content (SITEC)	424, 540
sudden ionospheric disturbances (SIDs)	27, 424, 538, 656, 661
sudden phase anomaly (sudden phase advance) (SPA)	424, 659
sudden storm commencement (SSC)	496, 521
Sun	12, 19
active	110
atmosphere	94
convection zone	281
Earth L1 libration point	126
echo	120
mean rotational period	12
optical	117, 120
quiescent	94, 109, 120
radio	117, 120
sunspots	10, 25, 29, 39
Sun-in-view	120
sunquake wave	421
high-energy particles	421
sunspot cycle	10, 400, 411, 439, 442
sunspot groups	426
sunspot maximum	10
sunspot minimum	10
sunspot number	410, 438, 520
sunspots	25, 97, 101, 107, 110, 116, 123, 125, 157, 172, 282, 283, 400, 402, 404, 405–407, 410, 411, 413, 416, 417, 421, 427, 429, 439
cycle	104, 125
number	116
pressure balance	282, 283
Sun-synchronous orbit	621
Super Dual Auroral Radar Network (SuperDARN)	579
supergranule network	111
supergranules, supergranulation	105–107, 121
cells	107, 113
magnetic field	107
network	111
supernova explosions	436
supernovae	436–437, 441
supersonic flow	208
supersonic region	214
supersonic stream	227
superstorms	621
suprathermal electrons	458–462
suprathermal particles	247
surface charging	32, 616
Swedish railways	642
sweep frequency burst	422, 423
shock wave	423
Swift satellite	679
Sym-H index	523
synchrotron radiation	88
system-generated EMP (SGEMP)	682, 683

T

tachocline	95, 101, 102, 106, 400
tail current	522, 526
tail lobes	312
TEC Units (TECUs)	662
telecommunication cable systems	640
temperature	46
function of energy, as a	54–55, 58
temperature-thermal velocity relationship	615
Temporary Sleep Station (TeSS)	606
termination shock	15, 234–236, 238, 436, 471
shock nose	598
terminator	18
signal propagation effects on	655
Terra spacecraft	296, 647
terrestrial gamma ray flashes (TGFs)	359
terrestrial weather	2, 6, 34
drag	32
thermal convection	104
thermal cycling	618
thermal energy	209, 212, 230, 296
density	230, 401, 404
thermal environment	620

thermal plasma	310
thermal pressure	69, 237, 324, 407, 498
thermal speed	209
thermal turbulence	238
thermalized	593
thermosphere	21, 136, 206, 349–355, 527, 557, 560, 563, 565, 566
Thermosphere Ionosphere Electrodynamics General Circulation Model	567
Thermosphere Ionosphere Mesosphere Energetics and Dynamics (TIMED)	559, 568
thermospheric cooling	567
thermospheric density	620
thermospheric heating	558, 563
thermospheric tides	558
thermospheric winds	352, 558–560
threshold displacement energy	594
tides	
Sun-synchronous	558
topside ionosphere	371
toroidal magnetic fields	101
torsional oscillations	402
total electron content (TEC)	367, 369, 516, 566, 574, 579, 581, 658, 662, 667, 668, 672
total magnetic field vector F	301
total solar irradiance (TSI)	133, 413, 501
Tracking and Data Relay Satellite (TDRS-1)	610
trailing-polarity flux	412
transformer failure	
alternating current (AC)	641
flux densities	641
geomagnetically induced currents (GICs)	641
harmonics	641
magnetic flux	641
nonlinear regime	641
out-gassing	641
reactive power losses	641
solar cycle	641
transient coronal hole (TCH)	470
transient geomagnetic storms	492
transient luminous events (TLEs)	358, 441
trans-ionospheric wave	376
transition region	112, 405
transmission, energy	134
transpolar arcs (TPAs)	548
trapped energetic protons	516
trapped particles	511
trapping regions	310, 339, 341, 508, 511, 520
traveling atmospheric disturbances (TADs)	561, 566, 579, 580, 582, 620
traveling ionospheric disturbances (TIDs)	572, 583, 666, 686
traveling shocks	473–475, 477
type II radio bursts	420, 476
type III burst	435

U

ultraviolet (UV) region	616
ultraviolet light	337
Uluru, Mount	13
Ulysses	200, 221, 225, 226
umbra	406
undershielding	529, 531
unipolar	124
unipolar regions	124, 400
unit of energy	
electron volt (eV)	593
universal gravitational constant	100
unsigned flux	401
upper atmosphere	311
Upper Atmosphere Research Satellite	647
upper plasmaspheric boundary	308
upper-atmospheric storms *See also* magnetic storms, geomagnetic storms	536, 548, 559
upwelling	563
Uranus	15

V

Van Allen radiation belts *See also* radiation belts	17, 316, 599, 608
Van Allen, James	17, 309
velocity loss cone	320
Venus	200
vertical guidance approach (APV)	673
very high- to extremely high-frequency (EHF) systems	667
very-low-energy particles	305
viscous interaction	327, 342
visible spectrum	71
von Humboldt, Baron Alexander	25
Voyager	200, 234, 237, 455
Voyager 1	15, 235, 598
Voyager 2	219, 235, 236, 471, 598

W

water equivalent material	603
wave energy dissipation	401
wave fronts	70
wave number k	284
wavelength λ	70
wave-particle interactions	515
Wide Area Augmentation System (WAAS)	671, 673, 685
Wien's Displacement Law	76
Wolf, Rudolph	410
Wolf sunspot counts	410
Zurich Sunspot Number	410
work	45
work functions	616
work-kinetic energy theorem	50
World Data Center-C2 for Geomagnetism	495

X

X rays	225, 401, 416, 418, 419, 428, 431, 437, 441
flares	428, 434
impulsive	428
hard X rays	419
image	225
long-duration	428
soft X rays	419
X-class flares	424, 517
X-line	329, 333, 427
X-ray bremsstrahlung	593
X-ray emissions	405, 521
X-ray flux	421
soft X rays	421
X-ray glow	593
X-ray region	616

Y

Yohkoh	113, 225, 434, 465

Z

zero momentum	637
electric coils	637
reaction jets	637
torque	637
zonal circulation	108
Zvezda Service Module	606